D1743159

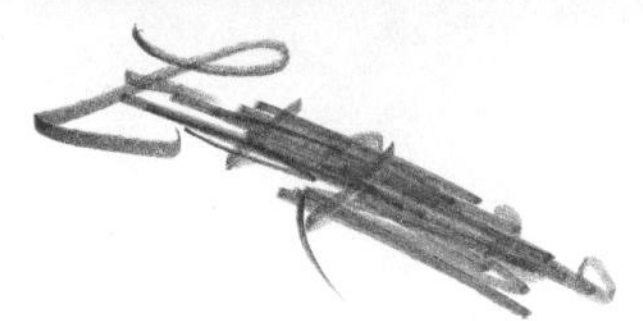

COMPOSITE CONSTRUCTION MATERIALS HANDBOOK

COMPOSITE CONSTRUCTION MATERIALS HANDBOOK

ROBERT NICHOLLS

Department of Civil Engineering

University of Delaware

PRENTICE-HALL, INC., *Englewood Cliffs, New Jersey*

Library of Congress Cataloging in Publication Data

Nicholls, Robert
Composite construction materials handbook.

Includes bibliographical references.
1. Composite materials. I. Title.
TA418.9.C6N52 624′.18 75–34921
ISBN 0–13–164889–6

Cover illustration reproduced with permission from Newman, K. and J. B. Newman, "Failure Theories and Design Criteria for Plain Concrete," in Structure, Solid Mechanics and Engineering Design, Proceedings of the Southampton 1970 Civil Engineering Materials Conference, *Part II, M. Te'eni, ed., New York: John Wiley & Sons, Inc. (Interscience Division), p. 963.*

10 9 8 7 6 5 4 3 2 1

Printed in the United States of America

PRENTICE-HALL INTERNATIONAL, INC., *London*
PRENTICE-HALL OF AUSTRALIA, PTY. Limited, *Sydney*
PRENTICE-HALL OF CANADA, LTD., *Toronto*
PRENTICE-HALL OF INDIA PRIVATE LIMITED, *New Delhi*
PRENTICE-HALL OF JAPAN, INC., *Tokyo*
PRENTICE-HALL OF SOUTHEAST ASIA (PTE.) Limited, *Singapore*

— To R, D, J, C —

Symbiotic Composites

with

joie de vivre

CONTENTS

PREFACE

This book is for undergraduate and beginning graduate students who have studied general chemistry and mechanics, and for engineers involved with materials.

Introductory materials science courses, based on the chemistry and internal structure of materials, have become accepted in all branches of engineering, analogous to physics, mathematics and chemistry.

A significant field of application for materials science is the construction industry, which accounts for more than 10 percent of the U. S. national product, and where diminishing supplies of raw materials and spiraling labor costs motivate examination of the chemistry and production procedures for processing raw materials into structures for housing, transportation, and public works.

Due to the profusion of raw materials and processes used in construction, it is impossible to optimize the materials selection, chemical processing, and structural design without a rudimentary knowledge of composite materials, based on fundamentals of chemistry and mechanics. This book presents unifying principles applicable to composite construction materials. It also seeks to bridge the gap between materials research and structural design and to provide a handy reference for those interested in modifying construction materials or developing new ones.

The text's principal aim is to integrate the (a) chemical structure, (b) mechanical behavior, and (c) design optimization of composite materials used in the construction industry. The emphasis throughout is on the dependence of mechanical properties upon chemical structure. Examples and problems are used as the essential teaching tool.

Following an introductory chapter which surveys the major classes of composite materials used in construction, Chapters 2 and 3 describe the chemical basis for the mechanical behavior of homogeneous materials. These two chapters emphasize general phenomena rather than their manifestations in any specific material system.

Chapters 4 to 6 describe the chemical and mechanical properties of specific materials used as the continuous and discontinuous phases of two-phase construction composites.

Chapters 7 to 12 apply systems optimization techniques to the design of aggregate-binder, structural foam, fiber-matrix, and laminate systems. These chapters are grouped according to the geometry of the discontinuous phase; e.g., bulky aggregates or voids—Chapters 7 to 10, elongate fibers—Chapter 10, and laminates—Chapter 12. Chapter 13 describes useful instrumental methods for structural determination, permitting the book to be used with laboratory demonstrations or student projects, where the equipment is available.

For students having a basic materials science course, Chapters 2 and 3 can be omitted, or treated as review topics. Student projects employing some of the instrumental techniques of Chapter 13 can then be included in a one semester course. For students not having an introductory materials science course, Chapters 2 and 3 are essential and may preclude the use of student laboratory projects.

I am indebted to many workers in various fields. Materials were gathered from diverse sources and illustrations were used with permission from many texts and papers. Special thanks are due to the reviewers of the manuscript, especially Professor Albert G.H. Dietz of the Massachusetts Institute of Technology, for helpful suggestions for improving the text, to the University of Delaware Civil Engineering Department for clerical assistance, and to my wife and family for patience while the manuscript was gestating. To the shade of Oliver Wendell Holmes, who, in "The Wonderful One-Hoss Shay," perceived the synergism of composite behavior before such catch words were in vogue, I offer humble recognition.

ROBERT NICHOLLS

Newark, Delaware

1

COMPOSITE MATERIALS IN CONSTRUCTION

Have you heard of the wonderful one-hoss shay
That was built in such a logical way
It ran a hundred years to a day,—

—HOLMES

Composite materials are mixtures of two or more phases. Typical natural composites are wood and bone. Wood is a composite of cellulose fibers, strong in tension but flexible, cemented together with lignin, which provides stiffness. Bone is a composite of the strong but soft protein collagen and the hard but brittle mineral apatite.

Most engineering materials are composites, for example, the addition of wood flour to plastics to increase their rigidity (Fig. 1), of carbon black to rubber, of fiber glass to resins, and of asphalt or portland cement to sand and gravel. Man-made composites may result from mechanical mixing, as for fillers in plastics and rubbers or for aggregates in cement, or they may result from solution and subsequent phase transformations, as in heat-treating processes. In these *eutectics* (from the Greek for "easily melted") part of the alloy develops into parallel whiskers and part becomes a matrix for the whiskers. The result is a strong whisker-reinforced composite. An

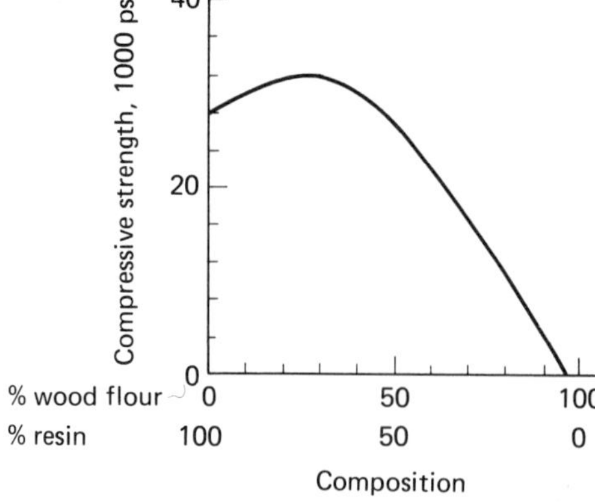

Fig. 1. Strength of wood flour—formaldehyde. From Van Vlack, L. H.: *Elements of Materials Science, First Edition*, Reading, Mass.: Addison-Wesley Publishing Company, 1959 [1].

example is niobium carbide whiskers in a niobium matrix, which has high strength at temperatures up to 1650°C. Additional methods of manufacturing composites include the infiltration of a solid discrete phase by a continuous phase initially in liquid form and the hardening of structural foams from a mixture of gas and liquid phases. All such composites have strength, rigidity, toughness, heat resistance, or some combination of desirable properties which is superior to the corresponding properties of the individual components. For example, glass is brittle, and small surface scratches cause failure at low tensile stresses, but it is superior to plastics in strength and stiffness, does not expand significantly, and can be made from inexpensive raw materials. Glass fibers can be bound together with a soft plastic matrix so that continuous cracks do not easily develop through the mixture. The beauty of the composite is that small cracks which occasionally develop in individual fibers will be unimportant.

Certain properties of composites are insensitive to their microstructure, while other properties are sensitive. Density and heat capacity, for example, depend only on the volume fractions of the phases. But thermal and electrical conductivity, Young's modulus, and strength are sensitive to the geometry and orientation of the discrete and continuous phases.

Many composites can be adequately described as a homogeneous isotropic matrix in which particles of a second homogeneous isotropic phase are dispersed. Assuming a uniform volume concentration, it seems possible, in principle, to calculate the microstructure-sensitive properties of the composite from the properties of the composing elements. Unfortunately, there are few cases where rigorous calculations can be made. For some special cases, however, in this book, the deformation and fracture properties of two-phase composites as a function of (1) the geometry of the discrete phase and (2) the strength and elastic moduli ratios between the continuous and discrete phases, are described. We shall consider three idealized, but bounding, geometries of the discrete phase: elongate (fiber), platelets (laminar), and bulky (aggregate). The study of bulky geometry will include structural foams as a limiting case representing aggregate with zero strength.

A. CONSTRUCTION COMPOSITES

The entire production of some industries and large portions of others are used in construction—crushed rock, cement, asphalt, timber and steel, as well as over one fourth of the plastics and most of the glass produced each year. Nearly 300 billion lb of building materials were used in the United States in 1970. Approximately 165 billion lb of cement were used, 30 billion lb of steel, 15 billion lb of asphalt, 4 billion lb of plastic, 2 billion lb of aluminum, 35 billion fbm of wood, 25 billion sq ft of glass, and 7.5 billion bricks.

The in-place yearly volume of construction is currently about $126 billlion, representing slightly over 10% of the gross national product and employing approximately 10% of the labor force. The major categories include

Housing	$ 48 billion
Highway and transportation	15
Maintenance (all categories)	15
Heavy industrial	10
Pollution control	13
Commercial building	7
Power generation	7
Marine and miscellaneous	6
Underground	3
Light industrial	2
	$126 billion

Although the techniques for processing raw materials into finished structures are continually being improved, most current building materials still require extensive curing or conditioning. This slows production and frequently requires a large investment in storage space or forms. The building industry still relies heavily on labor, and the costs of building construction continue to rise at a higher rate than the costs of other industrial products. There is thus a growing need to adapt materials drived from abundant mineral sources—clay, sand, and gravel—to the mass production of building elements.

All research and development activities related to construction, including materials research, is estimated at 0.1%, or about $126 million/yr, and basic engineering research at 0.02%, or $25 million/yr. Productivity rises about 1%/yr, compared with 8% for the agricultural industry, which is more highly research-and-development oriented. Although construction annually uses over one third of the total output value of the basic materials industries (metals, polymers, and minerals, including processed aggregates), the annual cost of applied materials science research related to construction applications has been less than applied research expenditures for materials used in space, military, and air transport applications. The reasons may stem from several factors:

1. Improvements in construction materials have not constituted a national priority to the extent that improvements in materials for applications in transportation and space have.
2. Because most construction materials are, of economic necessity, not highly processed, they tend to be chemically and structurally more variable and

are therefore less capable of precise characterization than more uniform materials, such as metals and synthetic polymers, obtained by extensive processing.

3. The chemical structures are usually significantly more complex (aside from their variability) than those of more highly processed materials. For example, the chemistry of asphalt is vastly more complicated than that of synthetic polymers, and greater effort is required to establish the phase diagrams of silicate cements than those of metals, or even the simpler ceramics, due to the numbers of components and the durations required to reach thermodynamic equilibrium.

Building materials research is frequently accomplished by engineers with little background in materials science or by chemists with little background in design and construction. Progress in construction materials science should improve with efforts by those grounded both in the requirements of the construction industry and in the aspects of materials science which bear most directly upon construction materials, i.e., silicate chemistry, with somewhat less emphasis upon organic chemistry and metallurgy.

B. OBJECTIVES

This book differs from most materials science texts in its greater emphasis on the deformation and fracture mechanics of two-phase systems. The structure and properties of metals (dislocation theory, corrosion, fatigue, forming processes) described in most materials texts, and the properties of solids used for magnetic and electronic applications, are omitted here. This emphasis stems from the responsibilities of the construction materials engineer. In structural metals he usually selects a premanufactured material, based upon behavioral properties developed by the metallurgist. In aggregate-binder systems such as portland cement or bituminous concretes, he is more frequently involved with several phases of the production process—exploration for raw materials, mix design, chemical processing, and field application. Also, since the annual cost of aggregate composites used in construction is several times that of structural metals, these materials deserve careful attention.

The book does not include the structural design of finished building or pavement components or the field construction technique used with different materials. Techniques for mix design optimization are emphasized, but quality-control procedures for composite materials are only briefly outlined. We shall emphasize chemistry and two-phase mechanics rather than laboratory and construction procedures, already well described by books on construction materials. This emphasis seemed appropriate to spark interest in developing and studying new composite construction materials.

Emphasis is placed on those properties and levels of structure which appear to offer significant promise for system improvement. Mechanical properties of multiphase composites normally depend less on the nature of primary chemical bonds or

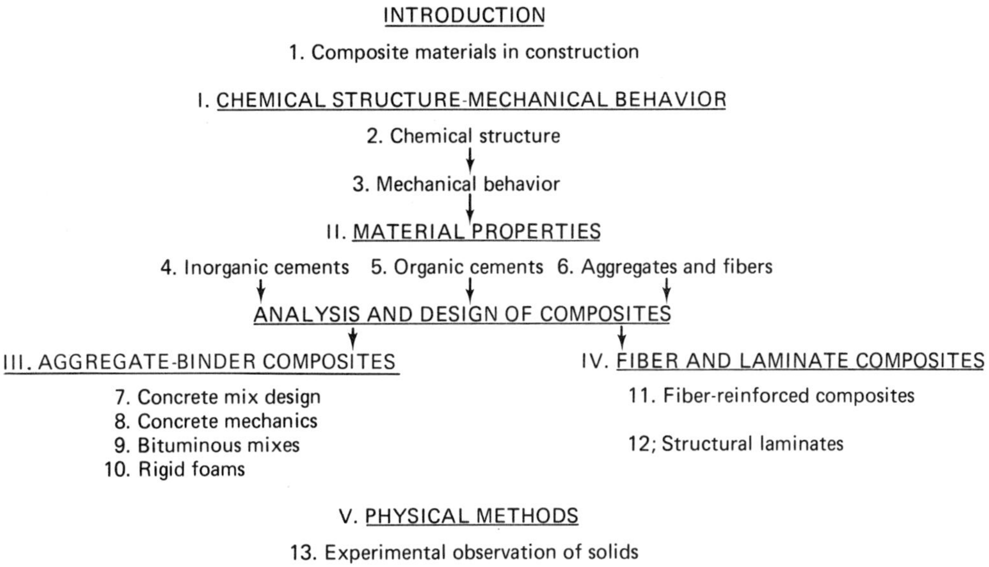

INTRODUCTION

1. Composite materials in construction

I. CHEMICAL STRUCTURE-MECHANICAL BEHAVIOR

2. Chemical structure

↓

3. Mechanical behavior

↓

II. MATERIAL PROPERTIES

4. Inorganic cements 5. Organic cements 6. Aggregates and fibers

↓ ↓ ↓

ANALYSIS AND DESIGN OF COMPOSITES

↓ ↓

III. AGGREGATE-BINDER COMPOSITES

7. Concrete mix design
8. Concrete mechanics
9. Bituminous mixes
10. Rigid foams

IV. FIBER AND LAMINATE COMPOSITES

11. Fiber-reinforced composites

12; Structural laminates

V. PHYSICAL METHODS

13. Experimental observation of solids

Fig. 2. Chapter relationships.

the dislocation in ductile crystals than on phase geometry, the nature of the bond between phases, and the ratios of elastic moduli between phases.

Figure 2 shows relationships among the chapters in the text. Chemical structure (Chap. 2) and mechanical behavior (Chap. 3) provide the tools for analyzing and designing composite materials, Chaps. 7–12, which are identified according to the gometry of the discrete phase* and the material class of the continuous phase. In each of Chaps. 7–12 the deformation and failure modes of the respective composites as a basis for the corresponding design methods is presented. The recurring theme is the way material properties affect structural behavior and the extent to which they can be modified in manufacture and accounted for in structural design.

C. CLASSES OF COMPOSITE MATERIALS IN CONSTRUCTION

The remainder of this chapter contains some representative applications of construction composites in each of the major classes shown in Fig. 2. Then after studying the dependence of mechanical behavior on chemical structure in Chaps. 2 and 3, in subsequent chapters we shall take up the analysis and design of the major classes of construction composites.

1. Aggregate-Binder Composites (Bulky Discrete Phase)

Examples of aggregate-binder composites in construction include

*One can visualize a bulky aggregate elongated to a fiber or flattened to a laminar shape.

	Material	Illustrations	Analysis and Design
1.	Portland cement concrete, with various normal, heavy-, or lightweight aggregates, either inorganic or organic	Figs. 3–7	Chaps. 7, 8
2.	Autoclaved calcium-silicate concretes, using normal, heavy-, or lightweight aggregates		Chaps. 7, 8
3.	Bituminous mixes		Chap. 9
4.	Synthetic polymer-aggregate mixes	Fig. 8	
5.	Rigid foams, including aerated concretes having densities to below 25 pcf, autoclaved aerated calcium-silicates having densities to below 12 pcf, and foamed polymers having densities to below 1 pcf		Chap. 10
6.	Sintered products, including structural and nonstructural ceramics		
7.	Stabilized soils, using inorganic or organic binders		

Fig. 3. Precast prestressed concrete. 100 West Harrison, Seattle; office building. (*Courtesy of the Prestressed Concrete Institute, Chicago, Ill.*)

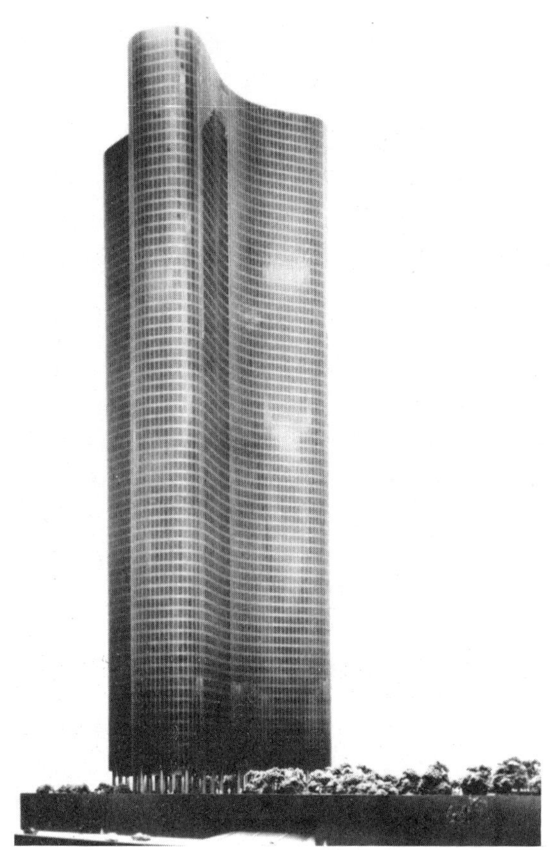

Fig. 4. Lightweight aggregate concrete. Lake Point Tower, Chicago; 70 stories. (*Courtesy of American Concrete Institute, Detroit, Mich.*)

Fig. 5. Polystyrene beads as concrete aggregate. (a) Unexpanded beads, bulk density about 50 pcf; (b) steam-expanded beads, bulk density about 1 pcf; (c) expanded beads coated with epoxy resin to improve aggregate strength and adhesion with portland cement. From Kohling, K., and F. Hohwiller, "Styropor Concrete—A New Building Material," *Rev. Badische Anilin-&Soda-Fabrik AG*, 20 (Sept. 1970) 69.

Fig. 6. Polystyrene bead aggregate concrete insulation for bottom, walls and roof of liquid ammonia storage tanks, Antwerpen. From Kohling, K. and F. Hohwiller, "Styropor Concrete—A New Building Material," *Rev. Badische Anilin-&Soda-Fabrik AG*, 20 (Sept. 1970) 69.

Fig. 7. Glass, ceramic, and epoxy bubbles, available in diameters up to 0.5 in., bulk densities as low as 8 pcf. Used as fillers in cast synthetic wood (polystyrene and polyurethane) furniture, plaster, concretes, and glass. (a) Model cut-a-way. From Beck, W. R., et al., "Two and Three Phase Glass Bubble Composites," *Soc. Plastics Engineers, J.*, 25, No. 4, (1969) 83. [5]; (b) in a marine buoy. (*Courtesy of Emerson and Cuming, Inc.*)

(a)

(b)

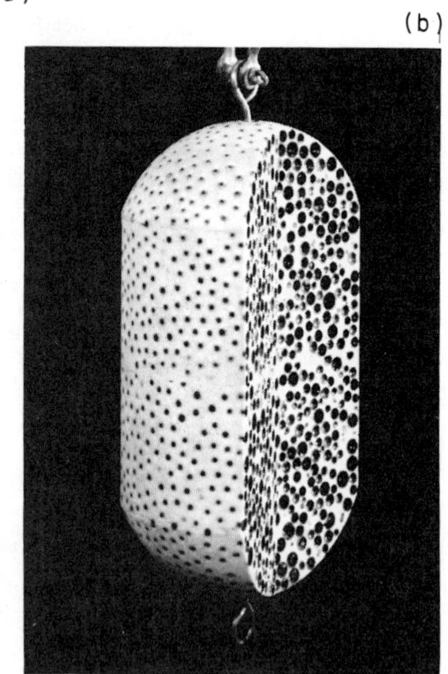

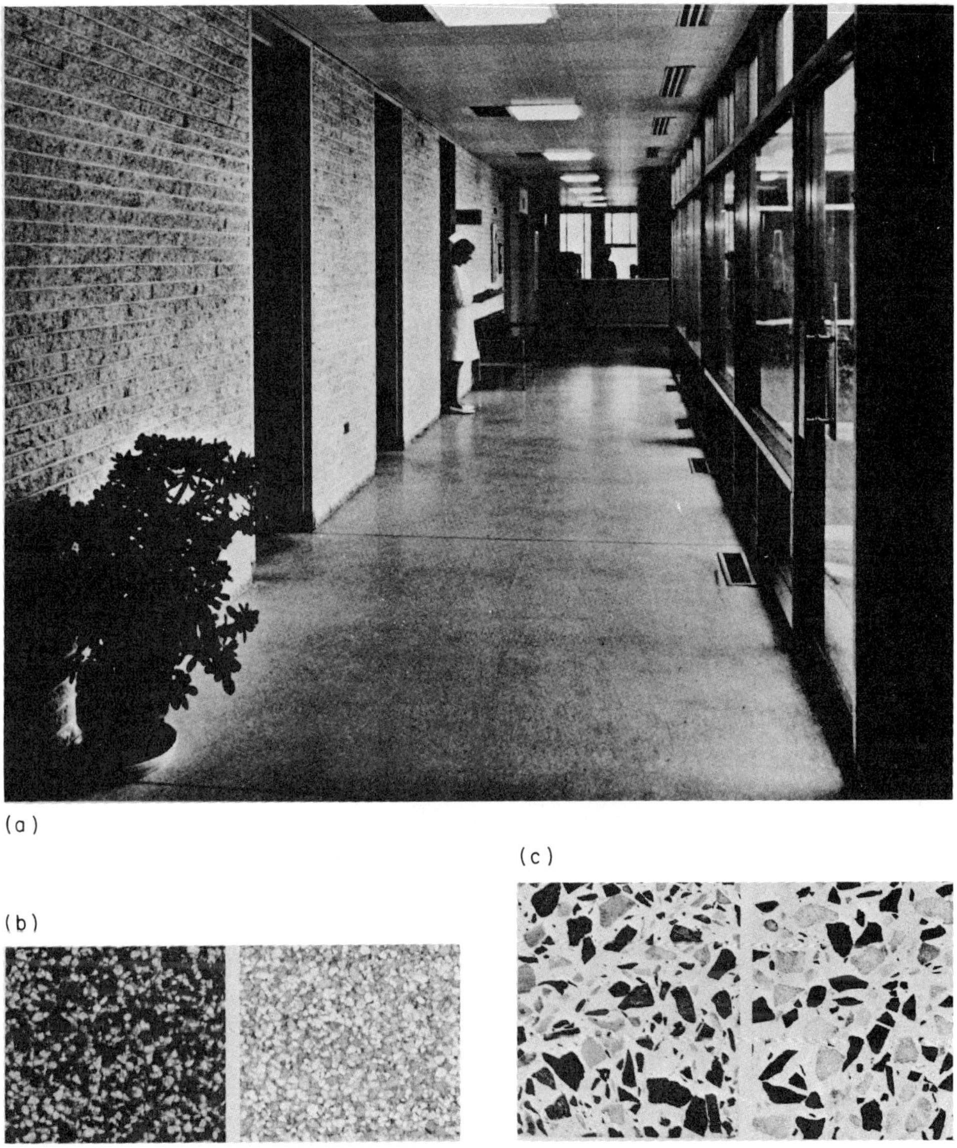

(a)

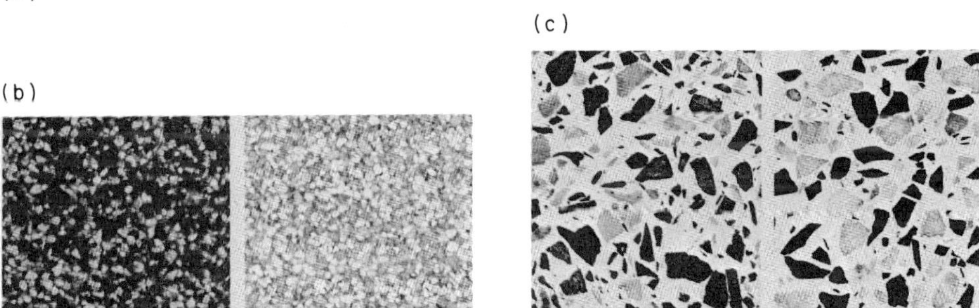

Fig. 8. Polyester-aggregate mixes for floors. (a) Spread and troweled; (b) tile; (c) terrazzo. (*Courtesy of Koppers Company, Inc., Pittsburgh, Pa.*)

2. *Fiber-Reinforced Composites (Elongate Discrete Phase)*

Examples of fiber-reinforced composites in construction include

Material	*Illustrations*	*Analysis and Design*
1. Asbestos cement products	Fig. 9	Chap. 11
2. Inorganic cements (portland cement and autoclaved calcium silicates) reinforced with other fibrous materials, including metal wires, polymers, or glass fibers		Chap. 11
3. Bitumen-aggregate mixes containing fiber reinforcement		Chap. 11
4. Plastics reinforced with glass, asbestos, hemp, or other fibers	Figs. 10–12	Chap. 11

(a)

Custom extruded Corspan mullions and panels provide for direct insertion of thermal pane glass into mullion slot using glazer tape and a neoprene wedge gasket. Stainless steel core clips tie Corspan to concrete slab.

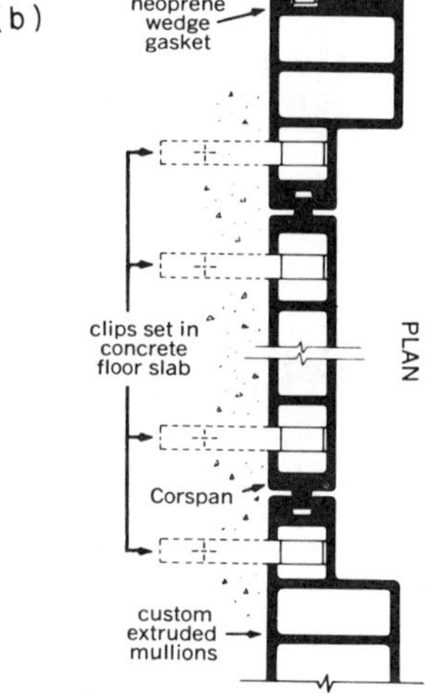

(b)

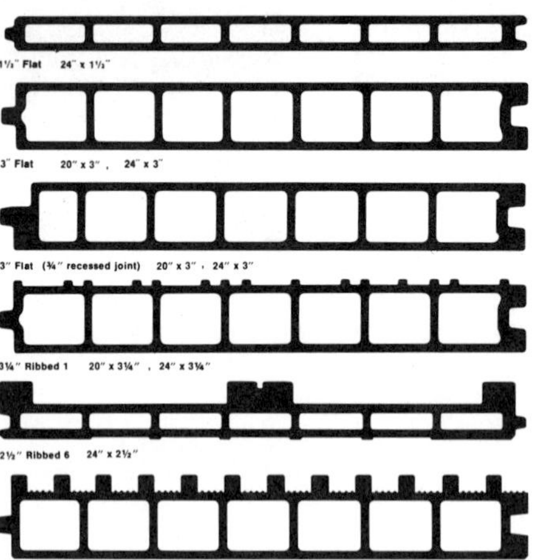

(c)

Fig. 9. Hydrothermally cured calcium and aluminum silicate reinforced with asbestos fiber. (a) For building walls, 200 Walnut St. Building, Wellesley, Mass.; (b) construction detail; (c) standard tongue and groove hollow core wall sections. (*Courtesy of Johns-Manville Corp.*)

Fig. 10. Radomes may be made of premolded fiber-reinforced plastic sections with integral edge ribs. Parts are bolted together. Larger radomes may have metal ribs, or sandwich panels having reinforced plastic facings and honeycomb core. A few are made of inflated coated fabrics. Reprinted from *Plastics for Architects and Builders* by A. G. Dietz, by permission of the MIT Press, Cambridge, Mass. [6].

Fig. 11. Glass fiber-reinforced plastic underground storage tank. *(Courtesy of Owens-Corning Fiberglas Corp.)*

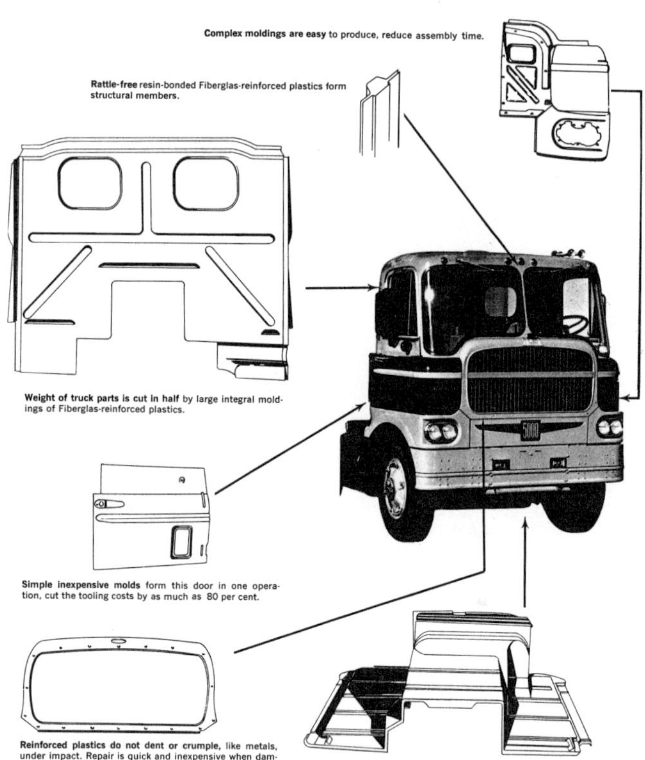

Fig. 12. Glass fiber-reinforced plastic parts, White Truck, *(Courtesy of Owens-Corning Fiberglas Corp.)*

Unreinforced plastics also find numerous applications in various types of construction. Figures 13–15 illustrate several examples. Figure 16 shows the major markets for all plastics, the largest being in construction, and the second largest in transportation, including boat hulls, tank cars, and automobile, truck, and aircraft parts.

(a)

(b)

Fig. 13. Rigid polyvinyl chloride. (a) Extruded interlocking clapboards in various profiles and colors. Color is integral and material dent-resistant; (b) gutter, downspout, and accessories. (*Courtesy of Bird and Son, Inc.*); Reprinted from *Plastics for Architects and Builders* by A. G. Dietz, by permission of the MIT Press, Cambridge, Mass. [6].

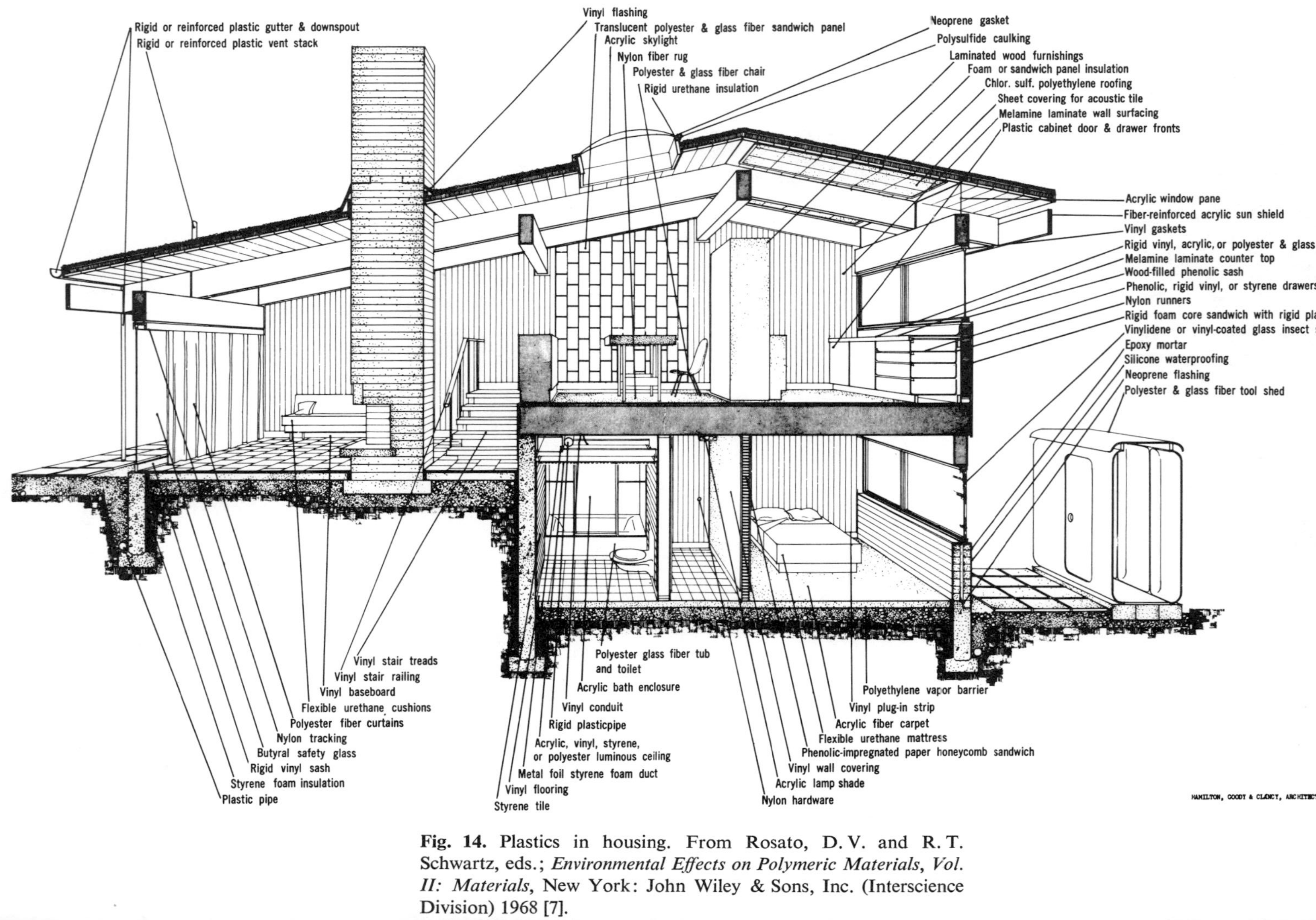

Fig. 14. Plastics in housing. From Rosato, D. V. and R. T. Schwartz, eds.; *Environmental Effects on Polymeric Materials, Vol. II: Materials*, New York: John Wiley & Sons, Inc. (Interscience Division) 1968 [7].

(a)

(b)

Fig. 15. Vacuum-formed acrylic bubbles on steel frame, Houston Astrodome. (a) Exterior; (b) interior. Reprinted from *Plastics for Architects and Builders* by A. G. Dietz, by permission of the MIT Press, Cambridge, Mass. [6].

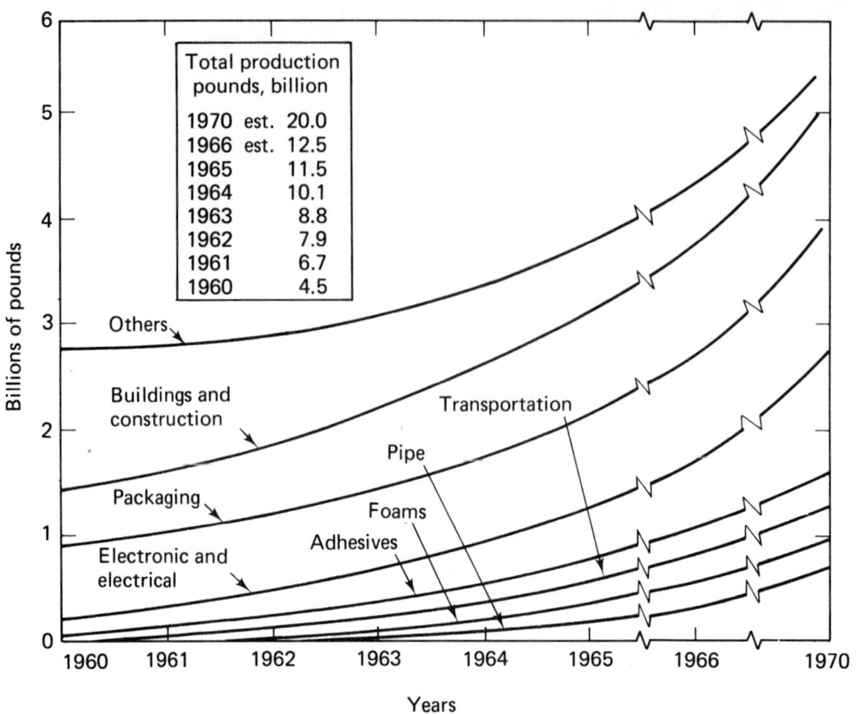

Fig. 16. Estimated U.S. consumption for all plastics, by major markets. From Rosato, D. V. and R. T. Schwartz, eds., *Environmental Effects on Polymeric Materials, Vol. II: Materials*, New York: John Wiley & Sons, Inc. (Interscience Division) 1968 [7].

Percentage consumption by application for all types of plastics in construction is currently about 23% in surface coatings, 20% in flooring, 20% in wire insulation, 10% in heat insulation and vapor barriers, 9% in plywood, 8% in pipe, and 10% in all other applications [7]. Major growth applications are anticipated in pipes and in composite building walls, roofs, and floors.

Chemical and physical properties of specific plastics groups and their suitability for various construction applications are described in Chap. 5.

3. Laminate Composites

Examples of structural laminates include

Material	*Illustrations*	*Analysis and Design*
1. Laminated timber and plywoods	Figs. 17–18	Chap. 12
2. Laminated plastics and fiber-reinforced plastics		Chap. 12
3. Structural sandwich panels	Figs. 19–23	Chap. 12

Fig. 17. Laminated timber arches. (*Courtesy of Koppers Company, Pittsburgh, Pa.*)

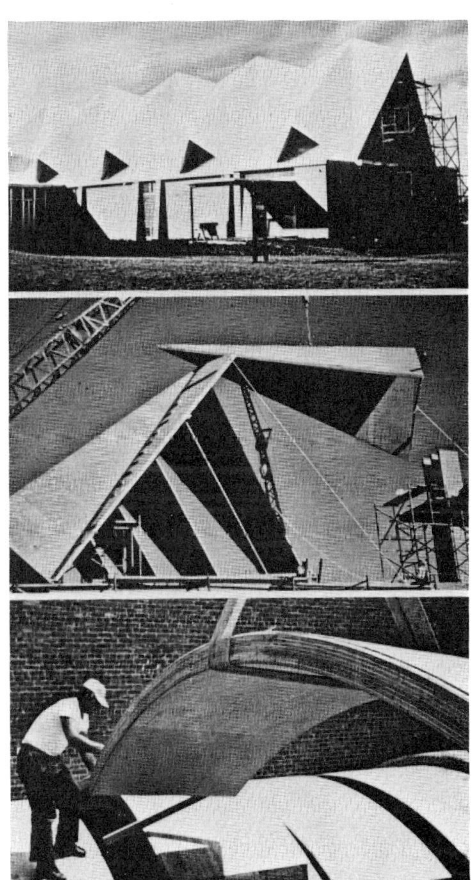

Fig. 18. Glue laminated preformed plywood panels. (*Courtesy of Koppers Company, Inc., Pittsburgh, Pa.*)

Fig. 19. Sandwich panel walls, 5 in. thick, of concrete faces and foamed polystyrene cores, University of Pittsburgh student residence towers. (*Courtesy of Koppers Company, Inc., Pittsburgh, Pa.*)

(a)

Fig. 20. Long span prestressed hollow core floors bolted to precast concrete end and interior cross walls. Exterior walls are sandwich panels; concrete faces with foamed polymer core. Bison System, developed by Concrete Ltd. and used extensively in England, eliminates need for column and beam framing, and scaffolding. Dormitories, University of Delaware; (a) exterior.

(b)

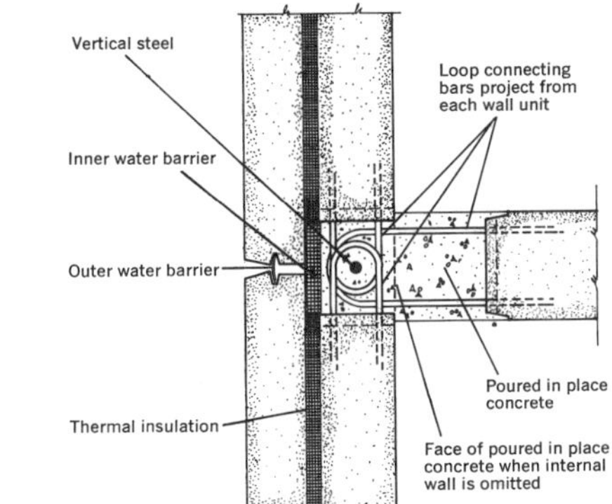

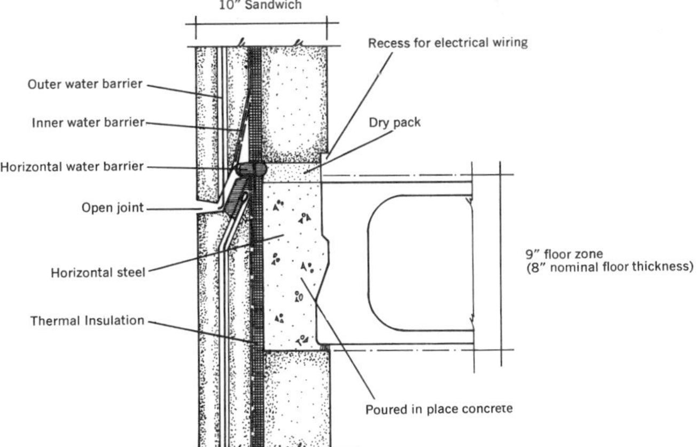

Fig. 20. (*cont.*) (b) construction details. (*Courtesy of Strescon Industries.*)

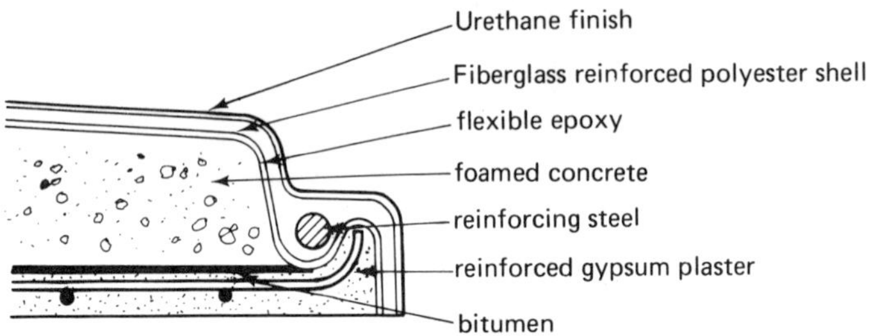

Fig. 21. Sandwich panels having polymer exterior face and concrete core, i.e., opposite from the materials usage illustrated in Figs. 19 and 20. Molded polyester shells are heavily loaded with mineral filler to provide zero flame spread, and coated with baked-on polyurethane finish. Shells are filled with 4-inch thick foamed concrete, bonded to the shell with a flexible epoxy-urethane layer. Bitumen, used to bond the inner face gypsum board to the concrete, serves also as a vapor barrier. Greater London Council high-rise flats. Reprinted from *Plastics for Architects and Builders* by A. G. Dietz, by permission of the MIT Press, Cambridge, Mass. [6].

(a)

Weather resistant exterior facing in various thicknesses.

Available in patterned or unpatterned surfaces.

Exceptional dent resistance.

Panel sizes to 5 x 18 ft.

Superior insulation and vapor barrier core in ¾ to 8-in. thicknesses.

Insulating core has fire protective characteristics.

(b)

Fig. 22. Sandwich panels with aluminum faces of various finishes, textures, and colors bonded to foamed polystyrene or polyurethane core. (a) Exterior; (b) connection detail. (*Courtesy of ALCOA.*)

Fig. 23. Translucent sandwich panels of 1/16 inch glass fiber-reinforced polyester faces bonded to extruded aluminum I beam grid cores, Newark Airport tower. Reprinted from *Plastics for Architects and Builders* by A. G. Dietz, by permission of the MIT Press, Cambridge, Mass. [6].

Several of the materials in Secs. D1–D3 have at least two discrete phases, for example, asbestos fiber used synergetically with fine aggregate in concretes, bituminous mixes, or plastics. In addition, various composites contain two or more continuous phases. Examples include creosote-impregnated timber and polymer-impregnated concretes.

Appendix E lists some frequently used standard test methods for construction materials in each of the preceding classes: aggregate-binder, fiber-reinforced, and laminate composites.

D. COMPOSITE STRUCTURES

Composite materials can be combined in various geometries to form composite structures (Figs. 24 and 25, for example). For purposes of analysis and design, the distinction between a composite material and a composite structure is not usually essential, because optimization of the structure and of the material cannot usually be

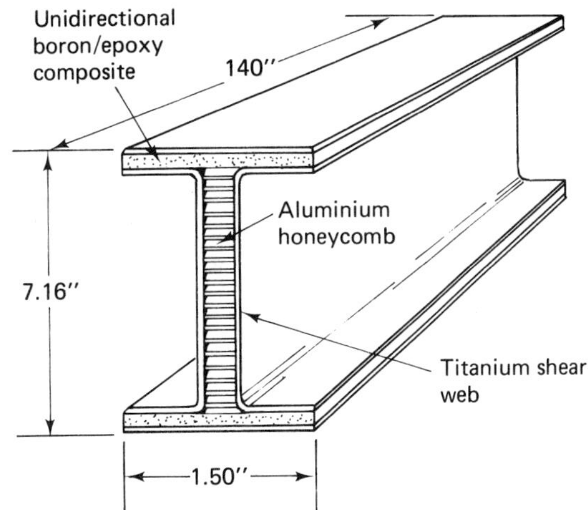

Fig. 24. Composite floor beam. High modulus epoxy resin in the compression (top) flange increases compressive strength and stiffness of the boron composite. A lower modulus epoxy resin in the tension flange enables full tensile strength of the boron fiber to be developed. From Lager, J. R. and R. R. June, "Design, Analysis, Fabrication and Test of a Boron Composite Beam." *J. Composite Mater.*, (1968) 128. Technomic Publishing Co., Inc., Westport, Conn.

treated independently anyway. With composite materials, where material properties and proportions may be continuously varied, optimization requires simultaneous variation of the material properties (strength, modulus, etc.) and the structure (size, shape, etc.). Effective use of composite materials requires that material design be included as an integral part of structural design.

Several of the examples and problems in Chaps. 7–12 illustrate combined material-structural design concepts for various aggregate-matrix and fiber-reinforced composites. In some instances constitutive equations are used to satisfy several physical property requirements simultaneously. Many of the optimization examples include material component costs as well as physical properties.

Fig. 25. Composite helicopter rotor blade. Torsion and cordwise bending carried by the $\pm 45°$ fiber-reinforced laminates. Beam bending carried by the laminates and the unidirectional fiber-reinforced span. Reprinted from McCullough, R. L., *Concepts of Fiber-Reinforced Plastics*, 1971, p. 15. (Courtesy of Marcel Dekker, Inc.)

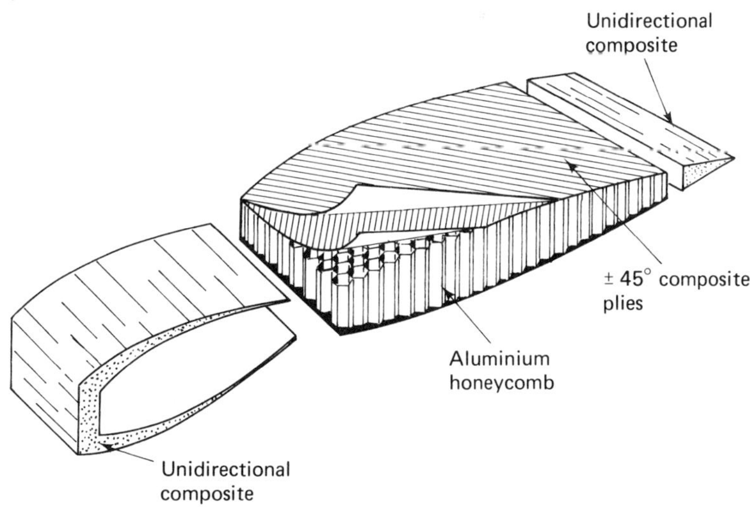

In formulating design problems we shall speak of *preassigned parameters* and *design variables*. Each set of mechanical, thermal, or other loads is referred to as a *load condition*, and the several distinct load conditions will be called a *loading system*. The various potential *failure modes*, and the criteria which constitute failure in each case, must be carefully designated by the engineer. Finally, the *objective function* is a function of the design variables, the value (maximum or minimum) of which can be used as a basis for choosing between alternative design. Examples and problems in Chaps. 7–12 illustrate these concepts.

REFERENCES

1. VAN VLACK, L. H., *Elements of Materials Science*. Reading, Mass.: Addison-Wesley Publishing Company, Inc., 1959.
2. *Prestressed Concrete Items*. Chicago: Prestressed Concrete Institute, 1972.
3. *American Concrete Institute Manual of Concrete Practice*, Part 3. Detroit: American Concrete Institute, 1970.
4. KOHLING, K., and F. HOHWILLER, "Styropor Concrete—A New Building Material," *Rev. Badische Anilin-& Soda-Fabrik AG*, 20 (Sept. 1970) 69.
5. BECK, W. R., et al., "Two and Three Phase Glass Bubble Composites," *Soc. Plastics Engrs. J.*, 25, No. 4 (1969) 83.
6. DIETZ, A. G., *Plastics for Arhitects and Builders*. Cambridge, Mass.: The M.I.T. Press, 1969.
7. ROSATO, D. V., and R. T. SCHWARTZ, eds., *Environmental Effects on Polymeric Materials*, Vol. II: Materials. New York: John Wiley & Sons, Inc. (Interscience Division), 1968.
8. LAGER, J. R., and R. R. JUNE, "Design, Analysis, Fabrication and Test of A Boron Composite Beam," *J. Composite Mater.*, 2 (1968) 128.
9. MCCULLOUGH, R. L., *Concepts of Fiber-Reinforced Plastics*. New York; Marcel Dekker, Inc., 1971.

I

CHEMICAL STRUCTURE–MECHANICAL BEHAVIOR

2

CHEMICAL STRUCTURE

I'll tell you what happened without delay,
Scaring the parson into fits,
Frightening people out of their wits,—
Have you ever heard of that, I say?
—HOLMES

A chemical structural basis for understanding and interpreting most of the mechanical behavioral properties described in Chap. 3* and in the subsequent chapters, which all deal with multiphase materials is provided in this chapter. Only those topics in chemistry which seem quite essential to the construction materials engineer are included.

**Useful constants:*		
	1 dyne:	1 g-cm/sec^2
	1 erg:	1 dyne-cm
	electronic charge, e:	4.8×10^{-10} statcoulomb
	1 statcoulomb:	1 dyne-cm^2
	1 electron volt, eV:	1.6×10^{-12} erg
	1 Angstrom unit, A:	10^{-8} cm
	speed of light, c:	3×10^{10} cm/sec
	Planck's constant, h:	6.625×10^{-27} erg-sec

For example, a brief introduction to solid-state bonding (Sec. A) provides a basis for *all* mechanical behavior of materials.

Section B includes a rudimentary basis for studying crystalline materials, which comprise over 85% by volume of all construction materials. These concepts will help explain the anisotropic physical properties of solids (Chap. 3) and the mechanical behavior of aggregates and fibers (Chap. 6) and of partially crystalline inorganic cements (Chap. 4). This section also includes the basis for using diffraction techniques to study crystalline materials (Chap. 13).

In Sec. C we shall relate the principal mechanical properties of polymers to their chemical structure. Applications in construction include all classes of sealants and adhesives, plastic pipes and conduits, surface finishes, foamed insulating materials, bituminous materials, and numerous others.

The few widely useful concepts and techniques of surface chemistry contained in Sec. D are important because *all* composite materials contain large interfacial areas between the various components. Interfacial properties usually differ drastically from those of the bulk phases and are typically a governing factor in determining the composite mechanical behavior. Concepts from Sec. D will help explain the mechanical behavior of composites described in Chaps. 7–12.

Material properties depend on *bonding* and *structure*, i.e., the types of chemical bonds which hold atoms together and the geometrical arrangement of atoms. Bonding is described in Sec. A, and the remaining sections are concerned with both bonding and structure.

A. SOLID-STATE BONDING

Several concepts of atomic structure and chemical bonding required for understanding the engineering properties of materials are presented in this section. The logical starting point is the basic building block, the atom.

1. Atomic Models

Of the various atomic structural models which have been proposed, only two will be reviewed: the Bohr-Sommerfeld orbital model and the Schrödinger-Heisenberg-Pauli wave mechanical model. The former provides a useful though inaccurate representation of atomic structure, and the latter provides the basis for explaining much about bonding theory.

a. BOHR-SOMMERFELD MODEL

Because hydrogen is the simplest atom, it is the subject of much study by physicists. Neils Bohr in 1913 visualized the hydrogen atom as an electron orbiting a nucleus. To avoid a dilemma raised by classical electrodynamics which requires accelerated charges to emit radiation, Bohr suggested that an electron would emit radiation only when it changed from one allowed orbit to another. The energies associated with the allowed orbits were called the quantum energy states of the electron. An electron can jump to a higher energy level (larger orbit) by absorbing radiant energy or to a lower energy level by emitting radiant energy.

The energy of a hydrogen atom at rest, E, is the sum of the potential energy E_p due to the electrical attraction between its proton and electron and the kinetic energy E_k of the orbiting electron:

$$E = E_p + E_k \tag{1}$$

The potential energy may be taken as the work required to move the electron from some datum, say at an infinite distance, to its orbit radius, r, about the proton

$$E_p = \int_{\infty}^{r} F\,dr' \tag{2}$$

where the force F, in dynes, between two charged particles is given by Coulomb's law:

$$F = \frac{-e_1 e_2}{r'^2}$$

where $-e_1$ and e_2 are the particle charges in electrostatic units and r' is the separation distance in centimeters. Positive F means that particles attract each other. If the two charges are of equal magnitude, $-e_1e_2 = -e^2$, substituting into Eq. (2),

$$E_p = \int_{\infty}^{r} \frac{e^2}{r'^2}\,dr' = \frac{-e^2}{r} \tag{3}$$

The kinetic energy for circular motion, from Newton's second law, is

$$E_k = \frac{mv^2}{2} = \frac{e^2}{2r} \tag{4}$$

where m and v are the electron mass and velocity, respectively. Substituting Eqs. (3) and (4) into (1),

$$E = \frac{-e^2}{r} + \frac{e^2}{2r} + \frac{-e^2}{2r} \tag{5}$$

Bohr's treatment of the hydrogen atom uses Eq. (5) except that the electron orbit radius can take only certain values, related by

$$r_n = n^2 r_1$$

where n = 1, 2, 3, 4, . . . , the quantum number,
r_1 = 0.528 A, the smallest radius that the electron can have,
r_n = allowed radius corresponding to the integer n.

The corresponding energies are then given by

$$E_n = \frac{-e^2}{2n^2 r_1}$$

Small discrepancies between Bohr's theory and experimental measurements led Sommerfeld to investigate elliptical as well as circular orbits.

b. WAVE MECHANICAL MODEL

de Broglie postulated in 1924 that matter, particularly electrons, possesses wave properties, and he derived the wavelength of an electron as

$$\lambda_m = \frac{h}{p}$$

where h is Planck's constant and p is the linear momentum of the electron. This equation provides, for example, the wavelengths of x-rays used in diffraction work (Chap. 13). Schrödinger and Heisenberg in 1926 developed a wave-mechanics model for hydrogen in which only the probability of finding an electron within a given range of radii could be computed, in contrast to the well-defined orbits in Bohr's model. According to this theory the description of electron energy distributions required three quantum numbers, designated n, l, and m, which assume values

$$n = 1, 2, 3, 4, \ldots$$

$$l = 0, 1, 2, 3, \ldots, n - 1$$

$$m = -l, -(l - 1), \ldots, 0, \ldots, (l - 1)l$$

The *principal quantum numbers*, n, indicate orbital radii and correspond to the Bohr quantum numbers. The l numbers are *azimuthal quantum numbers*, which describe the orbital ellipticity recognized by Sommerfeld. A circular orbit corresponds to $l = 0$, and higher values of l indicate more eccentric ellipses. The m numbers are *magnetic quantum numbers*, and the energy states which they designate are observed only where the atom is in a magnetic field.

Another feature of quantum mechanics is the quantization of electron spin vectors. An electron spins on its axis as it orbits the nucleus somewhat as the earth rotates on its axis while orbiting the sun. A spinning electron has an electromagnetic *spin angular momentum* represented by a vector pointing along the spin axis. Pauli [1] concluded that no more than two electrons in an atom can simultaneously occupy the same energy level and that these two must have opposite spins. This condition, called the *Pauli exclusion principle*, is indicated by giving an electron a spin quantum value

$$s = \pm\tfrac{1}{2}$$

The Pauli exclusion principle implies that no two electrons of an atom can simultaneously have the same set of four quantum numbers (n,l,m,s). In a later section we shall see that this quantization of electron energies together with the exclusion principle is responsible for the different bonding tendencies among elements.

c. QUANTUM THEORY AND THE PERIODIC TABLE

The four quantum numbers and the Pauli exclusion principle can be used to derive the periodic table (Appendix A): This has allowed scientists to predict unobserved elements and then seek them experimentally. Electrons in an atom tend to arrange themselves in the lowest energy levels (smallest orbits). The periodic table is derived by simply finding all the allowed energy levels for electrons in atoms and then filling these levels with electrons, starting with the lowest level. Figure 1(a) depicts the possible energy levels for the first four values of n. Since l can take values $l = 0, 1, 2, \ldots, n - 1$, there are n possible values of l corresponding to each n value. The usual notation [Fig. 1(b)] is to indicate n values by numbers and l values by letters, i.e.,

Value of l:	0	1	2	3	4
Designation:	s	p	d	f	g

so that the energy level corresponding to $n = 2$ and $l = 3$ would be called the $2f$ level, etc. Since m can take values $m = -l, \ldots, +l$, there are $2l + 1$ possible values of m. Figure 1(c) includes the third quantum number. For example, the heavy line has $n = 4, l = 1, m = -1$. Figure 1(d) includes energy levels with all four quantum numbers, with $s = +\frac{1}{2}$ and $s = -\frac{1}{2}$ indicated by upward and downward arrows, respectively. The periodic table contains an element for each of the energy levels in Fig. 1(d), which shows only n values through 4. The $1s^1$ level represents hydrogen. Adding a second electron with opposite spin gives helium, $1s^2$—etc., throughout the table.

The electron configuration is frequently described in terms of shells and subshells, corresponding to n and l, respectively. Table 1 shows the configuration for the first four shells.

TABLE 1. Electron Shell Configuration

Main shell		*Number of electrons, subshell (l)*				
n	*Shell notation*	*s*	*p*	*d*	*f*	*Total*
1	*K*	2				2
2	*L*	2	6			8
3	*M*	2	6	10		18
4	*N*	2	6	10	14	32

d. IONIZATION ENERGY

Ionization energy is the energy required to subtract a valence electron from an atom to form a positive ion. This energy is important to both ionic and metallic bonding. Ionization energy can be measured and in some instances calculated.

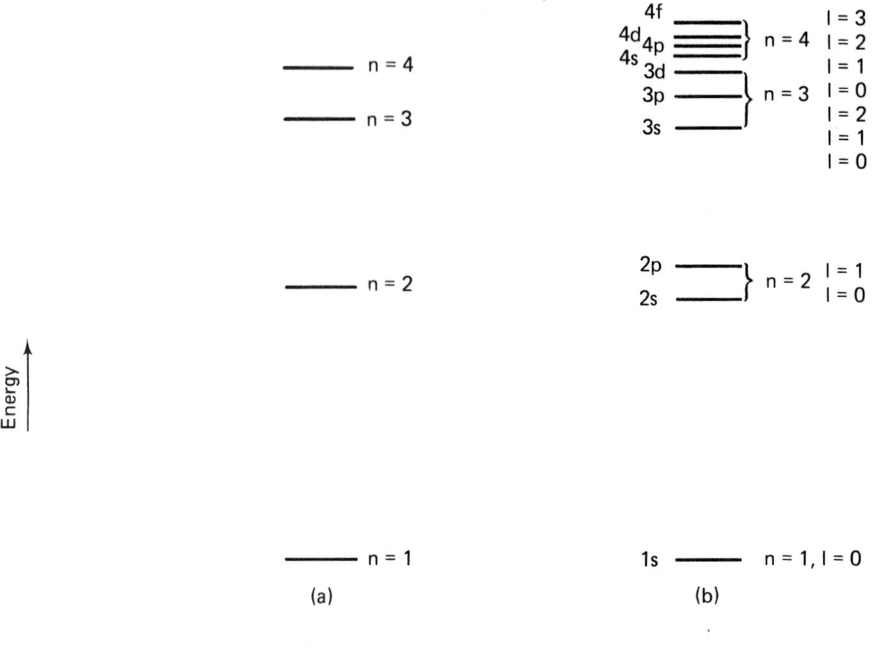

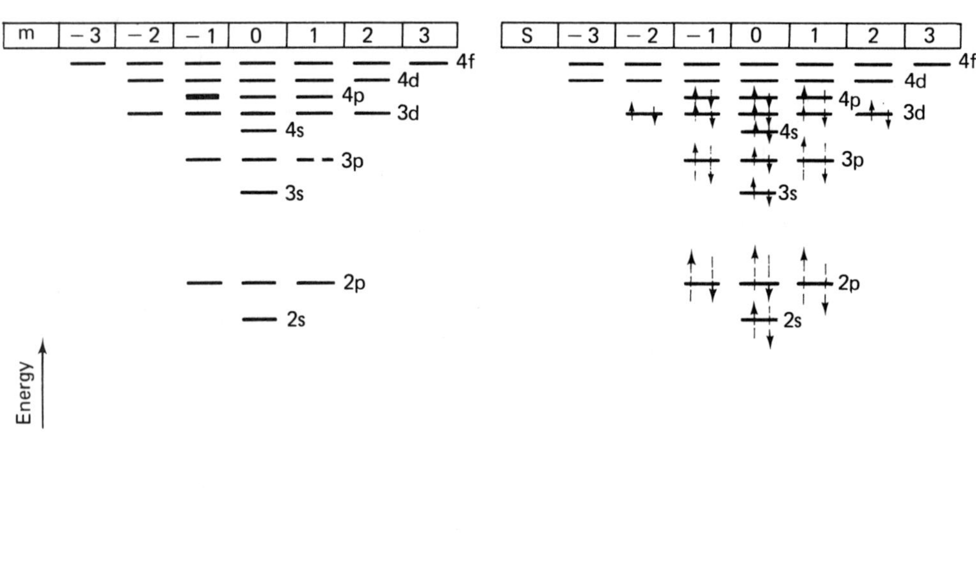

Fig. 1. Allowed energy levels in atoms. (a) n only; (b) n,l; (c) n,l,m; (d) n,l,m,s. From Eisenstadt, Melvin M. *Introduction to Mechanical Properties of Materials*, New York: Macmillan Company, © 1971 [2].

EXAMPLE 1: Calculate the ionization energy of hydrogen (the energy required to remove the electron from $r = 0.528$ A to an infinite distance).

Solution:

$$E_{\text{ionization}} = E_{\text{final}} - E_{\text{initial}}$$

Using Eq. (5),

$$E_{\text{ionization}} = \frac{-e^2}{2r_{\text{final}}} + \frac{e^2}{2r_{\text{initial}}} = \frac{e^2}{\infty} + \frac{(4.8 \times 10)^2}{2(0.528 \times 10^{-8})}$$
$$= 2.18 \times 10^{-11} \text{ erg} - 13.6 \text{ eV}$$

where 4.8×10^{-10} is the electron charge in electrostatic units. This agrees well with the experimental value 13.53 eV, where 1 eV $= 1.602 \times 10^{-12}$ erg is the energy acquired by an electron as it passes through a potential field of 1 V. The ionization energy is positive, because energy must be added to form an ion. As the equation suggests, large atoms should generally have lower ionization energy than small ions, because the valence electrons are farther from the nucleus. This is confirmed by reading ionization energy values down the columns in the periodic table (Appendix A). Elements in the lower left-hand corner of the table are easiest to ionize. More energy is always required to remove a second electron to form a doubly charged ion.

2. *Primary Bonds*

Four types of interatomic bonds are generally recognized. The first three—ionic, covalent, and metallic—are called *primary bonds* because they are relatively strong. Each depends on a sharing or exchange of valence electrons which lie in the *s* and *p* orbitals. The fourth type, *secondary bonds*, includes several varieties of weaker but still important attractive forces.

a. IONIC BONDS

Ionic bonds result from electrostatic attraction between positive and negative ions. The ease with which an atom forms a postive or negative ion depends on its position in the periodic table; e.g., sodium easily loses an electron to form a positive ion, and chlorine readily accepts an electron to form a negative ion. Ions of unlike charges attract each other with a force given by Coulomb's law:

$$f_a = e^2 \frac{Z_1 Z_2}{r_{12}^2} \tag{6}$$

where e is the electron charge (4.80×10^{-10} electrostatic unit), Z_1 and Z_2 are the valences of the two ions, and r_{12} is the interionic distance.

By integrating Eq. (6) the potential energy of the ions (work required to separate the ions to an infinite distance) is

$$V_a = \int_{\infty}^{r_{12}} f_a \, dr = e^2 Z_1 Z_2 \int_{\infty}^{r_{12}} \frac{1}{r^2} \, dr = \frac{-e^2 Z_1 Z_2}{r_{12}} \tag{7}$$

When the ions approach each other closely, the repulsive force due to interaction of their electron clouds is

$$f_r = \frac{-ne^2B_{12}}{r_{12}^{n+1}} \tag{8}$$

where B_{12} is a repulsion coefficient and n is the Born exponent [3]. Table 2 shows some typical Born exponent values.

TABLE 2. Experimental Values of Born Exponent, n

Type of ion	n
He	5
Ne	7
Ar, Cu^+	9
Kr, Ag^+	10
Xe, Au^+	12

SOURCE: PAULING, L., *The Nature of the Chemical Bond*, 3rd ed. Ithaca, N.Y.: Cornell University Press, 1960 [4].

The repulsion potential energy is

$$V_r = \int_{\infty}^{r_{12}} f_r\,dr = -ne^2B_{12}\int_{\infty}^{r_{12}} \frac{1}{r^{n+1}}\,dr = \frac{e^2B_{12}}{r_{12}^n} \tag{9}$$

The ion pair achieves equilibrium at the separation for which $f_a + f_r = 0$. The total energy $V = V_r - V_a$ is then minimum and is called the *bond energy*. This is the energy required to rupture the bond. Figure 2 shows the variations of force and energy with ion separation. The net bond force curve is quite informative. For example, Young's modulus of the material is related to the slope of the curve at r_0, and theoretical strength depends on the peak value of this curve (see Chap. 3).

Whether a compound is primarily ionic or primarily covalent depends on which type of bonding will lower total energy more. The more electropositive the metal and more electronegative the nonmetal, the greater is the ionic character of the bond.

EXAMPLE 2: A pair of oppositely charged divalent ions has an equilibrium spacing of 2.50 A. Assuming that n in Eq. (9) is 9, what energy is required to separate the ions?

Solution: At equilibrium spacing the sum of attractive and repulsive energies is minimum. Combining Eqs. (7) and (9) and setting the derivative to zero,

$$\frac{dV}{dr} = -Z_1Z_2\frac{e^2}{r^2} - \frac{ne^2B_{12}}{r^{n+1}} = 0$$

Therefore,

$$\frac{4e^2}{r^2} = \frac{9e^2B_{12}}{r^{n+1}} \qquad \text{or } B_{12} = \frac{4r^{n-1}}{9}$$

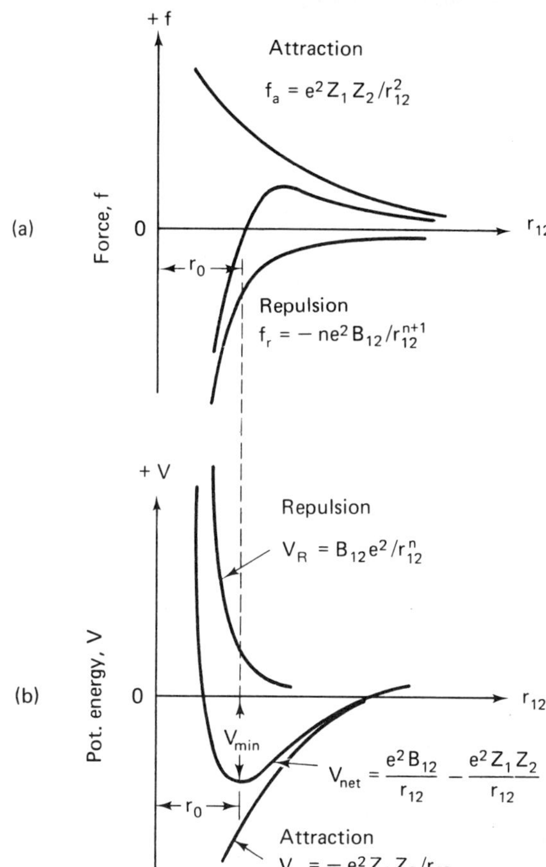

Fig. 2. (a) Force, and (b) potential energy of an ion pair as a function of distance of separation. At equilibrium separation, r_0, attraction and repulsion forces are equal and potential energy is minimum.

$$
\begin{aligned}
V_0 - V_{\min} &= 0 - \left(\frac{4e^2}{r} - \frac{4e^2 r^{n-1}}{9r^n}\right) \\
&= \frac{4e^2}{r}\left(1 - \frac{1}{9}\right) = \frac{32}{9}\frac{(4.80 \times 10^{-10})^2 \text{ esu}^2}{2.5 \times 10^{-8} \text{ cm}} \\
&= 32.8 \times 10^{-12} \text{ erg} \\
&= 7.87 \text{ eV}
\end{aligned}
$$

b. COVALENT BONDS

Covalent bonds result from the sharing of electrons by adjacent atoms, rather than by a transfer of electrons, as in ionic bonding. The hydrogen molecule, H_2, is an example. Consistent with the Pauli exclusion principle (Sec. A1), two electrons may have the same energy values if they have opposite spins. Two such electrons have a high instantaneous probability of occurring between the atoms. The covalent bond in a hydrogen molecule may be viewed, in simplified fashion, as an attraction of positive ions to the intervening pair of electrons with opposite spins.

Bond angles. Unlike ionic bonds, covalent bonds can form only in relatively fixed directions. This directional nature profoundly affects all the properties of covalent solids. For example, organic substances have much smaller densities than do minerals because directional bonding of the carbon atom does not permit the close

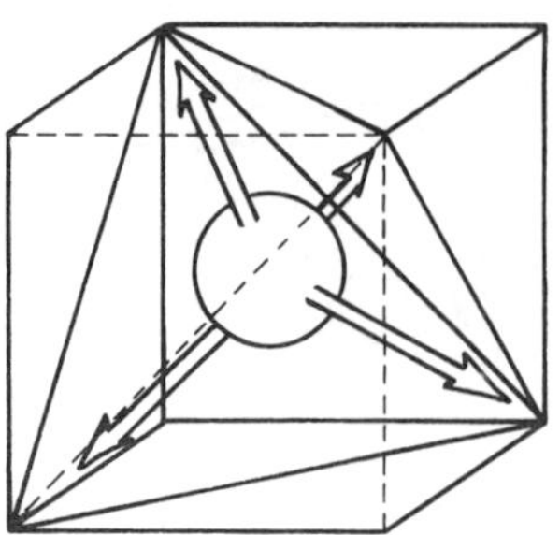

Fig. 3. Tetrahedral orientation of carbon bonds resulting from s-p^3 hybridization.

packing arrangements possible with ionic or partially ionic bonding. Therefore, coordination numbers for covalent bonding are not controlled by radii ratios, as they are for ionic bonding (see Sec. C).

The bond directionality produces characteristic bond angles. The most stable covalent bonds are formed between orbitals which can overlap the most. Therefore an element will use p electrons for bonding rather than s electrons in the same shell if a choice exists. The resulting bond angles are approximately 90° because this is the angle which p orbitals make with each other. These orbitals seek orbitals from other atoms to pair with, thus providing shared electrons for carbon. Table 3 lists some observed bond angles thought to be produced by p electron bonding. Bond angles larger than 90° in H_2O and NH_3 may be due to the partially ionic character of these bonds, causing mutual repulsion between hydrogen atoms [5].

Deviations from 90° bond angles can also be caused by hybridization of the s and p wave functions. For example, in carbon four orbitals are available to the four electrons in the valence shell: one $2s$ and three $2p$ orbitals. But instead of three bonds at 90° to one another plus a weaker spherically symmetric bond, carbon often has four bonds of equal strength directed toward the corners of a regular tetrahedron (Fig. 3). These bonds result from s-p^3 hybridization, which permits a lower energy configuration of orbitals, allowing more overlap in their spatial orientation. Other hybrid combinations exist for elements whose bonding orbitals differ from those of Group IV elements. Table 3 includes several bond angles for elements in other groups.

TABLE 3. Bond Angles

Water	H_2O	H—O—H	104°
Ammonia	NH_3	H—N—H	107°
Diamond	C	C—C—C	109°
Carbonate ion	CO_3^{-2}	O—C—O	120°
Methane	CH_4	H—C—H	109°
Ethane	C_2H_6	H—C—H	109°
Chloromethane	CH_3Cl	H—C—H	110°
Hydrogen sulfide	H_2S	H—S—H	93°
Sulfur	S	S—S—S	107°

The covalent bonds of carbon are particularly important because they largely determine the physical properties of polymers (plastics, elastomers, etc.), which constitute an important group in engineering construction materials (see Chaps. 1 and 5). Carbon bonding may be pictured as two tetrahedra joined corner to corner. The tetrahedra may also be joined edge to edge by double bonds between adjacent carbons or face to face by triple bonds, giving rise to the large variety of structures in hydrocarbons. Examples are ethylene ($C_2{=}H_4$) and acetylene ($C_2{\equiv}H_2$).

The effect of bonding structure on the physical properties of materials is well illustrated by the two forms of pure carbon, diamond and graphite. In diamond, each carbon is surrounded by four nearest neighbors at the tetrahedral angles, but not all the covalent bonds are filled. The entire diamond is a single large molecule. The strength and hardness of diamond reflect the strength of the covalent bond. In graphite, the more common form of carbon, three equivalent orbitals exist at 120° orientations in a plane, requiring the carbon atoms to bond together in sheets. The sheets are held together only by weak secondary forces, described in the next section. These weak forces permit the sheets to slide over one another easily, accounting for graphite's lubricating ability.

Bond energies. Representative bond energies and lengths of covalent bonds are shown in Table 4. Notice that the increased bond strength of multiple bonds reduces the C—C spacing from 1.54 A for single bonds to 1.33 A and 1.20 A for double and triple bonds, respectively. Covalent bond lengths are frequently additive, the bond length of A—B being approximately the average of the A—A and B—B bond lengths. For example the C—Cl bond length (1.77 A) approximates the average of 1.54 A and 1.99 A, the bond lengths of C—C and Cl—Cl, respectively.

TABLE 4. Bond Energies and Lengths*

Bond	*Bond energy (kcal/mole*)*	*Bond length (A)*	*Bond*	*Bond energy (kcal/mole*)*	*Bond length (A)*
C—C	88	1.5	O—O	52	1.5
C=C	162	1.3	O—H	119	1.0
C≡C	213	1.2	O—Si	90	1.8
C—O	86	1.4			
C=O	128	1.2	H—H	104	0.74
C—H	104	1.1			
C—N	73	1.5	N—O	60	1.2
C—F	108	1.5	N—H	103	1.0
C—Cl	81	1.8			

*Values vary with the type of adjacent bonds; e.g., the C—H bond is 104 kcal/mole in methane, 98 kcal/mole in ethane, and 90 kcal/mole in trichloromethane.

SOURCE: VAN VLACK, L. H., *Materials Science for Engineers*, Reading, Mass.: Addison-Wesley Publishing Co., 1970 [6].

Bond energies are usually calculated from thermochemical or spectroscopic data (Chap. 13). From known bond energies it is possible to calculate the approximate heat of a given reaction. For example, the reaction heat when ethylene reacts with hydrogen to form ethane ($H_2C{=}CH_2 + H_2 \rightarrow H_3C{-}CH_3$) is easily obtained by writing the energy change due to bonds which are either broken or formed in the reaction:

$$\underset{162}{C{=}C} + \underset{104}{H{-}H} \longrightarrow \underset{2 \times 104}{2\,C{-}H} + \underset{88}{C{-}C}$$

The difference is $296 - 266 = +30$ kcal/mole; i.e., the bonds on the right-hand side constitute a more stable combination by 30 kcal/mole.

Resonance. Whenever more than one electronic structure can be written for a molecule or ion, a more stable (higher-energy) bond exists than could be attained by any of the electronic arrangements taken individually. In fact, the greater the possible number of electronic arrangements, the more stable is the bonding. This condition is called resonance, and the structure is said to resonate among the different electronic forms. For the carbonate ion, for example, three electronic forms can be written:

```
   O            O            O
O  C  O      O  C  O      O  C  O
```

Whenever a bond can resonate between single and double bonds, or between double and triple bonds, the observed bond length is closer to the smaller value. Table 5 indicates several resonating structures and their bond lengths.

TABLE 5. Carbon-Carbon Distances in Resonating Molecules

Molecule	*Bond*	*Bond length (A)*	*Example of alternative structure*
$HC\equiv C—CH_3$	$C—CH_3$	1.46	$^-CH=C=CH_2H^+$
$HC\equiv C—C\equiv CH$	C—C	1.36	HC=CH=CH=CH
$H_2C=CH—CH=CH_2$	HC—CH	1.46	$H_2C—CH=CH—CH_2$
$HC\equiv C—CH_2Cl$	C—C	1.47	$^+HC=C=CH_2Cl$
$H_3C—C\equiv C—CH=CH_2$	$H_3C—C$	1.47	$^+HCH_2=C=C^-—CH=CH_2$
	$\equiv C—CH=$	1.42	$H_3C—C=C=CH—CH_2$

c. METALLIC BONDS

Most elements are metals; e.g., they possess loosely held valence electrons. Typically, the valence electrons are so disperse that their mean orbit radius in a free atom is greater than the interatomic distance in solid metal. Therefore the valence electrons in the solid are always closer to one or more neighboring nuclei than they are to the parent nucleus, which lowers the total energy.

Unlike ionic bonds, which may exist between several oppositely charged ions, and covalent bonds, which exist between only two atoms, metal bonds exist only in a large aggregate of identical atoms or between chemically similar atoms in alloys. Metallic bonds, like ionic bonds, are essentially nondirectional.

Unlike nonmetals, metals are good electrical and heat conductors, plastically deformable, and optically opaque and lustrous. These properties all depend on the metallic bond. Since the valence electrons lose their associations with particular nuclei, they are free to migrate throughout the solid as an *electron gas.* This free electron gas permits high rates of energy transfer under relatively low thermal and electrical gradients. It accounts for the plasticity of metals because it permits permanent rearrangement of lattice structures without internal fracturing. These "free" electrons cause metals to be opaque by absorbing energy from light photons and produce luster by reemitting this energy as they return to lower energy levels.

The fewer the valence electrons an atom has and the more loosely they are held, the more metallic the bonding. As the number of valence electrons and the energy with which they are held to the nucleus increase, the covalency of the bonding increases. Transition metals (those with incomplete *d* shells, such as iron and nickel) exhibit appreciable covalent bonding with hybridized inner shell electron orbitals, which partially explains their high melting points. The covalency of metallic bonds is strongest in the fourth column of the periodic table. Carbon in the form of diamond exhibits almost pure covalent bonding, silicon is more metallic, and lead is metallic.

d. MIXED PRIMARY BONDS

Although bonds have been categorized into three types, "pure" bonds are seldom encountered. Silicon, like carbon in diamond, covalently shares electrons except that a few electrons can escape from the covalent positions between adjacent atoms. This trend continues through germanium and tin, providing progressively more metallic behavior, and the basis for semiconductivity. Ionic-metallic and ionic-covalent bond combinations also are observed in numerous compounds. Primary bond combinations are illustrated schematically in Fig. 4.

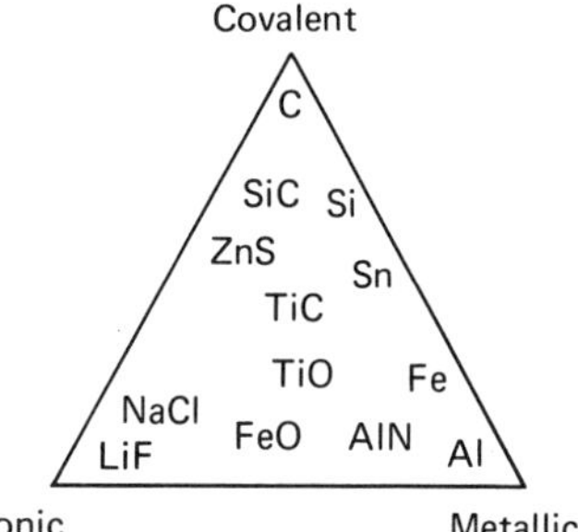

Fig. 4. Primary bond combinations. From Van Vlack, L. H.: *Materials Science for Engineers*, Reading, Mass.: Addison-Wesley Publishing Company, 1970 [6].

A bond may be purely covalent if the two atoms are alike in *electronegativity*, e.g., BrCl, while it possesses ionic character if the atoms are dissimilar in electronegativity, e.g., HF. Electronegativity is a measure of electron attracting power, which differs from ionization potential (the energy required to remove an electron from an atom), although the two are related. Pauling [4] defined an electronegativity scale by the relation

$$X_A - X_B = \sqrt{\frac{\Delta_{AB}}{30}}, \qquad X_{\text{fluorine}} = 4.0 \tag{10}$$

where X_A and X_B are the electronegativities of elements A and B and Δ_{AB} is the A—B bond energy in kilocalories per mole. Table 6 lists values of X from Eq. (10) for most elements, using $X = 4.0$ as a datum. The alkali metals are the most electropositive, fluorine is the most electronegative, and boron, at 2.0 on the scale, is the point where neither metallic nor nonmetallic properties predominate. From values in the table, one can conclude that there is little ionic character in a C—C bond, some in a C—H bond, and more in C—O and C—Cl bonds. Also, carbon is negative compared to hydrogen but positive compared to nitrogen or oxygen.

A simple formula was also proposed by Pauling to calculate the percentage ionic character of a bond:

$$\% \text{ ionic character} = 100\{1 - \exp[-\tfrac{1}{4}(X_A - X_B)^2]\} \tag{11}$$

Figure 5 compares percentage ionic character from Eq. (11) with experimental data.

TABLE 6. Pauling's Electronegativity Values of Elements

H	2.1	Si	1.8	Mn	1.5	Br	2.8	Rh	2.2	Ba	0.9	Hg	1.9
Li	1.0	P	2.1	Fe	1.8	Rb	0.8	Pd	2.2	La-Lu	1.1–1.2	Tl	1.8
Be	1.5	S	2.5	Co	1.8	Sr	1.0	Ag	1.9	Hf	1.3	Pb	1.8
B	2.0	Cl	3.0	Ni	1.8	Y	1.2	Cd	1.7	Ta	1.5	Bi	1.9
C	2.5	K	0.8	Cu	1.9	Zr	1.4	In	1.7	W	1.7	Po	2.0
N	3.0	Ca	1.0	Zn	1.6	Nb	1.6	Sn	1.8	Re	1.9	At	2.2
O	3.5	Sc	1.3	Ga	1.6	Mo	1.8	Sb	1.9	Os	2.2	Fr	0.7
F	4.0	Ti	1.5	Ge	1.8	Tc	1.9	Te	2.1	Ir	2.2	Ra	0.9
Na	0.9	V	1.6	As	2.0	Ru	2.2	I	2.5	Pt	2.2	Ac	1.1
Mg	1.2	Cr	1.6	Se	2.4			Cs	0.7	Au	2.4	Th	1.3
Al	1.5											Pa	1.5
												U	1.7
												Np-No	1.3

SOURCE: PAULING, L., *The Nature of the Chemical Bond*, 3rd ed. Ithaca, N.Y.: Cornell University Press, 1960 [4].

3. Secondary Bonds

The types of primary bonds discussed depend on the differing behaviors of valence electrons. Secondary bonds are also electrostatic in nature but depend on the distribution of the centroids of *all* electrons in neighboring atoms, rather than on the behavior of just valence electrons. Secondary bonds are weaker, having energies ranging from less than 1 to about 10 kcal/mole, compared with 20 to 50 kcal/mole for

Fig. 5. Comparison of experimental values of percent ionic character with values computed from Eq. (11). Reprinted from Linus Pauling: *The Nature of the Chemical Bond*, Third Edition, Ithaca, N.Y.: Cornell University Press, 1960. Used by permission of Cornell University Press.

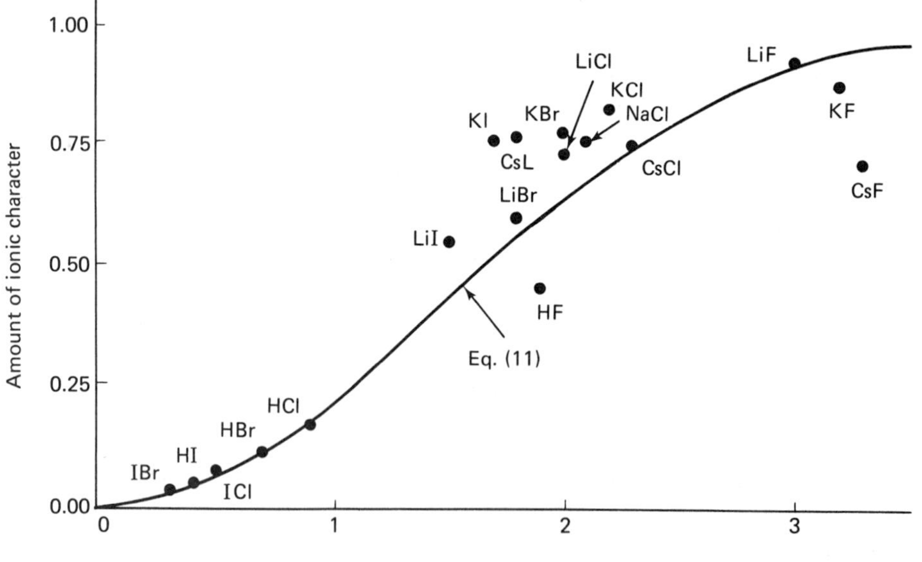

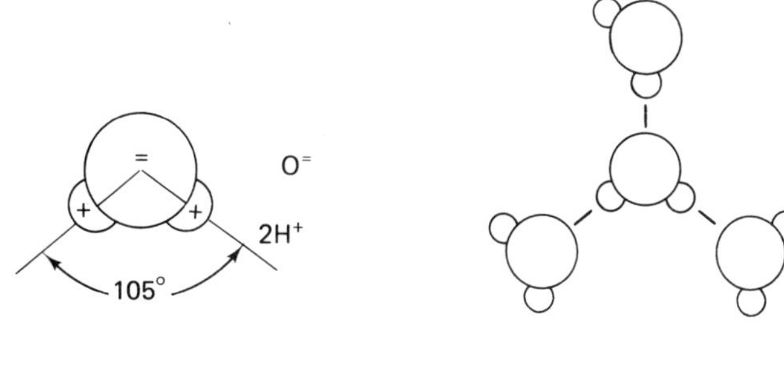

Fig. 6. (a) Water dipole; (b) hydrogen bonding of water molecules.

primary bonds. Nonetheless, the structures of more solids are determined by secondary bonds than by primary bonds. They are especially important in determining the properties of many polymers and may even give rise to crystalline structures if the molecules are not too large.

The major types of secondary bonds include *permanent dipole*, *induced dipole*, and *dispersion bonding*.

Molecules form permanent dipoles if their centers of positive and negative charges do not coincide. Water is the classical example for a permanent dipole [Fig. 6(a)]. Since the shared electrons spend their time between the hydrogen and oxygen atoms, the hydrogen atoms show a net positive charge, and the oxygen is negative. The positive end of one molecule attracts the negative end of another, forming a dipole bond [Fig. 6(b)]. Dipoles containing hydrogen are called *hydrogen bonds* and are the strongest of the secondary bonds because of hydrogen ion's small size. The hydrogen bond accounts for the fact that water and HF have higher boiling points than other molecules of comparable weights. The attraction is defined quantitatively as the dipole moment,

$$\mu = qs \tag{12}$$

where q is the charge on the particles in electrostatic units, s the distance between charge centers in centimeters, and μ the dipole moment in esu-cm. Because of the small numerical values, the Debye (1 Debye $= 10^{-18}$ esu-cm) is a more convenient unit. Table 7 shows dipole moments of various compounds in Debyes.

TABLE 7. Dipole Moments of Several Compounds in Debyes

H_2	0.00	H_2O	1.84
N_2	0.00	H_2O_2	2.10
NH_3	1.46	H_2S	1.10
		SO_2	1.60
HI	0.38	CO_2	0.00
HBr	0.78	CS_2	0.00
HCl	1.03	CCl_4	0.00
		CH_3Cl	2.00
CH_3OH	1.68	CH_2Cl_2	1.60
C_2H_5OH	1.70	$CHCl_3$	1.10
C_6H_5OH	1.70	C_6H_5Cl	1.73

It is apparent that molecules with no dipole moments have electrical symmetry. Therefore, CO_2, CS_2, etc., must be linear molecules. In the presence of an external electric field, however, the centroidal distribution of electrons may shift to one side, forming an induced dipole. In general, the magnitude of this centroidal shift increases with the number of electrons, so that heavy molecules can be more strongly bonded by induced dipole moments than light molecules. The electric field may be provided, for example, by the presence of permanent dipole molecules.

Symmetrical molecules can, however, form electrostatic bonds even without an external field. These are the *fluctuating*, *dispersion*, or *van der Waals* bonds. These bonds exist because at any one time there are a few more electrons on one side of the nucleus than on the other. The centers of negative and positive charge do not coincide at this instant, and a weak dipole is formed. An attraction can then exist between dipoles in adjacent atoms, which is maintained by the centers of charge resonating in unison. The resulting bond is nondirectional and typically an order of magnitude weaker than the hydrogen bond.

The equation describing bonding due to the various secondary effects is lengthy but reduces to the form

$$f = \frac{am}{r^{m+1}} - \frac{bn}{r^{n+1}} \qquad (n > m) \tag{13}$$

where a, b, m, and n are constants and r is the distance between atoms. The first term is an attractive term and the second term a repulsive one. The force equation can be

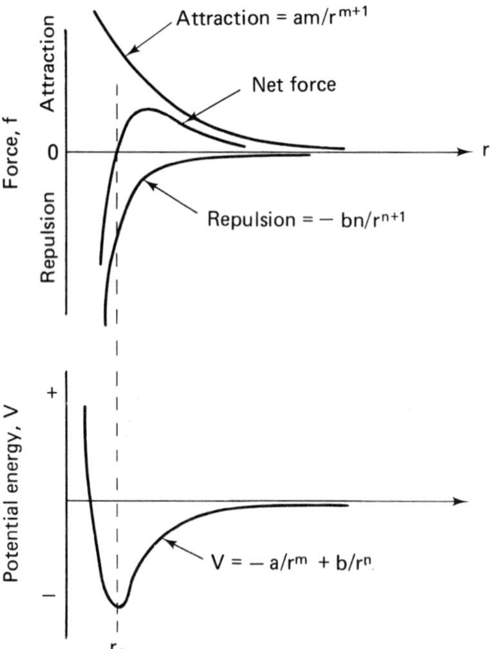

Fig. 7. Force and energy curves for Eqs. (13) and (14).

integrated to obtain an energy equation of the form

$$V = -\frac{a}{r^m} + \frac{b}{r^n} \qquad (n > m) \tag{14}$$

The two equations are analogous to Eqs. (6)–(9) for ionic bonding and have similar characteristic curves (Fig. 7). The equations are reasonably accurate for molecular vapors, less accurate for liquids, and still less applicable for solids.

EXAMPLE 3: Compare the percentage ionic character of the C—O bond as determined from Eq. (11) with the value obtained by making use of Eq. (12), given that the dipole moment of the C—O group is 0.7 Debye.

Solution: Using values of $X_C = 2.5$ and $X_O = 3.5$ from Table 6 in Eq. (11),

$$\begin{aligned}\%\ \text{ionic character} &= 100\{1 - \exp[-\tfrac{1}{4}(3.5 - 2.5)^2]\} \\ &= 100[1 - \exp(-\tfrac{1}{4})] = 22.1\%\end{aligned}$$

Using the C—O bond length value of 1.4 A from Table 4 in Eq. (12),

$$\begin{aligned}\mu &= (4.80 \times 10^{-10}\ \text{esu})(1.4 \times 10^{-8}\ \text{cm})(10^{-18}\ \text{Debye/esu-cm}) \\ &= 6.72\ \text{Debyes}\end{aligned}$$

Then

$$\%\ \text{ionic character} = 100\left(\frac{0.7}{6.72}\right) = 10.4\%$$

The difference arises in the choice of definition of percentage ionic character.

4. Summary

The preceding sections have described principal types of primary (ionic, covalent, metallic) and secondary (permanent and induced dipoles, dispersion force) bonds. Just as combinations of the three types of primary bonds were illustrated by a triangular chart (Fig. 4), mixed bonding including secondary bonds (represented as a single class) can be represented three-dimensionally by a tetrahedron (Fig. 8). Metals phase from pure metallic bonds, such as in alkali metals, to less perfect metals, such as Te and As, and finally to the pure covalent bonding of C diamond. The transition from covalent to the secondary (molecular) bonding apex ranges through graphite and high polymers such as cellulose to the rare gases at pure secondary bonding. The mixed bonding between pure ionic and covalent bonds ranges from the alkali halides such as NaCl through silicates and silica to covalent bonds. Layer structures fall between the ionic and secondary apexes, and certain defect structures, not discussed, lie between the metallic and ionic apexes. Generally there are no mixed bonds between metallic and secondary bonding, but a wide range of compounds can be represented by points lying throughout the interior of the tetrahedron. This figure emphasizes the continuity of the solid state and the lack of definitive demarcations between the various bond types.

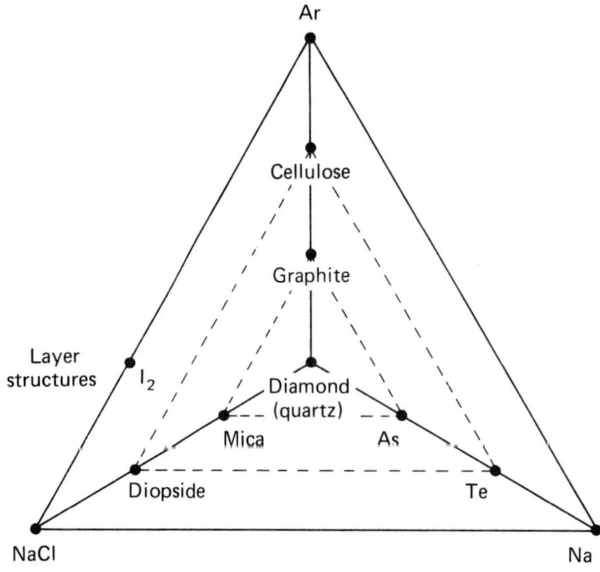

Fig. 8. Grimm tetrahedron showing four principal types of bonding at the corners and combinations of two bond types along the edges. From H. G. Grimm, *Naturwissenschaften*, 17, 535 (1929); *Agnew. Chem.*, 47, 53 (1934), by Sinnott [7].

PROBLEMS

1. Write the electron configurations of the following elements, using their atomic numbers and Table 1 as a guide: Na ($Z = 11$), Al ($Z = 13$), Si ($Z = 14$), Cu ($Z = 29$).
2. A divalent ion pair has values of $n = 7$ and $B_{12} = 2.22 \times 10^{-46}$ cm^6. What is the bond energy in ergs and the equilibrium spacing?
3. What are the values of the attractive and repulsive forces on the preceding ion pair at the equilibrium spacing?
4. Ethylene, C_2H_4, and hydrogen peroxide, H_2O_2, react to form ethylene glycol, $C_2H_4(OH)_2$:

$$\begin{array}{c} \mathrm{H}\ \ \mathrm{H} \\ \mathrm{C{=}C} \\ \mathrm{H}\ \ \mathrm{H} \end{array} + \mathrm{HO{-}OH} \longrightarrow \mathrm{HO}{-}\begin{array}{c} \mathrm{H}\ \ \mathrm{H} \\ \mathrm{C{-}C} \\ \mathrm{H}\ \ \mathrm{H} \end{array}{-}\mathrm{OH}$$

   What is the change in bond energy?

5. What energy is liberated when 1 mole of ethylene ($CH_2{=}CH_2$) polymerizes to form $(C_2H_4)_n$ with no double bonds?
6. Why do lead atoms bond metallically while carbon atoms in diamond bond covalently, even though both have four valence electrons? What resulting differences should be expected in strength, ductility, and conductivity?

7. Elements 21–28, 39–46, and 57–78 and elements from 89 on are the *transition elements*. From an inspection of Table 3 in Appendix A, what electron configuration characteristic is common to all transition elements? What effects might the electron configuration of transition metals have on their bonding characteristics? Compare their bond strengths with those of alkali metals.
8. Using Eq. (11), calculate the percentage ionic character of C—Cl and C—N bonds.
9. An O—H bond has a dipole moment of 1.52 Debyes. What is the dipole moment of a water molecule where the H—O—H bond is 104.5°?

B. CRYSTAL STRUCTURES

Solids are of two types: crystalline, in which atoms are stacked more or less in regular arrays, and amorphous, in which they are not. Since the mechanical properties of crystals and the diffraction techniques used for studying crystals depend on crystalline structure, we shall review the classes of crystal structure. Crystals whose mechanical, thermal, optical, or electrical properties differ along different crystal axes are anisotropic. Isotropic crystals have properties that are the same in all directions.

1. Coordination Theory

The most stable atomic arrangement in a crystal will be that which minimizes energy per unit volume, that is, one that

1. Satisfies electrical neutrality.
2. Satisfies the directionality and discreteness of any covalent bonds.
3. Minimizes strong ion-ion repulsion; i.e., cations of high valency tend to be far apart, producing open chain or framework structures instead of close three-dimensional packing.
4. Packs the atoms as closely as possible, consistent with properties 1–3.

In crystals having pure ionic bonding, where bond angles are not restricted to specific values, the number of anions surrounding each cation (coordination number) is a function of the radius ratio of the ions. A given coordination is predicted to be stable between that radius ratio at which the anions touch each other as well as the central cation (the *critical radius ratio*) and that radius ratio at which the next higher coordination becomes possible. The critical radius ratio is so named because below this ratio the cation-anion distance becomes greater than the equilibrium interionic distance (Chap. 3). For example, a coordination of 4 (tetrahedral coordination) is stable between radius ratios of 0.225 and 0.414. Below the lower limit, anions and cations would be farther apart than their equilibrium distance, and triangular coordination would provide a lower energy configuration. Above the upper limit, octahedral coordination would permit lower total energy because more anions would touch the cation. Calculating the critical radius ratio for each coordination is quite simple (Table 8). Notice that the only values of coordination possible in three-dimensional

TABLE 8. Computation of Critical Radius Ratios
(R = Anion Radius, r = Cation Radius)

Threefold (equilateral) coordination:

$$BO = R + r, \cos 30° = \frac{BE}{BO} = \frac{R}{R+r} = \sqrt{\frac{3}{2}}$$

$$\therefore r = \frac{(2 - \sqrt{3})R}{\sqrt{3}} = 0.155R$$

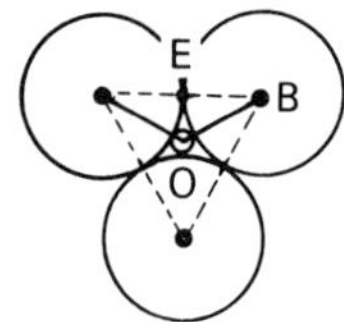

Fourfold (tetrahedral) coordination (central angle of tetrahedron – 109°28′).

$\angle CAO = 35°16'$ for AOC is an isosceles triangle

$$\cos \angle CAO = \frac{AE}{AO} = \frac{R}{R+r} = 0.8166$$

$$\therefore R - 0.817R = 0.817r, \quad r = 0.225R$$

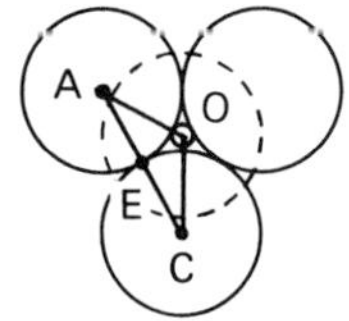

Sixfold (octahedral) coordination (each large sphere at corner of regular octahedron, 4 in plane of paper 1 above, 1 below):

$$BC = 2R, \quad AB = 2R + 2r$$

$$\cos \angle ABC = \cos 45° = \frac{BC}{AB} = \frac{2R}{2R + 2r} = \frac{1}{\sqrt{2}}$$

$$\therefore r = (\sqrt{2} - 1)R = 0.414R$$

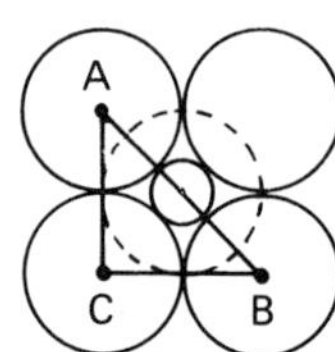

regular arrays are 1 (which is trivial), 2, 3, 4, 6, 8, and 12. Triangular coordination (Table 9) is rarely observed in ionic solids because it requires such a large difference in the sizes of the ions. An exception is the rare crystalline boron oxide, B_2O_3. Twelve-fold coordination occurs when both "ions" are the same size, as in the case of pure elements. Figure 9 shows the two possible packing arrangements in 12-fold coordination: hexagonal close packing (HCP) and face-centered cubic (FCC). In each arrangement, six spheres surround a central sphere in a midplane with three more spheres above and three below. The only difference between the two is that in the HCP arrangement the three spheres above the midplane are directly above the three spheres below the midplane, while in FCC they are in the alternate set of positions.

Because both HCP and FCC provide efficient packing and therefore relatively low energy, these two structures are encountered in approximately half of the elemental solids.

It was mentioned that a stable atomic arrangement requires electrical neutrality, and therefore the ionic valences are also significant in determining coordination. For

TABLE 9. Coordination as a Function of Radius Ratio

Coordination number	*Anion/cation radius ratio for which coordination is expected to be stable*	*Packing (disposition of anions)*
2	0–0.155	Linear
3	0.155–0.255	Triangular
4	0.255–0.414	Tetrahedral
6	0.414–0.732	Octahedral
8	0.732–1.0	Cubic
12	1.0	Hexagonal close packing
12	1.0	Face-centered cubic

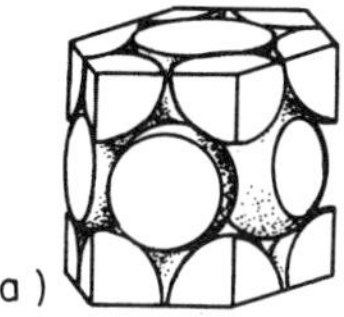

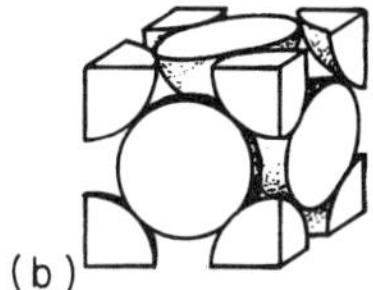

Fig. 9. 12-fold coordination packing arrangements. (a) Hexagonal close-packed structure; (b) face-centered cubic structure.

example, the sodium/chlorine radius ratio is 0.54, giving sodium chloride crystals a coordination number in the sixfold range (Table 9). There is one Cl^- for each Na^+. The calcium/chlorine radius ratio is similar (0.59). But since calcium chloride does not have equal numbers of calcium and chlorine ions because only half of the possible positive positions are filled, the rest are left vacant, causing some secondary shifting of the remaining atoms.

2. *Space Lattice and Unit Cell*

Crystal geometry is described by means of the space lattice and unit cell. A space lattice is a three-dimensional array of points, infinite in extent, in which every point has identical surroundings. A crystalline structure is found by placing an atom or group of atoms at each point in the space lattice. Space lattices can be generated by translating groups of lattice points called unit cells. Figure 10 shows several unit cells in a two-dimensional lattice. The unit cell is not unique, and the choice of unit cell used in describing a space lattice is normally taken as the one which best describes the lattice symmetry.

Fig. 10. Unit cell representations of a two-dimensional lattice.

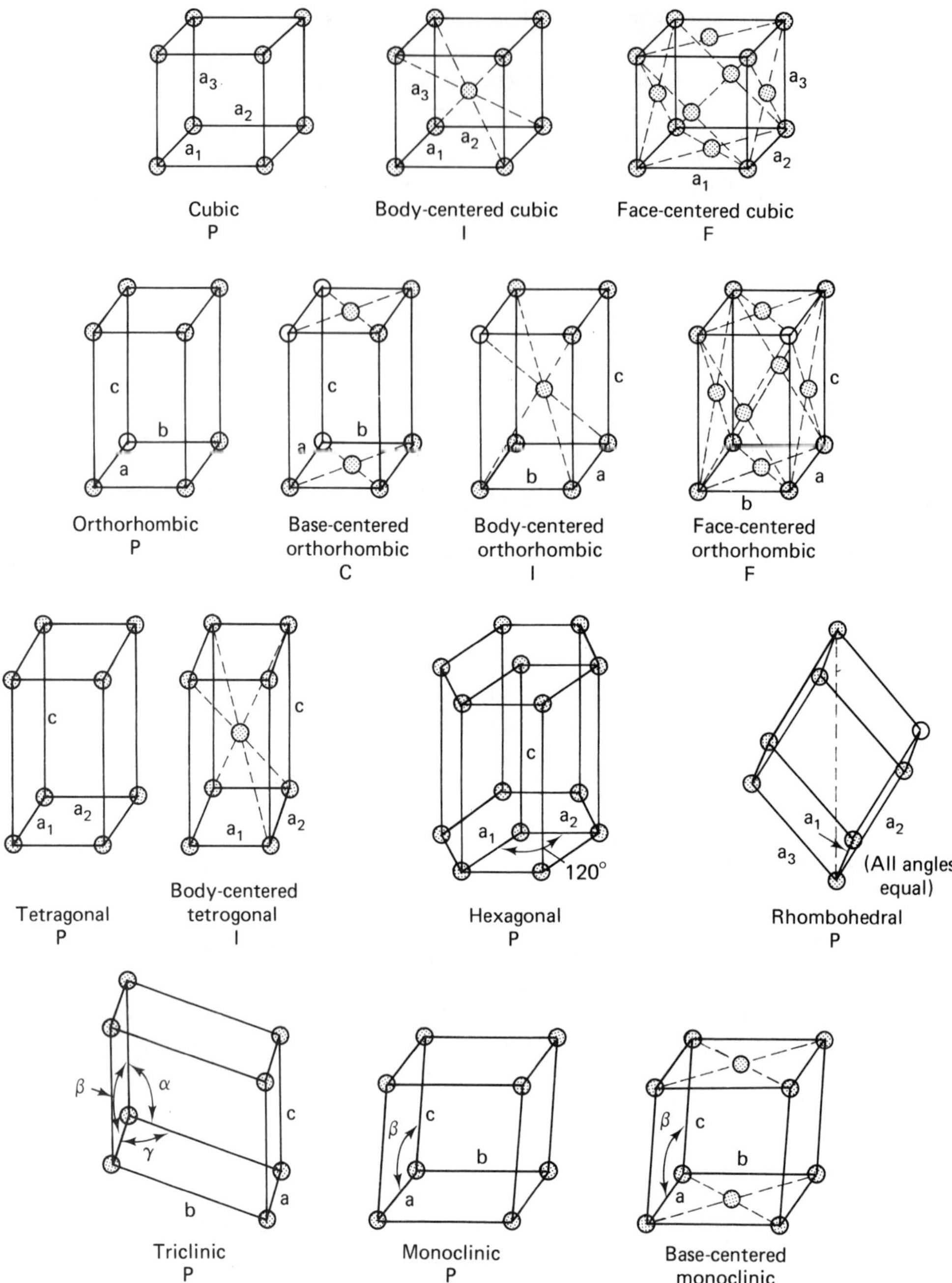

Fig. 11. The seven crystal systems and 14 Bravais lattices. P: primitive cell; C: cell with lattice point at center of two opposite faces; F: cell with lattice point at center of each face; I: cell with lattice point at center of interior.

Crystallographers have shown that space lattices in three dimensions have only 14 possible arrangements of points in space. These are called the Bravais lattices, for which the unit cells are shown in Fig. 11. They are classed in seven crystal systems, based upon the relative lengths of the three cell edges and the angles between the edges. Table 10 indicates the relationship between edge lengths and angles (the *lattice parameters*) for the seven systems.

TABLE 10. Lattice Parameters and Bravais Lattices of the Seven Crystal Systems

Crystal system	*Axial lengths and angles*	*Bravais lattice*
Cubic	$a = b = c, \quad \alpha = \beta = \gamma = 90°$	P, F, I
Tetragonal	$a = b \neq c, \quad \alpha = \beta = \gamma = 90°$	P, I
Orthorhombic	$a \neq b \neq c, \quad \alpha = \beta = \gamma = 90°$	P, C, F, I
Rhombohedral	$a = b = c, \quad \alpha = \beta = \gamma \neq 90°$	P
Hexagonal	$a = b \neq c, \quad \alpha = \beta = 90°, \gamma = 120°$	P
Monoclinic	$a \neq b \neq c, \quad \alpha = \gamma = 90° \neq \beta$	P, C
Triclinic	$a \neq b \neq c, \quad \alpha \neq \beta \neq \gamma \neq 90°$	P

Since lattice points at the faces, edges, and corners of a unit cell are shared by other unit cells, the number of lattice points, N, associated with one unit cell is

$$N = i + \frac{f}{2} + \frac{e}{4} + \frac{c}{8} \tag{15}$$

where i = number of lattice points in the interior of the cell,
f = number at the faces,
e = number at the edges,
c = number at the corners.

Cells containing only one lattice point are called primitive cells (P). Other possibilities include body-centered (I), lattice points at the centers of two parallel faces (c), and lattice points at the centers of all faces (F). Note that one primitive cell is shown for each of the crystal systems.

3. Lattice Coordinates, Directions, and Planes

Lattice coordinates are used to describe positions within a unit cell. A lattice coordinate is a fraction of the unit length of the a, b, or c axis with the origin taken at the corner of the cell. The coordinates of atoms in a body-centered cubic cell are, for example, 000 and $\frac{1}{2}\frac{1}{2}\frac{1}{2}$. The 000 refers to the lattice origin, which is identical to the other seven corners of the cube, each being worth one eighth, and whose sum therefore is equivalent to one lattice point with coordinates 000. The $\frac{1}{2}\frac{1}{2}\frac{1}{2}$ represents the body-centered atom.

Similarly, the equivalent lattice points for the 14 Bravais lattices in Fig. 11 can be represented as

Bravais Lattice	*Equivalent Lattice Points*
Primitive (*P*)	000
Face-centered (*C*)	000, $\frac{1}{2}\frac{1}{2}0$
Face-centered (*F*)	000, $\frac{1}{2}\frac{1}{2}0$, $\frac{1}{2}0\frac{1}{2}$, $0\frac{1}{2}\frac{1}{2}$
Body-centered (*I*)	000, $\frac{1}{2}\frac{1}{2}\frac{1}{2}$
Hexagonal	000, $\frac{1}{3}\frac{2}{3}\frac{1}{2}$

The indicated numbers of equivalent lattice points can be confirmed by using Eq. (15). The hexagonal lattice is a special case, since the simplest repeating unit is one third of a hexagon, subtended by an angle of 120° (Fig. 11).

Since one or more atoms may be associated with each lattice point, the number of atoms per unit cell in any crystal is partially dependent on its Bravais lattice. For example, the number of atoms per unit cell in a body-centered crystal must be a multiple of 2, since there must be, for any atom in the cell, a corresponding atom of the same kind at a translation of $\frac{1}{2}\frac{1}{2}\frac{1}{2}$ from the first. The reverse of this proposition is, of course, not true. It would be a mistake to assume, for example, that if the number of atoms per cell is a multiple of 4, then the lattice is necessarily face-centered.

Lattice directions are indicated as vector components of the direction resolved along the coordinate axes and reduced to smallest integers. For example, in a cubic unit cell, the body diagonal is represented as [111].

Lattice planes are indicated by their Miller indices, obtained by the following steps:

1. Determine the intercepts of the desired plane, expressed as fractions of the unit cell lengths along the *a*, *b*, and *c* crystallographic axes.
2. Take the reciprocals of these three numbers, reduce them to the least common denominator, and clear of fractions.
3. Enclose in parentheses (h, k, l). h is the reciprocal of the intercept on the *a* axis, k on the *b* axis, and l on the *c* axis.

EXAMPLE 4: Write the Miller indices of the plane indicated in the accompanying figure.

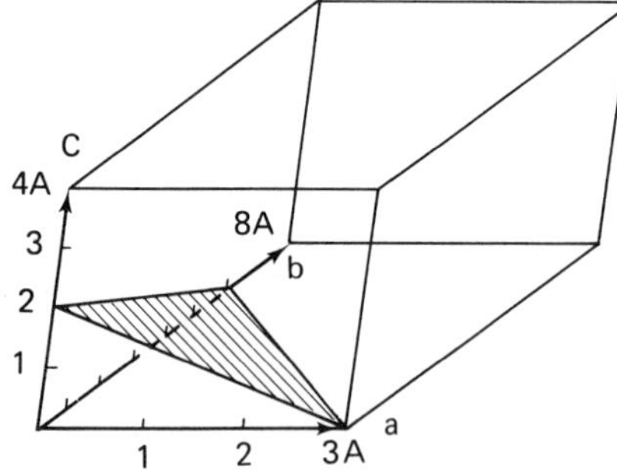

	a	*b*	*c*
Axial lengths	3	8	4
Intercept lengths	3	6	2
Fractional intercepts	1	$\frac{3}{4}$	$\frac{1}{2}$
Reciprocal	1	$\frac{4}{3}$	2
Clear of fractions	3	4	6

Write Miller indices as $(hkl) = (346)$.

The reciprocal is taken to avoid the computationally awkward occurrence of infinity in the indices of planes parallel to any of the crystallographic axes. Note that the same plane may belong to two parallel sets of planes, the Miller indices of one set being multiples of those of the other; thus, the same plane belongs to the (102) and (204) set. In fact, planes of the (102) set form every second plane of the (204) set.

In using indices to specify a plane with respect to a definite origin, note that negative indices can be obtained. These are specified by placing the negative sign above the index $(\bar{h}\bar{k}\bar{l})$. Planes whose indices are negative of one another are parallel and lie on opposite sides of the origin, e.g., $(\bar{1}02)$ and $(10\bar{2})$.

As an exercise, we write Miller indices for the planes in Fig. 12.

Note that the origin (shown as O) may be placed at any corner of the unit cell. Although the choice of origin may change the Miller indices, it does not change the results of x-ray diffraction computations, which depend only on the spacing between planes. One must not place the origin in the plane being indexed, however, since a Miller index of (000) is meaningless. This rule is not violated in Figs. 12(a) and 12(f),

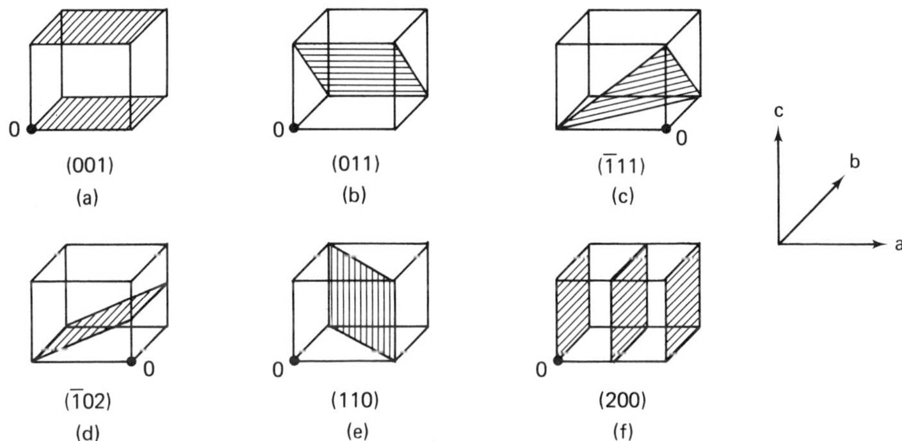

Fig. 12. Miller indices of lattice planes.

since the planes containing the origin are identical (in neighboring cells, not shown) to other planes in the indicated cells.

Any plane can be identified by three Miller indices. In hexagonal systems, however, four indices (*hkil*) are sometimes used to better depict symmetry relationships. The four axes include one along the axis of the hexagonal prism (*l*) and three in its hexagonal base 120° apart (*hki*). Miller indices are then computed in the same manner as for other crystal classes. Since only three noncoplanar axes are necessary to specify a plane in space, the four indices cannot be independent. The additional condition which their values must satisfy is $h + k = -i$.

Expressions for the *interplanar spacing*, the perpendicular distance between any parallel adjacent planes, can be developed for each of the seven crystal systems. For the cubic system, for example, consider the plane HKL, nearest the origin (Fig. 13). There is a plane of the same family passing through the origin (but not shown). Therefore, if OD is normal to the plane HKL, it is also an interplanar distance d of the family. Calling angles HOD, KOD, and LOD α, β, and γ, respectively,

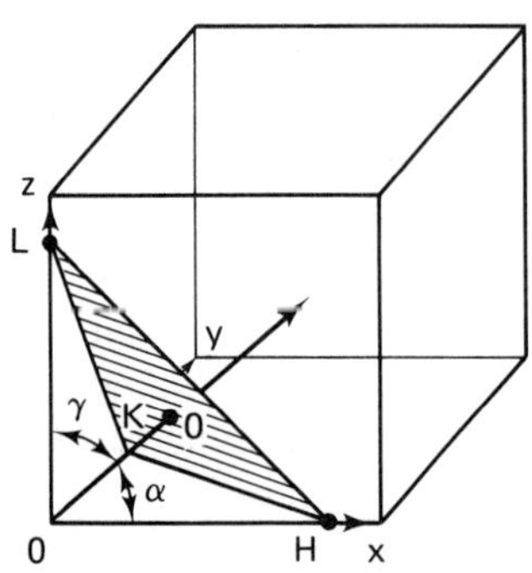

Fig. 13. Determination of interplanar spacing d for the cubic system.

$$\cos\alpha = \frac{OD}{OH}$$

$$\cos\beta = \frac{OD}{OK}$$

$$\cos\gamma = \frac{OD}{OL}$$

and from analytical geometry,

$$\cos^2\alpha + \cos^2\beta + \cos^2\gamma = 1$$

Hence

$$\left(\frac{OD}{OH}\right)^2 + \left(\frac{OD}{OK}\right)^2 + \left(\frac{OD}{OL}\right)^2 = 1$$

or

$$\frac{1}{OH^2} + \frac{1}{OK^2} + \frac{1}{OL^2} = \frac{1}{OD^2} \tag{16}$$

But

$$OH = \frac{a}{h}, \qquad OK = \frac{a}{k}, \qquad OL = \frac{a}{l} \tag{17}$$

and

$$OD = d$$

Substituting Eq. (17) into Eq. (16) and simplifying,

$$d_{hkl} = \frac{a}{(h^2 + k^2 + l^2)^{0.5}}$$

where a is the cubic unit cell dimension.

Similar equations can be derived for interplanar spacings in the remaining crystal systems. For tetragonal crystals, for example, since unit cell dimensions a and c are not equal,

$$d_{khl} = \frac{a}{[h^2 + k^2 + l^2(a^2/c^2)]^{0.5}}$$

Appendix B contains plane spacing equations for the remaining crystal systems.

The plane spacing, the relative packing densities of atoms on various crystallographic planes, and the average packing density of atoms within the unit cell are all important in determining physical properties (strength; modulus; optical, electrical, and thermal transmission; etc.) along various crystallographic axes of crystalline materials. Some of these properties are evaluated in terms of crystal structure in Chap. 3. Familiarity with plane spacing and packing density computations is also important in x-ray diffraction work (Chap. 13). The following example illustrates very simple calculations of this type.

EXAMPLE 5: Given the lattice parameters abc for the body-centered orthorhombic unit cell of an element having atomic weight of M, compute

(a) The density of the element.
(b) The number of atoms per unit area on the (110) plane.
(c) The distance between adjacent (101) planes.

Solution: (a) A body-centered cell contains two atoms, and

$$\text{density} = \frac{2M}{6.027 \times 10^{23}} \frac{1}{abc(10^{-8})^3} \text{ g/cm}^3$$

where 6.027×10^{23} is Avogadro's number and 10^{-8} converts angstroms to centimeters.

(b) The area of the (110) plane within the unit cell is

$$c(a^2 + b^2)^{0.5} \text{ A}^2$$

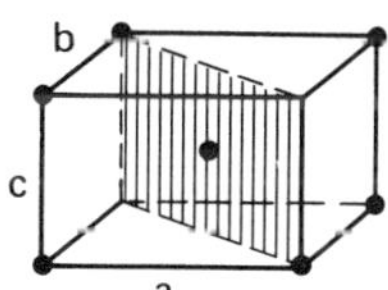

Atoms at the corners are each shared by four similar planes in adjacent unit cells. Therefore, there are 1 (at center) plus $4\frac{1}{4}$ atoms in this area. The number of atoms per unit area is then

$$\frac{2 \times 10^{16}}{c(a^2 + b^2)^{0.5}} \text{ atoms/cm}^2$$

(c) For the (101) plane,

$$d = a \sin\theta = a \sin\left(\tan^{-1}\frac{c}{a}\right)$$

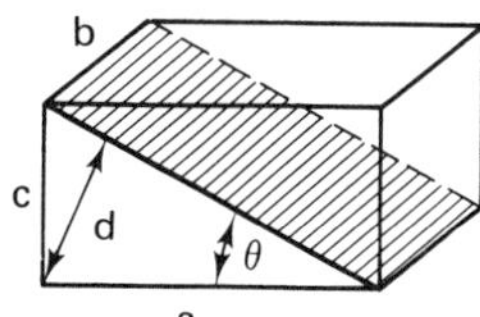

Planes of a form have different indices but the same spacing. This relationship will later be used to obtain multiplicity factors (Appendix B) for x-ray diffraction work. In cubic, for example, the six sides (100), (010), (001), ($\bar{1}$00), (0$\bar{1}$0), and (00$\bar{1}$) are planes of a form. In tetragonal, however, only four of the six belong to one form, the other two having different spacing.

EXAMPLE 6: Cite the planes which fall in the [022] form of (a) a cubic crystal and (b) an orthorhombic crystal.

Answer: (a)

(022) (0$\bar{2}$2) (02$\bar{2}$) (0$\bar{2}\bar{2}$)
(202) ($\bar{2}$02) (20$\bar{2}$) ($\bar{2}$0$\bar{2}$)
(220) ($\bar{2}$20) (2$\bar{2}$0) ($\bar{2}\bar{2}$0)

(b) Since in the orthorhombic $a \neq b \neq c$,

(022) (0$\bar{2}$2) (02$\bar{2}$) (0$\bar{2}\bar{2}$)

PROBLEMS

1. Calculate the critical radius ratio for a coordination number of 8. For this structure, the centers of the anions form the corners of a cube and the cation is at the center. Calculate the fraction of total space occupied by ions (packing factor) in this structure.
2. What is the packing factor for a face-centered cubic structure?
3. Calculate the c/a ratio for hexagonal close packing of equal-sized spheres and the corresponding packing factor.
4. Sketch the following planes:
   (a) (110) in an orthorhombic cell.
   (b) (100) in a monoclinic cell.
   (c) (111) in a tetragonal cell.
5. The unit cell is shown in the accompanying figure for an element whose atomic weight is 36. Compute
   (a) The density.
   (b) The spacing of adjacent 101 planes:

   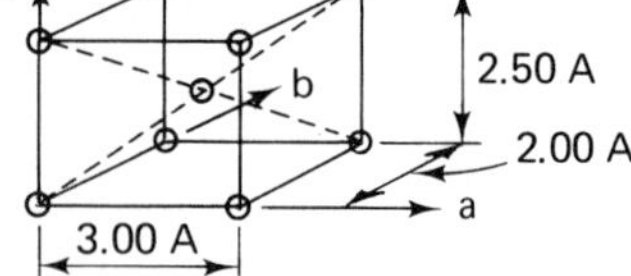

   1 A = 10^{-8} cm
   Avogadro's number = 6.027×10^{23}
6. Tungsten has a body-centered cubic structure and a lattice parameter of 3.165 A. Find the number of atoms per square centimeter on the (110) planes.
7. Calculate the size of the largest sphere which fits interstitially in a body-centered cubic

crystal structure, as a fraction of the radius of the BCC atoms. The coordinates of the largest holes are of the type $0\frac{1}{2}\frac{1}{4}$.

C. POLYMER STRUCTURES

Polymers are long molecules comprised of simpler molecular units called monomers (poly—many, mono—one, mer—unit). Monomers are commonly carbon compounds but may also be inorganic compounds such as silicates and silicones. Polymers are so widely used in construction that a rudimentary knowledge of the relationship between polymer structure and mechanical properties is extremely useful. Applications include all classes of construction sealants and adhesives, pipes and conduits, surface finishes, foamed insulating materials, bituminous binders and adhesion-improving additives for asphalts, various additives and foaming agents for concretes, soil stabilizing agents, and numerous others. It is estimated that by 1980 the use of synthetic polymers alone will outstrip metals on a volume basis.

Compared with other construction materials, polymers are generally more resistant to a wide range of salts, bases, and acids than metals, concrete, and wood. However, they oxidize to varying degrees in the presence of ultraviolet light and/or strong oxidizing agents.

Polymers also have many properties useful in structural applications. They are lightweight, easily transported and installed, and easily repaired. But they are often inadequate in structural strength, long-life durability, and, above all, resistance to high temperature and fire. Polymers can be modified to meet many of these demands, but the cost is often prohibitive.

1. Organic Materials

Since most polymers are organic materials, the relevant classes of organic materials will be considered before discussing polymer structures.

a. SATURATED HYDROCARBONS

Saturated hydrocarbons are those in which each carbon atom is surrounded by a full complement of four neighboring atoms. Since there is no way to add additional atoms, the molecule is called *saturated*. Methane, CH_4, is the simplest hydrocarbon, and the smallest member of the *paraffin* family, C_nH_{2n+2} (Table 11). The word paraffin indicates a chain of carbon atoms, as opposed to a ring. Paraffin molecules have strong *intra*molecular covalent bonds but weak *inter*molecular van der Waals bonds. Hence, the molecules behave quite independently and have low melting points, expressed approximately as

$$\frac{1}{T_m} = \frac{2.395 \times 10^{-3} + 17.1 \times 10^{-3}}{n}$$

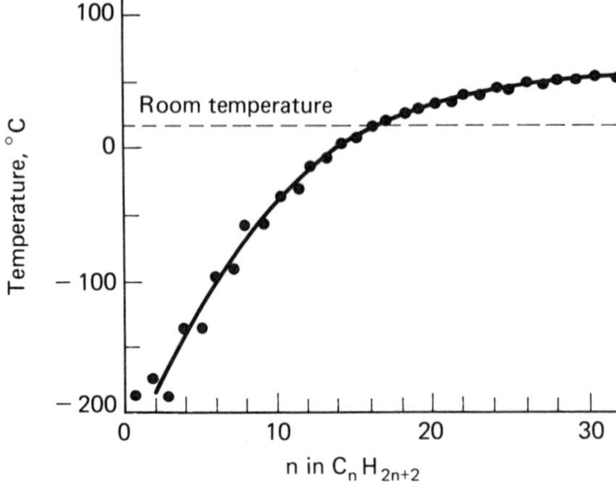

Fig. 14. Melting temperature vs. molecular size of paraffins. From Van Vlack, L. H.: *Elements of Materials Science, Second Edition.* Reading, Mass.: Addison-Wesley Publishing Company, Inc., 1964 [8].

where T_m is the absolute melting temperature (°K) for a molecule with n carbon atoms (Fig. 14).

TABLE 11. Paraffin Series

Name	*Composition*	*Structure*
Methane	CH_4	H | H—C—H | H
Ethane	C_2H_6	H H | | H—C—C—H | | H H
Propane	C_3H_8	H—H—H | | | H—C—C—C—H | | | H H H
Butane	C_4H_{10}	Etc.
Pentane	C_5H_{12}	
Hexane	C_6H_{14}	
Etc.	C_nH_{2n+2}	

Polyethylenes are paraffin chains several hundred carbon atoms long. Because of their high molecular symmetry, they are nonpolar and therefore insoluble in polar solvents such as water.

If the carbon chains are branched, the compounds are *isoparaffins,* and if ring structures, *cycloparaffins* or *napthenes.* Compounds of the same chemical composition but different arrangements of atoms are *isomers* (Fig. 15). Isomerization, which corresponds to polymorphism in crystalline phases (Sec. B), can profoundly affect properties such as melting and boiling points.

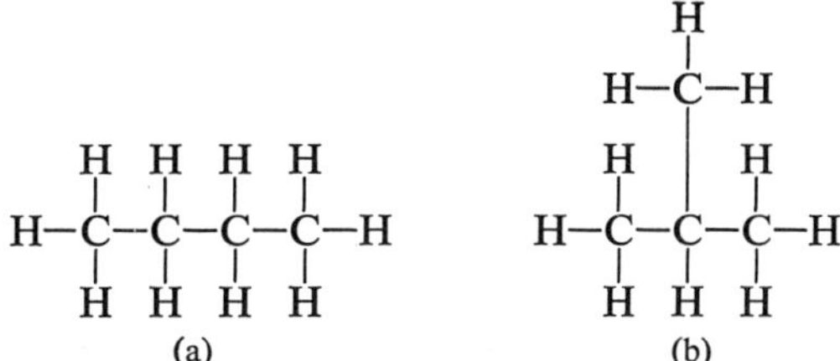

Fig. 15. Isomers: (a) butane, (b) isobutane.

Several other hydrocarbon functional groups are illustrated in Fig. 16. The halides are less combustible because hydrogen atoms are replaced with halogens (usually F or Cl), imparting combustion resistance by dilution. Halogens also form hydrogen halogenides as they burn and further inhibit combustion by preventing complete oxidation of the hydrogen (an important advantage of polyvinyl chloride over polyethylene). The alcohols are soluble in water due to the polar OH groups.

Halogen	R—Hal
Alcohol	R—OH
Ether	R—O—R′
Ketone	R >C=O R
Aldehyde	R—C=O
Acid	R<(=O)(—H)
Formate	O=C<(R)(OR)
Ester	O=C<(H)(OR)
Nitrile	R≡N
Aromatic Hydrocarbons	R | C HC CH HC CH C | H

Fig. 16. Hydrocarbon functional groups.

b. UNSATURATED HYDROCARBONS

In unsaturated hydrocarbons, two or three bonds (electron pairs) exist between some of the carbons. For example, ethylene ($H_2C{=}CH_2$) and acetylene ($HC{\equiv}CH$), unlike ethane, are not saturated with the maximum number of hydrogens. Unsaturation facilitates polymerization by the transfer of one of the double bonds to an adjacent monomer (Fig. 17). Each transfer involves breaking one double bond and forming two single bonds. Some additional double bond functional groups were

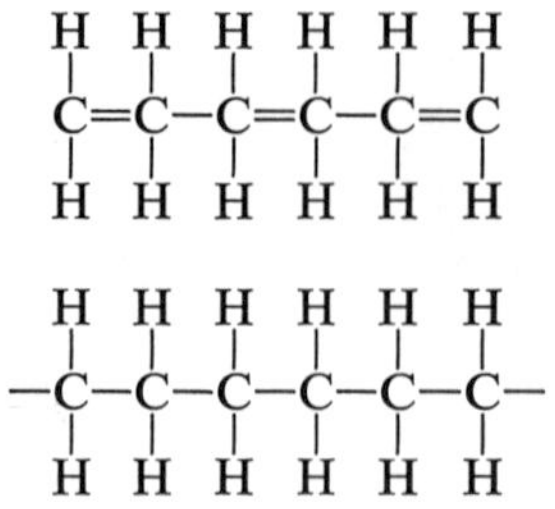

$$
\begin{array}{cccccc}
H & H & H & H & H & H \\
| & | & | & | & | & | \\
C{=}&C{-}&C{=}&C{-}&C{=}&C \\
| & | & | & | & | & | \\
H & H & H & H & H & H
\end{array}
$$

$$
\begin{array}{cccccc}
H & H & H & H & H & H \\
| & | & | & | & | & | \\
-C{-}&C{-}&C{-}&C{-}&C{-}&C- \\
| & | & | & | & | & | \\
H & H & H & H & H & H
\end{array}
$$

Fig. 17. Ethylene polymerization.

included in Fig. 16. The simplest aldehyde, for example, is formaldehyde (HCHO), a component of polymeric phenol-formaldehyde (Bakelite). The aromatic ring occurs in styrene, which is ethylene with one of the hydrogens replaced with the benzene ring. Styrene reacts at the ethylenic double bonds, so the rings appear as appendages to the chains in polystyrene.

2. *Polymer Synthesis*

Most polymers are produced by one of two familiar types of organic reactions, condensation and addition.

a. CONDENSATION

In condensation polymerization small molecules (such as water) are split out as successive monomer units combine to form the polymer chain. A common example is the reaction of a diol with a diacid:

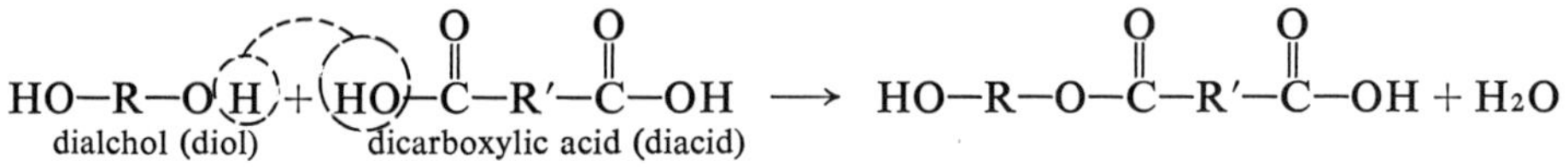

$$
\underset{\text{dialchol (diol)}}{HO{-}R{-}O\boxed{H}} + \underset{\text{dicarboxylic acid (diacid)}}{\boxed{HO}{-}\overset{\overset{\displaystyle O}{\|}}{C}{-}R'{-}\overset{\overset{\displaystyle O}{\|}}{C}{-}OH} \longrightarrow HO{-}R{-}O{-}\overset{\overset{\displaystyle O}{\|}}{C}{-}R'{-}\overset{\overset{\displaystyle O}{\|}}{C}{-}OH + H_2O
$$

The resulting molecule is still *difunctional;* i.e., its left end can react with another diacid molecule and its right end with another diol molecule. Continuing in this fashion, n moles of diol react with n moles of diacid to form a *polyester,*

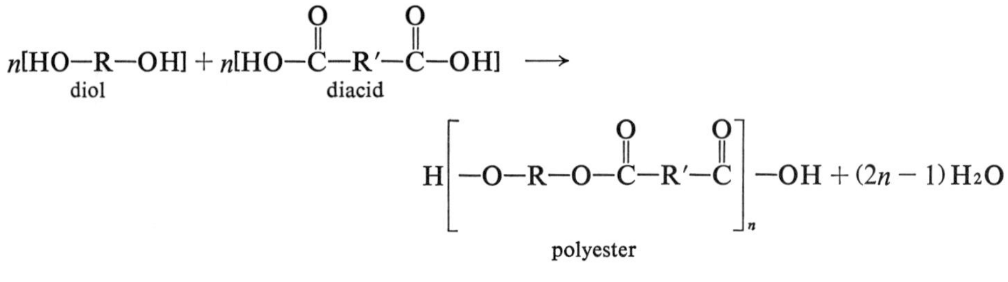

$$
n\underset{\text{diol}}{[HO{-}R{-}OH]} + n\underset{\text{diacid}}{[HO{-}\overset{\overset{\displaystyle O}{\|}}{C}{-}R'{-}\overset{\overset{\displaystyle O}{\|}}{C}{-}OH]} \longrightarrow
$$

$$
\underset{\text{polyester}}{H\left[{-}O{-}R{-}O{-}\overset{\overset{\displaystyle O}{\|}}{C}{-}R'{-}\overset{\overset{\displaystyle O}{\|}}{C}\right]_n{-}OH} + (2n-1)H_2O
$$

The number of repeating units, n, is called the *degree of polymerization.* The generalized organic groups R and R′ can vary widely, and their structures can strongly influence the mechanical properties of the resulting polymer.

Another example of polycondensation is the reaction of a diamine with a diacid to form a polyamide, or nylon:

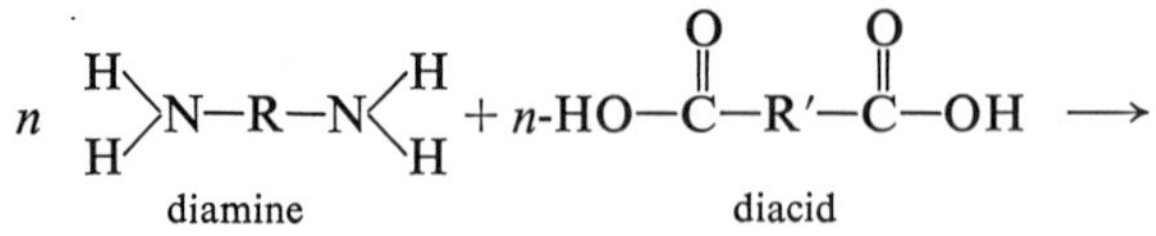

$$
n\ \underset{\text{diamine}}{\overset{H}{\underset{H}{>}}N{-}R{-}N\overset{H}{\underset{H}{<}}} + n\text{-}\underset{\text{diacid}}{HO{-}\overset{\overset{\displaystyle O}{\|}}{C}{-}R'{-}\overset{\overset{\displaystyle O}{\|}}{C}{-}OH} \longrightarrow
$$

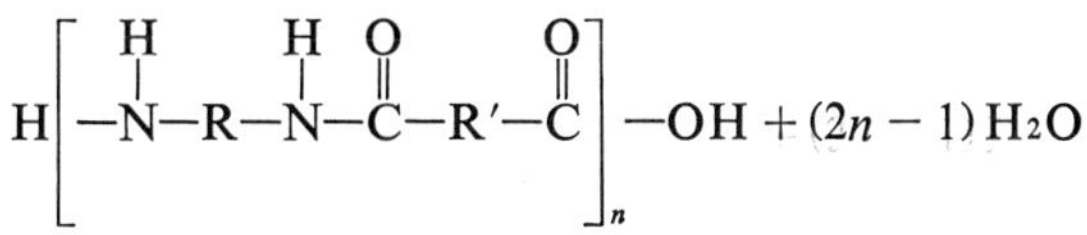

$$H\left[-\overset{\overset{\displaystyle H}{|}}{N}-R-\overset{\overset{\displaystyle H}{|}}{N}-\overset{\overset{\displaystyle O}{\|}}{C}-R'-\overset{\overset{\displaystyle O}{\|}}{C}\right]_n-OH+(2n-1)H_2O$$

polyamide (nylon)

A third example is the condensation of phenol and formaldehyde to form polyphenol-formaldehyde, or Bakelite:

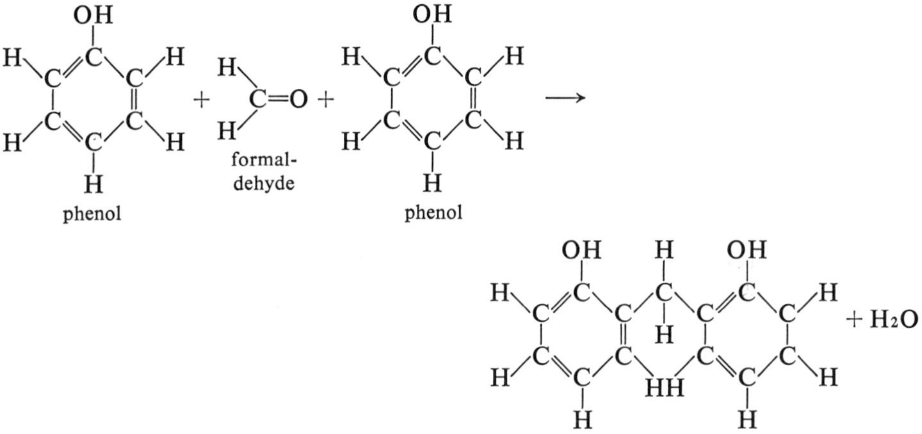

$$\text{phenol} + \underset{\text{formaldehyde}}{H_2C{=}O} + \text{phenol} \longrightarrow \text{Bakelite} + H_2O$$

phenol formal-dehyde phenol

Bakelite

b. ADDITION

Addition polymerization involves opening of a double bond, and no molecule is split out; i.e., the repeating unit has the same formula as the monomer. Monomers of the general type

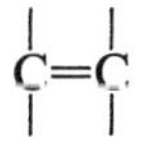

$$\begin{array}{cc} | & | \\ C & = C \\ | & | \end{array}$$

undergo addition polymerization:

$$n[\overset{|}{\underset{|}{C}}=\overset{|}{\underset{|}{C}}] \longrightarrow -\left[\overset{|}{\underset{|}{C}}-\overset{|}{\underset{|}{C}}\right]_n-$$

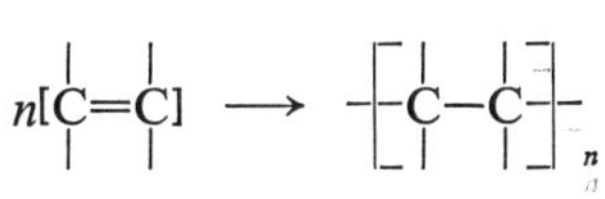

The double bond opens to form bonds to adjacent monomers. Vinyl monomers,

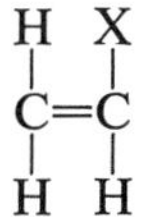

$$\begin{array}{cc} H & X \\ | & | \\ C & = C \\ | & | \\ H & H \end{array}$$

are an important group of double bond monomers, and addition polymerization is sometimes called vinyl polymerization. Acrylic acid is a vinyl monomer and undergoes addition polymerization to give polyacrylate,

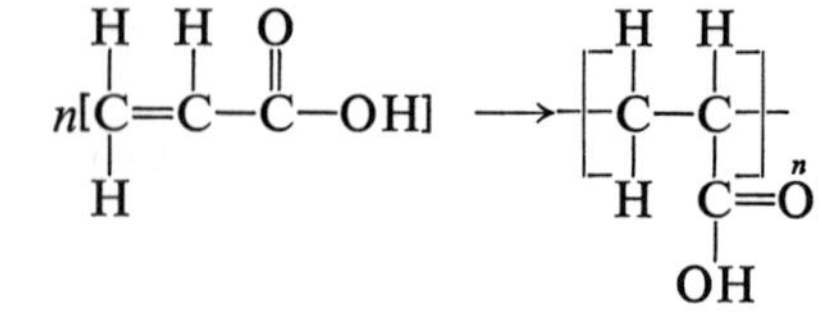

```
   H  H  O               H  H
   |  |  ||            [ |  |  ]
n[C=C—C—OH]  ——→     —[—C—C—]—
   |                   [ |  |  ]n
   H                     H  C=O
                            |
                            OH
```

Table 12 lists some common members of three classes of addition polymers: the vinyls, vinylidenes, and dienes.

TABLE 12. Common Addition Polymers

Vinyl polymers

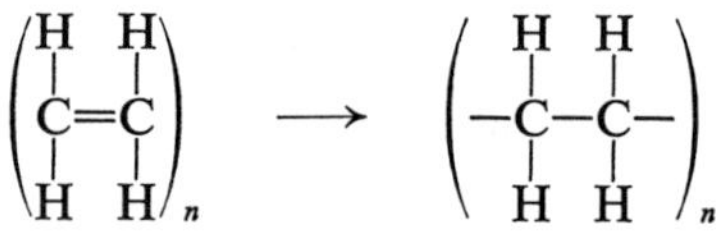

```
 / H  H \          /    H  H    \
 |  |  | |         |    |  |    |
 |  C=C  |   ——→   |  —C—C—     |
 |  |  | |         |    |  |    |
 \ H  H /n         \    H  H    /n
```

		Structures	*Name*	*Typical use*
R_1 →	 —Cl	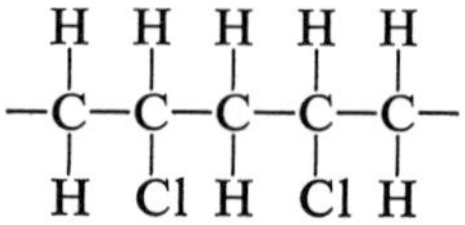 —CH(H)—CH(Cl)—CH(H)—CH(Cl)—CH(H)— (H H H H H above; H Cl H Cl H below)	Polyvinyl chloride (PVC)	Plastic pipes, liquid containers
R_1 →	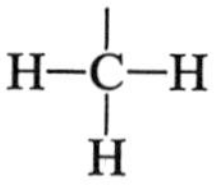 H—C(H)—H	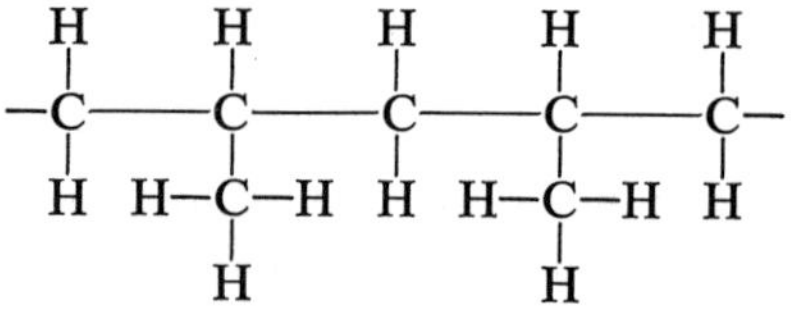 —C—C—C—C—C— (H above each; below: H, H—C—H (H), H, H—C—H (H), H)	Polypropylene	Steering wheels, radiator fans
R_1 →	 —C≡N	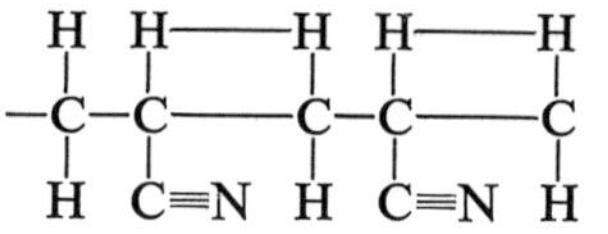 —C—C—C—C—C (H above each; below: H, C≡N, H, C≡N, H)	Acrylonitrile (orlon)	Orlon fiber
R_1 →	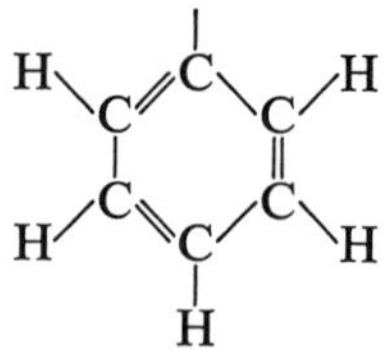 benzene ring (C with H's in six-membered ring)	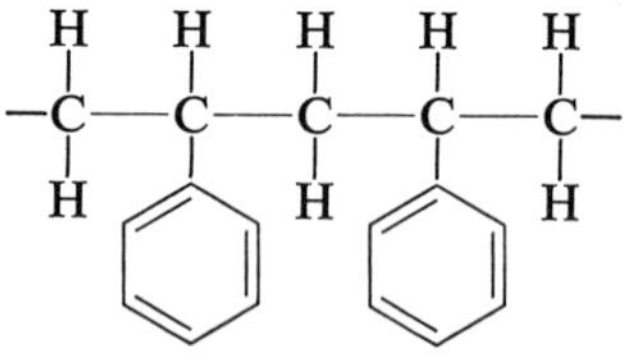 —C—C—C—C—C— (H above each; below: H, benzene ring, H, benzene ring, H)	Polystyrene	Kitchen appliances, food containers, battery cases

	Structures	Name	Typical use
$R_1 \longrightarrow$ 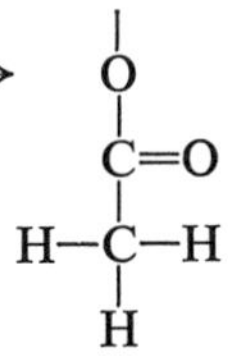 acetate: $-O-C(=O)-CH_3$	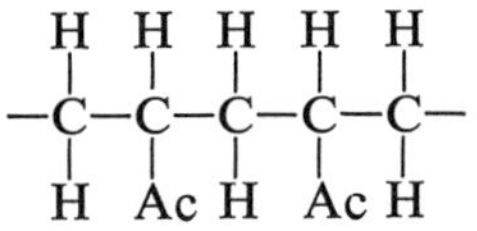 $-[CH_2-CH(Ac)]-$: H H H H H / —C—C—C—C—C— / H Ac H Ac H	Polyvinyl acetate	Adhesives, paints, flashbulb lining

Vinylidene polymers

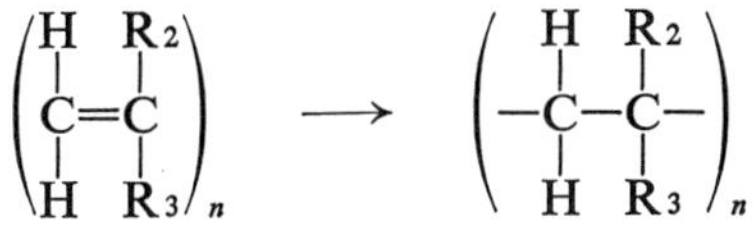

$$\left(\begin{matrix} H & R_2 \\ C = & C \\ H & R_3 \end{matrix}\right)_n \longrightarrow \left(\begin{matrix} H & R_2 \\ -C- & C- \\ H & R_3 \end{matrix}\right)_n$$

	Structures	Name	Typical use
$R_2 \longrightarrow Cl$; $R_3 \longrightarrow Cl$ 	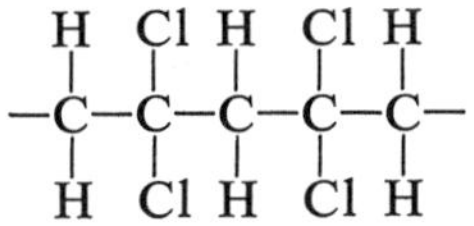 H Cl H Cl H / —C—C—C—C—C— / H Cl H Cl H	Polyvinylidene chloride (Saran)	Saran wrap, upholstery fabric
$R_2 \longrightarrow CH_3$ (H—C—H with H); $R_3 \longrightarrow -C(=O)-O-CH_3$ 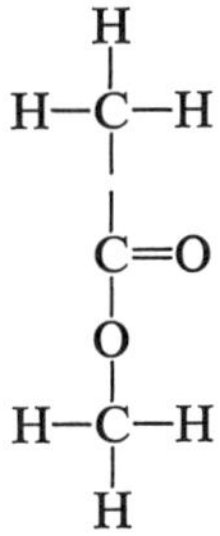	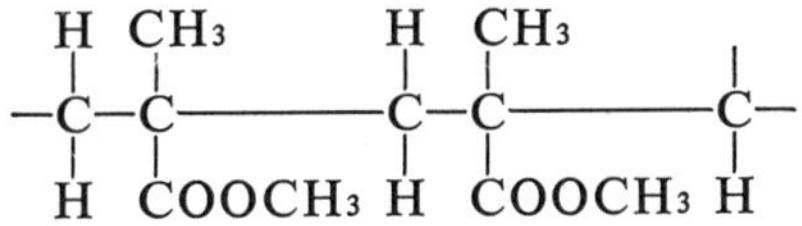 H CH_3 H CH_3 / —C—C———C—C———C— / H $COOCH_3$ H $COOCH_3$ H	Polymethyl methacrylate (Lucite, Plexiglas, Perspex)	Goggles, light pipes

Other

Structures	Name	Typical use
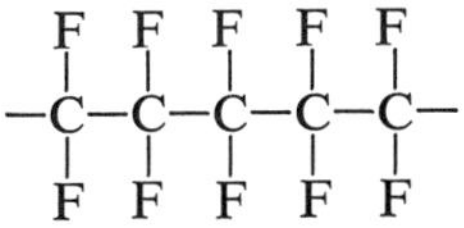 F F F F F / —C—C—C—C—C— / F F F F F	Polytetrafluoroethylene (Teflon)	Cookware, bearings

Diene polymers

```
 /H  R  H  H\          /  H  R  H  H  \
 |  |  |  |  |         |   |  |  |  |   |
 | C=C—C=C |   ——>    | —C—C=C—C—  |
 |  |        |  |        |   |        |   |
 \H        H/n         \  H        H  /n
```

	Structure	Name
R ——> H	H H H H H H H H / —C—C=C—C—C—C=C—C— / H H H H	Polybutadiene
R ——> Cl	H Cl H H H Cl H H / —C—C=C—C—C—C=C—C— / H H H H	Polychlorophene (neoprene)
R ——> H—C(H)—H	H CH_3 H H H CH_3 H H / —C—C=C—C—C—C=C—C— / H H H H	Polyisoprene (natural rubber, gutta percha)

SOURCE: EISENSTADT, MELVIN M., *Introduction to Mechanical Properties of Materials,* New York: The MacMillan Company, 1971 [2].

Numerous catalysts have been developed to initiate addition polymerization. For example, free-radical formers such as organic and hydrogen peroxides activate the double bond to form an intermediate free radical, which in turn activates another monomer which attaches itself to the chain.* The free radical then passes to the end of the chain to continue the propagation. Nuclear radiation and ultraviolet light are also used to initiate polymerization reactions.

3. Structure

The major structural classes include linear, branched, and cross-linked polymers.

a. LINEAR

If every monomer unit is difunctional (can form only two bonds), a linear polymer is formed. The polymer chain is not linear, however, and assumes a random configuration due to the freedom of rotation about the C—C—C single bonds (having $\sim 109.5°$ bond angles). The random configuration is also time-varying due to the Brownian motion of the chain segments.

*A free radical is an organic ion having a free orbital which can overlap with another, such as in a double bond, to form a covalent bond.

If two monomers A and B are used instead of a single monomer, the resulting polymer is a *copolymer* and may have a random arrangement A and B in the chain:

BAABABAABBB

The randomness depends on the relative amounts and reactivities of A and B. Figure 18 shows a copolymer of butadiene and styrene, the basis for many artificial rubbers.

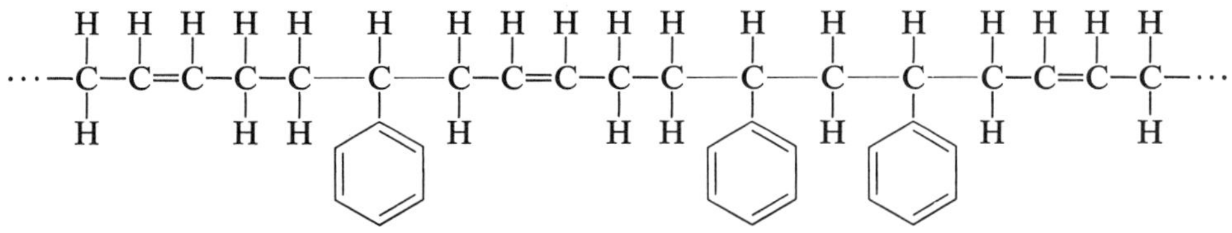

Fig. 18. Copolymerization of butadiene and styrene.

Copolymers can sometimes be used to obtain mechanical properties which vary greatly from polymers produced with either of the two monomers separately. Condensation polymers, which require two different monomers to provide the required functional groups, are not copolymers.

b. BRANCHED

If a few monomer units along the chain are tri- or tetrafunctional, branched polymers can be formed. Branching can profoundly affect physical properties through steric (geometric) effects.

c. CROSS-LINKED

As the frequency and length of chain branches increases, the probability of branches connecting between two chains also increases. If approximately equal molar quantities of di- and trifunctional monomers are used, tight, three-dimensional networks can be produced. When all chains become connected with such cross-links, the entire material becomes one large molecule, a *cross-linked* polymer. A bowling ball is one example of such a single large molecule.

Cross-linked polymers may be formed either by starting with reactants containing sufficient tri- or higher functional monomers or by chemically producing cross-links between previously formed linear or branched polymers, as in the vulcanization of rubber with sulfur. In vulcanization, double bonds in the chain are activated and cross-linked through S—C bonds (Fig. 19).

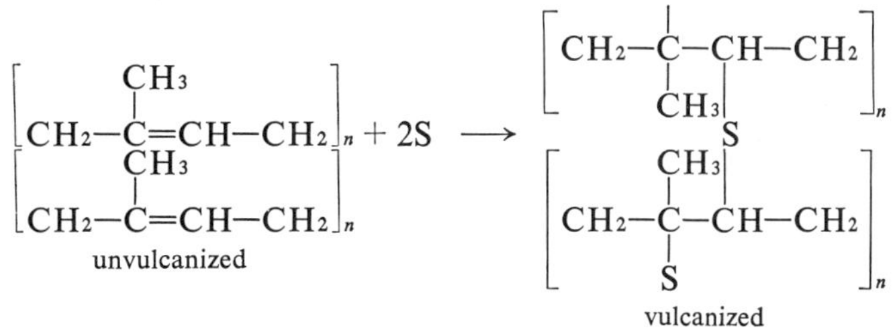

$$\left[CH_2{-}\overset{\displaystyle CH_3}{\overset{|}{C}}{=}CH{-}CH_2\right]_n \left[CH_2{-}\overset{\displaystyle CH_3}{\overset{|}{C}}{=}CH{-}CH_2\right]_n + 2S \longrightarrow \left[CH_2{-}\overset{|}{\underset{\underset{\displaystyle CH_3}{|}}{C}}{-}CH{-}CH_2\right]_n \overset{S}{} \left[CH_2{-}\overset{\overset{\displaystyle CH_3}{|}}{\underset{\underset{\displaystyle S}{|}}{C}}{-}CH{-}CH_2\right]_n$$

unvulcanized — vulcanized

Fig. 19. Vulcanization of polyisoprene (natural rubber).

EXAMPLE 7: Calculate the percentage of sulfur by weight of final product required to fully vulcanize polyisoprene.

Solution: One sulfur atom is required per double bond, i.e., per monomer unit. The mer weight of isoprene is

$$8(1.00) + 5(12.01) = 68.11$$

and

$$\% \mathrm{S} = \frac{32}{68 + 32} = 32\% \text{ sulfur} \qquad \text{for } 100\% \text{ vulcanization.}$$

By comparison, automobile tires are about 4–6% vulcanized.

The terms *thermoplastic*, *thermosetting*, and *elastomeric* are frequently used to designate the thermal behavior of polymers or their degree of cross-linking. Thermoplastic polymers are not cross-linked. They may be reformed repeatedly upon application of heat and pressure. Thermosetting polymers remain relatively strong until temperatures at which chemical decomposition sets in. They are highly cross-linked, or in some cases the chains may be joined by strong hydrogen bonds, as in cellulose. Thermosetting resins have better thermal and mechanical properties, but thermoplastic resins are more easily formed in intricate shapes, and the scrap resin may frequently be recycled, which is not possible with thermosets.

Elastomers are polymers which exhibit large and reversible deformations at ordinary temperatures; they can be stretched at least 100% and often 1000% and snap back to the original dimensions when load is removed. The structural properties which promote this behavior include

1. Long, coiled, and tangled chains with occasional built-in kinks, such as provided by carbon-carbon double bonds and bulky side groups which interfere with each other (called *steric hindrance*) and prevent freedom of rotation.
2. Room temperature must provide enough thermal energy for chain segments to be in a constant state of motion. Otherwise, quick recovery on removal of load is not possible.
3. *Infrequent* cross-links must exist between chains, sufficient to prevent permanent deformation due to chains slipping past one another under load but not sufficient to make the material inflexible.

Upon loading, elastomer chains uncoil until the infrequent cross-links prevent further deformation without rupture of covalent bonds. Natural rubber (polyisoprene,

$$\begin{array}{cc} -\mathrm{C}{=}&\mathrm{C}- \\ | & | \\ \mathrm{H} & \mathrm{CH_3} \end{array}$$

is the best known elastomer. The chains are bent because the methyl group interferes with hydrogen in the repeating unit. An isomeric form, gutta percha, has hydrogen and the methyl group on opposite sides of the chain so they do not interfere with each other. Gutta percha is not an elastomer, and in fact it crystallizes because the chains are free to line up more easily than in natural rubber.

EXAMPLE 8: Starting with monomers shown below, show the corresponding repeating polymer unit, and indicate the method of synthesis (condensation or addition) and structure (linear, branched, or network).

Monomer	*Repeating Polymer Unit*	*Synthesis and Structure*
(a) Ethylene $H_2C{=}CH_2$	Polyethylene 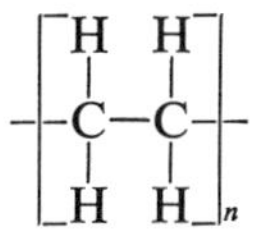 $\left[\text{-CH}_2\text{-CH}_2\text{-}\right]_n$	Addition Linear
(b) Isoprene 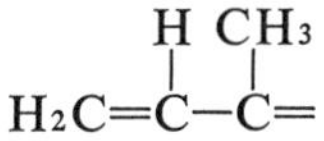 $H_2C{=}C(H){-}C(CH_3){=}$	Polyisoprene $CH_2\text{-}\left[\text{-CH}_2\text{-CH}{=}\text{C(CH}_3\text{)-CH}_2\text{-}\right]_n$ 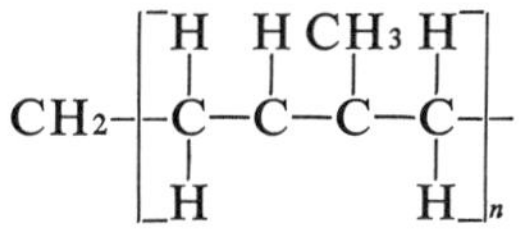	Addition Network
(c) Styrene 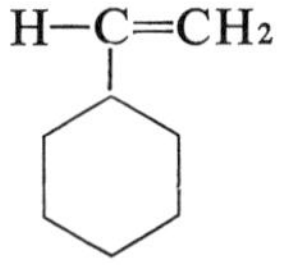 $H{-}C(C_6H_5){=}CH_2$	Polystryene $\left[\text{-CH}_2\text{-CH(C}_6\text{H}_5\text{)-}\right]_n$ 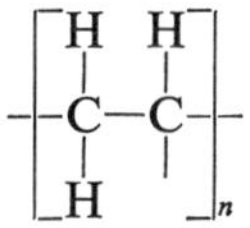	Addition Linear
(d) Divinylbenzene 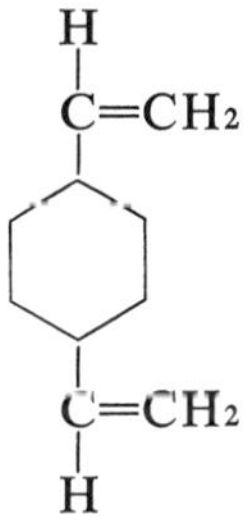 $CH_2{=}CH{-}C_6H_4{-}CH{=}CH_2$	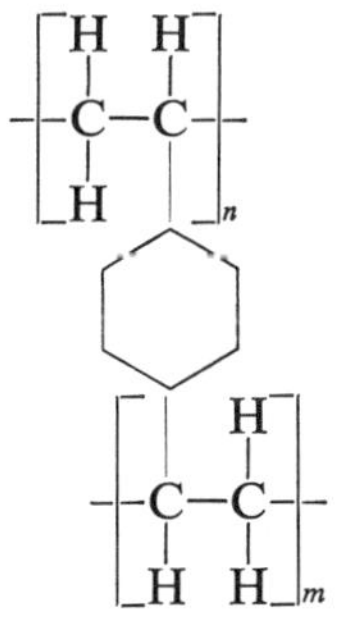 $\left[\text{-CH}_2\text{-CH-}\right]_n$ linked through C_6H_4 to $\left[\text{-CH-CH}_2\text{-}\right]_m$	Addition Network (monomer is tetra-functional)
(e) Hexamethylenediamine $H_2N(CH_2)_6NH_2$ plus 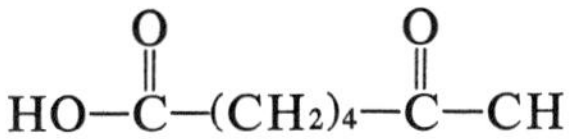 $HO{-}C(=O){-}(CH_2)_4{-}C(=O){-}CH$ adipic acid	Nylon 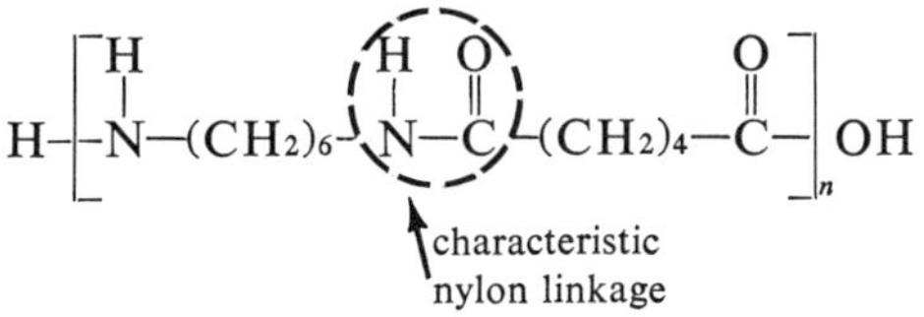 $H\text{-}\left[\text{-NH-(CH}_2\text{)}_6\text{-NH-C(=O)-(CH}_2\text{)}_4\text{-C(=O)-}\right]_n OH$ (circled N(H)–C(=O)): characteristic nylon linkage	Condensation Linear

d. POLYELECTROLYTES

Polymers are held together by covalent bonds within the chains and either covalent bonds (e.g., in cross-linking), ionic bonds, or the various secondary bonds (see Sec. A) between chains. This contrasts with metals, for example, where only one type of bonding exists. Polyelectrolytes are polymers containing side groups which permit ionic or polar bonding between the chains. They combine the properties of a water-soluble polymer and an electrolyte. Polyelectrolytes may be *anionic*, *cationic*, or *amphoteric*. Figure 20 shows examples of the three classes. Amphoteric polyelectrolytes contain both positive and negative ions. The largest group of naturally occurring amphoterics are proteins.

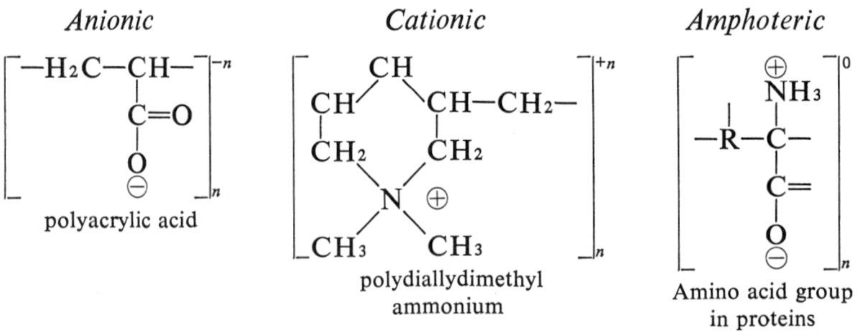

Anionic

$$\left[-H_2C-\underset{\underset{\underset{\ominus}{O}}{|}}{\underset{C=O}{\underset{|}{CH}}}-\right]_n^{-n}$$

polyacrylic acid

Cationic

$$\left[\begin{array}{c} CH \\ CH \diagup \quad \diagdown CH-CH_2- \\ | \qquad\qquad | \\ CH_2 \qquad CH_2 \\ \diagdown \quad \diagup \\ N^{\oplus} \\ \diagup \quad \diagdown \\ CH_3 \qquad CH_3 \end{array}\right]_n^{+n}$$

polydiallydimethyl ammonium

Amphoteric

$$\left[\begin{array}{c} \overset{\oplus}{N}H_3 \\ | \\ -R-C- \\ | \\ C= \\ | \\ O \\ \ominus \end{array}\right]_n^{0}$$

Amino acid group in proteins

Fig. 20. Classes of polyelectrolytes.

Polyelectrolytes are used extensively as ion exchange resins for water softening and for coagulating fine particles in mineral separation, sewage treatment, and soil stabilization operations. When polyacrylic acid in aqueous suspension is ionized by addition of a base, such as NaOH, it expands from a randomly kinked configuration to an elongated configuration due to osmotic pressure caused by electrostatic attraction between the solvated Na^+ ions and the carboxyl radicals (Fig. 21). Cations of higher valence and smaller size tend to provide ionic cross-linking between carboxyl groups on the same or on different chains (Fig. 22). The resulting polymer, though ionically cross-linked, is thermoplastic rather than thermosetting because the ionic bonds can be reversibly broken.

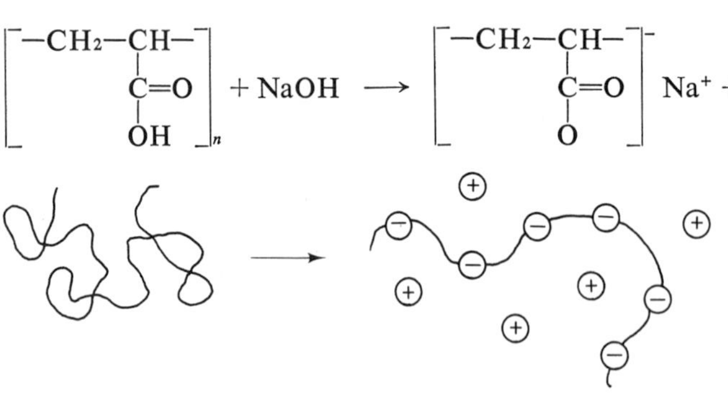

$$\left[-CH_2-\underset{\underset{OH}{|}}{\underset{C=O}{\underset{|}{CH}}}-\right]_n + NaOH \longrightarrow \left[-CH_2-\underset{\underset{O}{|}}{\underset{C=O}{\underset{|}{CH}}}-\right]^- Na^+ + H_2O$$

Fig. 21. Polyelectrolyte chain expansion due to ionization.

e. SOLUBILITY

Solubilities of each of the preceding structural classes of polymers depend on similarity in structure between solvent and polymer. Like dissolves like; i.e., polar solvents dissolve polar polymers, and nonpolar solvents dissolve nonpolar polymers. For example, water tends to dissolve polyvinyl alcohol, and toluene tends to dissolve polystyrene:

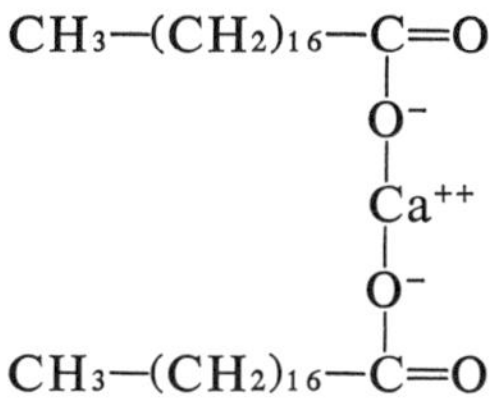

$$\left[-\underset{H}{\overset{H}{C}}-\underset{OH}{\overset{H}{C}}- \right]_n + HOH$$

and

$$\left[-\underset{H}{\overset{H}{C}}-\underset{C_6H_5}{\overset{H}{C}}- \right]_n + C_6H_5CH_3$$

$$CH_3-(CH_2)_{16}-C(=O)-O^- \; Ca^{++} \; {}^-O-C(=O)-(CH_2)_{16}-CH_3$$

Fig. 22. Ionic crosslinking; calcium ion to carboxyl groups.

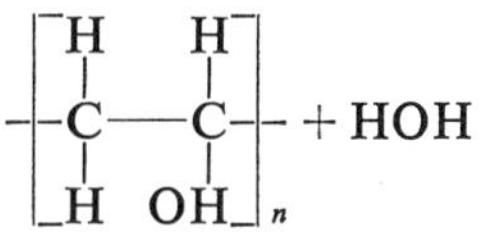

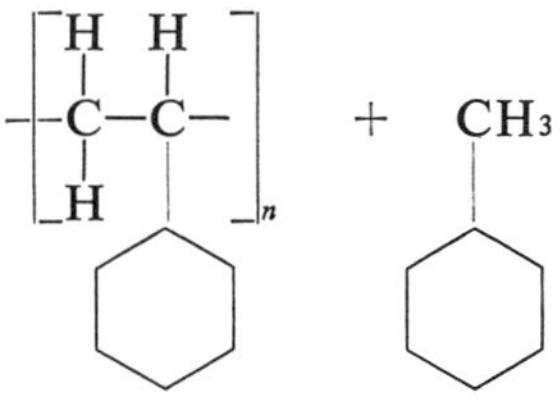

Solubility decreases with decreasing frequency of the "like" or soluble units on the polymer chain, with increasing molecular weight, with crystallinity, and, of course, with cross-linking.

4. Mechanical Properties

Several mechanical analogs and mathematical models for the viscoelastic stress-strain behavior of polymers are described in the next chapter. In this section we shall describe two ways, in addition to cross-linking, to increase polymer strength: *crystallization*, which improves strength by increasing both the packing density of polymer chains and the secondary bonding forces between chains, and *chain stiffening*, which improves strength by reducing the tendency of chains to uncoil and untangle from one another under stress.

a. CRYSTALLIZATION

(1) Requirements for crystallinity. Although crystallinity of polymers is not yet well understood, the requirements for crystallinity are well recognized:

1. Regularity in chain structures. If the chain contains side groups, their orientation should be continuous along the chain. Irregular spacing or orientation of side groups interfere with the arrangement of chains in regular patterns and hinder crystallinity.

2. In addition to regularity of structure, the secondary forces holding chains together in the crystal lattice must be strong enough to overcome the disordering effect of thermal energy. Strong secondary forces such as hydrogen bonding promote crystallinity.

Thus, polymers which lack side groups and which can form strong hydrogen bonds between adjacent chains tend to be crystalline. In spite of intensive efforts, no one has produced a completely crystalline polymer. There always remains some amorphous material (sometimes as little as 2 or 3%) connecting the various *crystallites*, or small crystalline zones. In the fiber industry, crystallinity is often increased by drawing or stretching the fibers or by cooling very slowly from the melt. When nylon is cooled slowly, it can be almost 100% crystalline, whereas if it is quenched from the melt, it is almost 100% noncrystalline.

In applications requiring flexible materials, it may be desirable to *prevent* crystallinity in otherwise crystalline polymers. Properties which favor noncrystallinity are

(a) Long, branched, molecular chains.
(b) Random arrangements of large side groups on the chains.
(c) Copolymer chains.
(d) *Plasticizers* (low-molecular-weight additives) which separate chains from one another.

The effect of random arrangement of side groups is illustrated by polystyrene, for example, which has bulky phenyl groups normally attached randomly along the molecule:

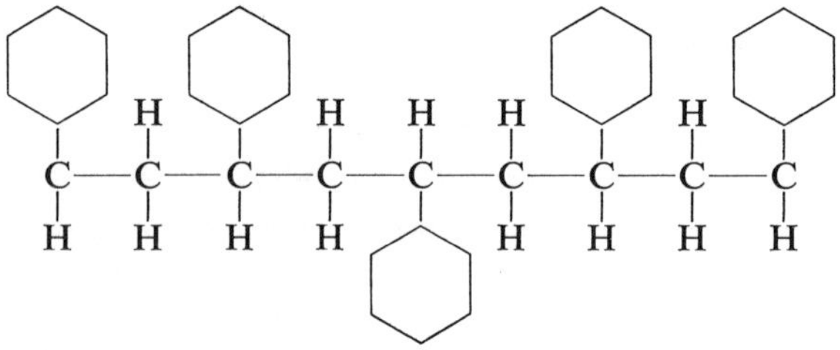

Structures with randomly attached side groups are called *atactic*. They do not allow close packing and therefore are not so strong as the same materials having *isotactic* or *syndiotactic* arrangements (all groups on the same side and alternating regularly on opposite sides, respectively).

Copolymerization always decreases the regularity of polymer chains and therefore promotes the formation of noncrystalline structures. For example, polyvinylidene chloride, which is normally crystalline and not very pliable, can be copolymerized with a small amount of polyvinyl chloride to make it noncrystalline and more flexible.

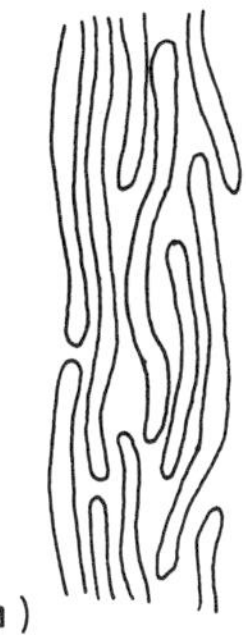

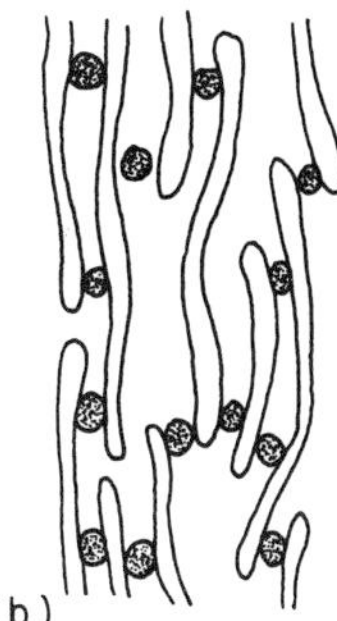

Fig. 23. Plasticizers reduce attraction between polymer chains, permitting flexibility and slipping. (a) No plasticizer; (b) with plasticizer (schematic).

Plasticizers prevent crystallization by separating chains (Fig. 23). For example, cellophane is cellulose chains prevented from crystallizing by the addition of glycerol as a plasticizer. Polyvinyl chloride may also be varied for use from a stiff floor tile to a flexible shower curtain by addition of plasticizers. A plasticizer must be a material which does not volatilize, or the plastic becomes brittle with time. Plasticizers lower the modulus and strength but improve impact resistance, ductility, and low-temperature processing.

(2) Effects of crystallinity on polymer properties. A crystallite may have up to 10% higher density than the corresponding amorphous polymer because chains are packed together more efficiently; density, in fact, is sometimes used as a measure of crystallinity. Since chains are more efficiently packed in crystalline areas, there is a greater density of chains to support stress. Also, since they are in close and regular contact over long distances, the secondary bonding forces holding them together are much greater than in amorphous regions, increasing strength and rigidity significantly.

Crystalline polymers can sometimes display unexpected behavior. For example, a weight may be suspended from a polyvinyl alcohol fiber and immersed in a beaker of boiling water without signs of distress. When the fiber is lowered so that the weight rests on the bottom of the beaker, however, the fiber dissolves. The stress caused by the weight maintains alignment of polymer chains in a crystal lattice, resisting the disordering effects of thermal action and solvent (water) penetration. When the stress is removed the polymer returns to an amorphous state and dissolves.

b. CHAIN STIFFENING

Polymer chains can be stiffened by attaching bulky side groups which restrict carbon bond rotation by steric hindrance, by introducing (nonrotating) double bonds in the carbon chain, or by various bonding arrangements, such as oxygen bonding between the hydrocarbon ring side groups in cellulose, which keeps the chains straight. Synthetic polymers have been developed which employ various combinations of cross-linking, crystallization, and chain stiffening, and some of the results are shown schematically in Fig. 24 and in Table 13.

TABLE 13. Examples of Mechanisms Shown in Fig. 24

Location	*Polymer characteristics*	*Examples*	*Uses*
1	Flexible and crystallizable chains	Polyethylene	Pails, pipes, thin films
		Polypropylene	Steering wheels
		Polyvinyl chloride	Plastic pipes and sidings
		Nylon	Stockings, shirts, dresses, coats
2	Cross-linked, amorphous networks of flexible chains	Phenol-formaldehyde	Television casings and receivers
		Cured rubber	Tires, transport belts, hoses
		Styrenated polyester	Finish on automobiles and appliances
3	Rigid chains	Polyimides	High-temperature insulation
		Ladder molecules	Heat shields
A	Crystalline domains in a viscous network	Terylene (dacron)	Fibers and films
		Cellulose acetate	Fibers and films
B	Moderate cross-linking with some crystallinity	Neoprene	Oil-resistant rubber goods
		Polyisoprene	Particularly resilient rubber goods
C	Rigid chains, partly cross-linked	Heat-resistant materials	Jet and rocket engines and plasma technology
D	Crystalline domains with rigid chains between them and cross-linking between chains	Materials of high strength and temperature resistance	Buildings and vehicles

SOURCE: MARKS, H. F., "The Nature of Polymeric Materials" *Scientific American*, Sept. 1967 [9].

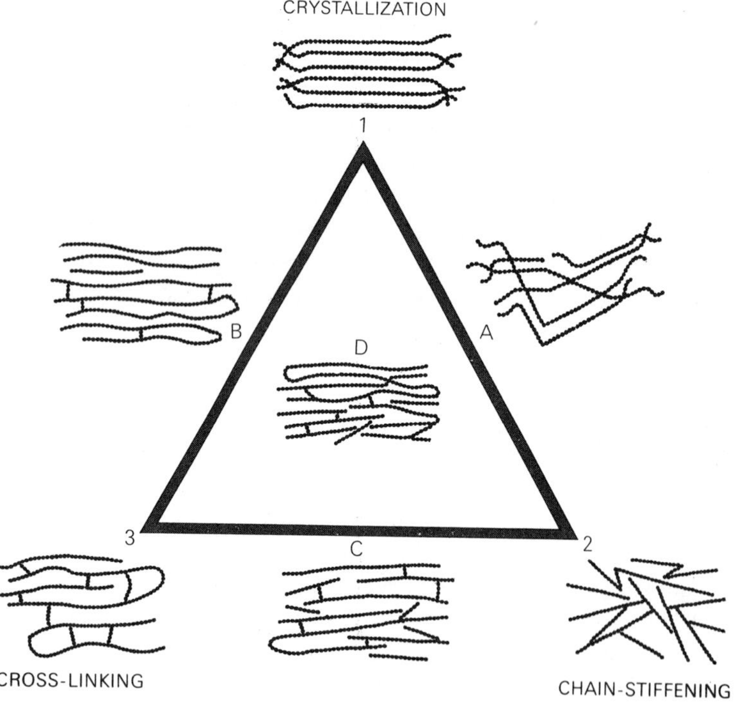

Fig. 24. Schematic of polymer strengthening mechanisms. Corners represent basic strengthening mechanisms and lettered areas represent combinations of mechanisms. From Marks, H. F.: "The Nature of Polymeric Materials," *Scientific American*, Sept. 1967.

PROBLEMS

1. How does each of the following affect the strength of a polymer?
   (a) Increased molecular weight.
   (b) Increased branching.
   (c) Increased crystallinity.
   (d) Increased cross-linking.
2. A rubber contains 93% polymerized chloroprene and 7% sulfur. What fraction of the possible cross-links are joined in vulcanization, assuming full utilization of the sulfur?
3. Will increased cross-linking increase or decrease the Young's modulus of a polymer? Why?
4. What monomer structural properties are required for
   (a) Addition polymerization.
   (b) Condensation polymerization.
5. Which would you expect to be a stronger, tougher polymer, polyethylene $[—CH_2—CH_2—]_n$ or polyvinyl chloride $[—CH_2—CHCl—]_n$? Why?
6. Which is more soluble in water, polystyrene or a polyacid? Why?
7. For polyethylene, polystyrene, and polyisoprene of equal chain lengths, which is likely to have
   (a) The lowest softening point? Why?
   (b) The highest softening point? Why?
   (c) The highest Young's modulus? Why?

D. SURFACE PROPERTIES

Unlike the atoms or molecules within a condensed phase, which are coordinated on all sides, those at the surface have a lower coordination because they lack a similar arrangement of neighbors on one side. Equilibrium bonding arrangements are disrupted, and unstable bonds abound, leading to an excess energy associated with the surface. This *surface energy*, γ, per unit surface area, is commonly expressed in ergs per square centimeter.

The magnitude of γ may be estimated for metallic and covalent materials by considering the number and energy of bonds which must be broken to form the surface. For ionic materials one determines the work which must be done against coulombic forces. If at least one of the phases is a fluid, a variety of methods is also available to measure surface energy experimentally.

Like surface energy, Young's modulus can also be approximately predicted on the basis of bond energies. In this section we shall derive the pertinent relationships and describe some of the engineering applications of surface energy, and in Chap. 3 we shall provide the bond energy basis for Young's modulus.

Surface behavior, i.e., the physical and chemical behavior of interfaces between two or more phases, is fundamental to the manufacture and use of all composite construction materials. This is especially true of air-entrained concretes (Chap. 7), asphalt emulsion and asphalt-aggregate systems (Chap. 9), structural foams (Chap. 10), and fiber-reinforced composites (Chap. 11). Surface energy is also the key to predicting theoretical tensile strength and explaining the brittle fracture of materials (Chap. 3). Since some familiarity with surface chemistry is helpful in the solution of all composite materials problems, in this section we shall summarize some key principles and techniques of surface chemistry.

1. Equivalence of Surface Energy and Surface Tension

Surface energy and surface tension are often described in units of ergs per square centimeter and dynes per centimeter, respectively. The units are equivalent, since an erg is 1 dyne-cm. The equivalence of energy per unit area and force per unit length can be illustrated by the behavior of a soap film on an expandable wire frame (Fig. 25) [10]. If γ is the surface tension of the soap-air interface, the force required to keep the movable wire stationary is $F = 2\gamma l$ (F must resist two soap-air interfaces). Surface energy is the energy required to increase the surface area. If the wire is moved

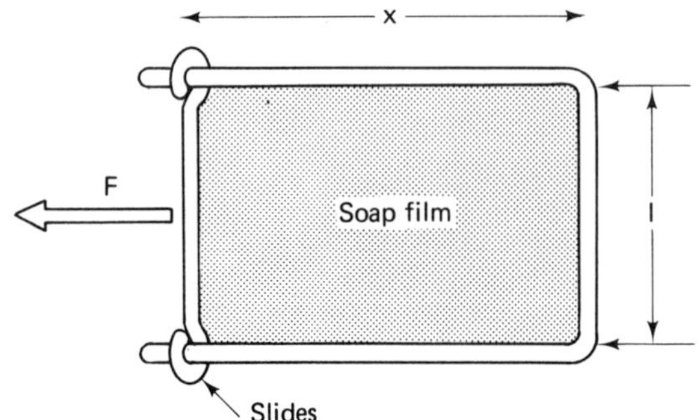

Fig. 25. Surface tension in a soap film [10].

to the left a distance dx, the area $2l\,dx$ will be created at the expense of energy $F\,dx$. The surface energy is

$$\frac{\text{energy}}{\text{area}} = \frac{F\,dx}{2l\,dx} = \frac{2\gamma l\,dx}{2l\,dx} = \gamma$$

Thus, the surface tension and energy are equivalent if the new surface is created with only mechanical work. The value of γ depends on the two phases which the surface separates. The relationship is true for liquids, and for solids near their melting points. In lower-temperature solids the determination of solid-vapor, solid-liquid, and solid-solid interfaces is usually complicated by crystallinity and/or the lack of atomic mobility, compared to that in liquids.

2. Calculations of Surface Energy

Surface energy can be calculated from bond energy considerations. It is convenient to consider short-range (metallic and covalent) and longer-range (ionic) bonding separately [2].

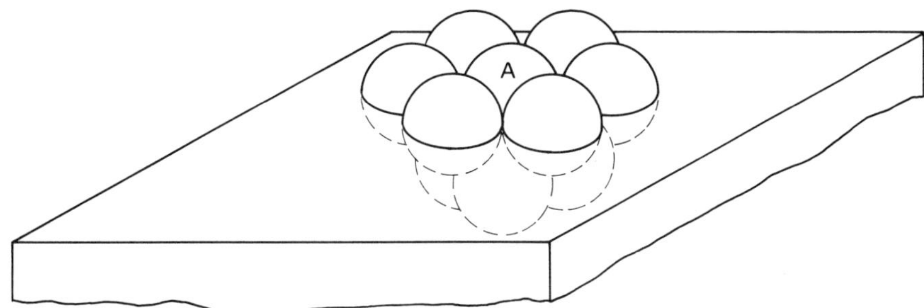

Fig. 26. Surface plane of atoms in a close-packed solid.

a. SURFACE ENERGY OF METALLIC AND COVALENT MATERIALS

Consider a close-packed solid in which each atom has 12 adjacent neighbors, corresponding to either the hexagonal close packing or face-centered cubic structure (Sec. B). An atom in the surface of such a structure, as atom A in Fig. 26, will have only 9 neighbors and therefore have a higher potential energy (the surface energy) due to the unsatisfied bonds. To compute the surface energy, let us assume that the potential energy of each atom depends only on adjacent neighbor interactions; i.e., assume that only short-range forces need be considered. To leave the solid surface (to sublimate), atom A must break 9 bonds, increasing the potential energy of A and each of its nearest neighbors. For a given metal the sublimation energy can be determined. For example, the sublimation energy of copper [2] is

$$H_s = 75{,}900 \text{ cal/g-mole} = 5.26 \times 10^{-12} \text{ erg/atom}$$

Since this is the energy required to break 9 bonds and there is a difference of 3 bonds (12 − 9) between an interior atom and a surface atom, the surface energy per atom is

$$(\tfrac{3}{9})\,(5.26 \times 10^{-12})(\tfrac{1}{2}) = 8.74 \times 10^{-13} \text{ erg/atom}$$

The factor $\frac{1}{2}$ accounts for the fact that the energy gained in sublimation is shared equally by atom A and its nearest neighbors, each of which is also left in a higher energy state. The surface energy per unit surface area can be calculated as

$$\gamma - (8.47 \times 10^{-13})(1.77 \times 10^{15}) = 1{,}550 \text{ ergs/cm}^2$$

where 1.77×10^{15} is the number of atoms per square centimeter on any close packed plane in copper. Surface energies of metals computed in this manner often agree reasonably well with measured values. Most metals have surface energies ranging above 500 ergs/cm^2, compared with 700 ergs/cm^2 for water at room temperature. Experimentally determined energies of fusion and vaporization can similarly be used to obtain surface energies at solid-liquid and liquid-vapor interfaces, respectively.

Generally, more densely packed planes have *lower* surface energies because an atom in such a plane has more nearest neighbors; therefore the number of bonds more closely approximates that of an interior atom. Solids attempt to form their most stable state (lowest free energy) and therefore tend to have larger portions of close-packed planes aligned with their surfaces than would be predicted statistically.

This tendency has been observed in metals, for example, with the field ionizing microscope. Also, a single crystal held at elevated temperatures will always assume a shape bounded by crystallographic planes of minimum surface energy.

b. SURFACE ENERGY OF IONIC MATERIALS

Since coulombic forces are essentially long-range in nature, the simple assumption of nearest neighbor bond breaking, used for metallic and covalent materials, is no longer acceptable. Consider, however, the creation of two surfaces by mechanically separating an ionic crystal into halves [10]. The surface energy created must equal the work of separation,

$$\gamma = \int_0^\infty \sigma \, dr$$

where σ is the stress (force per area) and r the separation of the two halves. The potential energy of an ionic lattice is periodic, repeating itself every interionic distance, $2r_0$. The stress varies with distance of separation, as shown in Fig. 27, which can be approximated as a sine curve. The stress required to separate the two halves increases from zero at equilibrium spacing to a maximum and reduces to zero again at spacing r_0. Therefore

$$\begin{aligned} \sigma &= \sigma_{\max} \sin \frac{2\pi r}{2r_0} \qquad && 0 \le r \le 2r_0 \\ \sigma &\sim 0 && r_0 < r \end{aligned} \tag{18}$$

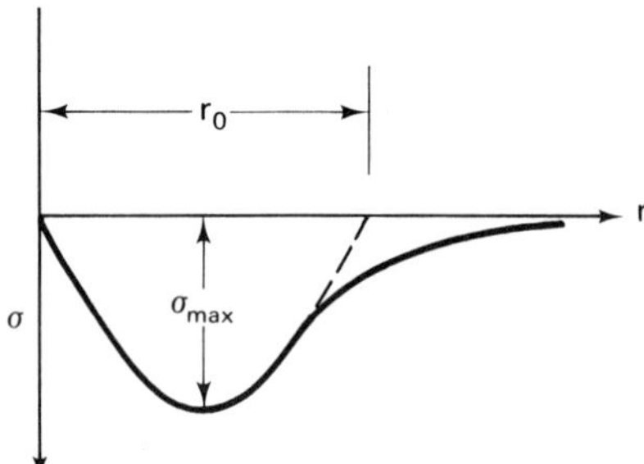

Fig. 27. Sine curve approximation of stress σ as a function of separation r between adjacent planes in an ionic crystal. Ordinate axis shown at equilibrium spacing.

At equilibrium spacing the sine curve has zero curvature, representing linear elasticity, and Young's modulus is

$$E \equiv \left[\frac{\partial \sigma}{\partial \epsilon}\right]_{\epsilon=0} = \left[\frac{d\sigma}{d(r/r_0)}\right]_{r=0}$$

Also from Eq. (18)

$$\left[\frac{d\sigma}{d(r/r_0)}\right]_{r=0} = \left[\pi\sigma_{\max} \cos\left(\frac{\pi r}{r_0}\right)\right]_{r=0} = \pi\sigma_{\max}$$

so that

$$\sigma_{\max} = \frac{E}{\pi}, \qquad \sigma = \frac{E}{\pi} \sin \frac{\pi r}{r_0}$$

and therefore

$$\gamma = \int_0^{r_0} \sin \frac{\pi r_0}{\pi^2}\, dr = \frac{E r_0}{\pi^2} \tag{19}$$

Surface energy can be calculated from Eq. (19) after having values for E (i.e., from mechanical test data) and r_0 (i.e., from x-ray or other data). Table 14 compares calculations of γ by Eq. (19) with values from direct measurements, described in the next section.

TABLE 14. Calculated and Measured Surface Energy Values for Several Ionic Materials

Material	γ *(Expt.) (erg/cm²)*	γ *(Calc.) (erg/cm²)*
NaCl	300	310
MgO	1,200	1,300
LiF	340	370
CaF_2	450	540
BaF_2	280	350
$CaCO_3$	230	380

SOURCE: BROPHY, J. H., R. M. ROSE, and J. WULFF, *The Structure and Properties of Materials*, Vol. II: Thermodynamics of Structure. New York: John Wiley & Sons, Inc., 1967 [10].

3. Measurements of Surface Tension

Only a few of the numerous techniques for measuring surface tension will be described.

The height to which a liquid rises in a capillary tube of known radius is a classical illustration. Liquid rises until the force of surface tension supporting it balances the force of gravity on the liquid column. Equating vertical forces [Fig. 28(a)],

$$\pi r^2 h\, dg = 2\pi r \gamma \cos \theta$$

or

$$\gamma = \frac{h\, dg\, r}{2 \cos \theta}$$

where d is the liquid density, g the acceleration due to gravity, and θ the contact angle between liquid and tube. If h and r are in centimeters, d in grams per cubic centimeter, and g in centimeters per square second, then surface tension has the units

$$\gamma = \text{cm}\left(\frac{\text{g}}{\text{cm}^2}\right)\left(\frac{\text{cm}}{\text{sec}^2}\right)\text{cm} = \frac{\text{g} - \text{cm}}{\text{sec}^2}\left(\frac{1}{\text{cm}}\right) = \frac{\text{dynes}}{\text{cm}}$$

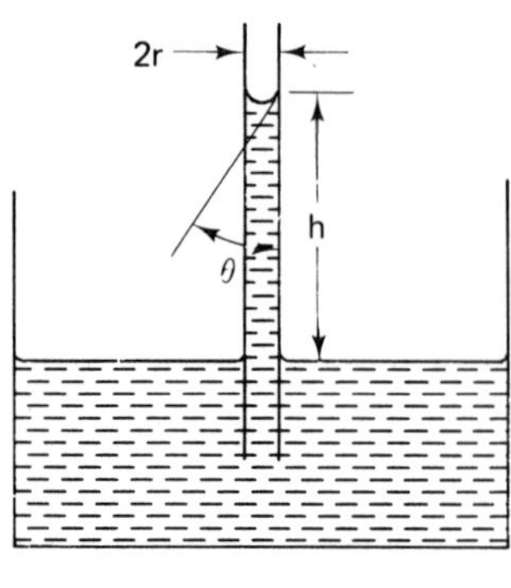

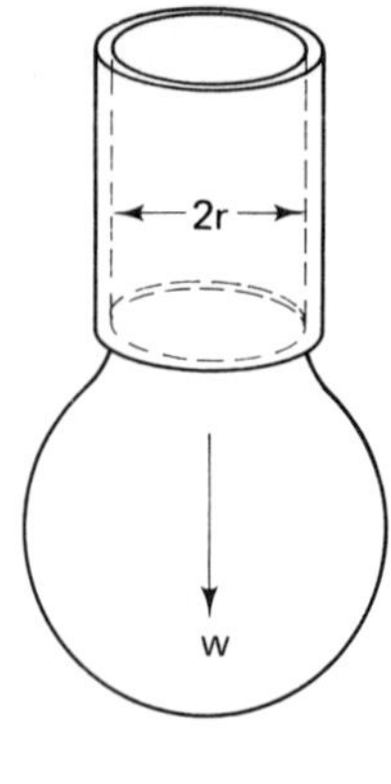

Fig. 28. Surface tension measurement by: (a) capillary rise; (b) weight of a liquid drop.

The height h may also be negative, as when a glass capillary tube is inserted into mercury. Capillarity accounts for the presence of soil moisture above the groundwater table, the transfer of moisture from roots to the tops of trees, and the frequently harmful moisture absorptions in cracks of portland cement concrete.

When a liquid wets a surface, the attraction of liquid molecules to foreign molecules is greater than the attraction to other liquid molecules. Therefore, when a drop of liquid breaks away from a thin-walled tube [Fig. 28(b)] the break occurs within the liquid, not at the liquid-solid contact. If w is the drop weight at the moment of breaking and r the tube radius, then

$$(2\pi r)\gamma = w, \quad \text{or} \quad \gamma = \frac{w}{2\pi r}$$

A very direct way to determine surface tension (surface energy) is to measure the heat of solution (or some reaction) of fine particles of known size. The heat of solution differs from that of the bulk material due to the greater surface areas of the small particles. The difference between the two heats, together with the known difference in initial surface areas, enables the calculation of surface tension.

Wetting of a solid surface by a liquid can also be used. A liquid spreads on a solid surface rather than forming a spherical drop when

$$\gamma_{SL} + \gamma_{LV} < \gamma_{SV}$$

That is, when the net free energy is lowered by replacing an S-V surface by an S-L and an L-V surface together. The subscripts indicate contacts between the solid, liquid, and gas phases [Fig. 29(a)]. No wetting will occur if

$$\gamma_{SV} + \gamma_{LV} < \gamma_{SL}$$

In the intermediate case of partial wetting, force equilibrium must be maintained between components of the three surface tensions in the plane of the solid surface. Hence

$$\gamma_{SV} = \gamma_{SL} + \gamma_{LV} \cos \theta \tag{20}$$

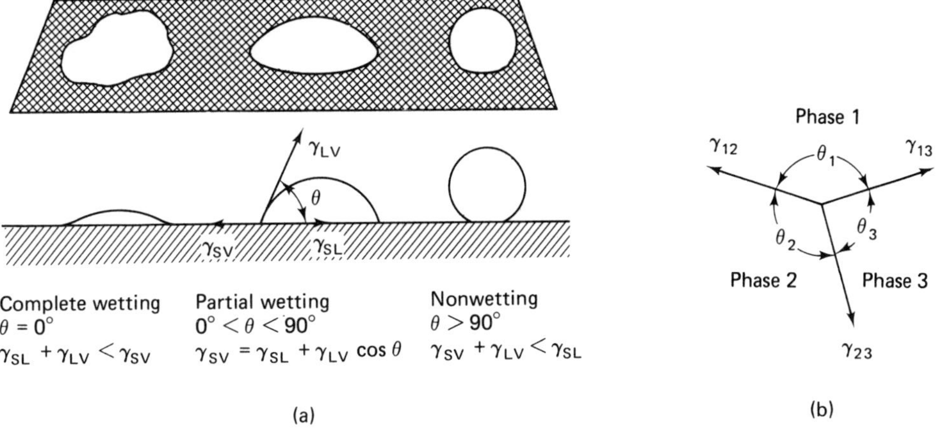

Fig. 29. (a) Wetting of a solid by a liquid; (b) surface tension equilibrium in fluid phases.

Force equilibrium may also be used when all three phases are fluid [Fig. 29(b)]:

$$\frac{\gamma_{12}}{\sin \theta_3} = \frac{\gamma_{23}}{\sin \theta_1} = \frac{\gamma_{31}}{\sin \theta_2} \tag{21}$$

Force balance equations (20) and (21) yield only ratios, not absolute values of γ, and one absolute value must be determined independently to use Eq. (21) and two values to use Eq. (20).

The Du Nouy tensiometer is a platinum ring of radius r attached to a torsion balance [Fig. 30(a)]. The force f required to raise the ring from a liquid surface is measured directly. Since two new surfaces are formed around the circumference of the ring,

$$f = 4\pi r \gamma$$

Obtaining and maintaining contamination-free surfaces is a serious problem in all the above experimental methods.

In blowing bubbles beneath a liquid surface [Fig. 30(b)], the work done in increasing the surface area of the expanding bubble, $\gamma\, dA$, equals the work of expansion at constant pressure ($P\, dV$), so that

$$\gamma(8\pi R\, dR) = 4\pi R^2\, dR\, \Delta P$$

and

$$\Delta P = \frac{2\gamma}{R}$$

This result applies to liquid-liquid as well as gas-liquid interfaces. For the more general curved surface with radii R_1 and R_2 it becomes

$$\Delta P = \gamma\left(\frac{1}{R_1} + \frac{1}{R_2}\right)$$

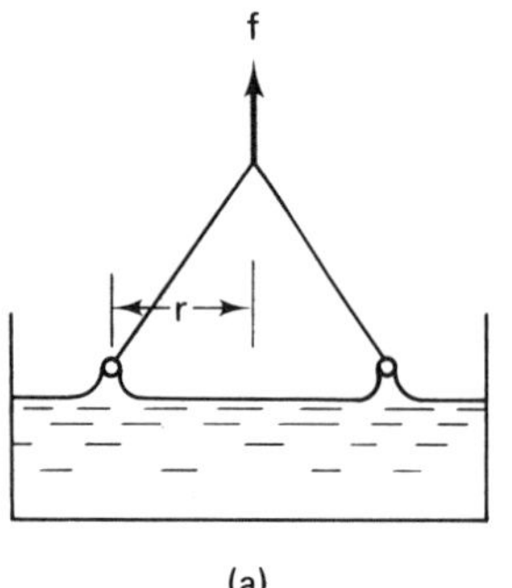

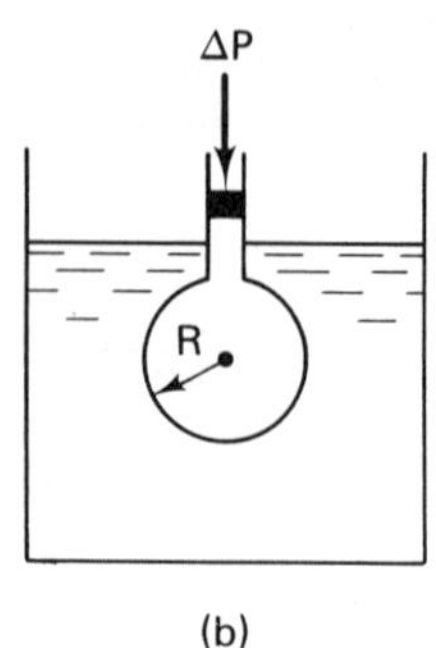

Fig. 30. (a) Du Nouy tensiometer; (b) pressure difference across a curved interface.

4. Adsorption

In any multiphase system nature tends to minimize total surface energy. There are two ways to do this: minimize surface area and minimize surface energy per unit area. Gas bubbles and liquid drops tend to assume spherical shapes in order to minimize surface areas. Similarly, minimal surface areas tend to exist in single-phase solids between grains of different crystallographic orientation. Surface energy per unit area is most noticeably minimized by *adsorption*, i.e., the segregation of various components to and from the surface. Those impurities or components which reduce γ will segregate to the surface; those which decrease γ will segregate away from the surface. Those solutes which concentrate in the surface layer (e.g., soap in water) have a large effect on surface phenomena and cause a great *decrease* in surface tension. Only tiny amounts of such components are sufficient to saturate the surface. Those solutes which concentrate in the bulk of the solution (e.g., NaCl in water) have little effect on surface phenomena and therefore can *increase* the surface tension only slightly.

Gibbs equation. The effect of concentration of a solute on the surface tension of a solution is given by the Gibbs equation:

$$\mu = -\frac{c}{RT}\frac{d\gamma}{dc}$$

where μ = excess concentration of solute in the surface layer per square centimeter of surface (as compared to the concentration in the bulk of the solution),
c = concentration of solute in bulk of solution,
R = molar gas constant,
T = absolute temperature,
$d\gamma/dc$ = rate at which surface tension changes with concentration.

From measurements of surface energy and free energy, the segregation of any component to or from a surface may be predicted with this equation.

Chromatography. Chromatography is a fractionation technique useful in the study of some construction materials, notably asphalts and various mineral and organic colloidal systems. Chromatography uses the principle of selective adsorption combined with the fact that the adsorption rate varies with a given adsorbent for different substances. Since much of the work has been with colored substances, we have the word *chromatography*.

The mixture to be separated is dissolved in a solvent and poured through a column containing the adsorbent. The various components are adsorbed at different rates. The most easily adsorbed substances concentrate at the top of the column and the least easily adsorbed substances at the bottom. Since the various components may overlap, pure solvent is poured through the column to continually desorb and resorb each component until complete separation is accomplished. This process is called *elution.*

5. Surface-Active Agents

Surface-active agents are essential components in the formulation of air-entrained concretes, foamed polymer insulating materials, asphalt emulsions, and fiber-reinforced composites. They are also used in various mineral beneficiation processes and occasionally as pretreatments to improve the bonding characteristics of aggregates.

Surface-active agents (contracted to *surfactants*) resemble soaps in that they contain a polar group which is hydrophilic (water-attracting) and an organic group which is hydrophobic (water-repelling) and are attracted to certain organic solvents. The combination of these two groups gives a molecule which is attracted to an oil-water interface and reduces the surface tension.

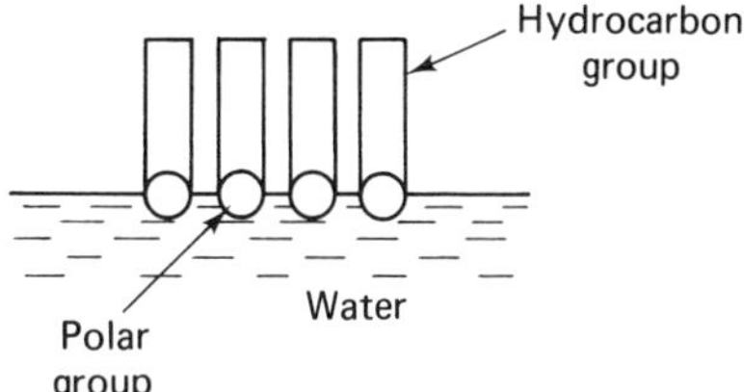

Fig. 31. Surfactant monolayer on water.

For example, if stearic acid ($C_{17}H_{35}COOH$) is dissolved in benzene and placed on water, the benzene evaporates, leaving a film of acid. The film spreads out to form a layer one molecule thick, or *monolayer*, with the polar carboxyl group (COOH) dissolved in the water and the insoluble hydrocarbon chain projecting from the surface (Fig. 31). The length and cross-sectional area of one molecule of acid can be calculated as follows:

$$\text{molecular length} = \text{monolayer thickness} = \frac{\text{acid volume}}{\text{monolayer area}}$$

$$\text{molecular cross-sectional area} = \frac{\text{monolayer area}}{\text{number of molecules}} = \frac{\text{monolayer area}}{N\,(\text{moles of acid})}$$

where N is Avogadro's number.

Surfactant properties are varied by changing the chemical structure and size of either the polar group or the hydrocarbon group, and surfactants may be classified according to the type of either of these two groups. Tables 15 and 16 indicate alternative classifications by these two criteria. They are not intended as complete classifications but indicate the more important types of surfactants.

TABLE 15. Classification of Surfactants by Type of Polar Group

Type	*General formula*	*Name or description*
A	$R\cdot CO\cdot O^{-+}Na$	Anionic
B.1	$R\cdot NH_2$	Cationic: Primary amines
2	$R\cdot NH\cdot R'$	Secondary amines
3	$R\cdot N\cdot R_2'$	Tertiary amines
4	$R\cdot N\cdot R_3'^{+-}Cl$	Quaternary ammonium salts
5	$R\cdot CO\cdot NH\cdot CH_2\cdot CH_2NR_2'$	Diamines
C	$R\cdot CO\cdot O\cdot CH_2\cdot CHOH\cdot CH_2\cdot OH$	Nonionic (various alcohols)
D		Amphoteric (positive or negative charge, depending on pH of solution)

In Table 16, Type A.1 is illustrated by ordinary soap. Type A.3 corresponds to type C in Table 15 and is distinctive in that its members are nonionic and may be used over a wide pH range. Type A.6 (B.5 in Table 15) is effective in both acid and neutral solutions and may be made active in the presence of alkali by further alkylation of the nitrogen atom to give the corresponding substituted alkyl trialkylammonium salt.

Areas of application for several classes of surfactants will be pointed out throughout the remainder of the text, particularly in the two chapters on bituminous systems.

TABLE 16. Classification of Surfactants by Type of Organic Group

Type	*General formula*	*Name or description*
A.1	$R\cdot CO\cdot O^{-+}Na$	Fatty acid salt
2	$C_nH_{2n-2}(O\cdot SO_2\cdot O^{-+}Na)CO\cdot O^{-+}Na$	Disodium salt of sulfated fatty acid
3	$R\cdot CO\cdot O\cdot CH_2\cdot CHOH\cdot CH_2\cdot OH$	Glyceryl ester of a fatty acid
4	$R\cdot CO\cdot O\cdot CH_2\cdot CH_2\cdot SO_2\cdot O^{-+}Na$	Sodium fatty acid ester sulfonate
5	$R\cdot CO\cdot NH\cdot CH_2\cdot CH_2\cdot SO_2\cdot O^{-+}Na$	Sodium fatty acid amide sulfonate
6	$R\cdot CO\cdot NH\cdot CH_2\cdot CH_2\cdot NR_2'$	Fatty acid amido ethyl dialkyl amine
B.1	$R\cdot O\cdot SO_2\cdot O^{-+}Na$	Sodium alkyl sulfate where R is large
2	$R_2'CH\cdot O\cdot SO_2\cdot O^{-+}Na$	Sodium secondary alkyl sulfate
3	$R'\cdot O\cdot OC\cdot CH_2$ | $R'\cdot O\cdot OC\cdot CH\cdot O\cdot SO_2\cdot O^{-+}Na$	Sodium salt of the bisulfate of a dialkyl dicarboxylate
C.1	$R\cdot SO_2\cdot O^{-+}Na$	Sodium alkyl sulfonate where R is large
2	$Ar\cdot SO_2\cdot O^{-+}Na$	Sodium aryl sulfonate
3	$R'\cdot O\cdot OC\cdot CH_2$ | $R'\cdot O\cdot OC\cdot CH\cdot SO_2\cdot O^{-+}Na$	Sodium salt of the sulfonic acid derivative of a dialkyl dicarboxylate
D	$R\cdot N^+$ (pyridine ring)	An alkyl pyridinium salt where R is a long aliphatic chain

Legend: R represents a long hydrocarbon chain of the aliphatic type,
Ar represents an aryl or aromatic nucleus, and
R′ represents a primary or secondary alkyl group of the short chain type.

SOURCE: DEGERING, E. F., *Organic Chemistry*. New York: Barnes & Noble, Inc., 1951 [11].

6. Colloidal Behavior

Colloidal systems contain two phases, one dispersed in the other. The dispersed phase is discontinuous, and the dispersion medium is continuous. Dispersed particle sizes typically range from 1 to 1,000 $m\mu$ (1 $m\mu = 10^{-7}$ cm). Below 1 $m\mu$ the system is a true solution; above 1,000 $m\mu$ it is a coarse suspension. The important characteristic on which all properties of colloidal systems depend is the large surface area of the dispersed phase. A 1-cm^3 cube, for example, has a surface area of 6 cm^2. If such a cube is divided into 1 $m\mu^3$ cubes, the total surface is 6,000 m^2 (about $1\frac{1}{2}$ acre).

Table 17 summarizes the classification of colloids. A few properties of emulsions and sols will be discussed.

TABLE 17. Classes of Colloidal Systems

Dispersion medium	*Dispersed phase*	*Class*	*Example*
Gas	Liquid	Liquid aerosol	Fog
Gas	Solid	Solid aerosol	Smoke
Liquid	Gas	Foam	Soap foam
Liquid	Liquid	Emulsion	Asphalt in water (Chap. 5)
Liquid	Solid	Sol	Cement paste (Chap. 7)
Solid	Gas	Solid foam	Foamed polystyrene (Chap. 10)
Solid	Liquid	Solid emulsion	Polymer-modified cement (Chap. 7)
Solid	Solid	Solid sol	Pigmented glass

Emulsions. Emulsions are dispersions of one immiscible liquid in another. Ordinarily two types are possible, i.e., an oil dispersed in water and water dispersed in an oil. Oil mixes readily with a water-in-oil emulsion and water mixes readily with an oil-in-water emulsion.

Emulsions may be prepared with a homogenizer in which oil, water, and an emulsifying agent (surfactant) are forced through small openings under high pressure to impinge against a hard surface. The surfactant forms films around the droplets and provides emulsion stability by preventing their coalescence.

Emulsions may be broken (the disperse phase coalesced) by neutralizing or deactivating the surfactant. For example, an emulsion having sodium palmitate can be broken by adding an acid. This converts sodium palmitate to palmitic acid, which has no emulsifying action. Emulsions may also be broken by mechanical or thermal means, such as centrifuging, freezing, or heating. Emulsion formulations which break within an appropriate time range after application are essential to the hardening of water-based paints and the coating of stone by asphalt emulsions.

Sols. Sols are classified according to the liquid used as a dispersion medium. Water is the dispersion medium in hydrosols, alcohol in alcosols, etc. If the dispersion medium attracts the dispersed phase, the sol is *lyophilic* (or *hydrophilic* if water is the dispersion medium). If little or no attraction exists, the sol is *lyophobic* (*hydrophobic*

TABLE 18. Distinctions Between Hydrophilic and Hydrophobic Sols

Property	*Hydrophilic*	*Hydrophobic*
Example	Protein in water	Metal in water
Viscosity	Higher than water	About same as water
Concentration of dispersed phase	Can be high	Low
Amount of electrolyte required for stability	None	Very little
Amount of electrolyte required for precipitation	Not easily precipitated by electrolyte	Small
Precipitation reversible	Usually	Usually not
Ease in varying electrical charge on particles	Particles may be +, −, or neutral, depending on pH	Charge not easily changed
Particles migrate under applied potential	May or may not migrate	Yes

in the case of water). Table 18 lists some distinguishing characteristics of hydrophilic and hydrophobic sols.

The pH at which a hydrophilic sol has no charge is called the *isoelectric point.* The stability is minimum at the isoelectric point. Sols are precipitated by ions whose charge is opposite that of the sol particles, and effectiveness of precipitation increases with the valence of the ion. Sols may also be precipitated by adding a sol of opposite charge. For example, a negative arsenious sulfide sol added to a positive ferric oxide sol causes the precipitation of both.

Sols and emulsions obey similar laws with respect to their sedimentation behavior and electrical properties.

Sedimentation behavior. In colloidal systems particle settling due to gravity is opposed by Brownian movement so that finally a state of equilibrium is reached in which settling and diffusion are just balanced. For particles of uniform size the governing equation is

$$2.303 \log \frac{n_2}{n_1} = \frac{Nvg(d - d')(x_1 - x_2)}{RT} \qquad (22)$$

where subscripts 1 and 2 refer to two different levels and

n = number of particles per unit volume at a given level,
N = Avogadro's number,
v = volume of one particle (cm^3),
g = acceleration due to gravity = 981 cm/sec^2,
d = particle density,
d' = solvent density,
x = distance of particle from bottom of container (cm),
R = molar gas constant (ergs/deg-mole),
T = absolute temperature.

Equation (22) can be used to determine the volume of a particle, v, from which the radius can be computed if the particle is spherical.

Electrical properties. Hydrophobic particles maintain fairly fixed electrical charge (Table 18) and migrate toward the electrode of opposite charge when placed in an electric field. This migration is called *electrophoresis*. Since all particles have the same charge, they repel each other and prevent precipitation. The charge usually results from adsorbed ions. These ions attract ions of opposite charge, resulting in an electrical double layer. Figure 32 gives a simplified picture of the double layer for a

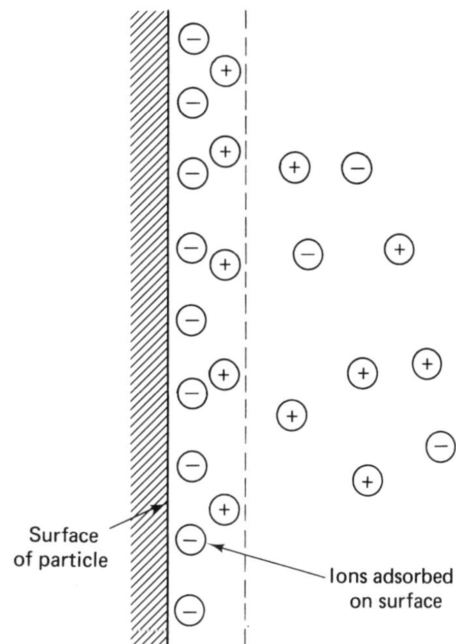

Fig. 32. Electrical double layer.

negatively charged particle. When placed in an electrical field, the particle and a film of ions between the particle surface and the dashed line migrate toward the positive electrode. The potential drop between the dashed line and a point in the bulk of the dispersion medium is called the *zeta potential.* Colloid systems having a high zeta potential have high stability. If the fluid is free to move instead of the particles, it will move to the opposite electrode. Such fluid motion is called *electroosmosis.*

PROBLEMS

1. What is the origin of surface energy?
2. What surface energy in kcal/g-mole does BCC iron have on the (110) plane? The energy of sublimation is 97 kcal/g-mole.

3. Determine the plane in an FCC crystal which has the minimum surface energy. [*Hint:* Find the number of unsatisfied bonds per unit area on the (111) plane and on several other low index planes.]

4. Estimate the height to which a column of water will rise in a glass tube of (a) 1-μ diameter, (b) 10-μ diameter. Assume zero contact angle and $\gamma = 70$ dynes/cm.

5. What pressure in dynes/cm^2 is required to force water out of a sintered glass filter having a uniform pore diameter of 0.10 μ? Assume zero contact angle and $\gamma = 70$ dynes/cm.

6. What is the weight of a water droplet breaking away from a thin-walled glass tube 0.4 cm in diameter? Assume that $\gamma = 70$ dynes/cm.

7. What type of wetting occurs when a drop of benzene is placed on a water surface? How much energy is released if the benzene spreads over 2 m^2 of water? Assume negligible surface area of the initial benzene drop. The surface tensions are

Interface	γ (dynes/cm)
Water-benzene	35
Water-air	73
Benzene-air	29

8. Find the contact angle of a drop of water on a mercury surface. The surface tensions are

Interface	γ (dynes/cm)
Mercury-water	375
Mercury-air	380
Water-air	74

9. An oiled steel needle floats on water. Derive the relationship between the oil-water surface tension, specific gravities of steel and water, and the diameter of the needle if it is immersed to one third of its diameter.

10. What is the usual effect of impurities on the surface energy of a material. Why?

11. 2×10^{-4} cm^3 of stearic acid (dissolved in benzene) placed on a water surface forms a monolayer covering 800 cm^2. Calculate the length of a stearic acid molecule.

12. State five differences in properties between hydrophilic and hydrophobic sols.

13. Show by a sketch the electrical double layer distribution at a negatively charged sol particle. Define zeta potential.

14. The following data were obtained in conjunction with a centrifuge test on an oil-in-water asphalt emulsion:

$$\begin{aligned} x_1 - x_2 &= 0.01 \text{ cm} \\ n_1 &= 1{,}500 \\ n_2 &= 300 \\ d &= 0.94 \text{ g/cm}^3 \\ d' &= 1.0 \text{ g/cm}^3 \\ T &= 295^\circ\text{K} \\ \text{centrifuge acceleration} &= 100 \text{ g} \end{aligned}$$

Calculate the volume and radius of one asphalt droplet. Note that the droplets are less dense than the dispersion medium.

REFERENCES

1. PAULI, W., *Zeit. Physik.*, **31** (1925) 765.
2. EISENSTADT, M. M., *Introduction to Mechanical Properties of Materials.* New York: The MacMillan Company, 1971.
3. BORN, M., and J. E. MAYER, *Zeit. Physik.*, **75** (1932) 1.
4. PAULING, L., *The Nature of the Chemical Bond*, 3rd ed. Ithaca, N.Y.: Cornell University Press, 1960.
5. MOFFATT, W. G., G. W. PEARSALL, and J. WULFF, *The Structure and Properties of Materials*, Vol. 1: Structure. New York: John Wiley & Sons, Inc., 1967.
6. VAN VLACK, L. H., *Materials Science for Engineers.* Reading, Mass.: Addison-Wesley Publishing Company, Inc., 1970.
7. SINNOTT, M. J., *The Solid State for Engineers.* New York: John Wiley & Sons, Inc., 1958.
8. VAN VLACK, L. H., *Elements of Materials Science*, 2nd ed. Reading, Mass.: Addison-Wesley Publishing Company, Inc., 1964.
9. MARK, HERMAN F., "The Nature of Polymeric Materials," in *Materials, A Scientific American Book.* San Francisco: W. H. Freeman and Company, Publishers, 1967.
10. BROPHY, J. H., R. M. ROSE, and J. WULFF, *The Structure and Properties of Materials*, Vol. II: Thermodynamics of Structure. New York: John Wiley & Sons, Inc., 1967.
11. DEGERING, E. F., *Organic Chemistry.* New York: Barnes & Noble, Inc., 1951.

3

MECHANICAL BEHAVIOR

Now in building of chaises, I tell you what,
There is always somewhere a weakest spot,
In hub, tire, felloe, in spring or thill,
In panel, or crossbar, or floor, or sill,
. . .
And that's the reason, beyond a doubt,
That a chaise breaks down, but doesn't wear out.
—HOLMES

Chapter 2 provided a chemical structural basis for studying the continuous and discrete phase materials described in Chaps. 4–6. In this chapter basic skills in the deformation and fracture mechanics of materials are presented. In Chaps. 7–12 we shall integrate our understanding of the chemistry of various continuous and discrete phases with basic skills in mechanics to study the deformation and fracture of two-phase systems having various geometries of the discrete phase.

A. DEFORMATION MECHANICS

All materials deform under changes in stress or temperature. The deformation is called *elastic* if it is completely recovered after the material returns to its original stress or temperature. Nonrecoverable deformation is called *plastic* deformation. Deformation (stress-strain) relationships can sometimes be predicted directly from the type of structure and atomic bonding.

1. Elastic Deformation

Elastic deformation is linear in some materials and nonlinear in others. Stress-strain relationships for typical materials are shown in Fig. 1, where the shaded regions indicate elastic deformation.

Crystalline materials exhibit linear elasticity [Fig. 1(a)]. The stress-strain ratio is high because the applied stress is resisted by primary bonds (ionic, covalent, metallic). Noncrystalline materials such as glass and cross-linked polymers may also exhibit linear elasticity if deformation is opposed from the start by primary bonds.

Elastomers, which are long chain polymers with little or no cross-linking, exhibit highly nonlinear elasticity [Fig. 1(b)]. Low tensile stress can produce large deformation simply by the straightening of kinked chains. But once the chains have been aligned, further elongation is resisted by secondary bonding forces between chains and by primary bonding forces within them. Similarly, low compressive stress can produce large deformation simply by causing a more efficient filling on space by the kinked chains. As available space decreases, further compression is resisted by primary bonding forces within the chains. The stress-strain slopes in both tension and compression thus increase with increasing stress.

Fig. 1. Elastic behavior: (a) crystalline materials; (b) elastomers; (c) cellular materials. (Adapted from Hayden, H. W., W. G. Moffatt, and J. Wulff, *The Structure and Properties of Materials, Vol. III: Mechanical Behavior.* New York: John Wiley and Sons, Inc., 1965 [1], by permission.

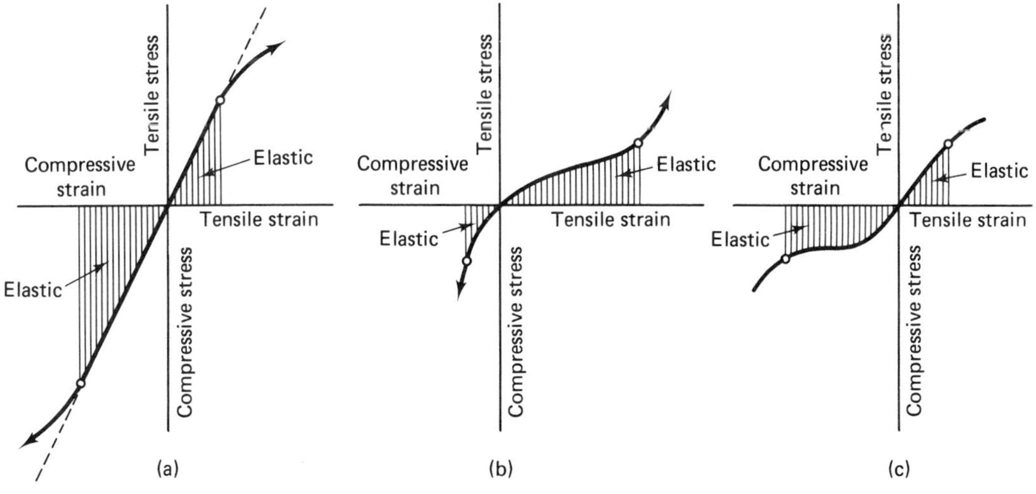

Cellular materials, such as wood, may be stiff in compression until elastic buckling of the cell walls occurs, beyond which additional strain is easily produced until stiffness again increases due to compaction of the cells [Fig. 1(c)]. If the cell walls crush instead of buckle elastically, the deformation is no longer elastic. In tension, the cell walls obviously do not buckle in the same manner, and thus elastic behavior is more nearly linear.

The slope of the linear portion of a stress-strain curve is

$$E = \frac{\sigma}{\epsilon} \tag{1a}$$

where σ is the uniaxial tensile or compressive stress (psi), ϵ is the strain (in./in.), and E is Young's modulus (psi), the elastic modulus of a material under uniaxial stress. Additional elastic moduli relate to different types of stress:

$$G = \frac{\tau}{\gamma} \quad \text{(shear modulus)} \tag{1b}$$

$$K = \frac{\sigma_t}{\epsilon_v} \quad \text{(bulk modulus)} \tag{1c}$$

where τ is the shear stress (psi), γ is the shear strain [the tangent of the shear angle β, Fig. 2(b)], σ_t is the triaxial or hydrostatic tensile or compressive stress (psi), and ϵ_v is the fractional volume expansion or contraction. The Poisson ratio, also an elastic constant, is the ratio of lateral to axial strain:

$$\nu = -\frac{\epsilon_x}{\epsilon_y} \quad \text{(Poisson ratio)} \tag{2}$$

where the normal load is applied in the y direction. Figure 2 illustrates the elastic moduli, and Table 1 gives Poisson ratios of some common materials.

TABLE 1. Poisson Ratios

Material	ν
Aluminum	0.34
Copper	0.35
Iron	0.28
Lead	0.40
Glass	0.25
Polyethylene	0.41

For isotropic materials (having properties at a point identical in all directions), if any two of the four constants E, G, K, and ν are known, the other two may be derived (see Problems 1 and 2):

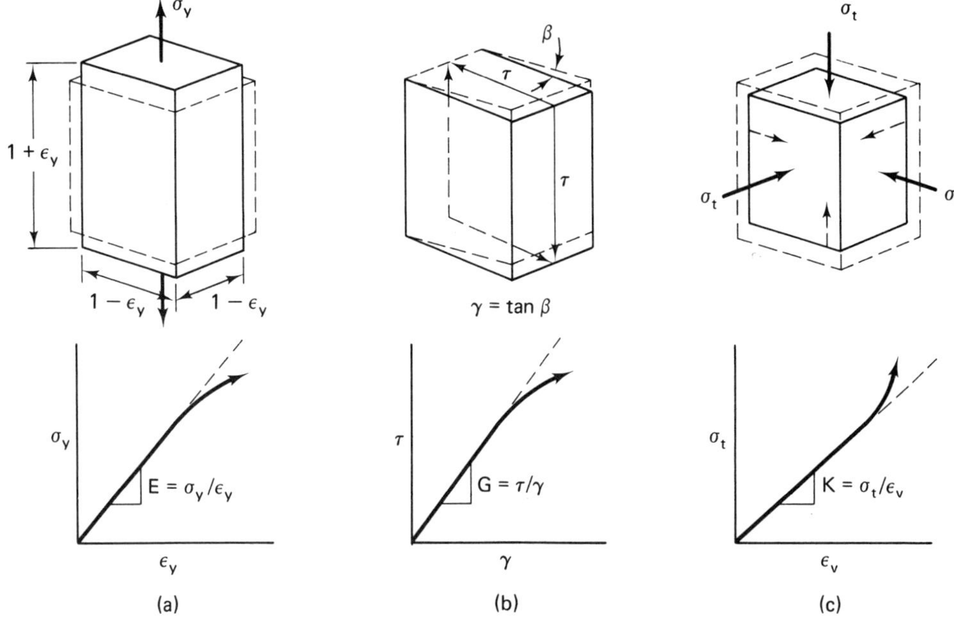

Fig. 2. Elastic moduli. (a) Young's modulus; (b) shear modulus; (c) bulk modulus. Dashed lines represent initial stress-free cube of edge length l.

$$G = \frac{E}{2(1 + \nu)} \tag{3}$$

$$K = \frac{E}{3(1 - 2\nu)} \tag{4}$$

Under uniaxial stress, normal strains are found by *Hooke's law*:

$$\epsilon_x = \frac{\sigma_x}{E}, \qquad \epsilon_y = \epsilon_z = -\frac{\nu\sigma_x}{E} \tag{5}$$

where the normal stress is in the x direction. The negative sign indicates that extension in one dimension is usually accompanied by compression in the other two. Under biaxial or triaxial stresses, the strains are calculated by superposition, giving the generalized Hooke's law for isotropic materials:

$$\begin{aligned} \epsilon_x &= \frac{1}{E}[\sigma_x - \nu(\sigma_y + \sigma_z)] \\ \epsilon_y &= \frac{1}{E}[\sigma_y - \nu(\sigma_z + \sigma_x)] \\ \epsilon_z &= \frac{1}{E}[\sigma_z - \nu(\sigma_x + \sigma_y)] \end{aligned} \tag{6}$$

The corresponding equations relating shear stress and shear strain can be shown to be

$$\gamma_{xy} = \frac{\tau_{xy}}{G}, \qquad \gamma_{yz} = \frac{\tau_{yz}}{G}, \qquad \gamma_{zx} = \frac{\tau_{zx}}{G}$$

where the subscripts identify the plane in which shear occurs.

For small strains, the volume strain (dilatation) under generalized stresses is

$$\epsilon_v = 3\bar{\epsilon} = \frac{\bar{\sigma}}{K} \tag{7}$$

where $\bar{\epsilon}$ is the mean normal strain and $\bar{\sigma}$ is the mean normal stress.

Strains may also be defined in terms of the displacement components of a point in a stressed body. If an element length is *dx* before deformation (Fig. 3), it becomes

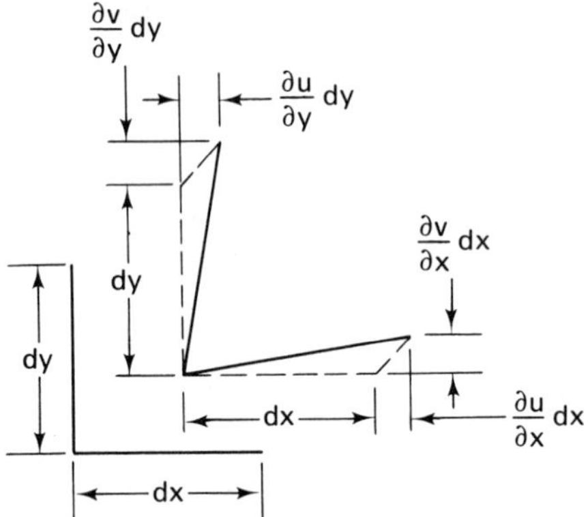

Fig. 3. Displacement components for defining strains.

approximately $dx + (\partial u/\partial x)\,dx$ after deformation, assuming small strain and hence small strain angle. The normal strain is then

$$\epsilon_x = \frac{\left(dx + \dfrac{\partial u}{\partial x}\,dx\right) - dx}{dx} = \frac{\partial u}{\partial x} \tag{8}$$

Similarly,

$$\epsilon_y = \frac{\partial u}{\partial y}, \qquad \epsilon_z = \frac{\partial u}{\partial z}$$

The sum of the tangents of the angular changes (Fig. 3) is

$$\frac{(\partial v/\partial y)\,dx}{dx} + \frac{(\partial u/\partial y)\,dy}{dy} = \frac{\partial v}{\partial x} + \frac{\partial u}{\partial y}$$

Since the shear strain was defined as the tangent of the angular change and since the sum of tangents of two small angles approximately equals the tangent of their sum,

$$\gamma_{xy} = \frac{\partial u}{\partial y} + \frac{\partial v}{\partial x} \tag{9}$$

Similarly,

$$\gamma_{yz} = \frac{\partial v}{\partial z} + \frac{\partial w}{\partial y}, \qquad \gamma_{xy} = \frac{\partial u}{\partial z} + \frac{\partial w}{\partial x}$$

where the double subscripts identify the plane of strain.

a. ATOMIC BASIS

The elastic properties of crystalline materials can be related directly to their atomic structure. Figure 2 of Chap. 2 indicates that the equilibrium spacing, r_0, between two ions occurs where the net force is zero or where the potential energy is a minimum. A displacement in either direction will increase the potential energy and call restoring forces into play. Similar behavior describes three-dimensional arrays of ions in crystals, where elastic strain results in a change in interionic spacing. It follows that Young's modulus E is proportional to the slope at r_0 of the force curve, or to the curvature at r_0 of the energy curve. To evaluate Young's modulus, we can first consider a pair of ions, then a linear arrangement of ions, and finally an ionic crystal.

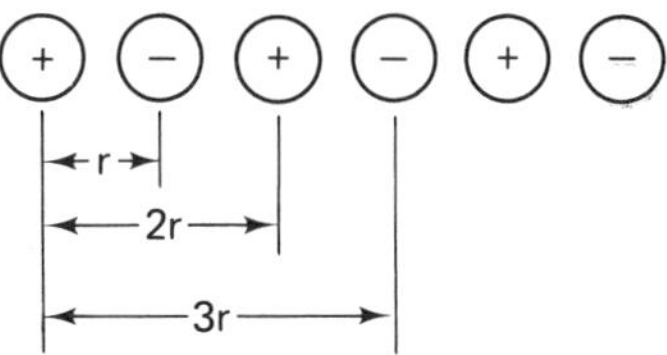

Fig. 4. Attraction of an ion by a chain of alternating anions and cations.

For a linear arrangement of cations and anions of equal valence (Fig. 4), the net force on the end ion, according to Eq. (6) of Chap. 2, is

$$f = \frac{e^2Z^2}{r^2} - \frac{e^2Z^2}{(2r)^2} + \frac{e^2Z^2}{(3r)^2} - \cdots$$

a rapidly converging series which approaches the value $0.80e^2Z^2/r^2$. The computation may be extended to a two-dimensional array, and finally to a three-dimensional cubic array of alternating charges to give

$$f = \frac{0.29e^2Z^2}{r^2}$$

a value less than one third the force extерted by a single ion of opposite charge. The coefficient 0.29 varies with the crystal geometry.

Repulsion predominates when r is so small that the electron shells impinge upon each other. To determine the value of B_{12} in Eq. (8) of Chap. 2, we write for the equilibrium spacing of two ions as

$$f = \frac{0.29e^2Z^2}{r^2} - ne^2\frac{B_{12}}{r_{12}^{n+1}} \tag{10}$$

Since $f = 0$ when $r = r_0$,

$$B_{12} = 0.29e^2Z^2r_0^{n+1}$$

Substituting into Eq. (10),

$$f = 0.29\frac{e^2Z^2}{r^2}\left[1 - \left(\frac{r_0}{r}\right)^j\right] \tag{11}$$

where for convenience we have put $j = n - 1$. The quantities e, Z, r_0, and j are constants for a given crystal; e is a universal constant, Z depends on the ion species, r_0 may be determined from x-ray diffraction analysis, and j (the Born exponent n, Table 2 of Chap. 2) can be calculated on the basis of theoretical considerations which we shall omit [2]. In case of different types of anions and cations, the corresponding values of j are averaged.

Thus the resistance of ionic crystals to change in dimension derives directly from ionic bonding. The same is true for changes in shape, although the derivations are more difficult. Extensions to other types of bonding may be made, using techniques of modern physics.

Equation (11) may be used directly to obtain Hooke's law. Since most materials have only small elastic deformations, we substitute

$$r = r_0 + \Delta r \tag{12}$$

into Eq. (11), where $\Delta r/r_0 \ll 1$. By the definition of a derivative,

$$\left|\frac{f(r_0 + \Delta r) - f(r_0)}{\Delta r}\right|_{\Delta r \to 0} = f'(r_0)$$

or approximately

$$f(r_0 + \Delta r) - f(r_0) = f'(r_0)\Delta r \tag{13}$$

since $\Delta r/r_0 \ll 1$. We note that $f(r_0) = 0$ and that

$$f'(r) = 0.29e^2Z^2\left[-\frac{2}{r^3} + (j + 2)\frac{r_0^j}{r^{j+3}}\right] \tag{14}$$

Substituting $r = r_0$ in Eq. (14) and using Eqs. (12) and (13),

$$f(r) = f(r_0 + \Delta r) = \frac{0.29e^2Z^2j}{r_0^3}\Delta r = k\,\Delta r \tag{15}$$

where k is a constant of the material dependent on the values e, Z, j, and r_0 and r_0 is temperature-dependent. Equation (15) represents Hooke's law ("extension is proportional to force"), where k is the spring constant. Since Young's modulus, E, is defined as stress, σ, per *unit* strain ϵ (e.g., in./in.), we may divide both sides of Eq. (15) by r_0^2 to obtain

$$\sigma = \frac{0.29e^2Z^2j}{r_0^4}\epsilon = E\epsilon \tag{16}$$

which is Hooke's law.

b. ELASTIC MODULI FROM THE PERIODIC TABLE

The periodic table provides a helpful clue to the values of elastic moduli of elements. For the bulk elastic modulus, for which much experimental data are available, we write

$$K = \frac{Ce^2}{r_0^4} \tag{17}$$

in analogy to Eq. (16) for Young's modulus. C is a constant whose value depends on the periodic group. Taking logarithms of both sides,

$$\log K = \log Ce^2 - 4 \log r_0$$

giving a linear relationship between $\log K$ and $\log r_0$, with a slope of 4. Several such lines are plotted (Fig. 5) using experimental points for elements from the first four groups of the periodic table. The values of C for the periodic groups are also included. Inspection of the data reveals that [3]

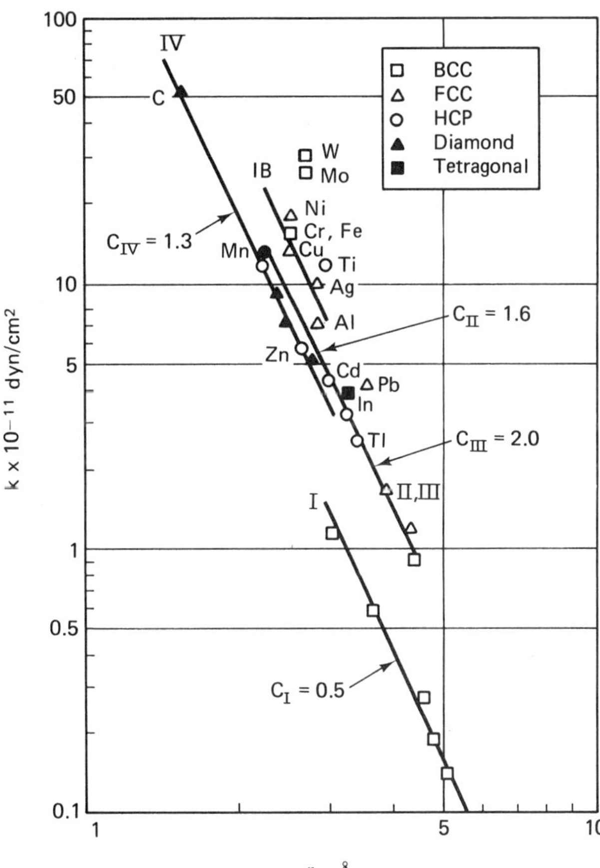

Fig. 5. Log K versus log r_0 for selected elements. From Rosenthal, D., *Introduction to Properties of Materials.* New York: Van Nostrand Reinhold Company (Litton Educational Publishing Division), 1964 [3].

1. Within the same periodic group the bulk modulus varies inversely with r_0^4, as predicted by Eq. (17).
2. For Groups I–IV, the coefficient C is little affected by the number of closed electron shells of an atom. This differs from the case for ionic crystals, where the value of j is sensitive to the number of closed shells (Table 2 of Chap. 2).
3. The value of C increases to a maximum in Group III: then decreases in Group IV, reflecting the transition of bond type from ionic to metallic; and then goes to covalent in Group IV.
4. Heavier elements display less regularity in behavior. Transition elements, especially, are characterized by high values of C and small values of r_0 [2], both of which contribute to high modulus.

From the relationships between K, E, and G [Eqs. (3) and (4)], the preceding observations also apply to E and G, although trends are less regular, possibly in part because E and G are vector quantities, dependent on crystallographic orientation, whereas K is a scalar quantity, independent of direction.

Relationships from Eq. (17) can sometimes be useful for predicting properties of new materials such as the synthetic modification of boron nitride. Its diamond crystal structure, together with a value $r_0 = 1.58$ A, predicted from the covalent radii of B and N, suggest a value of bulk modulus approximately equal to that of diamond. Unfortunately, it is not yet possible to predict the properties of molecular solids, held together by secondary rather than primary bonds, by any similar set of simple relationships.

c. THERMOELASTICITY

Thermoelasticity is the interdependence between elastic strain energy and thermal energy. At absolute zero temperature, there is almost no thermal motion, and the atomic spacing is r_0 (Fig. 6). As the material is heated, atomic spacings vary from r_a to r_b due to thermal motion, with mean spacing of r_m. Due to the asymmetry of the

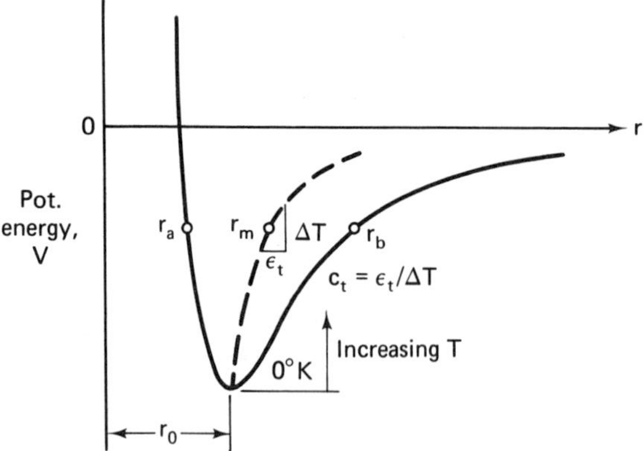

Fig. 6. Coefficient of thermal expansion.

potential energy curve, $r_0 \neq r_m$. The linear thermal expansion coefficient, c_t, is

$$c_t = \frac{\epsilon_t}{\Delta T}$$

where ϵ_t is the thermal strain (in./in.), resulting from temperature change ΔT. Since the dashed line is nonlinear, c_t is temperature-dependent. In crystalline materials, c_t may also vary, as does Young's modulus, with crystallographic direction.

Thus, the Young's modulus, thermal expansion coefficient, and melting temperature of a material are all related to its bond strength, as indicated by the potential energy curve (Fig. 7). Young's modulus is related to the curvature ($\partial^2 v/\partial r^2$) at the bottom of the trough, thermal expansion to the trough symmetry, and melting temperature to trough depth. Figure 8 shows the inverse relationship between expansion coefficient and melting temperature for several classes of materials.

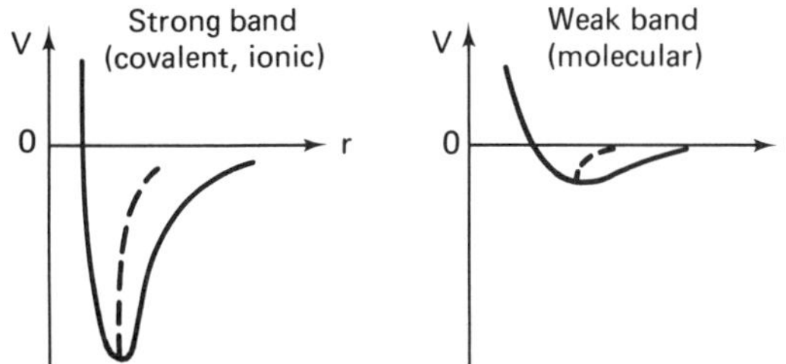

Fig. 7. Potential energy curves for strong and weak materials.

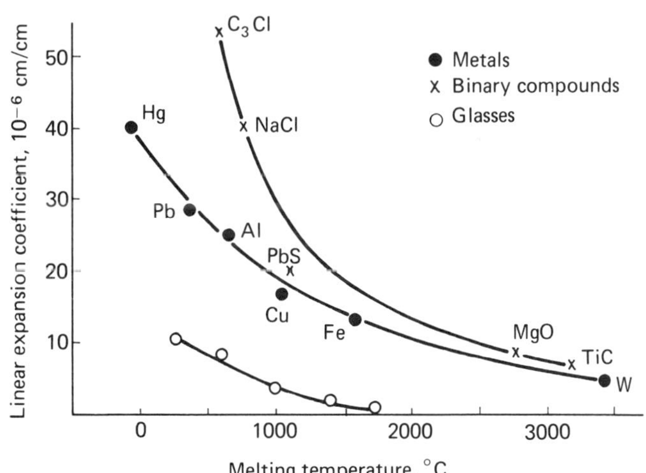

Fig. 8. Expansion coefficient at 20°C versus melting temperature for comparable materials. From Van Vlack L. H.: *Materials Science for Engineers*. Reading, Mass.: Addison-Wesley Publishing Company, Inc., 1970 [4], by permission.

2. Plastic Deformation

Elastic deformation is terminated by fracture in some solids and by nonrecoverable or *plastic* deformation in others. The ability to undergo plastic deformation allows machine and structural components to yield instead of break when loaded beyond the elastic range. The forming of metals into wires, bars, beams, plates, and doubly curved shapes by drawing, rolling, and stamping also depends on plastic deformation. The stress at which deformation changes from elastic to plastic is called the *yield stress* (σ_y in Fig. 9). The increase in stress beyond the yield stress is called strain-hardening or *work hardening*. Since work hardening is attained only at temperatures less than about 40% of the absolute melting temperature, it is sometimes called cold working. Work hardening is exploited in the metal industries to increase the yield stress. If deformation is continued to *b* (Fig. 9), the new elastic range is *a-b*, and the yield stress is increased to σ_y'.

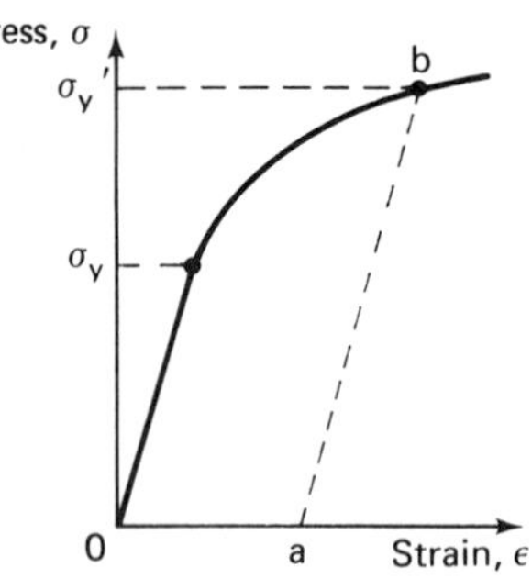

Fig. 9. Elastic-plastic deformation.

Several theories of plastic deformation, so-called *yield theories*, have been advanced. Probably the most widely used is the *maximum shear stress criterion*, proposed by Tresca in about 1865. It states that plastic deformation begins when shear stress reaches the value of shear at the yield point in a simple tension or compression test, that is, when the difference between the maximum and minimum principal stresses reaches σ_y. The principal stress combinations which cause plastic deformation lie outside of the hexagonal stress prism whose axis is the {111} vector in Fig. 10(a), where it is assumed that any of the stresses σ_1, σ_2, or σ_3 may be the maximum or minimum principal stress. Principal stress combinations within this yield surface produce elastic deformation. The prism is infinitely long, in accordance with the fact that very high normal stresses can produce purely elastic deformation if their differences are small. The trace of the prism on the σ_1-σ_2 plane is obtained by letting $\sigma_3 = 0$ [Fig. 10(b)], and similarly for the σ_1-σ_3 and σ_2-σ_3 planes.

a. PLASTIC STRESS-STRAIN RELATIONS

It was possible to define the *elastic* volume strain ϵ_v in Eq. (7) as the sum of the three component strains because, for small strains, their product can be ignored. This

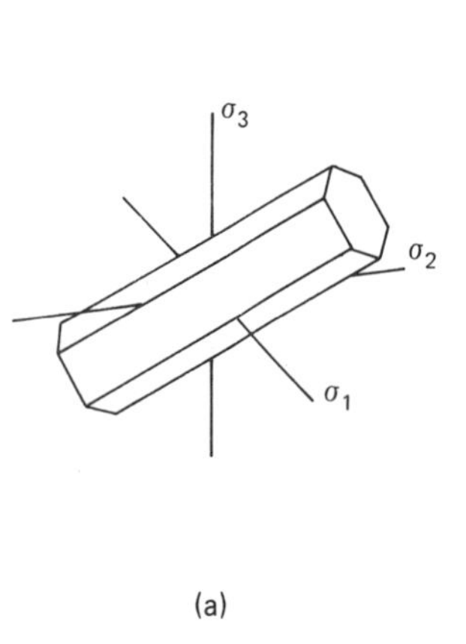

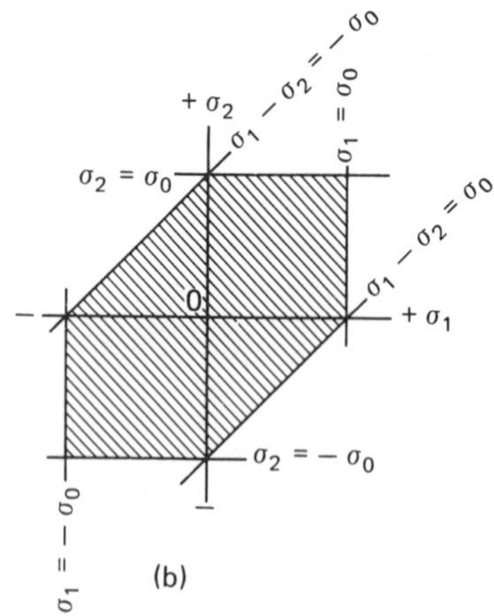

Fig. 10. (a) Yield surface according to the maximum shear stress criterion; (b) maximum shear stress yield boundary for plane stress. From Polakowski, N. H. and E. J. Ripling, *Strength and Structure of Engineering Materials.* Englewood Cliffs, N.J.: Prentice-Hall, Inc. 1966 [5]. Adapted by permission.

approximation is not suitable for plastic deformation, where component strains may be large and the value of their product significant. For plastic deformation it is more convenient to define strain in terms of the variable instantaneous length rather than the constant initial length. Integrating strain in terms of an instantaneous length L between two arbitrary lengths L_a and L_b gives

$$\bar{\epsilon} = \int_{L_a}^{L_b} \frac{dL}{L} = \ln\left(\frac{L_b}{L_a}\right)$$

where $\bar{\epsilon}$ is called the *logarithmic* or *natural* strain.

For a cube of sides L_0 being plastically deformed into a parallelepiped with sides L_1, L_2, and L_3, we can write

$$\frac{L_1}{L_0}\frac{L_2}{L_0}\frac{L_3}{L_0} = 1$$

since materials undergo practically no volume change during plastic deformation. Taking logarithms of both sides,

$$\ln\left(\frac{L_1}{L_0}\right) + \ln\left(\frac{L_2}{L_0}\right) + \ln\left(\frac{L_3}{L_0}\right) = 0$$

or

$$\bar{\epsilon}_1 + \bar{\epsilon}_2 + \bar{\epsilon}_3 = 0$$

For a uniaxial stress σ_1, the axial strain is $\bar{\epsilon}_1$, and the two lateral strains are $\bar{\epsilon}_2 = \bar{\epsilon}_3 = -\nu\bar{\epsilon}_1$. For zero volume change, $\bar{\epsilon}_1 - 2\nu\bar{\epsilon}_1 = 0$, making the Poisson ratio, ν, equal to $\frac{1}{2}$ for plastic deformation.

For an isotropic substance with negligible elastic strain, Levy, and later von Mises, expressed the relation between the plastic principal stresses and *strain increments*, $d\bar{\epsilon}$, in the form

$$\frac{d\bar{\epsilon}_1 - d\bar{\epsilon}_2}{\sigma_1 - \sigma_2} = \frac{d\bar{\epsilon}_2 - d\bar{\epsilon}_3}{\sigma_2 - \sigma_3} = \frac{d\bar{\epsilon}_3 - d\epsilon_1}{\sigma_3 - \sigma_1} = d\lambda \tag{18}$$

where $d\lambda$ is an instantaneous but otherwise variable proportionality factor. Equation (18) assumes that the principal stress axes coincide with the axes of principal strain increment at all times. For a Poisson ratio of $\frac{1}{2}$, one can obtain the following equations for any set of orthogonal components:

$$\begin{aligned}
d\bar{\epsilon}_x &= \frac{2}{3}\, d\lambda\left[\sigma_x - \frac{1}{2}(\sigma_y + \sigma_z)\right] \\
d\bar{\epsilon}_y &= \frac{2}{3}\, d\lambda\left[\sigma_y - \frac{1}{2}(\sigma_z + \sigma_x)\right] \\
d\bar{\epsilon}_z &= \frac{2}{3}\, d\lambda\left[\sigma_z - \frac{1}{2}(\sigma_x + \sigma_y)\right] \\
d\gamma_{xy} = 2\tau_{xy}\, d\lambda, &\qquad d\lambda_{yz} = 2\partial_{yz}\, d\lambda, \qquad d\gamma_{zx} = 2\tau_{zx}\, d\lambda
\end{aligned} \tag{19}$$

where τ and γ indicate shear stress and strain, respectively. Notice that Eqs. (19) resemble Eqs. (6) except for the replacement of strain by a logarithmic strain increment and Young's modulus by the proportionality factor $d\lambda$.

b. ATOMIC BASIS

Plastic deformation occurs by a variety of mechanisms in different materials (see Sec. E in Chap. 7, for example). In metals and some ionic crystals, plastic deformation involves *slip*, the sliding of parallel layers of atoms within or between crystals, so that atoms in one layer become aligned with new neighbors without substantially changing the crystal unit cells. Slip occurs most easily on those *planes* which are most densely populated with atoms and in the *directions* on those planes along which atoms are most closely spaced. For a simplified understanding of the reason, consider the shear force T required to displace atom A one atomic distance (Fig. 11), as presented by

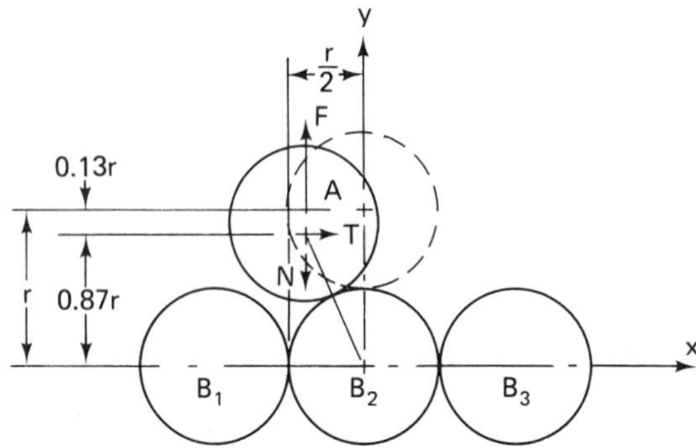

Fig. 11. Slip in a perfect two-dimensional close-packed crystal.

Rosenthal [3]. Suppose the atoms to be hard, frictionless spheres. Since they are actually deformable, this assumption increases our estimate of T but still gives a useful approximation. The force T must act over only the distance $r/2$, since once atom A attains a position atop atom B_2, it slides spontaneously into the new trough between atoms B_2 and B_3. The work done is

$$W = \int_0^{r/2} T\,dx$$

and equals the work by a force F (opposing attraction N) required to pull atom A from level $0.87r$ up to level r, because the resultant of T and N does no work, being always perpendicular to the path taken by atom A, and therefore the sum of works done by its components T and N is zero. Therefore,

$$W = \frac{r}{2} T_{\text{avg}} = \int_0^{0.13r} F\,dy$$

If the attractive force obeyed Hooke's law, Eq. (5), then, in analogy with the work done by a linear spring,

$$\frac{r}{2}T_{\text{avg}} = \frac{ES(0.13)^2}{2(0.87)}$$

where E is Young's modulus in the y direction and A is the area over which F acts. Since T_{avg} acts over the same area, the required shear stress is

$$\tau_{\text{avg}} = \frac{T_{\text{avg}}}{A} \simeq 0.02E \tag{20}$$

Since Hooke's law is not valid over a distance so large as $0.13r$ (net attraction decrease with distance) and since atoms are slightly deformable, this value might be taken as an upper bound for τ_{avg}.

Fig. 12. (a) Linear dislocation; (b) atomic deformation adjacent to a dislocation. From Rosenthal, D., *Introduction to Properties of Materials*. New York: Van Nostrand Reinhold Company, (Litton Educational Publishing Division) 1964 [3].

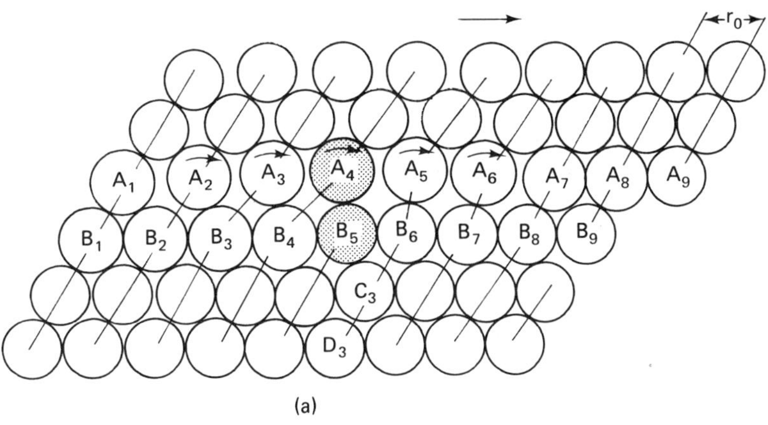

(a)

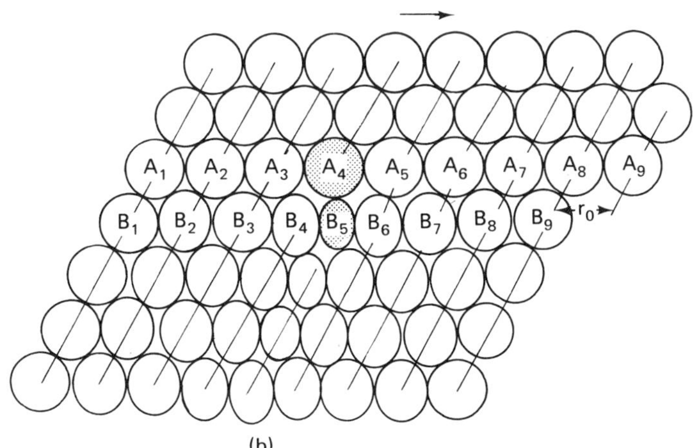

(b)

Due to crystal imperfections and local onset of plastic deformation, slip in metals actually begins at shear stresses on the order of 10^{-2} of the value indicated by Eq. (20). Figure 12(a) shows one type of crystal imperfection, a linear dislocation. Note that if all atoms A were atop atoms B, as is the crosshatched pair, no stress would be required to produce slip. In fact, the slip process is likely to be progressive and local, rather than occurring simultaneously across the shear surface. Also, slip phenomena are believed to be complicated by the fact that attractive and repulsive forces tend to close the gaps, causing differences in atomic radii in the neighborhood of the slip surface, as shown by the crosshatched atoms in Fig. 12(b).

In *polycrystalline* materials, plastic deformation may involve slip mechanisms in addition to the one just described. The subject has been under intensive study and is still not well understood [3]. As a rule, the smaller the grain size, the higher the stress at which plastic deformation begins. This occurs because the grain boundaries presumably offer resistance to the motion of dislocations.

3. Time-Dependent Deformation

In the preceding two sections, strain was assumed to be a function of stress alone. In reality, strain can lag behind stress, because the migration of atoms and thermal energy are time-dependent processes. Time-dependent deformation at constant stress is called *anelasticity* in the elastic range and *creep* in the plastic range. A reduction of stress at constant strain in either the elastic or plastic range is called *stress relaxation.* If a nonelastic deformation rate is exactly proportional to the applied stress, the deformation is called *viscous flow. Newtonian liquids* are idealized viscous materials, which, in addition, are assumed incompressible and without elastic deformation. Many materials at high temperatures, including glass, approach the behavior of a Newtonian liquid.

a. ANELASTICITY

Anelastic materials exhibit *elastic aftereffect*, or asymptotic approach of strain to constant value under static load, and *damping*, or dissipation of internal energy under vibrational load. The time-dependent component of elastic strain is sometimes defined by its *relaxation time*, θ. Upon loading or unloading, a material may undergo an immediate strain, ϵ_i, and a time-dependent strain, ϵ_t (Fig. 13). The time-dependent strains of many materials are approximately exponential functions of time. They may be expressed by

$$\epsilon_l = (\epsilon_i + \epsilon_t)\left[1 - \left(\frac{\epsilon_t}{\epsilon_i + \epsilon_t}\right)e^{-t/\theta}\right] \tag{21a}$$

and

$$\epsilon_u = \epsilon_t e^{-[(t-t_1)/\theta]} \tag{21b}$$

where ϵ_l is the strain after loading, ϵ_u the strain after unloading, θ (relaxation time) the time required for the time-dependent component of strain to rise (or decrease) to

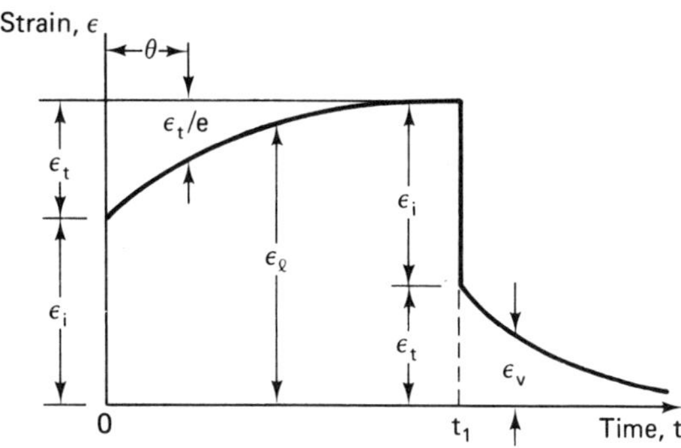

Fig. 13. Elastic aftereffect.

within $1/e$ of its final value on loading (or unloading), and the remaining notation is shown in Fig. 13.

b. VISCOUS FLOW

The same relationships that exist between strains and displacements for elastic deformation [Eqs. (8) and (9)] exist between *strain rates* and *velocities* for viscous flow. Hence, by analogy with Eqs. (8) and (9) the six strain rate components are

$$\dot{\epsilon}_x = \frac{\partial \dot{u}}{\partial x}, \qquad \dot{\epsilon}_y = \frac{\partial \dot{v}}{\partial y}, \qquad \dot{\epsilon}_z = \frac{\partial \dot{w}}{\partial z}$$

$$\dot{\gamma}_{xy} = \frac{\partial \dot{v}}{\partial x} + \frac{\partial \dot{u}}{\partial y}, \qquad \dot{\gamma}_{yz} = \frac{\partial \dot{w}}{\partial y} + \frac{\partial \dot{v}}{\partial z}, \qquad \dot{\gamma}_{zx} = \frac{\partial \dot{u}}{\partial z} + \frac{\partial \dot{w}}{\partial x}$$

where the dots represent differentiation with respect to time. The dimension of strain rate, i.e., $\dot{u} = \partial u/\partial t$, is inches per inch per time, or 1/time. By analogy with Eqs. (1a) and (1b), the relationships between the strain rates and the corresponding stresses are

$$\dot{\epsilon}_x = \left(\frac{1}{\xi}\right)\sigma_x, \qquad \dot{\epsilon}_y = \left(\frac{1}{\xi}\right)\sigma_y, \qquad \dot{\epsilon}_z = \left(\frac{1}{\xi}\right)\sigma_z$$

$$\dot{\gamma}_{xy} = \left(\frac{1}{\eta}\right)\tau_{xy}, \qquad \dot{\gamma}_{yz} = \left(\frac{1}{\eta}\right)\tau_{yz}, \qquad \dot{\gamma}_{zx} = \left(\frac{1}{\eta}\right)\tau_{zx}$$

where ξ and η are the tensile and shear viscosity, respectively, which are seen to have dimensions of stress times time. A common unit of measurement is the poise (1 dyne-sec/cm^2). Table 2 gives values of η for some common materials [6].

TABLE 2. Representative Viscosities

Substance	*Shear viscosity, η (poise)*
Air (20°C)	1.86×10^{-4}
Water (20°C)	0.010
Castor oil (20°C)	7.2
Glass (575°C)	1.1×10^{13}
Glass (20°C)	10^{22}

SOURCE: JAEGER, J. C., *Elasticity, Fracture and Flow*, 2nd ed. London: Methuen & Co. Ltd., 1962 [6].

The assumption of incompressibility ($\nu = \frac{1}{2}$) transforms Eq. (3), $E = 2(1 + \nu)G$, into $\xi = 2(1 + \frac{1}{2})\eta$, or $\xi = 3\eta$. Since the assumption of incompressibility was also used for plastic deformation, the generalized strain rate equations for Newtonian liquids are of the same form as Eqs. (19):

$$\dot{\epsilon}_x = \frac{1}{\xi}\left[\sigma_x - \frac{1}{2}(\sigma_y + \sigma_z)\right]$$

$$\dot{\epsilon}_y = \frac{1}{\xi}\left[\sigma_y - \frac{1}{2}(\sigma_x + \sigma_z)\right]$$

$$\dot{\epsilon}_z = \frac{1}{\xi}\left[\sigma_z - \frac{1}{2}(\sigma_y + \sigma_x)\right]$$

The correspondence between elastic, plastic, and viscous deformation can now be summarized by tabulating the appropriate terms for the analogous equations (Table 3).

TABLE 3. Corresponding Elasticity, Plasticity and Viscosity Terms

Elasticity	E	G	ν	ϵ	γ	u
Plasticity	$\frac{3}{2}d\lambda$	$\frac{1}{2}d\lambda$	$\frac{1}{2}$	$d\bar{\epsilon}$	$d\gamma$	u
Viscosity	ξ	η	$\frac{1}{2}$	$\dot{\epsilon}$	$\dot{\gamma}$	$\dot{u}$

Viscosity is temperature-sensitive, decreasing exponentially with the reciprocal of absolute temperature, T:

$$\eta = \eta_0 e^{Q/RT}$$

or

$$\ln = \ln \eta_0 + \frac{Q}{RT} \tag{22}$$

where Q is the activation energy (cal/mole-°K) for viscous shear of the material, R is the gas constant (1.987 cal/mole), and η_0 is a constant for the material.

Deformation which contains both elastic and viscous components is called *viscoelastic*. It differs from anelastic deformation in that the time-dependent component is viscous and hence not recoverable. Long chain polymers with no cross-linking are typically viscoelastic. They deform elastically due to the unkinking of polymer chains and undergo viscous flow due to the time-dependent wriggling of chain segments into adjacent holes.

c. RHEOLOGICAL MODELS

Many materials exhibit complex mixtures of deformation modes. The equations describing their stress-strain behavior can be developed with the aid of mechanical models made by combinations of several basic *model elements*. Either normal or shear stresses can be represented by equations based upon the models, although physically they represent only forces and extensions. The basic model elements and their cor-

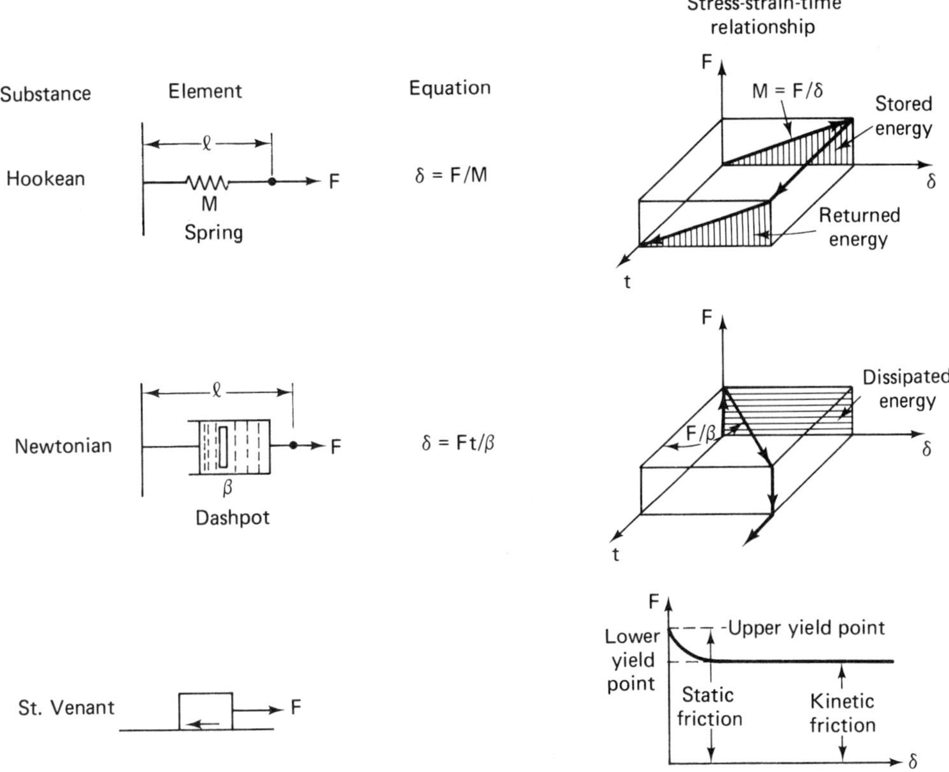

Fig. 14. Basic rheological elements. From Polakowski, N. H. and E. J. Ripling, *Strength and Structure of Engineering Materials.* Englewood Cliffs, N.J.: Prentice-Hall, Inc. 1966 [5]. Adapted by permission.

responding equations and stress-strain-time relations are shown in Fig. 14. The *Hookean* element stores all the energy supplied to it as strain energy and displaces instantaneously under an applied force. The *dashpot* dissipates all the energy supplied to it, and the strain rate is proportional to stress. The *St. Venant* element, pictured as a sliding block on a plane in Fig. 14, is used to represent substances which have zero displacement up to a certain stress level.

The deformation of various materials is represented by combinations of the basic elements and by varying the constants of the springs, dashpots, and friction blocks. The simplest combination is a *Maxwell model*, a spring and dashpot in series [Fig. 15(a)], for which

$$\delta = \delta_1 + \delta_2 = \frac{F}{M} + \frac{Ft}{\beta}$$

Differentiating with respect to time,

$$\dot{\delta} = \frac{\dot{F}}{M} + \frac{F}{\beta} \tag{23}$$

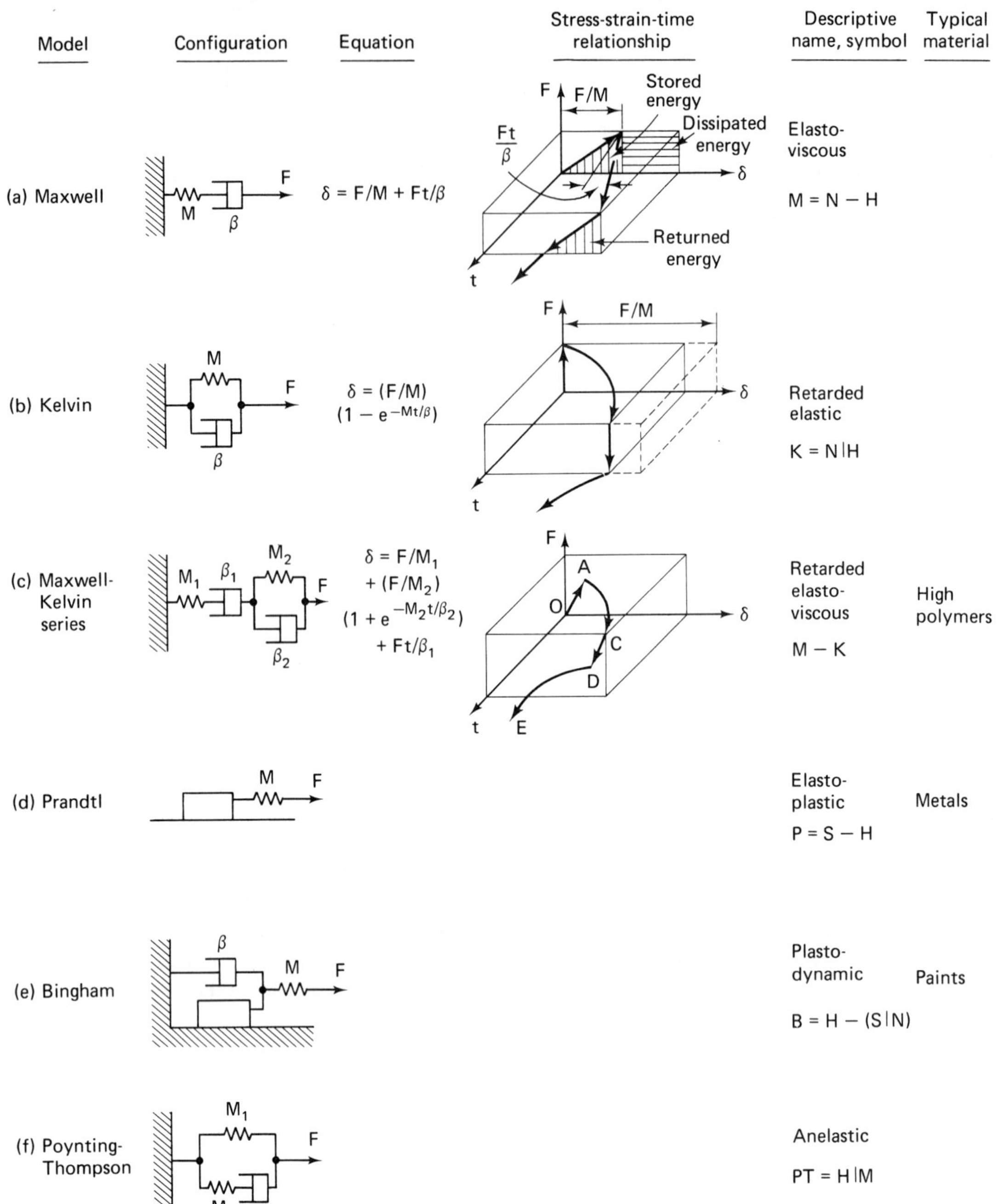

Fig. 15. Rheological models. From Polakowski, N. H. and E. J. Ripling, *Strength and Structure of Engineering Materials*, Englewood Cliffs, N.J.: Prentice-Hall, Inc. 1966 [5]. Adapted by permission.

where δ, δ_1, and δ_2 are the total, spring, and dashpot extensions, respectively, F is an instantaneously applied force, t is time, and M and β are the spring and dashpot constants, respectively.* On release of the load, the specimen retracts, and the dashpot extension remains. If, instead of applying a constant load, the model is suddenly extended to δ_1, the time-varying force required to maintain this extension is found from Eq. (23). For this case $\dot{\delta}$ is zero. On integrating Eq. (23) and using the limits

$$t = 0, \qquad F = M\delta_1; \qquad t = t, \qquad F = F$$

one obtains

$$F = M\delta_1 e^{-Mt/\beta} \tag{24}$$

a form similar to that for relaxation time of the anelastic model [Eq. (21b)]. This is the equation for stress relaxation in a Maxwell model.

Since η is a logarithmic function of temperature [Eq. (22)], the viscous relaxation time, β/M, or η/G, also decreases logarithmically with temperature:

$$\frac{\eta}{G} = \left(\frac{\eta_0}{G}\right) e^{Q/RT} \tag{25}$$

Another simple combination is the *Kelvin* or *Voight model* [Fig. 15(b)], used in representations of rubber elasticity and of high polymers. For this model the spring and dashpot extensions are equal,

$$\delta = \delta_1 = \delta_2$$

and their forces are additive,

$$F = F_1 + F_2 = M\delta + \beta\dot{\delta} \tag{26}$$

Note that in series coupling the element forces are equal and their intensities are additive, while the opposite is true in parallel coupling. In the Kelvin model, part of the deformation is stored in the spring, and the remainder is dissipated in the dashpot. The extension as a function of time for an instantaneously applied force is found by integrating Eq. (26) and using the limits

$$t = 0, \qquad \delta = 0; \qquad t = t, \qquad \delta = \delta$$

for which

$$\delta = \left(\frac{F}{M}\right)(1 - e^{-Mt/\beta}) \tag{27}$$

Deformation models of high polymers can be approximated by Kelvin and Maxwell models in series [Fig. 15(c)]. If a constant force F is applied to this system at

---

*The symbols M, β, and δ are used instead of E, θ, and ε (or G, η, and Υ), respectively, because of differences in units. For example, E was used earlier with units of psi, whereas M is expressed as force per unit length.

$t_0 = 0$ and removed at t, then

$$\underset{\substack{\text{Hookean}\\\text{displacement}}}{\delta = \frac{F}{M_1}} + \underset{\substack{\text{Kelvin}\\\text{displacement}}}{\left(\frac{F}{M_2}\right)(1 + e^{-M_2 t/\beta_2})} + \underset{\substack{\text{Newtonian}\\\text{displacement}}}{\frac{Ft}{\beta_1}}$$

where subscripts 1 and 2 refer to the Maxwell and Kelvin models, respectively. A series of one Maxwell model and two or more Kelvin models having different moduli and viscosities permits fairly accurate modeling of most any viscoelastic material.

Many materials, particularly metals, show a combination of elastic-plastic deformation and can be represented by a *Prandtl model*, a Hookean and St. Venant element in series [Fig. 15(d)]. The rheology of paint can be described by a *Bingham model*, a St. Venant-Newtonian combination in series with a Hookean element [Fig. 15(e)]. Bingham substances suffer negligible deformation at light loads (as governed by the spring constant) but undergo large deformations at slightly greater loads. Paint, for example, can be easily brushed onto a wall, and it then flows enough under the action of gravity to obliterate the brush marks but not enough to run off the wall.

Rheologists frequently use a shorthand in which each element is referred to by its initial; i.e., *H* for Hookean, *N* for Newtonian, and *S* for St. Venant. A vertical line (|) indicates a parallel connection, and a horizontal line (–) indicates series. The

TABLE 4. Compound Rheological Models

Model	*Symbol*	*Descriptive name*
Maxwell	$M = N\text{–}H$	Elasticoviscosity
Kelvin	$K = N \mid H$	Retarded elasticity
Bingham	$B = H\text{–}(St\ \mathrm{V} \mid N)$	Plasticodynamic
Letherich	$L = K\text{–}N$	Elastic sols
Jeffrey	$J = N \mid M$	Relaxing gels
Schwedoff	$\mathrm{Schw} = H\text{–}(St\ \mathrm{V} \mid M)$	Plastic gels
Poynting-Thomson	$P T\mathrm{h} = H \mid M$	Anelasticity (standard linear solid)
Burgers	$\mathrm{Bu} = M\text{–}K$	
Trouton-Rankine	$TR = N\text{–}PT\mathrm{h}$	
Schofield-Scott-Blair	$\mathrm{Sch\ Sc\ B} = \mathrm{Schw}\text{–}K$	

SOURCE: POLAKOWSKI, N. H., and E. J. RIPLING, *Strength and Structure of Engineering Materials*. Englewood Cliffs, N. J.: Prentice-Hall, Inc., 1966 [5].

Maxwell body is written $M = N\text{–}H$ and the Kelvin body is $K = N \mid H$. Several models are shown in Table 4. More complex models can be developed to account for *nonlinear* behavior in the various spring and dashpot elements.

d. VISCOSITY AND STRUCTURE

The mechanical behavior and corresponding molecular structural distortion of a typical linear *amorphous* polymer are shown in Fig. 16 as a function of strain rate at constant temperature. Increasing the test temperature moves the curve to the left. The polymer undergoes elastic deformation, due to the kinking or unkinking of polymer chains or, in dense materials, to slight changes in the length of valence bonds and changes in valence angles. This essentially Hookean response corresponds to the glassy range in Fig. 16. At a lower strain rate *retarded* or *configurational* elasticity manifests itself, due to the time-dependent wriggling of chain segments into adjacent holes. This anelastic response is sometimes called rubbery deformation, and the intermediate transition range in which the modulus is very sensitive to strain rate is called leathery deformation. At still lower strain rates the time-dependent wriggling of chain segments accumulates along the entire chain length, so that the molecule, in effect, wanders through the body to produce irreversible viscous flow. The regions of Fig. 16 can be represented by the following rheological elements and models:

Glassy	Hookean
Leathery	Kelvin
Rubbery	Kelvin
Viscous flow	Newtonian

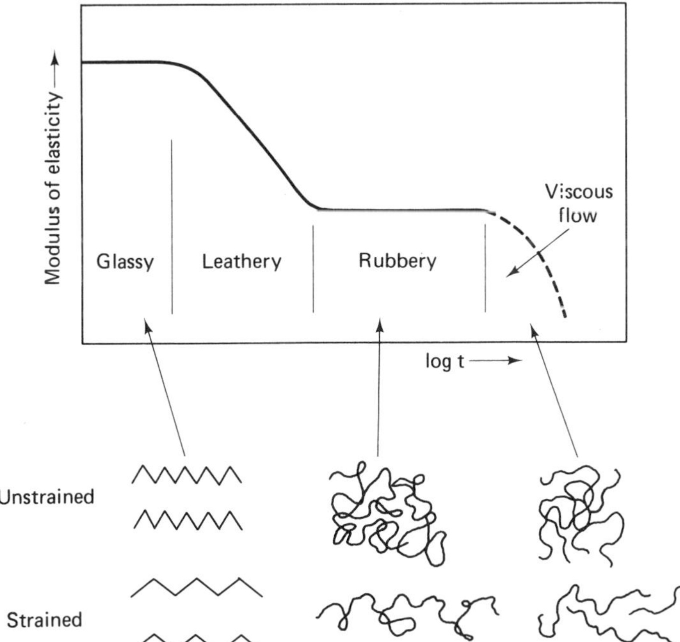

Fig. 16. Typical mechanical behavior and molecular structural behavior of a linear amorphous polymer as a function of strain rate at constant temperature. From Polakowski, N. H. and E. J. Ripling, *Strength and Structure of Engineering Materials*, Englewood Cliffs, N.J.: Prentice-Hall, Inc. 1966 [5]. Adapted by permission.

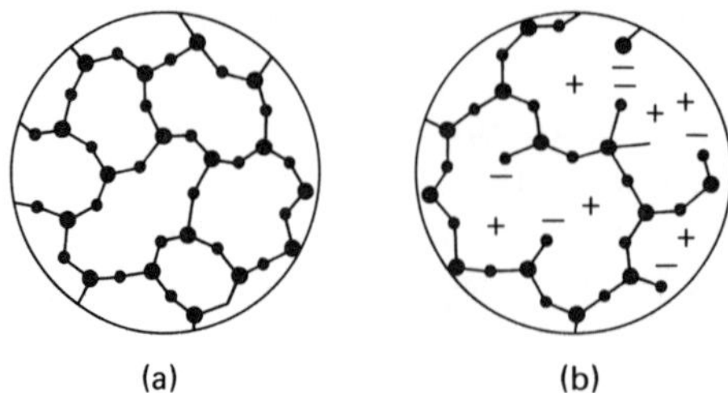

Fig. 17. (a) Fused silica glass, with crosslinking saturation; (b) Lime-soda glass, with a more open network and less specific bonds. Adapted from Van Vlack, L. H.: *Materials Science for Engineers*. Reading, Mass.: Addison-Wesley Publishing Company, Inc., 1970 [4], by permission.

Figure 16 is representative of linear amorphous polymers such as polyvinyl chloride or polystyrene. Its shape is sensitive to polymer structure. *Elastomers*, such as cis-isoprene, are polymers which have very low elastic moduli (10^2 to 10^4 psi) due to extreme coiling or kinking. *Cross-linking* of polymers, such as of phenol-formaldehyde or vulcanized rubber, restricts the possibility of viscous flow. The same is true of cross-linked glasses, such as fused silica glass, where every SiO_4 group is coordinated with four neighbors, compared with soda glass (Fig. 17). *Crystalline* polymers, such as crystalline polyethylene, show little or no rubbery behavior but gradually decrease in elastic modulus until an abrupt drop at the melting temperature. This behavior is typical of metals and crystalline materials in general. Crystalline polymers are desired for stress-carrying applications near the melting temperature. Additional structural factors which influence the shape of the curve in Fig. 16 include molecular shape and polarity.

The stress-strain-time relationships of nonlinear viscoelastic polymers can be represented in another way, shown in Fig. 18. From the constant temperature

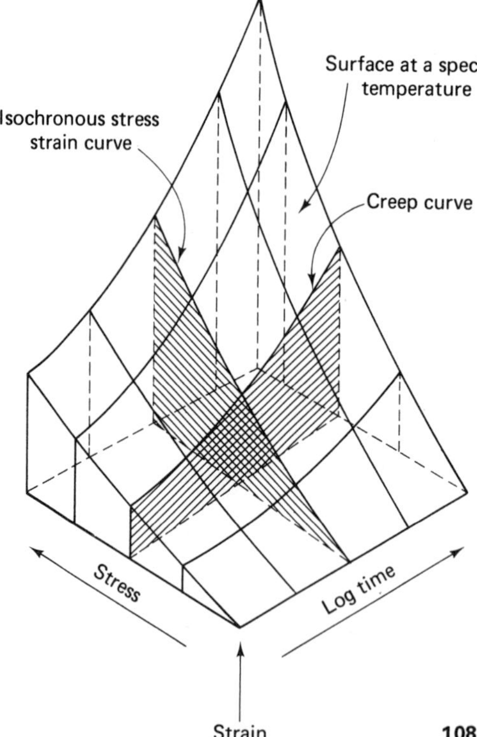

Fig. 18. Stress-strain-time-temperature relationship for nonlinear viscoelastic polymer. From Benjamin, B. S., *Structural Design with Plastics*. New York: Van Nostrand Reinhold Company (Litton Educational Publishing Division), 1969 [7].

(isochronous) stress-strain surface, it is possible to determine the typically required stress-strain and creep characteristics (crosshatched) and the corresponding nonlinear elastic and creep moduli.

B. FRACTURE MECHANICS

Fracture is the breaking of a solid into two or more parts and is usually characterized as either *brittle* or *ductile*, depending on the amount of plastic deformation which accompanies failure. Amorphous materials such as glass or glassy polymers nearly always exhibit brittle fracture, whereas crystalline materials may fail by either brittle or ductile fracture. Brittle fracture in polycrystalline materials usually occurs on

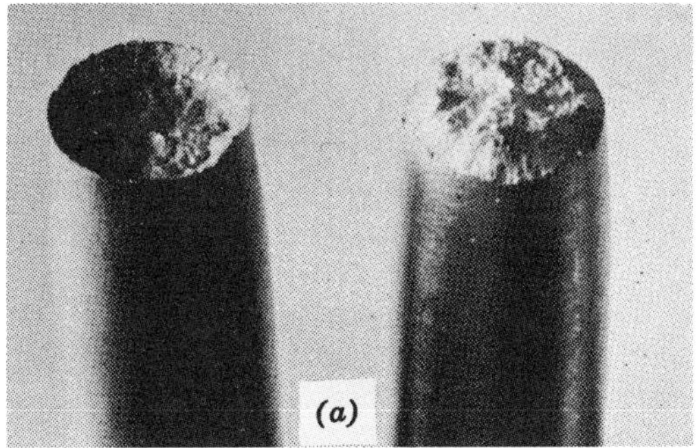

Fig. 19. (a) Ductile (cup and cone) fracture in aluminum; (b) brittle fracture in mild steel. From Hayden, H. W., W. G. Moffatt, and J. Wulff, *The Structure and Properties of Materials, Vol. III: Mechanical Behavior*, New York: John Wiley and Sons, Inc., 1965 [1], by permission.

crystallographic planes called *cleavage* planes, and the fracture surface appears granular with highly reflective spots, due to the difference in orientation of these planes in adjacent crystals [Fig. 19(a)]. Brittle fracture may also occur along *grain boundaries* (intergranular fracture), due to a soft or brittle phase at the boundaries or due to the combined action of stress and a corrosive environment at the boundaries.*
In both cases, brittle fracture occurs normal to the maximum tensile stress.

*Fracture due to a soft phase at grain boundaries may occur, for example, in magnesia brick if silica-rich liquid forms intergranular films and prevents solid-solid contact between (Mg, Fe)O grains or in copper where small amounts of bismuth can form a weak phase between grains of copper.

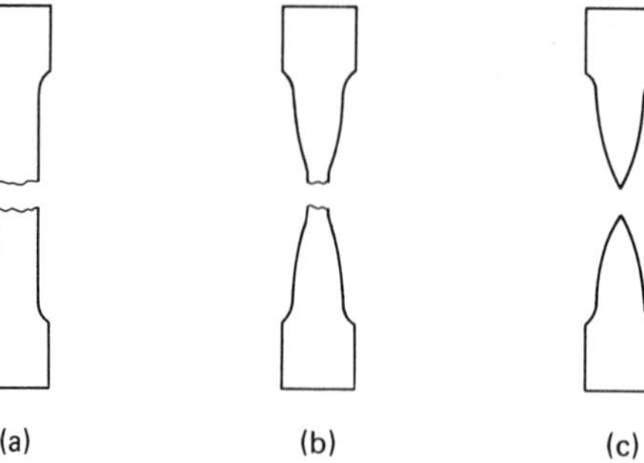

Fig. 20. Tensile failure. (a) Brittle fracture; (b) ductile fracture; (c) rupture. From Van Vlack L. H.: *Materials Science for Engineers*, Reading, Mass.: Addison-Wesley Publishing Company, Inc. 1970 [4], by permission.

Ductile fracture results from extensive deformation along *slip* planes. The cleavage and slip planes in a crystal may or may not be identical. Typical specimen deformation shapes for brittle and ductile tensile fractures are shown in Fig. 20(a) and (b). A ductile fracture appears dull and fibrous. Ductile tension fracture involves necking and the growth of cavities in the neck. The cavities then coalesce into a crack near the center of the cross section which extends outward toward the surface in directions 45° to the tensile axis, resulting in the "cup and cone" appearance in Fig. 19(a).

Rupture [Fig. 20(c)] occurs if the specimen necks to zero cross-sectional area without fracture and is found only at high temperatures where strain hardening is minimal or in crystalline materials where there are very few heterogeneities to nucleate the growth of cavities.

Any specific fracture may, of course, be a mixture of brittle and ductile, and large or small deformation may precede either of these fracture modes. For example, a metal may become sufficiently embrittled after large or repeated plastic deformation to undergo brittle fracture. Such *ductile-brittle transition* may also be induced in many materials by increasing the strain rate, by lowering the temperature, or by cutting a notch in the materials. Asphalt provides an example of the strain rate effect. It can be shattered by a hammer but will deform slowly under small stress. Most materials (face-centered cubic metals excepted) become more ductile at higher temperatures, because the yield stress ∂y decreases with increasing temperature, while the fracture strength ∂f is much less sensitive to temperature change (Fig. 21). The temperature T at which the two stresses are equal (for a specific loading rate) is called the ductile-brittle transition temperature. The transition temperature is sometimes taken to be that at which 50% ductile (shear) fracture is observed on a failure surface. This temperature is not sharply defined for most materials, and considerable experimental scatter can be observed in one lot of test specimens. The yield stress can be increased to σ_y' by altering the stress pattern, such as by making a notch, for example, thereby increasing the transition temperature to T' (Fig. 21).

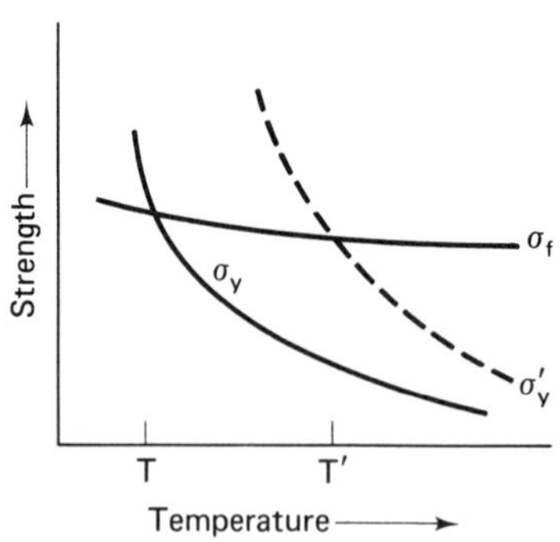

Fig. 21. Sensitivity of transition temperature to notching.

Ductile fracture is preferred to brittle fracture for most structural engineering applications, because ductile deformation provides a visible warning prior to failure. Also, plastic flow at the tip of a crack will blunt the crack, reducing the stress concentration—possibly to a level insufficient to continue the crack propagation. In commercial metals, fracture properties can be highly anisotropic because the flaws which control fracture are oriented by such processes as rolling and drawing. Engineers must specify both the strength and ductility of materials and be mindful of their anisotropic properties in design.

A material's ability to retain strength in the presence of cracks is called its *work of fracture*, which is the energy required to break it. Glass has a small work of fracture compared to steel. For comparison, the length of a transverse crack or machined notch in a sheet of material which causes immediate fracture if the material is stretched to 100,000 psi is up to 1 in. in strong steel, 1/64 in. in aluminum, and only 1/10,000 in. in glass.

The inherently strong materials such as graphite, boron, and silicone carbide all behave like glass; i.e., they are vulnerable to the presence of cracks and therefore have small work of fracture. But with metals and most polymers the engineer obtains moderately high strength without having to eliminate all but the finest cracks. The chemical reason that metals and polymers are more resistant to cracks than ceramics is that the interatomic forces in metals and intermolecular forces in polymers do not depend on particular directional alignments for strength. Also, the chemical bonds of metals and polymers are unsaturated, so that the atoms or molecules can more easily form new bonds. These two factors make it possible for atoms or molecules to slide over one another at the leading edge of a crack. In ceramics, however, the interatomic forces are highly directional, and the bonds are saturated so that a crack, once initiated, spreads easily.

1. Fracture Strengths

The theoretical strengths of materials are useful standards against which to measure actual performance. The theoretical tensile strength (as in brittle fracture) and shear strength (as in ductile fracture) of ionic crystals can be computed by extensions of the methods used in Secs. A1a and A2b for elastic deformation. In order of magnitude, these theoretical strength relationships will also apply to solids held together by other primary bonds.

a. THEORETICAL TENSILE STRENGTH

The force of attraction between ions on opposite sides of a break in a cubic ionic crystal were given by Eq. (11):

$$f = 0.29\frac{e^2Z^2}{r^2}\left[1 - \left(\frac{r_0}{r}\right)^j\right] \tag{11}$$

As shown in Fig. 2(a) of Chap. 2, f increases to a maximum at r_1 and then reduces to zero as r approaches infinity. To find f_{max}, the force required for breaking, we set the derivative of f with respect to r to zero,

$$\frac{df}{dr} = 0.29Z^2e^2\left[-\frac{2}{r^3} + \frac{(j+2)r_0^j}{r^{j+3}}\right] - \frac{2}{r_1^3} + \frac{(j+2)r_0^j}{r_i^{j+3}} = 0$$

from which

$$\frac{r_1}{r_0} = k = \left(\frac{j+2}{2}\right)^{1/j} \tag{28}$$

Substituting into Eq. (11),

$$f_{\max} = \frac{0.29Z^2e^2}{r_1^2}\left(1 - \frac{2}{j+2}\right) = \frac{0.29Z^2j}{k^2(j+2)}\frac{e^2}{r_0^2} \tag{29}$$

If the distance between ions is r_0, the number per unit area in the layer on one side of the cleavage plane is $1/r_0^2$, and the stress (force per unit area) is

$$\sigma_{\max} = \frac{f_{\max}}{r_0^2} = \frac{0.29Z^2j}{k^2(j+2)}\frac{e^2}{r_0^2} \tag{30}$$

where e is the electron charge, Z is the valency, j the repulsive exponent (Table 2 of Chap. 2), and k the coefficient defined by Eq. (28). Substituting Young's modulus, E, from Eq. (16),

$$\sigma_{\max} = \frac{E}{k^2}(j+2) \tag{31}$$

Taking values of j from Table 2 of Chap. 2 and k from Table 5 for alkali halides as an example, $\sigma_{\max}$ falls in the range

$$0.06E \leq \sigma_{\max} \leq 0.09E$$

Since we have neglected the contraction due to the Poisson ratio, actual values are somewhat larger, and we can write

$$\sigma_{\max} \cong 0.1E \tag{32}$$

TABLE 5. Values of $r_1/r_0 = k$ for Some Na^+Cl^- Type Ionic Crystals

Crystal	*k*
NaF	1.26
NaCl	1.25
NaBr	1.24
NaI	1.22
KF	1.25
KBr	1.22
RbF	1.24
RbI	1.20

SOURCE: ROSENTHAL, D., *Introduction to Properties of Materials.* New York: Van Nostrand Reinhold Company, 1964 [3].

Thus, the theoretical strengths of ionic crystals are on the order of one tenth of their Young's moduli. This ratio also provides a rough approximation for the strengths of many other solids. Table 6 shows approximate Young's moduli for several materials. Zinc and graphite are layer structures, with small Young's modulus for stretching normal to the cleavage planes [8].

Other estimates of tensile strength can be made based upon known values of heats of sublimation, interatomic forces, and surface energies. Although each approach differs, they all predict theoretical tensile strengths on the order of 10^6 to 10^7 psi. These strengths are several orders of magnitude greater than those usually observed.

TABLE 6. Approximate Young's Moduli

Material	*Crystallographic plane*	*Young's modulus* (10^6 *psi*)
Copper	(111)	27.5
Copper	(100)	9.6
Diamond	(111)	173
Gold	(111)	158
Graphite	(0001)	1.4
α-Iron	(100)	18.9
α-Iron	(111)	37.3
Tungsten	(100)	56.0
Silica glass	—	10.5
Zinc	(0001)	5.0

b. BRITTLE FRACTURE; GRIFFITH THEORY

To account for the usually observed discrepancy between theoretical and observed strengths, Griffith [9] proposed that fracture occurs by the spreading of preexisting cracks in a brittle material and that fracture strength is related to the maximum size of the random cracks existing in the material before being loaded. For a crack to spread, the work done by the external forces (strain energy) must equal the increase in surface energy due to crack growth.

The surface energy of a crack of length $2c$ [Fig. 22(a)] which extends through a plate of unit thickness is $4cV_\alpha$, where V_α is the unit surface energy. To calculate the strain energy released as the crack propagates requires a knowledge of the stress at the crack tip. Griffith used the tip stress for an ellipitical crack previously obtained by Inglis to calculate the released strain energy, V_s:

$$V_s = \frac{\pi(1 - \nu^2)\sigma^2 c^2}{E}$$

The energy available for crack propagation is the difference between the surface energy and strain energy,

$$V = \frac{4cV_\alpha - \pi(1 - \nu^2)\sigma^2 c^2}{E}$$

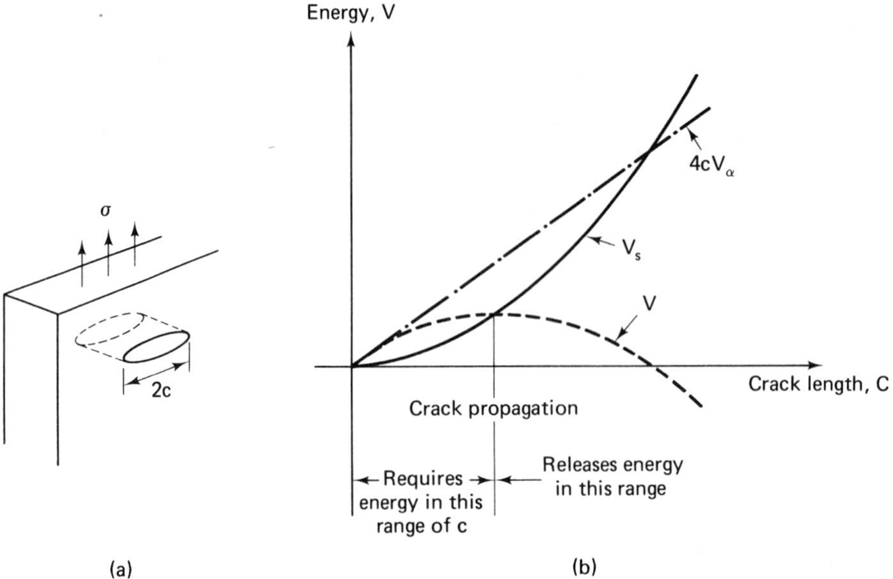

Fig. 22. (a) Elliptical crack; (b) energy V required to propagate a crack as a function of crack length. From Polakowski, N. H. and E. J. Ripling, *Strength and Structure of Engineering Materials*, Englewood Cliffs, N.J.: Prentice-Hall, Inc., 1966 [5]. Adapted by permission.

and the crack will spread spontaneously when $dV/dc = 0$, giving

$$\sigma = \left[\frac{2EV_\alpha}{\pi(1-\nu^2)c}\right]^{1/2} \tag{33}$$

which shows the stress required for crack growth to be proportional to $c^{-1/2}$. Hence the tensile strength of a completely brittle material is determined by the length of the largest crack prior to loading.

The foregoing analysis applies to a crack in a thin plate under uniaxial tension. Others have analyzed the behavior of an oblate spherical crack in a volume of material. For biaxial stress, including tension and/or compression, it has been shown that the criterion for brittle failure is

$$\sigma_1 = \left[\frac{2EV_\alpha}{\pi(1-\nu^2)c}\right]^{1/2} \qquad \text{if } 3\sigma_1 + \sigma_2 > 0 \tag{34a}$$

$$(\sigma_1 - \sigma_2)^2 + 8\left[\frac{2EV_\alpha}{\pi(1-\nu^2)c}\right]^{1/2}(\sigma_1 + \sigma_2) = 0 \qquad \text{if } 3\sigma_1 + \sigma_2 < 0 \tag{34b}$$

where σ_1 and σ_2 are the two principal stresses, with σ_1 larger than σ_2 (tensile stress being positive). These failure conditions are shown in Fig. 23, where brittle failure will occur only for stress combinations outside of the shaded region. Notice from Fig. 23 that the strength in uniaxial compression is eight times the strength in uniaxial tension (when either σ_1 or σ_2 is zero). Many brittle materials such as glass do have compression-tensile strength ratios of this order.

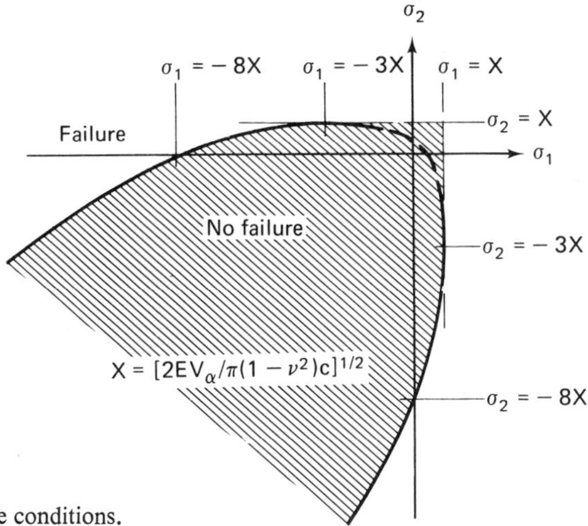

Fig. 23. Biaxial Griffith theory failure conditions.

The Griffith theory applies to completely brittle materials, for which Orowan [10] represented the crack appearance as in Fig. 24, having atomic bonds at every stage of elongation. If the crack advances one interatomic distance, each bond takes the strain previously held by its predecessor, giving a sum of displacements just equal to the complete breaking (surface energy) of one bond. Although there is general agreement that the stress required to propagate a crack through a truly brittle material is given by Eq. (33), there is no way to determine what solids can be regarded as truly brittle.

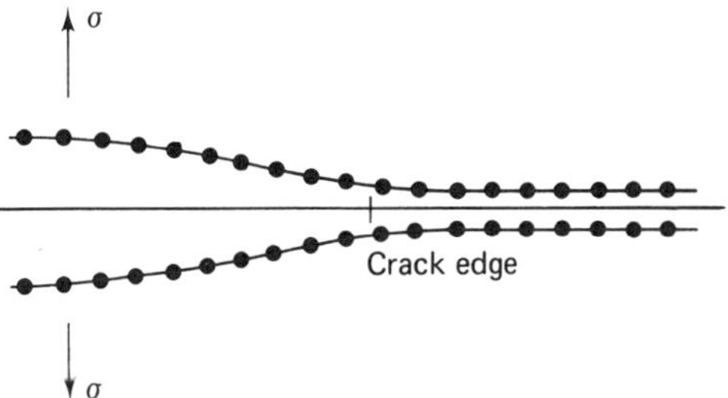

Fig. 24. Displacement of atomic planes at a crack.

Careful experiments on the fracture of most apparently brittle materials reveal some plastic flow near the crack tip. Orowan and Irwin independently proposed that the energy per area consumed in plastic flow, V_p, be added to the surface energy V_α. Since V_p is many times larger than V_α, V_α can be neglected, and Eq. (33) becomes

$$\sigma = \left(\frac{EV_p}{c}\right)^{1/2} \tag{35}$$

Using this equation for crystalline materials which exhibit some ductility, it is found that the crack length necessary for brittle fracture is on the order of a few millimeters, instead of the few microns necessary for amorphous materials. The stress concentration at the tip of a crack can initially be accommodated by plastic deformation, with spontaneous propagation occurring when Eq. (35) is satisfied.

The foregoing relationships have been extensively evaluated by experiment, including correlations of strength with crack and notch size. Griffith obtained strengths of glass approaching the theoretical strength by using freshly drawn fine fibers to reduce crack size and atmospheric corrosion. It has been found that glass fiber strengths increased by factors of 2 or 3 when broken repeatedly to eliminate the largest cracks. The strengths of mica sheets have been shown to increase 10-fold when tension grips narrower than the sheet were used to reduce stress in the vicinity of microcracks at the sheet edges.

EXAMPLE 1: Compute the brittle uniaxial fracture strength of KF, assuming an ionic spacing of 1.33 A, a Poisson ratio of 0.3, and maximum incipient crack length of 5,000 A. Obtain necessary data from Table 2 of Chap. 2 and Table 5 of Chap. 3.

Solution: From Eq. (33)

$$\sigma = \left[\frac{2EV_\alpha}{\pi(1-\nu^2)c}\right]^{1/2}$$

in which E can be obtained from Eq. (16),

$$E = \frac{0.29e^2Z^2j}{r_0^4} = \frac{0.29(4.77\times10^{-10})^2(1)^2\times 7}{(1.33\times10^{-8})^4} = 14.76^{12}\times 10 \text{ g/cm}^2$$

where $j = (6+8)/2 = 7$ (from Table 2, Chap. 2), $e = 4.77\times10^{-10}$ (cgs units), $\text{A} = 10^{-8}$ cm, and surface energy

$$V_\alpha = \int_{r_0}^{r_1} f\,dr$$

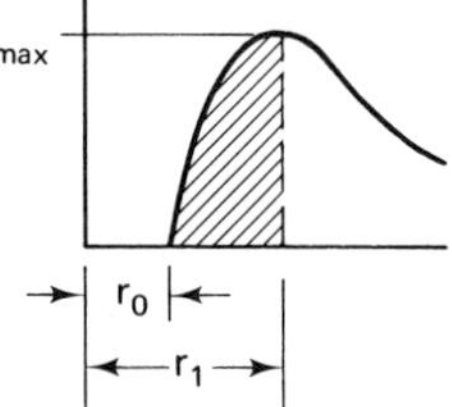

Substituting f from Eq. (11) and integrating,

$$V_\alpha = \frac{0.29Z^2j}{j+1}\frac{e^2}{r_0}P$$

where

$$P = 1 - \frac{j+3}{k(j+2)} \quad \text{and} \quad k = \frac{r_1}{r_0} = 1.25 \qquad \text{(from Table 5)}$$

$$P = 1 - \frac{10}{1.25(9)} = 0.11$$

$$V_\alpha = \frac{0.29(1)^2\times 7}{8}\frac{(4.77\times10^{-10})^2}{1.33\times10^{-8}}(0.11) = 0.477\times10^{-12}\text{ g-cm}$$

Then

$$\sigma = \left[\frac{2(14.76\times10^{12})(0.477\times10^{-12})}{\pi(1-0.09)(5\times10^{-5})}\right]^{1/2}$$

$$= 314 \text{ g/cm}^2$$

c. THEORETICAL SHEAR STRENGTH

The value of $\sigma_{\max} \simeq 0.1E$ [Eq. (32)] compares with the upper bound value $\tau_{\text{avg}} \simeq 0.02E$ [Eq. (20)]. The 5 : 1 ratio between theoretical tensile and shear strengths reflects approximately the ratio between the value of f_{max} [Eq. (29)] and the force required to just lift one atom over its neighbor in an adjoining layer to permit slipping. In shear failure, the bonds with neighbors are renewed periodically, and no new surfaces are created, except for steps at the ends of the plane. The process is therefore less drastic than tensile failure, and lower values of shear strength should be expected.

Alternative approximate solutions for theoretical shear strength have been developed by Frenkel [11] and McKenzie [12]. Frenkel's solution, the simpler of the two, assumes two neighboring planes which remain individually undistorted as shear stress is applied. The shear stress for one atomic displacement is represented as a sine function

$$\tau = c \sin \frac{2\pi x}{b}$$

where x is the shear displacement, b is the repeat distance in the direction of shear (Fig. 25), and c is a constant. For small x, $h(d\tau/dx)$ equals G, the shear modulus, so c equals $Gb/2\pi h$. The maximum restraining force therefore occurs when $x = b/4$, and hence

$$\tau_{max} = \frac{Gb}{2\pi h} \tag{36}$$

For the simple cubic crystal used for previous discussions, $b/h = 1$ for the planes having the highest packing density [the (100) planes]. Therefore $\tau \simeq G/6$. Assuming a Poisson ratio of 0.3, $G \simeq 0.4\,E$ [Eq. (3)] and $\tau_{max} \simeq 0.06E$, compared with the value $\sim 0.02E$ obtained in Eq. (20). For a face-centered cubic crystal, $b/h = 0.67$ for planes with highest packing density, for which $\tau_{max} \simeq 0.04E$. For a layer structure such as graphite, $b/h = 0.4$, for which $\tau_{max} \simeq 0.025E$.

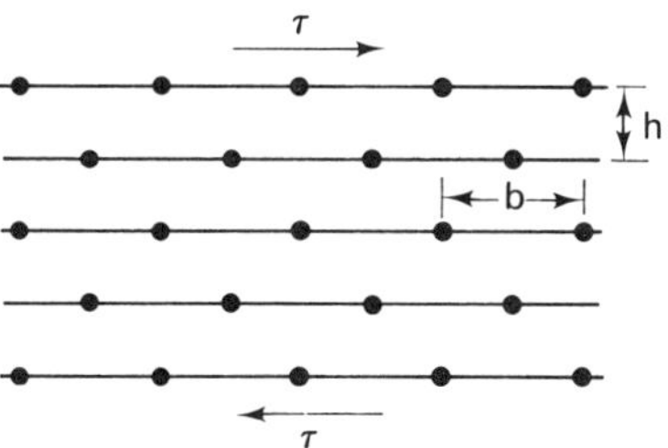

Fig. 25. Dimensions for Frenkel's solution of shear strength, τ.

Metals, as a class, are distinguished by low ratios of τ_{max}/σ_{max}, due primarily to their dense packing (most are face- or body-centered cubic or hexagonal) and to the nondirectional character of the metal bond. Hence metals are ductile; they often fail plastically. The ratio for body-centered cubic metals is larger than for face-centered cubic metals but smaller than for most covalent and ionic solids. For the noble metal copper, $\tau_{max}/\sigma_{max} \simeq \frac{1}{30}$, while diamond and rock salt are characterized by ratios approaching 1. If τ_{max}/σ_{max} is less than $\frac{1}{4}$, crystals under uniaxial tension will fail in shear rather than in tension. The value of $\frac{1}{4}$ comes from the fact that the maximum ratio of σ/τ, where σ is an applied tensile stress and τ the corresponding resolved shear stress on any slip system, is 3.7 when τ is on a slip system of high multiplicity (say $\{111\}$ or $\{11\bar{1}\}$ in a cubic crystal) [8]. Since $\tau_{max} < \sigma_{max}/4$ for metals, the strength of perfect metal crystals is limited by τ_{max}.

d. DUCTILE FRACTURE

As with brittle fracture, observed ductile fracture strengths may be several orders of magnitude below theoretical shear strengths. Ductile fracture theory has not been reduced to equations as simple as those of Griffith [Eq. (33)] and Orowan-Irwin [Eq. (35)] for brittle fracture, although much is known about various causes of ductile fracture. Atomic dislocation and the localized nature of shear failures were mentioned in Sec. A (Fig. 13). A variety of phenomena above the atomic scale also reduces ductile fracture strengths. For example, ductile fracture in alloys commonly initiates at inclusions, such as oxides trapped during solidification, or at precipitated brittle phases. When a metal undergoing deformation flows past a rigid inclusion, dislocations pile up near the inclusion and form a void. The void formation depends partly on the adhesion between the metal matrix and the rigid particle. Neighboring cavities eventually coalesce to initiate the ductile fracture. On the other hand, metal strengths are sometimes improved by *diffusion hardening* or *dispersion strengthening*, a process in which a fine dispersion of rigid particles is introduced to restrain dislocation motion. An example is the diffusion hardening of martensite steel with ammonia gas at 500 to 600°C for 12 to 36 hr. The ammonia reacts with alloying elements such as Al or Cr to form rigid nitrides. The beneficial effect in this case is due to the fact that the rigid particles are very fine and densely distributed so that dislocations do not pile up excessively at one inclusion, and therefore cavities are less likely to form. To obtain effective diffusion hardening the spacing of the obstacles to dislocation must be less than about 10^4 atom spacings, or 2–3 μm, and they are usually distributed within the metal crystals. They can greatly increase the rate of work hardening of a metal, as well as increasing its strength.

Several *plastic failure theories* have been advanced, corresponding to those for plastic yield (Sec. A2). The law which appears most representative of a wide variety of experimental data for *plane stress* of ductile materials is an extension of the Tresca law for plastic yielding. The failure stress surface resembles that in Fig. 10(b), except that fracture stress σ_f replaces yield stress σ_0. Failure stress surfaces for three-dimensional stress have not been well established, due largely to experimental limitations.

2. Chemical Properties of Strong Solids

Strong solids possess large values of both $\sigma_{\max}$ and $\tau_{\max}$. From Eqs. (16) and (32), $\sigma_{\max}$ depends on E, i.e., high valency (z) and small atoms, to permit close spacing (r_0). From Eqs. (20) and (36), $\tau_{\max}$ depends on G, also related to E by the Poisson ratio [Eq. (3)], and on b/h for the planes of highest atomic packing density and direction. In addition, large shear strength is found in solids with interatomic forces of a directed nature, i.e., covalent or strongly polarized ionic bonds. For covalent bonding, a covalence of 3 or 4 is required to produce a three-dimensional molecular crystal instead of separate molecules bound only by secondary forces. Considering tensile and shear strengths together, strong materials are simply those with the highest density of strong directional bonds. The elements possessing these properties are

I	II	III	IV	V	VI	VII	VIII
Li	Be	B	C	N	O	F	Ne
Na	Mg	Al	Si	P	S	Cl	A

Fig. 26. Elements which form strong but brittle materials. Most of these elements are abundant and hence potentially cheap.

beryllium, boron, carbon, nitrogen, oxygen, aluminum, and silicon (Fig. 26). The strongest materials contain one of these elements. Kelly [8] points out that the strongest materials will have low densities since the requirement of small atoms ensures the presence of the lighter elements and directional bonding implies non-close-packed crystal structures. Furthermore, high bond energy implies high Young's modulus [Eq. (16)], which in turn is normally accompanied by high melting point (see Fig. 8). Hence, the strongest solids will possess low density, high elastic moduli, and high melting point, all attractive properties in most engineering applications.

Notice that the discussion of fracture mechanics has pertained entirely to primary bonding: ionic, covalent, and metallic. The reasons for discrepancies between observed and theoretical strengths of polymers, where secondary (molecular) bonding is important, are understood less quantitatively.

PROBLEMS

1. Derive Eq. (3) by using Eq. (1b), the generalized Hooke's Law [Eq. (6)] in the principal stress directions, and the following relationships for principal stresses and strains:

$$\tau_{12} = \frac{\sigma_1 - \sigma_2}{2} \tag{a}$$

$$\gamma_{12} = \epsilon_1 - \epsilon_2 \tag{b}$$

2. Derive Eq. (4) by considering a hydrostatically stressed cube and using the generalized Hooke's law for isotropic materials [Eq. (6)] to compute the volume change.
3. Derive the relationship between elastic volume change and the algebraic sum of the three principal stresses for an isotropic material.
4. What uniaxial stress on an isotropic material produces the same volume change as the three-dimensional stresses $\sigma_1 = -2{,}000$ psi, $\sigma_2 = 4{,}000$ psi, $\sigma_3 = 1{,}000$ psi?
5. Derive the relationship between G and K using Eqs. (3) and (4).
6. Compute the theoretical bulk and Young's moduli of zinc, which has an atomic spacing of 2.66 A and a Poisson ratio of 0.32. Use Eqs. (4) and (17) and Fig. 5.
7. What theoretical uniaxial stress is required to produce a compressive strain of 0.002 in $CaCl_2$? The Ca-Cl equilibrium spacing is 2.81 A.
8. What value of the Poisson ratio results in zero volume change under deformation?
9. A metal cube has an edge length of 1 in. when unstressed. Compressive stresses of 20,000 psi and 15,000 psi are applied in two orthogonal directions. If the Young's modulus is

30×10^6 psi and the Poisson ratio is 0.30, what is the resulting length in the third (unstressed) direction of the cube?

10. Iron has a bulk modulus of 2.46×10^7 psi. What is the volume of a piece of iron 10,000 ft under the ocean if its volume at atmospheric pressure is 1 in.^{3}? Assume constant temperature and seawater density of 64 lb/cu ft.

11. An aircraft component 3 in. long must not elongate more than 0.002 in. under a tensile load of 700 lb. For the given data, which of the following three materials will perform satisfactorily with least weight:

Material	*Young's Modulus (psi)*	*Density (lb/cu ft)*
Steel	30×10^6	490
Aluminum alloy	10×10^6	168
Nylon	4×10^5	53

What ratio of physical properties allows the determination of the best material without doing a complete calculation for each material?

12. A steel oil well drill pipe of uniform cross section having a density of 480 lb/cu ft and Young's modulus of 30×10^6 psi is to be hoisted from its hole for inspection of the drill bit. The pipe initially rests with its full weight on the hole bottom, and 12 ft of pipe are pulled out of the ground before the load on the hoisting cable stops increasing. Assuming zero friction between the side of the pipe and the hole, how deep is the hole?

13. Compute the theoretical Young's modulus of Na^+F^- if $r_0 = 2.31$ A. See Eq. (16) and Table 2, Chap. 2.

14. Determine the ratio of the bulk moduli of tin (Group IV, $r_0 = 3.24$ A) and magnesium (Group II, $r_0 = 2.72$ A). Use Eq. (17) and Fig. 5.

15. With reference to the conclusion drawn from Fig. 5, which of the elements in each of the following pairs is likely to have the higher bulk modulus:
    (a) Cu or Ag?
    (b) Tl or In?
    (c) Li or Na?

16. Thirty percent of the anelastic strain in a structural polymer is recovered in 12 sec after the stress is removed.
    (a) What is the relaxation time, θ?
    (b) How much anelastic strain remains at 60 sec after stress removal?

17. A wire tendon to be used for concrete prestressing had an immediate elongation of 0.082 in. after applying a fixed load. Ten minutes later the elongation was 0.089 in., and several days later it was 0.091 in. Estimate the remaining elongation in the wire 3 min after the load is removed. Assume that the relaxation time θ is the same for loading and unloading.

18. After a fixed load was applied to a sample of steel strand the immediate strain was 0.021. The elastic strain was 0.26 after 1 min of loading, and 0.28 after 10 min, at which time the load was removed. How long would it take for the strain to drop to 0.005? Assume relaxation time θ to be the same for loading and unloading.

19. The Maxwell model provides a simplified representation of the behavior of asphalt. Determine the viscosity and shear modulus of an asphalt under a shear stress of 2×10^3

dynes/cm² if the shear strains are

30 sec	0.010
300 sec	0.100

20. If an asphalt has a viscosity of $10^{0.5}$ poise at 200°C and $10^{0.1}$ poise at 350°C, what is its viscosity at 300°C?

21. What are the shear modulus and viscosity of a polymer which behaves as a Kelvin model if under a shear stress of 5×10^6 dynes/cm² the shear strains are

1 hr	0.0050
5 hr	0.010
10 hr	0.010

22. What is the viscous relaxation time, $\theta_v = \eta/G$, for the polymer in Problem 22? If the stress is removed after 10 hr, how long will it take for the strain to return to 0.002?

23. The stress in a concrete prestressing tendon at 20°C relaxes from 90 ksi to 85 ksi in 200 days.
    (a) Do you think steel should behave more like a Maxwell model or a Kelvin model?
    (b) What is the relaxation time?
    (c) How long would it take to relax to 80 ksi?
    (d) How long would it take to relax to 80 ksi at 200°C, assuming an initial stress at this temperature of 85 ksi?

24. A stress of 1,000 psi produces an initial strain in an elastomer of 0.5. After maintaining that strain for 20 days the stress is 600 psi. What stress would maintain the same strain at 100 days?

25. Calculate the ideal fracture strength of NaBr.

26. What uniaxial stress is required to enlarge a crack 0.2 mm long in steel which has a Young's modulus $E = 40 \times 10^6$ psi (2.8×10^{12} dynes/cm²) and an energy per area consumed in plastic flow $V_p = 5 \times 10^5$ ergs/cm²?

27. In cutting plate glass, a scratch 0.010 cm deep is made in a plate 0.25 in. thick. What bending moment per linear inch of scratch is required to break the glass? Assume that the glass has a Young's modulus $E = 5.6 \times 10^{11}$ dynes/cm², a Poisson ratio $\nu = 0.25$, and a specific surface energy $V_\alpha = 280$ ergs/cm².

28. Why does the fracture strength of brittle materials decrease with specimen size?

29. A 3-ft-OD glass cylinder with 1-in.-thick wall is used for submarine research. For a cable load, submerged, of 2,000 lb (Fig. 27), determine the safe theoretical depth of submergence below sea level, in feet. Neglect the weight of the cylinder ends. Assume that seawater weighs 65 pcf and that the properties of the glass are

$$E = 6 \times 10^{11} \text{ dynes/cm}^2$$

$$V_\alpha = 250 \text{ ergs/cm}^2$$

$$\nu = 0.25$$

$$c = 0.002 \text{ cm} \quad \text{(controlled)}$$

Refer to Eq. (34) and Fig. 23.

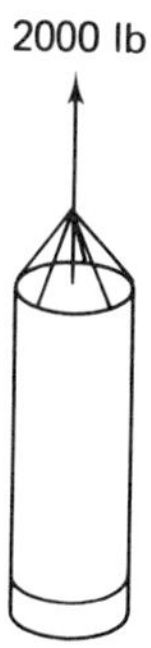

2000 lb

Fig. 27. Submarine research chamber.

30. Two methods have been proposed for prestressing the glass in Problem 29: one by placing the cylinder in a metal shroud to provide circumferential compressive prestress and the second by tie rods between flanges on the cylinder ends to provide longitudinal compressive prestress. Which method can permit use of the chamber to the greater depth? Why?

31. What maximum compressive prestress could be applied in the direction indicated in Problem 29 before the cylinder would fail at zero water depth? To what theoretical depth of submergence would the cylinder by safe with this magnitude of compressive prestress?

REFERENCES

1. HAYDEN, H. W., W. G. MOFFATT, and J. WULFF, *The Structure and Properties of Materials*, Vol. III: Mechanical Behavior. New York: John Wiley & Sons, Inc., 1965.
2. PAULING, L., *The Nature of the Chemical Bond*, 3rd ed. Ithaca, N.Y.: Cornell University Press, 1960.
3. ROSENTHAL, D., *Introduction to Properties of Materials.* New York: Van Nostrand Reinhold Company, 1964.
4. VAN VLACK, L. H., *Materials Science for Engineers.* Reading, Mass.: Addison-Wesley Publishing Company, Inc., 1970.
5. POLAKOWSKI, N. H., and E. J. RIPLING, *Strength and Structure of Engineering Materials.* Englewood Cliffs, N.J.: Prentice-Hall, Inc., 1966.
6. JAEGER, J. C., *Elasticity, Fracture and Flow*, 2nd ed. London: Methuen & Co., Ltd., 1962.
7. BENJAMIN, B. S., *Structural Design with Plastics.* New York: Van Nostrand Reinhold Company, 1969.
8. KELLY, A., *Strong Solids.* New York: Oxford University Press, Inc., 1966.
9. GRIFFITH, A. A., *Phil. Trans. Roy. Soc.*, A221 (1920) 163.
10. OROWAN, E., *Zeit. Physik.*, 82 (1933) 235.
11. FRENKEL, J., *Zeit. Physik.*, 37 (1926) 572.
12. MCKENZIE, J. K., Ph.D. thesis, Bristol, 1949.

II

MATERIAL PROPERTIES

4

INORGANIC CEMENTS

It should be so built that it *couldn'* break daown:
"Fur," said the Deacon, "'t's mighty plain
Thut the weakes' places mus' stan' the strain;
'N' the way t' fix it, uz I maintain,
    Is only jest
T' make that place uz strong uz the rest."
—HOLMES

Cement denotes any kind of adhesive, inorganic or organic. In construction it generally implies a binder for sand, aggregate, or fibrous materials to produce various concretes, mortars, or fibrous cement products. It may be a single chemical compound, but more often it is a mixture. In this chapter we shall describe mechanical properties of inorganic cements in terms of their chemical structure. It provides the basis, together with Chap. 6, for studying concretes in Chaps. 7 and 8.

A. CLASSES OF INORGANIC CEMENTS

Inorganic cements may be hydraulic or nonhydraulic. *Hydraulic* cements harden by reaction with water and give water-resistant products. As a group, silicate cements are hydraulic because they can harden under water. The most widely used hydraulic cement is portland cement. Others include aluminous, pozzolanic, slag, and natural cements.

Hydraulic cements having small interfacial areas between the reacting components may harden very slowly. The reaction can be accelerated by curing under steam pressure in an autoclave, as for mixtures of lime and fine sand or fly ash. Such cements, called *hydrothermal* cements to distinguish the necessary curing conditions, will be described later.

Nonhydraulic cements are those which either do not harden by reaction with water or which do not give water-insoluble products. Gypsum and lime are nonhydraulic cements. Gypsum hardens by reacting with water, but the product is water-soluble. Although hardened lime is water-resistant, its hardening results from reaction with atmospheric carbon dioxide to form $CaCO_3$, rather than from reaction with water.

Table 1 classifies typical inorganic cements, and Fig. 1 shows the composition of those in the lime-silicate-aluminate system [1]. Several nonportland cements will first be described briefly, and then the portland and hydrothermal cements will be studied in more detail.

Fig. 1. Approximate compositions of various cements by weight.

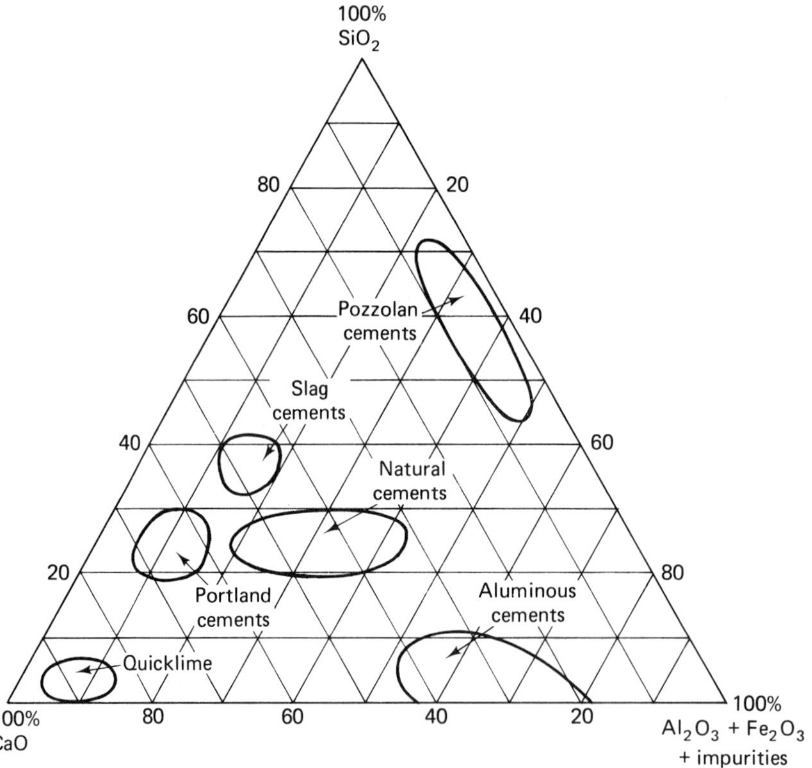

TABLE 1. Classes of Inorganic Cements

Hydraulic cements		*Nonhydraulic cements*
1. Portland cements		1. Gypsum
2. Other lime-silicate-aluminate cements		2. Lime (quick, hydrated)
a. Aluminous		3. Magnesium oxychloride (sorel cement)
b. Pozzolanic		
c. Slag		
d. Natural		
e. Expansive		
f. Lime—silica sand	Hydrothermal	
g. Lime—fly ash	Hydrothermal	

1. Hydraulic Cements

Aluminous cement, first produced in France in 1913, attains a 24-hr strength equivalent to the 4-week strength of portland cement and an ultimate strength about twice that of portland cement. It is more resistant to high temperatures and to attack by sulfate and seawater than is portland cement. It develops optimum strength at a high water/cement ratio, thus giving a very workable concrete mix (Fig. 2). These advantages compensate, in some applications, for its cost of about three times that of portland cement. A very serious limitation of aluminous cement is a conversion reaction which degrades its strength after hot wet exposure and therefore limits its use-

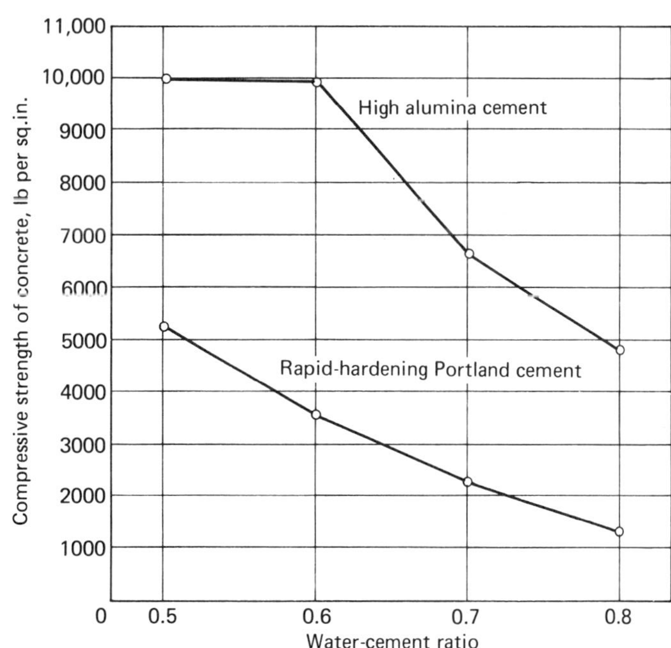

Fig. 2. Strengths of concretes made with aluminous and portland cements as a function of water: cement weight (w/c) ratio. From Lea F. M. and C. H. Desch; *The Chemistry of Cement and Concrete*, Second Edition. London: Edward Arnold Co., 1956 [2].

fulness in structural applications. For example, this has been identified as the cause of some roof beam failures in England. Aluminous cements are used in mortar for fire brick. A refractory concrete useful to 1600°C (2919°F) can be made, using fire clay or corundum as an aggregate. Much of the aluminous cement output is used in industrial furnaces and kilns.

Although hardened aluminous cement withstands high temperature, its strength is critically dependent on the temperature during curing. High curing temperature is much more damaging to aluminous cement than to portland cement. Therefore, aluminous cement, which has a high rate of heat liberation during curing, must usually be placed in quite thin sections. Figure 3 compares temperature rise curves for adiabatically-cured aluminous and portland cements.

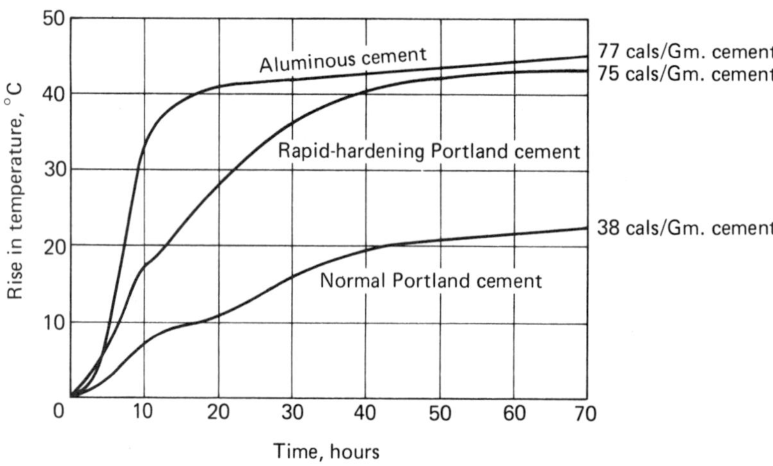

Fig. 3. Temperature rise in adiabatically cured 1 : 2 : 4 concrete, w/c 0.6. From Davey, N. and E. N. Fox, *Building Research Technology Paper No. 15*, London: H. M. Stationery Offices, 1933 [3]. Crown copyright. Reproduced by permission of the Directors, Building Research Establishment, UK.

The chemical composition of aluminous cement is more complicated than that of portland cement, and the mineralogical composition cannot be so easily computed. The approximate composition is [4]

SiO_2	3–11%
Al_2O_3	33–44%
CaO	35–44%
Fe_2O_3	4–12%
FeO	0–10%

Aluminous cements are made by fusing limestone and bauxite in an electric furnace or blast furnace and grinding the resulting clinker. Clays, as used for portland

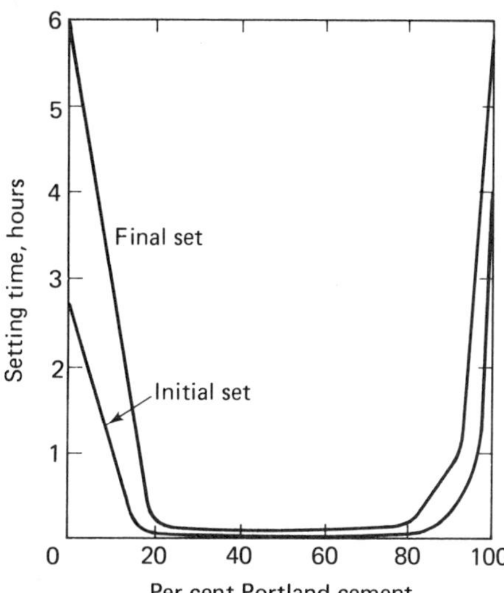

Fig. 4. Setting time of mixtures of aluminous and portland cements. From Lea, F. M.: *The Chemistry of Cement and Concrete*. London: Edward Arnold & Co., 1970 [5].

cement, are unsuitable because they contain too little alumina and too much silica. Bauxite-limestone mixtures melt completely within a narrow temperature range, especially in the presence of ferric oxide. For this reason aluminous cement cannot be clinkered in a rotary kiln as portland cement is but must be melted in a metallurgical smelting furnace.

A very quick setting cement may be obtained by mixing aluminous cement with either lime or portland cement (Fig. 4) but always with a reduction in strength below that of the individual cements. Such mixtures can be used for emergency repairs of water leaks, etc., provided the resulting strength is sufficient. Both the gypsum and calcium hydroxide present in hydrating portland cement are known to accelerate the setting of aluminous cement. On the other hand, the removal of gypsum from portland cement by combination with hydrated calcium aluminate formed in aluminous cement is probably responsible for the accelerated set of portland cement with small additions of aluminous cement [5].

Strontium aluminate and barium aluminate refractory cements have been developed which withstand still higher temperatures than calcium aluminate cements [6].

Pozzolans are noncrystalline (glassy) siliceous materials which react with $Ca(OH)_2$ and water to form cements. The word comes from the Roman "pozzuolana," a volcanic ash found in Italy. Pozzolanic cements are mixtures of pozzolan with portland cement and consist principally of natural pozzolan of volcanic origin or artificial pozzolan inthe form of fly ash or calcined clays or shales.

Slag cement is made by rapidly cooling blast furnace slag with cold water, causing it to break up into fine particles, which are dried, partially ground, mixed with hydrated lime, and ground again.

Iron ores contain clay impurities, consisting of silica and alumina, which must be separated from the molten iron. Since the fusion temperature of silica-alumina mixtures is much higher than the melting point of iron, lime is added as a fluxing agent. The resulting molten slag, consisting of lime, silica, and alumina, floats on top of the molten iron and can be drawn from the blast furnace in liquid state.

The composition of blast furnace slag resembles that of portland cement, except that it contains practically no ferric oxide, since the purpose of the blast furnace is to remove iron as efficiently as possible. Slag also contains less lime than does portland cement. The eutectic (lowest melting) temperature in iron production is obtained at lime contents considerably below those of portland cements, and it is therefore uneconomical to add higher lime contents.

Although slag cements harden more slowly than portland cement, there is little difference in final strength values, and the hydration products are very similar. Since slag cements are usually ground finer than portland cements due to their lower hardening rate, their shrinkage is also apt to be higher. Slag cement is cheaper than portland cement and is sometimes used in bulk concrete not requiring high initial strength but where low heat of hydration is desired, as in massive structures such as dams. Slag cements are also sometimes added to portland cement to increase plasticity. Slag cements are more widely used in Europe than in North America. In the Trief process, granulated slag is wet-ground in ball mills and then fed with portland cement and aggregate directly into the concrete mixer. The wet grinding requires less energy to obtain a given fineness than dry grinding.

Natural cements are made by calcining clayey limestone at about 2300°F. This temperature is lower than required for the sintering of portland cement. The clay gives the product a high alumina content. The clinker is slaked and then crushed and ground. Natural cements set quickly, and gypsum is often added as a retarder. Strengths are commonly $\frac{1}{3}$ to $\frac{1}{2}$ those of portland cements. Hence, natural cements, like slag cements, are used where bulk rather than strength or weather resistance is required.

2. Nonhydraulic Cements

Gypsum manufacture begins with the mining, crushing, and dehydration of the mineral gypsum. The dehydration, performed in large heated kettles or a rotary kiln, is accomplished in two stages:

$$\underset{\text{dihydrate}}{\overset{\text{gypsum}}{CaSO_4 \cdot 2\,H_2O}} \xrightarrow{130^\circ C} \underset{\text{half-hydrate}}{CaSO_4 \cdot \tfrac{1}{2}\,H_2O} + 1\tfrac{1}{2}H_2O$$

$$\underset{\text{half-hydrate}}{CaSO_4 \cdot \tfrac{1}{2}\,H_2O} \xrightarrow{205^\circ C} \underset{\text{anhydrite}}{CaSO_4} + \tfrac{1}{2}\,H_2O$$

The half-hydrate is quite stable and is packaged in bags for use in plaster. The anhydrite absorbs water rapidly and is converted to the half-hydrate. Calcium sulfate is unusual in that its hydrate is less soluble in water than its half-hydrate. Thus, when the half-hydrate is mixed with water it slowly reacts to precipitate needles of dihydrate:

$$1\tfrac{1}{2}\,H_2O + CaSO_4 \cdot \tfrac{1}{2}\,H_2O \longrightarrow CaSO_4 \cdot 2\,H_2O$$

The formation of interlocking dihydrate crystals causes the hardening of plaster of Paris. The rate of hardening depends on the fineness of grinding of the half-hydrate and the water temperature. It is easily workable, and setting occurs typically in about 10 min, without the use of animal glues or other organic retarding agents. The strength of the dihydrates is normally 1,000–2,000 psi. It provides excellent fire resistance. Upon exposure to fire the insulating value of the dihydrate increases due to loss of water of hydration and a corresponding increase in porosity. Since it is slightly water-soluble, it is not used for exterior applications.

Lime manufacture and use as a cement is based upon the following three reactions:

Burning:

$$\underset{\text{limestone}}{CaCO_3} \xrightarrow{1000°C} \underset{\text{quicklime}}{CaO} + CO_2$$

Slaking:

$$CaO + H_2O \longrightarrow \underset{\text{slaked lime}}{Ca(OH)_2}$$

Hardening:

$$Ca(OH)_2 + CO_2 \longrightarrow CaCO_3 + H_2O$$

When heated to about 1000°C limestone rapidly decomposes to quicklime and carbon dioxide. Although about 50% of the weight is lost as CO_2, the volume of the CaO is almost the same as that of the original limestone. The high porosity resulting from the expelled CO_2 molecules enable quicklime to react almost violently with water to produce $Ca(OH)_2$. Because the volume of $Ca(OH)_2$ (slaked lime) is 20% greater than the volume of CaO, the expansive force due to the absorption of water causes the product to disintegrate to a white powder having particle sizes on the order of 2 μ.

Slaked lime mixed with water and sand to form mortar develops a strength of 400–500 psi upon slowly drying. Any finely divided material, clay, for example, develops similar strength due to cohesive forces developed during drying. However, further hardening of the lime mortar results from slow reaction with atmospheric CO_2, which reconverts the slaked lime to limestone. The resulting product is quite porous and therefore much weaker than most natural limestones.

Magnesium oxychloride (sorel cement) is made by treating a mixture of MgO and aggregate with a concentrated solution of $MgCl_2$ to produce $Mg_2(OH)_3Cl \cdot 4H_2O$ crystals. The MgO is obtained by calcining $MgCO_3$ at a low temperature. The product is sometimes used as a flooring material. It is hard but does not resist water and can be protected by polishing with wax in turpentine.

B. PORTLAND CEMENTS

1. Manufacture

Portland cement takes its name from the color of a natural stone quarried on the Isle of Portland, off the British Coast. Joseph Aspdin, a Yorkshire bricklayer, is regarded as the discoverer of portland cement, having used limestone dust and clay as raw materials. Aspdin's patent for portland cement in 1824 was followed by manufacture in the United States about 50 years later.

a. RAW MATERIALS

Two classes of raw materials are used for portland cement, argillaceous (clayey) and calcareous (lime-bearing). The argillaceous materials, which may be clay, shale, or slate, provide alumina and silica. The calcareous materials, which may be limestone, marl, or oyster shells, provide lime. Cement rock, which is a clayey limestone, is much easier to grind than other limestones and contains nearly the correct proportions of the ingredients, although usually small amounts of pure limestone must be added. Blast furnace slag is also mixed with limestone and ground to form a true portland cement. This is distinguished from the previously described slag cement, produced by grinding slag and hydrated lime together without calcination.

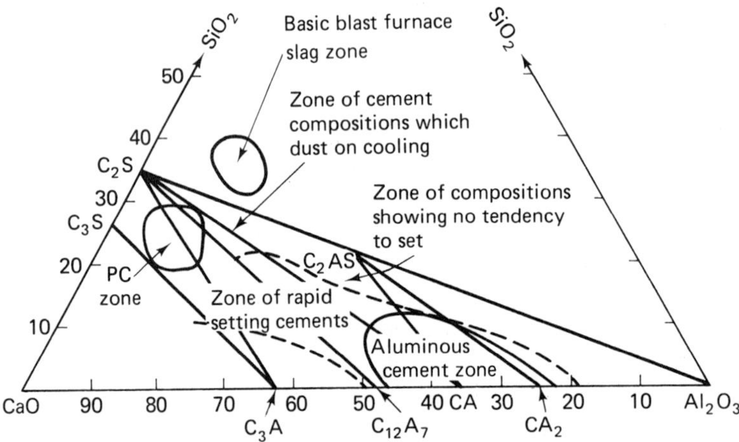

Fig. 5. Cement zones in the CaO-Al_2O_3-SiO_2 system. From Lea, F. M.: *The Chemistry of Cement and Concrete*. London: Edward Arnold & Co., 1970 [5].

The major raw component of portland cement is lime, followed by smaller percentages of silica, alumina, and finally iron oxide. Figure 5 shows the compound composition ranges of some of the cements in Fig. 1. Compositions immediately above the upper dashed line in Fig. 5 have little cementing value, and those below the lower dashed line exhibit flash set due to the large proportion of tricalcium aluminate (C_3A). A weight ratio of lime to alumina and silica of about 1.7 is frequently specified for portland cements. This corresponds to a lime content in the CaO-Al_2O_3-SiO_2 system of 63%. If one allows for 2–3% ferric oxide, the minimum lime content is about 60%, which is seen to mark the lower lime limit of portland cement in Fig. 5. This lower limit results from decreasing cementing ability and from increasing trouble due to dusting caused by the transformation from the β to the γ form of dicalcium silicate on cooling. This dusting is more pronounced when the low lime content is combined with low alumina content and correspondingly high silica content. The high lime content is limited by the requirement that nearly all the lime should combine with the silica-alumina clinker. The low alumina limit is fixed by the added cost of burning due to high clinkering temperatures rather than by any compositional requirements. White portland cements have lower alumina contents. The high alumina content limit is determined by too rapid setting of the hydration products. The setting rates of cements having SiO_2/R_2O_3 rations much below 1.5:1 cannot be controlled by the addition of gypsum due to their large tricalcium aluminate contents.* If the R_2O_3 contains a significant proportion of ferric oxide, somewhat lower $SiO_2 : R_2O_3$ is acceptable because $(CaO)_4 \cdot Al_2O_3 \cdot Fe_2O_3$ does not set as rapidly as $3CaO \cdot Al_2O_3 \cdot Fe_2O_3$.

Modern portland cements fall in the triangle C_3S-C_2S-C_3A in Fig. 5 [5]. The general trend in manufacture has been toward higher lime contents, resulting in higher tricalcium silicate contents.

The kiln reaction products of portland cement, in weight percent, can be determined approximately from the chemical analysis data. The abbreviations in general use are

$$C = CaO, \quad S = SiO_2, \quad A = Al_2O_3, \quad F = Fe_2O_3, \quad U = SO_3 \text{ (sulfur trioxide)}$$

and the composition equations are

Tricalcium silicate:

$$C_3S = 4.07\ C - (7.60\ S + 6.72\ A + 1.43\ F + 2.85\ U)$$

Dicalcium silicate:

$$C_2S = 2.87\ S - 0.754\ C_3S$$

Tricalcium aluminate:

$$C_3A - 2.65\ A - 1.69\ F$$

Tetracalcium aluminate-ferrite:

$$C_4AF = 3.04\ F$$

These equations were derived by Bogue [7], based upon the known reaction sequence and the molar ratios and molecular weights of the reactants. For reasons which are only partially understood, the mineralogical composition of commercial portland cements can vary appreciably from the calculations, although they provide useful

*R = Fe or Al.

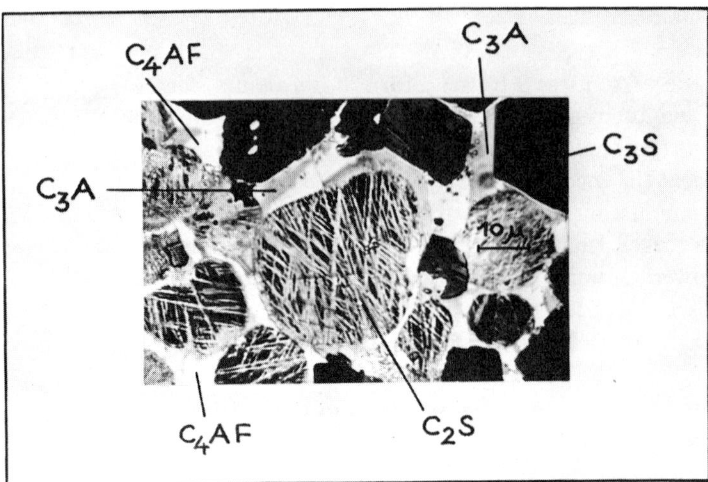

Fig. 6. Photomicrograph of polished and etched portland cement clinker. Photograph by Dr. F. Trojer, Österr., Amerikan. Magnésit A. G., Radenthein. From Czernin, W.: *Cement Chemistry and Physics for Engineers*. New York: Chemical Publishing Company, Inc., 1962 [4].

approximations. Various corrections to the Bogue equations have since been proposed.

These major reaction products can be identified microscopically on a piece of clinker which is cut, polished, and etched, due to different reflecting powers of the minerals. Figure 6 shows dark polygonal crystals of C_3S and rounded crystals of C_2S with a characteristic line network embedded in strongly reflecting (hence white) iron compounds of C_4AF as well as gray C_3A crystals [4].

The chemical composition can be evaluated by several ratios, of which the *lime saturation factor* is the most important. The lime saturation factor is the ratio, in weight percent, of the lime present in the cement to the quantity required to form the three major clinker components (C_3S, C_2S, and C_3A):

$$\text{lime saturation factor} \quad \text{LSF} = \frac{\text{CaO} - 0.7(\text{SO}_3)}{2.8(\text{SiO}_2) + 1.2(\text{Al}_2\text{O}_3) + 0.65(\text{Fe}_2\text{O}_3)}$$

If the LSF is too low, there is not enough tricalcium silicate (the major component responsible for early hardening) in the cement. If the LSF exceeds 1.0, free lime is present in the cement, and the hydrated cement is not volume stable. Free lime in finished cement behaves entirely differently from hydraulic lime. Hydraulic lime is not sintered, is soft and porous, and slakes quickly on contact with water. Free lime in portland cement is trapped within dense shells of solidified melt and is not easily accessible to slaking water, even if the clinker is finely ground. The increase in volume which accompanies final slaking occurs after the concrete has hardened to an extent where differential volume changes cause severe internal stressing and cracking. This process is called "unsoundness due to free lime" in portland cement. The acceptable lime contents shown in Fig. 5 can be expressed in terms of the LSF approximately as $0.66 < \text{LSF} < 1.02$ [8].

LSF > 1.02	Unsoundness due to free lime
LSF = 1.00	Practically all silica in the form of tricalcium silicate
LSF < 1.00	Silica forms a mixture of tricalcium and dicalcium silicates
LSF < 0.66	Excessive dicalcium silicate

The LSF of commerical portland cements is in the range of 0.85 to 0.95.

Apart from the major components, the following minor components also affect the properties of portland cement.

Magnesia (MgO) is present in small quantities in the raw materials of almost all portland cements. It does not combine with alumina or silica during kilning and remains free in the finished product. As with free lime, slaking of MgO may occur largely after the concrete has hardened and is accompanied by volume increase and cracking. The allowable magnesia content is usually limited to 5%.

Sulfuric anhydrate (SO_3) is ground with portland cement clinker in the form of gypsum ($CaSO_4 \cdot 2\,H_2O$) to regulate the setting time. When mixed with water the tricalcium and dicalcium silicate remain soft and plastic for several hours before setting, but tricalcium aluminate sets almost immediately. Although the weight fraction of tricalcium aluminate is small, it can cause flash setting of concrete. A small amount of gypsum can prevent flash setting by quickly combining with tricalcium aluminate to form voluminous needle-shaped crystals of calcium sulfoaluminate hydrate. The resulting expansion is not harmful but can cause internal stressing and cracking if it continues after sufficient hardening has occurred (usually after about 1 day). The SO_3 content is therefore normally limited to about 3%.

Alkalies (K_2O and Na_2O) are usually present in small amounts in cement raw materials. If the concrete aggregate contains reactive (noncrystalline) silica in such forms as opal, chalcedony, cristobalite, or tridymite, alkaline hydroxides derived from hydrating alkalies in the cement can react with the silica to imbibe water and form a violently expanding gel. If the gel is surrounded by hardened cement paste, internal stresses and cracking develop. With nonreactive aggregates, small amounts of alkali in the cement are harmless, but with reactive aggregates cements with low alkali content are required.

b. PLANT OPERATION

Portland cement manufacture, after raw materials have been obtained, involves

1. Grinding the raw materials.
2. Proportioning and mixing raw materials.
3. Pulverizing the mixture (to under 100 μ).
4. Burning the mixture to form a clinker.
5. Cooling the clinker.
6. Adding gypsum to reduce the rate of set.
7. Final grinding (to under 100 μ).

The raw materials may be fed into the kiln dry (the dry process) or in an aqueous slurry (the wet process). The two processes are illustrated in Fig. 7. In the wet process a cement of more uniform composition can be manufactured and the grinding costs less, but there is increased expense in kiln-drying the slurry. Most plants now use the wet process. Anionic and cationic floatation techniques (see Sec. B in Chap. 6) are used at a few plants to separate undesired components from the pulverized raw

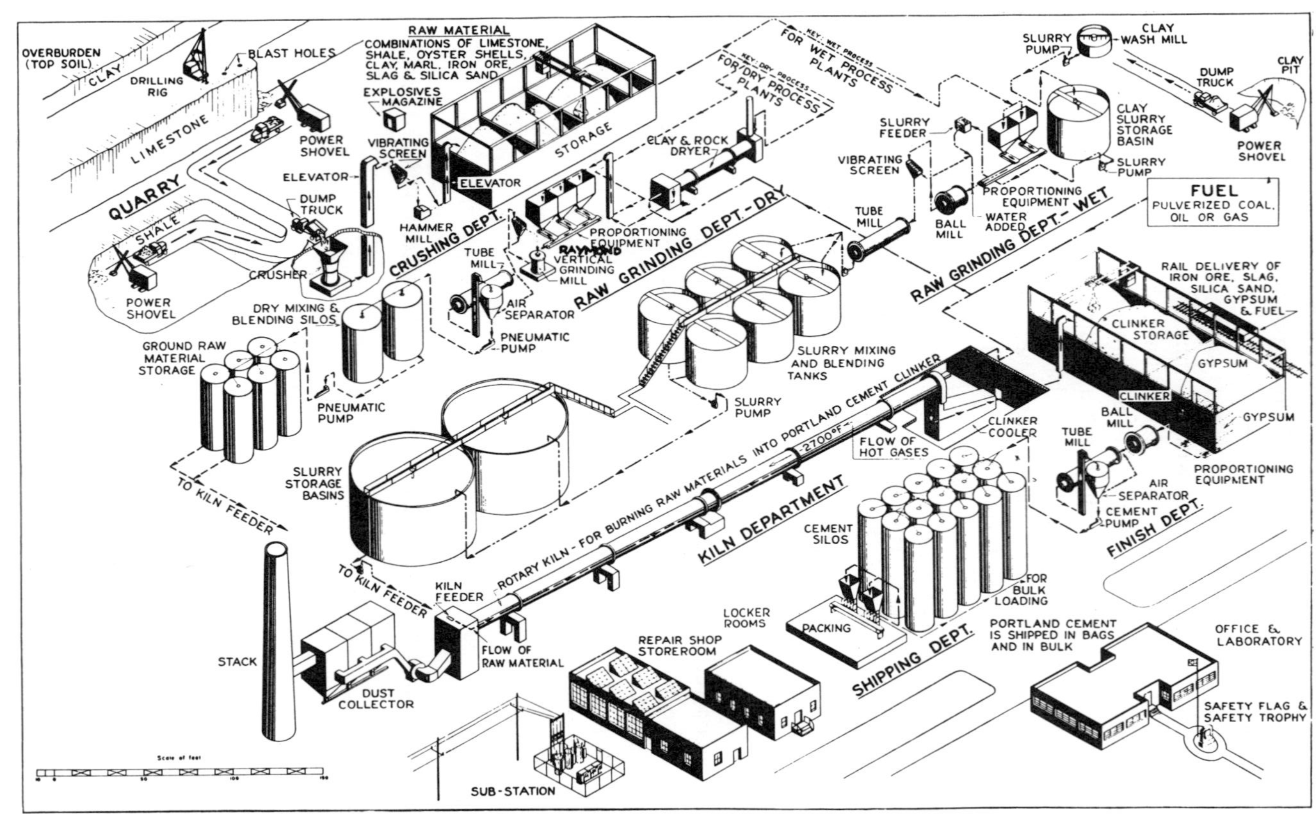

Fig. 7. Flowchart of portland cement manufacture, dry and wet processes. *Courtesy of Portland Cement Association.*

materials. A detailed account of portland cement manufacturing processes and descriptions of chemical tests for component analysis is presented by Witt [9].

The properties of finished cement depend primarily on four factors:

1. Chemical composition of the mix.
2. Physiochemical state of the raw components.
3. Temperature and duration of burning.
4. Clinker cooling rate.

To proportion the raw materials the plant chemist first determines the composition of each material. In all probability the first proportioning of raw materials will not produce quite the desired composition, due to variability of the raw materials. After blending and grinding, the raw mix is transferred to large blending bins and analyzed again. The final kiln feed is obtained by mixing materials from several such blending bins, each closely approximating the desired composition. In the wet process, slurry tanks replace the blending bins. Careful blending control is quite critical even with fairly uniform raw materials. For example, an increase of 1% limestone in a typical limestone-shale mix can cause an increase of up to 12% C_3S and a corresponding reduction of C_2S in the resulting cement.

The mixed raw material from step 3 (pulverizing) is fed into a slightly inclined, rotating kiln in which coal, oil, or gas fuel is used to produce a calcining temperature of about 1400°C. Such kilns, up to 12 ft in diameter and 450 ft long, are lined with special fire brick and rotate at about 1 rpm to produce slow movement of material down the incline. The time of passage through the kiln may be $2\frac{1}{2}$ to 6 hr, of which 20 min or less is at a temperature required for clinkering. The calcined material is burned to incipient fusion and discharged at the lower end as clinkers about the size of walnuts or smaller. The cooled clinker is ground to a fine powder, mixed with 2–3% gypsum to retard the setting time, and stored for shipment.

c. KILN REACTIONS

The major reactions occurring in the kiln are

Reaction	*Temperature*	*Heat Change*
Evaporation of free water	Above 100	Endothermic
Evolution of combined water from clay	Above 500	Endothermic
Evolution of CO_2 from magnesium carbonate	Above 600	Endothermic
Crystallization of amorphous dehydration products of clay	Above 900	Exothermic
Evolution of CO_2 from calcium carbonate	Above 900	Endothermic
Initial reaction between lime and clay	900–1200	Exothermic
Evolution of SO_3 from sulfate	Above 1000	Endothermic
Commencement of liquid formation	1250–1280	Endothermic
Continued liquid formation and formation of cement compounds	Above 1280	Probably net endothermic

Approximately one third of the original weight is lost due to volatilization, and 20–30% of the remaining mass is melted, forming new crystalline and glass products on cooling.

Under equilibrium conditions above 1280°C, the solid and liquid state reactions of the five major raw components of portland cement (CaO, MgO, Al_2O_3, Fe_2O_3, and SiO_2) proceed in approximately the following sequence:

1. Fe_2O_3 reacts with $Al_2O_3 + CaO$ to form C_4AF.
2. Remaining $Al_2O_3 + CaO \longrightarrow C_3A$.
3. Remaining $CaO + SiO_2 \longrightarrow C_2S$.
4. Remaining $CaO + C_2S \longrightarrow C_3S$.
5. Remaining $CaO \longrightarrow$ uncombined.
6. $MgO \longrightarrow$ uncombined.

In C_3S, silica is combined with the maximum possible amount of lime, about 75% of weight. Thus, although none of the raw mix components possess cementing properties, they form calcium silicates and aluminates during clinkering which are cementitious when mixed with water.

Equilibrium conditions are not completely obtained in commercial clinker due to various factors, including incomplete solid-liquid interaction of the coarser particles during the short clinkering period. If the clinker is underburned, the resulting cement may be unsound due to an excess of free, or unreacted, lime. If the clinker is overburned, it will be difficult to grind and may yield a less reactive cement due to the formation of larger crystals or ones containing fewer defects [10].

Since clinkering is dependent on liquid formation, the minimum temperature at which liquid forms is important in terms of fuel economy. Hansen [11] has reported the following minimum eutectic melting temperatures:

Components	*Minimum Eutectic Temperature (°C)*
C_3S-C_2S	2065
C_3S-C_2S-C_3A	1455
C_3S-C_2S-C_3A-N	1430
C_3S-C_2S-C_3A-M	1375
C_3S-C_2S-C_3A-F	1340
C_3S-C_2S-C_3A-F-M	1300
C_3S-C_2S-C_3A-N-F-M	1280

where M = MgO and N = Na_2O or K_2O.

Alumina and iron oxide are the main fluxes in cement clinkering. In fact this is their primary contribution to cement manufacture, since neither contribute significantly to ultimate strength of the hydrated cement. To lower the clinkering temperature still further, fluxing agents such as 1–3% of calcium fluoride and various fluoride wastes are sometimes used [12]. The fluoride forms solid solutions with tricalcium

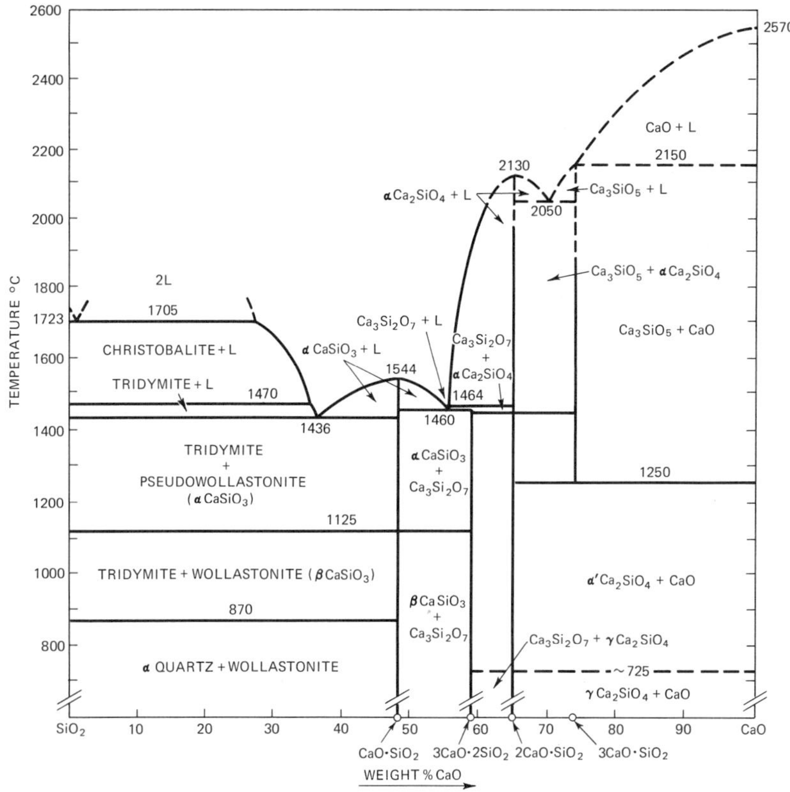

Fig. 8. The CaO-SiO_2 system. From Lea, F. M.: *The Chemistry of Cement and Concrete*. London: Edward Arnold & Co., 1970 [5].

silicate, which give lower strength than pure C_3S [13]. Tricalcium aluminate does not form in the presence of CaF_2 and is replaced by $C_{12}A_7$, making some additional lime available for the formation of C_3S [14]. Thus, the net effect depends on the balance between the increase in C_3S and the decrease in its cementing value.

Some notion of the complexity of the kiln reactions can be gained from the phase diagrams for combinations of the reacting components. For example, there are four distinctive compounds in the binary system CaO-SiO_2 (Fig. 8): (1) the metasilicate $CaO \cdot SiO_2$, (2) the compound $3\,CaO \cdot {}_2\,SiO_2$, (3) the orthosilicate $2\,CaO \cdot SiO_2$, and (4) the compound $3\,CaO \cdot SiO_2$. Each compound has several crystal forms which are stable in different temperature ranges, although not all are found in portland cement.

The ternary system containing the three oxides CaO-Al_2O_3-SiO_2 comprises about 90% of portland cement. The equilibrium diagram for this system is shown in Fig. 9, where the range of portland cement compositions is also indicated.

The quaternary system CaO-Al_2O_3-SiO_2-Fe_2O_3 comprises about 95% of portland cement. A knowledge of the phase relations in this system is fundamental to all portland cement studies. Just as the ternary system can be illustrated by an equilateral triangle with each apex representing one pure component (Fig. 9), a quaternary system can be illustrated by a regular tetrahedron, with each of the four apices

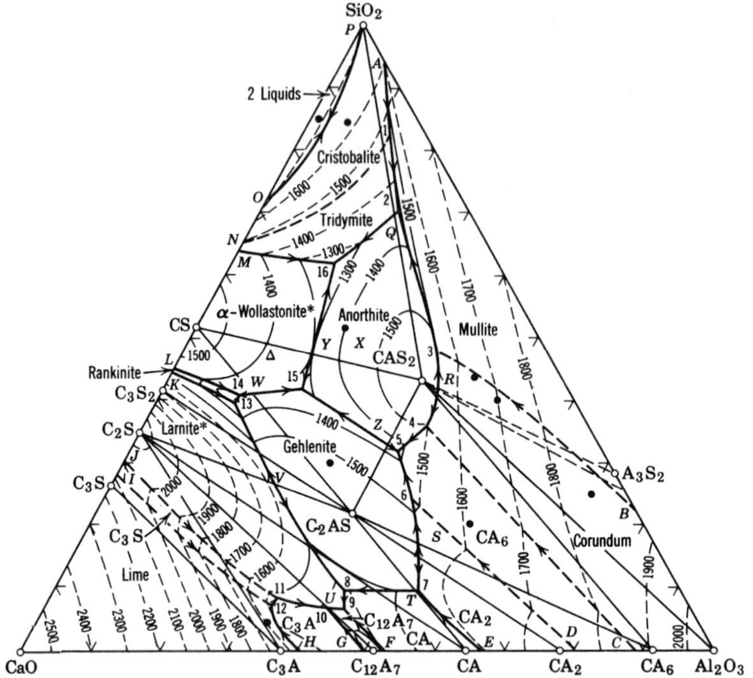

Fig. 9. The CaO-SiO_2-Al_2O_3 system. From Osborn, E. F. and A. Maun: *Equilibrium Diagrams of Oxide Systems.* Columbus, Ohio: *American Ceramics Society*, 1960 [15].

representing one pure component. The six edges of the tetrahedron in Fig. 10 each represent a binary system, and the four equilateral faces each represent a ternary system. The range of portland cement compositions is indicated by the smaller tetrahedron having apices CaO-C_4AF-C_2S-$C_{12}A_7$.

Cooling rate. The rate at which the clinker is cooled controls the amount of liquid which is frozen as a glass, rather than crystallizing. Figure 11 is a schematic

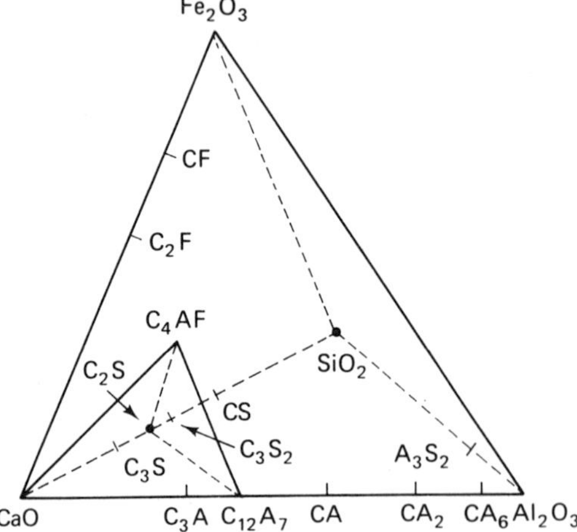

Fig. 10. The CaO-SiO_2-A_2O_3 system. From Lea, F. M.: *The Chemistry of Cement and Concrete.* London: Edward Arnold & Co., 1970 [5].

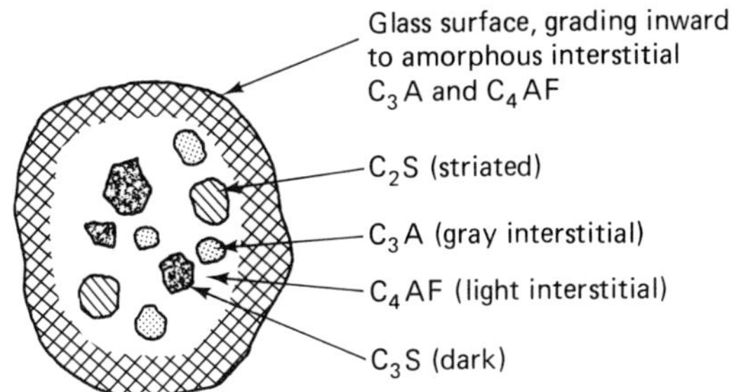

Fig. 11. Clinker nugget cross section.

view of a typical clinker nugget cross section, as it might appear after being cracked open. The smaller crystals, and finally the glassy phase, are concentrated at the surface of the nugget where cooling occurs most rapidly. Glass contents of commercial clinkers usually range from 2 to 12%.

The cooling rate, within practical commercial limits, seems to have little effect on strengths up to 7 days, but fast cooling rates tend to increase the strength somewhat at 28 days. Whether this is simply a manifestation of the normally higher reaction rates of glasses compared with crystalline phases of the same materials has not been determined.

The most important effects of cooling rate are on the soundness and sulfate resistance of the resulting cement. The presence of periclase (MgO) crystals in clinker contributes to long-term unsoundness due to the expansion which accompanies its slow hydration, but it is not harmful when present as part of the glass. Small grains of periclase or glass hydrate more rapidly than large ones and have less tendency to cause delayed expansion. The clinker melt can dissolve about 5% MgO, or about 1.5–2%, based upon total clinker weight. Rapid cooling is therefore very useful in offsetting the harmful effects of MgO, and higher MgO contents can be tolerated in clinkers which are cooled more rapidly [15].

Fineness of grinding. Fine grinding produces a large surface area for hydration, and more gel will form at an early age, giving higher strength. However, the hydration rate is retarded, in turn, by the presence of cement gel, since the gel provides a diffusion

Fig. 12. Effect of cement fineness on concrete strength. From Price, W. H.: *Proc. Am. Concrete Inst.*, 47 (1951) 417 [17].

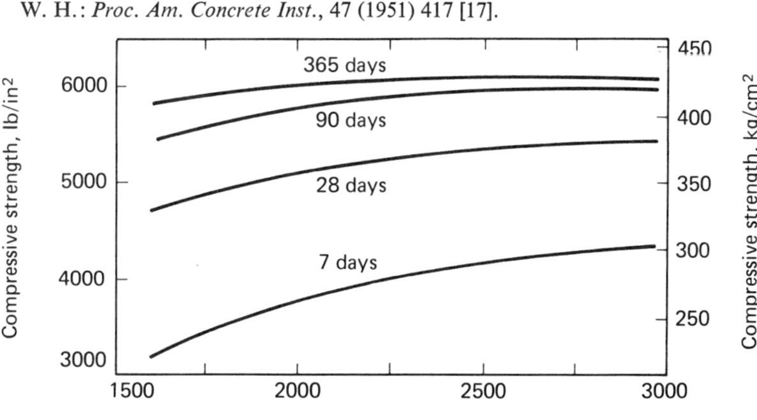

barrier between water and the unreacted cement grains. Therefore, extra fine grinding increases the cement strength only up to about 7 days, with little effect noticeable after 28 days and none at later ages. Figure 12 shows these relationships. Furthermore, certain disadvantages may arise with extra fine grinding, such as increased shrinkage. The following properties are frequently associated with high cement fineness [8]:

Rapid set
High early strength
High water requirement
Reduced bleeding
High shrinkage, cracking
Increased deterioration on exposure to atmosphere
Increased reaction with reactive aggregates

d. KILN REACTION PRODUCTS

Some characteristics of the principal kiln reaction products in portland cement are shown in Table 2. C_3S gives early strength and heat liberation, and hydration is virtually complete after about 20 days. C_2S is responsible for late strength and heat liberation. Hydration becomes significant about 14 days after mixing with water, and the strength equals that of C_3S after about 1 yr. C_2S may slowly continue to hydrate and increase in strength for 10 yr or more. C_3A reacts most rapidly, with rate of heat liberation about twice that of C_3S. C_3A gains about half its maximum strength in 1 day and almost its maximum strength in 1 month. Its maximum strength is about 15% that of C_3S and C_2S. C_3A accounts for much of the shrinkage in cement. C_4AF accounts for little insofar as strength, heat liberation, and volume change are con-

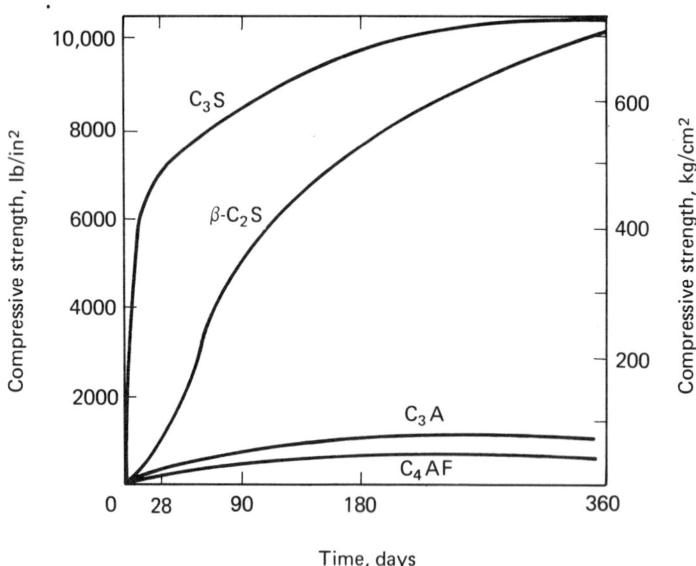

Fig. 13. Strengths developed by portland cement compounds. From Bogue, R. H. and W. Lerch, *Ind. Eng. Chem.*, 26 (1934) 837 [18].

cerned. Figure 13 shows strengths developed by these four principal cement compounds.

TABLE 2. Some Characteristics of Principal Portland Cement Components

Property	*Relative behavior*			
	C_3S	C_2S	C_3A	C_4AF
Reaction rate	Medium	Slow	Fast	Slow
Heat liberated per unit wt	Medium	Small	Large	Small
Cementing value per unit wt				
Early	High	Low	High	Low
Ultimate	High	High	Low	Low

e. PORTLAND CEMENT TYPES

The American Society for Testing and Materials (ASTM) specifies five types of portland cement. Figure 14 shows their chemical compositions, and Tables 3 and 4 indicate the range of chemical compositions and typical phase compositions for the five types. Table 5 and 6 give chemical and physical specifications for the five types according to ASTM Designation C150-61. In Table 6, soundness tests consist es-

Fig. 14. Approximate weight compositions of portland cement types.

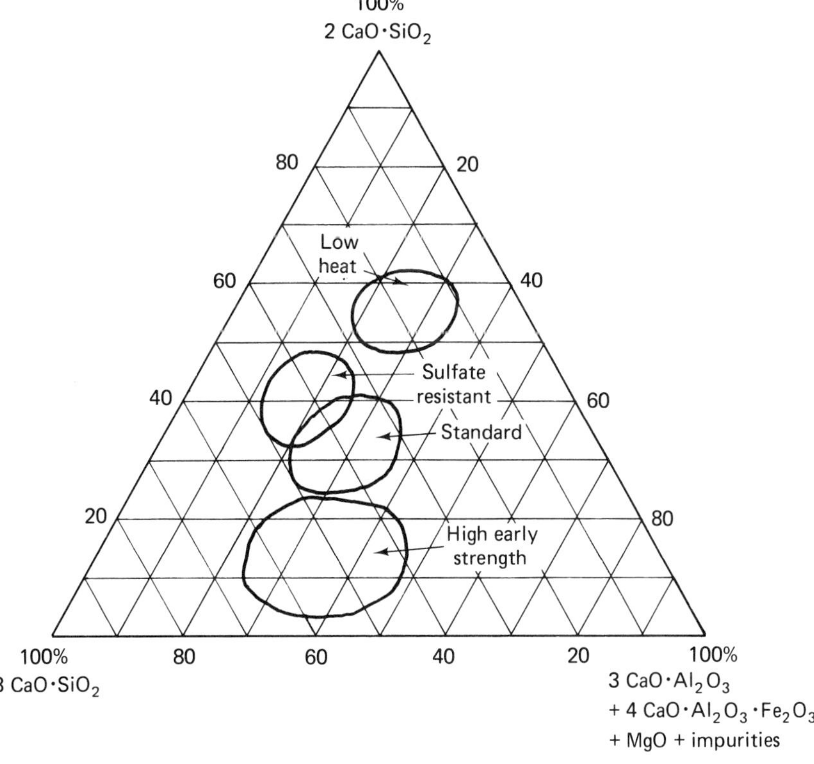

sentially of placing a pat of neat cement in a steam bath for 24 hr and measuring the expansion. Initial set can be described as the condition at which a cement paste cannot be worked but can be marked, while final set signifies the time required for the sample to become solid.

TABLE 3. Composition Ranges of Portland Cement

Component	CaO	SiO_2	Al_2O_3	Fe_2O_3	MgO	$Na_2O + K_2O$	SO_3
% Weight	60–66	20–26	3–7	1–5	1–5	0.5–1.3	1–3

TABLE 4. Typical Potential Phase Compositions by Percent Weight of Cement Types

	ASTM type	*Characteristics*	C_3S	C_2S	C_3A	C_4AF	MgO	*Free* CaO	$CaSO_4$	*Total**
I	Standard		45	27	11	8	2.9	0.5	3.1	98
II	Moderate heat		44	31	5	13	2.5	0.4	2.8	99
III	High early strength	High C_3A, C_3S	53	19	11	9	2.0	0.7	4.0	99
IV	Low heat	Low C_3A, C_3S	28	49	4	12	1.8	0.2	3.2	98
V	Sulfate resistant	Low C_3A, C_3S	38	43	4	9	1.9	0.5	2.7	99

SOURCE: BOGUE, R. H., *Chemistry of Portland Cement*, 2nd ed. New York: Van Nostrand Reinhold Company, 1955 [19].
*The remaining 1–2% consists primarily of insoluble residue and alkali oxides combined in various ways.

TABLE 5. Chemical Requirements (ASTM C150-72)

	Type I	*Type II*	*Type III*	*Type IV*	*Type V*
		General Chemical Composition			
SiO_2, min		21.0			
Al_2O_3, max		6.0			
Fe_2O_3, max		6.0		6.5	
MgO, max	5.0	5.0	5.0	5.0	4.0
SO_3:					
When $C_3A < 8\%$, max	2.5	2.5	3.0	2.3	2.3
When $C_3A < 8\%$, max	3.0		4.0		
Ignition loss, max	3.0	3.0	3.0	2.3	3.0
Insoluble residue, max	0.75	0.75	0.75	0.75	0.75
		Compound Composition			
C_3S, max				35	
C_2S, max				40	
C_3A, max		8	15	7	5
$C_3S + C_3A$, max		58			
$C_4AF + 2C_3A$					20

TABLE 6. Physical Requirements (ASTM C150-72)

	Type I	*Type II*	*Type III*	*Type IV*	*Type V*
Fineness, specific surface, cm^3/g (alternative methods)					
Turbidimetric test:					
Avg. value, min	1,600	1,600		1,600	1,600
Min value, one sample	1,500	1,500		1,500	1,500
Air-permeability test:					
Avg. value, min	2,800	2,800		2,800	2,800
Min value, one sample	2,600	2,600		2,600	2,600
Soundness: autoclave expansion, max, %	0.50	0.50	0.50	0.50	0.50
Time of setting (alternative methods)					
Gillmore test:					
Initial set, minutes, not less	60	60	60	60	60
Final set, hours, not more	10	10	10	10	10
Vicat test:					
Set, minutes, not less	45	45	45	45	45
Air content, max %	12.0	12.0	12.0	12.0	12.0
Compressive strength, psi, min:					
1 day			1,700		
3 days	1,200	1,000	3,000		
7 days	2,100	1,800		800	1,500
28 days	3,500	3,500		2,000	3,000
Tensile strength, psi, min:					
1 day			275		
3 days	150	125	375		
7 days	275	250		175	250
28 days	350	325		300	325

Type I (standard) is used where special properties of the other types are not required.

Type II (moderate heat of hydration) is used in more massive structures, such as large retaining walls. To reduce the heat of hydration, and thereby the potential danger due to cracking caused by volume changes, the faster-reacting components, C_3A and C_3S, are reduced, and the slower-reacting component, C_2S, is increased. The reduction of C_3A also improves the sulfate resistance of Type II cement, but it is not intended for severe sulfate conditions. Type II is intermediate in composition between Types I and IV (low heat).

Type III (high early strength) is used where forms must be removed early or in cold weather construction. In road repairs it permits traffic to resume within a day or two. The percentage of C_3S and sometimes C_3A is increased in Type III, while C_2S is decreased. Type III is also more finely ground, which accelerates the hydration. Because of the fineness and the higher percentage of faster-reacting compounds, Type III cement requires more gypsum than Types I and II. The pronounced expansion and contraction caused by the high heat of hydration may be a disadvantage.

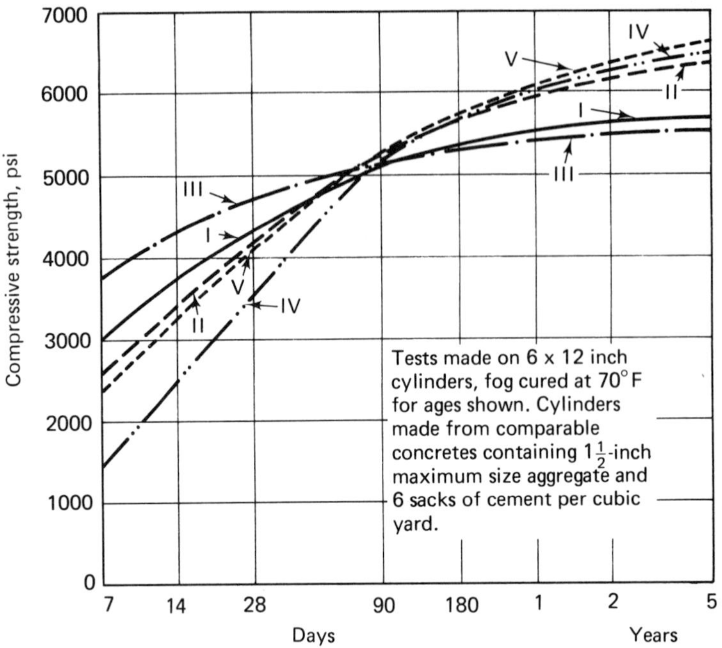

Fig. 15. Rates of strength increase for concretes made with five cement types. From U. S. Bureau of Reclamation Concrete Manual, Seventh Edition, 1966 [20].

Type IV (low heat of hydration) is used in massive structures such as large dams to avoid harmful volume changes due to the higher temperatures liberated by standard cement.

Type V (sulfate-resistant) is used for exposure to sulfate-bearing waters. The sulfates in seawater and some groundwaters react with C_3A in portland cement to form ettringite ($C_3A \cdot 3\ CaSO_4 \cdot 32\ H_2O$), a highly hydrated structure whose formation may be accompanied by expansion (over 200%) and gradual disintegration of the concrete. Harmful expansive sulfate reactions also occur with MgO and free CaO. Sulfate attack is greatly accelerated by alternate wetting and drying, as in the tidal zone on marine structures. Since C_3A is the compound most susceptible to sulfate attack, C_3A is limited to 5% in Type V cement. The C_3S is also limited to 50%

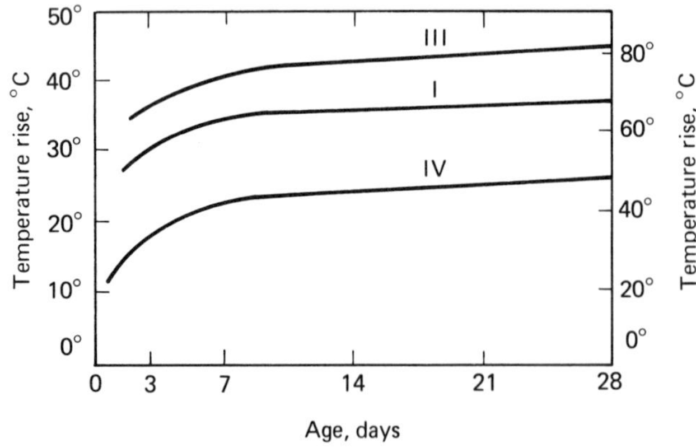

Fig. 16. Temperature rise in concrete samples (10% cement by weight) cured adiabatically. From Lea, F. M., and C. H. Desch: *Chemistry of Cement and Concrete*, Second Edition. London: Edward Arnold & Company, 1970 [2].

because C_3S releases a considerable amount of $Ca(OH)_2$ during hydration, which in turn reduces the resistance to sulfate attack.

Figure 15 compares rates of strength increase of the five types, and Fig. 16 compares the adiabatic temperature rises for Types I, III, and IV.

An unlimited range of additional cements are available by intergrinding portland cement clinker in various proportions with other cements such as aluminous, natural, or slag cements or with additives such as pozzolans and air-entraining agents. Other cements and pozzolans were mentioned previously, and air-entraining cements will be described subsequently.

2. *Hydration Reactions*

The hydration reactions of portland cement are even more complex and less understood than the high-temperature phase reactions which occur in the kiln. The important questions are (1) how do individual components hydrate, and (2) how do the various reaction products affect strength, durability, and volume stability of the concrete? To answer these questions we can consider first the hydration reactions and kinetics and then their physical manifestations.

Hydration reactions and reaction kinetics have been studied by observing the reaction products after various curing periods by optical and electron microscopy, the electron microprobe, x-ray diffraction, infrared absorption, and differential thermal analysis [21, 22, 23]. Colorimetry and electroconductivity studies have also been used.

Using the additional notation $Ch = Ca(OH)_2$, $Cs = CaSO_4$, and $H = H_2O$, the principal hydration reactions and end products are

$$2\,C_3S + 6\,H \longrightarrow \underset{\text{tobermorite}}{C_3S_2H_3} + 3\,Ch \tag{1}$$

$$2\,C_2S + 4\,H \longrightarrow C_3S_2H_3 + Ch \tag{2}$$

$$C_3A + 10\,H + CsH_2 \longrightarrow \underset{\text{calcium alumino monosulfate hydrate}}{C_3ACSH_{12}} \tag{3}$$

$$C_3A + 12\,H + Ch \longrightarrow \underset{\text{tetracalcium aluminate hydrate}}{C_3AChH_{12}} \tag{4}$$

$$C_4AF + 10\,H + 2\,Ch \longrightarrow \underset{\text{calcium alumino ferrite hydrate}}{C_6AFH_{12}} \tag{5}$$

Tricalcium silicate (C_3S) and β-dicalcium silicate (β-C_2S) constitute about 75% of portland cement by weight. The end hydration product of both compounds at atmospheric pressure is tricalcium silicate hydrate ($C_3S_2H_3$), called tobermorite because it resembles the rare mineral tobermorite ($C_5S_6H_5$) [24, 25]. Tobermorite has a layer structure analogous to the clay mineral vermiculite. Hardened portland cement is mostly tobermorite having a low degree of crystallinity. It plays a dominant role in cement setting and is the major adhesive component in hardened cement, determining both its strength and dimensional stability. Although C_3A and C_4AF hydrates alter the setting rates of cements, it is known that their reaction products, as well as $Ca(OH)_2$, contribute little to cement strength.

When tobermorite is dried so that only chemically bound water remains, it has one less water molecule ($C_3S_2H_2$) [24]. In saturated tobermorite the third water molecule is probably intercrystalline water.

Although Eqs. (1) and (2) give little difference in the weight percentages of water combined by C_3S and C_2S, C_3S hydration releases much more free lime. Since free lime can be undesirable from the standpoint of long-term unsoundness and sulfate attack, a low C_3S/C_2S ratio is preferred in portland cement, consistent with obtaining the required early strength provided by the faster-hydrating C_3S.

Complete hydration of C_3S can be obtained in 1 to 2 days, and of C_2S in 40 to 50 days, by grinding the pure components with excess water in a laboratory ball mill [26]. This treatment continually removes reaction products from grain surfaces, allowing hydration at fresh surfaces so that water does not have to diffuse through the fairly impervious and ever-thickening shell of reaction products. These intermediate reaction products consist of calcium silicate hydrates having various ratios of C/S/H and various degrees of crystallinity [5]. In portland cements, which contain C_3S and C_2S in combination with other components, strength increases for over 20 yr have been observed. The two silicate hydration reactions are therefore sometimes written in a more general form applicable to any curing age or water solids ratio as

$$C_3S + (2.5 + n)\,H \longrightarrow C_{1.5+m}SH_{1+mn} + (1.5 - m)\,CH$$

and

$$C_2S + (1.5 + n)\,H \longrightarrow C_{1.5+m}SH_{1+m+n} + (0.5 - m)\,CH$$

Fig. 17. Tobermorite hydrates. From Brunauer, S., *Am. Sci.* 50 (1962) 210 [25].

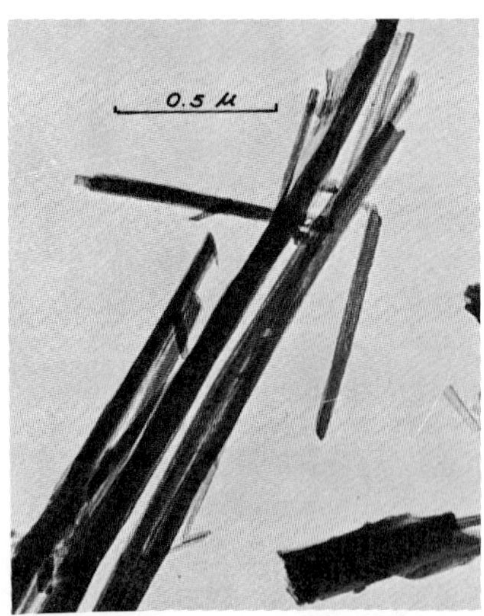

where *m* and *n* are variables. The early hydration products typically have a C/S molar ratio of 0.8 to 1.5 and are denoted CHS(I). Upon additional hydration in a solution with increasing lime concentration these transform to CSH(II), having molar ratios of 1.5 to 2. Electron micrographs [24, 25, 27] show that CSH(I) hydrates are crumpled sheets or crinkled foils only one or two molecular units thick [28] which gradually transform to fibrous or needle-like crystals up to 1 μ long and 500 A wide with the formation of CSH(II). As the lime solution concentration increases, it appears that the sheets and foils twist and roll into long interlocking tubes or fibers (Fig. 17) [29]. The average width of pores between the fibers is about 20 A, or five times the width of a water molecule, and the average porosity about 28% by volume [30]. Table 7 indicates identifying characteristics of the various CSH hydrates.

TABLE 7. Identifying Characteristics of CSH Hydrates

Primary subdivisions	*Type of x-ray powder pattern*	*Secondary subdivisions*	*Composition*	*Appearance in electron microscope*
Crystalline tobermorites	Full pattern showing many *hkl* reflections; often 40–50 lines given adequate experimental technique	14-A Tobermorite 11.3-A Tobermorite 9.3-A Tobermorite 12.6-A Tobermorite 10-A Tobermorite	$C_5S_6H_9$ $C_5S_6H_5$ $C_5S_6H_{0-2}$ (?) (?)	Flat plates or laths, usually euhedral; rarely fibers
Semicrystalline tobermorites	Patterns of about 6–12 lines, including mainly *hk* or *hk*0 reflections	C-S-H(I)	Ca/Si $<$ 1.5	Crumpled foils
	and usually a basal reflection at 9–14 A	C-S-H(II)	Ca/Si $\geq$ 1.5	Usually fibers
Near-amorphous tobermorites	Weak patterns of 1–3 *hk* lines or bands (3.05, 2.8, and 1.8 A approx.)	Tobermorite gel [predominant constituent(s)], etc.	Ca/Si probably usually $\geq$ 1.5	Irregular platelets or foils, fibers (?)

The initials C-S-H denote "calcium silicate hydrate"; hyphens are used to show that the composition $CaO \cdot SiO_2 \cdot H_2O$ is not necessarily indicated.
SOURCE: TAYLOR, W. H., *Concrete Technology and Practice*, 2nd ed. London: Angus & Robertson Ltd., 1967 [31].

The aluminate compounds hydrate more rapidly than C_3S or C_2S. Finely ground C_3A reacts with water in a few minutes to produce hexagonal crystals, and C_4AF hydrates somewhat less rapidly than C_3A, also as hexagonal crystals, mostly in foliated masses [5].

Calcium sulfate (gypsum) is added to portland cement clinker (3–4% by weight) to prevent flash set. Gypsum goes into solution almost immediately and reacts with calcium aluminates to precipitate either of two water-rich insoluble compounds: calcium alumino trisulfate hydrate ($C_3ACS_3H_{31}$: ettringite) or calcium alumino monosulfate hydrate (C_3ACsH_{12}). With sufficient gypsum, these reactions reduce the C_3A concentration in solution below the point where C_3A hydrate can precipitate to cause flash set. The ettringite appears as extremely fine needles on a microscope slide (Fig. 18). The reaction proceeds until all gypsum is used, usually within 24 hr. With continued cement hydration, ettringite also tends to convert, by reaction with

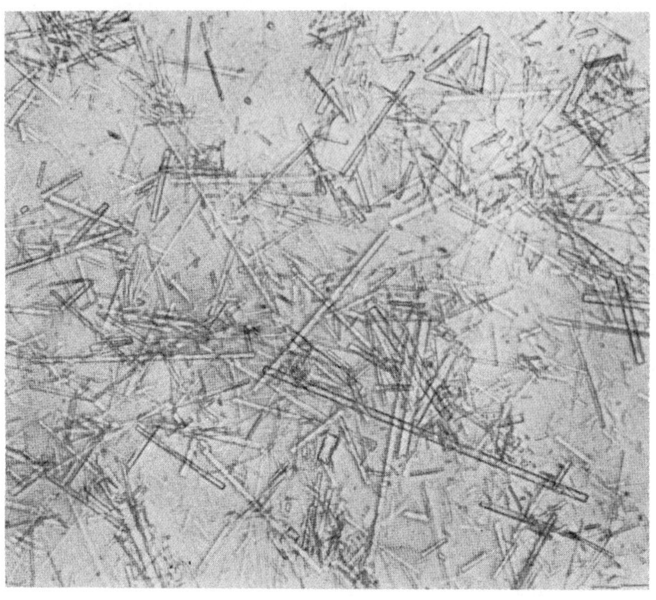

Fig. 18. Ettringite needles. From Bogue, R. H.: *Chemistry of Portland Cement*, Second Edition. New York: Litton Educational Publishing, Inc., a division of Van Nostrand Reinhold Company [19].

additional C_3A, to the monosulfate hydrate. The aluminate hydrations [Eqs. (4) and (5)] are completed only after gypsum has completed its reactions [32]. The C_3S and C_2S hydrations are only slightly modified by gypsum, some sulfate entering the silicate hydrate gels and changing their morphology.

In spite of the wide variation in water-binding capacities of the different clinker minerals, the water-binding capacities of commercial cements having significant differences in chemical composition usually lie in the narrow range $28 \pm 1\%$ by weight. This is primarily because the two dominant components, C_3S and C_2S, both bind approximately the same ratio of water. In practice, however, water cement ratios of 0.35 to 0.70 are used, for reasons described later.

a. HYDRATION MECHANISMS

The hydration rates of individual cement components can be related to provide some understanding of the hydration behavior of cement itself. The following stages of cement hydration can be distinguished, based upon major changes in reaction rate and physical structure:

Hydration Stage	*Typical Exothermic Reaction Rate*	*Typical Duration*	*Principal Chemical or Physical Changes*
Initial reaction	40 cal/g/hr	5–10 min	Initial solution and hydration
Dormant period	1 cal/g/hr	1 hr	Growth of reaction product membrances around cement particles
Setting	Increasing gradually to 5 cal/g/hr at 6 hr	6 hr	Membrane rupture, renewed hydration of particles

Hydration Stage	*Typical Exothermic Reaction Rate*	*Typical Duration*	*Principal Chemical or Physical Changes*
Hardening	Decreasing to 1 cal/g/hr at 24 hr; less thereafter	6 hr to years	Hydration products exude into and fill capillary space
Destructive hydration	Negligible	1–5 years	Disruption of surrounding hardened paste by expansion due to late hydration

Initial reaction. During initial reaction C_3S starts hydrating, releasing $Ca(OH)_2$ [Eq. (1)], which immediately dissolves in the mix water and increases the pH to about 13. The solution becomes supersaturated, and $Ca(OH)_2$ precipitates as hexagonal crystals. Aluminates exposed at the grain surfaces also dissolve and quickly react with dissolved gypsum to precipitate ettringite. Only about 1% of the cement hydrates during initial reaction. Most of the reaction products are insoluble and adhere to the cement grains.

C_2S hydrates much more slowly than C_3S. The crystal structure of C_3S is more open and porous to water molecules, and C_3S is also less stable thermodynamically than C_2S [33].

In rare cases a premature stiffening may occur during or shortly after initial reaction. If the temperature of the grinding mill is too high, gypsum added to the clinker may dehydrate to plaster of Paris ($CaSO_4 \cdot \frac{1}{2} H_2O$) or soluble anhydride ($CaSO_4$) in place of the original dihydrate ($CaSO_4 \cdot 2\,H_2O$). The dehydrated gypsum quickly hydrates and precipitates as a delicate skeleton rigid enough to resemble setting. This is called *false set* because vigorous mixing breaks down the skeleton and the paste regains its workability.

Dormant period. During this period the hydration of all components slows as their surfaces become coated with gels of the reaction products. Specific surface measurements indicate that the sizes of cement grains hardly change until the end of the dormant period [34].

Setting. During hydration gelatinous layers of calcium silicate hydrate form around the cement grains (Fig. 19). Water diffuses inward through the gel membrane and the hydration products diffuse outward, as governed by the concentration gradients across the membrane and the diffusion constants of each species. Since the membrane is more permeable to calcium ions than to the larger SiO_2 ions, $Ca(OH)_2$ crystals precipitate primarily outside the membrane, while silicate gel forms primarily on the inside, producing a supersaturated transition zone between the grain and surrounding membrane. The rate-limiting step is probably the diffusion of SiO_2 ions through the gel layer. The gel layer gradually grows inward as the grain dissolves, and pseudomorphs can sometimes be observed under the microscope. The mass transfer of water into the membrane exceeds the transfer of diffusion products out, producing osmotic pressure. The membrane eventually bursts, and the supersaturated solution of reaction products fingers outward to fill any voids and gradually sets to provide bonding between partially hydrated cement grains (Fig. 20). Rupture of the gel layer permits access to the grain surface by the less highly saturated interparticle solution, thus

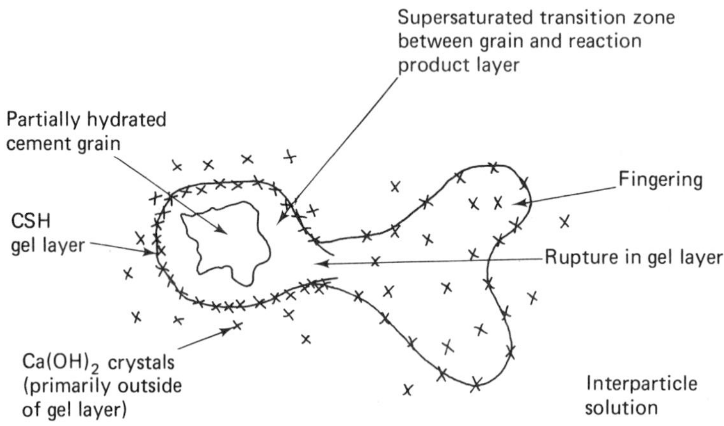

Fig. 19. Schematic of a cement grain and its reaction membrane.

allowing increased reaction rate, until new gel material slowly mends the rupture again. Ruptures occur randomly in time and location, and the increased exothermic reaction marking the initiation of setting probably identifies a period during which numerous ruptures first occur. Part of the hydration may also occur without the cement compounds going into solution, by direct topochemical or solid-state reaction [36]. The solid-state reactions may predominate during the late stages of hydration when diffusion becomes more difficult.

The initial set is due primarily to C_3S hydrates, if sufficient gypsum is present to reduce the concentration of C_3A in solution below the level which would cause flash set by C_3A hydrate. At the time of initial set C_3S has hydrated sufficiently that laths of tobermorite begin to interlock. At the end of setting (typically 6 hr), about 15% of the cement is hydrated, and the interlocking of tobermorite particles is pronounced (see Fig. 17).

Hardening. After 24 hr the capillary space is mostly filled with cement gel, but unhydrated cement cores remain for many months. Hydration typically proceeds to a depth of 5–9 μ in 9 months, and normally over 50% of cement by weight consists of particles exceeding 10 μ in diameter and ranging to 100 μ [5]. Figure 21 illustrates the

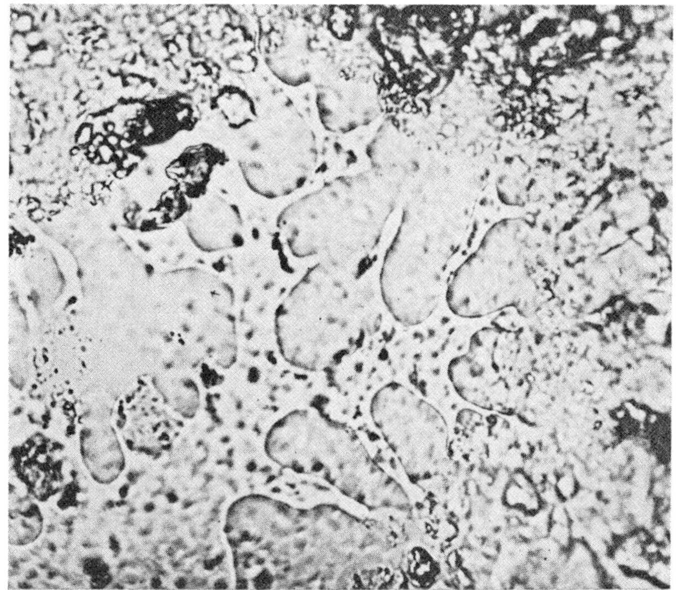

Fig. 20. Thin section showing gel exuding or fingering into capillary space (× 900). From Bogue, R. H.: *Chemistry of Portland Cement*, Second Edition. New York: Litton Educational Publishing, Inc., a division of Van Nostrand Reinhold Company [19], and Ward, G., Res. Report, PCAF (April 1944) [35].

Fig. 21. Schematic representation of hydration and gel formation. From Powers, T. C., *Chemistry of Cement, Proceedings of the 4th International Symposium*, Washington, 1960. National Bureau of Standards Monograph 43, 1962, p. 577 [30].

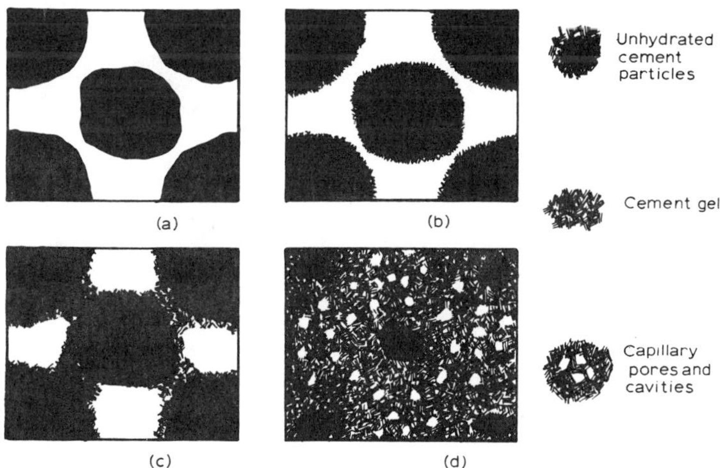

process schematically, and Fig. 22 illustrates the hydration sequence in a series of electron micrographs selected by Czernin [4]. The cement, prepared in the laboratory, consisted almost exclusively of C_3S grains from 6–9 microns. After one hour, acicular or worm-like growths are observed extending from the grains into the capillary space. After 28 days it is denser and fills almost all available capillary space to bond the original cement grains, which now appear joined at points of contact.

Fully hydrated cement contains about 25% $Ca(OH)_2$ and 50% tobermorite by volume. Tobermorite accounts largely for both the strength and shrinkage. It has

Fig. 22. Cement structure after various periods of hydration. (a) shortly after mixing with water; (b), one hour; (c) 24 hours; (d) 28 days. From Czernin, W., *Cement Chemistry and Physics for Engineers.* New York: Chemical Publishing Company, 1962 [4].

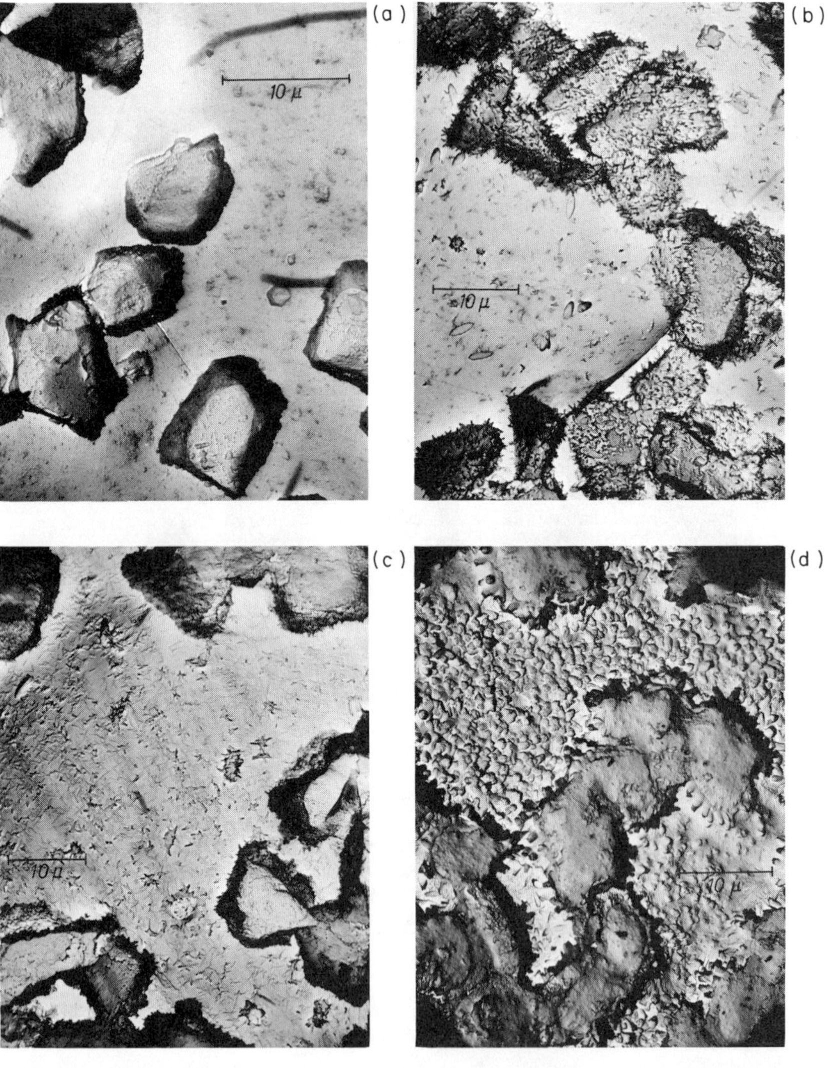

been suggested that compressive strength is greater than tensile strength because valence forces within the crystals must be overcome to cause compressive failure, whereas only van der Waals forces between the interlocking laths must be overcome to cause tensile failure [25]. Shrinkage is caused by the desorption of interlaminar water in the crystal structure of tobermorite (normally only 2 or 3 molecular layers thick), and by evaporation of water from the gel capillaries.

Destructive hydration. Any hydration reaction which occurs after cement hardening and which causes expansion can be destructive. Among these are the long-term hydration of free lime (CaO) and magnesia (MaO), of excess gypsum with C_3A to produce expanding calcium sulfoaluminate, of alkalies (Na_2O and K_2O) in the cement with glassy silicates in some aggregates to produce expanding alkali silicate hydrates (called alkali-aggregate reaction), and the reaction of sulfates in seawater (particularly $MgSO_4$) with C_3A.

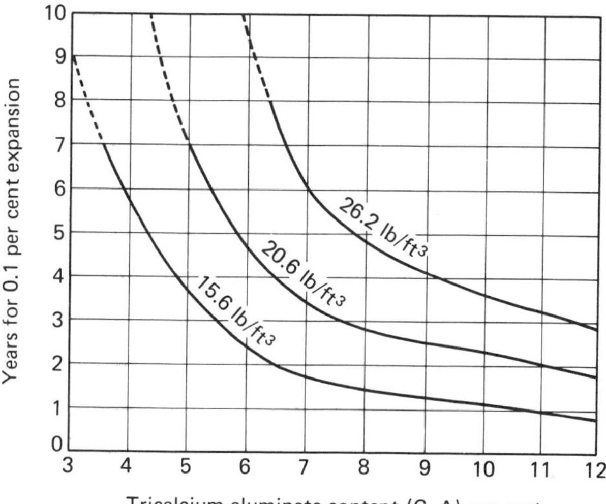

Fig. 23. Sulfate resistance of concrete as a function of C_3A content in the cement. From Lea, F. M., *The Chemistry of Cement and Concrete*. London: Edward Arnold & Co., 1970 [5].

The preventive measures in each case include limiting one of the reactive components. For example, the MgO content in cements is limited to about 5%, the SO_3 content to about 3%, and alkali contents to about 2%, and in Type IV (sulfate-resistant) cement the C_3A content is limited to 7% (Fig. 23). C_3A can be reduced by increasing the amount of Fe_2O_3 in the raw mix so that most of the alumina is converted during burning to C_4AF, which has much greater resistance to sulfate attack.

To summarize the hydration mechanisms simply, setting is normally caused by rapidly hydrating compounds, ultimate strength by compounds which hydrate at intermediate rates, and destructive expansion by very slowly hydrating compounds.

b. STRENGTH DEVELOPMENT

The strengths of different hydraulic cements may be associated with the formation of collidal products, crystalline products, or both. In portland cement cured at normal temperatures the strength is due almost entirely to a gel (tobermorite). In sorel cement and gypsum plaster, hardening is due almost entirely to interlocking crystals, observed under the microscope. In normally cured aluminous cements, crystalline CAH_{10} and C_2AH_8 and colloidal $Al(OH)_3$ apparently all contribute to strength [37], and in autoclaved lime-silicate and portland cements apparently crystalline and colloidal products also act synergetically. Other things being equal, highly crystalline cements tend to undergo less volume change with wetting and drying and tend to be more resistant to chemical attack.

Aside from crystalline-colloidal character, the surface structure and specific surface area of cement pastes appear extremely important in strength development. Although the forces responsible for adhesion between cement particles are little understood, it is assumed they are greater for materials having irregular coordination of surface atoms or marked separations of electric charge, i.e., high surface energy. For example, 11-A tobermorite and α-C_2S hydrate ($Ca_2(HSiO_4)$ (OH)) are both platy submicroscopic crystals of similar specific surface, but tobermorite, having higher surface energy, also has much higher strength [37]. Surface area is important because adhesion between particles depends on the extent of interparticle contacts, which increases with the state of subdivision. In the course of hydration the specific surface of cement increases about 1,000 times [38].

An additional factor important to strength development is the time at which hydration occurs, mentioned earlier. So long as hydration does not disrupt preexisting structures, combinations of many sizes and shapes of crystalline and gel structures can yield high strength. For example, ettringite can contribute to strength by early hydration in supersulfate or expansive cements but is disruptive due to late hydration in portland cement when it is exposed to sulfate solutions.

Strength development depends on the compound composition of portland cement (Fig. 13), with somewhat higher ultimate strengths being attained by cements poor in lime, i.e., rich in C_2S, because C_2S produces more tobermorite and less $Ca(OH)_2$ than does C_3S.

A more important factor than chemical composition, however, is the paste porosity, as expressed by the water/cement (w/c) ratio rule. Expressed in a general form to account for both water and air voids, it is the Feret equation [Eq. (24), Chap. 7]. It permits one to predict the strength of any concrete mix from limited experimental data or to provide a mix of specified strength (Fig. 24).

Powers and Brownyard determined an analogous relationship between strength and the gel/space ratio [40]. This is defined as the ratio of the volume of the hydrated cement paste to the sum of the volumes of the hydrated cement and the capillary pores. Let

$$c = \text{weight of cement}$$

$$v_c = \text{specific volume of cement}$$

$$w_0 = \text{volume of mix water}$$

$$a = \text{fraction of cement which has hydrated}$$

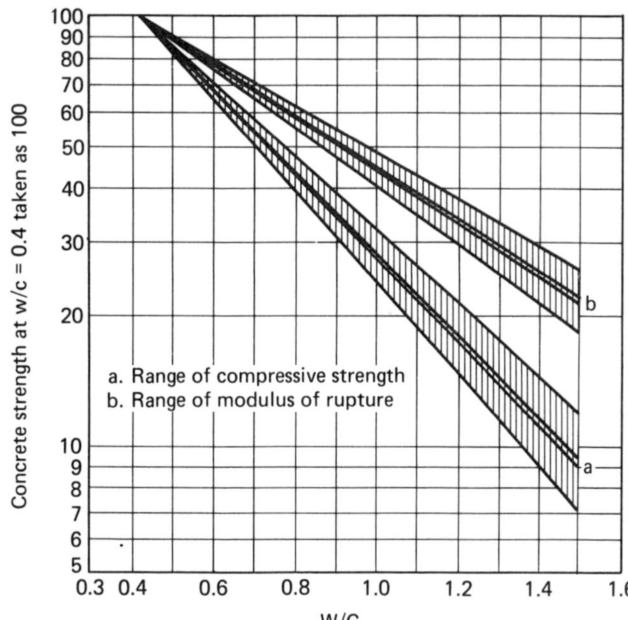

Fig. 24. Compressive strength variation with water:cement ratio. From Hummel, A., *The ABC of Concrete*, Twelfth Edition. Berlin: W. Ernst and Sohn, 1959 [39].

Since 1 cm³ of cement hydrates to initially occupy approximately 2.06 cm³ of gel, the gel volume $= 2.06cv_c a$, and the total space available to the gel, assuming zero air voids, is $cv_c a + w_0$. The gel/space ratio is therefore

$$x = \frac{2.06v_c a}{v_c a + w_0/c}$$

Taking the specific volume of dry cement as 0.319 cm³/g, the gel/space ratio is

$$x = \frac{0.647a}{0.319a + w_0/c}$$

If A cm³ of air are present in the paste, the term w_0/c is replaced by $(w_0 + A)/c$.

Powers found the compressive strengths of concretes to be $34{,}000x^3$ psi and to be independent of concrete age and mix proportions [41]. Thus compressive strength varies approximately as the cube of the gel/space ratio, with 34,000 psi as the intrinsic gel strength for the particular cement and specimen types used (Fig. 25). Numerical values differed little for various portland cement compositions, except that higher C_3A contents gave lower strengths at a given gel/space ratio.

Additional factors affecting concrete strength are the curing temperature (Fig. 26) and the availability of moisture during curing (Fig. 27). The results in Fig. 27 are for small samples, which were easy to resaturate after air curing.

As illustrated by Fig. 26, high curing temperature increases the initial hydration rate and therefore increases early strength. This rapid initial hydration is detrimental to ultimate strength, however, because it produces a nonuniformly distributed hydration product. Nonuniformity is believed detrimental to strength because the less concentrated zones of hydration product reduce overall strength and the more con-

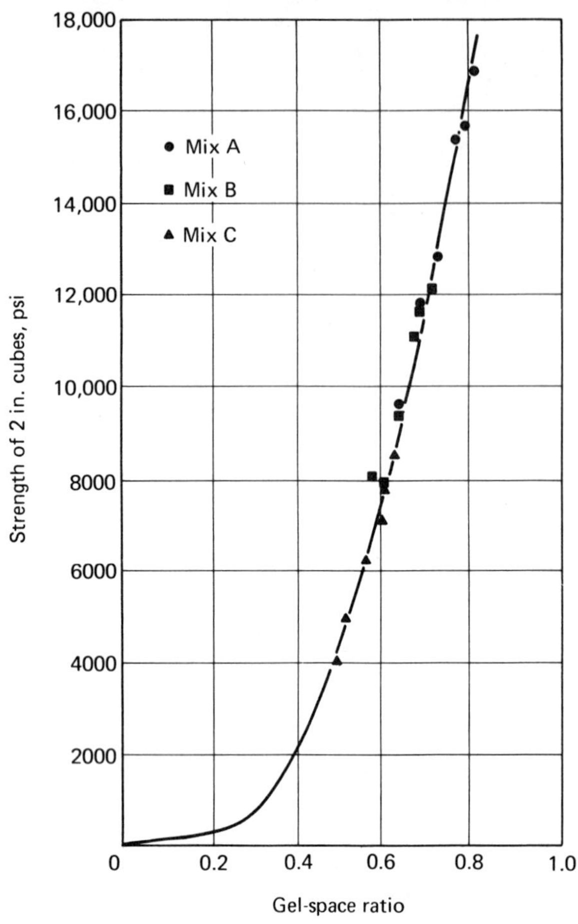

Fig. 25. Mortar compressive strength as a function of gel-space ratio. From Powers, T. C., *Port. Cem. Assoc. Res. Develop. Dept. Bull.* 90 (July 1958) 1, [42].

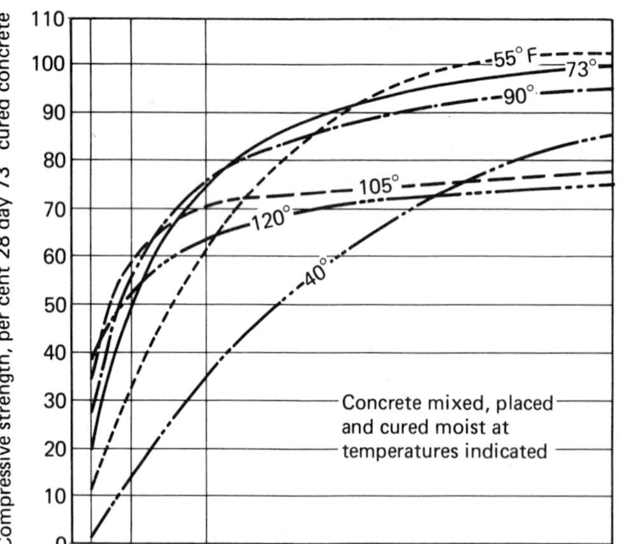

Fig. 26. Compressive strength variation with curing temperature. From *Design and Control of Concrete Mixtures*, Eleventh Edition. Portland Cement Association, 1968 [43].

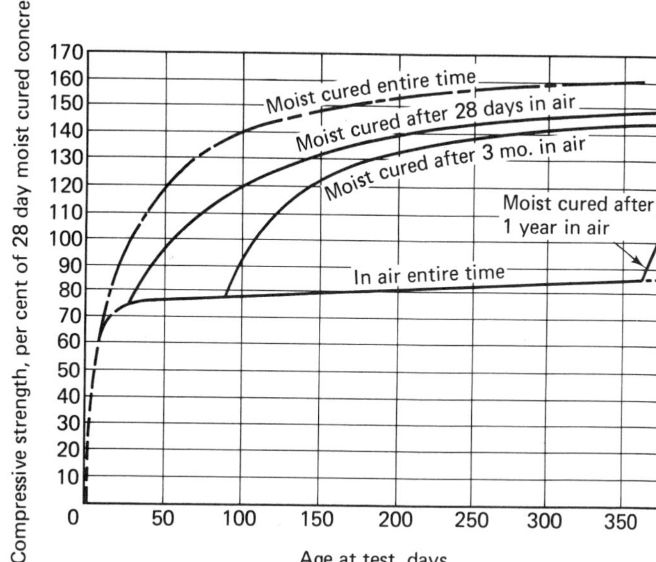

Fig. 27. Compressive strength variation with moisture availability during curing. From *Design and Control of Concrete Mixtures*, Eleventh Edition. Portland Cement Association, 1968 [43].

centrated zones retard subsequent hydration by encapsulation of hydration product immediately surrounding the cement grain [44]. Figure 28 shows the mechanism schematically. At lower temperatures, such as 40–60°F, the hydration product is uniformly distributed throughout the paste because the slow hydration allows ample time for diffusion of the product.

Therefore the more rapidly a cement undergoes early hydration, the greater the probability of reduced ultimate hydration and ultimate strength, due to nonuniform distribution of the hydration product. This relationship applies only to cements cured near atmospheric pressure. In autoclaved products (Sec. C) the increased curing

Fig. 28. Effect of hydration temperature upon amount and uniformity of distribution of hydration product (Schematic). From Verbeck, G. and L. E. Copeland: *Menzel Symposium on High Pressure Steam Curing*, 1969. Detroit: American Concrete Institute Publication SP-32, 1972, p. 1 [44].

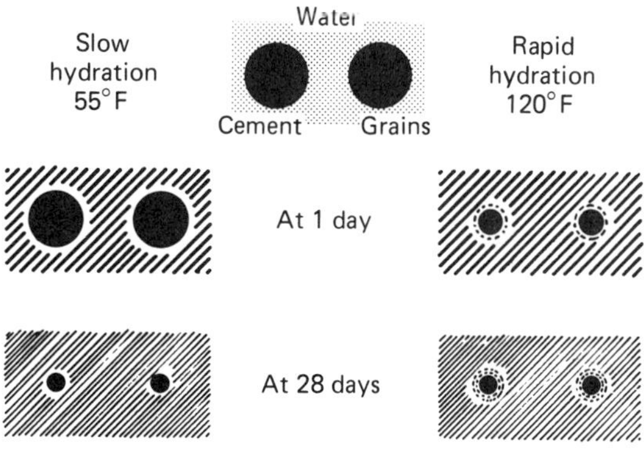

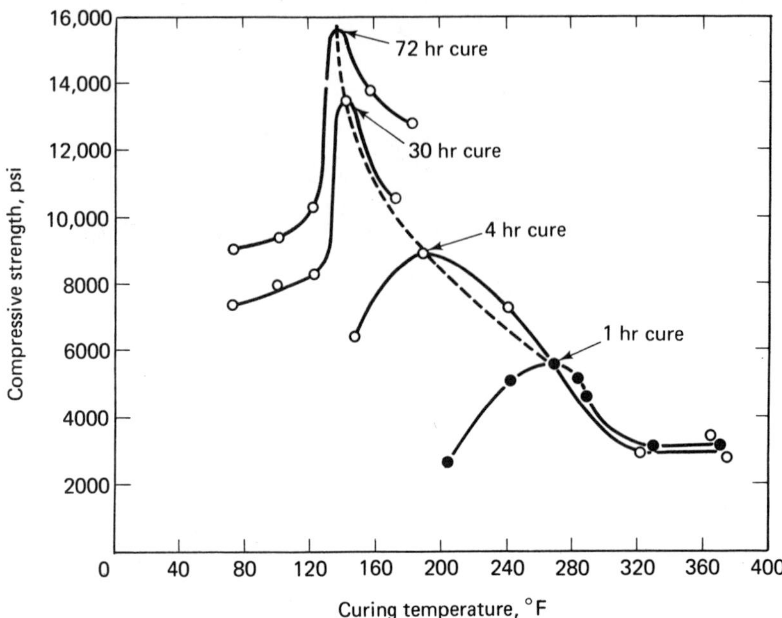

Fig. 29. Optimum curing temperatures for maximum strength as a function of curing time. From Verbeck, G. and L. E. Copeland: *Menzel Symposium on High Pressure Steam Curing*, 1969. Detroit: American Concrete Institute Publication SP-32, 1972, p. 1 [44].

pressure and temperature affect the fundamental structure of the hydration products, as well as the hydration rates.

Figure 29 shows the general relationship of Fig. 26 plotted to indicate optimum curing temperatures for maximum strength as a function of curing time.

c. VOLUME RELATIONSHIPS

Water is held in hardened cement paste in at least three ways [40]:

1. *Chemically bound water*, i.e., water of hydration or nonevaporable water.
2. *Gel pore water*, absorbed in the cement gel pores.
3. *Capillary pore water*, in larger continuous cavities in the order of 1,000 A wide.

Completely hydrated portland cement contains about 25% by weight of chemically bound water plus 15% of gel water, which can be removed by dessication or heating to 100°C. Due to the volatility of gel water, dessicated cement contains about 25% by volume of gel pores. However, the average gel pore size is small ($\sim$ 20 A), and the water permeability of hardened paste ($\sim 10^{-12}$ cm/sec) is less than for most rocks [45].

The volume change upon transforming 100 g of cement (32 cm³) and 40 g of water into cement gel is

$$32 \text{ cm}^3 \text{ of cement} + 40 \text{ cm}^3 \text{ of water} = [32 + (25 - 6)] \text{ cm}^3 \text{ of hydrate} + 15 \text{ cm}^3 \text{ of W} = 66 \text{ cm}^3 \text{ of cement gel}$$

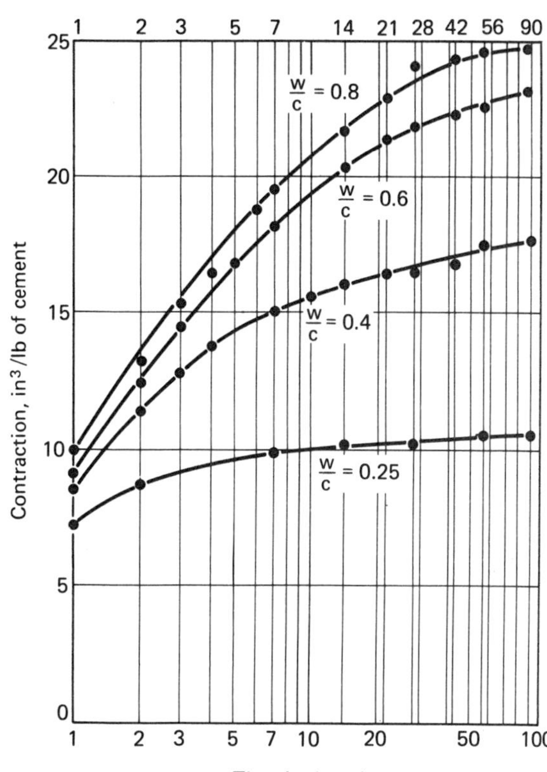

Fig. 30. Variation of hydration (contraction) rate with w/c ratio of cement pastes. From Czernin, W., *Cement Chemistry and Physics for Engineers.* New York: Chemical Publishing Company, Inc., 1962. [4].

where W = gel water [4]. Since 1 cm^3 of cement produces about 2.2 cm^3 of gel, 45% of the gel might be considered to form within the original boundary of a grain and 55% in the surrounding pore space. Thus the gel volume exceeds twice that of the original cement grains, and the mix volume is reduced by 6 cm^3 because the chemically bound water occupies less space than free water. The 6 cm^3 becomes capillary space which may be occupied either by air or by water absorbed from the atmosphere. But if the specimen is sealed from the atmosphere, or is so large that water absorbed from the atmosphere cannot penetrate the interior, the relative vapor pressure drops as water is chemically bound, and if it falls below 0.80, hydration ceases [46]. Unless the initial w/c ratio exceeds about 0.48, this self-dessication effect will retard hydration in large concrete structures [47]. Figure 30 shows the pronounced effect of the w/c ratio on hydration rate, as determined by measurements of volume change due to change in volume of the chemically bound water.

The penalty paid for complete hydration is larger capillary volume, as illustrated in Fig. 31 for 48% mix water. The 7.5% capillary air represents the case where no water enters the specimen during curing. The larger capillary volume decreases the paste strength and makes a fresh concrete mix more susceptible to *bleeding*, whereby aggregate tends to settle and mixing water rises to the surface, carrying the finest cement particles with it, removing cement from the mortar and also creating flow channels which cause increased permeability of the hardened concrete.

The lack of hydration due to low w/c ratios, on the other hand, does not necessarily produce weaker cement pastes. The highest reported cement strength, cured at

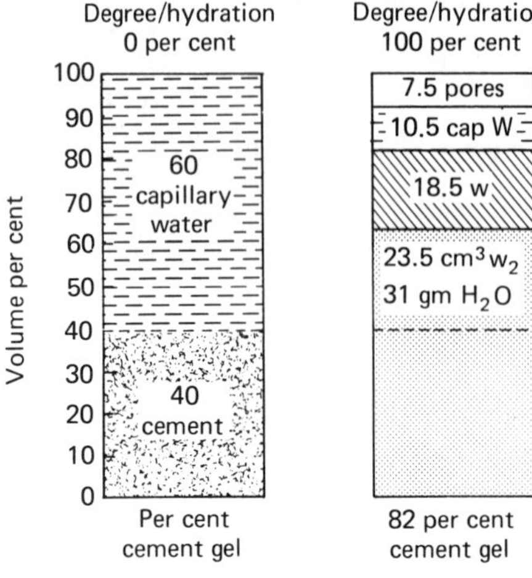

Fig. 31. Volume proportions in cement paste with w/c ratio = 0.48. From Czernin, W., *Cement Chemistry and Physics for Engineers.* New York: Chemical Publishing Company, Inc. 1962. [4].

normal temperatures, is about 42,660 psi, at a w/c ratio of 0.08 [4]. With adequate compaction, the unhydrated grains permit reduction in the total porosity and also are probably much more strongly bonded by cement gel than is any type of aggregate.

Aside from the irreversible self-dessication shrinkage due to chemical binding of water during hydration, reversible volume change due to moisture changes in hardened cement paste are also important in concrete structures. As water evaporates from the gel and capillary pores, increased hydrostatic tension draws the paste structure together, causing it to shrink. As minute capillaries become smaller, increased van der Waals attraction may also contribute to shrinkage. Removal of all water from a hardened cement having a porosity of 28% results in a shrinkage of 3% [5]. If the surface tension of water is eliminated, no shrinkage occurs, as has been demonstrated by freeze-drying the cement, allowing water to pass directly from the solid to vapor phase, without surface tension becoming effective [4].

A third type of cement shrinkage is caused by atmospheric carbon dioxide. In the presence of CO_2 and moisture, the large $Ca(OH)_2$ crystals in hardened cement slowly convert to $CaCO_3$. The reaction is accompanied by an irreversible *carbonation shrinkage*, which may equal 50% of the initial drying shrinkage [29]. Powers suggested that carbonation shrinkage may be caused by solution of $Ca(OH)_2$ crystals which are under triaxial compression due to drying shrinkage, followed by precipitation of $CaCO_3$ in stress-free configurations within the paste [48]. The consequent reduction of restraint would allow irreversible shrinkage. Carbonation shrinkage is greatest at relative vapor pressures around 0.5 and negligible at pressures above 1.0 or below 0.25 [49].

Concrete shrinks less than cement paste due to restraint by the aggregate, and the concrete shrinkage can be calculated if the properties and fractional volumes of aggregate and paste are known. Concrete and mortar shrinkage calculations are described in Sec. A in Chap. 8.

d. ACCELERATED CURING METHODS

The primary purposes of accelerated curing are to increase production rate for a given investment in field formwork, precasting forms, or prestressing beds. Concrete curing can be accelerated by

1. Mechanical treatment (vibration, application of load or vacuum).
2. Heating (by electrical resistance or low- or high-pressure steam).
3. Additives (described in the following section).

Any process which reduces water and air voids in cement can potentially increase strength. Mechanical vibration, the application of load, the removal of excess water by vacuum filter pressing or electrosmosis, or combinations of these methods after initial set has occurred can increase concrete strengths. Vibration serves additionally to disrupt and disperse the reaction product layers surrounding cement grains and to hasten hydration and obtain better distribution of cement gel throughout capillary pores and adjacent to aggregates. Strength increases of 70% have been obtained with ultrasonic treatment at 20–25-kH_z frequencies and $(20–60 \times 10^{-3})$-mm amplitudes [8].

Electrical heating can be accomplished somewhat more efficiently and uniformly than steam heating, by embedding small-diameter (high-resistance) wires directly in the fresh concrete. The influence of heating and cooling rates and maximum temperature upon strength are factors which required further study.

Steam curing may be accomplished either at atmospheric pressures (160–200°F) or at autoclave temperatures (325–375°F) and pressures (80–175 psi).

During low-pressure curing approximately the same hydration products are formed as at 20°C, but the reactions are accelerated. Steam is used to avoid moisture loss. Typical curing cycles are [8]

	Short Cycle	*Long Cycle*
At 20°C	2 hr	2 hr
Uniform rise to 80°C	90 min	5 hr
At 80°C	3 hr	3 hr
Uniform fall to 40°C	90 min	10 hr

Low-pressure curing gives 24-hr strengths that are 2–3 times that of concrete cured at 20°C, with 5–10% reduction in long-term strength. The long cycle gives somewhat higher strengths than the short cycle. Since it is the temperature at the time of setting which most influences long-term strength, a delay in steam curing is advantageous. Delays after mixing of 2, 3, 5, and 6 hr, respectively, for 40, 55, 70, and 85°C are recommended. Satisfactory results can be obtained up to 100°C, provided the initial heating rate is not so fast as to interfere with the setting process [50, 51]. The adverse effect of quicker heating is more pronounced at the higher w/c ratios.

Practical curing cycles are chosen as a compromise between early and late strength requirements and time requirements, such as shift length. Economic factors dictate to what extent the curing cycle is adapted to fit the mix or the mix modified to fit the curing cycle.

Autoclave steam curing is described in Sec. C.

C. HYDROTHERMAL CEMENTS

1. Autoclaved Products: Properties and Economics

The annual production of tobermorite and related calcium silicate hydrates exceeds that of any other chemical compound. Aside from hydrated portland cement, large amounts are produced by direct synthesis from various lime-silicate mixtures by high-pressure steam curing (autoclaving). Mixtures of portland cement with less expensive sources of lime (quick or hydrated) and silica (fly ash or fine silica sand) are often autoclaved.

Autoclaving is used in North America in the asbestos cement industry and in about 20% of the concrete block industry, and in Europe for cellular precast concrete units, including wall and roof panels and insulating materials.

In the asbestos cement industry, most large products such as pipe and corrugated or flat sheets are made by a wet process similar to papermaking. A thin slurry containing about 10% solids and 90% water is applied in layers and vacuum-filtered on a screen, laminated to obtain desired thickness, pressed on a form, and then stored for a presteaming period of several hours. The solids portion may typically contain 20% chrysotile asbestos fiber, 45–50% portland cement, and 30–35% ground quartz flour which reacts with lime formed during the cement hydration to provide additional binder material.

In more recent extrusion techniques for intricate shapes, small amounts of various additives are used which provide the required consistency of the asbestos-cement mixture at low moisture contents which make the dewatering unnecessary [52].

A typical curing sequence provides 2 hr at room temperature to obtain initial set (if portland cement is used), increases the saturated steam pressure for $1\frac{1}{2}$ hr to 150 psi at 320°F, maintains pressure for 8 hr, and reduces to atmospheric pressure in 20 min. Peak pressures vary from 100 psi for 20 hr to 370 psi for 2 hr. Changing the steam pressure and temperature too rapidly causes cracking due to pressure differences between the interior and surface of the product, particularly in foamed concretes, due to differential expansion of the entrained air cells during setting. Advantages of autoclaving include [53]

1. A strength after autoclaving normally equal to 28-day moist-cured strength.
2. A reduction of about 50% in drying shrinkage.
3. Increased resistance to sulfate attack.
4. Practical elimination of efflorescence, unless the aggregate contains sulfates or other salts.
5. Freedom from popping or spalling in service.
6. Lighter color.
7. Lower moisture content after manufacture.
8. The freedom to substitute for portland cement cheaper sources of lime and silicates, which also give higher strengths.

Major disadvantages include about a 50% reduction in bond strength with reinforcing steel, increased permeability and brittleness, and the added initial and operating costs of autoclaving. The major cost trade-off therefore is between the cost of raw materials and the cost of autoclaving.

Autoclaving produces primarily gel-like hydrosilicates at the lower pressure (100 psi) and cyrstalline hydrosilicates at the higher pressures (370 psi). Tobermorite gel is a high-lime silicate and tobermorite is a low-lime silicate. To transform tobermorite gel to tobermorite requires more silica than available in portland cement, plus high-temperature curing to force the silica to combine with the high-lime silicates and the $Ca(OH)_2$ liberated during the hydration of C_3S and C_2S. High-pressure autoclaving produces a portland cement paste having only 5% of the internal surface of normally cured paste. Because it is coarser and more crystalline, shrinkage and creep are reduced and permeability is increased. In general, crystalline tobermorite shrinks about half as much as tobermorite gel and is two to three times stronger. In foamed portland cement concretes, autoclaving typically reduces the drying shrinkage by four times, halves the reversible shrinkage due to moisture changes, and triples the strength [54, 55, 56]. Increased sulfate resistance results primarily from the greater crystallinity and sulfate stability imparted to the aluminate hydrates by autoclaving. Efflorescence is reduced because autoclaving reduces the amount of unreacted lime.

Strengths of autoclaved products depend on the raw materials, molding pressure, and autoclaving conditions.

a. RAW MATERIALS

Although the long-term strength of autoclaved cement is lower than that of normally cured cement, this can be corrected by substituting finely ground silica for 30–40% of the cement. Since any finely ground mixture of lime and reactive silica hardens when autoclaved, the expensive preliminary manufacture of portland cement is not essential. Cement does, however, allow shorter autoclaving periods by providing a more reactive source of lime and silica than is available in cheaper materials. The reactivity is due both to the solid solution of Ca and Si in portland cement and to their thermodynamic instability caused by a low order of crystallinity due to rapid cooling of the clinker.

The strength of portland cement autoclaved at 350°F has been found to decrease with the addition of silica flour up to about 10%, to increase markedly with increases of 10–30%, and to decrease again with silica additions above 30% [Fig. 32(a)]. In these tests no additional lime was added. DTA and microscopic examinations revealed the following changes in reaction products [57]

0–10% silica	Decreasing $Ca(OH)_2$, increasing α-C_2SH
10–30% silica	Decreasing α-C_2SH, increasing tobermorite
30–40% silica	Tobermorite
40–100% silica	Decreasing tobermorite, increasing unreacted silica

Figure 32(b) shows that increase in compressive strength is accompanied by increase

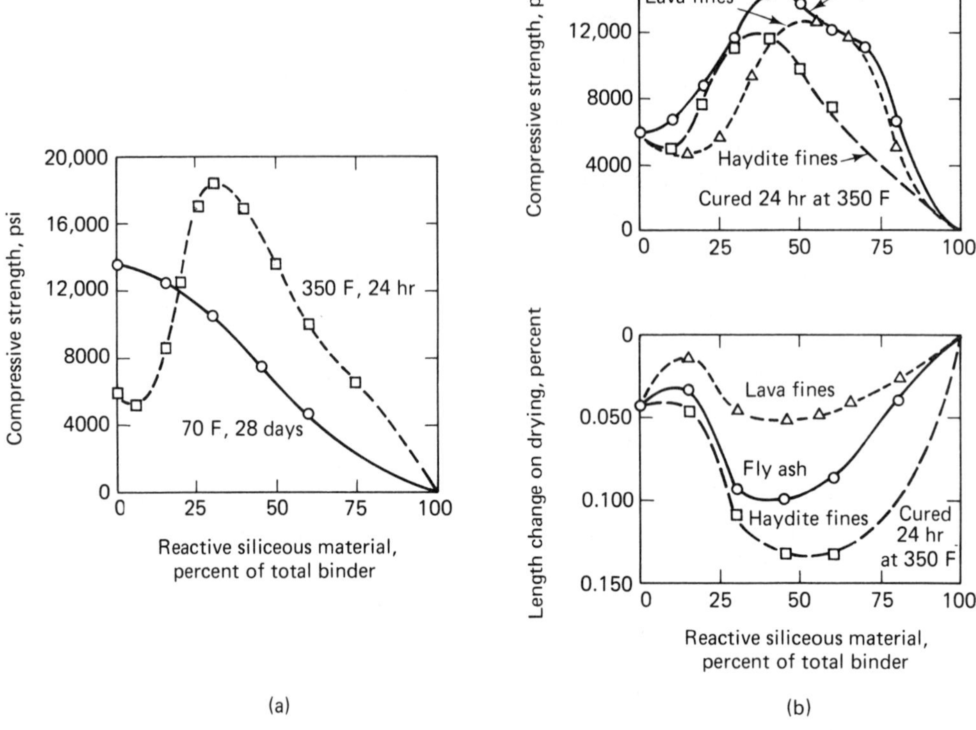

Fig. 32. Compressive strengths and shrinkage of portland cement pastes containing (a) silica flour; and (b) other siliceous fines. From Menzel, C. A.: *Am. Concr. Inst. J. Proc.*, 31, No. 2. (1934) 125 [58].

in drying shrinkage, since both depend largely on crystallinity of the hardened paste. Generally, variations in silica and moisture contents or autoclaving duration and temperature which increase strengths of autoclaved pastes also increase their drying shrinkage [49, 58].

In lime-silicate products containing little or no portland cement, autoclaving has been shortened by using highly reactive silica, such as silica flour, diatomaceous earths, or small amounts of sodium silicate. In the manufacture of cellular calcium silicate insulating products, asbestos fiber and up to 5% bentonite or other colloidal clay serve to keep the reactants in suspension prior to gelling. The mix is then filter pressure-molded to remove excess water prior to autoclaving.

Figure 33 shows a typical strength variation for a fine silica sand-lime mix (no portland cement). Taylor and Moorehead found the optimum Ca/Si ratio for strength of lightweight quartz-lime blocks to rise linearly with specific surface of the quartz:

$$R = 0.14 + 0.00009s$$

where R is the weight ratio of $Ca(OH)_2$ to SiO_2 and S is the specific surface of the quartz (cm^2/g) [59]. The coefficients vary somewhat with autoclaving conditions. Strengths also increase with the fineness, silica content, and particle size distribution (density) of the sand.

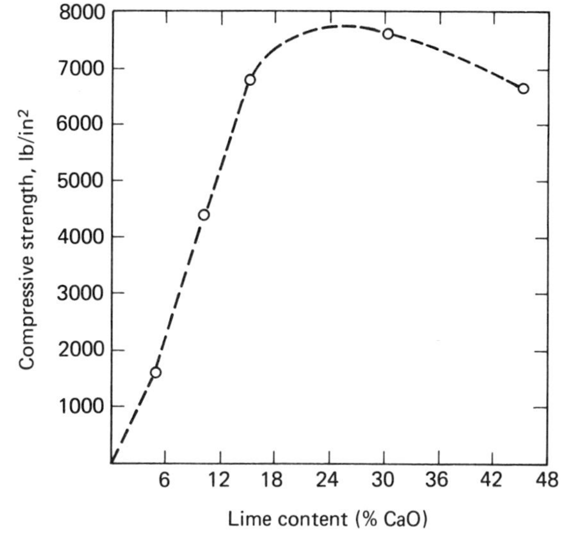

Compressive strength, lb/in²

Lime content (% CaO)

Fig. 33. Typical strength variation of autoclaved sand-lime brick with lime-silica ratio. Data from The Chalk Lime and Allied Industries Research Association, Welwyn, England. From Bessey, G. H., "Hydrated Calcium Silicate Products Other than Hydraulic Cements," in *The Chemistry of Cements*, H. F. W. Taylor, ed., New York: Academic Press, Inc., 1964, Chap. 16 [3].

High-calcium limes are preferred for autoclaving. MgO tends to hydrate late during the autoclave cycle and to expand, producing unsoundness in the same way it does after a period of several years in normally cured cements. Three percent MgO is usually taken as a limit, although higher percentages are acceptable in lightly burned limes [60]. Other impurities in limestone or chalk, i.e., silica, alumina, and iron oxide, reduce the usefulness of the lime only to their extent of dilution. Quicklime permits slightly shorter autoclaving than hydrated lime and is also cheaper. Well-sealed lime gives noticeably higher autoclaved strengths than the same product used after partial recarbonation due to exposure to atmospheric CO_2.

Industrial fly ash is an attractive silica source where locally available because of its low cost, fine particle size, and reactivity. Fly ash consists typically of about 75% glassy particles formed from the clay minerals originally present in the coal and 25% crystalline components (quartz, mullite, hematite, and magnetite). Typical chemical compositions of fly ash are

SiO_2	40–50%	MgO	1–3%
Al_2O_3	22–35%	Na_2O	0.5–3%
Fe_2O_3	6–13%	K_2O	2–4%
CaO	2–8%	C	2–12%

Chemical reactivity and strength in lime-fly ash products appears to increase with (1) the SiO_2-Al_2O_3 content, (2) the fineness, and (3) the glass content of the fly ash [61].

Figure 34 shows the distribution of fly ash production in the United States and Table 8 compares the percentage utilization of fly ash in several nations. Other major uses of fly ash are as a pozzolan in normally cured concretes, as a filler in asphalt paving mixes, as lightweight aggregate after being pelletized and sintered, and for soil stabilization. In some populated areas of the United States the cost of transporting and disposing of unused fly ash exceeds $2/ton.

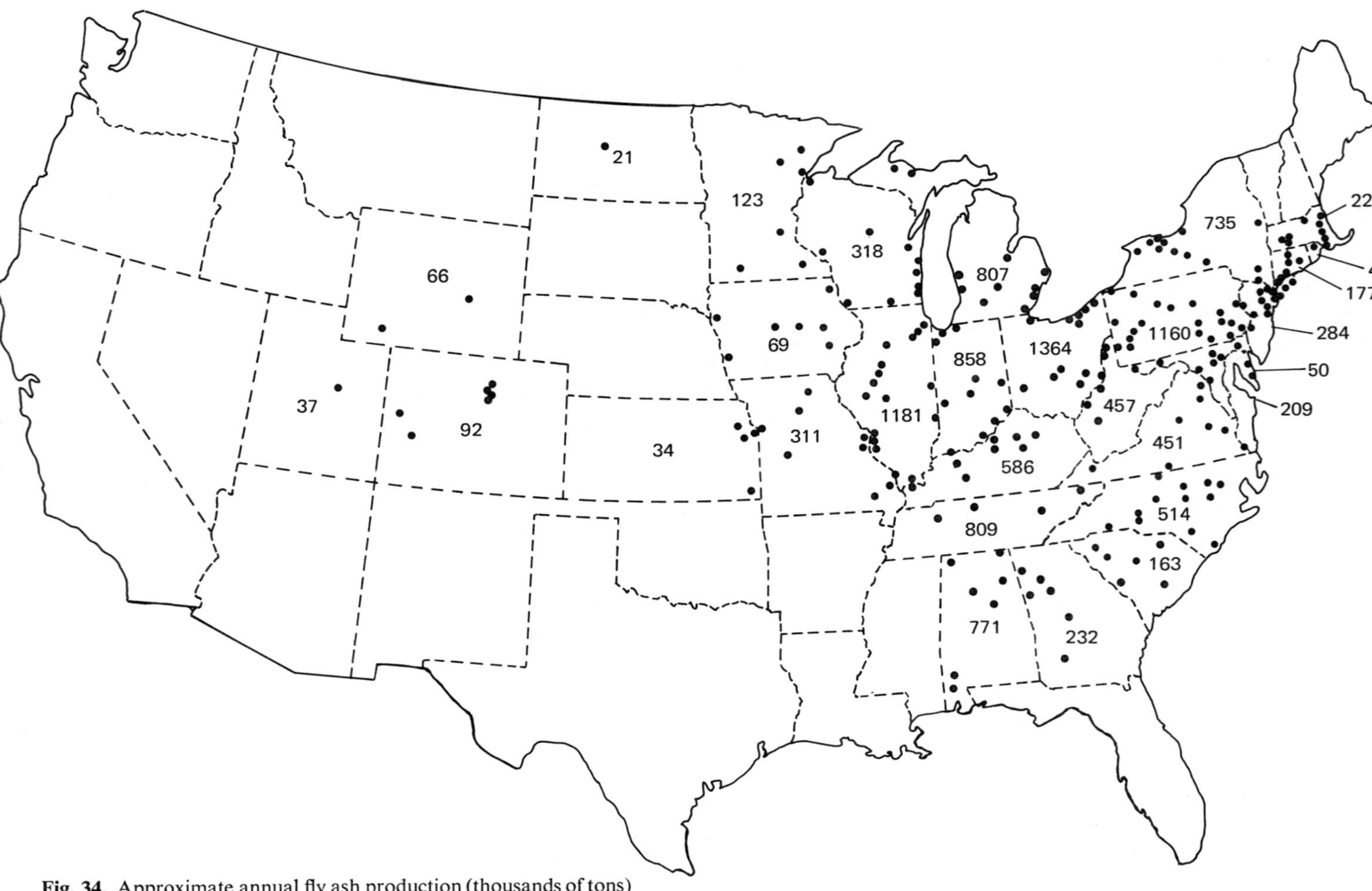

Fig. 34. Approximate annual fly ash production (thousands of tons) by state, 1965.

TABLE 8. Fly Ash Utilization, 1965
(Includes Fly Ash, Hearth Ash, and Molten Ash)

		Utilization (1,000 tons)												
						Compacted concrete								
Country	*Production in 1,000 tons*	*Cement (addition or replacement of hydraulic binder)*	*Cement kilns*	*Roads*	*Cellular concrete*	*Blocks*	*Prepared concrete*	*Dams*	*Lightweight aggregate*	*Bricks*	*Filler on construction sites*	*Miscellaneous*	*Total*	*Percent of production used*
France	4,022	832	124	726	—	102	—	—	—	1	286	53	2,124	53
Greece	250	—	—	—	—	—	—	—	—	—	—	—	—	—
Netherlands	391	—	—	57	—	10	—	—	—	77	95	9	248	63
Poland	4,990	15	—	25	525	175	—	—	—	—	200	—	940	19
Rumania	1,500	22	—	—	33	—	—	—	—	—	—	23	78	5
United Kingdom	9,440	18	—	1,647	—	938	55	—	65	31	1,013	12	3,779	40
U.S.S.R.	30,000	—	—	—	—	—	—	—	—	—	—	—	—	—
United States	20,000+	—	—	—	—	—	—	—	—	—	—	—	2,400	12
West Germany	10,400	5	—	550	40	850	800	—	—	—	400	35	2,730	26

SOURCE: FABER, J. H., *Proceedings of the 1st Mineral Waste Utilization Symposium*, M. A. Schnarty, ed. Chicago: U.S. Bureau of Mines, March 1968, p. 35 [62].

Fig. 35. Relation of compressive strength to curing time and temperature of portland cement pastes containing optimum amounts of siliceous material (data from Menzel). From *ACI Manual of Concrete Practice, Part 3: Products and Processes*. Detroit: American Concrete Institute, 1968 [53].

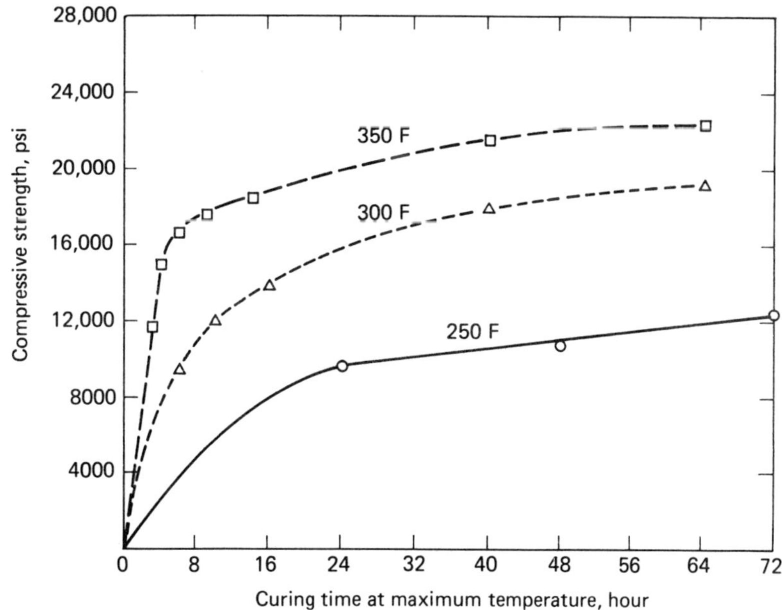

b. MOLDING AND AUTOCLAVING

In the manufacture of sand-lime bricks moisture contents of 4–7% are common, as are molding pressures of 2–5 tons/sq in., or about 100 tons/brick. Autoclaved lime-silica compressive strengths as high as 20,000 psi at 100-pcf density and 36,000 psi at 130 pcf have been obtained, using finely ground lime and silica, low moisture content, and high molding pressure. Portland cement paste strengths as high as 95,000 psi have been obtained by autoclaving under 50,000 psi pressure at 250°C to produce very low-porosity hydrated cement gel surrounding residual unhydrated cement grain cores [63]. Strength has been commonly observed to be more sensitive to density of the hardened paste than to variations in chemical composition or any other property [10, 64].

The optimum autoclaving period of lime-silica and portland cement-silica mixes depends on the temperature, on the fineness and reactivity of the siliceous material, and also on the specimen size. Figure 35 shows data from Menzel for 2-in. cubes made with silica passing the No. 200 sieve [58].

2. *Hydrothermal Reactions*

a. HYDRATION MECHANISMS

At least 17 crystalline calcium silicate hydrates are known to form by autoclaving. Intermediate reaction products of indefinite composition, such as the poorly crystalline calcium silicate hydrate gels [CSH(I) and CSH(II)] also contribute to strength. The crystalline hydrates can be classified by crystal structure into five groups [65]:

Wollastonite Group	*Tobermorite Group*	*Gyrolite Group*
1. Nekoite	6. 9-, 11-, and 14-A tobermorites	7. Gyrolite
2. Okenite		8. Truscottite
3. Xonotlite		9. Reyerite
4. Foshagite		10. *Z* Phase
5. Hillebrandite		

γ-C_2S Group	*Other Compounds*
11. γ-C_2S hydrate	14. Afwillite
12. Calciochondrodite	15. α-C_2S hydrate
13. Kilchoanite	16. Phase *Y*
	17. C_3S hydrate

Representative molecular structures of these silicates are shown in Fig. 36, corresponding to the silicate structures described in Chap. 6. The α-C_2S hydrate is an isolated silica tetrahedron (monomer). The C_3S hydrate is a dimer (Si_2O_7 groups). Tobermorite and hillebrandite are infinite silicate chains. Xonotlite is a double chain, and gyrolite appears to have sheet-like and three-dimensional silicate networks.

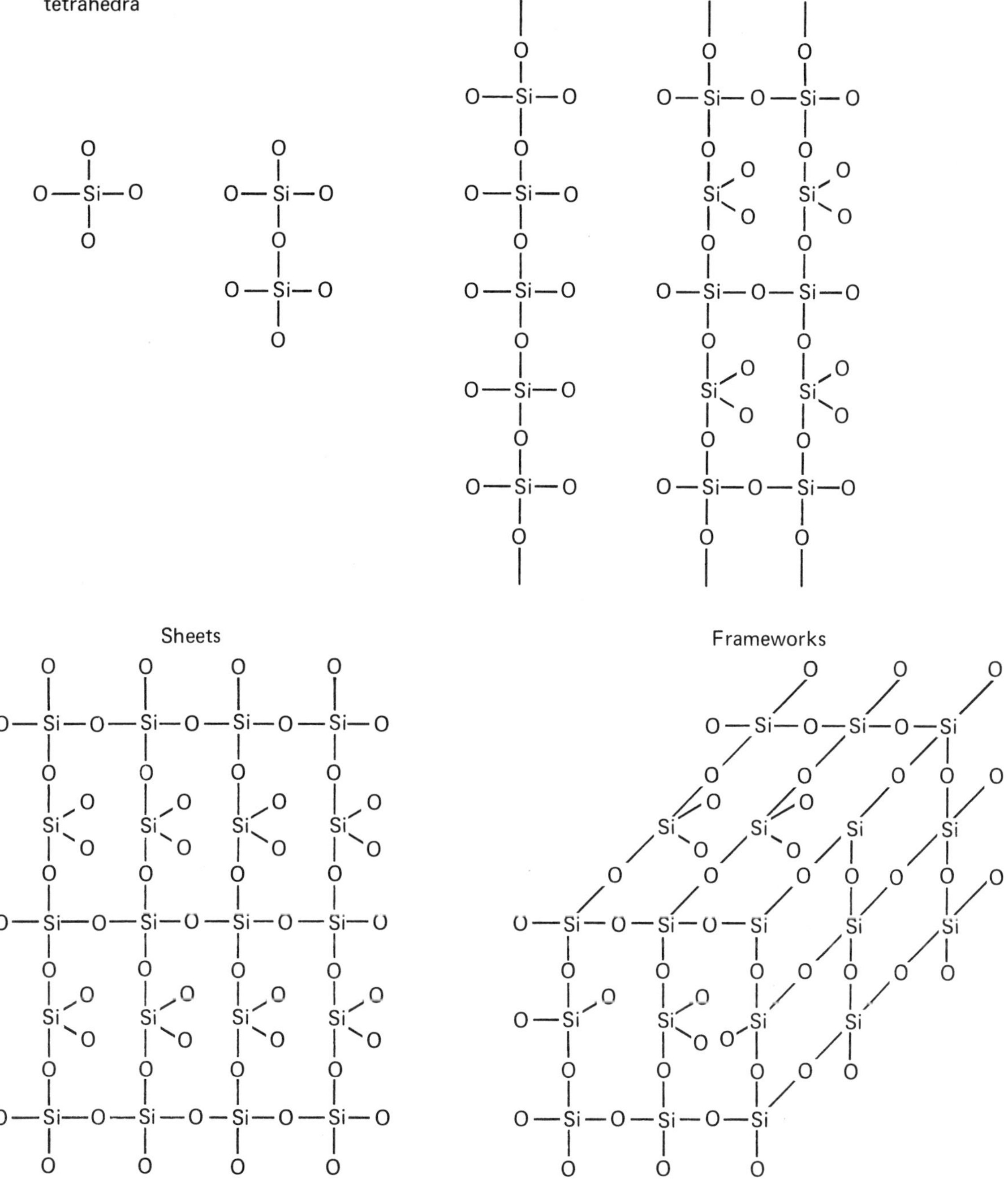

Fig. 36. Molecular structures of hydration products of various lime-silica mixtures at elevated temperatures. From Verbeck, G. and L. E. Copeland: *Menzel Symposium on High Pressure Steam Curing, 1969*. Detroit: American Concrete Institute Publication SP-32, 1972, p. 1 [44].

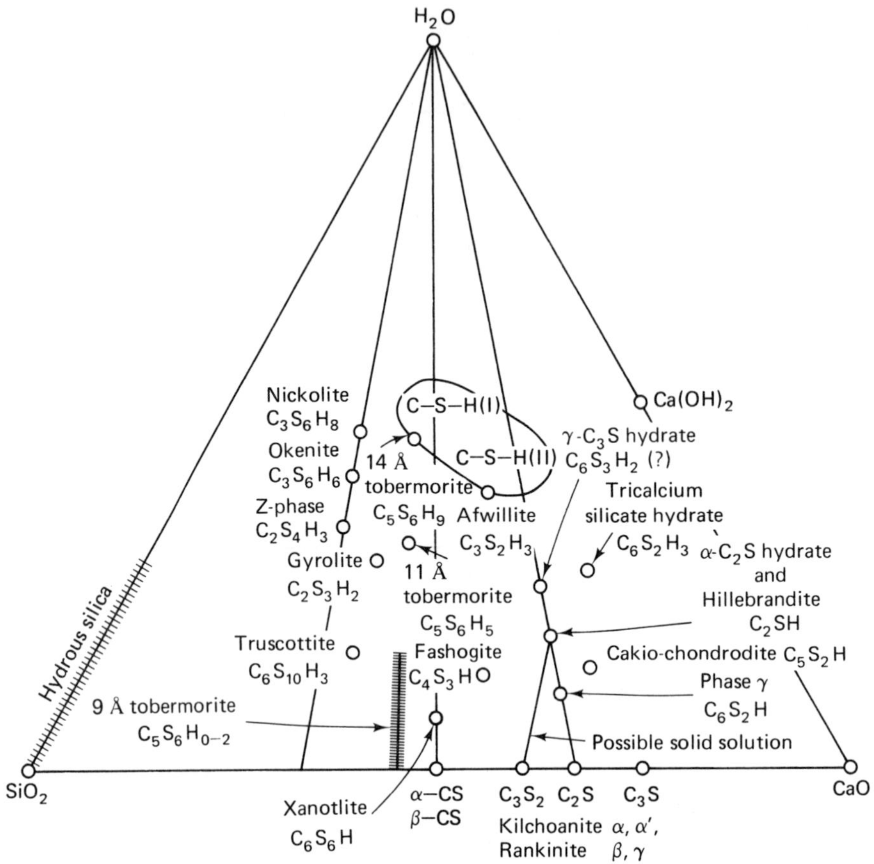

Fig. 37. Compounds in the CSH system. From Taylor, W. H., *Concrete Technology and Practice*, Second Edition. London: Angus & Robertson, Ltd., 1967 [31].

Figure 37 shows the phase compositions and Fig. 38 the equilibrium temperature ranges in which the various silicates form. As would be expected, the more hydrous compounds tend to form at the lower temperatures. In Fig. 38 the vertical line lengths indicate the stable temperature ranges. Line lengths are representative of saturated steam pressures below 374°C (the critical point) and about 5,000 psi for higher temperatures. Displacement of the name to the left or right of the line indicates that the compound tends to form also below or above the indicated C/S ratio. Wavy underlines denote variable composition. At temperatures above the line *ab* anhydrous compounds are formed. Compounds below the line *cd* generally contain Si-OH groups and may also contain molecular water and ionic hydroxyl.

The components generally believed to contribute most to compressive strength are shaded in Figs. 37 and 38 [66, 67]. It is probable that some blend of crystalline and gelatinous materials is desirable [68], and the reductions in strength observed after prolonged autoclaving may be due to too large a crystalline/gel ratio [65]. It is generally believed that high compressive strength is obtained from phases of high specific

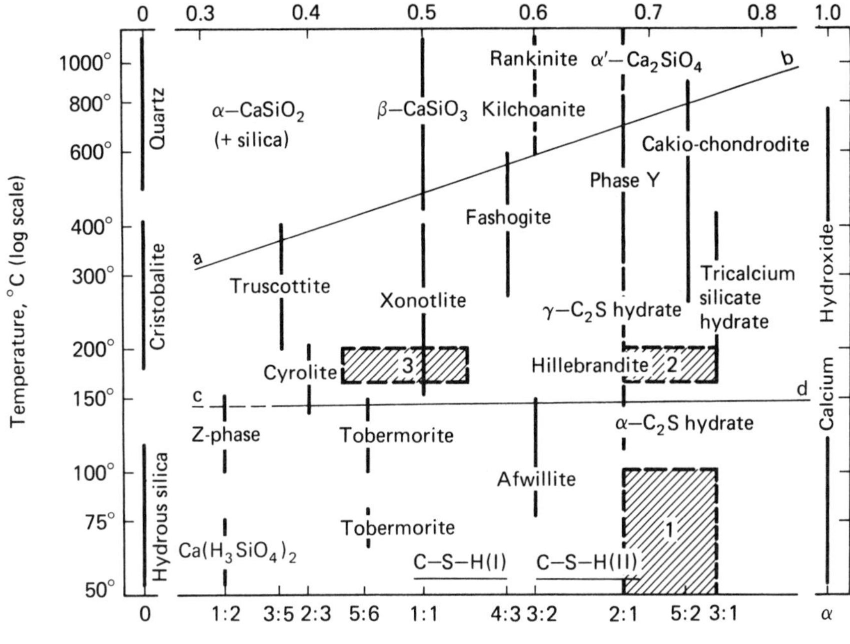

Fig. 38. Equilibrium temperature ranges for hydrothermal calcium silicate hydrates. Adapted from Verbeck, G. and L. E. Copeland: *Menzel Symposium on High Pressure Steam Curing, 1969*. Detroit: American Concrete Institute Publication SP-32, 1972, p. 1 [44] and Taylor, W. H.: *Concrete Technology and Practice*, Second Edition. London: Angus & Robertson, Ltd., 1967 [31].

surface which provide many contact points and that high tensile strength is obtained with needle-shaped crystals (xonotlite or C_3S hydrate) [65]. Low strengths are observed for α-C_2S hydrate, which has coarse crystals with few points of contact.

The areas 1, 2, and 3 in Fig. 38 delineate the usual temperature and composition ranges used in commercial practice. Area 1 represents steam curing at atmospheric pressure of normal portland cements. It has a line-silica ratio in the range 2–3 (since cements contain both dicalcium and tricalcium silicate) and extends up to the boiling point of water (100°C). The products formed in area 1 are calcium silicate hydrates with good strength but unfortunately also with the tendency to shrink when dried or carbonated.

Area 2 represents autoclaved portland cement (160 to 190°C). The products are predominately α-C_2S hydrate and hillebrandite, which have little tendency to shrink when dried or carbonated, but unfortunately also have little strength at practical/ratios [44].

The low strength in area 2 can be improved (but at the expense of increased shrinkage) by adding finely divided silica to portland cement, reducing the lime/silica ratio to area 3. In commercial practice the aim is to obtain a product intermediate between the low-strength, low-shrinkage α-hydrate of area 2 and the high-strength, high-shrinkage tobermorite products of area 3.

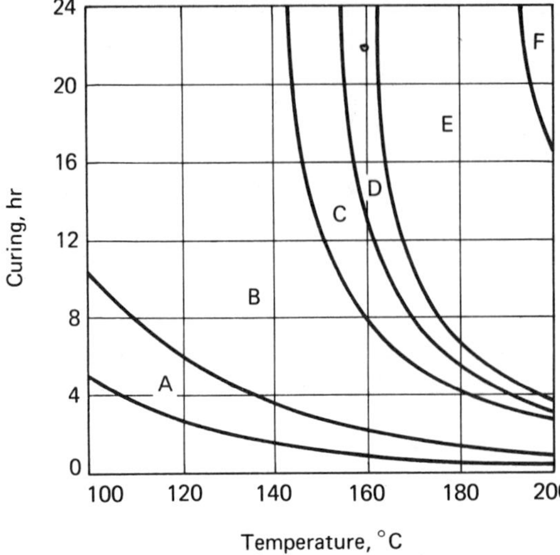

Fig. 39. Reaction products in autoclaved lime-quartz pastes: (a) $C_{1.75}SH_n$, (b) $C_{1.25}SH_n + Ca(OH)_2$, (c) $C_{1.25}SH_n$, (d) $C_{1.0}SH_n$, (e) $C_{1.0}SH_n$ + tobermorite, (f) tobermorite. From Neese, H., *Tonindustr. Zg.*, 83 (1959) 124 [69].

Figure 38 shows only the end products at the indicated temperatures. Hydrothermal lime-silicate reactions proceed through a series of products starting with the formation of ill-crystallized CSH(II) with a high C/S ratio, followed by CSH(I) with a lower C/S ratio and crystalline tobermorite [66]. Figure 39 shows the reaction products detected by chemical, optical, x-ray, DTA, and thermogravimetric studies as a function of autoclaving temperature and time. Generally, in industrial practice where autoclaving is below 200°C and for less than 24 hr, the predominant hydration products are the poorly crystalline CSH(I) and (II), 11-A tobermorite, and α-C_2S hydrate. The remaining products below 200°C in Fig. 38 apparently form too slowly to be obtained in industrial practice. Xonotlite is the major additional phase at somewhat higher temperatures [65].

Reaction mechanisms. Hydrothermal lime-silicate reactions are believed to proceed approximately in the following sequence [70]:

1. Chemisorption of $Ca(OH)_2$ on the silica surface.
2. Solution of the silica: $H_4SiO_4 \longrightarrow H_3SiO_4^- \longrightarrow H_2SiO_4^{--} \longrightarrow$ increasing pH.
3. Reaction of silicic acid with Ca^{++} ions in solution.
4. Formation of CSH nuclei, growth and precipitation of CSH crystals.

Figure 40 depicts the mass transport relationships for the case where the lime and

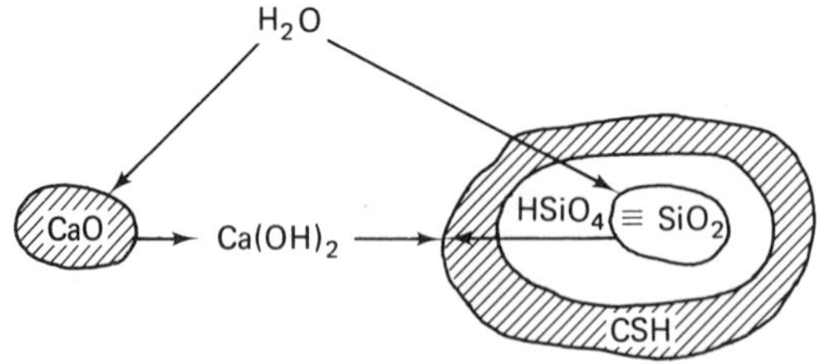

Fig. 40. Mass transport relationships in in lime-silicate hydration.

silicate exist initially as separate particles, as in lime-fly ash or lime-silicate sand systems [71]. The reaction mechanisms have been studied by the use of optical and electron microscope observations, diffusion measurements using radioactive isotopes, and x-ray diffraction traces of samples progressively sectioned from the lime solution boundary of the reaction product layers. The results indicated that

1. Initially, step 2, above, is rate-limiting. But very shortly the rate-limiting step becomes the diffusion of SiO_2 hydrates through the CHS reaction product layer at temperatures below approximately the critical point of steam (374°C).
2. At temperatures below approximately 374°C, CSH formation occurs on the outside of the product layer (SiO_2 hydrates diffuse outward through the layer; Ca^{++} species do not diffuse inward).
3. Water diffuses inward through the reaction product layer according to the H_2O concentration gradient and swells it by osmotic pressure until it bursts. SiO_2 hydrates contained at high concentration within the reaction product layer then exude outward into the neighboring solution space and react with $Ca(OH)^+$ to form fingered configurations of CSH reaction products which can be observed in photomicrographs similar to Fig. 20, which showed the analogous behavior for portland cement. New reaction product fills the break in the initial reaction layer, and the process repeats itself. Reaction product layers grow to a thickness of the order of 5 mm.
4. At temperatures considerably above 374°C the product layer forms adjacent to the lime particle rather than adjacent to the silica particle. The reason may lie primarily in their relative solubilities; lime solubility decreases with increasing temperature, whereas silica solubility increases exponentially.

b. ADMIXTURES

The foregoing results suggest that industrial autoclaving periods might be shortened by processes which increase the specific surface of the silicate, increase its reactivity, or remove the reaction product layers more effectively. Significant strength increases and reductions in autoclaving times of lime-fly ash mixes have been obtained in laboratory studies by the author with caustics such as NaOH. Upon adding up to 5% NaOH by weight of lime plus fly ash and vigorously mixing for 5 min at 95°C, the paste becomes very sticky and can be compacted with equivalent pressure to 8–12% higher densities than similar compositions mixed at room temperature without caustic.

It is believed that some initial hydration and perhaps stripping of reaction products is obtained during the mixing, permitting a partially hydrated and denser product prior to autoclaving. With regard to sodium hydroxide, most silicate glasses are more rapidly attacked by alkalies than by neutral or acid solutions, because alkali supplies hydroxyl ions for reaction with the silica network [72]. Also, cement and lime-silicate hydrations are favored by alkaline environments. The beneficial effect of NaOH may also be partially due to the great capacity of sodium species ranging from the Na^+ ion to the hydrate $Na_2Si(OH)_6 \cdot 6\,H_2O$ to imbibe water [73], which would increase osmotic pressures and cause more frequent rupture of the CSH reaction product layers. A quantity of silica being converted to solid $Na_2Si(OH)_6 \cdot 6\,H_2O$, for example, expands much more than the same quantity being converted to $K_2Si(OH)_6$. KOH was also found to be somewhat less effective than NaOH in these studies.

PROBLEMS

1. What properties of hydrated aluminous cement distinguish it from portland cement? Name three areas of application in which aluminous cement might be preferred. Describe the composition, characteristics, and particular applications of two additional kinds of hydraulic cements.

2. Lime, alumina, and silica are all intersoluble in the liquid state. Which of the following would you expect to have the lowest melting point? Why?
   (a) 40% A, 60% S.
   (b) 40% S, 60% C.
   (c) 40% C, 40% A, 20% S.
   (d) 40% C, 60% A.

3. How might the following characteristics of portland cement composition affect the behavior of concrete?
   (a) Too high lime.
   (b) Too low lime.
   (c) Too high alumina.
   (d) Too high gypsum.
   (e) Too low gypsum.
   (f) Too high magnesia.
   (g) Too high alkalies.

4. Briefly describe the chemical and physical changes and the reaction sequence which occur in the portland cement kiln. Which components are given off as gases? What are the major reactions and reaction products?

5. What are the effects of (a) underburning, (b) overburning the clinker?

6. In what ways does fineness of grinding of the clinker affect portland cement performance?

7. In terms of portland cement hydration, mention two potential advantages of rapid cooling of the clinker.

8. Of the four major compounds in portland cement (C_3S, C_2S, C_3A, C_4AF), which can contribute primarily to
   (a) Fast setting.
   (b) Early strength.
   (c) Ultimate strength.
   (d) Reducing kilning temperatures, by acting as fluxing agents.

9. Briefly describe the sequence of chemical and physical changes during portland cement hydration. Which components hydrate rapidly, slowly; which contribute most to early set, early hardening, long-term strength, and eventual deterioration, and why?

10. Describe the mechanisms involved in the following portland cement concrete behaviors: (a) setting and hardening, (b) dimensional change, (c) autogenous healing.

11. Describe how the following properties might be obtained in concrete: (a) high strength, (b) high durability, (c) impermeability, (d) minimum of cracks.

12. What factors might affect the water tightness of reinforced concrete pressure pipe?

13. What reactions tend to occur in portland cement concrete under moist sulfate conditions? What methods can be used to improve the durability of concrete exposed to a marine environment?

14. The vicat needle tests on portland cement mortar bars indicate flash set. What possible changes would you recommend be made in the manufacturing process?

15. Why, in terms of hydration, does low-temperature curing (say 40°F) produce concrete having greater ultimate strength than high-temperature curing (say 100°F)?

16. Explain how the statement "Lack of hydration due to low water/cement ratios does not necessarily produce weaker cement pastes" can be true.

17. Why is it that lime, $Ca(OH)_2$, serves admirably as a cementing material in brick mortar but can be harmful if present in more than small quantities as a component in portland cement?

18. Formation of the mineral hydrate ettringite ($C_3A \cdot 3\ CaSO_4 \cdot 32\ H_2O$) is the key factor in preventing flash set during initial hydration. It is also the key component in nonshrinking cements, in which most of the hydration occurs during the first 30 days. It is also the component primarily responsible for the destructive expansion of concrete exposed to seawater, where most of the hydration may occur after several years. Explain how such differing hydration rates can be associated with the same reaction product.

19. Briefly explain and differentiate three principal types of cement shrinkage. Propose at least two methods for reducing each type of shrinkage.

20. Describe four possible causes of expansion (unsoundness) of portland cement concrete.

21. How should concrete mixes be adjusted for autoclave curing? What effect does autoclaving have on the paste structure and its physical properties.

22. Why, in autoclaved cements, is an increase in compressive strength generally associated with an increase in drying shrinkage? What structural characteristics of the paste are generally believed to account for high compressive strength?

23. Briefly describe the similarities and differences between the mass transport relationships in the hydration of portland cement cured at room temperature and the hydration of a lime-silicate mixture cured by autoclaving.

24. In terms of the mass transport of reactants and reaction products through the gel membrane of reaction product surrounding partially hydrated particles, what difference exists between the hydration of portland cement and the hydration of a lime-fly ash mixture?

REFERENCES

1. MURPHY, G., *Properties of Engineering Materials*, 2nd ed. Scranton, Pa.: International Textbook Company, 1947.
2. LEA, F. M., and C. H. DESCH, *Chemistry of Cement and Concrete*, 2nd ed. London: Edward Arnold & Co., 1956.
3. DAVEY, N., and E. N. FOX, *Building Research Technology Paper No. 15*. London: Stationery Office, 1933.
4. CZERNIN, W., *Cement Chemistry and Physics for Engineers*. New York: Chemical Publishing Company, Inc., 1962.
5. LEA, F. M., *The Chemistry of Cement and Concrete*. London: Edward Arnold & Co., 1970.
6. FARRAN, J., *Rev. Matér. Constr.*, 491 (1956) 155; 492 (1956) 191.
7. BOGUE, R. H., *Ind. Eng. Chem., Anal. Ed.*, 1 (1929) 192.
8. HANSEN, T. C., "Notes from a Seminar on Structure and Properties of Concrete," *Stanford University Civil Engineering Department Tech. Report No. 71*, 1966.
9 WITT, J. C., *Portland Cement Technology*. New York: Chemical Publishing Company, Inc., 1966.
10 MIDGLEY, H. G., and S. K. CHOPRA, *Mag. Concr. Res.*, 12 (1960) 73.
11. HANSEN, W. C., *J. Res. Natl. Bur. Stand.*, 4 (1930) 55.
12. NAGY, N., *Rev. Matér Constr. Trans.*, Publ. 638 (1968) 428.
13. WELCH, J. H., and W. GUTT, *Chemistry of Cement, Proceedings of the 4th International Symposium*, Washington, 1960. National Bureau of Standards Monograph 43, 1962, p. 59.
14. EITEL, W., *Zement*, 30 (1949) 29.
15. OSBORN, E. F., and A. MAUN, *Equilibrium Diagrams of Oxide Systems*. Columbus, Ohio: American Ceramics Society, 1960.
16. LERCH, W., and W. C. TAYLOR, *Concrete*, 45 (1937) 217.
17. PRICE, W. H., *Proc. Am. Concr. Inst.*, 47 (1951) 417.
18. BOGUE, R. H., and W. LERCH, *Ind. Eng. Chem.*, 26 (1934) 837.
19. BOGUE, R. H., *Chemistry of Portland Cement*, 2nd ed. New York: Van Nostrand Reinhold Company, 1955.
20. *U. S. Bureau of Reclamation Concrete Manual*, 7th ed., 1966.
21. STEINOUR, H. H., *Port. Cem. Assoc. Res. Lab. Bull.*, 34 (1951).
22. MIDGLEY, H. G., *Chemistry of Cement, Proceedings of the 4th International Symposium*, Washington, 1960. National Bureau of Standards Monograph 43, 1962, p. 479.
23. LEHMANN, H., and H. DUTZ, *Chemistry of Cement, Proceedings of the 4th International Symposium*, Washington, 1960. National Bureau of Standards Monograph 43, 1962, p. 513.
24. BRUNAUER, S., and S. A. GREENBERG, *Chemistry of Cement, Proceedings of the 4th International Symposium*, Washington, 1960. National Bureau of Standards Monograph 43, 1962, p. 135.
25. BRUNAUER, S., *Am. Sci.*, 50 (1962) 210.
26. BRUNAUER, S., D. L. KANTRO, and L. E. COPELAND, *J. Amer. Chem. Soc.*, 80 (1958) 761.
27. GRUDEMO, A., *Chemistry of Cement, Proceedings of the 4th International Symposium*, Washington, 1960. National Bureau of Standards Monograph 43, 1962, p. 615.
28. KANTRO, D. L., S. BRUNAUER, and C. H. WEISE, *Adv. Chem. Ser.*, 33 (1962) 199.
29. NEWMAN, K., "Concrete Systems," in *Composite Materials*, L. Holliday, ed. Amsterdam: Elsevier Publishing Company, 1966, Chap. 8.
30. POWERS, T. C., *Chemistry of Cement, Proceedings of the 4th International Symposium*, Washington, 1960. National Bureau of Standards Monograph 43, 1962, p. 577.

31. TAYLOR, W. H., *Concrete Technology and Practice*, 2nd ed. London: Angus & Robertson Ltd., 1967.

32. SELIGMANN, P., and N. R. GREENING, *Hwy. Res. Record*, No. 62 (1964) 80.

33. BERNAL, J. D., J. W. JEFFERY, and H. F. TAYLOR, *Mag. Concr. Res.*, 4 (1952) 49.

34. POWERS, T. C., *Port. Cem. Assoc. Res. Lab. Bull.*, 2 (1939).

35. WARD, G., *Res. Report, PCAF* (April 1944).

36. HANSEN, W. C., *Chemistry of Cement, Proceedings of the 3rd International Symposium*, London: Concrete and Cement Association, 1954, p. 318.

37. BESSEY, G. H., "Hydrated Calcium Silicate Products Other Than Hydraulic Cements," in *The Chemistry of Cements*, H. F. W. Taylor, ed. New York: Academic Press, Inc., 1964, Chap. 16.

38. POWERS, T. C., and T. L. BROWNYARD, *Port. Cem. Assoc. Res. Lab. Bull.*, 22 (1948).

39. HUMMEL, A., *The ABC of Concrete*, 12th ed. Berlin: W. Ernst und Sohn, 1959.

40. POWERS, T. C., and T. L. BROWNYARD, *Proc. Am. Concr. Inst.*, 43 (1947) 101, 249, 469, 549, 669, 845, 933.

41. POWERS, T. C., *Port. Cem. Assoc. Res. Develop. Dept. Bull.*, 90 (July 1958) 1.

42. POWERS, T. C., *J. Am. Ceram. Soc.*, 41, No. 1 (Jan. 1958) 1.

43. *Design and Control of Concrete Mixtures*, 11th ed. Portland Cement Association, 1968.

44. VERBECK, G., and L. E. COPELAND, *Menzel Symposium on High Pressure Steam Curing*, 1969. American Concrete Institute Publication SP-32, 1972, p. 1.

45. POWERS, T. C., et al., *Proc. Am. Concr. Inst.*, 51 (1955) 285.

46. POWERS, T. C., *Proc. Hwy. Res. Bd.*, 27 (1947) 178.

47. COPELAND, L. E., and R. H. BRAGG, *Am. Soc. Testing Mater. Bull.*, 204 (1955) 34.

48. POWERS, T. C., *J. Port. Cem. Assoc. Res. Labs*, 4, No. 2 (1962) 40.

49. VERBECK, G. J., *Am. Soc. Testing Mater. STP*, No. 205 (1958) 17.

50. SAUL, A. C., *Mag. Concr. Res.*, 1, No. 6 (1951) 127.

51. NURSE, R. W., *Proc. Bldg. Res. Cong.* (1951) 86.

52. YANG, J. C., *Menzel Symposium on High Pressure Steam Curing*, 1969. American Concrete Institute Publication SP-32, 1972, p. 117.

53. *ACI Manual of Concrete Practice*, Part 3: Products and Processes. American Concrete Institute, 1968.

54. *ACI Manual of Concrete Practice*, Part 1: Materials and Properties of Concrete. American Concrete Institute, 1970.

55. MIRONOV, S. A., and L. A. MALININA, *Autoclave Cured Concrete.* Moscow: State Publishing Office of Structural Materials, 1958.

56. GRANHOLM, H., ed., *Proceedings of the Symposium on Autoclaved Cellular Concrete*, Göteborg, 1960. Paris: RILEM, 1961.

57. KALOUSEK, G. L., and M. ADAMS, *Am. Concr. Inst. J., Proc.*, 48, No. 1 (1951) 77.

58. MENZEL, C. A., *Am. Concr. Inst. J., Proc.*, 31, No. 2 (1934) 125.

59. TAYLOR, W. H., and D. R. MOOREHEAD, *Mag. Concr. Res.*, 8 (1956) 145.

60. WUHRER, J., W. STEYER, and G. RODERMACHER, *Zement-Kalk-Gips*, 14 (1961) 566.

61. WATT, J. D. and D. J. THORNE, *J. Appl. Chem.*, 15 (Dec. 1965) 585.

62. FABER, J. H., in *Proceedings of the 1st Mineral Waste Utilization Symposium*, Chicago: U. S. Bureau of Mines, March 1968, p. 35.

63. ROY, D. M., and G. R. GOUDA, *Cem. Concr. Res.*, 3 (1973) 807.

64. SATAVA, V., *Autoclaved Calcium Silicate Building Products, Proceedings of the International Symposium*, London, (1965) 148.

65. TAYLOR, H. F., *Autoclaved Calcium Silicate Building Products, Proceedings of the International Symposium*, London, (1965) 195.

66. KALOUSEK, G. L., and A. F. PREBUS, *J. Am. Ceram. Soc.*, 41 (1958) 124.

67. SANDERS, L. D., and W. J. SMOTHERS, *J. Am. Concr. Inst.*, 54 (1957) 127.

68. BOZHENOV, P. I., et al., *Chemistry of Cement, Proceedings of the 4th International Symposium*, Washington, 1960. National Bureau of Standards Monograph 43, 1962, p. 327.

69. NEESE, H., *Tonindustr. Ztg.*, 83 (1959) 124.

70. GREENBERG, S. A., *J. Phys. Chem.*, 65 No. 1 (1961) 12.

71. MOOREHEAD, D. R., *Autoclaved Calcium Silicate Building Products, Proceedings of the International Symposium*, London, (1965) 86.

72. HOLLAND, L., *The Properties of Glass Surfaces*. New York: John Wiley & Sons, Inc., 1964.

73. HANSEN, W. C., *J. Mater., Am. Soc. Testing Mater.*, 2, No. 2 (1967) 408.

74. VALORE, R. C., *Am. Concr. Inst. J., Proc.*, 50, No. 9 (1954) 773; 50, No. 10 (1954) 817.

75. MENZEL, C. A., *Am. Concr. Inst. J., Proc.*, 39, No. 3 (1943) 165.

5

ORGANIC CEMENTS

Spring, tire, axle, and linchpin too,
Steel of the finest, bright and blue;
Thoroughbrace bison-skin, thick and wide;
Boot, top, dasher, from tough old hide.

—HOLMES

In Sec. C in Chap. 2 we described the synthesis of polymers and introduced fundamental relationships between polymer structure and mechanical behavior. In Sec. A in Chap. 3 we presented mathematical models for describing the viscoelastic behavior of polymers, their properties, and applications in construction. In this chapter we shall describe the two major categories of polymers used in construction: synthetic polymers and bitumens. The synthetic polymers, which have more controllable and uniform chemical structure and physical properties, will be described first.

TABLE 1. Applications of Plastics in Buildings

Exterior	*Interior*
Adhesives	Acoustical panels
Air support structures	Adhesives*
Air vents	Baseboards
Cables	Cabinets
Caulkings	Ceilings
Coating—metal, wood	Conduits
Concrete forms	Counter tops
Concrete mixes	Coverings
Curtain walls	Decorative panels
Doors (prime and storm)	Drawers
Expansion joints	Ducts
Facings	Electrical fixtures
Flashings	Floorings*
Gaskets*	Gaskets*
Glazings	Graphic arts
Grilles	Grilles
Hardwares	Hardwares
Illuminating panels	Insulations*
Lighting fixtures	Light diffusers
Louvers	Molding, trims*
Moisture barriers	Paints
Mortar mixes	Panelings
Paints	Partitions
Panels	Pipe fittings*
Pipes*	Plastic backings
Railings	Plumbing fixtures*
Rain system—gutters, downspout, etc.	Railings
Roof edging, panels*	Sealants
Safety and thermal glasses	Shower stalls
Screens	Stair treads
Sealants	Tanks
Sheathings	Tile—floor, wall, ceilings
Shingles	Vapor barriers*
Shutters	Wall coverings*
Sidings	Wire insulations
Signs	
Skylights	
Stuccos	
Sun shields	
Swimming pools	
Tapes	
Tool sheds	
Topping—walk, driveways	
Vent stacks	
Water proofings	
Weather strippings	
Window panes	
Window sash (prime and storm)	
Wire insulations	

*Applications listed in Table 2.

SOURCE: ROSATO, D. V., and R. T. SCHWARTZ, eds., *Environmental Effects on Polymeric Materials*, Vol. II: Materials. New York: John Wiley & Sons, Inc. (Interscience Division), 1968, [1].

TABLE 2. Classes of Plastics Frequently Used for Several Applications from Table 1

Adhesives	
Plywood, laminated timber, and particle boards	*Phenolics*, or *resorcinol formaldehydes*, which have the water resistance of phenolics but can be cured at ordinary temperatures
Wood impregnation	Acrylics, which are polymerized in place to provide hard surfaces to some depth
Concrete repairs	Epoxies
Floorings	
Resilient	*Polyvinyl chloride*, compounded with fillers such as asbestos
Hard floorings	*Epoxies* and *polyesters* with fine aggregate fillers; especially applicable where chemical attack on concrete would occur
Gaskets, sealants	*Polysulfides, silicones, polyvinyl chloride, elastomers, synthetic rubbers*; for window glazes, fillers between curtain wall panels, etc.; must weather well, be watertight, undergo extreme tension and compression without cracking, be capable of application under adverse field conditions
Molded bathrooms	*Polymethyl methacrylate*, i.e., including floor, half-height walls, and fixtures in one piece, upper walls and ceiling in second piece
Piping	
Interior	*Polyvinyl chloride* and *ABS*; most common for water and drainage
Exterior	*Polyethylene*; for service from street to building, irrigation systems
Rooflights, domes	*Polymethyl methacrylate*
Thermal insulation	*Foamed polystyrene, polyurethane*, or *polyurethane* blown with *fluorocarbons*; in planks or foamed in place
Vapor barriers	*Polyethylene*
Wall coverings	*Polyvinyl chloride, polyvinyl fluoride, polyvinylidene fluoride*; for toughness, resistance to wear, weather resistance

A. SYNTHETIC POLYMERS

Tables 1 and 2 indicate various uses of synthetic polymers in building construction (see also Chap. 1, Figs. 13–15). The most commonly used plastics in construction are polyesters, because of their relatively low cost and ease of handling and molding and because their physical properties are adequate for a wide range of construction uses. Epoxies, and, to a much less extent, phenolics, are employed where superior mechanical and weathering properties are required.

Although plastics are used extensively in construction, the types suitable specifically for structural applications are limited. The following properties differentiate plastics from many other structural materials:

1. Plastics generally have nonlinear stress-strain curves, even in the elastic deformation range.
2. The modulus of elasticity is usually low and often differs between tension and compression.
3. Plastics exhibit significant viscoelasticity and creep.
4. The ultimate strength under static loading often reduces with time.
5. They have large thermal expansion coefficients.
6. Ultraviolet (sun) light is harmful to most plastics, although certain fillers and additives can retard deterioration.
7. Most seriously, they have low fire resistance. The fire resistance may be improved with various formulations but usually at considerably extra cost and often with reduced mechanical properties and weathering resistance.

Table 3 lists additional properties important in various construction applications. Table 4 compares selected properties of the major plastics groups, and Table 5 indicates typical construction applications for each group.

TABLE 3. Properties of Plastics Important in Various Construction Applications*

- A. *Physical properties*
  1. *Mechanical properties*
     a. Tensile, compressive, shear, and flexural strengths and moduli
     b. Creep and viscoelastic moduli
     c. Impact strength
     d. Surface hardness: indentation and abrasion resistance
     e. Brittleness (notch sensitivity)
     f. Stress crazing
  2. *Thermal properties*
     a. Coefficients of thermal expansion and conductivity, specific heat
     b. Flammability and ignition
  3. *Optical properties*
     a. Spectral transmission, refractive index, optical uniformity, and distortion
     b. Crazing resistance
  4. *Electrical properties*
     a. Insulation resistance, arc resistance
     b. Dielectric strength and breakdown voltage
  5. *Machinability*
- B. *Chemical properties*
  1. Resistance to acids, bases, solvents, fuels, fungi, salt spray
  2. Water absorption
  3. Effects of sunlight
  4. Toxicity

*See Appendix E for appropriate ASTM tests.

TABLE 4. Selected Properties of Plastics

Property	*ASTM Test method*	*Acrylonitrile-butadiene-styrene*	*Acrylic*	*Cellulosics*	*Epoxies*	*Fluoro-plastics*
Tensile strength, psi	D638–D651	4,000–8,000	7,000–11,000	2,000–9,000	4,000–30,000	2,000–7,000
Elongation, %	D638	2–300	2–10	5–100	0.5–70	80–300
Tensile modulus, 10^6 psi	D638	0.23–1.03	0.35–0.50	0.065–0.60	0.001–3.04	0.05–0.30
Compressive strength, psi	D695	7,000–22,000	11,000–19,000	2,000–36,000	1,000–40,000	1,700–10,000
Compressive modulus, 10^6 psi	D695	0.17–0.39	0.37–0.46	—	—	None to 0.12
Flexural yield strength, psi	D790	5,000–27,000	12,000–17,000	2,000–16,000	1,000–60,000	7,400–9,300
Flexural modulus, 10^6 psi	D790	0.20–1.30	0.39–0.47	—	—	None to 0.20
Hardness, Rockwell	D785	R75–M100	M80–M105	R34–R125	M80–M120	R25–95 (Shore) D50–D80
Impact strength, ft-lb/in. notch	D256	1.0–10	0.3–0.5	0.4–8.5	0.2–10	3.0 to no break
Thermal conductivity, Btu/sq ft/in./hr, °F	C177	1.3–2.3	1.2–1.7	1.1–2.3	1.2–8.7	0.9–1.7
Thermal expansion, 10^{-6}/°F	D696	39–73	28–50	44–111	3–55	25–66
Resistance to heat, continuous, °F		140–230	140–200	115–220	200–550	300–550
Burning rate, in./min	D635	Slow to self-extinguishing	Slow	Self-extinguishing to very fast	Slow to non-burning	None to self-extinguishing
Effect of sunlight	—	None to slight yellowing	None	Slight to discoloration, embrittlement	None to slight	None to slight bleaching
Clarity		Translucent to opaque	Excellent to opaque	Transparent to opaque	Transparent to opaque	Transparent to opaque
Machining qualities		Good to excellent	Fair to excellent	Good to excellent	Poor to excellent	Excellent
24-hr water absorption, ⅛-in. thickness, %	D570	0.2–0.45	0.3–0.4	0.8–7.0	0.08–4.0	0.00–0.04

TABLE 4. (Continued)

Property	*Melamine-formaldehyde*	*Nylon polyamide*	*Phenol-formaldehyde*	*Polycarbonate*	*Polyesters*	*Polyethylene*
Tensiles strength, psi	5,000–13,000	7,000–35,000	3,000–18,000	8,000–20,000	800–50,000	1,000–5,500
Elongation, %	0.30–0.90	10–320	0.13–2.25	0.9–1.30	0.5–310	15–1,000
Tensile modulus, 10^6 psi	1.2–2.4	0.11–1.80	0.25–5.00	0.35–1.85	0.3–2.0	0.014–0.18
Compressive strength, psi	20,000–45,000	6,700–24,000	10,000–70,000	12,500–19,000	12,000–50,000	None to 5,500
Compressive modulus, 10^6 psi	—	0.185–0.248	—	0.3–0.45	—	None to 0.15
Flexural yield strength, psi	9,000–23,000	No break to 17,500	4,000–60,000	13,500–30,000	8,000–80,000	None to 7,000
Flexural modulus, 10^6 psi	—	0.14–1.14	None to 2.4	0.34–1.20	None to 2.0	None to 0.35
Hardness, Rockwell	M110–M125	R108–E75	M37–E101	M70–R118	60(Barcol)–E98	D30(Shore)–R15
Impact strength, ft-lb/in. notch	0.24–6	1.0–5.5	0.2–18	1.2–17.5	0.2–16.0	0.5–2.0 to no break
Thermal conductivity, Btu/sq ft/in./hr/°F	1.9–4.9	1.5–2.5	0.9–6.4	0.7–1.5	1.2–7.2	2.3–3.6
Thermal expansion, 10^{-6}/°F	11–25	7–83	14–33	10–37	7–56	56–195
Resistance to heat, continuous, °F	210–400	175–400	200–550	250–275	250–450	180–275
Burning rate, in./min	Nonburning to very slow	Self-extinguishing to slow burning	None to slow	Self-extinguishing	Slow to nonburning	Slow to self-extinguishing
Effect of sunlight	Slight to darkening	Slight discoloration	Darkens	Slight color change	None to slight yellowing, embrittlement	Unprotected crazes fast, weather resistance available
Clarity	Translucent to opaque	Translucent to opaque	Transparent to opaque	Transparent to opaque	Transparent to opaque	Transparent to opaque
Machining qualities	Fair to good	Fair to excellent	Poor to good	Fair to excellent	Poor to excellent	Fair to excellent
24-hr water absorption, ⅛-in. thickness, %	0.08–0.80	0.4–1.5	0.1–2	0.07–0.20	0.01–1.0	<0.01–0.06

TABLE 4. (Continued)

Property	*Poly-propylene*	*Polystyrene*	*Silicones*	*Urea-form-aldehyde*	*Urethanes*	*Polyvinyls*
Tensile strength, psi	2,900–9,000	1,500–20,000	800–35,000	5,500–13,000	175–10,000	500–9,000
Elongation, %	2–700	0.75–80	None to 100	0.5–1.0	10–1,000	2–450
Tensile modulus, 10^6 psi	0.1–0.9	0.15–1.4	0.0009–3.0	1.0–1.5	0.01–1.0	0.05–0.6
Compressive strength, psi	3,700–8,000	4,000–22,000	100–18,000	25,000–45,000	20,000	1,000–22,000
Compressive modulus, 10^6 psi	None to 0.3	None to 0.53	—	—	0.004–0.1	None to 0.6
Flexural yield strength, psi	5,000–11,000	5,000–26,000	None to 35,000	10,000–18,000	None to 9,000	None to 17,000
Flexural modulus, 10^6 psi	0.125–0.825	None to 1.8	—	1.3–1.6	0.01–0.35	None to 0.4
Hardness, Rockwell	R30–R110	R50–E60	40(Shore)–M95	M110–M120	20A(Shore)–M28	10A(Shore)–M85
Impact strength, ft-lb/in. notch	0.5–20.0	0.25–11.0	None to 15	0.25–0.40	5 to flexible	0.4–20; impact strength varies with type and amount of plasticizer
Thermal conductivity, Btu/sq ft/in./hr/°F	0.6–1.2	0.3–1.0	1.0–3.8	2.0–2.9	0.5–2.1	0.9–20
Thermal expansion, 10^{-6}/°F	16–57	19–117	4–167	12–20	56–112	28–195
Resistance to heat, continuous, °F	190–320	140–220	400–> 600	170	190–250	120–210
Burning rate, in./min	Slow to non-buring	Slow to non-burning	None to slow	Self-extin-guishing	Slow to self-extinguishing	Slow to self-extinguishing
Effect of sunlight	Unprotected crazes fast, weather resis-tance available	Slight yellowing	None to slight	Pastels, gray	None to yellowing	Slight
Clarity	Transparent to opaque	Excellent to opaque	Clear to opaque	Transparent to opaque	Clear to opaque	Transparent to opaque
Machining qualities	Fair to good	Fair to good	Fair to good	Fair	Fair to excellent	Poor to excellent
24-hr water absorption, ⅛-in. thickness %	< 0.01–0.05	0.03–0.6	None to 0.2	0.4–0.8	0.02–1.5	0.02–3.0

SOURCE: DIETZ, ALBERT G. H., *Plastics for Architects and Builders*. Cambridge, Mass.: The MIT Press, 1969 [2].

TABLE 5. Typical Construction Applications of Plastics Groups in Table 4

Acrylonitrile-butadiene-styrene	—
Acrylic	Skylights, light diffusers, glazing, facing panels, domes
Cellulosics	Hardware, handrails, moldings, piping; not for exterior use except in special formulations
Epoxies	Flooring and reinforced plastics, bonding and coating agents
Fluoroplastics	Gaskets and packings, pipe and vessel linings, bearing pads, pipe joints and sleeves, antifriction devices
Melamine-formaldehyde	Laminate surfaces and adhesives for laminates
Nylon polyamide	For fabric, often with other plastics, in air-supported structures
Phenol-formaldehyde	Electric insulation, hardward and adhesives, protective coatings
Polycarbonate	Lamp diffusers, light globes, shatter-resistant glazing
Polyesters	Building panels, skylights, translucent roofing, pipe; usually with glass or other fiber reinforcing
Polyethylene	Laminates, moldings, pipes, film for moisture and weather barriers
Polypropylene	Pipe, fittings, laminates, wire insulation, hard hats
Polystyrene	Foamed sandwich cores, handrails, hardware
Silicones	Moistureproof and thermal-resistant coatings
Urea-formaldehyde	Hardware and handles, laminates and laminate adhesives
Urethanes	Foamed insulation in sandwich panel cores, protective and decorative coatings
Polyvinyls	—

The major groups will be briefly described.

1. Polyesters

Polyesters comprise about 90% of the resins used in reinforced plastics. Advantages include low cost, low viscosity, and good handleability; disadvantages include high curing shrinkage, poor bonding to glass fiber, and limited pot life of the catalyzed resin.

A typical unsaturated polyester may be prepared by reacting an unsaturated dibasic acid, such as fumaric acid, with a glycol, such as ethylene glycol, to give

$$HO(CH_2-CH_2-O-\overset{\overset{\displaystyle O}{\|}}{C}-CH=CH-\overset{\overset{\displaystyle O}{\|}}{O}-O)_n-H$$

The chain structure may be varied by using over 30 monomer acid and glycol reactants [3]. Cross-linking may be accomplished if the acid used is unsaturated, and the degree of cross-linking can be controlled by using various combinations of saturated and unsaturated acids. The temperature at which the resin cures can be controlled by the choice of catalyst.

Fillers, such as clays and carbonates, are generally used to reduce cost and to reduce shrinkage and the tendency to crack and craze during curing. Fillers may be varied from 10 to 70% by weight of final product, 20 to 30% being common. Other additives may include ultraviolet absorbers and flame retarders (see the next section).

The various additives must be compatible with the catalyst used and must not increase the viscosity of the monomer adversely.

2. Epoxies

Epoxy polymers contain two or more epoxy groups

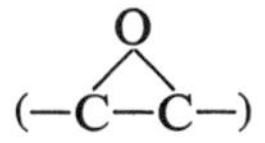

$$(-\overset{\overset{O}{\diagup\ \diagdown}}{C-C}-)$$

Epoxies may be either thermoplastic or thermosetting, depending on the availability of double bonds for cross-linking. Over 10 chemical structures of epoxies exist which may be cured with over 20 curing agents, plus a variety of other additives to modify physical properties. Many epoxies can be handled at room temperature, and the pot life, viscosity, and required cure temperature and time can be widely altered without seriously affecting the resulting properties. They have low shrinkage and adhere well to many fiber reinforcements, especially glass. A typical monomer is diglycidyl ether of bisphenol:

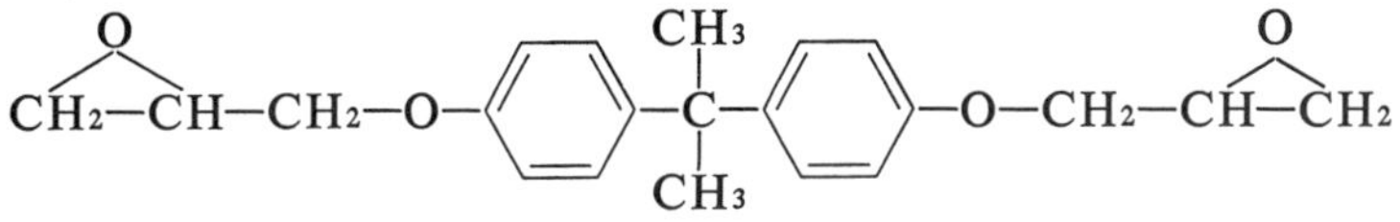

$$\overset{\overset{O}{\diagup\ \diagdown}}{CH_2-CH}-CH_2-O-C_6H_4-\underset{CH_3}{\overset{CH_3}{\overset{|}{\underset{|}{C}}}}-C_6H_4-O-CH_2-\overset{\overset{O}{\diagup\ \diagdown}}{CH-CH_2}$$

It may be polymerized and cross-linked with diethylene triamine (an aliphatic amine) as the curing agent, which becomes part of the network:

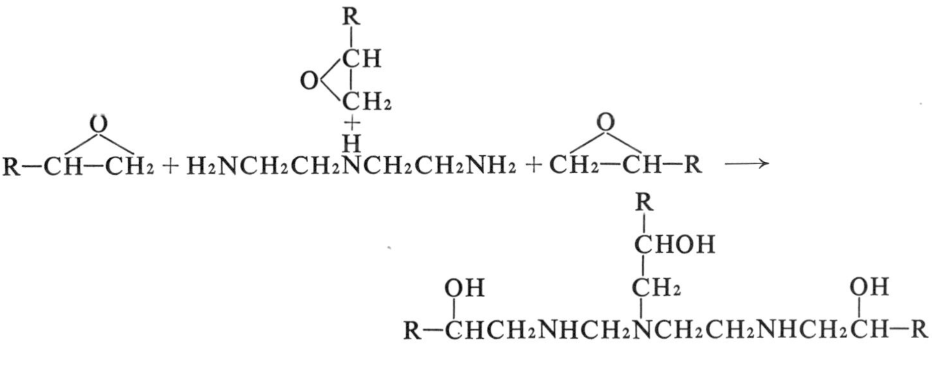

$$R-\overset{\overset{O}{\diagup\ \diagdown}}{CH-CH_2} + H_2NCH_2CH_2\overset{\overset{R-CH(O)CH_2\ +}{H}}{N}CH_2CH_2NH_2 + \overset{\overset{O}{\diagup\ \diagdown}}{CH_2-CH}-R \longrightarrow$$

$$R-\overset{OH}{\overset{|}{C}}HCH_2NHCH_2\overset{\overset{\overset{R}{|}}{CHOH}}{\overset{|}{\overset{CH_2}{\overset{|}{N}}}}CH_2CH_2NHCH_2\overset{OH}{\overset{|}{C}}H-R$$

The reaction occurs at room temperature and is highly exothermic. Other curing agents include aromatic amines, amides, and anhydrides.

Applications of epoxies in construction include all types of high-strength adhesive bonds between similar and dissimilar materials. In building construction, they are used to bond vinyl and steel tile to concrete floors, as a mortar for concrete and glass block walls and windows, and for epoxy terrazzo and exposed aggregate floors and wall panels—using a wide variety of aggregate including glass or marble chips. In highway construction, epoxies are used to repair spalls and potholes, as a bonding agent between old and fresh concrete, for bonding traffic markers to roads, as bridge waterproofing membranes, etc.

As adhesives, epoxies offer the following advantages:

1. Outstanding adhesion to concrete, metals, wood, glass, masonry, and ceramics.
2. A wide range of curing times and temperatures can be achieved.
3. Pressure is often unnecessary, because no volatiles are evolved during curing.
4. They can usually provide high-strength joints between dissimilar materials—and without galvanic corrosion where metals are involved.

Numerous mineral fillers can be used with epoxies in quantities up to about 85% of the product. Frequently used fillers include quartz sand, ground quartz, glass flake, metal powders, graphite, and ground slate.

3. Phenolics and Silicones

Phenolics are the third most widely used composite polymer binder. They are derived from the condensation of phenols (C_6H_5OH) and aldehydes such as formaldehyde, HCHO. Many combinations of these two reactive groups are possible. Phenolics have the advantage of low cost but are extremely brittle, and properties are usually measured for filled (wood flour, cotton flock) polymers. Since, unlike polyesters and epoxies, phenolics are condensation polymers and liberate water during polymerization, they are not easy to handle. Phenolic moldings and laminates are generally cured in high-pressure autoclaves. Phenolics are useful where high strength and heat resistance (to 600°F) are required. Working temperatures to 1000°F are possible with asbestos-reinforced phenolics.

Silicone polymers have chains of the form

$$-\overset{|}{\underset{|}{\mathrm{Si}}}-\mathrm{O}-\overset{|}{\underset{|}{\mathrm{Si}}}-$$

They have excellent electrical resistance, water repellancy, and retention of properties at high temperatures, although room temperature mechanical properties are not comparable to many organic polymers. Both rigid and elastomeric polymers can be prepared. Three methods are used to prepare rigid (highly cross-linked) silicone polymers [4]:

1. Hydrolyzing dimethyldichlorosilane followed by air oxidation.
2. Cocondensing the product of a mixture of dimethyldichlorosilane and silicon tetrachloride.
3. Methylation of silicon tetrachloride followed by hydrolysis.

The first two methods provide better control of the final composition (carbon-oxygen side chains) and are therefore more used. A typical reaction to produce a silicon is

$$x(CH_3)_2SiCl_2 + yCH_2SiCl_3 + zH_2O \longrightarrow \left[\begin{array}{c} -CH_2\ CH_2-CH_2-CH-CH_3 \\ -O-Si-O-Si-O-Si-O- \\ | \\ O \\ | \\ -O-Si-O-Si-O-Si-O- \\ CH_3CH_2-CH_2-CH_2 \end{array}\right]_v + wHC$$

The —O—Si—O— linkage provides heat resistance, while the C—H— linkages provide practical working characteristics of the polymer [4].

4. *Heat-Resistant Polymers*

An important disadvantage of most polymers used in construction is their low heat resistance. Upon heating, polymer integrity may be destroyed either by oxidation or by rupture of carbon-carbon bonds by thermal scission. Oxidation typically occurs at much lower temperatures than thermal scission [1]. The most heat-resistant polymers have been found to be those which have

1. A minimum of readily oxidizable hydrogen atoms and a maximum of rigid structural units, such as aromatic and/or heterocyclic rings, making up the backbone of the polymer chain.
2. No mechanisms by which the chains can readily undergo thermal depolymerization.

An infinite variety of polymers containing stable ring chains can be synthesized, a few of which are shown in Fig. 1. The polyimides are currently one of the most heat-resistant commercially available polymers. They withstand long-term aging at 600°F and are still stable but soften at 700°F.

The predominantly or completely aromatic cyclic structures provide extremely rigid chains, which resist thermoplastic behavior at elevated temperatures. Such structures allow most of the benefits of highly cross-linked polymers without the extreme brittleness caused by cross-linking. In addition, aromatic chains permit higher packing densities and hence greater strengths. Most importantly aromatic C—H and N—H bonds are much less prone to oxidation than aliphatic C—H or N—H bonds. It has been postulated that the resonating double bonds in aromatic structures increase oxidative stability by serving as energy sinks [4].

A major difficulty in using aromatic chain polymers is in obtaining formable or soluble intermediates which can be easily fabricated into useful shapes. All aromatic polymers are characterized by high melt temperatures and low solubilities. Another difficulty is that all polymerizations are condensation reactions and evolve volatile by-products which must be removed to prevent porous structures which would allow more rapid oxidation.

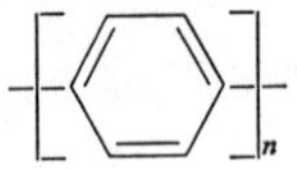

polyphenylene

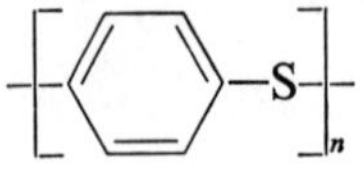

polyphenylene sulfide

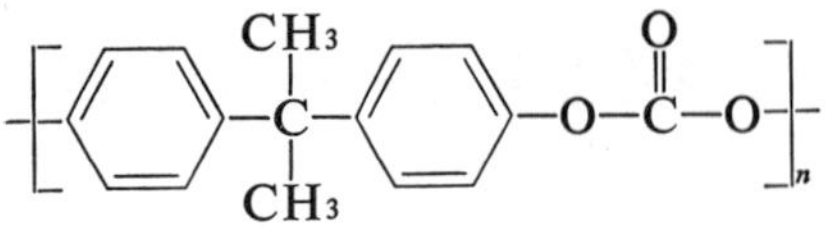

polycarbonate

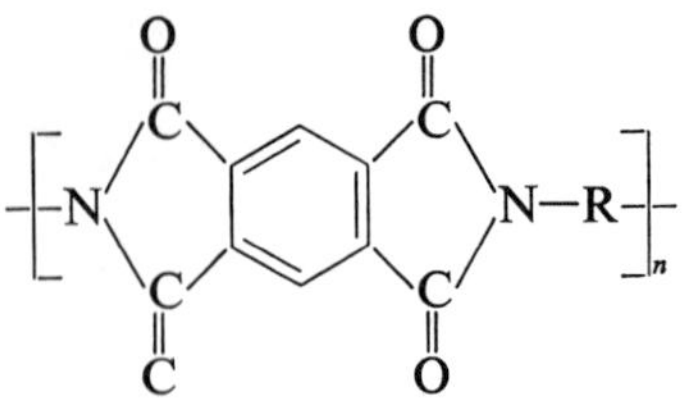

polyimide

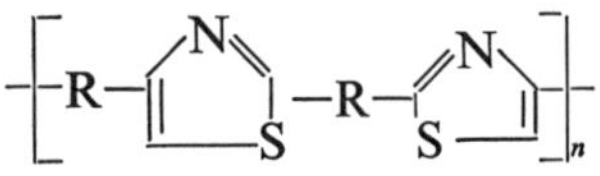

polythiazole

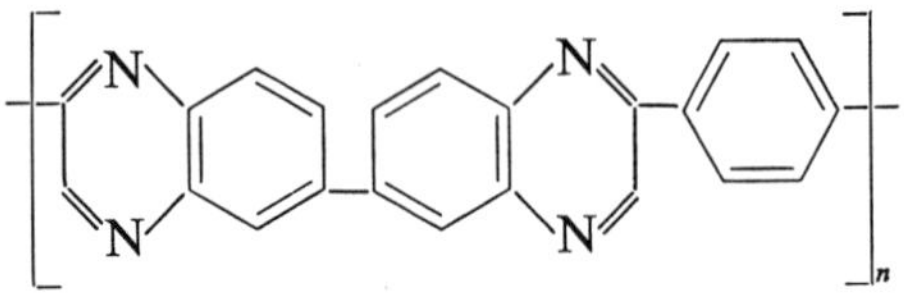

polyquinoxaline

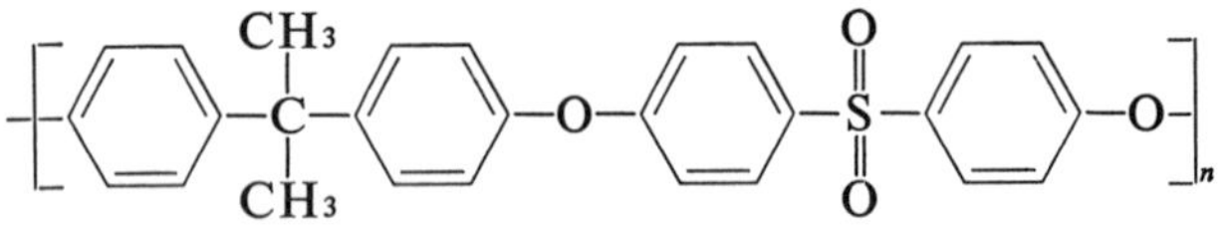

polysulfone

Fig. 1. Examples of heat-resistant polymer structures.

B. ASPHALTS AND TARS

Bitumens are hydrocarbons which are soluble in carbon disulfide (CS_2), including materials ranging from hydrocarbon gases to bituminous coal. They are characterized primarily by their molecular weight distributions and by the carbonhydrogen ratios in the various molecular weight fractions. Bitumens used in construction include tar and asphalt. Tar is a condensate from the destructive distillation of coal, petroleum, or wood. Most asphalt is obtained from petroleum refining, although some is processed directly from bitumen-bearing rocks (rock asphalt) and from surface deposits (lake asphalt).

The asphalt content of crude oils normally ranges from 10 to 60%. Asphalt-bearing rocks such as the limestones and sandstones found in Kentucky, Oklahoma, Utah, and California contain from trace amounts to 20% of asphalt by weight. The rocks are usually crushed, mixed with other asphalt or aggregate, and used for paving. Asphalt extracted from rocks cannot currently compete economically with petroleum asphalts. In the United States approximately 75% of the asphalt is used for paving and most of the remainder for roofing and building construction.

Chemically, tar molecules tend to be more aromatic and unsaturated, and asphalt molecules more aliphatic and saturated. Physically, the chief difference is that tars are more temperature-sensitive than asphalts, becoming softer in summer and more brittle in winter. But tars generally coat mineral aggregates more readily than asphalts do, especially when the aggregates are either wet or dusty. Tar coatings are also more permanent in the presence of water. Hence tars are widely used for applications below ground. Tars are also generally more chemically resistant than asphalts. For example, they are less affected by jet fuel spillage at airports, though more susceptible to damage by jet heat blast. Tars generally undergo hardening in service more rapidly than asphalts, a disadvantage when used in exposure to atmosphere and sunlight. Factors which contribute to the hardening of both asphalts and tars include evaporation of lighter-molecular-weight volatiles, polymerization, and oxidation. Table 6 summarizes the major chemical and physical differences of asphalts and tars.

TABLE 6. Major Chemical and Physical Differences of Asphalts and Tars

Asphalt	*Tar*
Chemical Differences	
More aliphatic and saturated	More aromatic and unsaturated
Poor resistance to many organic solvents	Good resistance to most organic solvents
Less temperature-susceptible	More temperature-susceptible
Physical Differences	
Poor coating and retention on wet or dusty aggregate	Good coating and retention on wet or dusty aggregate
Good weathering in thin films	Tendency to crack in thin films

1. Asphalt Chemical Composition and Physical Properties

Asphalts are composed of hydrocarbons having the general formula $C_nH_{2n+b}X_d$, where X represents such elements as sulfur, nitrogen, oxygen, or trace metals; d is usually small, and b may be negative. The elemental composition generally lies within the following ranges:

Carbon	80–85%
Hydrogen	9–10%
Oxygen	2–8%*
Nitrogen	0–1%
Sulfur	0.5–7%
Trace metals (iron, nickel, vanadium, calcium)	0–0.5%

The components vary from low-molecular-weight to high-molecular-weight polymers and condensation products, with isomers, ring structures, condensed rings, and chain hydrocarbons. Typical asphalts contain approximately one half of the carbon in ring structures. The rings may be condensed aromatic nuclei or five- or six-member rings connected by side chains. Sulfur, nitrogen, and oxygen tend to concentrate in the aromatic and higher-molecular-weight fractions, where they commonly form bridges between rings or between rings and side chains [5]. Purely aliphatic components

*Depends on production method, as well as source.

unattached to rings are quite rare. The rings may be either saturated (napthenes) or unsaturated (aromatics), and the chains may also be saturated (paraffins) or unsaturated (olefins). Olefins do not occur in significant quantities, however, except in cracked asphalts.

For explaining the physical behavior of asphalts, the various molecular structures are typically classified as oils, resins, and asphaltenes, shown on the triangular chart in Fig. 2. The three groups are distinguished by the following elemental composition ranges:

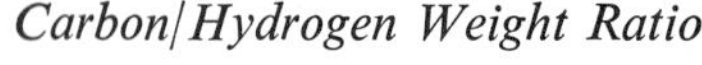
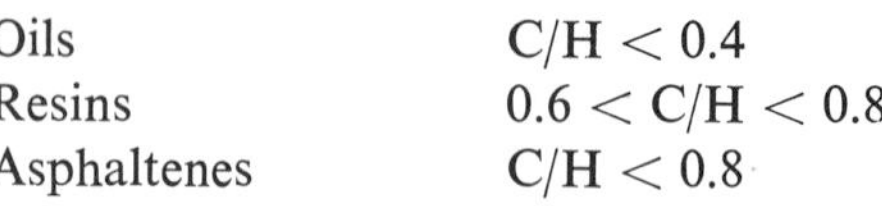

	Carbon/Hydrogen Weight Ratio
Oils	C/H < 0.4
Resins	$0.6 <$ C/H < 0.8
Asphaltenes	C/H < 0.8

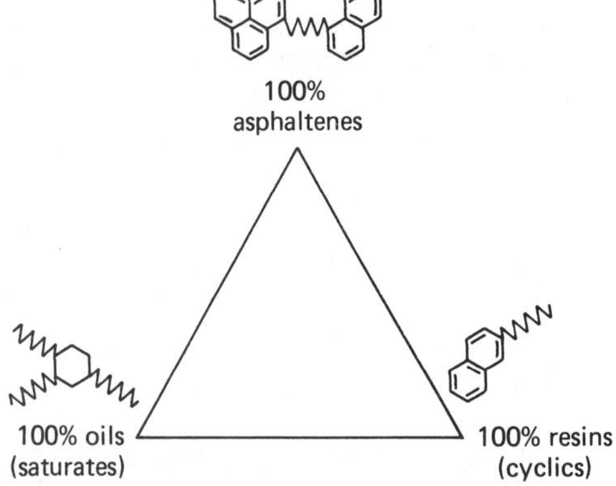

Fig. 2. Asphalt composition.

Structurally, the oils consist primarily of naphthenic rings saturated with long side chains. Resins contain primarily aromatic rings with many double bonds and a minimum of side chains. Asphaltenes have the highest molecular weight and are the most aromatic of all, containing many condensed rings.

The molecular species can be separated and analyzed by a variety of methods and reconstituted to study the effects of chemical composition on physical behavior [6]. Commonly used separation techniques include

Process	*Purpose*
Distillation	Separation by molecular size
Silica gel chromatography:	To separate saturates and resins:
With methyl cyclohexane washing	To remove saturates
With 1-butanol washing	To remove resins
Urea-complex formation	To separate long chain paraffins
Alumina chromatography	To separate monocyclic aromatics
Peroxide oxidation followed by chromatography	To remove theophene analogs
Thermal diffusion in the liquid state	To separate naphthenes on the basis of ring number

Refractive indices and ultraviolet spectrography are the two principal methods used to study the fractions obtained.

Although the dependence of physical properties on chemical structure of asphalts is still poorly defined, generalizations can be made (Table 7). Unsaturation favors adhesion because most aggregates are polar; saturation favors ductility and temperature susceptibility due to the absence of double bonds to restrict rotational freedom of the carbon chains. In general, low-molecular-weight fractions contribute to adhesiveness and ductility but are temperature-susceptible. Intermediate-molecular-weight fractions are more plastic and less temperature-susceptible, and the highest-molecular-weight fractions may no longer be colloidally dissolved but exist like finely divided filler, contributing rigidity and shear strength but not adhesiveness.

TABLE 7. Dependence of Physical Properties on Chemical Structure of Asphalt

Physical property	*Causative chemical structure*
Adhesion to mineral aggregates	Low molecular weight and unsaturation (resins)
Ductility (deformation without cracking)	Low-molecular-weight saturates (oils)
Temperature susceptibility (viscosity change with temperature)	Low-molecular-weight saturates
Rigidity and shear resistance	Asphaltenes

Rheological properties, important during mixing and coating of aggregates, depend primarily on the relative proportions of the two lower-weight fractions, resins and oils. Resins are chemically more similar to asphaltenes than are oils (due to the predominance of unsaturated ring structures) and, therefore, dissolve asphaltenes better than do oils. Resins tend to form sols, and oils tend to form gels. In sols the asphaltenes are mostly dissolved in the lower-molecular-weight fraction. In gels the asphaltene micelles form a continuous rigid structure through which the smaller molecules may flow with relative freedom. Sols are also favored by a smooth grading curve in molecular weights and gels by gap grading between the larger and smaller size molecules. For example, a gel-type asphalt may be produced by blending a hard (high-molecular-weight) asphalt with light oil, which allows only limited mutual

TABLE 8. Dependence of Physical Properties on Oil/Resin Ratio

Composition and rheology	*Physical properties*
Oil predominates (gel structure)	Elasticity and high resiliency (recoverable deformation) which may be further increased by the presence of many long side chains on the asphaltene ring structures; less temperature-susceptible and less ductile than sols
Resin predominates (sol structure)	More viscoelastic than gels, or almost purely Newtonian

solubility. Table 8 shows the physical manifestations of oils or resins predominating to produce gels or sols. Most asphalts exhibit properties intermediate between gels and sols and become more sol-like at higher temperatures.

2. Property Changes During Refining

Asphalt is a by-product of the refining of petroleum for gasoline, kerosene, fuel oils, lubricating oils, and related products. Figure 3 shows the fractional distillation processes by which this is usually accomplished. The refining processes for producing various asphalts include

1. Straight-run distillation.
2. Vacuum and steam reduction.
3. Solvent extraction.
4. Air blowing.
5. Cracking.
6. Blending of products from the preceding operations.

No sharp separation of these methods is made in asphalt manufacture in general practice.

When distillation is adjusted so that overheating and subsequent chemical change do not occur, the residue is a *straight-run asphalt.* This is accomplished by heating the crude in a tube still to 550°F and releasing it into the bottom of an atmospheric fractionating column. Traps catch the various boiling temperature fractions after rising to a position in the column where the temperature causes condensation.

Most crudes contain significant volatiles in molecular size ranges sufficiently close to those of asphalt that they cannot be volatilized at atmospheric pressure without heating to temperatures which cause chemical changes in the asphalt. For this reason a second fractionation is often performed under partial vacuum to lower boiling temperatures (Fig. 3). Steam introduced at the bottom of the column further lowers boiling temperatures since the boiling point of each fraction becomes a combination of the boiling points of that fraction and of water. The residue is called a *steam-* or *vacuum-reduced asphalt.* Asphalt quality is controlled by varying temperature, vacuum pressure, and proportion of steam. Harder asphalts are obtained by increasing any of these three variables, so that greater amounts of volatiles are removed.

Solvent-extracted asphalt is produced by adding selective solvents such as proponal or furfural to separate paraffinic oils from the asphaltic napthenes and aromatics. The oils are dissolved by the solvent and come overhead in the fractionating chamber. The process, therefore, accomplishes separation by chemical type, rather than by boiling point alone, as in straight-run and vacuum distillations. The residual asphalt is then steam-stripped to remove any remaining solvent.

Air-blown asphalt is produced by blowing air through a perforated pipe into the bottom of a still containing hot asphaltic stock. The stock may be anything from a

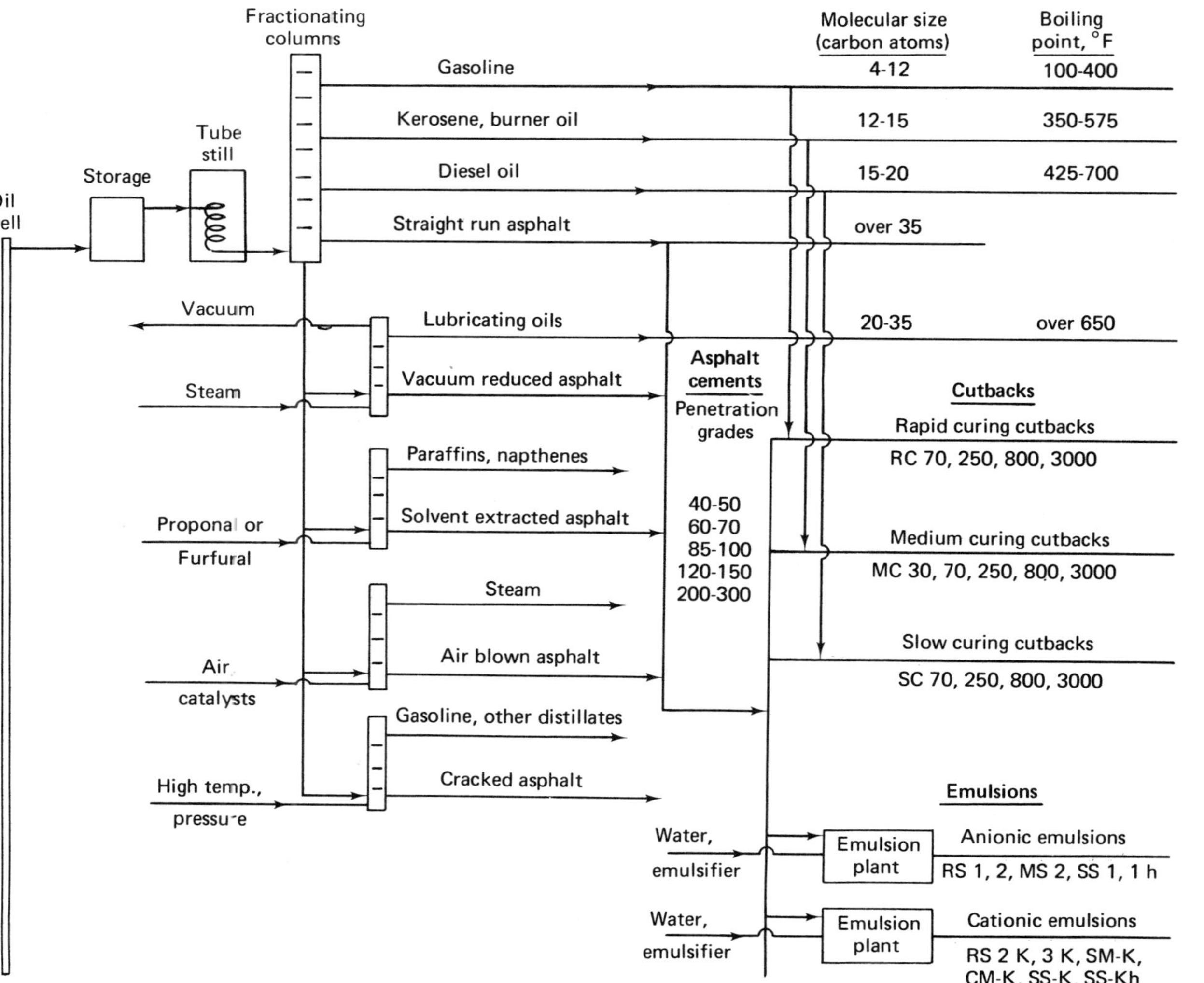

Fig. 3. Asphalt refining flow chart.

residual asphalt to a heavy distillate oil. Properties are controlled by the charging stock, temperature, and blowing duration. Oxygen from the air combines exothermically with hydrogen in the hydrocarbons to form steam, which is bled off the top of the still. The hydrocarbons condense through the resulting double bonds to form larger molecules. The main advantage of air blowing is that the asphalt is less temperature-sensitive. This is especially desirable for roofing asphalts, which are exposed to temperature extremes. By adding catalysts such as ferric chloride or phosphorous pentoxide the asphalt can be further modified with the following advantages:

1. Higher softening point for a given penetration (hardness).
2. Tougher and more cohesive.
3. More ductile at low temperatures, for a given softening point.
4. Materials with desired penetration-softening point relationships, with higher flash points and less tendency to exude oils, can be blown from more viscous residual stocks [7].

Air-blown asphalts, with or without catalysts, are generally less adhesive than straight-run asphalts, however, and are, therefore, less suitable for paving, where bonding to aggregates under traffic loads is essential. They are used extensively for roofing and waterproofing and coatings and are sometimes blended with other asphalts for paving uses.

The yield of higher-priced light distillates such as gasoline can be increased by *cracking* or degrading larger molecules primarily by splitting off the aliphatic side chains which become gasoline. Cracking operations involve either the liquid or gaseous phase at high pressure and temperature (450–600°C). The remaining primarily aromatic residue condenses and polymerizes to form *cracked asphalt*. It is less weather-resistant than straight-run asphalt, has properties more resembling tar (see Table 6), and is used primarily in applications for tars: roofing, subgrade stabilization, fiber soil pipes, insulation board, and building products.

Comparison of asphalts produced by different methods. Asphalt selection for a particular use requires comparing chemical and physical properties resulting from the various refining methods. Table 9 summarizes these properties, and Fig. 4 illustrates their effect on the triangular composition chart for a hypothetical charge stock. The exact slopes and lengths of arrows indicating compositional changes depend in each case on the charge stock and operating conditions. Table 10 compares softening point, penetration, and ductility for asphalts produced by different methods from the same crude [6].

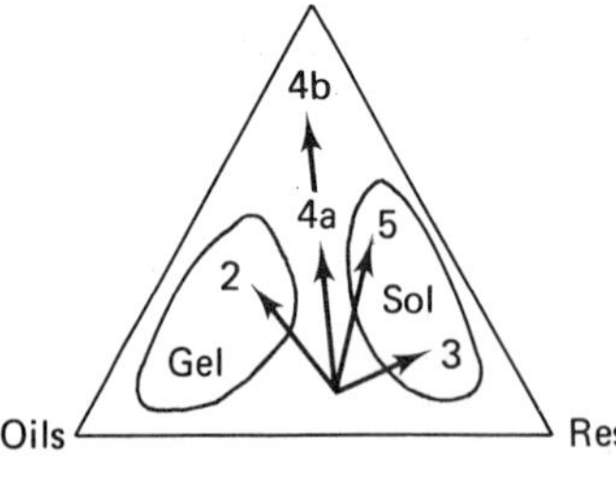

1. Straight run
2. Vacuum reduced
3. Solvent extracted
4a. Air blown
4b. Air blown with catalyst
5. Cracked

Fig. 4. Effect of refining operations on asphalt composition.

TABLE 9. Effect of Refining Process on Chemical and Physical Properties of Asphalts

Refining process	*Chemical change*	*Physical effect*	*Typical uses*
1. Straight-run	None	None	Paving
2. Steam or vacuum reduction	Increases unsaturation and molecular weight due to condensation	Leaves asphalt ductile but temperature-susceptible	Paving
3. Solvent extraction	Removes paraffins	Produces sol-type asphalt	When mixing and coating of fine aggregates is a problem
4. Air blowing	$O_2^{+} + (—CH_2—CH_2—) \longrightarrow H_2O + (—CH{=}CH—)$; Resulting hydrocarbons condense, polymerize to form harder asphalt	Less temperature-susceptible; lower ductility and tendency to exude oils; higher flash point; tendency to crack rather than knead under traffic loads	Roofing
5. Cracking	Degradation and condensation to form larger, more aromatic molecules	More resembles tars than asphalts	All typical uses for tars

TABLE 10. Physical Properties of Asphalts Produced by Different Methods

Refining process	*Softening point (°C)*	*Penetration at 25°C (0.1 mm)*	*Ductility at 22°C (cm)*
Vacuum reduction	98	8	0
Blowing	103	22	3
Solvent extraction	94	2	0

SOURCE: ZAKAR, PAL, *Asphalt*. New York: Chemical Publishing Company, Inc., 1971 [6].

Due to the costs incurred in changing operating conditions and in storing different grades of asphalts, the usual refinery practice is to produce only two or three standard asphalt grades and to blend these to meet customer specifications.

3. Asphalt Classification and Uses

In this section we shall outline the more important aspects of asphalt selection. Selecting the best type and grade of asphalt for a specific application requires careful consideration of numerous factors related to materials handling, mixing with aggregate, transporting, placing and compaction, and service performance. As a general summary, Table 11 shows typical uses of the three asphalt types—cements, cutbacks, and emulsions—shown in Fig. 3. Tables 12a and b show relative production of the three asphalt types in the United States and relative consumption of all asphalts used in paving construction in the United States [6].

TABLE 11. Recommended Uses of Asphalt Cements, Cutbacks, and Emulsions

Type of construction	*Asphalt cements*					*Cutbacks*													*Emulsions*										
						RC				*MC*					*SC*				*Anionic*					*Cationic*					
	40–50	*60–70*	*85–100*	*120–150*	*200–300*	*70*	*250*	*800*	*3000*	*30*	*70*	*250*	*800*	*3000*	*70*	*250*	*800*	*3000*	*RS-1*	*RS-2*	*MS-2*	*SS-1*	*SS-1h*	*CRS-1*	*CRS-2*	*CMS-2S*	*CMS-2*	*CSS-1*	*CSS-1h*
Asphalt plant mix, hot-laid																													
Highways		x	x	x										x				x											
Airports, parking areas, driveways, curbs		x	x																										
Industrial floors, asphalt block	x																												
Plant mix, cold-laid																													
Road mix													x				x					x						x	x
Open-graded aggregate							x	x					x	x							x						x		
Dense-graded aggregate							x	x				x	x			x	x				x	x				x		x	
Clean sand							x	x				x	x										x					x	x
Sandy soil						x	x	x				x	x			x	x					x				x		x	x
Penetration macadam																													
Large voids			x					x	x											x					x				
Small voids				x			x												x					x					
Surface treatments																													
Aggregate seal				x	x		x	x	x			x	x	x					x	x		x		x	x			x	
Sand seal							x	x				x	x						x			x	x	x				x	x
Prime coat, open surface							x					x																	
Prime coat, tight surfaces						x				x	x				x														
Tack coat						x													x			x	x	x				x	x
Dust laying										x	x				x							x						x	
Patching mix																													
Immediate use												x				x													
Stock pile												x	x		x	x										x	x		
Hydraulic structures																													
Canal and reservoir linings, dam facings	x	x																											
Crack filling							x												x			x	x	x				x	x
Expansion joints, roofing	Blown asphalts																												

NOTATION: *Asphalt cements.* 40–50, 60–70, etc.: penetration grade (range of penetration values).
Cutbacks. RC, MC, and SC: rapid, medium, slow curing, respectively.
70, 250, etc.: lower bound of kinematic viscosity, with upper bound double the figure indicated.
Emulsions. RS, MS, SS: rapid, medium, slow setting, respectively.
C: cationic.
1, 2: viscosity grades.

SOURCE: *The Asphalt Handbook* (Series MS-4). College Park, Md.: The Asphalt Institute, 1965 [8].

TABLE 12a. U.S. Asphalt Production

Type	*Approximate amount* (%)
Cements	36
Cutbacks and emulsions	32
Blown asphalts	24
Other types	8

TABLE 12b. U.S. Asphalt Consumption for Paving

Type	*Approximate amount (%)*
Cements	66
Cutbacks	23
Emulsions	11

Asphalt cement comes primarily from petroleum refining, described in Sec. 2, and classification is based upon penetration grade. The *penetration* is the distance in tenths of a millimeter that a standard needle penetrates the asphalt in 5 sec under a 100-g load at 77°F (Fig. 5).* Specifications for the various penetration grades used in paving have been prepared by the American Association of State Highway Officials and the ASTM (Table 13). Lower penetration grades are usually used to avoid soften-

TABLE 13. ASTM Requirements for Asphalt Cement Used in Paving

	Penetration grade									
	40–50		*60–70*		*85–100*		*120–150*		*200–300*	
	Min	*Max*	*Min*	*Max*	*Min*	*Max*	*Min*	*Max*	*Min*	*Max*
Penetration at 77°F (25°C) 100 g, 5 sec	40	50	60	70	85	100	120	150	200	300
Flash point, °F (Cleveland open cup)	450	—	450	—	450	—	425	—	350	—
Ductility at 77°F (25°C) 5 cm/min, cm	100	—	100	—	100	—	100	—	—	—
Retained penetration after thin-film oven test, %	55+	—	52+	—	47+	—	42+	—	37+	—
Solubility in trichloroethylene, %	99.0	—	99.0	—	99.0	—	99.0	—	99.0	—

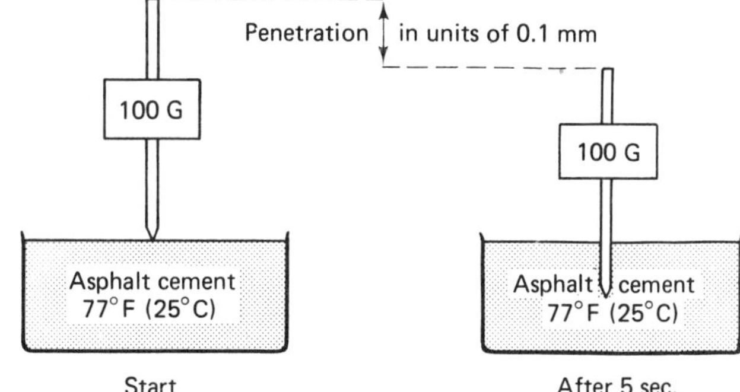

Fig. 5. Standard penetration test. From the Asphalt Handbook (Series MS-4). College Park, Maryland: The Asphalt Institute, 1965 [8].

*Appendix E contains the ASTM designations for asphalt tests.

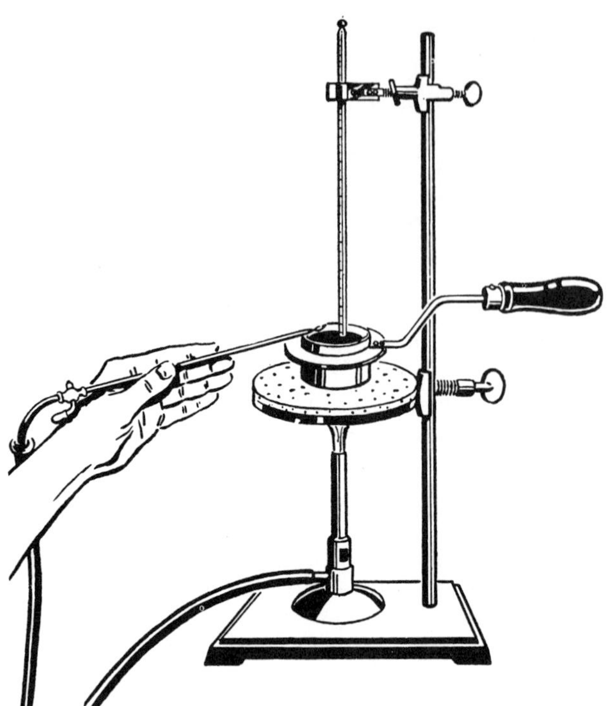

Fig. 6. Cleveland open cup flash point test. From The Asphalt Handbook (Series MS-4). College Park, Maryland: The Asphalt Institute, 1965 [8].

ing in warmer climates and under heavier traffic loads. Conversely, higher penetration grades are used in northern climates to reduce brittle cracking in cold weather. Most asphalt cements are used in hot mix bituminous concrete, but the highest penetration grade is also used for spray applications.

In Table 13, the *flash point* is the temperature at which asphalt gives off sufficient hydrocarbon vapors to ignite when exposed to an open flame immediately above the sample. Figure 6 illustrates the Cleveland open cup type of flash point test, in which the container of asphalt is heated directly with a Bunsen burner. Flash points are normally above 350°F for native asphalts and soft petroleum asphalts and above 450°F for harder petroleum asphalts. It is considered good practice not to heat asphalt cements above about 50°F below their flash points.

Ductility is the distance in centimeters to which a 1-cm^2 cross-section briquet of asphalt can be elongated at a specified rate in a constant temperature bath before breaking. The test is normally conducted at 77°F and 5 cm/min elongation rate. High ductility indicates greater adhesiveness but also greater temperature sensitivity, and the usefulness of the results obtained is controversial. Where asphalts are used in thin films, such as a matrix for paving aggregate, adhesiveness is more important, while in thick films, such as crack and joint filling, low-temperature susceptibility is more important.

The *retained penetration* after a thin film oven test indicates the hardening tendency of asphalt when heated for application. Specifications prescribe a minimum ratio, in percent, of the penetration values before and after the oven treatment. The *solubility* in trichlorethylene specification guards against the inclusion of undesirable amounts of foreign materials, particularly in native asphalts.

In addition to the above requirements, some states specify a maximum temperature susceptibility, T_s, for paving asphalts, by the relationship

$$T_s = \frac{\text{penetration, 77°F}}{\text{penetration, 32°F}}$$

or some similar quotient of penetration values. Low-temperature susceptibility assures that the asphalt will withstand temperature extremes without undue softening at high temperatures and brittleness at low temperatures.

Asphalt cutbacks are liquid asphalts made by adding lighter petroleum distillate as solvents for asphalt cement. The advantage is that cutbacks can be mixed with cold aggregate and at lower asphalt temperatures. After the mix is in place, the cutback reverts to asphalt cement by curing, i.e., by evaporation of the volatile solvent. *Rapid curing* (RC) cutbacks are made by dissolving asphalt cement of about 85–100 penetration in naptha or gasoline. Careful control of the solvent keeps the flash point of the cutback above about 80°F. *Medium curing* (MC) cutbacks contain approximately 100–200 penetration asphalt dissolved in kerosene. Thus RC cutbacks consist of hard asphalt in a solvent which evaporates rapidly at low temperature. MC cutbacks consist of softer asphalt in a less volatile solvent and cure more slowly.

Either the asphalt or solvent might be the primary basis for selection. For example, in cold climates the softer asphalt in an MC cutback will be less brittle and subject to cracking in winters. Also, when asphalt deteriorates to a penetration value less than 40, it becomes very susceptible to cracking, and an MC cutback takes longer to deteriorate to this condition. Conversely, in warm climates an RC cutback may be used to provide a harder asphalt to give a mix with better high-temperature stability. In these situations the difference in curing rate was ignored [10]. In other situations it may be important that the cutback cure rapidly in order to carry traffic quickly or not be damaged by a rainstorm or that it cure slowly to allow sufficient time to roll the stone, for example, in a surface treatment.

Slow curing (SC) cutbacks are made by dissolving asphalt cement in higher boiling point distillates such as lubricating oils. *Road oils*, resembling SC cutbacks, are produced by straight-run distillation in the manner of asphalt cements but with lower-temperature distillation so that some of the lubricating oils remain in the residue. SC cutbacks and road oils are used primarily in the drier western states for dust laying, road mixes, stockpiled patching mixes, and for secondary roads where an oil mat is periodically scarified and relaid. SC cutbacks do not have the fire and toxicity hazards of RC and MC. Also, gasoline and kerosene solvents are expensive and add nothing to the finished product.

Cutbacks are graded by viscosity. The number in the grade designation (Table 11) signifies the lower limit of kinematic viscosity. The upper viscosity limit is twice the lower limit. For example, RC 250 is an RC cutback with viscosity in the range 250–500 centistokes at 140°F. Low-viscosity materials are required for priming and for fine or dusty aggregate and more viscous materials may be used with coarser aggregate.

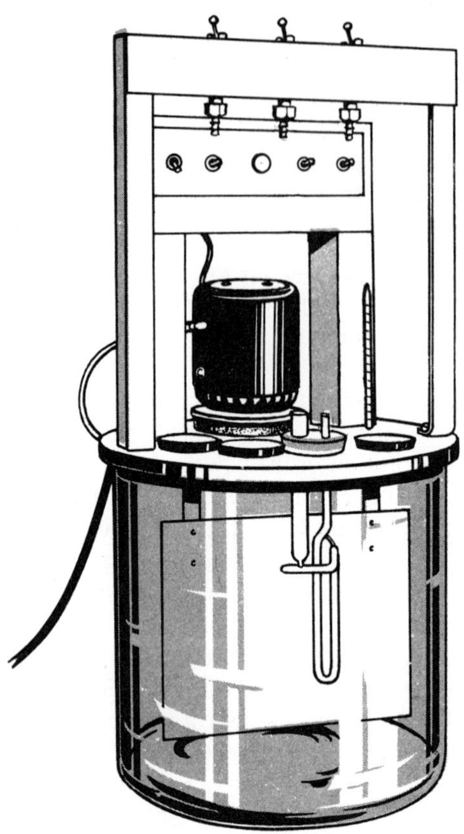

Fig. 7. Kinematic capillary viscosity test. From The Asphalt Handbook (Series MS-4). College Park, Md.: The Asphalt Institute, 1965 [8].

Table 14 lists the specifications for RC cutback. Specifications for the MC and SC cutbacks differ in minor respects.

Additional tests not described for asphalt cements include kinematic viscosity and distillation. Kinematic viscosity, in stokes, is the absolute viscosity divided by density, both at the same temperature. The test is performed with a gravity flow capillary viscometer in a constant temperature bath (Fig. 7).

In the distillation test (Fig. 8) cutback is slowly heated to 680°F, and the amount of distillate driven off at various temperatures is recorded. The results indicate the proportions of asphalt and solvent, and the distillation temperature distribution of the solvent, which provides a clue to the relative field curing rate. The residual asphalt is usually tested for penetration, ductility, and solubility, although these values generally differ somewhat from those of the initial asphalt cement due to the high final distillation temperature.

Asphalt emulsions are made by dispersing asphalt droplets in water, usually with a colloid mill or homogenizer. A high-speed rotor moving within the fixed stator of a colloid mill provides the high shear required to break asphalt into 1–2-μ diameter droplets which become stabilized by an emulsifier added to the water phase.

Asphalt emulsions "break" or demulsify on contact with aggregate surfaces, and the water phase then erodes and evaporates. Emulsions have the added advan-

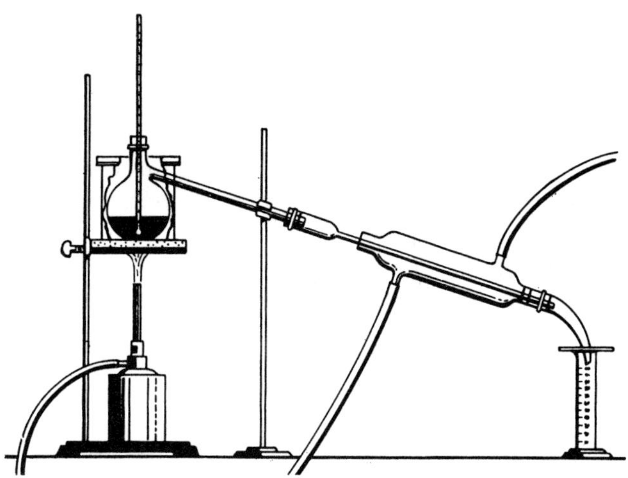

Fig. 8. Distillation test. From The Asphalt Handbook (Series MS-4). College Park, Md.: The Asphalt Institute, 1965 [8].

TABLE 14. ASTM Requirements for RC Cutbacks

	RC-70		*RC-250*		*RC-800*		*RC-3000*	
Designation	*Min*	*Max*	*Min*	*Max*	*Min*	*Max*	*Min*	*Max*
Kinematic viscosity at 140°F (60°C), cSt	70	140	250	500	800	1,600	3,000	6,000
Flash point (Tag open cup), °F (°C)	—	—	80+ (27+)	—	80+ (27+)	—	80+ (27+)	—
Distillation test:								
Distillate, volume % of total distillate to 680°F (360°C):								
To 374°F (190°C)	10	—	—	—	—	—	—	—
To 437°F (225°C)	50	—	35	—	15	—	—	—
To 500°F (260°C)	70	—	60	—	45	—	25	—
To 600°F (316°C)	85	—	80	—	75	—	70	—
Residue from distillation to 680°F (360°C). % volume by difference	55	—	65	—	75	—	80	—
Tests on residue from distillation:								
Penetration at 77°F (25°C), 100 g, 5 sec	80	120	80	120	80	120	80	120
Ductility at 77°F (25°C), cm	100	—	100	—	100	—	100	—
Solubility in trichloroethylene, %	99.0	—	99.0	—	99.0	—	99.0	—
Water, %	—	0.2	—	0.2	—	0.2	—	0.2

SOURCE: *Specifications for Asphalt Cements and Liquid Asphalts* (SS-2). College Park, Md.: The Asphalt Institute, 1974 [9].

tages over cutbacks that they require no heating of the asphalt, they can be used with wet aggregates as well as cold aggregates, fire and toxicity hazards are less, and they generally set faster and more completely than cutbacks, water being volatized more readily than the light distillate solvents. A fresh emulsion-aggregate mix may, however, be damaged by washing due to rain.

Rapid-, medium-, and slow-setting (RS, MS, SS) emulsions are produced primarily by varying the amount and type of emulsifier used and secondarily by the asphalt/water ratio and the penetration grade of asphalt. RS grades break easily and are recommended for penetration macadam, seal coats, and surface treatments. The lower-viscosity grade, RS-1, is used where thinner applications are required. MS grades are used for clean, coarse aggregate mixtures (they can also be used in hot mixes). SS grades are used where high mixing stability is required, such as for aggregates with large amounts of dust and for soil stabilization.

Most emulsifying agents are surfactants (see Sec. D in Chap. 2), having a polar group attached to an organic molecule to make them soluble in both the oil and water phases. Some finely divided powders are also used as emulsifiers.

Some emulsions have water droplets in a continuous asphalt phase. To obtain sufficiently low viscosity these *inverted* emulsions require a cutback (usually an RC is used) as the oil phase rather than a soft asphalt cement, as for oil-in-water emulsions. Inverted emulsions are sometimes more effective in coating highly hydrophilic aggregates. Inverted emulsions are obtained with surfactants which have bulky and highly soluble hydrophobic groups, which induce concavity of the interface toward the water phase due to crowding of the surfactant monolayer in the oil phase. A larger proportion of asphalt to water also promotes the formation of inverted emulsions.

The hydrophilic portion of the surfactant may be either anionic or cationic (Fig. 9). Anionic surfactants are effective in causing asphalt droplets to demulsify

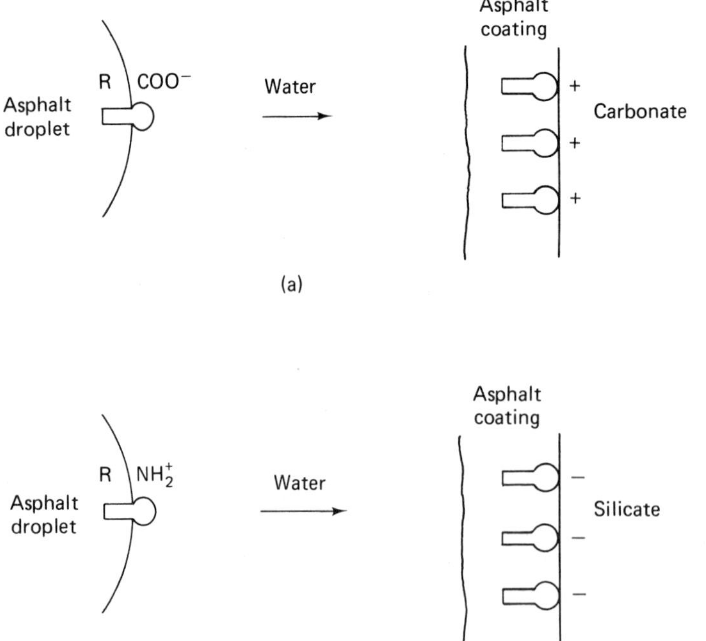

Fig. 9. Aggregate coating by asphalt emulsions. (a) Carbonate plus anionic emulsifier; (b) silicate plus cationic emulsifier.

TABLE 15. ASTM Requirements for Emulsified Asphalt

Type	Rapid-setting				Medium-setting						Slow-setting			
Grade	RS-1		RS-2		MS-1		MS-2		MS-2h		SS-1		SS-1h	
	Min	Max	Min	Max	Min	Max	Min	Max	Min	Max	Min	Max	Min	Max
Tests on emulsions:														
Viscosity, Saybolt Furol at 77°F (25°C), sec	20	100			20	100	100		100		20	100	20	100
Viscosity, Saybolt Furol at 122°F (50°C), sec			75	400										
Settlement, 5 days, %		5		5		5		5		5		5		5
Storage stability test, 1 day		1		1		1		1		1		1		1
Demulsibility, 35 ml, 0.02 *N* $CaCl_2$ solution, %	60		60											
Coating ability and water resistance:														
Coating, dry aggregate					Good		Good		Good					
Coating, after spraying					Fair		Fair		Fair					
Coating, wet aggregate					Fair		Fair		Fair					
Coating, after spraying					Fair		Fair		Fair					
Cement mixing test, %												2.0		2.0
Sieve test, %		0.10		0.10		0.10		0.10		0.10		0.10		0.10
Residue by distillation, %	55		63		55		65		65		57		57	
Tests on residue from distillation test:														
Penetration, 77°F (22°C), 100 g, 5 sec	100	200	100	200	100	200	100	200	40	90	100	200	40	90
Ductility, 77°F (25°C), 5 cm/min, cm	40		40		40		40		40		40		40	
Solubility in trichloroethylene, %	97.5		97.5		97.5		97.5		97.5		97.5		97.5	
Suggested uses	Surface treatment, penetration macadam and tack coat		Surface treatment and penetration macadam		Plant or road mixture with coarse aggregate substantially all of which is retained on a No. 8 (2.36-mm) sieve and practically none of which passes a No. 200 (75-μm) sieve; tack coat		Plant or road mixture with coarse aggregate substantially all of which is retained on a No. 8 (2.36-mm) sieve and practically none of which passes a No. 200 (75-μm) sieve				Plant or road mixture with graded and fine aggregates, a subtantial quantity of which passes a No. 8 (2.36-mm) sieve and a portion of which may pass a No. 200 (75-μm) sieve; slurry seal treatments			

SOURCE: *Specifications for Asphalt Cements and Liquid Asphalts* (SS-2). College Park, Md.: The Asphalt Institute, 1974 [9].

and coat positively charged minerals such as carbonates. Cationic emulsifiers are best with predominantly negative aggregates, such as the silicates. Typical emulsifiers are

Anionic $(CH_3R_nCOO)^-Na^+$
Cationic $CH_3R_nN^+H_2$

where R_n generally ranges from an 8 to 18 carbon atom chain. Others are illustrated in Chap. 2, Figs. 15 and 16.

Anionic emulsions are generally stable in the 8–12 pH range and cationic emulsions in the 2–6 pH range. The breaking time in the field can often be controlled by minor additions of a common acid or base. Emulsion *stabilizers* such as clay, sodium lignate, or cellulose derivatives are sometimes added to increase emulsion viscosity and thereby increase settling stability during storage.

Table 15 lists the ASTM specifications for anionic emulsions. Specifications for cationic emulsions are similar. Specifications for emulsified asphalts include viscosity, stability in storage and in mixing, aggregate coating ability, water content, and asphalt residue properties (Appendix E). Storage stability tests include sieve, miscibility with water, settlement of asphalt globules, and freezing tests. Mixing stability tests include stone-coating and water-stripping tests, cement mixing, and demulsibility. After distillation the asphalt is tested for penetration, ductility, and solubility in trichloroethylene.

4. Tars

In the United States asphalts have replaced tars almost exclusively for road paving, but tars are used extensively in roofing and underground waterproofing. Road tars are still used in Britain due to the large bituminous coal industry there.

a. PRODUCTION AND REFINING

The two common methods of producing tars are by the coke-oven and water-gas processes. Coke-oven tar is part of the volatile by-product removed in the production of coke from bituminous coal. The vapors removed from the oven are condensed to form liquid crude tar, which is then refined for various uses.

A smaller amount of tar is produced from the water-gas process for cracking petroleum to produce gas for heating purposes. The high temperatures used in this process cause destructive distillation (cracking) of petroleum to form hydrocarbon

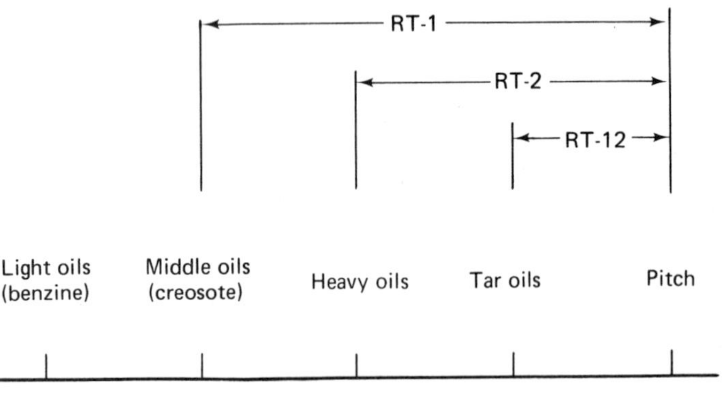

Fig. 10. Grades of road tars.

gases and the hydrocarbon residue crude called water-gas tar, which may also be further refined for various uses.

After removing any water by either settling or low-temperature heating, crude tar is refined by fractional distillation, much as petroleum is refined to produce asphalt. Twelve viscosity grades of road tar are produced by straight-run distillation (Fig. 10). Vacuum is sometimes used to obtain the more viscous grades at lower temperatures and hence with less decomposition. RT-1 is a very light fraction, and RT-12 is approximately as hard as a 200 penetration asphalt. Table 16 indicates typical uses for the various grades. Grades RT-1 to 6 can be used at temperatures below about 150°F.

TABLE 16. Uses of Road Tars

Grade	*Uses*
RT–1	Dust palliatives
RT–2, 3	Prime coats
RT–4	Prime coats and surface treatments
RT–5 to 7	Surface treatments and road mixes
RT–8, 9	Surface treatments, seal coats, road mixes
RT–10, 11	Seal coats, tar concrete, hot repairs
RT–12	Penetration macadam, tar concrete, hot repairs

SOURCE: ZAKAR, PAL, *Asphalt*. New York: Chemical Publishing Company, Inc., 1971, [6].

b. TESTS

Many of the standard tests for tars are analogous to those for asphalts, i.e., viscosity, specific gravity, distillation, and softening point tests. The apparatus and test procedures usually differ slightly, reflecting the differing development of the asphalt and tar industries. One test distinctive to tars is the sulfonation index, which permits determining approximately the percentages of coke-oven and water-gas tars in a blend. Also, specifying a maximum sulfonation index limits the amount of petroleum solvent which a refiner can use in blending operations, because the petroleum constituents do not react with sulfuric acid.

PROBLEMS

1. Distinguish between asphalts and tar in terms of their chemical structure, general adhesiveness, and temperature susceptibility.
2. If separate portions of a single crude were treated by each of the following refining processes, which product would you expect to have the greatest asphaltene content? The lowest asphaltene content? The lowest saturates content?
   (a) Vacuum or steam reduction.
   (b) Air blowing.
   (c) Air blowing with a catalyst.
   (d) Solvent extraction.

Which would you expect to be most temperature-susceptible? Least temperature-susceptible? Most adhesive? Least adhesive?

3. Of two asphalts, a vacuum-reduced paraffin stock and an asphalt produced from solvent extraction, both having approximately the same average molecular weight, which would you expect to be more gel-like and which more sol-like in colloidal behavior? Therefore, which would you expect to have a higher penetration value and, therefore, to mix more readily and to form thinner films when coating the aggregate?
4. Generally speaking, does an increase in average molecular weight of an asphalt increase or decrease adhesion? Increase or decrease temperature susceptibility?
5. Name some important classes of surface-active agents used as emulsifiers.
6. Assume that you have an oil-in-water (o/w) asphalt emulsion with an anionic surface-active agent such as sodium palmitate as the emulsifier.
   (a) Would you prefer to use this emulsion with a crushed limestone aggregate or with a quartz aggregate? Why?
   (b) The emulsion does not break as readily as you would like (i.e., asphalt and oil phases do not separate rapidly on mixing with the aggregate). Which would you recommend adding in small quantities to enhance breaking, calcium hydroxide (or a calcium salt) or hydrochloric acid?

REFERENCES

1. ROSATO, D. V., and R. T. SCHWARTZ, eds., *Environmental Effects on Polymetric Materials*, Vol. II: Materials. New York: John Wiley & Sons, Inc. (Interscience Division), 1968.
2. DIETZ, ALBERT G. H., *Plastics for Architects and Builders.* Cambridge, Mass.: The M.I.T. Press, 1969.
3. BROUTMAN, L. J., and R. H. KROCK, eds., *Modern Composite Materials.* Reading, Mass.: Addison-Wesley Publishing Company, Inc., 1967.
4. LUBIN, GEORGE, ed., *Handbook of Fiberglass and Advanced Plastics Composites.* New York: Van Nostrand Reinhold Company, 1969.
5. SCHWEYER, H. E., "Asphalt Composition and Properties," *Hw. Res. Bd. Bull.*, 192 (1958) 33.
6. ZAKAR, PAL, *Asphalt*, New York: Chemical Publishing Company, Inc., 1971.
7. GOETZ, W. H., and L. E. WOOD, "Bituminous Materials and Mixtures," in *Highway Engineering Handbook*, K. B. Woods, ed. New York: McGraw-Hill Book Company, 1960, Chap. 18.
8. *The Asphalt Handbook* (Series MS-4). College Park, Md.: The Asphalt Institute, 1965.
9. *Specifications for Asphalts Cements and Liquid Asphalts* (SS-2). College Park, Md.: The Asphalt Institute, 1974.
10. KREBS, R. D., and R. D. WALKER, *Highway Materials.* New York: McGraw-Hill Book Company, 1971.
11. BENJAMIN, B. S., *Structural Design with Plastics.* New York: Van Nostrand Reinhold Company, 1969.

6

AGGREGATES AND FIBERS

Colts grew horses, beards turned gray,
Deacon and deaconess dropped away,
Children and grandchildren—where were they?
But there stood the stout old one-hoss shay.
—HOLMES

Aggregates constitute a larger tonnage of materials used in the construction industry than all other materials combined. They also represent a large fraction of the total materials cost. In highway construction, for example, approximately 30 cents of the materials dollar is spent on aggregates, compared with 25 cents on steel, 19 cents on bituminous materials, 10 cents on cement, and lesser amounts on pipe, lumber, and petroleum products [1].

Most aggregates contain primarily silicon dioxide, which comprises 60% of the earth's crust either in the free form or, more commonly, combined with other oxides to form silicates. A familiarity with silicate structures is therefore essential to an understanding of aggregate behavior. In this chapter we shall describe the frequently encountered forms of silicate structures (Sec. A) and then classify and define properties of the major classes of aggregates and fibers used in construction (Sec. B).

A. SILICATE STRUCTURES

The basic building block of silicates is the $(SiO_4)^{-4}$ tetrahedron (Fig. 1). The Si—O bond is partly ionic and partly covalent, and tetrahedral coordination satisfies both the relative size ratio for ionic bonding and the directional requirements of covalent bonding. The variety of silicate structures depends on the various combinations of these tetrahedra, both with each other and with other ions. The tetrahedral $(SiO_4)^{-4}$ units are rarely joined edge to edge and never face to face, due to the high charge and low coordination number of the Si^{+4} ion. The remaining possible combinations may occur both in crystalline and amorphous form and will be presented in these two categories.

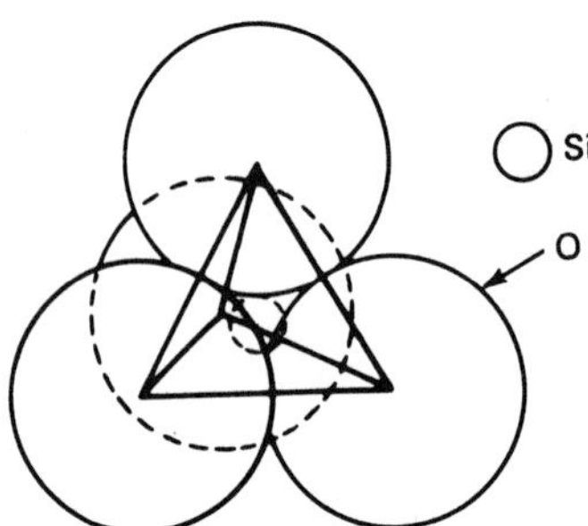

Fig. 1. Tetrahedral $(SiO_4)^{-4}$ group.

1. *Crystalline Silicates*

a. CRYSTAL STRUCTURE

The extent and type of polymerization or sharing between adjacent silica tetrahedra is used as the basis for classifying the crystalline silicates. Gillott [2] provides a concise summary of this classification, and much of the following presentation is based upon his summary. Table 1 summarizes the major classes. The crystals produced when adjacent tetrahedra share no oxygens with each other but are joined only through other positive ions in the structure are called *island* silicates. Minerals of this structure are called *nesosilicates* (Greek *nesos*, "island"). In olivine, the silicate tetrahedra are arranged so that the oxygen atoms form a distorted HCP structure with silica atoms filling one eighth of the tetrahedral cation sites. One half of the octahedral cation sites are then filled with Mg and Fe atoms.

In the *sorosilicates* (Greek *soros*, "group") two tetrahedra share a common oxygen to form a double tetrahedron $(Si_2O_7)^{-6}$ (Fig. 2).

The sharing of two corners of each tetrahedron results in either a *single chain* or a *ring* structure with a formula containing $(SiO_3)^{-2}$. The *cyclosilicates* (Greek *cyclos*, "ring") include minerals having silicate radicals such as the three-member ring $(Si_2O_9)^{-6}$ in benitoite and the six-member ring $(Si_6O_{18})^{-12}$ in beryl (Table 1). The *inosilicates* (Greek *inos*, "thread") or chain silicates display numerous structures

TABLE 1. Classification of Crystalline Silicates

Class	*Arrangement of tetrahedra*	*Structure*	*Silicate radical*	*Oxygen-to-silicon ratio*	*Examples*
Nesosilicates	Individual	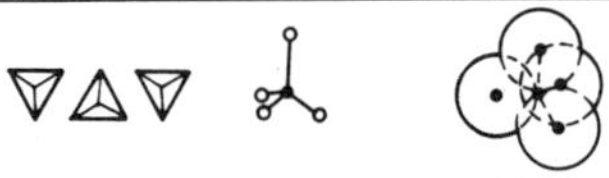	$(SiO_4)^{-4}$	4 : 1	Olivine $(Mg, Fe)_2(SiO_4)$
Sorosilicates	Double	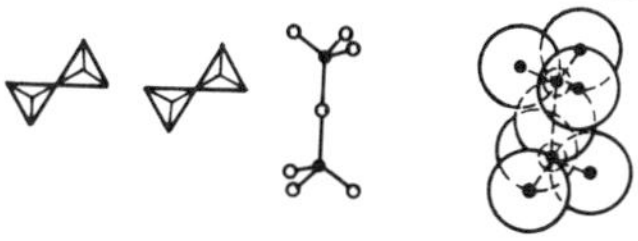	$(Si_2O_7)^{-6}$	7 : 2	Akermanite $(Ca_2MgSi_2O_7)$
Cyclosilicates	Three-member ring	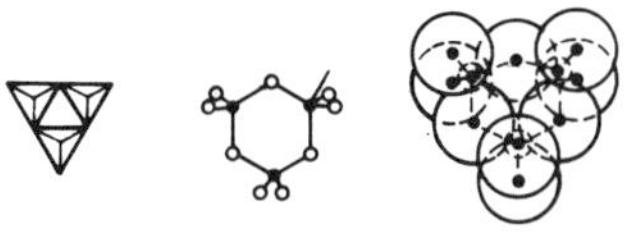	$(Si_3O_9)^{-6}$	3 : 1	Benitoite $(BaTiSi_3O_9)$
	Six-member ring	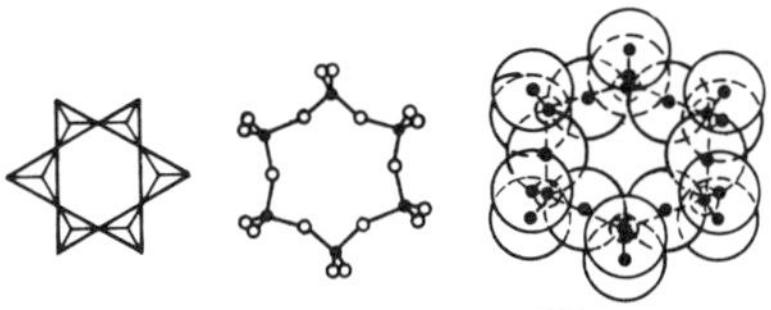	$(Si_6O_{18})^{-12}$	3 : 1	Beryl $(Be_3Al_2Si_6O_{18})$
Inosilicates	Chain	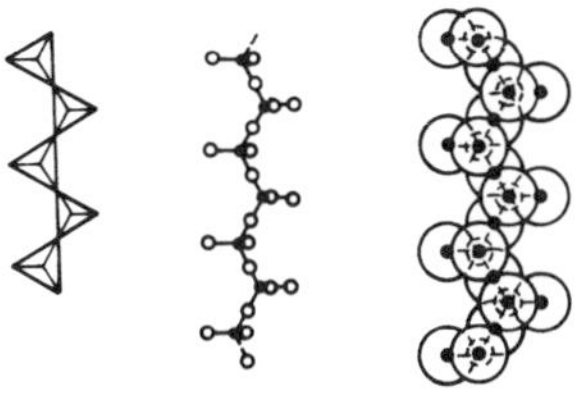	$(SiO_3)_n^{-2}$	3 : 1	Pyroxenes
	Band	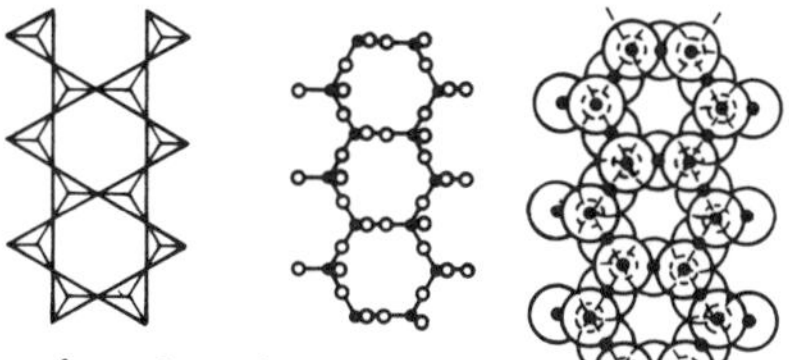	$(Si_4O_{11})_n^{-6}$	11 : 4	Amphiboles
Phyllosilicates	Sheet	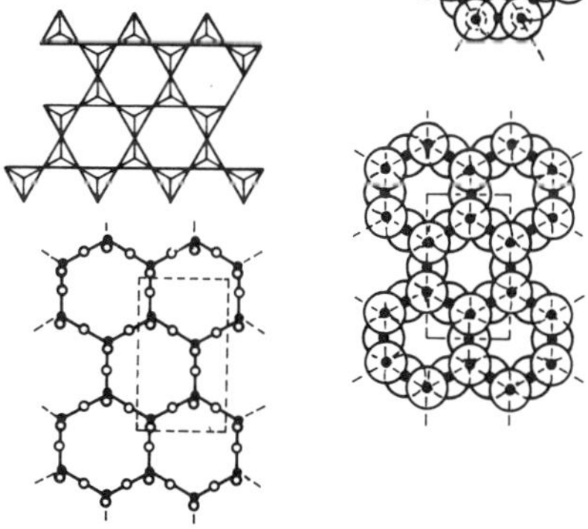	$(Si_4O_{10})_n^{-4}$	10 : 4	See Table 2
Tektosilicates	Framework	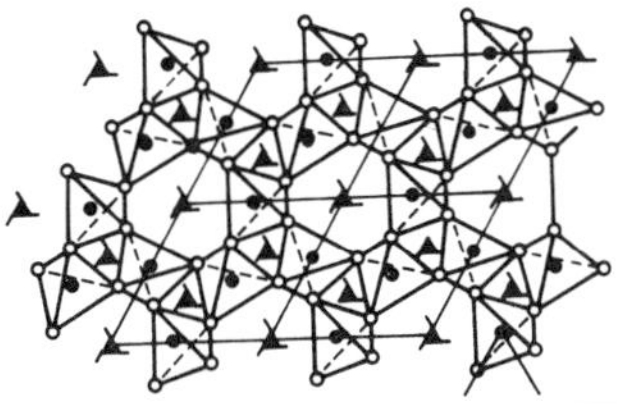	$(SiO_2)_n^0$	2 : 1	Quartz $(SiO_2)_n$ Feldspars

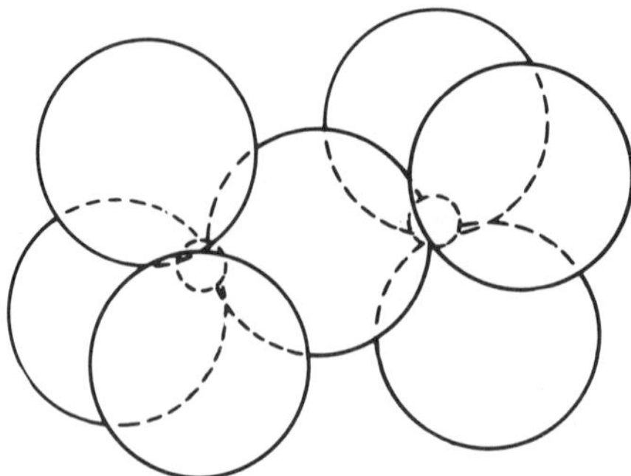

Fig. 2. Double tetrahedral silicate $(Si_2O_7)^{-6}$. The center oxygen receives an electron from each adjacent silicon.

depending on the arrangement of the linked tetrahedra and differences in relative positions of neighboring chains.

If, on the average, two and one-half corners of each tetrahedron are shared with each other, *double chains* of tetrahedra result, having the silicate radical $(Si_4O_{11})^{-6}$. The amphiboles are an example. The single and double chain structures can extend indefinitely and may be compared with organic polymers, except that adjacent chains are held together by ionic bonds (Fig. 3) rather than the weaker van der Waals forces which are more common in organic polymers. Since the ionic bonds between chains are not so strong as the partially covalent Si—O bonds within chains, cleavage occurs parallel to the chain. The pyroxenes and amphiboles, of which asbestos is one example, display this fibrous morphology.

```
 O⁻    O⁻    O⁻    O⁻    O⁻    O⁻    O⁻    O
 |     |     |     |     |     |     |     |
—Si—O—Si—O—Si—O—Si—O—Si—O—Si—O—Si—O—Si—O—
 |     |     |     |     |     |     |     |
 O⁻    O⁻    O⁻    O⁻    O⁻    O⁻    O⁻    O⁻
Mg⁺⁺  Mg⁺⁺  Mg⁺⁺ Na⁺Na⁺ Mg⁺⁺  Mg⁺⁺ Na⁺Na⁺Na⁺Na⁺
---------------------------------------------------
 O⁻    O⁻    O⁻    O⁻    O⁻    O⁻    O⁻    O⁻
 |     |     |     |     |     |     |     |
—Si—O—Si—O—Si—O—Si—O—Si—O—Si—O—Si—O—Si
 |     |     |     |     |     |     |     |
 O⁻    O⁻    O⁻    O⁻    O⁻    O⁻    O⁻    O⁻
```

Fig. 3. Ionic bonding between chains. The weaker ionic bonds permit cleavage parallel to the chains. From Van Vlack, L. H.: *Elements of Materials Science, First Edition.*, Reading, Mass.: Addison-Wesley Publishing Company, Inc., 1959 [3].

Sheet structures, or *phyllosilicates* (Greek *phyllos*, "sheet") result when three out of four oxygens in every tetrahedron are shared, with the resulting silicate radical $(Si_2O_5)^{-2}$. The micas, chlorites, serpentines, talc, and most clay minerals are built up from this structural unit. The layers are held together by different secondary forces in different minerals, but the interlayer bonding is weaker than that within layers. The cleavage of mica, the plasticity of clay, and the lubricating ability of talc all result from these weak secondary interlayer forces.

Three-dimensional framework structures, or *tektosilicates* (Greek *tektos*, "framework"), result when all four oxygens in the tetrahedron are shared. Silica, SiO_2, is the simplest example of a network silicate. Feldspars and zeolites are also framework silicates. In feldspars, for example, Al atoms substitute for some of the Si atoms in the silica structure. Additional alkali ions in the voids between tetrahedra balance the charge deficit resulting from the trivalent Al atoms substituting for tetravalent Si atoms. Network silicates alone account for over 50% of the earth's crust.

The sequence island-through-framework represents a spectrum of structures which have (1) a decreasing number of ionic bonds, (2) a decreasing silicon-to-oxygen ratio, and (3) a greater sharing of the oxygens by adjacent tetrahedra.

b. ISOMORPHOUS SUBSTITUTION

As described for feldspar above, one ion may sometimes replace another in a crystal lattice without altering the structure. Such *isomorphous* (*iso*, "same"; *morphous*, "form") substitutions occur extensively among the silicates and in varying degrees of replacement. When the ions involved are similar in size, the minerals, although different chemically, may exhibit few differences in properties. Cations frequently proxy for one another, as does oxygen for hydroxyl. The substitution of Al for Si is one of the most predominant substitutions. Because the Al/Si radii ratio (0.43) is very close to 0.414, the value for transition from fourfold to sixfold coordination, Al frequently occupies both tetrahedral and octahedral sites in coordination with oxygen. When two adjacent tetrahedra share an oxygen, it is believed that only one can be occupied by aluminum. Or if two aluminum ions do join a common oxygen, at least one of them must have a coordination number greater than 4. This rule would explain the 50% limit on substitution of aluminum for silicon observed in most silicates with linked tetrahedra [2]. The increased negative charge on the tetrahedron due to substitution of silicon with aluminum must be compensated for elsewhere in the structure. This might be accomplished by replacement of a univalent by a divalent cation or by the addition of cations in voids, as previously mentioned for feldspar.

c. THE PHYLLOSILICATES

The phyllosilicates, which comprise most of the clay minerals, are of particular importance to the construction industry because of their widespread occurrence at the earth's surface, their consequent availability and low cost, and their comparatively high hydroscopicity and chemical reactivity resulting from the plate-like structure. These minerals form the basis for most of the ceramics industry, including all brick and tile products, and their reactions are essential to various soil stabilization techniques. Clays are also used in portland cements, paints, and numerous other products manufactured for the construction industry.

The phyllosilicate clay minerals contain two types of structural units, the SiO_4 silicon tetrahedral sheet and the $Mg(OH)_2$ brucite or $Al(OH)_3$ gibbsite octahedral sheet.

In silica sheets, the corners of each tetrahedron lie in the same plane, with the fourth or apical oxygens all pointing the same direction. The sheets are built up from hexagonal units each containing six such tetrahedra.

Brucite and gibbsite sheets contain cations coordinated by six hydroxyls forming an octahedron. Adjacent octahedra share edges to form, again, hexagonal sheet structures. In the brucite structure, the cations lie between two planes of hydroxyls, each cation having three hydroxyl neighbors above and three below. One group of three hydroxyls is rotated relative to the other through a 60° angle in the plane of the sheet to form an approximately hexagonal close-packed arrangement of hydroxyls about each cation. In the brucite structure, all octahedral sites are occupied by divalent cations. In the gibbsite structure, two thirds of these positions are filled by trivalent cations, and the remainder are unfilled.

The various phyllosilicates contain different combinations of tetrahedral and octahedral sheets. Apical oxygens of the tetrahedral layer replace two thirds of the hydroxyls in one plane of the octahedral layer, and the remaining hydroxyls in this plane occupy the centers of hexagons formed by the apical oxygens in the tetrahedral layer [2].

The phyllosilicates are classified in Table 2. Kaolinites and serpentines consist of a combination of one tetrahedral and one octahedral layer, as just described, and are designated as a 1:1 structure.

When two tetrahedral sheets are joined similarly to each side of a central octahedral sheet, a 2:1 structure results. These include the montmorillonites, micas, vermiculite, and talc.

An additional octahedral layer may be joined between two successive 2:1 units, giving the 2:1:1 layer structure typical of chlorites. Table 2 indicates the sequence of tetrahedral and octahedral layers and the unit cell thickness in each structural class. The unit cell thickness is easily detected by x-ray analysis and is the primary criterion for classifying layer silicates.

TABLE 2. Proposed Classification Scheme for Phyllosilicates, Related to Clay Minerals*

Structural class	*Layer sequence*	*Unit cell thickness*† *(A)*	*Group* (*X* = *layer charge*)	*Subgroups*
1:1	OT‡	7	Kaolinite-serpentine ($X \sim 0$)	Kaolinites Serpentines
2:1	TOT	10	Pyrophyllite-talc ($X \sim 0$)	Pyrophyllites Talcs
			Montmorillonite-saponite ($X \sim 0.5$–1.0)	Montmorillonites (dioctahedral) Saponites (trioctahedral)
			Vermiculite ($X \sim 1.0$–1.5)	Dioctahedral vermiculite Trioctahedral vermiculite
			Micas ($X \sim 2$–4)	Dioctahedral micas Trioctahedral micas
2:1:1	TOTO	14	Chlorite (X variable)	Dioctahedral chlorites Trioctahedral chlorites

*Adopted from Gillott [2], after Brindley [4].
†Approximate thickness.
‡T = tetrahedral layer, O = octahedral layer.

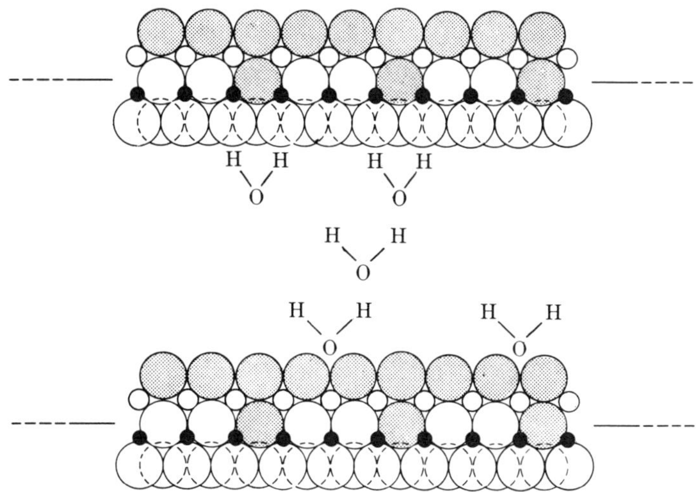

Fig. 4. Adsorption of H_2O by kaolinite. Adapted from Van Vlack, L. H.: *Elements of Materials Science*, Second Edition. Reading, Mass.: Addison-Wesley Publishing Company, Inc., 1964 [3].

Many of the differences in physical and chemical properties of the phyllosilicates can be explained by the structural distinctions shown in Table 2. For example, the nonsymmetrical nature of kaolinite, in which two thirds of the hydroxyl ions in only *one* side of the gibbsite sheet are replaced with tetrahedral oxygens, helps explain why the group attracts water so readily and thereby exhibits plasticity. The polar kaolinite crystal attracts polar water molecules by dipole forces (Fig. 4). Kaolinite typically occurs as thin platelets in the order of 5,000 A (0.00005 cm) wide and 300 A

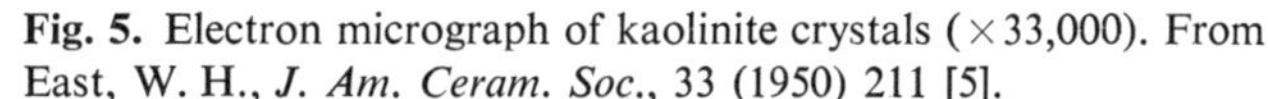

Fig. 5. Electron micrograph of kaolinite crystals (×33,000). From East, W. H., *J. Am. Ceram. Soc.*, 33 (1950) 211 [5].

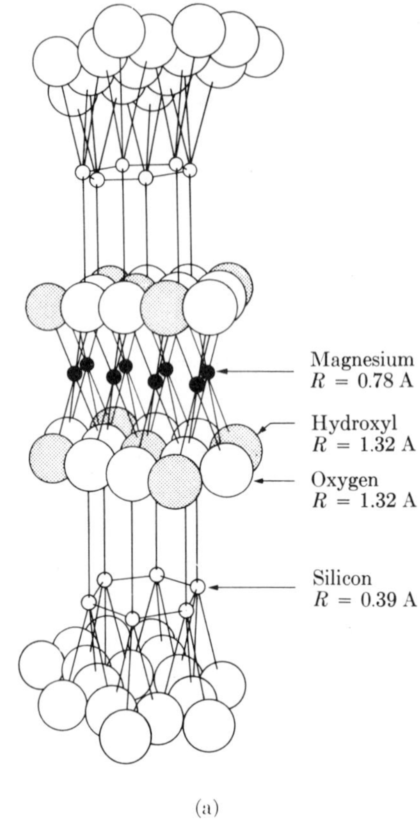

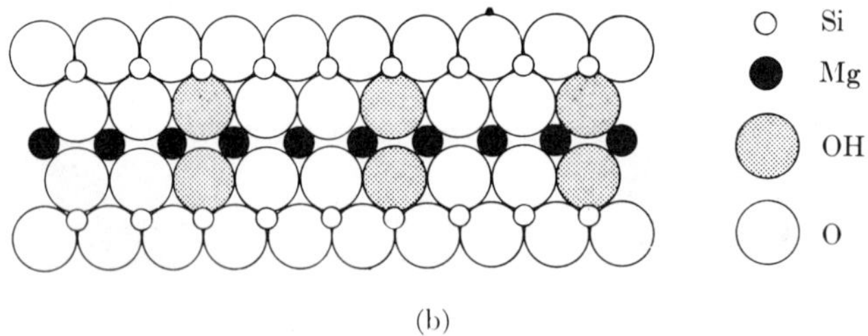

Fig. 6. Structure of talc; an $Mg(OH)_3$ sheet between two SiO_4 sheets, forming a symmetrical structure. (a) Exploded view. From Hauth, W. E., "Crystal Chemistry of Ceramics," *Bull. Am. Ceram. Soc.* 34 (1951) 61 [6]. (b) Cross sectional view. From Van Vlack, L. H.: *Elements of Materials Science*. Reading, Mass.: Addison-Wesley Publishing Company, Inc., 1959 [3].

thick and have an irregular hexagonal shape (Fig. 5). When wet, they cling together like wet sheets of paper, and with water as a lubricant, the crystals readily slide over one another, which accounts for the plasticity of kaolinite.

Talc does not absorb water so readily because the silicate sheets in talc are bonded on *both* sides of a $Mg_3(OH)_6$ sheet (Fig. 6). In both kaolinite and talc, bonding between unit cells is of the weak van der Waals type. In micas, however, one fourth of the silicon atoms are replaced by aluminum atoms, requiring an alkali such as K^+ for the structure to remain electrically neutral. These alkali ions fit in the *holes* between adjacent unit cells (adjacent tetrahedral sheets) and *bind* them together with ionic bonds, which are stronger than van der Waals bonds. The micas therefore tend to occur in nature as thicker particles than most other groups in the 2:1 structural class.

The variations in electrical charge indicated in Table 2 also account for significant differences in properties. When there is a balance of charges within sheets, the layers are held together either by van der Waals forces, as in pyrophyllite and talc, or by hydroxyls, as in kaolinite. A net negative charge results from unbalanced isomorphous substitution. Except for chlorites, this is neutralized by cations in the interlayer positions. The tightness of this interlayer cation bonding depends partly on the magnitude and location of the negative charge within the layer and partly on the size and valence of the interlayer cation. In micas, for example, the size of the K^+ ion almost equals the available hole size in adjacent tetrahedral layers, and therefore a close fit and short-range, strong, ionic bonding is possible. In chlorites, the net negative charge on the 2:1 layer is balanced by a positive charge in the additional octahedral layer due to such substitutions as replacement of some of the divalent magnesium by trivalent aluminum cations.

The mineral groups in Table 2 are divided into subgroups according to the distribution of cations in the octahedral layer. When the dominant cation in this layer is trivalent aluminum, only two thirds of the octahedral sites are occupied, and the structure is called dioctahedral because there are two octahedral cations per half unit cell. When the dominant cation in this layer is divalent, such as magnesium, all available sites are filled, giving three cations per half unit cell, and the structure is called trioctahedral.

A slight difference in the hexagonal dimensions between tetrahedral and octahedral sheets produces internal stresses when the two units are combined. The stresses might be relieved internally by ionic substitution or rotation of tetrahedra or externally by curling or twisting of the composite sheets [2]. Efforts have been made to relate the magnitude of this slight mismatch between layers in the various minerals to such properties as weathering stability, frequency of occurrence, and the known limits of solid solution in the various minerals.

The archetypal example of curling and twisting is observed in the halloysite minerals, which resemble kaolinites except that successive tetragonal and octahedral layers are displaced randomly in directions in the plane of the layers. Figure 7(a) and (b) shows halloysite particles curled into tubular shapes. Curling appears to occur in very thin flakes, where the number of layers is so few that the strain produced by the misfit in tetrahedral and octahedral layers is not overcome by the cumulative effect of interlayer bonds between a large number of successive layers [8]. Curvature can only occur in hydrated halloysite, due to the irregular stacking of layers, and the interlayer water molecules, which weaken the bond between successive layers. Upon dehydration, the tubes frequently collapse, split, or unroll. Curling does not occur in kaolinite because the regular stacking and close spacing of layers causes a relatively strong bond between successive layers.

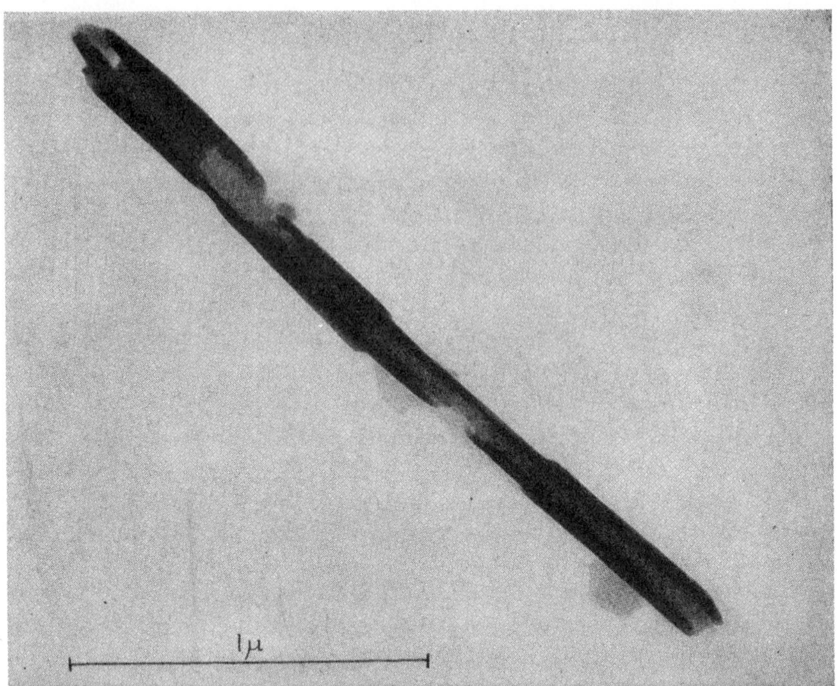

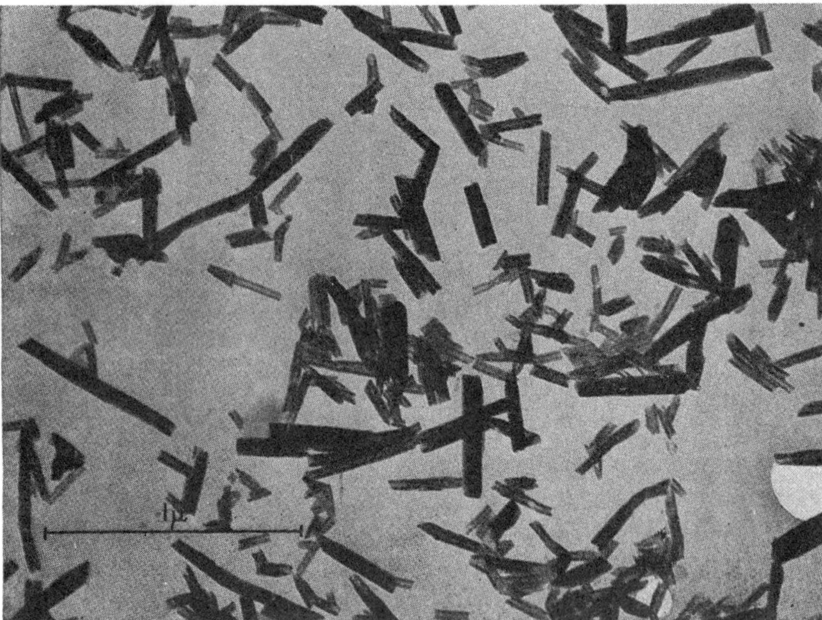

Fig. 7. Electron micrographs of halloysite. (a) from British Guiana; (b) from Wendover, Utah. From Bates, T. F., F. A. Hildebrand, and A. Swineford, *Am. Mineral.*, 35 (1950) 463 [7].

d. CATION EXCHANGE

The isomorphous substitution of lower-valence cations for higher-valence cations, such as Mg^{+2} for Al^{+3}, or Al^{+3} for Si^{+4} is most pronounced in the phyllosilicates, giving most clay minerals a residual negative charge (the layer charge, in Table 2). Sometimes such substitutions are canceled by other lattice changes, e.g., the substitution of OH^{-1} for O^{-2}, or by filling more than two thirds of the octahedral positions, but frequently they are balanced by cations which are adsorbed and held in an exchangeable state at the particle surface. Two other factors also contribute to this cation exchange capacity of clays. For example, broken (unsatisfied) positive and negative bonds at the edges of the layer structure may attract anions and cations, respectively. The hydrogen of exposed hydroxyls at the broken edges of clay minerals may also be replaced by cations which are exchangeable.

In the kaolinite group (1:1 minerals), since the layer charge is approximately zero, the cation exchange capacity is small and depends primarily on the broken bonds at layer edges. Cation exchange capacity of these minerals is quite dependent on particle size, e.g., the surface area of broken edges per unit volume of particles. In the montmorillonite group (2:1 minerals), the cation exchange capacity is much larger and depends primarily (typically around 80%) on the extent of isomorphous substitution. Thus, in the 1:1 minerals, most exchangeable cations can be expected near the particle edges, and in 2:1 minerals, on the cleavage faces.

Cation exchange capacity is normally measured in milliequivalents per 100 g of soil. For example, one equivalent of Na^+ expressed as Na_2O would have a combining weight of 31, and 1 meg/100 g would be 0.031% by weight. Table 3 shows typical ranges of cation exchange capacities for major clay mineral groups. Exchange capacity is usually determined at pH 7 and varies greatly with variations in pH, cation species, clay concentration in suspension, and other properties of the clay-water system. In clay minerals, the most common exchangeable cations are Ca^{+2}, Mg^{+2}, H^+, K^+, NH_4^+, and Na^+ in approximately that order.

TABLE 3. Cation Exchange Capacities and Specific Surface Areas of Clay Minerals

Mineral	*Cation exchange capacity (meg/100 g)*	*Typical specific surface (m^2/g)*
Kaolinite	3–15	15
Halloysite	10–50	40
Illite	20–40	80
Chlorite	20–40	80
Montmorillonite	80–150	800

Ion exchange properties of clays are profoundly important for widely varying reasons in almost every clay application: fertility and tilth in agriculture; plasticity of "green" clays in the brick, tile, and ceramic industries; thixotropic properties as sealants in well drilling; swelling characteristics as foundation soils; imperviousness as canal and reservoir linings; and chemical reactivity in soil stabilization applications.

2. Amorphous Silicates

Amorphous solids are those which are devoid of long-range order. Glasses and many plastics are prime examples. Inorganic compounds tend to be noncrystalline if (1) each anion is linked to not more than two cations, (2) no more than four anions surround a cation, (3) the anion polyhedra share corners but not edges or faces, and (4) the material has many constituents. Most materials must be cooled rapidly to form glasses. Natural amorphous silicates such as obsidian and rhyolite are frequently used as aggregates. In this section manufactured glasses are also described, because of their wide use in other construction applications. Most of the glass produced annually is used in the construction industry, in applications such as fiber-glass-reinforced composites and thermal insulating materials, structural glass units and windows, and corrosion-resistant pipes and fittings for chemical industries. Uses of glass cullet and waste glass products as aggregates in portland cement and bituminous concretes are also being developed.

a. GLASS MORPHOLOGY

Glasses do not exhibit long-range periodicity of atomic structure because they do not crystallize when cooled from a melt. Figure 8 depicts the difference, for example, between the short-range order of B_2O_3 glass and the long-range order of B_2O_3 crystals. The structure of glass is often inferred from the analysis by x-ray diffraction and other means of some crystalline modification of the material which forms it. Information on the degree of disordering in glass structure is usually obtained by thermodynamic measurements. Such information cannot describe disorder structure on an atomic scale but provides useful mass-average disorder characteristics.

Fig. 8. Short-range order of B_2O_3 glass (a), compared with long-range order of B_2O_3 crystal (b). (Schematic) From Van Vlack, L. H.: *Materials Science for Engineers.* Reading, Mass.: Addison-Wesley Publishing Company, Inc., 1970 [9].

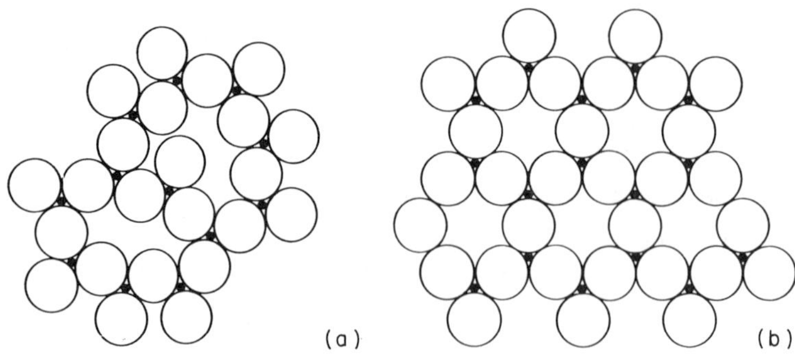

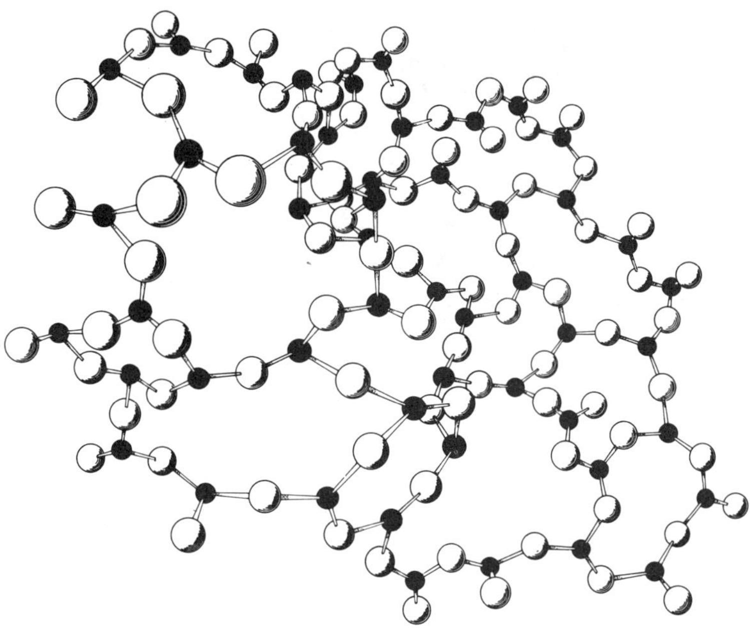

Fig. 9. Pure oxide glass; a random three-dimensional network with each oxygen (white) bonding two metal atoms such as boron. Here each metal atom bonds to three oxygen atoms. In some glasses, as in silica glass, each metal atom bonds to four oxygen atoms, producing a more rigid network. From Charles, R. C.: "The Nature of Glasses"; *Scientific American*, Sept. 1967.

In accordance with free energy considerations (Chap. 2), liquid glass has a freezing temperature below which a crystalline structure is more stable. But the activation energy required for atomic rearrangement from the disordered to the ordered state (Fig. 8) is sufficiently high and the heat of fusion released sufficiently low that crystallization proceeds very slowly at the freezing temperature, and practically not at all at room temperature, at which atoms so rarely attain the required energy to break existing bonds for rearrangement. Glasses therefore exist metastably for indefinite periods.

In glass making, oxygen forms stable bonds (mainly covalent) with small multivalent ions such as silica, boron, phosphorus, or germanium. Oxygen combined with these ions yields low-coordination polyhedra—primarily tetrahedra or triangles with oxygen atoms at the corners and the multivalent ion at the center [10]. The polyhedra connect flexibly through corner oxygen atoms.

In pure oxides, each oxygen shares electrons with two positive ions to form a completely cross-linked network (Fig. 9). Such glasses are stable and have high softening temperatures. Their cross-linking can be reduced and their softening temperature lowered by adding to the melt strong alkali metals such as sodium and potassium which relinquish electrons easily and form weak, nondirectional ionic bonds with oxygen. Such *fluxing ions* are used to inhibit crystallization and to control fluidity, in addition to reducing the melting temperature of glasses (Fig. 10).

Sufficient fluxing ions frequently cause a separation of components in a glass melt, on cooling, into two or more intermixed phases of different chemical composi-

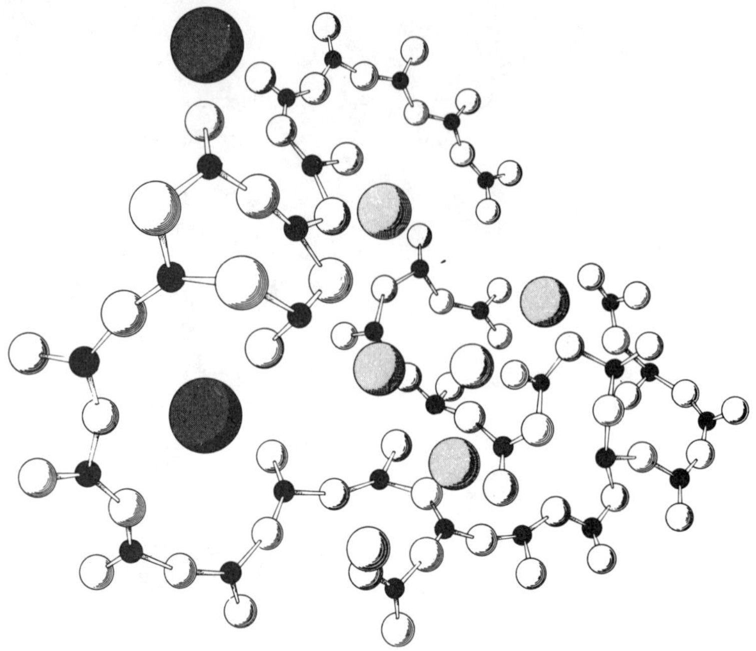

Fig. 10. Flux-containing glass; also a random three-dimensional network, but with flux atoms such as sodium (gray) which reduce the amount of cross-linking. Some atoms are now strongly bonded to only a single atom, with weaker attractions to one or more flux atoms. The melting point is reduced. From Charles, R. C.: "The Nature of Glasses," *Scientific American*, Sept. 1967 [10].

tion (Fig. 11). The nucleation and growth of one liquid out of the other is analogous in some respects to the nucleation and growth of crystals from solution. Such separation phenomena are important in glass technology. The distribution geometry of the phases, for example, can profoundly affect strength, electrical conductivity, corrosion resistance, and optical properties. The two-phase structure is also significant in influencing the crystallization sequence and final properties of crystallized glasses. These are glasses which, after melting, are transformed by prolonged heat treatment into durable ceramic glass-crystalline mixtures. Some of the typical trade names found in home and industry are Pyroceram, CERVIT, and Re-X.

Practically all elements can be dissolved to various degrees in glass melts. Many types of chemical reactions—oxidation and reduction, ion exchange, decomposition, and precipitation—can also be accomplished in glass melts and frozen at desired stages. Such process control has enabled the production of glasses which lighten or darken in response to light, transport ions selectively, fluoresce, and permit semiconduction or photoconduction [10]. New phenomena are sure to be encountered as progress in glass science continues.

b. MAJOR TYPES OF GLASSES

The properties of six important types of glass, described by Keyser [11], are outlined below and summarized in Table 4.

Soda-lime glass is the most common type. The raw materials, including silica sand, soda ash, and limestone, are proportioned to produce a product having approxi-

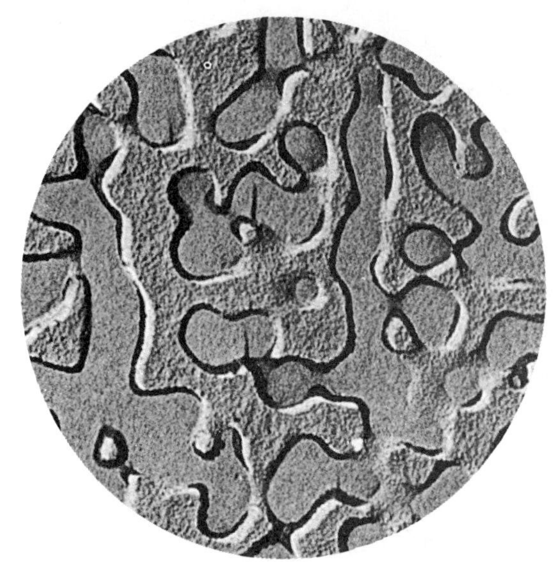

Fig. 11. Electron micrograph of two-phase sodium borate glass. Mechanical, electrical, and optical properties depend upon composition of the two phases and geometry of distribution. (*Courtesy of R. J. Charles.*) From Charles, R. J.: "The Nature of Glasses," *Scientific American*, Sept. 1967 [10].

mately the composition Na_2O-CaO-6 SiO_2. The addition of Na_2O to SiO_2 lowers the softening temperature but also reduces resistance to chemical attack. With addition of enough Na_2O the glass becomes water-soluble and is called waterglass. The addition of lime (CaO) with the Na_20 improves hardness and resistance to chemicals and water. Too much lime causes the glass to devitrify, or crystallize, causing loss in transparency and altering other properties. Variations in properties may be obtained by replacing portions of the lime and soda with alkaline earth oxides and K_2O, respectively. Soda glass is used for plate, containers, light bulbs, etc.

TABLE 4. Composition, Properties, and Uses of the Major Types of Glasses

Glass type	*Typical composition*	*Properties and uses*
Soda-lime	Na_2O, CaO, SiO_2	Fair strength, thermal and chemical resistance. Plate glass, containers, light bulbs, etc.
Lead alkali	Na_2O, PbO, SiO_2	High refractive index and dispersion. Radiation shielding, electrical insulation, and optical refraction applications.
Borosilicate	Na_2O, B_2O_3, SiO_2	High thermal shock resistance. Cooking ware, high-temperature industrial uses.
Alumina silicates	55% SiO_2, 20% Al_2O_3, lesser amounts of B_2O_3, MgO, CaO	High strength at high temperature. Combustion tubes, cooking ware, etc.
96% silica	Soda-boron oxide silicate	Very high strength and corrosion resistance, very low coefficient of expansion. Very high-temperature applications.
Fused silica	SiO_2, from pure silica sand	Slightly superior in those properties listed for 96% silica glass. Most stringent temperature applications, where properties justify extra manufacturing cost.

Lead alkali glass is similar to sodium silicate glass except that most of the lime is replaced by PbO. Lead oxide contents of 15 to 60% by weight are typical. Leaded glass is useful for absorbing gamma- and x-rays because of the high mass absorption coefficient of lead. The large lead ions also reduce ion movement and thus increase the electrical insulation value of the glass. The higher refractive index and dispersion power of leaded glasses are useful properties in compound lenses and for producing brilliance and sparkle in tableware.

Borosilicate glasses are made by replacing a large portion of the alkali and often all the lime in soda-lime glasses with B_2O_3. The use of B_2O_3 can reduce the coefficient of thermal expansion to about one third, thereby improving thermal shock resistance. The Pyrex used for cooking utensils is a common form of borosilicate glass.

Aluminosilicate glasses are used where high strength at high temperatures is required. The high alumina and silica contents increase melting temperature and reduce coefficient of expansion.

Ninety-six percent silica glass is a soda-boron oxide silicate glass. The melt is shaped into oversize objects, which are then leached with acid to remove nearly all the soda, leaving a porous glass. Upon firing the glass shrinks about 14% linearly and becomes dense, having a composite of about 96% silica. The high silica content gives excellent high-temperature strength and corrosion resistance and a coefficient of expansion less than one tenth that of soda-lime glass, resulting in excellent resistance to thermal shock.

Fused silica, made from pure silica sand, has physical properties slightly superior to those of 96% silica glass but is much more difficult to fabricate because of the high melting temperature. It is therefore used only in applications where its superior properties are worth the much higher manufacturing cost.

B. AGGREGATE AND FIBER PROPERTIES

Aggregates can be conveniently classed as dense or lightweight. Dense aggregates include sand, gravel, and crushed rock, as well as artificial products such as air-cooled blast furnace slag and broken brick. Lightweight aggregates include pumice, clinker, foamed slag, expanded shale and slate, exfoliated vermiculite, and expanded volcanic glasses such as perlite.

1. Dense Aggregates

a. SUITABILITY OF AGGREGATE DEPOSITS

Aggregate may be obtained from sand and gravel deposits or from quarries in bedrock outcropping. Geologic maps of the area in question will provide clues to the occurrence of suitable aggregate deposits, as well as rock formations where quarrying and plant operations might be possible [12]. In the western United States sand and gravel deposits are commonly the most accessible source. They are usually stream deposits but may also be glacial deposits or alluvial fans. Talus slopes can occasionally be processed for use, and fine blending sands can sometimes be obtained from windblown deposits.

Stream deposits are the most common and generally most desirable, because (1) streams exercise a sorting action which improves grading, and (2) weaker materials are partially removed in the processes of stream abrasion, transportation, and deposition. Sand and gravel deposits may occur in a stream channel, along its borders, or on adjoining terrace remnants of previously existing stream beds.

b. CLASSIFICATION AND IDENTIFICATION

Classification. On the basis of origin, rocks are igneous, metamorphic, or sedimentary. Table 5 indicates the morphologic distinctions among rocks in these three classes, and Table 6 shows the principal mineral constituents of igneous rocks which are often encountered in geologic reports of aggregate sources.

Most igneous rocks are excellent aggregate materials. Exceptions include tuffs and certain lavas which are porous due to inclusion of gas bubbles. These, however, are frequently used in lightweight concrete. Sedimentary rocks are extremely variable in their suitability as aggregate. Sandstones and limestones are suitable if hard and dense. The value of sandstones as aggregate depends principally on the type of rock fragments and cementing agents involved. For example, an unweathered siliceous quartzite (quartz particles cemented with silica) would normally provide a hard, tough, dense aggregate, but an argillaceous graywacke (particles containing significant amounts of feldspar cemented with consolidated clay) would be unsatisfactory. Limestones may also contain clay, grading into limey shales. Shales generally form poor aggregate, being soft, light, weak, and absorptive. Due to their thin bedding nature, they also tend to form flat, slabby shapes when reduced to sand and gravel. Metamorphic rocks, as a class, lie intermediate between igneous and sedimentary rocks as suitable aggregates. Table 7 summarizes engineering properties of the major rock types.

TABLE 5. Classification of Commonly Encountered Rocks

Igneous (solidified from a molten state)		
Coarse-grained crystalline	*Fine-grained crystalline (or crystals and glass)*	*Fragmental (crystalline or glassy)*
Origin: Deep intrusion slowly cooled	Origin: Quickly cooled volcanic or shallow intrusive	Origin: Explosive volcanic fragments deposited as sediments
Granite Diorite Gabbro (↑ Increasing quartz and light minerals)	(↓ Increasing dark minerals) Rhyolite Andesite Basalt	Ash and pumice (volcanic dust or cinders) Tuff (consolidated ash)
Note: Rock names are based upon mineral content. Color may be used as a rough index as noted above.	Essentially glass (suddenly chilled, few or no crystals) Obsidian, pitchstone, etc.	Agglomerate (coarse and fine volcanic debris)

TABLE 5. (*Continued*)

Sedimentary
(sediments transported by water, air, ice, gravity)

Mechanically deposited		*Chemically or biochemically deposited*	
A. Unconsolidated:		A. Calcareous:	
Clay	According to particle size	Limestone ($CaCO_3$)	
Silt	According to particle size	Dolomite ($CaCO_3 . MgCO_3$)	
Sand		Marl (calcareous shale)	
Gravel		Caliche (calcareous soil)	
Cobbles		Coquina (shell limestone)	
B. Consolidate:		B. Siliceous:	
Shale (consolidated clay)		Chert	
Siltstone (consolidated silt)		Flint	
Sandstone (consolidated sand)		Agate	Spring deposit, vein or cavity filling
Conglomerate (consolidated gravel or cobbles—rounded)		Opal	
		Chalcedony	
Breccia (angular fragments)		C. Others:	
		Coal, phosphate, salines, etc.	

Metamorphic
(igneous or sedimentary rocks changed by heat, pressure)

A. Foliated
- Slate: dense, dark, splits into thin plates (metamorphosed shale)
- Schist: predominantly micaceous, semiparallel lamellae
- Gneiss: granular, banded, subordinately micaceous

B. Massive
- Marble: coarsely crystalline, calcareous (metamorphosed limestone)
- Quartzite: dense, very hard, quartzose (metamorphosed sandstone)

SOURCE: *U.S. Bureau of Reclamation Concrete Manual*, 7th ed. Denver: U.S.B.R., 1966, [12].

TABLE 6. Principal Mineral Constituents of Common Igneous Rocks

Coarsely crystalline rocks	*Principal constituent minerals**	*Finely crystalline or porphyritic rocks*
Granite	Q + O + (P) + A	Rhyolite
Syenite	O + (P) + A	Trachyte
Quartz monzonite	Q + O + P + A	Dellenite
Monzonite	O + P + A	Latite
Quartz diorite	Q + (O) + P + A or B	Dacite
Diorite	(O) + P + A or B	Andesite
Gabbro	P + B	Basalt

*Mineral symbols:
Q—quartz (hard, shiny, conchoidal fracture).
O—orthoclase feldspar (commonly pinkish, unstriated, regular cleavage faces).
P—plagioclase feldspar (commonly white or nearly so, good cleavage faces which are often striated).
A—amphibole and/or biotite.
B—pyroxene.
( )—minerals in parentheses are subordinate in amount.

TABLE 7. Summary of Engineering Properties of Rocks

Type of rock	*Mechanical strength*	*Dura-bility*	*Chemical stability*	*Surface character-istics*	*Presence of undesirable impurities*	*Crushed shape*
Igneous						
Granite, syenite, diorite	Good	Good	Good	Good	Possible	Good
Felsite	Good	Good	Question-able	Fair	Possible	Fair
Basalt, diabase, gabbro	Good	Good	Good	Good	Seldom	Fair
Peridotite	Good	Fair	Question-able	Good	Possible	Good
Sedimentary						
Limestone, dolomite	Good	Fair	Good	Good	Possible	Good
Sandstone	Fair	Fair	Good	Good	Seldom	Good
Chert	Good	Poor	Poor	Fair	Likely	Poor
Conglomerate, breccia	Fair	Fair	Good	Good	Seldom	Fair
Shale	Poor	Poor		Good	Possible	Fair to poor
Metamorphic						
Gneiss, schist	Good	Good	Good	Good	Seldom	Good to poor
Quartzite	Good	Good	Good	Good	Seldom	Fair
Marble	Fair	Good	Good	Good	Possible	Good
Serpentinite	Fair	Fair	Good	Fair to poor	Possible	Fair
Amphibolite	Good	Good	Good	Good	Seldom	Fair
Slate	Good	Good	Good	Poor	Seldom	Poor

SOURCE: Federal Highway Administration Manual, *The Identification of Rock Types.* Washington D.C.: U.S. Govt. Printing Office, 1963.

Identification. Practice in identifying rock types rapidly leads to adequate proficiency for field reconnaissance work. Table 8 provides a system for the preliminary identification of common rocks.

TABLE 8. Preliminary Identification of Common Rocks

Group I. Glassy, wholly or partly.
Group II. Not glassy; dull or stony, homogeneous; so fine-grained that grains cannot be recognized.
Group III. Distinctly granular.
Group IV. Distinctly foliated; no effervescence with acid.
Group V. Clearly fragmental in composition; rounded or angular pieces or grains cemented together.

Group I. Glassy Rocks

1. Glassy luster; hard; conchoidal fracture; colorless to white or smoky gray; generally brittle. *Quartz.*
2. Solid glass; may have spherical inclusions; brilliant vitreous luster; generally black. *Obsidian.*
3. Cellular or frothy glass. *Pumice.*

Group II. Dull or stony, very fine-grained rocks

Subgroup IIA. Not scratched by fingernail but readily scratched with knife.
1. Particles almost imperceptible; dull luster; homogeneous; clay odor; little if any effervescence with acid; laminated structure; breaks into flakes. *Shale.*
2. Little if any clay odor; brisk effervescence with acid. *Limestone.*
3. Little if any clay odor; brisk effervescence with acid only when rock is powdered or acid is heated. *Dolomite.*

4. Soapy or greasy feel; translucent on thin edges; green to black; no effervescence. *Serpentinite.*

Subgroup IIB. Not scratched with the knife or scratched only with difficulty; no effervescence with acid.

1. Light to gray color; clay odor possible; may have a banded flow structure. *Felsite.*
2. Very hard; pale colors to black; no clay odor; conchoidal fracture; waxy or horny appearance. *Chert.*
3. Heavy; dark color; may have cellular structure; may contain small cavities filled with crystalline minerals. *Basalt.*

Group III. Granular rocks

Subgroup IIIA. Easily scratched with the knife.

1. Brisk effervescence with acid. *Limestone or marble.*
2. Brisk effervescence only with warm acid, or with powdered rock. *Dolomitic marble.*

Subgroup IIIB. Hard; not scratched with knife or scratched with difficulty; grains of approximately equal size.

1. Mainly quartz and feldspar; usually light colored, sometimes pinkish. *Granite.*
2. Mainly feldspar; little quartz (less than 5%); light colors of nearly white to light gray or pink. *Syenite.*
3. Feldspar and a dark ferromagnesian mineral.
   a. Major constituent feldspar; rock of medium color. *Diorite.*
   b. Ferromagnesian mineral equal to or in excess of feldspar; rock of dark color.
      (1) Grains just large enough to be recognized by the unaided eye. *Diabase.*
      (2) Coarse-grained rock. *Gabbro.*
4. Mainly ferromagnesian minerals; generally dark green to black.
   a. Predominant olivine with pyroxene or hornblende. *Peridotite.*
   b. Predominant augite. *Pyroxenite.*
   c. Predominant hornblende. *Hornblendite.*
5. Mainly quartz.
   a. Fracture around grains. *Sandstone.*
   b. Fracture through all or through an appreciable percentage of grains. *Quartzite.*

Subgroup III*C.* Hard; not scratched with knife or scratched with difficulty; large distinct crystals in finer groundmass.

1. Crystals of feldspar and quartz with some of a ferromagnesian mineral (generally biotite) in a light-colored groundmass. *Granite porphyry.*
2. Crystals of feldspar and usually a ferromagnesian mineral in a light-colored groundmass. *Syenite porphyry.*
3. Crystals of ferromagnesian minerals, or of striated feldspar, or both, in a medium-colored groundmass. *Diorite porphyry.*
4. Crystals of quartz, or feldspar, or both, generally with a ferromagnesian mineral, in a predominant, fine-grained groundmass of light color. *Felsite porphyry.*
5. Crystals of feldspar or of a ferromagnesian mineral or both, in a fine-grained, dark or black, heavy groundmass. *Basalt porphyry.*

Group IV. Foliated rocks

1. Medium to coarse grain; roughly foliated. *Gneiss.*
2. More finely grained and foliated. *Schist.*
   a. Consists mainly or largely of mica with some quartz. *Mica schist.*
   b. Medium green to black; consists mostly of a felted or matted mass of small, bladed or needle-like crystals arranged in one general direction. *Hornblende schist or amphibolite.*
   c. Glassy or silky luster on foliation surfaces; splits readily into thin pieces. *Sericite schist.*
   d. Soft, greasy feel; marks cloth; easily scratched with fingernail; whitish to light gray, or green. *Talc schist.*
   e. Smooth feel; soft; glimmering luster; green to dark green. *Chlorite schist.*
3. Very fine grain; splits easily into thin slabs; usually dark gray, green, or black. *Slate.*

Group V. Fragmental

1. Rounded pebbles embedded in some type of a cementing medium. *Conglomerate.*
2. Angular fragments embedded in a cementing medium. *Breccia.*
3. Fragments of volcanic (fine-grained or glassy) rocks embedded in compacted volcanic ash. *Volcanic tuff or volcanic breccia.*

4. Quartz grains, rounded or angular, cemented together. *Sandstone.*
5. Quartz and feldspar grains cemented together to resemble the appearance of granite. *Arkose* (feldspathic sandstone.)

SOURCE: Federal Highway Administration Manual, *The Identification of Rock Types.* Washington D.C.: U.S. Govt. Printing Office, 1963.

Detailed examination for aggregate acceptance requires microscopic and chemical analyses. ASTM 295 (Appendix E) recommends procedures for petrographic examination of concrete aggregate and has information usable by persons at all levels of ability in rock identification. Petrographic studies commonly consist of two parts: (1) identification of the material or component; (2) an attempt to predict performance from past records of similar materials in service or from theoretical considerations. The references in Chap. 13 provide the essential techniques for petrographic analysis.

c. AGGREGATE PROPERTIES AND SELECTION

Aggregate selection depends on both physical and chemical properties. Typical physical properties include

Specific gravity
Gradation
Particle shape
Surface texture
Compressive strength
Young's modulus
Thermal expansion
Abrasion resistance
Moisture absorption and volume change due to moisture absorption
Volume change due to freezing and thawing

Typical chemical properties include

Presence of water-soluble substances such as gypsum in the form of coatings or seam fillings
Susceptibility to oxidation, hydration, or carbonation when exposed to the atmosphere
Alkali reactivity and other expansive reactions with cement (see Sec. C in Chap. 7)
Effects of various clay, mineral, or organic surface coatings on the bond with cements, asphalts, or other binder phases
Degree of particle coating in bitumen-aggregate mixtures
Effect of water on stripping of bitumen-aggregate mixtures

Two characteristics especially important to bituminous aggregates include hydrophobic surface character (to enhance coating by asphalt and reduce stripping by water) and angular or blocky shape (to improve aggregate interlocking, essential with soft binders such as asphalt).

Appendix E lists frequently used ASTM tests for physical and chemical properties of aggregates used with cements and asphalts. A detailed description of aggregate properties and selection is provided by McLaughlin et al. [13].

d. AGGREGATE BENEFICIATION

It is often more economical to remove unwanted materials from locally available aggregates than to ship higher-quality aggregates from greater distances. Such *beneficiation* is accomplished by a variety of methods, similar to those used in mining industries. Process economy depends on the magnitude of differences in physical properties, such as hardness, density, and elasticity, between the desirable and undesirable constituents. Processes which have been used with variable success include

1. *Crushing*. Crushing reduces soft and friable particles in coarse aggregates, and the degraded material may be eliminated by screening. Certain impact crushers are particularly useful for *selective* crushing. There is always loss of sound material, which may become excessive.
2. *Washing*. Washing is used to remove particle coatings or to remove fine particles. Clay and organic coatings can be removed with water jets while the aggregate passes over screens or conveyor belts or through special wash tanks. Fine particles can be removed either by the use of water jets or by vertical or horizontal flow elutriators, in which the aggregate is submerged in flowing water and the separation of particle sizes depends on Stokes law (see Fig. 1 in Chap. 13).
3. *Specific gravity separation*. Often the more desirable aggregate has greater specific gravity. Advantage is taken of this difference in a variety of processes.
   a. *Heavy media separation*. Aggregate is introduced into a tank containing a controllable specific gravity suspension of water and finely ground magnetite and ferrosilicon. Deleterious lightweight material is removed by flotation, while heavy particles are drawn off from bottom pockets. The separation medium is reused, and the aggregates are washed and stockpiled.
   b. *Jigging*. Upward pulsations created by air or rubber diaphragms tend to hinder settlement of lighter particles, which are removed by skimming devices. Another technique consists of a box with a perforated bottom in which a separating layer is formed by a pulsating water current. Lightweight shales and cherts are often removed successfully by jigging.
4. *Elastic fractionation*. This process is of limited applicability. It involves dropping aggregate onto an inclined steel plate from which those with higher modulus of elasticity bounce farther than the presumably less desirable particles of lower modulus. Suitable collection devices are used to collect the separated fractions.

2. *Lightweight Aggregates*

Lightweight aggregates, used to produce concretes having densities usually below 100 pcf, can be divided into the following groups:

Unprocessed natural materials
Unprocessed by-products
Processed natural materials
Processed by-products
Organic materials

The properties of concrete produced with these various materials are described in Sec. D in Chap. 7.

a. UNPROCESSED NATURAL MATERIALS

Pumice is the only widely used natural lightweight aggregate. Pumice is lava swollen by escaping gases to form a glassy honeycombed mass. It is used chiefly in Greece, Germany, and the northwestern United States. Less frequently used natural lightweight aggregates include scoria, volcanic cinders, tuff, and diatonite.

b. UNPROCESSED BY-PRODUCTS

A classification of all lightweight synthetic aggregates (i.e., excluding only the unprocessed natural materials) recommended by the RILEM Lightweight Concrete Committee is shown in Table 9, and properties of selected synthetic aggregates are compared in Table 10.

TABLE 9. Classification of Synthetic Lightweight Aggregates

Unprocessed by-products

1. Air-cooled blast furnace slag
2. Ground blast furnace slag
3. Furnace clinker
4. Winkle clinker
5. Pulverized fuel ash (fly ash)
6. Furnace bottom ash
7. Crushed bricks
8. Crushed aerated concrete

Processed natural materials

1. Exfoliated vermiculite
2. Expanded clay
3. Expanded diatomite
4. Expanded obsidian
5. Expanded perlite
6. Expanded shale
7. Expanded slate
8. Expanded mixture of sand and limestone
9. Agglomerated clay
10. Sintered clay
11. Sintered desert rock-debris
12. Sintered fire clay
13. Sintered under clay
14. Sintered shale
15. Hollow ceramic particles
16. Ground burnt shale
17. Coated pumice

Processed by-products

1. Expanded blast furnace slag
2. Expanded pulverized fuel ash (fly ash)
3. Foamed blast furnace slag
4. Granulated blast furnace slag
5. Sintered colliery waste
6. Sintered pulverized fuel ash (fly ash)
7. Sintered household refuse
8. Agglomerated coal shale
9. Agglomerated furnace clinker
10. Expanded glass waste

Organic materials

1. Plastic particles
2. Chaff
3. Wood particles
4. Wood fibers

SOURCE: RILEM Lightweight Concrete Committee, *Materiaux et Construction*, 13, No. 3 (1970) 61. [14].

TABLE 10. Properties of Selected Synthetic Aggregates

Coarse synthetic aggregate	*Aggregate dry unit weight (pcf)*	*Concrete unit weight (pcf)*	*28-Day compressive strength (psi)*	*Thermal conductivity (K factor)*	*% Absorption by weight*	*Cost ($/cu yd)*
Expanded shale, clay, slate	35–65	70–115	2,000–6,000	1.9–4.5	5–15	5–12
Expanded slag	40–55	70–115	2,000–6,000	1.7–3.3	5–25	2–4.7
Sintered fly ash	37–63	85–120	2,000–6,000	1.7–5.0	14–24	4–7
Exfoliated vermiculite	4–12	25–60	95–420	0.5–1.1	20–35	4.3–13†
Expanded perlite	4–15	35–50	80–500	0.5–1.0	10–50	3.2–12‡
Expanded glass	15–30	75	1,300 [7-day]	2.8	5–10	
Expanded polystyrene beads	(2–10)*	20–55	100–1,800	(< 1.0)		7.5§
Carbonized wheat	(4–12)	40	800	(< 1.0)	—	2.7
Air-cooled slag	70–90	115–130	3,000–6,000	(6–8)	1–5	1.4–7.5
Synopal	(65–80)	(115–125)	(3,000–6,000)	(6–8)	(0.7–2.0)	(30)
Granulated slag	55–65				(4–15)	0.6–1.2
Brick rubble	~47	110–120	1,100–3,000	3.8–5.0	19–36	
Fused clinker	~50	65–100		1.8–3.8	16–23	
Steel shot	~300	340–375	3,000–6,000	~35	0	432
Crushed stone¶	90–110	140–150	3,000–7,000	~10	0.5–2.0	3.4–5.4

*Figures in parentheses are estimates.
†Based upon a price of $0.04/lb.
‡Based upon a price of $0.03/lb.
§Based upon a price of $0.28/cu ft
¶Natural aggregate for comparison.
SOURCE: *Proceedings of the Conference on New Materials in Concrete Construction*, S. P. Shah, ed., University of Illinois at Chicago Circle, Dec. 1971, pp. I–VIII, [15].

Unprocessed by-products, such as furnace clinker and cinders, are cheap but extremely variable in quality, compared with products which are processed from either natural or by-product materials. Only clinker and cinders from high-temperature combustion in industrial coal or coke furnaces is used, cinders from other sources being generally unsuitable. Waste glass and glass in combination with sand and asbestos fibers are being evaluated as aggregates in bituminous concrete [16, 17] and as aggregate in masonry block [18]. Cationic antistripping additives are used to maintain asphalt coatings on the glass particles.

Wastes such as mine tailings can sometimes be used directly as aggregate but are not used extensively because they generally occur in low construction areas.

c. PROCESSED NATURAL MATERIALS AND BY-PRODUCTS

Currently the United States produces about 1 ton of refuse per capita annually, plus 10 million tons of iron and steel scrapped each year, over 3 billion tons of waste rock and mine tailings, and large amounts of industrial wastes such as slag and fly ash. The incentive to use these wastes is particularly strong in construction, since most of the wastes are generated in regions of high population density, where large volumes of construction materials are required. Whereas natural aggregate sources are constantly being depleted, solid waste sources of aggregate are continually being replenished.

The raw materials in the manufacture of processed aggregates are either (1) these by-products, such as fly ash and blast furnace slag, or (2) natural deposits such as clays, shales, or slates.

Production processes. The manufacture of processed aggregates depends on one or more of the following techniques [19]:

1. Vaporization of components of the raw materials, coincident with incipient fusion of the mineral, causing expansion by the entrapped gases.
2. Mixing a molten mass with controlled amounts of water to produce a cellular structure by entrapped steam.
3. Burning off the combustible materials in a matrix.

The cells produced by the various processes and materials may very widely in size and be predominantly connected or discrete. The major manufactured aggregates and their corresponding production processes include

Expanded clay, shale, and slate. These aggregates are made by heating prepared materials in a rotary kiln or traveling sinter grate to the fusion point (1000–1200°C) where they become soft and expand because of entrapped expanding gases. In some cases the particles are coated with a material of higher fusion point to prevent agglomeration during heating.

Expanded slag. Expanded slag aggregates are produced by treating blast-furnace slag with water. The molten slag is run into pits containing water or is broken up by mechanical devices and subjected to water sprays. The products are fragments which have been vesiculated by steam. Blast furnace slags have been found to produce consistently higher concretes than steelmaking slags. Table 11 indicates the amounts of these two materials used for various other purposes in the construction industry [20].

TABLE 11. Slag Uses in the United States (in Thousands of Tons) (1968)

Use	*Blast furnace*	*Steel*	*Total*
Portland cement concrete	3,381	—	3,381
Bituminous concrete pavements	3,910	779	4,389
Concrete masonry block	1,833	—	1,833
Railroad ballast	4,223	492	5,015
Bases for cement & bituminous concrete	11,199	4,379	15,578
Cement manufacture	1,131	—	1,131
Roofing slag (cover and granules)	519	—	519
Mineral wool	416	—	416
Agricultural slag	61	85	146
Glass	181	—	181
Sewage trickling filter medium	24	—	24
Antiskid and ice control	107	93	200
Paths, driveways, and parking areas	40	54	94
Fill	1,461	70	1,531
Miscellaneous	258	258	516
Totals	28,744	6,210	34,954

Slag and expanded clay or shale are the dominant processed aggregates in the United States, with about 10 and 5.5 million tons, respectively, currently used in structural concrete and concrete products each year.

Sintered fly ash. The fly ash is first pulverized and aggregated in the desired shape either by pelletizing or extrusion. The pellets are then sintered at 1000–1200°C in a rotary kiln or on a traveling sinter grate. In some cases powdered coal is blended with the fly ash to facilitate sintering [21].

Exfoliated vermiculite. Vermiculite is a mica-like mineral which when heated to 700–900°C exfoliates, giving a product weighing 5–15 pcf, depending on grading.

Expanded perlite. Volcanic glasses such as perlite, obsidian, and rhyolite contain small amounts of moisture. Rapid heating to incipient fusion at about 1000°C causes the resulting steam to produce expansion, with densities comparable to those for exfoliated vermiculite.

The products mentioned so far account for over 95% of the synthetic aggregate used in concrete. Some newer synthetics include

Expanded glass, produced in a rotary kiln or fluidized bed at 700–800°C from a pelletized mixture of glass and an expanding component [22].

Synopal, produced in a double kiln process in which the materials react with one another in the first kiln and are annealed in the second. Synopal consists primarily of calcium silicate (pseudowollastonite). It is white and is used on highway surfaces for its high reflectivity and skid resistance [15].

d. ORGANIC MATERIALS

Organic materials used for aggregate include sawdust and wood shavings, foamed polystyrene beads, and expanded carbonized cereal grains.

Sawdust and wood shavings cannot be used without special treatments, because most woods contain soluble carbohydrates such as sugar, tannin, and aromatic oils which may delay or even prevent the setting of cement. Douglas fir and larch can, however, usually be used without special precautions.

Expanded polystyrene. Expanded polystyrene beads can be produced by steam-heating polystyrene granules containing a blowing agent which is usually a low-mocular-weight saturated hydrocarbon [23]. Upon expanding, the bulk density is reduced from about 50 pcf to 1 pcf. The beads are usually cylindrical or spherical, varying in size from about $\frac{1}{16}$ to $\frac{1}{4}$ in. The beads have negligible moisture absorption due to their closed cellular structure and the hydrophobic nature of polystyrene. They provide excellent thermal insulation and can be treated to make them self-extinguishing in a flame, although they melt instantly. Also the soft beads offer little resistance to curing shrinkage of polystyrene concrete (on the order of 0.1 to 0.2%), which is a limiting factor in some applications. Fine sand may be added to the cement paste to reduce shrinkage, but this also reduces compressive strength.

The expanded beads may be precoated with suitable wetting agents such as a water-soluble epoxy to ensure coating by the cement paste. Ordinary air entraining and/or plasticizing agents also serve this purpose. Typically, 28-day compressive strengths of 150 and 450 psi are achieved at densities of 25 to 40 pcf, respectively; 40 pcf is approximately the density for maximum packing of the beads with the voids completely filled with cement paste. Higher and lower densities represent separation of the beads with paste and closely packed beads with interstitial air voids, respectively. Polystyrene concretes are used for insulating ceiling and floor slabs, sandwich core for wall panels, and road and railway base courses in permafrost areas.

Carbonized cereal grains. Carbonized cereals such as wheat, oats, corn, and rice are first puffed, and then the volatiles are "distilled" off, leaving an inert carbon residue which maintains the shape of the original particles. Their use as concrete aggregate has been proposed for insulating foundations for buildings, roads, and railways in permafrost areas and for high-temperature insulating fire brick [24].

The grains are expanded to about 25 times their original volume by the conventional puffing or extrusion methods used for breakfast cereals, usually by extrusion. The expanded grains are then carbonized by passing through an oven and have a typical bulk unit weight after carbonization of 1.5 pcf. Up to 40% of the original weight is given off as volatile gases, which can be recycled as fuel to the oven burners. The resulting carbon beads have slightly lower thermal transmission than loose polystyrene beads, can be produced cheaply, and are stable to as high as 3000°F when embedded in concrete paste. Most matrix materials, including portland cement, disintegrate at lower temperatures than the carbon beads. Since the beads are almost pure carbon, they are very inert and resistant to biological attack and most chemicals. The carbon beads are hydrophobic and do not suffer the disadvantages of high moisture absorption experienced with most lightweight mineral aggregates. Various abrasion-resistant coatings for the beads have been evaluated. Coming from cereal grains, the carbon beads have a renewable source, in contrast to, say, plastic foams which derive from nonreplenishable petroleum supplies.

3. Fibers

Large volumes of fibers are used annually in construction composites. Asbestos is the principal natural fiber. Man-made fibers include a variety of polymers, glasses, and metals.

a. ASBESTOS

The most commonly used asbestos is a fibrous variety of serpentine known as chrysotile, a hydrated magnesium silicate (3 MgO-2 SiO_2-2 H_2O). Chrysotile accounts for 95% of the world production of asbestos, the remainder consisting of other fibrous minerals of the serpentine and amphibole groups.

Asbestos cement typically contains 1 part of asbestos to between 6 and 12 parts of weight of cement. An excess of water is used for mixing and is subsequently drawn off during the molding process. Figure 12 illustrates several other materials in which asbestos reinforcement is used.

Asbestos Cement Products

Asphalt Coatings and Cements

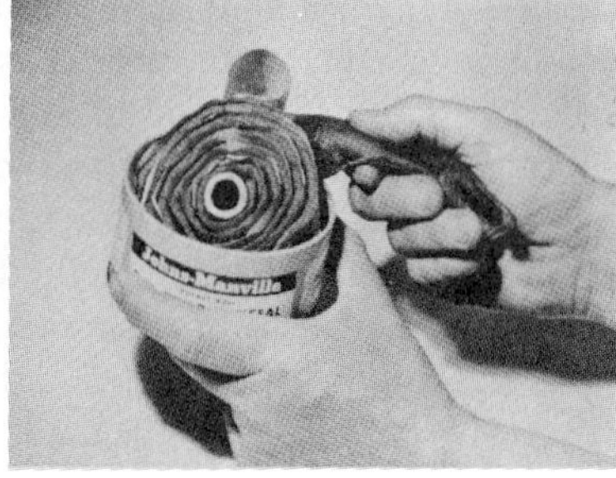

Asphalt Paving

Caulking Compounds

Cements—Insulating

Paper

Floor Tile

Friction Materials

Fig. 12. Some products containing asbestos fiber reinforcement. (*Courtesy of Johns Manville Corporation.*)

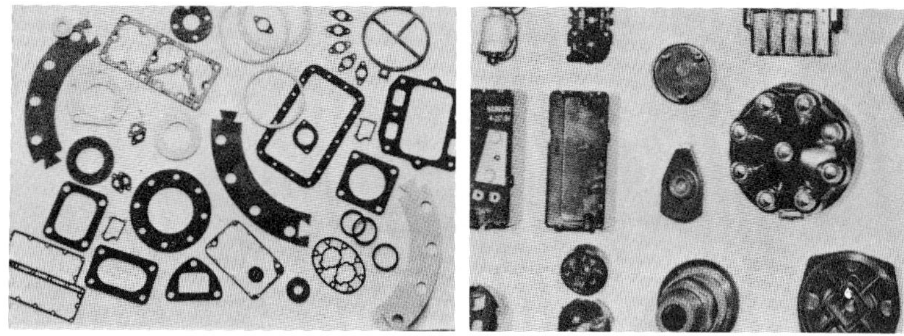

Gaskets

Plastics (Filler)

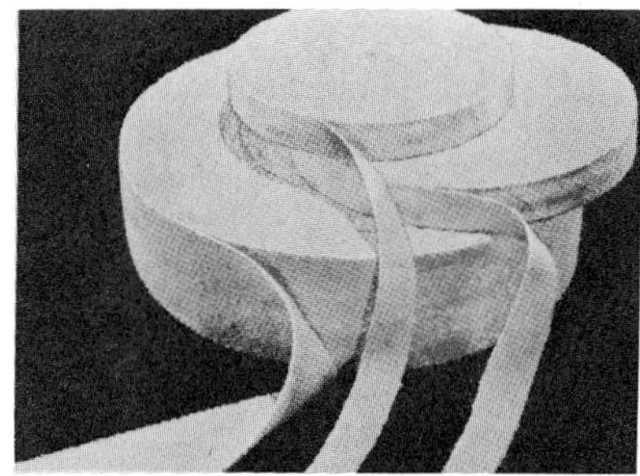

Textiles

Texture Paints

*Effect of temperature on loss in weight of asbestos fibers**

Temp. °F	*% of loss in weight chrysotile*
400	0.30
600	0.86
700	1.78
800	2.17
900	2.83
1000	3.99
1100	10.38
1200	12.75
1400	13.43
1600	13.62
1800	13.77

*Time: 2 hours

Effect of heat on tensile strength of Canadian chrysotile crude

		Tensile strength (lb/sq in.)	*% of original tensile strength*
Original crude— No heat		131,000	—
	600°F	120,000	91.6
Heated 3	800°F	96,000	73.3
min. at	1000°F	78,000	59.5
	1200°F	42,000	32.0

Comparison of approximate fiber diameters

Type of fiber	*Fiber diameter in inches*	*Fibers in one linear inch*
Human Hair	0.00158	630
Ramie	0.000985	1,015
Wool	0.0008 to 0.0011	910–1,250
Cotton	0.0004	2,500
Rayons	0.0003	3,300
Nylon	0.0003	3,300
Glass	0.00026	3,840
Rock Wool	0.000142 to 0.000284	3,520–7,040
Asbestos (Chrysotile)	0.000000706 to 0.00000118	850,000–1,400,000

Fig. 12 (Continued)

Crude asbestos is milled into fibrous form and sold in various grades which depend on the degree of milling (fiberization) and upon the length distribution (Fig. 13). Chrysotile, for example, is graded by the asbestos industry into groups 1 to 7. Group 1 contains the longest fibers ($\frac{3}{4}$ in. or greater) and group 7 the shortest fibers. Shorter fibers are much less expensive. Table 12 compares the tensile strength and surface area of chrysotile asbestos with several other commonly used fibers.

TABLE 12. Comparison of Tensile Strength and Surface Area of Chrysotile Asbestos with Other Fibers

Material	*Tensile strength (lb/sq in.)*	*Fiber*	*Surface area, by N_2 absorption (cm^2/g)*
Ingot iron	45,000	Nylon	3,100
Wrought iron	48,000	Acetate rayon	3,800
Carbon steel	155,000	Cotton	7,200
Ni-Cr steel	243,000	Silk	7,600
Cotton fiber	73,000–89,000	Wool	9,600
Glass fiber	100,000–200,000	Viscose rayon	9,800
Chrysotile asbestos	80,000–100,000	Asbestos (chrysotile)	130,000–220,000

Asbestos fiber is frequently used in conjunction with glass fiber to reinforce plastics. Finely divided short asbestos fiber can improve the bond between resin and glass, reduce shrinkage crazing of resin, make the resin tackier, reduce the thermal coefficient of expansion, and augment the glass in increasing mechanical properties of the composite structure [25]. A glass-asbestos reinforcement mix can also substantially reduce product costs.

b. MAN-MADE FIBERS

The properties of glass fibers and their production methods are summarized in Chap. 11. New organic fibers are now becoming available having strengths and Young's moduli on the order of 400×10^3 and 16×10^6 psi, respectively. The combination of high strength and low specific gravity (~ 1.47) gives them higher specific tensile strengths than currently available glass-, boron-, or graphite-reinforcing fibers, with stiffness intermediate between those of glass and graphite [26]. In common with organic fibers, they are much more handleable than mineral or metal fibers and do not cause balling in mixing operations. Unlike common organic fibers such as nylon, polypropylene, and polyethylene, however, their current costs limit their use primarily to aircraft and aerospace applications.

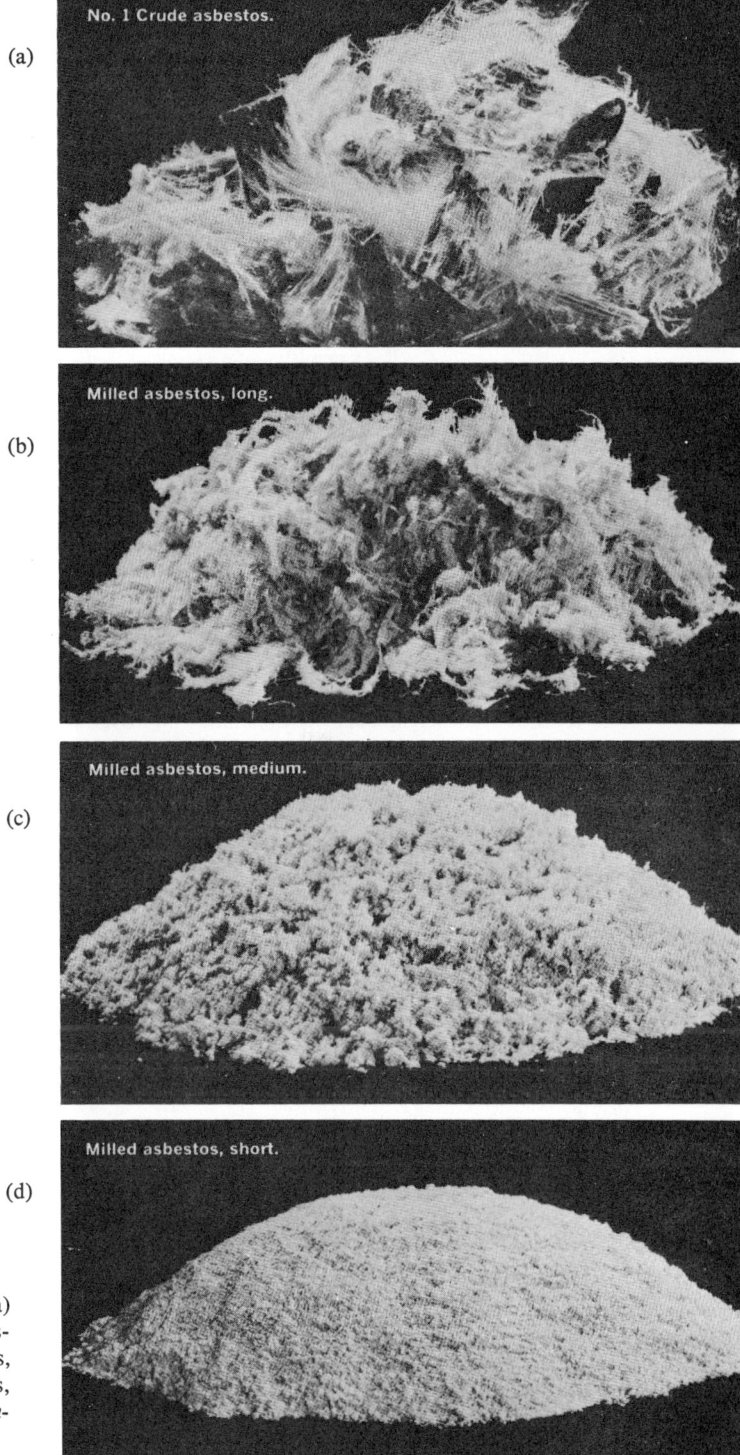

Fig. 13. Grades of asbestos. (a) Crude asbestos; (b) milled asbestos, long; (c) milled asbestos, medium; (d) milled asbestos, short. (*Courtesy of Johns Manville Corporation.*)

REFERENCES

1. KREBS, R. D., and R. D. WALKER, *Highway Materials.* New York: McGraw-Hill Book Company, 1971.
2. GILLOTT, J. E., *Clay in Engineering Geology.* New York: American Elsevier Publishing Company, Inc., 1968.
3. VAN VLACK, L. H., *Elements of Materials Science.* Reading, Mass.: Addison-Wesley Publishing Company, Inc., 1959.
4. BRINDLEY, G. W., *Proc. Natl. Conf. Clays, Clay Minerals,* 14 (1965) 27.
5. EAST, W. H., *J. Am. Ceram. Soc.,* 33 (1950) 211.
6. HAUTH, W. E., "Crystal Chemistry of Ceramics," *Bull. Am. Ceram. Soc.* (1951).
7. BATES, T. F., F. A. HILDEBRAND, and A. SWINEFORD, *Am. Mineral.,* 35 (1950) 463.
8. GRIM, R. E., *Clay Mineralogy,* 2nd ed. New York: McGraw-Hill Book Company, 1968.
9. VAN VLACK, L. H., *Materials Science for Engineers.* Reading, Mass.: Addison-Wesley Publishing Company, Inc., 1970.
10. CHARLES, R. J., "The Nature of Glasses," *Sci. Am.,* 217 (Sept. 1967). 127.
11. KEYSER, C. A., *Materials Science in Engineering.* Columbus, Ohio: Charles E. Merrill Publishers, 1968.
12. *U.S. Bureau of Reclamation Concrete Manual,* 7th ed., U.S.B.R., 1966.
13. MCLAUGHLIN, J. E., et al., "Distribution, Production, and Engineering Characteristics of Aggregates," in *Highway Engineering Handbook,* K. B. Woods, ed. New York: McGraw-Hill Book Company, 1960, Chap. 16.
14. RILEM Lightweight Concrete Committee, *Materiaux et Construction,* 13, No. 3 (1970) 61.
15. BERGER, R. L., *Proceedings of the Conference on New Materials in Concrete Construction,* S. P. Shah, ed. University of Illinois at Chicago Circle, Dec. 1971, pp. I–VIII.
16. MALISCH, W. R., D. E. DAY, and B. G. WIXSON, *Proceedings of the 2nd Mineral Waste Utilization Symposium,* M. A. Schwartz, ed. Chicago, U.S. Bureau of Mines, March 1972, p. 371.
17. BENNETT, W. R., "Crushed Glass in Asphalt Pavement Construction," *D.H.O. Report IR 41.* Ontario, Canada: Department of Highways, May 1971.
18. PHILLIPS, C., D. S. CAHN, and G. W. KELLER, *Proceedings of the 3rd Mineral Waste Utilization Symposium,* M. A. Schwartz, ed. Chicago, U.S. Bureau of Mines, March 1972, p. 385.
19. *ACI Manual of Concrete Practice,* Part 1: Structural Lightweight Aggregate Concrete. Detroit: American Concrete Institute, 1970.
20. EGGLESTON, H. K., *Proceedings of the 2nd Mineral Waste Utilization Symposium,* M. A. Schwartz, ed. Chicago, U.S. Bureau of Mines, March 1970, p. 15.
21. MINNICK, J. L., *Hwy Res. Rec.,* 307 (1970) 21.
22. LYNDON, F. D., *Concr. Mag.,* 9, No. 2 (1968) 355.
23. EICH, H., "Styropor-Concrete," *Zement-Kalk-Gips,* 6, No. 12 (1959) 253.
24. RILEY, V. R., and J. TIMUSK, "New Concepts for Panels," Toronto: University of Toronto Systems Building Centre Report, May 1972.
25. LUBIN, GEORGE, ed., *Handbook of Fiberglass and Advanced Plastics Composites.* New York: Van Nostrand Reinhold Company, 1969.
26. MOORE, J. W., "A New Organic High Modulus Reinforcing Fiber," Wilmington, Del.: *Du Pont Textile Fibers Department Report PRD-49,* 1973.

III

AGGREGATE–BINDER COMPOSITES

7

CONCRETE MIX DESIGN

> There are traces of age in the one-hoss shay,
> A general flavor of mild decay,
> But nothing local, as one may say.
> There couldn't be—for the Deacon's art
> Had made it so like in every part.
>
> —HOLMES

Concrete is a mixture of gravel, crushed rock, sand, slag, or other natural or man-made *aggregates* bound together by some form of cement. The aggregates are normally graded from about 100 μ (10^{-4} cm) up to a maximum size usually limited by the dimensions of the structure to be formed, but a particle size generally not exceeding 6 in. Particles larger than about $\frac{3}{16}$ in. are called coarse aggregate and smaller particles fine aggregate. A concrete mixture of only fine aggregate (such as sand) and cement is called mortar. Concrete normally contains about 7–14% cement, 15–20% water, and 66–78% aggregate by weight, plus any desired amount of entrained air.

Portland cement concrete is the world's most widely used manufactured building material. About 4 billion tons, one for each person on earth, are manufactured each year. In the United States and Canada, much of this is precast concrete, representing a billion dollar industry (since 1973) which grows almost 20% per year.

A. VOLUME RELATIONSHIPS

Optimal grading of concrete aggregates has been studied for many years [1, 2]. Some considerations include

1. Other factors being equal, for aggregate having higher strength than the cement paste and lower cost per unit volume, an aggregate gradation approaching the *minimum aggregate void ratio* is economically favored. The minimum void ratio is obtained from a smooth grading curve, using the widest possible spread in sizes between the largest and smallest particles. For given compactive effort, the minimum void ratio is also obtained with approximately spherically shaped particles.
2. The *maximum aggregate size* is governed only by the aggregate-size-handling limitations in mixing and placing the concrete and by the size of the structural member being formed. Other things being equal, a larger maximum aggregate size reduces the void ratio and therefore cement demand.
3. The *minimum size* limit of aggregate may be governed, among other factors, by the adverse effect of very fine particles upon the nucleation and growth of interlocking tobermorite crystals. For example, although adding extremely finely ground rock flour to a given aggregate may reduce its void ratio, thereby reducing the cement demand, the resulting concrete may be weaker.
4. A useful maximum size of the smallest aggregate may also exist, because the cement shrinkage strain fields around fine aggregate can be accommodated without cracking, whereas shrinkage strains about larger particles in the absence of finer aggregate result in cracking and therefore reduced strength.

Consider the first of these, the void ratio, in greater detail. The void ratio is the volume of aggregate voids divided by the volume of aggregate particles. It depends on three factors:

1. Particle geometry.
2. Particle gradation.
3. Nature and amount of compactive effort.

With regard to *particle geometry*, one can imagine various idealized shapes such as rectangular prisms (ranging from rectangular needles through cubes to plates) or hexagonal prisms of similar proportions. For a unary aggregate (all particles of one size and shape) having any of these shapes, ordered packing (as by hand) could produce a zero void ratio. Void ratios for the corresponding materials with random packing would depend on the particle shape and the nature and magnitude of compactive effort. Generally, the higher the sphericity (the more closely the particle resembles a sphere), the lower the void ratio under random packing, due to the ease with which spheres roll over and slide past one another during compaction.

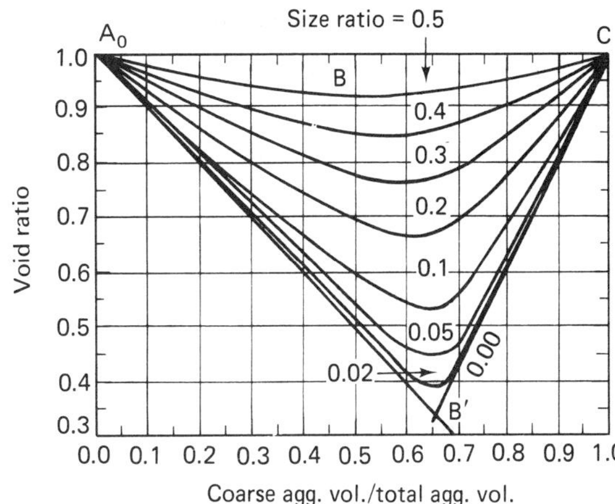

Fig. 1. Void ratio of binary mixture as a function of aggregate volume ratio and size ratio. From Furnas, C. C., Washington, D. C.: U. S. Bureau of Mines *Report of Investigation* No. 2894. 1928 [4].

With regard to *particle gradation*, consider a binary mixture, a mixture of two different but uniformly sized aggregates. Assume for ease of visualization that the two sizes differ greatly, as by a factor of 10. If we pour fine sand into a box of pebbles, the void ratio decreases, as the original void spaces become partially filled with sand particles. If we place pebbles in a box of sand, the void ratio also decreases, as certain spaces initially occupied by sand of a given void ratio become occupied by solid stone. Thus, any binary mixture will have a lower void ratio than a unary mixture of the same aggregate shape. But the reduction in void ratio is not so great as might be imagined, since the void ratio in the vicinity of particles packed against a surface is always greater than the void ratio in the interior of the aggregation of particles [3]. Similar *boundary voids* exist wherever small particles touch large ones, causing a dilation effect whenever particles of different sizes are mixed.

The thickness of the layer of increased void ratio in particles resting against a surface clearly increases with particle size. Therefore, the average void ratio within the fine aggregate of a binary mixture tends to decrease, the smaller the fine aggregate, other factors being similar. Stated differently, the greater the difference in size between fine and coarse aggregates in a binary mixture, the lower the minimum void ratio which can be obtained. This is illustrated in Fig. 1, where void ratios of binary mixtures are plotted against volume ratio of the coarse aggregate for various particle size ratios, assuming the two individual groups each have void ratios of 1. The straight lines intersecting at B' indicate hypothetical (but physically impossible) mixtures having zero dilation at contacts between the two aggregate sizes. The same general principles apply to mixtures of two aggregates each having a gradation in particle sizes, so long as their mean particle sizes differ, and also to mixtures of more than two aggregate sizes.

The general conclusion may be drawn that the minimum void ratio is obtained with a maximum difference in size ratio between the smallest and largest aggregate and a fairly uniform grading between these two sizes. For an idealized case of spherical aggregates with the largest spheres in hexagonal close packing, this can be visualized by assuming the selection of adequate numbers of successively smaller spheres to just fit snugly into the largest remaining voids (Fig. 2). In real mixes, randomness of packing dilates this representation, but the relationship remains.

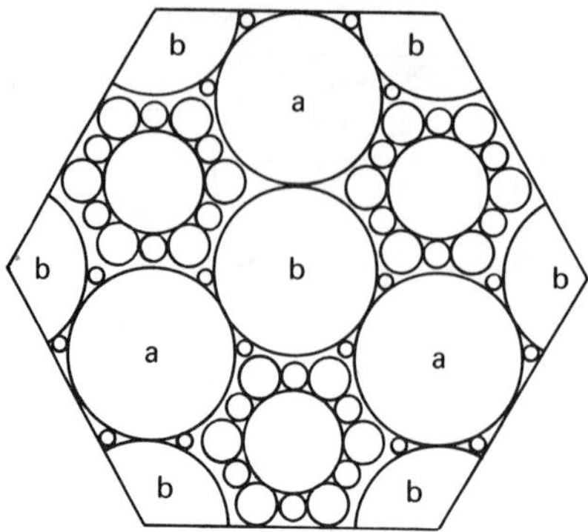

Fig. 2. Minimum void ratio with ordered packing of spherical aggregates. Cross section on the C plane with hexagonal close packing of the largest spheres. Largest spheres labeled "a" project above the plane of the paper, those labeled "b" project below.

The size distribution of an aggregate may be specified by the formula

$$p = 100\left(\frac{d}{D}\right)^n \tag{1}$$

where p = percentage weight of material passing a given sieve of opening width d,
D = maximum particle size in the aggregate,
n = an exponent related to the coarseness or fineness of gradation.

Studies by Fuller and Thompson indicated that $n \sim \frac{1}{2}$ for a minimum aggregate void ratio [5].

B. MIX OPTIMIZATION

Consistency, strength, and durability are the three essentials designed into a concrete mix. Consistency facilitates proper placement in the forms, strength provides structural performance, and durability ensures a reasonable service life.

A variety of tests is used to specify these three properties of concrete (see the listing under Chap. 7 in Appendix E). For example, consistency of fresh concrete is measured by the slump test or by the Kelley ball test. Strength is assured by specifying properties such as strength and gradation of aggregate, cement type and content, and water/cement ratio. Durability is the long-term resistance to disintegration of concrete by chemical or physical weathering. Durability may be assured by specifying properties such as resistance to sulfate attack of the cement paste, to harmful chemical reactions between cement and aggregate, and required air content to control cracking caused by freezing of unreacted water in the cement paste. The particular test methods and required levels of performance vary with the job requiremments, geographic locality, and the client or code-establishing agency.

1. Aggregates

"... For most purposes, aggregates should consist of clean, hard, strong, and durable particles free of chemicals or coatings of clay or other fine material that may affect coatings or as loose, fine material. Many of them can be removed by proper washing.

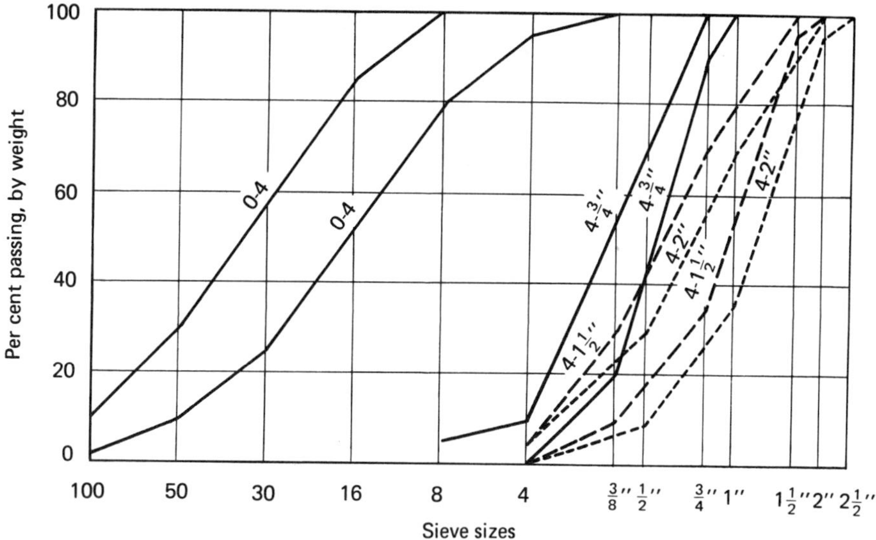

Fig. 3. Grading limit curves for fine aggregate and three sizes of coarse aggregate (ASTM C33).

". . . The commonly used aggregates are sand, gravel, crushed stone, and blast furnace slag. Cinders, burnt clay, expanded blast furnace slag, and other materials are also used.

"Very sharp and rough aggregate particles or flat and elongated particles require more fine material to produce workable concrete than aggregate particles that are more rounded or cubical. When the aggregates are made up largely of such particle shapes, more cement may therefore be required. Excellent concrete is made by using crushed stone and other crushed materials but the particles should be more or less cubical in shape. . . ." [6]*

Generally, aggregates with smooth grading curves produce the most satisfactory results. The ASTM C33 grading limits for fine aggregate and for No. 4 to 1-in. and No.4 to $1\frac{1}{2}$-in. coarse aggregate are shown in Fig. 3. Specifications for particular applications may be made more restrictive.

The *fineness modulus* of an aggregate is the sum of the cumulative percentages of material retained on the standard sieve sizes divided by 100. For example,

Sieve Size	*Cumulative % Retained*
No. 4	3
8	12
16	30
30	49
50	75
100	97
	266/100 = 2.66 (fineness modulus)

*Excerpts by permission of the Portland Cement Association.

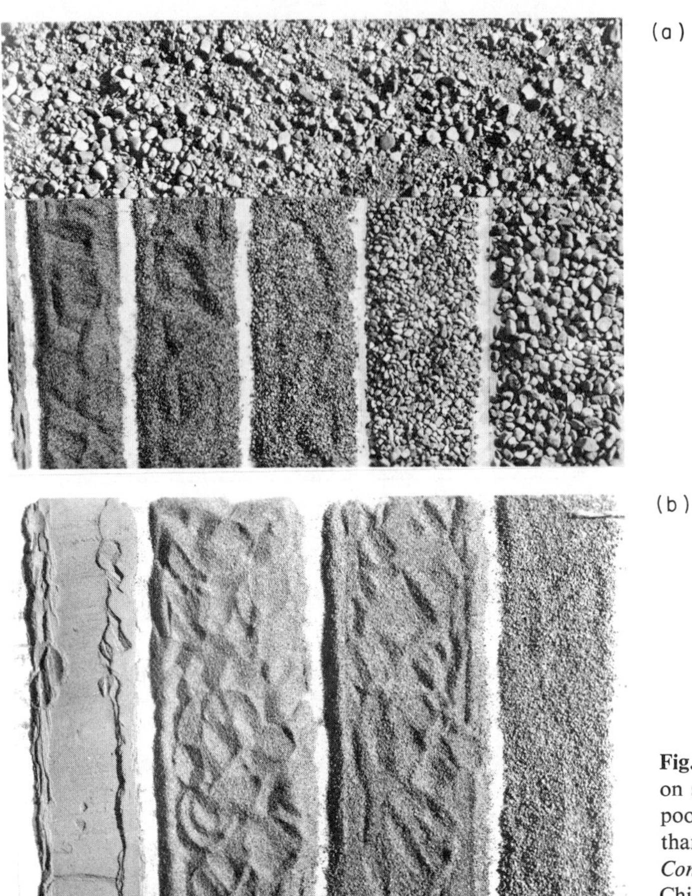

Fig. 4. Sands before and after separation on standard sieves. (a) well-graded; (b) poorly-graded; lacking particles coarser than $\frac{1}{16}$ in. From *Design and Control of Concrete Mixtures*, Twelfth Edition. Chicago: Portland Cement Association, 1970 [6].

Very coarse sands give harsh, unworkable mixes, and very fine sands require larger amounts of cement. Figure 4 shows examples of well-graded and poorly graded sands. For leaner mixes, or when using small-sized coarse aggregate, a grading approaching the maximum percentage on each sieve is desirable. For richer mixes, a grading approaching the minimum percentage passing each sieve is more economical. The Bureau of Reclamation requires a fineness modulus of sand between 2.50 and 3.00 and size distribution within the following limits:

Sieve Size	*Cumulative % Retained*
No. 4	0–5
8	10–20
16	20–40
30	40–70
50	70–80
100	92–98

Fig. 5. Schematic representation of bleeding. From Czernin, W.: *Cement Chemistry and Physics for Engineers.* New York: Chemical Publishing Company, Inc., 1962 [7].

For thin walls and smooth surfaces cast against forms, the fine aggregate should contain at least 15 and 4% passing the No. 50 and 100 sieves, respectively [6].

Concrete mixes containing insufficient fine sand are also susceptible to *bleeding*, a condition in which aggregate particles settle within the fresh concrete, forcing water to the surface, resulting in a lower-strength top layer which may eventually scale off. The upward flow of water may form channels in the concrete, making it vulnerable to freezing and thawing. Fine particles may settle from under larger particles, leaving a water-rich cement slurry of low strength (Fig. 5). Similar behavior under reinforcing steel may reduce bond strength and resistance to corrosion. Bleeding is also caused by a too-high water/cement ratio or by insufficient aggregate of any size in the grading curve, a condition called *gap grading*.

2. *Trial Mix Design*

The allowable maximum particle size, minimum strength, consistency, and sometimes air content of a concrete are usually specified, based upon its intended use. From these specifications, the mix designer calculates a trial mix using tables based upon extensive prior experience for preparation in the laboratory. The trial mix may then be modified one or more times to improve desired concrete properties or economy. The mix may be further modified in the field when the nature of available aggregates, weather, and other factors vary.

Certain physical properties of the fine and coarse aggregates and cement must be determined before the trial mix can be designed. These include the specific gravity, moisture content, and absorption of fine and coarse aggregates; fineness modulus of the fine aggregate; dry rodded unit weight of the coarse aggregate; and the specific gravity of cement, usually taken as 3.15.

Of several current trial mix design methods, the ACI method is most frequently used. Tables 1–7 contain the necessary data [8]. Two trial mix designs will be illustrated.

TABLE 1. Concrete Mixes for Small Jobs*

Maximum size of aggregate (in.)	Mix designation	Approximate bags of cement per cu yd of concrete	Aggregate, lb per l-bag batch: Sand†, Air-entrained concrete‡	Sand†, Concrete without air	Gravel or crushed stone	Iron blast furnace slag
½	A	7.0	235	245	170	145
	B	6.9	225	235	190	165
	C	6.8	225	235	205	180
¾	A	6.6	225	235	225	195
	B	6.4	225	235	245	215
	C	6.3	215	225	265	235
1	A	6.4	225	235	245	210
	B	6.2	215	225	275	240
	C	6.1	205	215	290	255
1½	A	6.0	225	235	290	245
	B	5.8	215	225	320	275
	C	5.7	205	215	345	300
2	A	5.7	225	235	330	270
	B	5.6	215	225	360	300
	C	5.4	205	215	380	320

Note: Air-entrained concrete should be used in all structures which will be exposed to alternate cycles of freezing and thawing.

*May be used without adjustment.

†Weights are for dry sand. If damp sand is used, increase weight of sand 10 lb for one-bag batch, and if very wet sand is used, add 20 lb for one-bag batch.

‡Air-entrained concrete can be obtained by the use of an air-entraining cement or by adding an air-entraining agent. If an agent is used, the amount recommended by the manufacturer will, in most cases, produce the desired air content.

Procedure: Select the proper maximum size of aggregate, and then, using mix B, add just enough water to produce a sufficiently workable consistency. If the concrete appears to be undersanded, use mix A, and if it appears to be oversanded, use mix C.

SOURCE: ACI 613–54.

TABLE 2. Maximum Sizes of Aggregate Recommended for Various Types of Construction

Minimum dimension of section (in.)	Maximum size of aggregate (in.)*: Reinforced walls, beams, and columns	Unreinforced walls	Heavily reinforced slabs	Lightly reinforced or unreinforced slabs
2½–5	½–¾	¾	¾–1	¾–1½
6–11	¾–1½	1½	1½	1½–3
12–29	1½–3	3	1½–3	3
30 or more	1½–3	6	1½–3	3–6

*Based upon square openings.

SOURCE: ACI 613–54.

TABLE 3. Recommended Slumps for Various Types of Construction*

Types of construction	Slump (in.)† Maximum	Minimum
Reinforced foundation walls and footings	5	2
Plain footings, caissons, and substructure walls	4	1
Slabs, beams, and reinforced walls	6	3
Building columns	6	3
Pavements	3	2
Heavy mass construction	3	1

*Adapted from Table 4 of the 1949 Joint Committee on Recommended Practice and Standard Specifications for Concrete and Reinforced Concrete.
†When high-frequency vibrators are used, the values given should be reduced about one third.
SOURCE: ACI 613–54.

TABLE 4. Approximate Mixing Water Requirements for Different Slumps and Maximum Sizes of Aggregates*

	Water (gal/cu yd of concrete for indicated maximum sizes of aggregate)							
Slump (in.)	$\frac{3}{8}$ *in.*	$\frac{1}{2}$ *in.*	$\frac{3}{4}$ *in.*	1 *in.*	$1\frac{1}{2}$ *in.*	2 *in.*	3 *in.*	6 *in.*
	Non-air-entrained concrete							
1–2	42	40	37	36	33	31	29	25
3–4	46	44	41	39	36	34	32	28
6–7	49	46	43	41	38	36	34	30
Approximate amount of entrapped air in non-air-entrained concrete, %	3	2.5	2	1.5	1	0.5	0.3	0.2
	Air-entrained concrete							
1–2	37	36	33	31	29	27	25	22
3–4	41	39	36	34	32	30	28	24
6–7	43	41	38	36	34	32	30	26
Recommended average total air content, %	8	7	6	5	4.5	4	3.5	3

*These quantities of mixing water are for use in computing cement factors for trial batches. They are maxima for reasonably well-shaped angular coarse aggregates graded within limits of accepted specifications. If *more* water is required than shown, the cement factor, estimated from these quantities, *should* be increased to maintain desired w/c ratio, except as otherwise indicated by laboratory tests for strength. If *less* water is required than shown, the cement factor, estimated from these quantities, *should not* be decreased except as indicated by laboratory tests for strength.
SOURCE: ACI 613–54.

TABLE 5. Compressive Strength of Concrete for Various Water/Cement Ratios*

Water/cement ratio (gal/bag of cement)	*Probable compressive strength at 28 days (psi)*	
	Non-air-entrained concrete	*Air-entrained concrete*
4	6,000	4,800
5	5,000	4,000
6	4,000	3,200
7	3,200	2,600
8	2,500	2,000
9	2,000	1,600

*These average strengths are for concretes containing not more than the percentages of entrained and/or entrapped air shown in Table 4.4. For a constant w/c ratio, the strength of the concrete is reduced as the air content is increased. For air contents higher than those listed in Table 4.4, the strengths will be proportionally less than those listed in this table.

Strengths are based on 6- by 12-in. cylinders moist-cured 28 days.

SOURCE: ACI 613–54.

TABLE 6. Maximum Permissible Water/Cement Ratios (Gal/Bag) for Different Types of Structures and Degrees of Exposure

Type of structure	*Exposure conditions**					
	Severe wide range in temperature or frequent alternations of freezing and thawing (air-entrained concrete only)			*Mild temperature rarely below freezing or rainy or arid*		
	In air	*At the water line or within the range of fluctuating water level or spray*		*In air*	*At the water line or within the range of fluctuating water level or spray*	
		In fresh water	*In seawater or in contact with sulfates†*		*In fresh water*	*In seawater or in contact with sulfates†*
This sections, such as railings, curbs, sills, ledges, ornamental or architectural concrete, reinforced piles, pipe, and all sections with less than 1-in. concrete cover over reinforcing	5.5	5.0	4.5‡	6	5.5	4.5‡
Moderate sections, such as retaining walls, abutments, piers, girders, beams	6.0	5.5	5.0‡	∥	6.0	5.0‡
Exterior portions of heavy (mass) sections	6.5	5.5	5.0‡	∥	6.0	5.0‡
Concrete deposited by tremie under water	—	5.0	5.0	—	5.0	5.0
Concrete slabs laid on the ground	6.0	—	—	∥		
Concrete protected from the weather, interiors of buildings, concrete below ground	∥	—	—	∥		
Concrete which will later be protected by enclosure or backfill but which may be exposed to freezing and thawing for several years before such protection is offered	6.0	—	—	∥		

*Air-entrained concrete should be used under all conditions involving severe exposure and may be used under mild exposure conditions to improve workability of the mixture.

†Soil or ground water containing sulfate concentrations of more than 0.2%.

‡When sulfate-resisting cement is used, maximum w/c ratio may be increased by 0.5 gal/bag.

∥Water/cement ratio should be selected on basis of strength and workability requirements.

SOURCE: ACI 613–54.

TABLE 7. Volumes of Coarse Aggregate per Unit of Volume of Concrete*

Maximum size of aggregate (in.)	*Volume of dry-rodded coarse aggregate per unit volume of concrete for different fineness moduli of sand*			
	2.40	2.60	2.80	3.00
$\frac{3}{8}$	0.46	0.44	0.42	0.40
$\frac{1}{2}$	0.55	0.53	0.51	0.49
$\frac{3}{4}$	0.65	0.63	0.61	0.59
1	0.70	0.68	0.66	0.64
$1\frac{1}{2}$	0.76	0.74	0.72	0.70
2	0.79	0.77	0.75	0.73
3	0.84	0.82	0.80	0.78
6	0.90	0.88	0.86	0.84

*Volumes are based upon aggregates in dry-rodded condition as described in Standard Method of Test for Unit Weight of Aggregate (ASTM C29). These volumes are selected from empirical relationships to produce concrete with a degree of workability suitable for usual reinforced construction. For less workable concrete such as required for concrete pavement construction, they may be increased about 10%.
SOURCE: ACI 613–54.

	EXAMPLE 1		EXAMPLE 2	
		Data		
Structure:	Lightly reinforced warehouse floor slab, 6 in. thick, requiring 4,000 psi 28-day compressive strength		Reinforced bridge abutment wall, 12 in. thick, requiring 3,000 psi 28-day compressive strength	
Aggregates	*Fine*	*Coarse*	*Fine*	*Coarse*
Specific gravity	2.60	2.72	2.64	2.70
Moisture content	3.5%	1.5%	4.0%	0.5%
Absorption	1.5%	1.0%	2.5%	1.5%
Fineness modulus	2.90		2.50	
Dry-rodded unit weight		110 pcf		104 pcf
		Solution		
Volumes				
Maximum aggregate size (Table 2)	say 2 in.		say 3 in.	
Slump (Table 3)	say 3 in.		say 3 in.	
Mix water (Table 4)	$\frac{34\text{ gal/cu yd}}{7.5}$	= 4.54 cu ft	$\frac{28\text{ gal/cu yd}}{7.5}$	= 3.73 cu ft
Air (Table 4)	0.5% × 27	= 0.13	3.5% × 27	= 0.94
Water/cement ratio (Tables 5, 6)	6 gal/bag		6 gal/bag	
Cement	34/6	= 5.67 bags	28/6	= 4.67 bags
	$\frac{5.67 \times 94}{3.15 \times 62.4}$	= 2.71	$\frac{4.67 \times 94}{3.15 \times 62.4}$	= 2.23
Coarse aggregate (Table 7)	$\frac{0.74 \times 110 \times 27}{2.72 \times 62.4}$	= 12.92	$\frac{0.83 \times 104 \times 27}{2.70 \times 62.4}$	= 13.82
Sum		19.30		20.73
Sand	27 − 19.30	= 7.70 cu ft	27 − 20.72	= 6.28 cu ft
Weights including moisture corrections				
Cement	5.67 × 94	= 532 lb	4.67 × 94	= 440 lb
Sand	7.70 × 2.60 × 62.4 = 1,250		6.28 × 2.64 × 62.4 = 1,036	
	1,250(1 + 0.035)	= 1,295	1,036(1 + 0.040)	= 1,076
Coarse aggregate	12.92 × 2.72 × 62.4 = 2,190		13.82 × 2.70 × 62.4 = 2,330	
	2,190(1 + 0.015)	= 2,200	2,330(1 + 0.005)	= 2,340
Water	4.54 × 62.4 = 283		3.73 × 62.4 = 233	
	283 + 1,250(0.015 − 0.035) + 2,190(0.010 − 0.015) = 247		233 + 1,036(0.025 − 0.040) + 2,330(0.015 − 0.005) = 241	

These proportions would then be used for the first trial batch.

(a)

(b)

(c)

Fig. 6. Concrete mixtures. (a) Insufficient cement-sand mortar to fill all spaces between coarse aggregate particles. Concrete will be difficult to place and be porous; (b) correct amount of cement-sand mortar; (c) excess of cement-sand mortar. Concrete is uneconomical of cement and will likely be porous. *From Design and Control of Concrete Mixtures*, Twelfth Edition. Chicago: Portland Cement Association, 1970 [6].

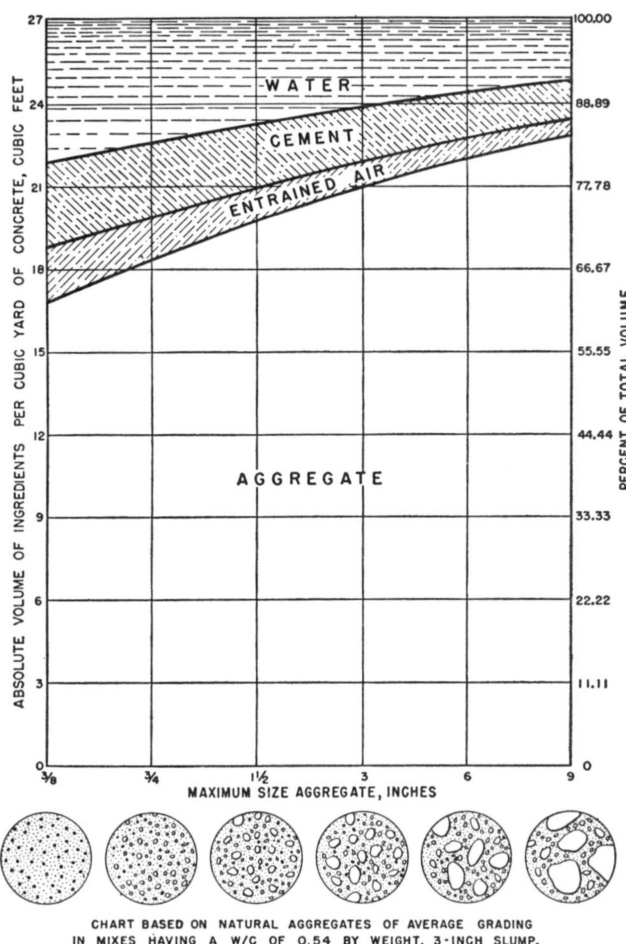

Fig. 7. Amounts of water, cement, and entrained air for various maximum sizes of aggregate. Mixes having larger coarse aggregate require less water and less cement per cu. yd. than do mixes with small coarse aggregate. From *U. S. Bureau of Reclamation Concrete Manual*, Eighth Edition. Washington, D. C.: U. S. Government Printing Office, 1975 [9].

Figure 6 shows concrete ranging from too lean to too rich in cement-sand mortar. Figure 7 indicates amounts of water, cement, and entrained air for various maximum aggregate sizes, suggested for exposed concrete by the U.S. Bureau of Reclamation.

3. *Minimum-Cost Mix Design*

a. GRAPHICAL METHODS

Various schemes are used to minimize the cost of a concrete mix while satisfying constraints on properties such as strength, consistency, aggregate gradation, and air content. An important subproblem is the minimum-cost blending of two or more aggregates having different unit costs in order to satisfy specified gradation limits on the combined aggregate. Several graphical methods are available. One method consists of the following steps [10]:

1. Plot the percentages of aggregate *A* passing various sieve sizes on line *a* and of aggregate *B* on line *b* (Fig. 8).

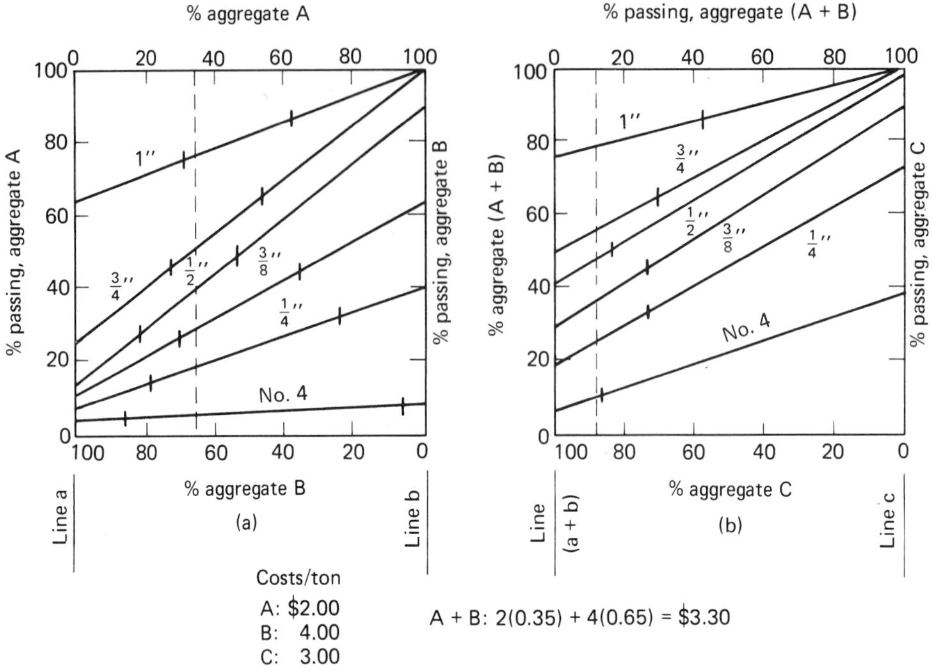

Fig. 8. Graphical method of blending aggregates. (a) Two aggregates; (b) three aggregates.

2. Connect the percentages passing various sieve sizes on lines *a* and *b* by straight lines.
3. Place tick marks on each sieve line indicating the specified limits for the particle size.
4. Draw a vertical line which falls between the two ticks on each line, if possible, and as close to 100% of the lower-priced aggregate as possible. The line drawn in Fig. 8(a) represents 35% of aggregate *A* and 65% of *B*.
5. Extend the method to three or more aggregates, as needed, by drawing additional graphs. Project the percentages passing each sieve size for the combined aggregate $A + B$ horizontally onto line $a + b$ [Fig. 8(b)]. The process is then repeated in the same manner for combined aggregate $A + B$ and aggregate *C*, plotted on line *c*. The answer is 12% $A + B$ and 88% *C*, or equivalently 4.2% *A*, 7.8% *B*, and 88% *C*. The intersection of the final vertical line with each sieve line indicates the particle gradation of the blend, i.e.,

Sieve Size	*Percentage Passing*
1	79
$\frac{3}{4}$	56
$\frac{1}{2}$	47
$\frac{3}{8}$	35
$\frac{1}{4}$	26
No. 4	9

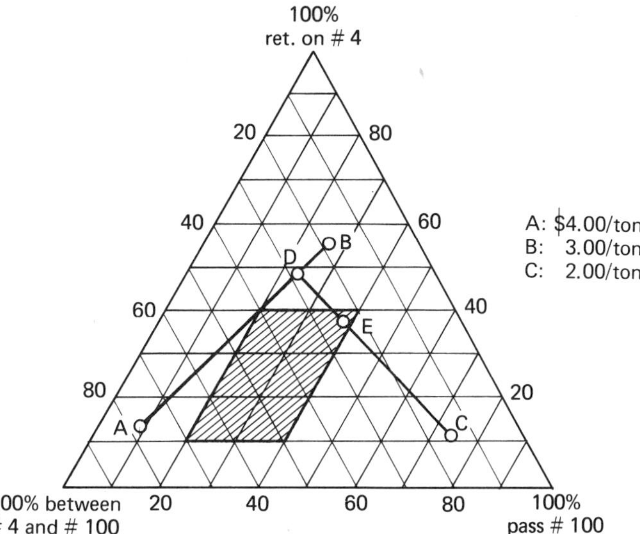

Fig. 9. Chart for combining three aggregates, specification limits on two sieves.

This is a trial-and-error method. The minimum-cost mix of two aggregates in Fig. 8(a) does not ensure the minimum cost of all three aggregates in Fig. 8(b). Also, a vertical line which falls outside of some of the tick ranges in Fig. 8(a) may lead not only to an acceptable but also to an optimal blend of three aggregates in Fig. 8(b).

Figure 9 shows an alternative scheme useful for blending three aggregates when the gradation is specified on only two sieves [11]. On the triangular chart are plotted the specified grading limits (crosshatched) and the gradations of the three aggregates, Points *A*, *B*, and *C*. Point *A*, for example, represents an aggregate having 13% retained on No. 4, 78% between Nos. 4 and 100, and 9% passing No. 100. The objective is to blend the three aggregates to obtain a minimum-cost mix within the crosshatched area. Points *A*, *B*, and *C* are connected by lines as shown such that point *E* lies within the crosshatched area and as close to the cheapest and/or as far from the most costly aggregate as possible. The mix proportions are obtained by the lever rule, using the lengths of line segments between the various points:

$$\% \text{ aggregate } A = \frac{CE}{CD} \cdot \frac{DB}{AB}, \quad \text{etc.}$$

$$\% \text{ aggregate } B = \frac{CE}{CD} \cdot \frac{AD}{AB}$$

$$\% \text{ aggregate } C = \frac{DE}{DC}$$

$$= 100 - \%A - \%B$$

Figure 10 shows a second scheme for blending three aggregates but with gradation limits specified on any number of sieves instead of just two [12]. The gradations for individual aggregates are shown at the three corners of the triangle. The three rectangular sides are graphs of two aggregate mixes, plotted identically to that in Fig. 8(a). The tick marks are projected onto the sides of the triangle (circles). The

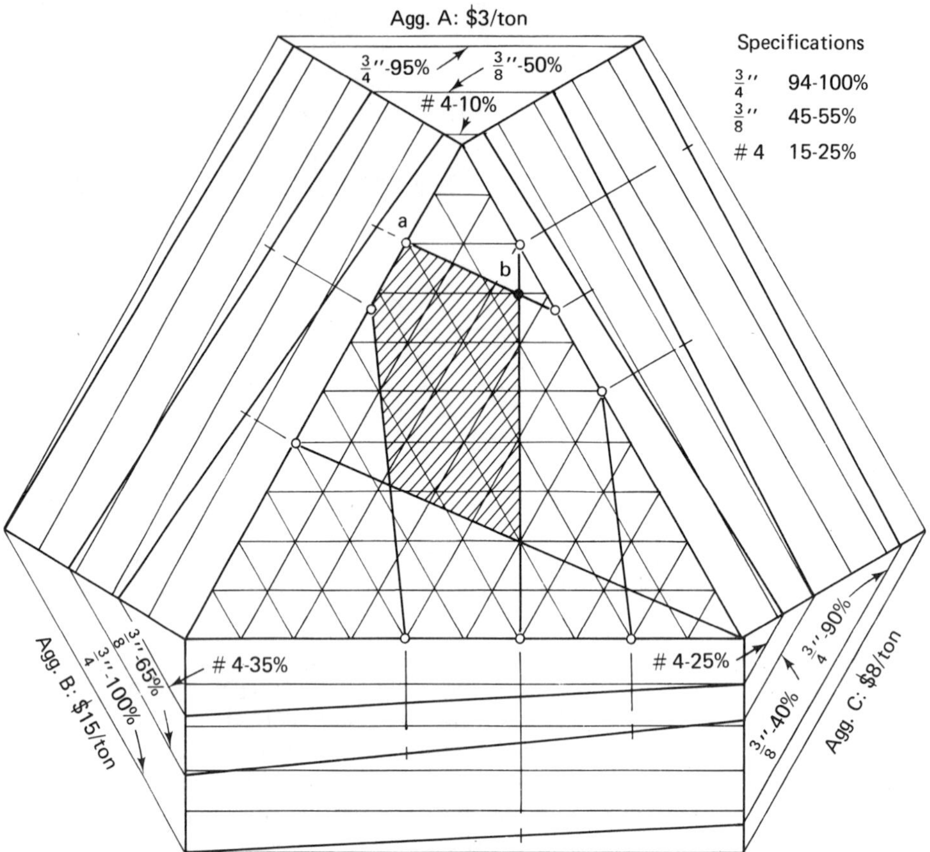

Fig. 10. Chart for combining three aggregates, specification limits on any number of sieves.

circles representing corresponding sieve limits on the three sides of the triangle are then connected by straight lines, forming an envelope of the allowable blend proportions (crosshatched). Since it is not readily apparent from the graph which of vertices *a* or *b* of the feasible region represents the minimum-cost blend, the cost is evaluated at each of these points, revealing the optimum mix at point *b*, for which $A = 70\%$, $B = 5\%$, $C = 25\%$, and cost $= 0.7(3) + 0.05(15) + 0.25(8) = \$4.85$/ton.

Note that if each of the aggregates were within specifications on any sieve size, no limits could be indicated on the triangular chart for that sieve size, and any combination of the three aggregates would satisfy specifications for that size. Note also that the upper and lower specification limit lines in the triangular region for any particle size are parallel. This can be shown to be a necessary condition of the graphical construction, and it provides a quick check of graphical accuracy.

The use of triangular charts can be extended to a wide range of composite material design problems where the mix percentages of *three components* are to be selected to satisfy *any number* of composite material property constraints. Consider an example:

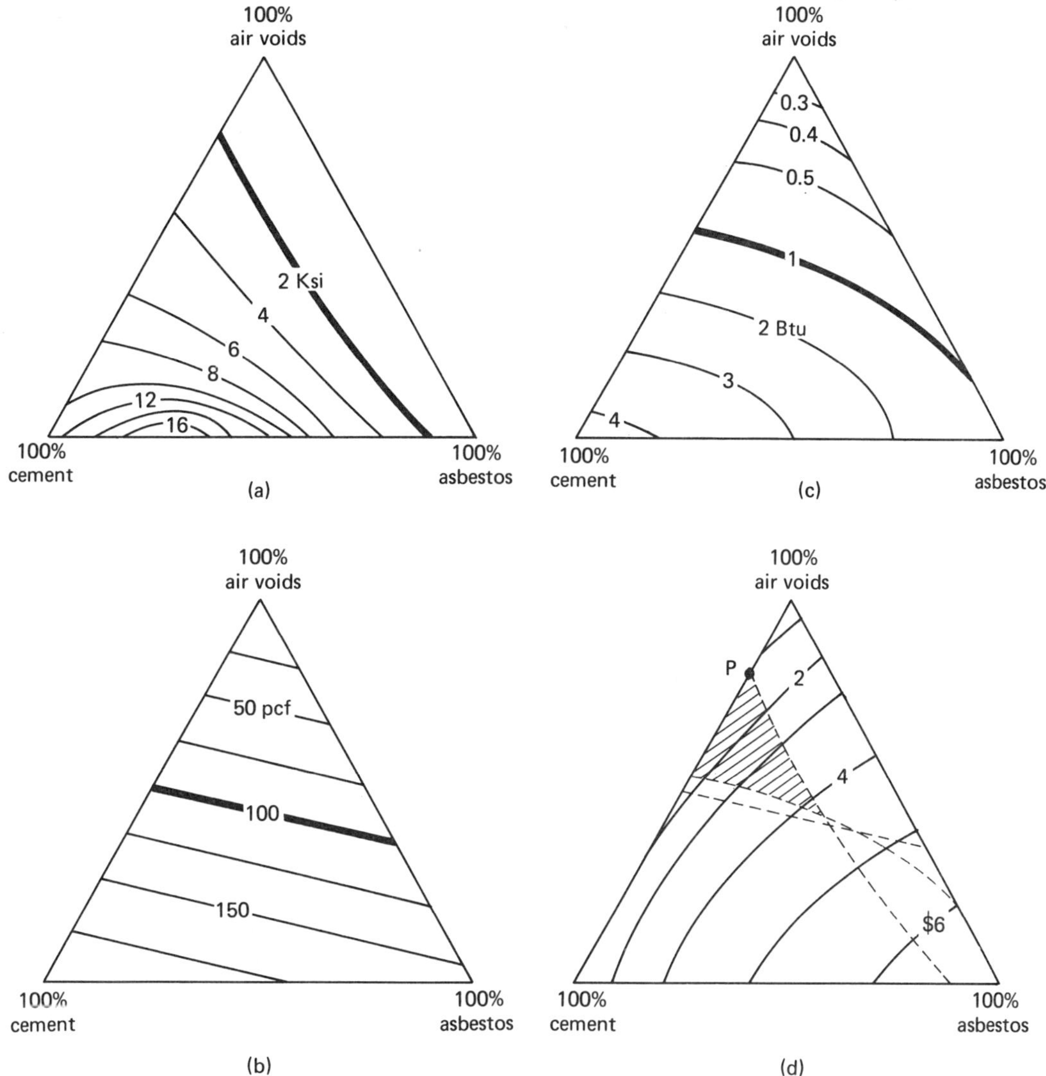

Fig. 11. Property maps for foamed cement-asbestos fiber composites, based on the volume percentages of cement paste, asbestos fiber, and air. (a) Compressive strength, Ksi; (b) density, pcf; (c) thermal conductivity, BTU/sq ft/hr/°F/in. thickness; and (d) manufacturing cost, \$/cu ft.

EXAMPLE 3: Figure 11 shows ternary property maps of (a) compressive strength, (b) thermal conductivity, (c) density, and (d) manufacturing cost per unit volume of an autoclaved foamed cement-asbestos fiber mixture to be used in load-bearing thermal insulation panels. Determine the minimum-cost volume percentages of cement, asbestos fiber, and air to satisfy the following constraints:

(a) Compressive strength $\geq$ 2,000 psi.
(b) Density $\leq$ 100 pcf.
(c) Thermal conductivity $\leq$ 1.0 Btu/sq ft/hr/°F/in. thickness.

The constraint values are shown as heavy lines in Fig. 11(a), (b), and (c). The three constraints are superposed on the cost graph in Fig. 11(d) (dashed lines), leaving the crosshatched portion as the feasible region, within which the minimum-cost design, point P, is selected by inspection, i.e., 21% cement, 79% air, and 0% asbestos, by volume.

The method can be used for any ternary composite, for example,

Components	*Property Constraints*
Cement-fine aggregate-coarse aggregate	Strength, durability, workability
Asphalt-aggregate-asbestos fiber	Air voids, Marshall stability, tensile strength
Cement-sand-polystyrene bead	Density, fire rating, compressive strength, thermal conductivity
Cement-fiber-polymer impregnate	Compressive strength, tensile strength
Resin-glass fiber-boron fiber, etc.	Tensile strength, modulus

The usual requirement would be to optimize one composite material property (or cost) subject to constraints on all remaining properties.

In the case of properties which all vary linearly with composition, the technique can, in principle, be extended to four-component composites by plotting property values within an isometric projection of a tetrahedron. At this level of complexity, however, linear programming becomes more attractive, or one of the nonlinear optimization techniques, if property values do not vary linearly with composition.

b. LINEAR-PROGRAMMING METHODS

The constitutive equations for many of the properties of multicomponent systems are linear in concentration terms. For such cases the material design problem reduces to linear programming, for which standard computer techniques are available [13]. Consider an example:

EXAMPLE 4: A concrete is required having a maximum aggregate size of $\frac{3}{4}$ in., a minimum compressive strength of 5,000 psi, a minimum water/cement weight ratio of 0.3, and the following aggregate gradation limits:

Sieve Size	*Percentage Passing by Weight*	
	Minimum	*Maximum*
#4	30	48
#30	9	27

(The ASTM gradation limits on only two sieve sizes have been selected to simplify the illustration.) Find the weight percentages of cement, water, and fine and coarse aggregates which minimize the mix cost, given the following data.

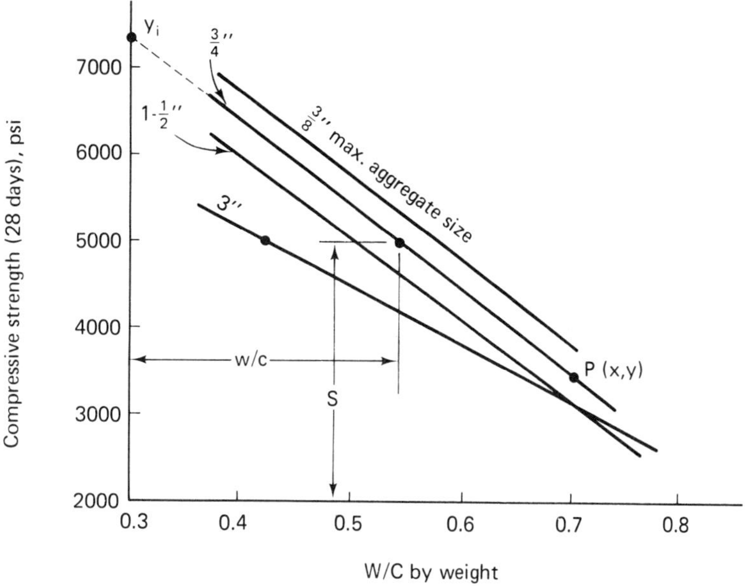

Fig. 12. Compressive strength as a function of water-cement ratio for various maximum aggregate sizes. From Cannon, J. D. and G. R. Murti, *J. Cement and Concrete Res.* 1, (1971) 353 [13].

1. Two aggregates are available, having the following gradations:

Sieve Size	*Percentage Passing by Weight* — *Coarse Aggregate*	*Fine Aggregate*
#4	10	92
#30	0	52

2. Unit costs, dollars per ton:

Cement	$C_c = 25$
Coarse aggregate	$C_a = 3.50$
Fine aggregate	$C_f = 4.50$
Water	$C_w = 0$

3. The compressive strength can be approximated as a linear function of the water/cement ratio, based upon maximum aggregate size, as averaged from numerous experimental results (Fig. 12) [14, 15, 16].

Solution: To ensure adequate strength, given the maximum aggregate size, we write, for the equation of the straight line,

$$\frac{w}{c} \leq \frac{x(y_i - s)}{y_i - y}$$

or

$$\frac{w - cx(y_i - s)}{y_i - y} \leq 0 \tag{2}$$

where w = weight ratio of water to concrete,
c = weight ratio of cement to concrete,
s = required compressive strength,
y_i = y intercept of the line for the appropriate maximum aggregate size,
x, y = coordinates of another point on the same line.

Arbitrarily selecting point P on the $\frac{3}{4}$-in. maximum aggregate size line, $(x,y) = (0.7, 3{,}400)$ and $y_i = 7{,}300$. Substituting into Eq. (2),

$$w - 0.413c \leq 0$$

The percentage of aggregate mix passing any sieve is

$$\frac{a_i a + f_i f}{a + f}$$

and the gradation limits are

$$l_i \leq \frac{a_i a + f_i f}{a + f} \leq u_i, \qquad \text{all } i$$

where a = weight ratio of coarse aggregate to concrete,
f = weight ratio of fine aggregate to concrete,
a_i = percentage coarse aggregate passing ith sieve,
f_i = percentage fine aggregate passing ith sieve,
l_i = lower gradation limit on ith sieve,
u_i = upper gradation limit on ith sieve.

For the No. 4 and No. 30 sieves,

$$30 \leq \frac{10a + 92f}{a + f} \leq 48$$

and

$$9 \leq \frac{0a + 52f}{a + f} \leq 27$$

respectively.

The concrete workability must also be specified. Workability depends on the angularity and gradation of the aggregate, the cement factor (bags per cubic yard), and the water/cement ratio. Various equations have been developed for concrete workability, or its equivalent, the compacting factor [13], which can be incorporated into a linear-programming model. Assume, for simplicity, that adequate workability for our particular concrete, given the previously stated constraints, is provided by the relationship

$$\frac{c}{a + f} \geq 0.17$$

The linear-programming problem is

$$\text{minimize} \quad Z = c_c c + c_a a + c_f f = 25c + 3.50a + 4.50f$$

$$\begin{aligned}
\text{subject to} \quad & w - 0.3c \geq 0 && \text{minimum water/cement ratio} \\
& w - 0.413c \leq 0 && \text{compressive strength} \\
& c - 0.17a - 0.17f \geq 0 && \text{workability} \\
& w + c + a + f = 1 && \text{summation constraint}
\end{aligned}$$

$$\left.\begin{aligned}
& \frac{10a + 92f}{a + f} \geq 30 \\
& \frac{10a + 92f}{a + f} \leq 48 \\
& \frac{52f}{a + f} \geq 9 \\
& \frac{52f}{a + f} \leq 27
\end{aligned}\right\} \quad \text{gradation}$$

where the gradation constraints have been separated according to lower and upper bounds.

Substituting $w = 1 - c - a - f$ into the first and third constraints, multiplying both sides of the gradation constraints by $a + f$ and transposing, and writing all constraints as "greater than" inequalities,

$$\begin{aligned}
\text{minimize} \quad & Z = 25c + 3.5a + 4.5f \\
\text{subject to} \quad & -1.3c - a - f \geq -1 \\
& 1.413c + a + f \geq 1 \\
& c - 0.17a - 0.17f \geq 0 \\
& -20a + 62f \geq 0 \\
& 38a - 44f \geq 0 \\
& -9a + 43f \geq 0 \\
& 27a - 25f \geq 0
\end{aligned}$$

The problem has seven constraints and three variables (c, a, and f). Computational work can be saved by solving its dual, which has only three constraints and seven variables. (A large number of variables can be solved more easily than a large number of constraints.)

The dual is

$$\text{maximize} \quad Z' + x_1 - x_2 = 0$$

subject to

$$\begin{aligned}
-1.3x_1 + 1.41x_2 + x_3 \qquad\qquad\qquad\qquad\qquad\qquad\qquad\qquad + x_8 \qquad\qquad &= 25 \\
-x_1 + x_2 - 0.17x_3 - 20x_4 + 38x_5 - 9x_6 + 27x_7 \qquad + x_9 \qquad &= 3.5 \\
-x_1 + x_2 - 0.17x_3 + 62x_4 - 44x_5 + 43x_6 - 25x_7 \qquad\qquad + x_{10} &= 4.5 \\
x_1 \text{ to } x_{10} \geq 0 &
\end{aligned}$$

in which x_8 to x_{10} are slack variables. Table 8 gives the solution of the dual. The absence of negative coefficients in the final Z' row indicates an optimal solution, for which $x_2 = 6.44$, $x_3 = 15.85$, and $x_4 = 0.0122$ (underlined in the final tableau). A convenient way to obtain the primal variables utilizes the final dual tableau. The ith primal variable (c, a, or f) equals the objective coefficient of the $(m + i)$th slack (or excess) variable (x_8,x_9,x_{10}) in the final dual tableau, where m is the number of primal constraints. Thus $c = 0.137$, $a = 0.609$, $f = 0.197$ (underlined in the final tableau), and $w = 1 - c - a - f = 0.057$. Also $Z' = -x_1 + x_2 = 6.44$, and $Z = 25(0.137) + 3.5(0.609) + 4.5(0.197) =$ \$6.44/ton of concrete. The technique is described in many references [17].

TABLE 8. Dual Simplex Linear-Programming Solution; Concrete Mix Design Problem

Iteration	Basic variable	Z'	x_1	x_2	x_3	x_4	x_5	x_6	x_7	x_8	x_9	x_{10}	Right-hand side	Upper bound on entering variable
				↓										
1	Z'	1	1	−1										
	x_8		−1.3	1.413	1					1			25	25/1.413
	x_9		−1	1	−0.17	−20	38	−9	27		1		3.5	3.5 ←
	x_{10}		−1	1	−0.17	62	−44	43	−25			1	4.5	4.5
						↓								
2	Z'	1			−0.17	−20	38	−9	27		1		3.5	
	x_8		0.113		1.24	28.3	−53.8	12.7	−38.2	1	−1.41		20	20/28.3
	x_2		−1	1	−0.17	−20	38	−9	27		1		3.5	
	x_{10}					82	−82	52	−52		−1	1	1	1/82 ←
					↓									
3	Z'	1			−0.17		18	3.7	14.3		0.756	0.244	3.74	
	x_8		0.113		1.24		−25.2	−5.3	−20.2	1	−1.07	−0.345	19.66	19.66/1.24←
	x_2		−1	1	−0.17		18	3.7	14.3		0.756	0.244	3.74	
	x_4					1	−1	0.634	−0.634		−0.0122	0.0122	0.0122	
4	Z'	1	0.0155				14.5	2.97	11.5	<u>0.137</u>	<u>0.609</u>	<u>0.197</u>	6.44	
	x_3		0.091		<u>1</u>		−20.6	−4.28	−16.3	0.806	−0.862	−0.278	<u>15.85</u>	
	x_2		−0.985	<u>1</u>			14.5	2.97	11.5	0.137	0.609	0.197	<u>6.44</u>	
	x_4					<u>1</u>	−1	0.634	−0.634		−0.0122	0.0122	<u>0.0122</u>	

Figure 13 shows the optimal aggregate grading relative to the fine and coarse aggregate grading limits. Notice that the optimal grading tends to approach the coarse grading limit, because the coarse aggregate has lower unit cost.

In the primal problem, the number of terms in each constraint increases with the number of aggregates, and the number of constraints increases with the number of sieves used to specify gradation limits. Additional constraints on properties such as air content and durability can be included whenever linear approximations of their values in terms of the component concentrations are acceptable [13].

C. CEMENT-AGGREGATE REACTIONS

In concrete, the bond strength between aggregate and cement paste is usually the weakest link in the system and plays an important role in determining the properties

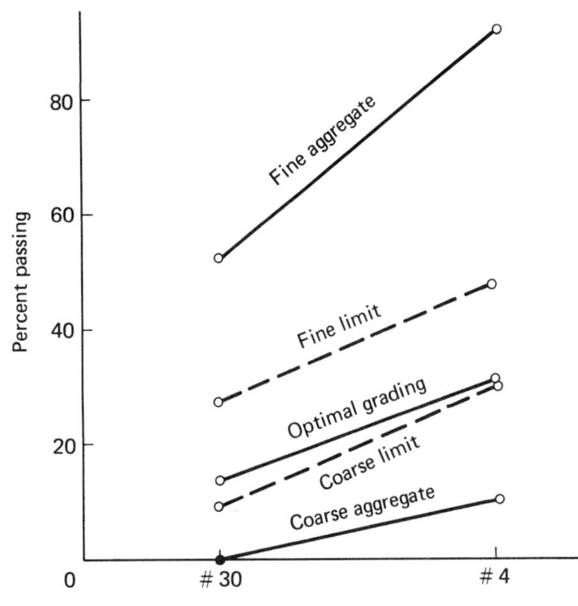

Fig. 13. Optimal aggregate grading relative to fine and coarse aggregate gradings and the grading limits.

of the composite material. Bonding has been studied by measuring the shear and tensile strengths of cement pastes placed on pure mineral substrates, by microhardness measurements across the paste-mineral interface, and by microscopic examination.

The forces which bind cement paste to an aggregate may be physical, due to adhesion and mechanical interlock, and chemical, due to the formation of surface reaction products.

1. Physical Bonding

The aggregate shape, surface texture, and stiffness are factors which influence the mechanical interlock between paste and aggregate [18, 19, 20]. The flexural and tensile strengths of concretes made with rough-textured granite and limestone aggregates may be up to 30% greater than those made with smooth-textured gravels, the strength being proportional to surface roughness [19, 21, 22]. The aggregate-paste bond strength may also be dependent on vibration of the fresh concrete. Upon drying, the restraint of paste shrinkage by stiff aggregate particles may cause large tensile stresses and cracking at the aggregate-paste interface [23, 24]. This cracking may decrease the aggregate-paste bond strength by up to 70%. Cracks initiate predominantly at the largest aggregate-paste interfaces, where shrinkage forces are most highly concentrated.

2. Chemical Bonding

Most aggregates show some degree of chemical reaction with cement paste. Here, aggregates fall into two classes [25]: (1) those which produce a strong contact

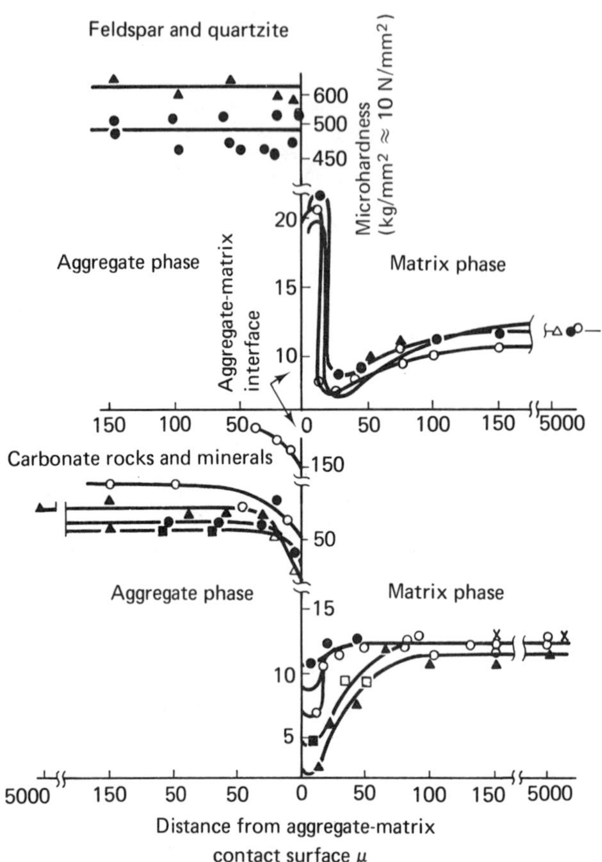

Top	Bottom
● Labradorite	● Calcite parallel to cleavage plane
○ Microcline	x Limestone (porous)
△ Quartzite	○ Marble
	□ Dolomite (porous)
	△ Calcite perpendicular to cleavage plane

Fig. 14. Variation of microhardness across the aggregate-cement interface. w/c = 0.25; age = 3 mo. From Lyubimova, T. Yu and E. R. Pinus, *Colloid J.*, 24 (1962) 491 [26].

layer in the paste together with an aggregate surface layer weaker than the rest of the aggregate (e.g., quartzite) and (2) those which produce weak contact layers in both the aggregate surface and the cement paste [e.g., many carbonate rocks [26] (Fig. 14)].

Siliceous (acidic) aggregates tend to absorb lime and form a strong hydrated calcium silicate reaction layer with the matrix. The removal of calcium hydroxide near the interface results in a greater degree of hydration and the formation of a more dense and isotropic gel, with a consequent increase in the paste strength [27]. With newly crushed sand, the reaction proceeds quickly, probably due to the formation of a highly disturbed layer on the surface of the aggregate. With unground sand, similar strengths can be achieved with much longer reaction times [28]. Figure 15 shows how the modulus of rupture in bond strength increases with increasing silica content of

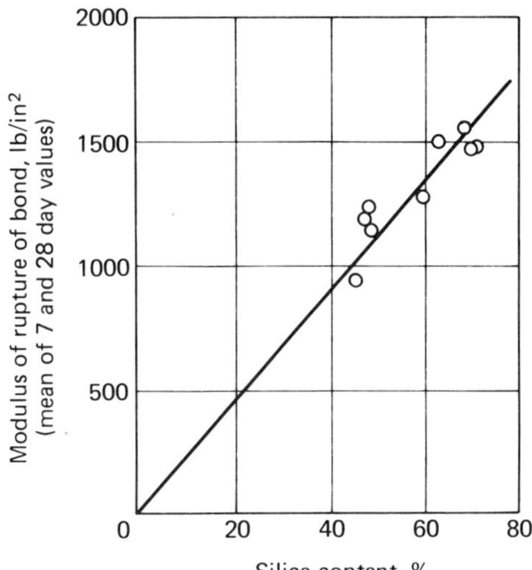

Fig. 15. Dependence of bond strength on silica content of extrusive rocks. From Alexander, K. M., J. Wardlaw and D. J. Gilbert, *Intern. Conf. Struct. Concr.*, London (1965) 58 [29].

several aggregates. The most effective rocks developed about twice the bond strength of the least effective, and bond strengths ranged from about 50 to 100% of paste strength. The larger the specific surface of the aggregate, that is, the smaller the particles, the more reactive is the silica per unit volume, and the higher is the concrete strength—although there is a limit of fineness beyond which little increase in strength can be obtained [28, 30].

Many carbonate rocks, on the other hand, including dolomites, cause a drop in cement paste hardness at the interface. The drop in hardness may be related to the formation of carbo-aluminates at the interface [26]. Dolomites frequently undergo surface exchange reactions with cement which are expansive and detrimental to strength.

A few carbonate rocks, however, are capable of producing high bond strength, and high concrete strength, even with relatively weak aggregates. Concrete containing chalk aggregate (compressive strength: 41–48 N/mm^2), for example, was found to have higher compressive strength than otherwise similar concrete having basalt aggregate (55–62 N/mm^2) [31]. The lower strength of basalt concrete was found to be due to the development of weak bonds with augite crystals in the basalt aggregate. The formation of strong bonds with the chalk aggregate was attributed to *epitaxial* overgrowth. Epitaxial overgrowth can arise when the lattice of the aggregate mineral is similar to that of some component of the set cement so that growing crystals can align themselves in an arrangement corresponding to that of the substrate. On calcium carbonate aggregates, for example, calcium hydroxide crystals may sometimes form a contact layer between the cement paste and the aggregate in a epitaxial arrangement [32]. Blast furnace slag aggregate is another material which appears capable of surface chemical reaction with cement, giving a bond of higher strength than with an inert aggregate.

There are several classes of substances occurring in mineral aggregates which can cause harmful reactions, beyond just the reduction of interfacial bond. They include

1. Certain siliceous minerals which produce deleterious expansion and cracking of concrete by reaction with alkalies released during the hydration of portland cement [33]. Alkali-aggregate reaction is a serious problem in concrete in many parts of the United States, including most of the western states and parts of the midwest, south, Atlantic seaboard, and portions of upper New York and Ontario, Canada. Table 9 lists several reactive rocks and minerals. Tests to evaluate potential alkali reactivity of aggregate and cement-aggregate combinations are listed in Appendix E.

TABLE 9. Reactive Rocks and Minerals

Reactive minerals	*Chemical composition*	*Physical character*
Opal	SIO_2NH_2O	Amorphous
Chalcedony	SIO_2	Cryptocrystalline fibrous
Tridymite	SIO_2	Crystalline

Reactive rocks	*Reactive component*
Siliceous rocks	
Opaline chert	Opal
Chalcedonic chert	Chalcedony
Siliceous limestone	Chalcedony and/or opal
Volcanic rocks	
Rhyolites and rhyolite tuff	Volcanic glass, devitrified glass, and tridymite
Dactites and dactite tuff	
Andesites and andesite tuff	
Metamorphic rocks	
Phyllites	Hydromica
Miscellaneous rocks:	
Any rocks containing veinlets, inclusions, or grains of the reactive rocks or minerals listed above	

SOURCE: MIELENZ, R. C., *Geol. Soc. Am.*, Bull. 57 (1946) 312 [35].

2. Soluble sulfates such as gypsum, and oxidation products of iron sulfides, which may cause expansion and disintegration of cement paste as a result of sulfate attack. These substances are rarely important, except for the iron sulfide oxidation products.
3. Inorganic compounds such as water-soluble salts of borates and zinc which retard hardening and strength development of portland cement. These substances are also rarely significant.
4. Organic substances such as carbohydrates and certain hydrocarbons which may occur on the surface of aggregate particles and retard hardening or cause staining or excessive air entrainment in concrete [34].

Control of alkali-aggregate reaction. Expansion due to the alkali-aggregate reaction may be controlled by (1) selecting a cement with low alkali content, (2) selecting the proper aggregate, (3) use of reactive pozzolans, (4) air entrainment, (5) reduction of the cement content, and (6) preventing access of water to the structure.

Some federal agencies specify maximum cement alkali contents of 0.4 to 0.6% by weight, depending on how severely reactive the aggregate is. The addition of pozzolans is often the most practical means of controlling expansion. Pozzolan reactions are described in Sec. B in Chap. 4. The effectiveness of a pozzolan may be expressed as the ratio between the expansion of specimens made with pozzolan and the expansion without pozzolan. Entrained air voids reduce expansion by providing storage reservoirs for the expansive siliceous gel formed during the alkali aggregate reaction.

D. LIGHTWEIGHT AGGREGATE CONCRETES

Lightweight aggregate concrete structures of historical interest include concrete ships and barges built during both World Wars I and II, the upper roadway of the San Francisco-Oakland Bay Bridge, and the increased number of roadway lanes placed on the original piers of the Tacoma Narrows Bridge after its failure. Since the 1950s innumerable multistory buildings have been built with lightweight concrete floors and walls. An early example was the 42-story Prudential Life Building in Chicago.

Advantages of lightweight concretes include the saving in structural support and decreased foundation sizes due to decreased loads and better fire resistance and thermal insulation. Disadvantages include greater cost (20 to 50%), greater porosity, more drying shrinkage, and the need for more care in placing.

Fig. 16. Densities of lightweight aggregate concretes. From *ACI Manual of Concrete Practice*, Part I: Materials and Properties of Concrete. Detroit: American Concrete Institute, 1973 [36].

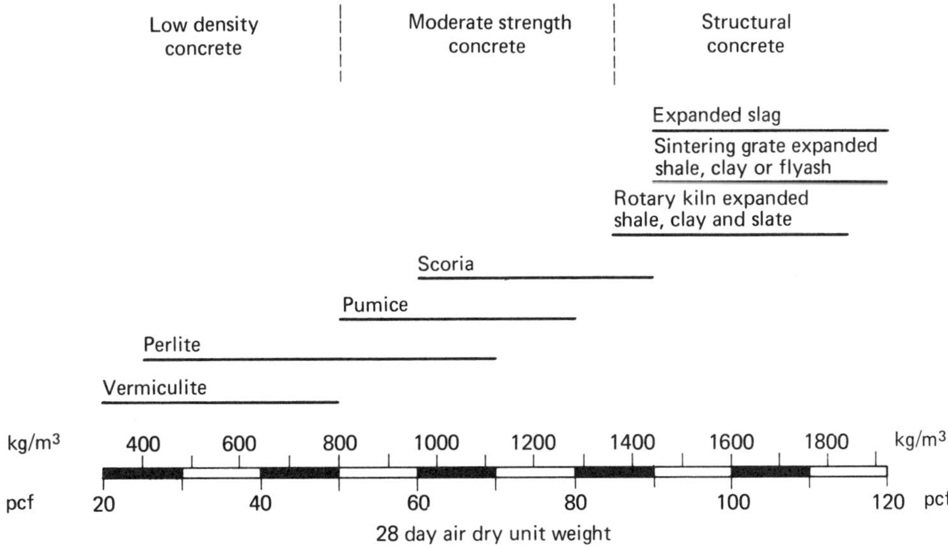

Figure 16 shows densities and uses of concretes produced with various lightweight aggregates. The aggregate properties are described in Chap. 6. Low-density concretes, made with exfoliated vermiculite and expanded perlite, range in density from 20 to 50 pcf, with thermal conductivities from 0.5 to 1.5 Btu/sq ft/hr/in./°F. Although in the higher densities these materials can be used for concrete block, their main use is for insulation, due to their low compressive strengths (up to 600 psi) and high drying shrinkage (0.2–0.4%).

Moderate-strength concretes range from 50 to 90 pcf with thermal conductivities of 1–4 Btu. Table 10 shows typical property ranges of moderate-strength aggregates made with various lightweight aggregates. All lightweight aggregates, with the exception of expanded shales and clays and scoria, produce concretes subject to high shrinkage.

Structural lightweight concrete aggregates are generally expanded shales, clays, slates, slags, or pelletized fly ash. Minimum compressive strength is 2,500 psi and maximum density is 115 pcf, by definition.

Concretes made with many lightweight aggregates are difficult to place and finish, because of the aggregate porosity and angularity. The aggregate may tend to separate from the cement mortar and float to the surface. This can generally be corrected by crushing the larger particles or otherwise adjusting the aggregate grading or by adding natural sand or other filler materials. Workability is also improved by adding an air entrainer.

TABLE 10. Properties of Some Medium Lightweight Aggregate Concretes

	Density of dry concrete		*Compressive strength*			*Thermal conductivity*	
Aggregate	*kg/m³*	*lb/ft³*	*N/mm²*	*lb/in.²*	*Drying shrinkage, %*	*SI units*	*Btu/sq ft/hr/in./°F*
Pumice	800	50	4.2	600	0.04–0.08	0.15–0.3	1–2
Clinker	960–1,520	60–95	2.1–7	300–1,000	0.04–0.08	0.35–0.6	2.5–4
Expanded clay or shale	960–1,200	60–75	5.6–8.4	800–1,200	0.04–0.07	0.3–0.45	2–3
Sintered pulverised fuel-ash	1,120–1,280	70–80	4.2–10.5	600–1,500	0.04–0.07	0.3–0.45	2–3
Foamed slag	960–1,520	60–95	2.1–7	300–1,000	0.03–0.07	0.2–0.45	1.5–3

Conversions rounded off: 1 N/mm² = 10 kg/m².
SOURCE: LEA, F. M., *The Chemistry of Cement and Concrete*. London: Edward Arnold & Co., 1970 [37].

Lightweight aggregate is often wetted 24 hr before use to ensure uniform moisture content during mixing. Unwetted aggregate may absorb water from the cement mortar and cause segregation and stiffening before placement is completed.

E. ADMIXTURES

Admixtures are defined by ASTM as materials other than water, aggregates, and portland cement (including air-entraining portland cement and portland blast furnace slag cement) which are added to concrete during mixing. Products added by the cement manufacturer are usually not regarded as admixtures. Some major classes of admixtures are

Accelerators
Retarders
Pozzolans
Air entrainers
Gas formers
Water reducers
Expansion admixtures
Surface hardeners
Coloring agents
Bonding admixtures

Commercial products often combine the effects of several classes. In each case the costs associated with using the admixture must be compared with alternative methods for obtaining the desired result.

Accelerators. Accelerators may be added to hasten either the rate of setting or hardening of concrete in order to permit early form removal or early use of the finished product or to compensate for low temperature on the rate of strength development. Accelerators for portland cement include soluble chlorides, carbonates, silicates, fluorosilicates, and hydroxides [38] and triethanolamine [39, 40]. Alumina cements and finely ground prehydrated portland cement have also been used.

The most used accelerator is calcium chloride, which acts as a catalyst in the hydration of C_3S and C_2S. As a catalyst, it is not removed by the reaction, but eventually reacts with C_3A in the cement to form $C_3A \cdot CaCl_2 \cdot 10\ H_2O$. Since $CaCl_2$ corrodes reinforcing steel, the cement should contain sufficient C_3A to remove all $CaCl_2$ after it has served its purpose. Corrosion can be reduced by using no more than 1.5% $CaCl_2$ by weight of cement and compacting the concrete well around the reinforcing steel. $CaCl_2$ is not used in prestressed concrete due to stress corrosion of the steel. 1.5% $CaCl_2$ has about the same effect on accelerating hydration as a temperature rise of 15°F. At 70°F, the 3-day strength increases by about 30%, the 28-day strength increase by 20%, and long-term strength is practically unaffected. The early strength increases are more pronounced at lower temperatures. The drying shrinkage of concrete tends to be increased with $CaCl_2$, and its resistance to sulfate attack is also lowered. $CaCl_2$ is added to the mix in solution or in flake, pellet, or granulated form. Sodium sulfate and stannochloride produce early strength gains similar to those of $CaCl_2$ but with much less corrosion of steel.

The *seeding* of portland cement with 2% of finely ground fully hydrated cement has an accelerating effect equivalent to that of 2% of $CaCl_2$ but also increases the 90-day strength 20 to 25% and causes no drying shrinkage [41].

Aluminum chloride is a strong accelerator, 1% increasing 10-day concrete strengths from 50 to 170%, but the effect on long-term strength is irregular, and strength is more sensitive to curing temperature than in the case of $CaCl_2$. Sodium silicate, the fluorosilicates, and triethanolamine act still more rapidly. Some ready-to-use mixtures of accelerator, cement, and sand provide initial set in 1 to 4 min and final set in 3 to 10 min. These are useful for stopping pressure leaks but usually have much lower ultimate strengths.

By modifying portland cement with the addition of certain ternary cementing compounds, mortars can be made with 15-min handling times and 1-hr compressive strengths of 1,000 to 3,000 psi. The resulting rapid heat liberation can be useful for winter concreting or *autogenous* steam curing of precast products but is disadvantageous in massive structures. Adipic or citric acids may be used as retarders to adjust handling times. Such *regulated-set* cements can be steam-cured without the holding period required for portland cements since at the end of handling the structure is sufficiently firm to withstand thermal shock. Shrinkage due to drying after hardening and creep under sustained load are both significantly greater for regulated-set cement concrete than for portland cement concrete.

Retarders. Retarders are used to counteract the damaging effects of quick setting in hot weather, in slip form operations where old and new concrete must be joined after an interruption in casting, where fresh concrete must be in transit for long periods, after an interruption in casting, where fresh concrete must be in transit for long periods, and for pouring bridge decks where it is desirable to obtain form deflection under the entire concrete weight before setting begins. Retarders are also used in long prestress members to keep the concrete plastic until vibrating is completed to ensure good bond strength between concrete and steel along the entire length of bed.

Commonly used retarders are the calcium, sodium, and ammonium salts of lignosulfonic acids and of hydroxylated carboxylic acid. The same compounds serve as water reducers, described later. Retarding is believed primarily due to the reduction of calcium ion concentration in solution by complexing with the aldehyde and ketone groups of the retarders. The lignosulfonate salts are typically used in liquid form, in fractions of a percent by weight of cement, and may retard the set by 1 to 4 hr. Lignosulfonate retarders normally also entrain 2 to 6% air in the concrete, and the mix generally requires 5 to 10% less water. Three-day strengths typically equal strengths without retarder, and 28-day strengths may be 10 to 20% higher.

In addition, a variety of water-soluble starches and sugars as well as sulfite lye act as retarders. Inorganic retarders include boron compounds, sodium phosphate, and sodium hexametaphosphate.

Specifications sometimes require that surface aggregate be exposed for architectural effect. Retarders applied in greater concentrations to the surface can prevent cement from bonding to surface aggregate without retarding cement in the remainder of the pour. After a few days of curing the retarded surface paste can be hosed or brushed off to expose the aggregate.

Pozzolans. Pozzolans are siliceous compounds which react with calcium hydroxide and water to form cementing materials. Pozzolans include

1. Volcanic ashes and tuffs containing rhyolite, dacite, or andesite.

2. Siliceous sedimentary rocks, as diatomaceous earth and opaline shales and cherts.
3. Industrial by-products, as blast furnace slag, fly ash, and ground brick.
4. Calcined clays and shales.

The clay pozzolans must be calcined at 1200 to 1800°F to activate the clay.

Finely ground pozzolans may be substituted for part of the cement to

1. Reduce the heat of hydration and resultant thermal cracking in massive concrete structures and to increase their ability to creep under load.
2. Reduce unsoundness.
3. Reduce the expansion due to alkali-aggregate reaction.
4. Increase resistance to attack by sulfate and seawater.

Reduced heat of hydration is due to the slower hydration rates of pozzolans than of portland cement compounds. Pozzolans share this property with low-heat portland cements. The substitution of fly ash having under 10% carbon for 20% portland cement reduces the strengths of concretes cured at 20°C by 10–30% in 7 days, 0–25% in 28 days, and 0% to slight strength increase in 1 year. For 30% substitution the corresponding strength reductions are 25–50%, 4–45%, and 0–15% [37]. Higher strengths are obtained with finer ashes.

Reduced unsoundness is attributed to the formation of hydration reaction products between the pozzolan and free lime or magnesium oxide which cause less swelling than do the pure hydrates of these compounds.

It may seem anomalous that pozzolans can reduce expansion due to alkali-aggregate reaction, since pozzolans are siliceous glasses, as are the opaline cherts and other aggregate minerals which are responsible for expansion. But since the pozzolans are finely ground, they have a large surface-reactive area uniformly distributed throughout the hydrating mortar in contrast to the more concentrated glassy inclusions exposed at the surfaces of aggregate particles. Therefore, the expanding alkali-silicate hydration product is produced somewhat earlier, before the cement paste is completely hardened, and is distributed more uniformly throughout the mass, causing less damage than the highly localized expansions due to hydrations at aggregate surfaces.

The increased sulfate and seawater resistance is attributed largely to the reduction in free lime by the pozzolan, preventing its reaction with sulfates.

Air entrainers. Air entrainers form finely dispersed air bubbles 0.01 to 2 mm in diameter in the cement paste which serve several useful functions. In fresh concrete they act as air cushions which improve the workability. Or conversely, the same workability is obtained with a lower water/cement ratio. The bubbles serve as barriers to the movement of water through fresh concrete and therefore reduce bleeding. They reduce the segregation of large aggregate when concrete is transported long distances, pumped through pipes, or poured through tremies. In hardened concrete the bubbles are pressure relief chambers into which the capillary water remaining in hardened cement paste can escape under pressure due to expansion by the formation of ice

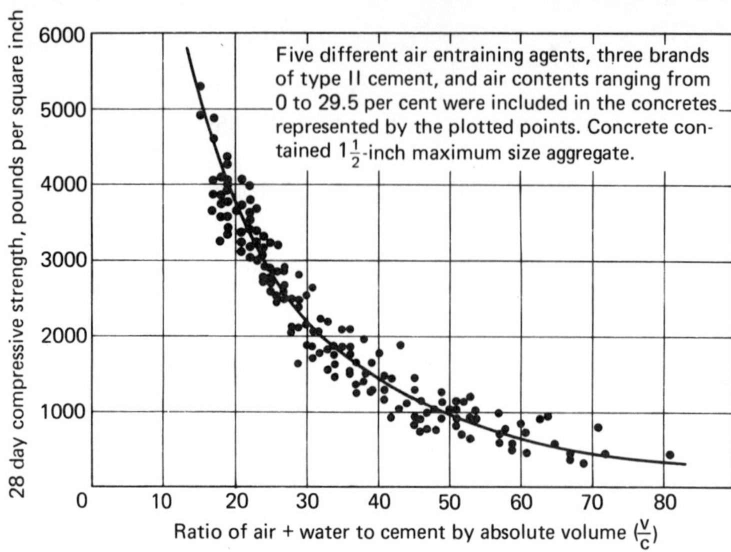

Fig. 17. Concrete compressive strength related to voids-cement ratio. From *U. S. Bureau of Reclamation Concrete Manual*, Eighth Edition. Washington, D. C.: U. S. Government Printing Office, 1975 [9].

crystals or the formation of salt crystals, particularly sulfate salts in seawater or chloride salts used for deicing roads. Deterioration is prevented by reducing expansive pressures to below the tensile strength of the paste. In addition, entrained air slightly increases the resistance of concrete to chemical attack, partly because the cement gel surrounding small air voids tends to break up continuous capillaries in the paste and therefore provide a barrier to water movement by capillary action.

The price paid for these advantages is reduced strength and increased drying shrinkage. The water/cement ratio law [Eq. (22), Chap. 8] states that concrete

Fig. 18. Drying shrinkage of concrete related to water and air contents. From *U. S. Bureau of Reclamation Concrete Manual*, Eighth Edition. Washington, D. C.: U. S. Government Printing Office, 1975 [9].

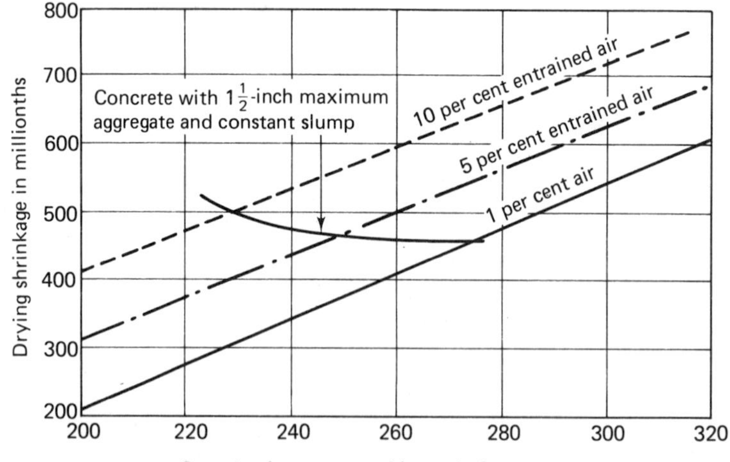

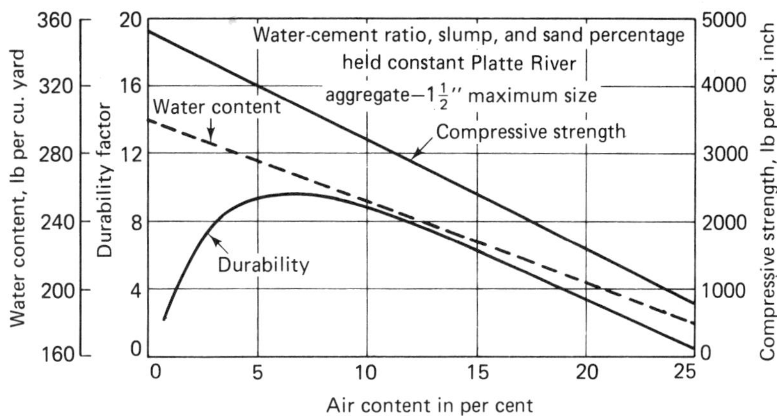

Fig. 19. Effects of air content on durability, compressive strength, and required water content. From *U. S. Bureau of Reclamation Concrete Manual*, Eighth Edition. Washington, D. C.: U. S. Government Printing Office, 1975 [9].

strength varies with the ratio of cement volume v_c to the water volume v_w plus air volume v_a. Figure 17 illustrates this trend of Type II cements. The air contents can be measured with an air meter, available in various styles.

Although strengths generally decrease about 5% per percent of entrained air, a concommitant reduction in water content to maintain constant workability offsets most of the strength reduction. It also offsets most of the increased drying shrinkage, as shown by the heavy line representing constant slump in Fig. 18. Each percent of entrained air permits a reduction in mixing water of 2–4%.

The optimum air content for freeze-thaw durability varies with maximum aggregate size:

Maximum Aggregate size (in.)	*Optimum Air Content (% by Volume of Concrete)*
No. 4	9 ± 1
½	7.5 ± 2
1	6 ± 2
1½	4.5 ± 1.5
3	3.5 ± 1

Figure 19 shows the variation in durability and compressive strength with air content for a concrete of 1½-in. maximum aggregate size. Although the theoretical air space required for expansion due to freezing of capillary water would be only about 1% for the concrete in Fig. 19, maximum durability is observed at 6% air space. The reason lies in the heterogeneous size and spacings of the air bubbles. Protection against freeze-thaw damage is obtained if the spacing is less than about 0.008 in. [42], as determined in accordance with ASTM C457. Some bubbles are coarser than required and not uniformly spaced. As air content is increased the 1% volume requirement is satisfied before the effective bubble spacing of 0.008 in. is satisfied. In fact, air-entrained concrete can be vibrated sufficiently to remove the larger bubbles, reducing air contents to 1 or 2%, with practically no reduction in durability.

Air entrainers are usually surface-active agents which reduce the surface tension between air and water and depend on the neutralization and saponification of a resin by alkalies in the cement. Typical entrainers are pine pitch resin, beef tallow, sodium lauryl sulfate, alkyl aryl sulfonate, soluble resin soaps, and a variety of other stearates. A few hundredths percent by weight of cement normally entrains 3 to 5% air, by volume of concrete. Their use requires close control because the volume of air entrained varies greatly with the clinker type and mixing temperature. Air entrainers may also be interground with the clinker and gypsum at the mill. The ASTM chemical requirements (C175) for the three types of air-entrained portland cement IA, IIA, and IIIA are the same as for non-air-entraining portland cements (C150). The physical requirements differ slightly from those for non-air-entraining cements (Table 6). There are no ASTM-specified Type IVA and VA cements.

Gas formers. To introduce larger quantities of air for lightweight (20- to 110-pcf) concrete, additional foaming agents and/or methods are used. Such concretes, called aerated, cellular, or gas concretes, are used for thermal insulation in ceilings, walls, and floors of buildings and cold storage structures and for grouting under machines and backfilling under horizontal surfaces. Cellular concrete is used in over 80% of the buildings in Sweden, and its use is widespread in many countries. Cellular concrete mix design is normally governed by three interrelated requirements: strength, density, and thermal insulating value. Neat cement mortar is usually used for concrete under 25 pcf, whereas fine lightweight aggregate or sand up to a cement/sand ratio of 1 to 4 may be used for heavier aerated concretes. Only concretes above about 50 pcf are used for structural applications.

The advantages of aerated concrete are

1. High thermal insulation.
2. Easily sawed and nailed.
3. Low density, which reduces dead load in structures and simplifies the handling of precast components.
4. Formation of a good key for exterior or interior finishing or plastering.

Major disadvantages include reduced strength (Fig. 20) and high drying shrinkage and permeability. In ordinary concrete, shrinkage and swelling are restrained by the aggregate, whereas foam provides little restraint. However, curing by steam autoclaving for 12 hr at 150–200 psi can increase the strengths two to three times (Fig. 21) and reduce volume changes on wetting and drying to one fourth to one sixth, in addition to slightly increasing the thermal insulation value [43]. Autoclaving also makes feasible the use of large proportions of cheap siliceous materials such as fly ash, or mixtures of lime and fly ash, in place of portland cement. These components provide additional strength increases associated with the formation of monocalcium silicate hydrate due to the reaction between lime and the finely divided siliceous aggregrate (see Fig. 21).

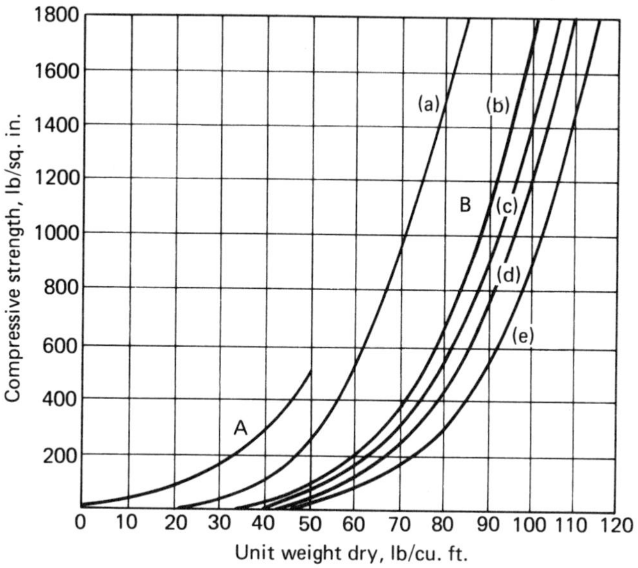

Curve	Mix proportions by weight	
	Cement: sand (FM 1.1)	Water/cement ratio
(a)	1 : 0	0.3
(b)	1 : 1	0.4
(c)	1 : 2	0.5
(d)	1 : 3	0.6
(e)	1 : 4	0.7

Fig. 20. Compressive strength of foamed concrete; moist-cured 14 days, tested at 56 days. Curves: A, prefoaming method; B, entraining method. From Taylor, W. H., *Concrete Technology and Practice*, Second Edition. London: Angus & Robertson Ltd., 1967 [44].

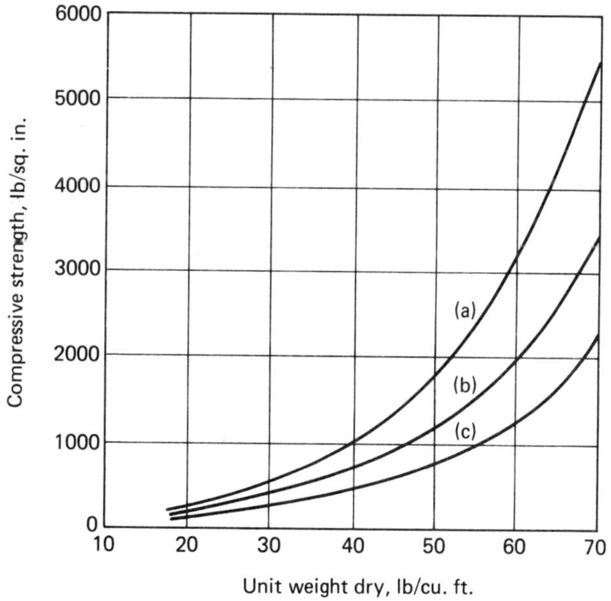

Fig. 21. Compressive strength at one day of autoclaved foamed concretes. (a) Lime concrete; (b) cement-lime concrete; (c) cement concrete. From Taylor, W. H., *Concrete Technology and Practice*, Second Edition. London: Angus & Robertson, Ltd., 1967 [44].

There are several processes for producing aerated concretes:

1. Air may be whipped into the mass by rapid agitation, using larger amounts of air entrainers than mentioned in the previous section. This technique is not useful for air contents over about 50%.
2. The foam may be produced separately, using hydrolyzed waste proteins or other foamers of the types used to smother gasoline fires. The premixed foam is then added to the cement slurry in the mixer in place of conventional aggregate.
3. Gas foamers such as aluminum powder, hydrogen peroxide (H_2O_2), or calcium carbide CaC_2 may be used. Aluminum powder reacts with calcium hydroxides in cement paste to produce hydrogen gas bubbles [53, 54]:

$$2\,Al + 3\,Ca(OH)_2 + 6\,H_2O \longrightarrow 3\,CaO_2Al_2O_3 \cdot 6\,H_2O + 3\,H_2$$

   Although calcium aluminates are formed as secondary products, the amounts are insufficient to affect the setting process. Normally 0.01 to 0.1% by weight of cement is used. Other metals such as magnesium and zinc react similarly, but only aluminum has been used extensively. Alkaline compounds such as sodium hydroxide, lime, or trisodium phosphate can be added to accelerate the generation of gas. Sodium benzoate is a useful stabilizer to prevent bubble coalescence and segregation. Aluminum powder methods are used more in European countries than in North America. A mixture of hydrogen peroxide and bleaching powder produces oxygen bubbles:

$$2\,H_2O_2 + Ca(OCl)_2 \longrightarrow CaCl_2 + 2\,O_2 + 2\,H_2O$$

4. Finely ground siliceous material, lime, and asbestos fiber can be mixed with excess water and autoclaved. Insulating materials as light as 10 pcf can be obtained by this method.

Water reducers. Cement tends to flocculate into clumps in water, causing incomplete hydration and reduced strength. The clumps have rough surfaces which reduce mobility in the paste and therefore reduce workability. Water trapped between the clumps may bleed to the surface due to the weight of overlying material, and the channels thus formed later provide passages for the movement of capillary water in the hardened concrete.

Water reducers are anionic agents which are adsorbed to cement surfaces, making them negatively charged and hydrophilic. Water dipoles then become oriented in sheaths around the charged and dispersed cement particles, giving better workability, more rapid hydration, and strength gain. Stated differently, for a given work-

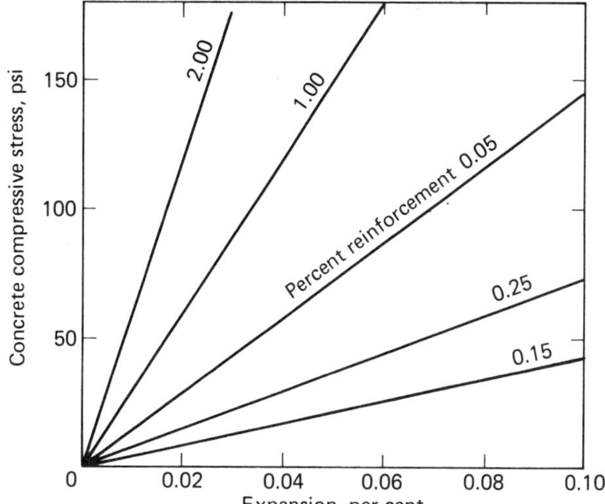

Fig. 22. Reinforcing requirements for expansive cement concrete, based on steel Young's modulus of 29 × 10^6 psi.

ability and strength, water reducers permit a reduction in water and cement contents and also reduce bleeding.

The most common water reducers are the same compounds used as retarders, i.e., salts of the lignosulfonic acids and of hydroxylated carboxylic acid. The anionic groups responsible for adsorption on the cement grains are HO—C—H, OH, COOH, and HO—C—C=O.

Expansion admixtures. Expansion admixtures minimize cracking due to drying shrinkage. They are used for machinery grouting, for patching, and to reduce cracking and permit increasing the spacing of construction joints in structures such as parking garages, large floor and pavement slabs, and tilt-up wall panels. Their use for producing self-stressing prestressed concrete is also being studied. Expansive cement concretes are classed as shrinkage-compensating (about 25–100 psi restrained expansive pressure) or self-stressing (150–500 psi expansive pressure). Expansion is normally resisted by adequate steel reinforcing, without additional strengthening of forms. Figure 22 indicates steel requirements, and Fig. 23 shows typical unrestrained length changes for shrinkage-compensating and self-stressing concretes.

Expansion admixtures either expand themselves or react with other cement constituents to expand. The amount and duration of expansion must be carefully controlled in some applications. The more common expansion admixtures are

1. Finely granulated cast iron plus a chloride salt to promote its oxidation to an expanding rust.
2. Magnesia [MgO; periclase which hydrates to magnesium hydroxide (Mg $(OH)_2$] [46]. Magnesia in powdered form hydrates earlier and does not cause the long-term unsoundness attributed to magnesia intercrystallized in portland cement clinker. Hydration of the magnesia is accelerated and controlled by steam curing.

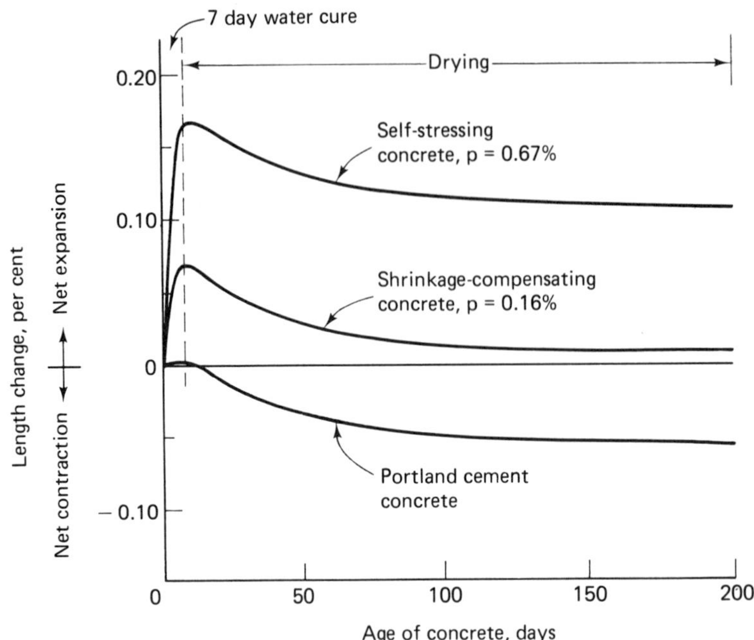

Fig. 23. Typical unrestrained length changes of concrete. From Polivka, M., *Proceedings of the Conference on New Materials.* Concrete Construction, Chicago Circle: University of Illinois, 1971, I–II [45].

3. Sulfoaluminous cement, made by burning gypsum, bauxite, and limestone [47]. This admixture is added to portland cement, 10 to 25% by weight, plus 15 to 20% of ground slag. The expansion rate and duration are controlled by varying the slag content and the fineness of both slag and sulfoaluminous cement [47] (Fig. 24). The slag is thought to combine eventually with the excess of calcium sulfate. The expansion is believed due to the formation of calcium aluminum trisulfate hydrate ($C_3A \cdot 3\ CaSO_4 \cdot 3$ O-32 H_2O; ettringite), the hydration occurring in the solid state. This compound, capable

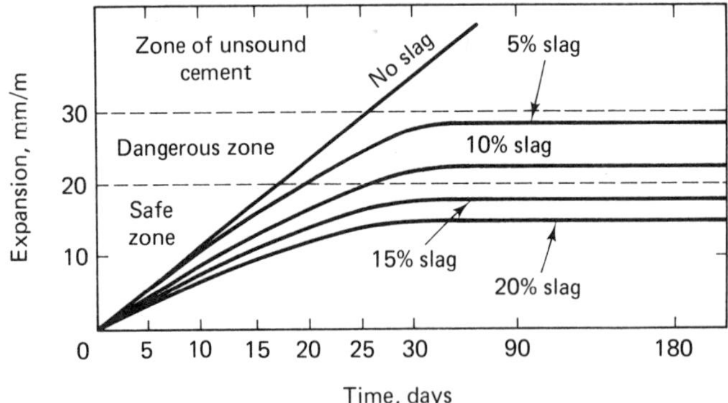

Fig. 24. Effect of slag in establishing volume stability of expanding cement. From Lafume, H.: Chemistry, *Proc. 3rd Intern. Symposium*, London 1952, 58 [47].

both of cementing and expanding, also causes the destructive expansion when high C_3A portland cements are exposed to solutions containing sulfate ions. Since the expansion accompanies water absorption, it may be terminated temporarily by air drying or permanently terminated by autoclaving.

4. Various portland cement-calcium sulfate combinations have been studied [48, 49]. Table 11 summarizes three commercially available types and their constituents. These also depend on the formation of ettringite [50]. Each contains portland cement, a source of aluminate and calcium sulfate. The reactive aluminate which forms ettringite in Type K cement is anhydrous calcium aluminosulfate (C_4A_3S); in Type M, calcium aluminate (CA and $C_{12}A_7$); and in Type S, tricalcium aluminate (C_3A) [45].

TABLE 11. Types of Sulfoaluminate Expansive Cements and Their Constituents

Type K	*Type M*	*Type S*
Portland cement	Portland cement	Portland cement, high in C_3A
+	+	+
Calcium sulfate	Calcium sulfate	Calcium sulfate
+	+	
Portland-like cement containing calcium sulfoaluminate	Calcium-aluminate cement	

The letter designations were assigned by the American Concrete Institute. The three types currently cost about 50% more than normal portland cement, the cost being offset in some applications by savings in construction, maintenance, and performance of the structure.

Type K is the most widely used. Figure 25 shows a representative relationship between the restrained expansion and compressive stress as a function of steel reinforcing in Type K self-stressing concrete.

Most of the ettringite must form after the cement has hardened; otherwise expansion only deforms the still plastic concrete. Most of the expansion should also be completed within the normal moist curing period, 4 to 7 days, since ettringite formation requires additional water, sufficiently available only during moist curing.

Cement properties which govern the rate and amount of expansion include the amounts of aluminate and calcium sulfate added and the fineness of grinding. Mix properties which influence expansion include the cement constant (high cement content gives greater expansion), aggregate properties (high modulus, nonshrinking aggregate gives greater expansion), and mixing time (ettringite formation during long mixing can slightly reduce expansion after placing).

Concretes containing K, M, and S admixtures have strength and modulus properties similar to those made with Types I and II cements. Mix design is similar, except for a slight increase in water with K and S admixtures, required for early ettringite formation, and to compensate for the corresponding loss in slump between the time of mixing and placing of the concrete.

Expansion admixtures are used successfully in combination with most air-entraining and water-reducing/retarding admixtures but not with most accelerators, which tend to restrict expansion and increase subsequent drying shrinkage.

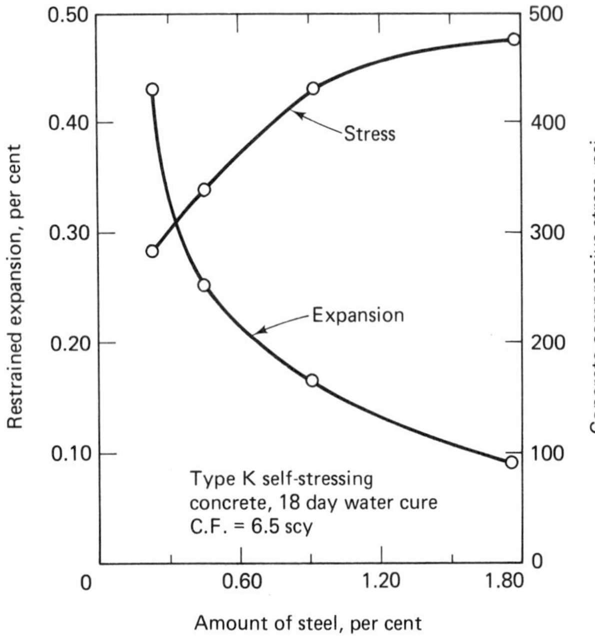

Fig. 25. Typical effect of steel reinforcing on compressive stress and restrained expansion in Type K self-stressing concrete (from Univ. of California studies). From Polivka, M.: *Proceedings of the Conference on New Materials.* Concrete Construction, Chicago Circle: University of Illinois, 1971, I–II [45].

Surface hardeners. To reduce surface abrasion of floors, metallic or chemical surface hardeners can be used.

Metallic hardeners are finely graded carborundum, fused alumina, or iron particles which are dry-mixed with portland cement, spread evenly over fresh concrete, and floated into the surface.

Chemical hardeners are water-soluble silicofluorides or fluorosilicates plus a wetting agent which reduces surface tension to allow greater penetration into concrete pores. These compounds convert lime into highly insoluble calcium fluoride (fluorspar) and densify the surface by causing colloidal silica to precipitate in the hardened cement pores. The protective layer is thin, however, because the densifying effect prevents deeper penetration of the solution. Greater penetration can be obtained with gaseous silicon fluoride (SiF_4). The thickness can be controlled by varying the gas pressure and treatment duration. The gas or *Ocrat* process provides good acid resistance and impressive compressive strength but is expensive.

Other chemical hardeners include sodium silicate, aluminum and zinc sulfates, and linseed and tung oils. Sodium silicate should have a silicate-to-soda molecular ratio of 3 or 4 to 1. The solution is usually diluted in water and applied in several coats.

Coloring agents. Pigments may be added to produce colored concrete. They should (1) not affect the strength development of concrete adversely and (2) be chemically stable in cement alkalinity and possible should (3) be colorfast in sunlight and (4) retain color stability during autoclaving [51].

Frequently used pigments are [36]

Color	*Pigment*
Gray, black	Black iron oxide, mineral black, carbon black
Blue	Ultramarine blue, phthalocyanine blue
Reds	Red iron oxide
Brown	Brown iron oxide, raw and burnt amber
Ivory, cream, buff	Yellow iron oxide
Green	Chromium oxide, phthalocyanine green
White	Titanium dioxide

Most of the pigments are available in natural or synthetic form. The synthetic materials are more expensive but brighter and more uniform in color due to their fineness and purity. Pigments are usually added in amounts of 2 to 10% by weight of cement or mixed with fine silica sand, spread evenly over the fresh concrete, and floated and troweled into the surface.

Bonding admixtures. Bonding of fresh to hardened concrete requires that the aggregate in the hardened surface be exposed and clean. Otherwise the fresh concrete shrinks and pulls away from the hardened surface during setting. The surface aggregate may be exposed by sandblasting or scrubbing with muriatic acid. A cement paste slurrying is often applied just before pouring new concrete to increase the paste available for bonding.

Alternatively, either metallic aggregate or a latex emulsion may be used as bonding agents. The metallic aggregate may be similar to the iron expansion admixture but with larger particles. Tiny fingers of iron oxide thrust out into the old and new concretes and bind them together.

The various latex bonding agents consist of synthetic resin polymers emulsified in water. When sprayed or painted on concrete, surface pores in the concrete absorb the water and allow the resin colloids to coalesce and bond. When mixed with cement paste the water is used in cement hydration, and the resin is left to bond to both surfaces.

Various polymer additives are also used to increase the strength of concrete by the combined effects of void filling, which increases strength in accordance with Eq. (22) of Chap. 8, and by serving as a ductile crack arrestor in an otherwise brittle material (Sec. B in Chap. 3). These benefits are more pronounced in concretes than in cement pastes, and polymer additives are described in Sec. F.

Table 12 summarizes the major types of cement admixtures and their properties.

TABLE 12. Cement Admixtures

Type	*Uses*	*Typical materials*	*Reactions*
Accelerators	Early form removal Early strength Accelerate cold weather curing	$CaCl_2$ Alumina cement Hydrated portland cement Other soluble chlorides, carbonates, silicates, hydroxides	Catalyze hydration of C_2S, C_3S Remove retarder $CaSO_4$ from solution in portland cement
Retarders	Retard hot weather curing Prevent joints between hardened and fresh concrete Bridge decks Long precast beds Long transit times	Lignosulfonate salts Hydroxylated carboxylic acid salts Water-soluble starches, sugars Sodium hexametaphosphate	Calcium ion concentration reduced by complexing with aldehyde and ketone groups
Pozzolans	Reduce heat of hydration, thermal cracking Reduce unsoundness Reduce alkali-aggregate reaction expansion Increase sulfate resistance	Volcanic ash Diatomaceous earth Calcined clay Fly ash	Silica slowly reacts with free lime in cement to produce additional C_2S hydrate
Air entrainers	Improve workability Reduce bleeding Increase freeze-thaw durability Increase resistance to attack by salts	Sodium lauryl sulfate Alkyl aryl sulfonate Other soluble resin soaps Hydrolyzed proteins	Reduce air-water interfacial tension to permit air entrainment
Gas formers	Produce cellular concrete	Aluminum powder Hydrogen peroxide Calcium carbide	React with hydroxides in cement to produce hydrogen bubbles
Water reducers	Increase workability Decrease water/cement ratio, increase strength	Same as retarders	Anionic groups HO—C—H, OH, COOH, HO—C—C=O adsorb at cement surfaces, deflocculate particles
Expansion admixtures	Machinery grouting Reduce shrinkage cracking	Finely granulated cast iron plus chloride Sulfoaluminous cement	Expansion due to rust formation Expansion due to formation of $C_3A \cdot 3CaSO_4$ hydrate
Surface hardeners	Abraision or acid-resistant surfaces	Carborundum, fused alumina, iron particles Silicofluorides, fluorosilicates Gaseous silicon fluoride	— Convert lime to insoluble calcium fluoride
Coloring agents	Surface colors	Natural and synthetic inorganic pigments	—
Bonding admixtures	Bonding fresh to hardened concrete	Fine iron aggregate plus chloride Latex emulsions	Bonding by iron oxide filaments Polymeric bonding

F. POLYMER-MODIFIED CONCRETES

Ductile materials may be strengthened by the use of brittle inclusions as a discrete phase. Examples include fiber-reinforced plastics (Sec. A in Chap. 11) and the diffusion hardening of ductile alloys (Sec. B in Chap. 3) in which rigid inclusions restrain dislocation along slip planes. In contrast, the strength and toughness of materials having a brittle continuous phase, such as concrete, can also be increased by the inclusion of more ductile materials such as polymers as the discrete phase. Ductile inclusions absorb the work of fracture of cracks propagating through the brittle matrix and retard crack propagation. They also provide some tensile strength across cracks already formed in the brittle phase.

In concretes, polymer additives are often observed to concentrate at the aggregate-paste interfaces and to fill continuous or discontinuous voids in the paste or to simply line such voids, depending on the relative volumes of voids and polymer. These void-filling functions can increase the bond between cement paste and aggregate and reduce water penetration into the concrete, thereby improving freeze-thaw durability and chemical resistance, in addition to increasing the brittle fracture strength of the paste and the work of fracture, already mentioned.

Polymer-modified concretes have been used in limited amounts for concrete pipe, tunnel support linings, desalting structures, concrete piles, precast bridge decks, and surface protection for cast-in-place bridge decks. The high costs of polymer additives are also often justified for repairing or resurfacing existing concrete structures such as pavements, bridge decks, industrial floors, and tank linings. The higher costs are justified because they may save replacement of the entire structure.

Polymer additives in common use can be classed in three groups: (1) water-dispersed polymers (latexes or latex emulsions), (2) water-dispersed resins, and (3) water-soluble polymers [52]. In addition, experimental work has been done with polymer-impregnated concrete: precast portland cement concrete impregnated with a liquid monomer which is subsequently polymerized in situ. Cement binder may also be totally replaced with polymer, as in epoxy-aggregate and polyester-aggregate flooring systems (Sec. B in Chap. 1).

1. Water-Dispersed Polymers

The most common polymer additives are dispersions of thermoplastic solids in water (latexes). The excellent bond of latex-modified mortars to existing concrete has resulted in their wide use for repair of concrete surfaces and for leveling concrete floors prior to laying tile or carpeting, where good bond to the concrete substrate and the ability to the troweled to a feather edge are essential. Lattices generally increase fluidity, allowing a reduction in the water/cement ratio and also serve to retain mix water and retard drying during curing.

Polymer/cement weight ratios of 0.10 to 0.20 are often found to give maximum strengths. At lower polymer contents, the polymer is found to exist primarily only as coatings on cement voids; at intermediate contents, it fills most of the voids; and at higher contents, the polymer creates discontinuities in the cement paste which reduce strength. Hence, a polymer volume near the natural void volume of the concrete gives maximum strength.

Polymer-modified mortars are predictably more susceptible to creep and reduction in strength and stiffness at high temperatures than mortars without polymer.

These effects are more pronounced with thermoplastic latexes than with thermosetting polymer additives. For example, a thermoplastic latex-modified mortar may lose up to 50% of its Young's modulus and flexural strength at about 45°C, a temperature which can be reached under exposure to natural sunlight [53].

Some latex-modified mortars also lose strength and stiffness when soaked in water. This is especially true of thermoplastics such as polyvinyl acetate which undergo hydrolysis in the presence of water [52].

Natural rubber latex, used in mortar as early as the 1930s, produced extremely flexible mortar systems. Polyvinyl acetate latexes, introduced in the 1940s, improved the bond strength between mortar and coarse aggregates but were water-sensitive, limiting their use to relatively dry climates or interior applications. Styrene-butadiene, acrylic, and Saran latexes, introduced in the 1950s and 1960s, have the advantages of polyvinyl acetate plus good water resistance.

Saran latex mortars have appreciably higher strengths than the others. Saran latex has also been used in brick masonry mortars and for prefabricated brick and concrete block structural panels where it provides both outstanding tensile and compressive strengths. The *Threadline* adhesive mortar, which can be applied with a caulking gun, contains four components: portland cement, stryene-butadiene latex, epoxy, and hardener. Table 13 shows typical mechanical properties of these two latex-modified mortars.

TABLE 13. Typical Mechanical Properties of Mortars

Property	*Control (psi)*	*Styrene-butadiene (psi)*	*Saran (psi)*
Shear bond	50–200	<650	<650
Compressive	4,480	5,500	8,430
Tensile	350	740	870
Modulus of rupture	820	1,620	1,820
Elastic modulus	3.4×10^6	1.92×10^6	2.52×10^6

SOURCE: EASH, R. D. and G. L. EMIG, *Concrete Construction, Proceedings of the Conference on New Materials*, S. P. Shah, ed. Chicago Circle: University of Illinois, 1971, p. XI [54].

2. Water-Dispersed Resins

Unlike latexes, which are water-dispersed solid thermoplastic polymer particles, resins are liquid monomers which are polymerized during hydration of the cement paste to form thermosetting polymers such as polyesters or epoxies. The resin can be mixed with the dry cement or added to the concrete mix as a water dispersion. The resin typically contains a catalyst, for example, a water-soluble redox catalyst, which becomes active in the presence of water.

Various liquid resin additives have been investigated, including those based upon epoxies, polyurethanes, and phenol-formaldehydes and chlorosulfonated polyethlenes.

Like latexes, the primary value of resins is to improve the bond of mortar to existing concrete. Resin mortars cure much faster, however, since cross-linking occurs rapidly compared to the drying action required for latexes. The high degree of cross-linking in thermosets also gives greater resistance to water and chemicals and much better thermal resistance than for latexes.

3. Water-Soluble Polymers

Water-soluble polymers are usually added to concrete as aqueous solutions and are primarily used as plasticizers in small amounts [52]. Most additives in this group are based upon water-soluble thermosetting resins such as phenol-formaldehyde and urea formaldehyde, melamine formaldehyde [55], and water-soluble epoxy resins [53]. The increased plasticity which is provided allows much lower water/cement ratios, higher strengths, and greater durability for the concretes. For example, a water/cement ratio below 0.3 can be used for mortars modified with a 0.02 polymer/cement ratio epoxy-carbamide resin additive, whereas a 0.5 water/cement ratio is normal for unmodified mortar [52]. Since only small amounts of polymer are present, the cement paste is the predominant matrix phase. The polymer may also enhance hydration of the cement paste by retarding evaporation of water.

4. Polymer-Impregnated Concrete

Limited use has been made of polymer-impregnated concrete: precast portland cement concrete impregnated with a liquid monomer which is subsequently polymerized. Methyl methacrylate, styrene, polyester-styrene, and epoxy-styrene are several polymers which have been investigated [56, 57]. A typical treatment consists of drying moisture from a concrete specimen, evacuating it in a chamber, and then introducing liquid monomer to the chamber at 25-psi pressure. In test cylinders of standard concrete, the monomer penetrates the outer $\frac{1}{2}$ in. or so. Polymerization may be accomplished thermal-catalytically or by Co^{60} gamma radiation, or by a combination of both.

The physical changes include

1. Increased compressive strength and Young's modulus. Strength increases of two to four times with 5 to 8% weight of polymer were typical. The strength and modulus increases were more pronounced in steam-cured specimens, where larger pore sizes allowed greater polymer loadings (up to 8% by weight).
2. Increased freeze-thaw durability, sulfate resistance, and acid resistance—all properties which are pore-volume-dependent.
3. A 5- to 10-fold reduction in creep.
4. Increased sonic velocity and Schmidt impact hammer rebound.
5. Cracks were more often found to pass through rather than around large aggregate, suggesting that the polymer may contribute to increased bond between mortar and aggregate.

Major limitations of polymer impregnation include its cost versus relative benefit in precast concrete applications and inadequate techniques for field applications.

PROBLEMS

1. What factors govern the maximum and minimum aggregate sizes used in a concrete mix?
2. What factors determine the aggregate void ratio?
3. Compute the optimum diameter ratio of the two spherical aggregates in a binary mixture which can give a minimum aggregate void ratio, assuming the larger spheres to be in a perfect hexagonal close-packing arrangement.
4. All concrete mixes are designed for at least what three essential properties? Briefly describe at least one test used to measure each property.
5. What difficulties may be caused by insufficient fine sand in a concrete mix? By too much fine sand?
6. What materials data are required in order to use the ACI trail mix design method? List an appropriate sequence of design steps.
7. Briefly describe three graphical methods of aggregate blending, and mention the capabilities and limitations of each.
8. Determine the optimum blending proportions of the following three aggregates to satisfy the indicated grading constraints on only two sieve sizes:

Sieve size	*Aggregate, % passing*			*Grading limit, % passing*	
	A	*B*	*C*	*Upper limit*	*Lower limit*
No. 4	40	75	90	60	80
No. 30	10	65	40	60	30
Cost per ton	$3	$5	$4		

9. Determine the optimum blending proportions of the following three aggregates to satisfy the indicated grading constraints on three sieve sizes:

Sieve size	*Aggregate, % passing*			*Grading limit, % passing*	
	A	*B*	*C*	*Upper limit*	*Lower limit*
No. 8	100	100	35	100	70
No. 30	85	30	10	60	20
No. 100	45	0	0	5	0
Cost per ton	$6	$3	$4		

10. Figure 26 shows the variations in (a) compressive strength (ksi), (b) tensile strength (ksi), (c) flexural work of fracture (10^2 lb-in./cu in.), and (d) production cost ($/cu ft)

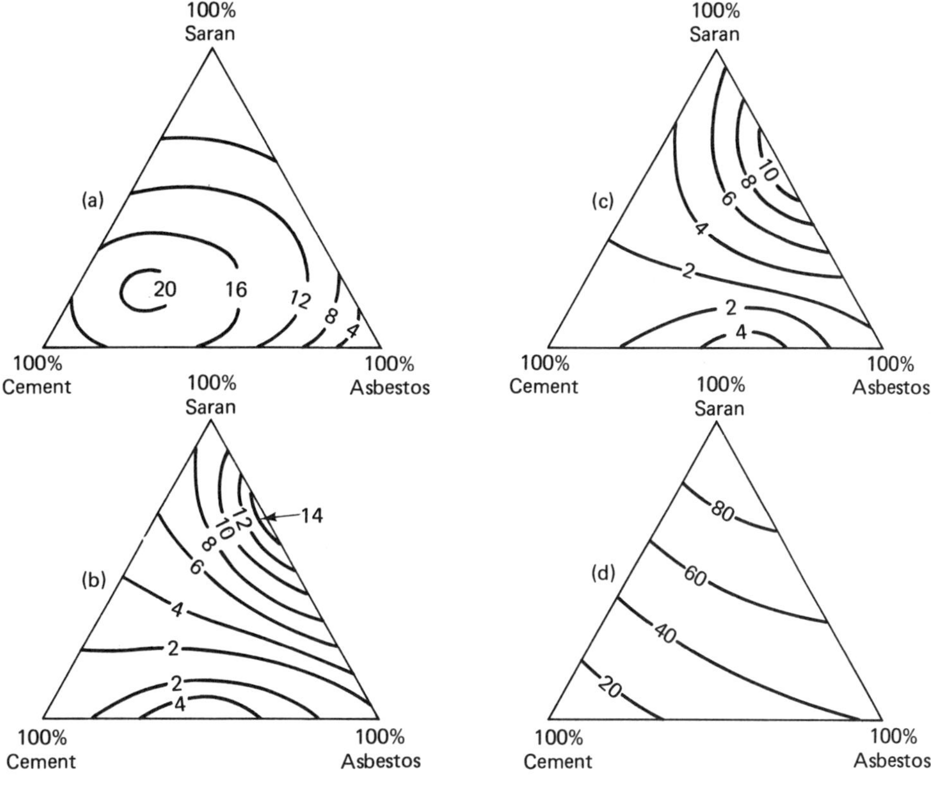

Fig. 26.

with composition of cement-asbestos fiber-Saran latex mixtures. Determine the minimum-cost volume percentages of cement, fiber, and latex to satisfy the following constraints:

(a) Compressive strength $\geq$ 15 ksi.

(b) Tensile strength $\geq$ 2.5 ksi.

(c) Work of fracture $\geq$ 300 lb-in./cu in.

11. Assuming that the two aggregates and cement paste in Example 4 have similar Young's moduli so that the concrete shrinkage can be defined by Eq. (19) of Chap. 8 and that the ultimate shrinkage of the cement paste is 2,500 microstrain, rewrite the primal linear-programming model to include an upper constraint on shrinkage of 500 microstrain. Assume the specific gravities to be 2.70 for both aggregates and 3.15 for the cement.

12. Assuming that the concrete compressive strength in Example 4 would vary linearly from the strength shown in Fig. 12 at approximately zero air content to 30% of that strength at 25% air content (see Fig. 35, Chap. 4), rewrite the primal programming model to include lower and upper constraints of 3 and 6% on the entrained air content.

13. What is the alkali-aggregate reaction, chemically? Why is it harmful? Name five methods for reducing expansion due to the alkali-aggregate reaction and state the physical-chemical basis on which each method depends.

14. Why is it that noncrystalline silica in such forms as opaline chert or siliceous limestone aggregate can be harmful to long-term strength (due to the alkali-aggregate reaction),

while pozzolans, which also contain large percentages of noncrystalline silica, are beneficial to long-term strength?

15. Into what main groups may lightweight aggregates be classified? What features distinguish their method of production and their performance in concrete? Include structural, insulating, and shrinkage characteristics.
16. Name two common classes of concrete admixture, and briefly describe the physical and chemical behavior responsible for their effectiveness.
17. A concrete paving mixture is found to set too rapidly to permit adequate finishing during the hot summer months.
    (a) What modifications might be made in manufacturing the cement? (See Chap. 4.)
    (b) What modifications might be made during mixing and placing of the concrete?
18. What materials are used to promote hardening of concrete placed in cold weather? Under what circumstances is it undesirable to use calcium chloride as a setting accelerator?
19. What adjustments in grading and proportioning should be made to suit the use of air entrainment?
20. The theoretical air space required for expansion due to freezing of capillary water in a given concrete is only about 1% by volume, yet maximum durability against freezing for the concrete is experimentally observed to be at 6% air voids. Why?
21. What basic mechanical function do polymer additives perform in terms of the microcracking of concrete? Distinguish between three major classes of polymer additives. What are the physical-chemical differences? What are the relative advantages, limitations, and areas of application of each?
22. Pose a question for yourself on some significant aspect of concrete technology with which you are acquainted and answer it with adequate detail in about 500 words.

REFERENCES

1. TALBOT, A. N., *ASTM Proc.* (1921).
2. GOLDBECK, A. T., and J. E. GRAY, *Natl. Crushed Stone Assoc. Bull.*, 11 (Dec. 1942), revised 1953, reviewed 1965.
3. POWERS, T. C., *J. Port. Cem. Assoc. Res. Lab.* **6**, No. 1 (1964) 2.
4. FURNAS, C. C., *Report of Investigations No. 2894*. Washington, D.C.: U.S. Bureau of Mines, 1928.
5. FULLER, W. B., and S. E. THOMPSON, *Trans. ASCE*, **59** (1907) 67.
6. *Design and Control of Concrete Mixtures*, 12th ed. Chicago: Portland Cement Association, 1970.
7. CZERNIN, W., *Cement Chemistry and Physics for Engineers*. New York: Chemical Publishing Company, Inc., 1962.
8. LARSON, T. D., *Portland Cement and Asphalt Concretes*. New York: McGraw-Hill Book Company, 1963.
9. *U.S. Bureau of Reclamation Concrete Manual*, 8th ed. Washington, D.C.: U.S. Printing Office, 1975.
10. WOODS, K. B., ed., *Highway Engineering Handbook*. New York, McGraw-Hill Book Company, 1960.

11. SPANGLER, M. G., *Soil Engineering*. Scranton, Pa.: International Textbook Company, 1960.

12. SARGENT, C., *Hwy. Res. Bd. Bull.*, **275** (1960), p. 1.

13. CANNON, J. P., and G. R. MURTI, *J. Cem. Concr. Res.*, 1 (1971) 353.

14. WALKER, S., and D. L. BLOEM, *J. ACI* **57** (Sept. 1960) 283.

15. CORDON, W. A., and A. GILLESPIE, *J. ACI* **60** (Aug. 1963) 1029.

16. BLOEM, D. L., and R. D. GAYNOR, *J. ACI* **60** (Oct. 1963) 1429.

17. HILLIER, F. S., and G. J. LIEBERMAN, *Introduction to Operations Research*. San Francisco: Holden-Day, Inc., 1967.

18. BACHE, H. H., and P. NEPPER-CHRISTENSEN, *Intern. Conf. Struct. Concr., London* (1965) 1.

19. KAPLAN, M. F., *Proc. Am. Concr. Inst.*, **55** (1959) 1193.

20. HSU, T. T., and F. O. SLATE, *Proc. Am. Concr. Inst.*, **60** (1963) 465.

21. KAPLAN, M. F., *Proc. Am. Concr. Inst.*, **60** (1963) 853.

22. JONES, R., and M. F. KAPLAN, *Mag. Concr. Res.*, **9** (1957) 89.

23. HSU, T. T., *Proc. Am. Concr. Inst.* **60** (1963) 371.

24. HSU, T. T., et al., *Proc. Am. Concr. Inst.*, **60** (1963) 209.

25. SWAMY, N., Structure, *Solid Mechanics and Engineering Design, Proc. Southampton 1969 Civil Engineering Materials Conference*. New York: John Wiley & Sons, Inc. (Interscience Division), 1970, p. 301.

26. LYUBIMOVA, T. YU, and E. R. PINUS, *Colloid J.*, **24** (1962) 491.

27. BROWN, L. S., and R. W. CARLSON, *Proc. Am. Soc. Test. Mater.*, **36** (1936) 332.

28. LOGGINOV, G. I., et al., *Colloid J.*, **21** (1959) 429.

29. ALEXANDER, K. M., J. WARDLAW, and D. J. GILBERT. *Intern. Conf. Struct. Concr., London* (1965) 58.

30. ALEXANDER, K. M., *Proc. Am. Concr. Inst.*, **57** (1960) 557.

31. GALWEY, A. K., K. A. JONES, and R. S. BECKETT. *J. Appl. Chem.*, **16** (1966) 159.

32. FARRAN, J., and J. C. MASO, *C. R. Hebd. Seanc. Acad. Sci.*, 260 (1965) 5195.

33. LEARCH, W., *ASTM Spec. Tech. Publ.*, **169** (1955) 334.

34. MACNAUGHTON, M. F. and J. B. HERBECH. Proc. Amer. Concrete Inst., **27** (1954) 273.

35. MIELENZ, R. C., *Geol. Soc. Am. Bull.*, **57** (1946) 312.

36. *ACI Manual of Concrete Practice*, Part 1: Materials and Properties of Concrete. Detroit: American Concrete Institute, 1973.

37. LEA, F. M., *The Chemistry of Cement and Concrete*. London: Edward Arnold & Co., 1970.

38. STEINOUR, H. H., *J. Port. Cem. Assoc. Res. Lab. Bull.*, **2**, No. 3 (1960) 32.

39. NEWMAN, E. S., et. al., *J. Res. U.S. Natl. Bur. Stand.*, **30** (1943) 281.

40. FORBRICK, L. R., *Am. Concr. Inst. J., Proc.*, **37**, No. 2 (1940) 161.

41. BALAZA, G., J. KELEMEN, and J. KILIAN. *Am. Concr. Inst. J., Proc.*, **57**, No. 2 (1960) 239.

42. POWERS, T. C., *Proc. Hwy Res. Bd.*, **29** (1949) 184.

43. DILNOT, S., *Rock Prod.*, **55** (1952) 110.

44. TAYLOR, W. H., *Concrete Technology and Practice*, 2nd ed. London: Angus & Robertson Ltd., 1967.

45. POLIVKA, M., *Concrete Construction, Proceedings of the Conference on New Materials*. Chicago Circle: University of Illinois, 1971, I-II.

46. SLATANOFF, V., and N. DJABAROFF, *Rev. Mater. Constr.*, **544** (1961) 30.

47. LAFUMA, H., *Chemistry, Proc. 3rd Intern. Symp. London* (1952) 581.

48. KLEIN, A., et al., *Am. Concr. Inst. J., Proc.*, **58**, No. 1 (1961) 59.

49. KLEIN, A., and TROXELL, *Proc. Am. Soc. Testing Mater.*, **58** (1958) 986.

50. American Concrete Institute, "Committee 223 Report," *J. Am. Concr. Inst.* (Aug. 1970) 583.

51. PAYNE, H. F., *Organic Coating Technology*, Vol. II. New York: John Wiley & Sons, Inc., 1961.

52. RILEY, V. R., and I. RAZL, *Polymer Additives for Mortars and Concretes.* Toronto: Departments of Civil and Chemical Engineering, University of Toronto, June 1972.

53. SOLOMATOV, V. L., *Izdatel'stuo Literatury po Stroitel'stva, Moscow* (1967). English translation by U.S. AEC Div. Tech. Info. Series AEC-tr-7147, Chemistry (TID-4500).

54. EASH, R. D., and G. L. EMIG, *Concrete Construction, Proceedings of the Conference on New Materials,* S. P. Shah, ed. Chicago Circle: University of Illinois, 1971, p. XI.

55. AIGNESBERGER, A., N. L. FASH, and T. REY, *J. Am. Concr. Inst.*, 68 (Aug. 1971) 608.

56. AUSKERN, A., *Concrete Construction, Proceedings of the Conference on New Materials.* Chicago Circle: University of Illinois, 1971, p. IX.

57. "Polymer-Impregnated Concrete: First Application," *Civil Eng. Mag.* (Jan. 1974) 36.

58. Detroit: "Polymers in Concrete," *Publication SP-40.* American Concrete Institute, 1973.

8

CONCRETE MECHANICS

For the wheels were just as strong as the thills,
And the floor was just as strong as the sills,
And the panels just as strong as the floor,
And the wipple-tree neither less nor more.

—HOLMES

Following the format of Chap. 3, we shall consider first the deformation and then the fracture mechanics of concrete materials.

A. DEFORMATION MECHANICS

In concrete technology, models are used to explain the deformation of hardened cement paste, mortar, and concrete. Concrete is an aggregate-matrix composite. It consists of aggregate inclusions in a cement-mortar matrix, while the mortar itself consists of sand embedded in cement paste. At a microscopic level, the paste consists of cement gel, which is a mixture of amorphous and crystalline fiber needles and crumbled and curled sheets and foils containing minute water-filled and air-filled voids. Strictly speaking, concretes are multiphase materials composed of at least seven components, viz., coarse aggregate, sand, unhydrated cement particles, cement gel, gel pores, capillary pores, and entrapped air voids [1]. The aggregate particles are also multiphase composites, containing various amorphous and crystalline minerals.

Two-phase models of mechanical behavior are acceptable approximations, however, as long as the dimensional level at which the behavior is being considered is large compared with the maximum particle size in the matrix. Thus, certain phases are assumed homogeneous and isotropic. For example, concrete may be treated as a two-phase composite of aggregate and cement mortar, mortar as a two-phase composite of sand and cement paste, paste as a two-phase composite of gel and gel pores, etc. Thus, at their respective dimensional levels, concretes, mortars, and cement pastes can all be considered as two-phase materials having a 0.3–0.6 volume fraction of coarse particles, roughly spherical in shape, embedded in a somewhat homogeneous matrix [2]. In each case we assume statistical homogeneity and isotropy for the two phases under consideration. Similar two-phase approximations are useful for other construction material composites such as bituminous mixes, asbestos-cement products, and mineral-filled polymers.

1. Elastic Deformation

The *transport* properties of composites, such as deformation, thermal and electrical conductivity, diffusivity, and magnetic permeability (Table 1), follow mixture rules [3]. The rules use weighting procedures which depend on the geometry of the phase distributions. For the phase distribution models in Fig. 1 the following relationships can be derived:

TABLE 1. Transport Properties
$q = kh$

Phenomena	*q*	*k*	*h*
Elastic deformation	Stress	Young's modulus	Unit strain
Thermal conduction	Heat flux	Heat conductivity	Temperature gradient
Electrical conduction	Current density	Electrical conductivity	Voltage gradient
Diffusion	Concentration flux	Diffusion coefficient	Concentration gradient
Electrostatics	Electric induction	Dielectric constant	Electric field intensity
Magnetostatics	Magnetic induction	Magnetic permeability	Magnetic field intensity

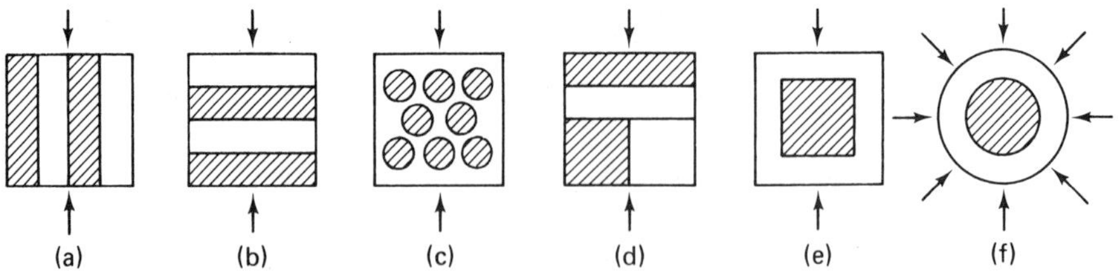

(a) (b) (c) (d) (e) (f)

Fig. 1. Phase distribution models. (a) Parallel phases; (b) series; (c) disperse; (d) Hirsch's model; (e) Counto's model; (f) Hashin's model.

For parallel phases,

$$k = \sum_{i=1}^{n} v_i k_i \tag{1}$$

For series phases,

$$\frac{1}{k} = \sum_{i=1}^{n} \frac{v_i}{k_i} \tag{2}$$

or for just two phases

$$k = \frac{k_1 k_2}{v_1 k_2 + v_2 k_1} \tag{3}$$

where k is a transport value of the composite (Table 1), k_i and v_i are the corresponding transport values and volume fractions of the individual components, respectively, and n is the number of components. All transport properties in Table 1 are vector properties except Young's modulus, which is generally a tensor property of second order. Young's modulus reduces to a first-order vector property for which Eqs. (1) and (2) are strictly valid only if all phases have zero Poisson's ratio. A deformation equation not having this restriction will be discussed shortly.

Equations (1) and (2) can be generalized to

$$k^m = \sum_{i=1}^{n} v_i k_i^m \tag{4}$$

where m is a constant, e.g., 1 in Eq. (1) and -1 in Eq. (2).

Bruggermann [4, 5] suggested that Eq. (4) also applies to the disperse model [Fig. 1(c)] by letting $m = 0$ for high-conductivity spheres embedded in a low-conductivity matrix and $m = \frac{1}{2}$ for the reverse case. For thin disks the values are $m = -\frac{2}{3}$ and $m = -\frac{1}{3}$, respectively.

Hansen [6] noted that Eqs. (1) and (2) represent the upper and lower bounds for elastic moduli of disperse systems such as concrete. Equation (1) applies when the matrix has a very high modulus compared with the disperse phase, and Eq. (2) applies for the reverse case. Therefore in a hard matrix composite the two phases undergo equal strain but have stresses proportional to their individual elastic moduli, while in a soft matrix composite the two phases have equal stress and undergo strains proportional to their individual moduli. Figures 1(a) and 1(b) therefore represent the boundary analogs for the moduli of the two-phase disperse composite [Fig. 1(c)]. Such bounds can also be derived rigorously from variational principles of minimum potential energy and minimum complementary energy [7] and from other methods [8]. Hansen [9] concluded that the moduli of concrete containing natural aggregate, i.e., hard particles in a soft matrix, approximated the lower bound solution [Eq. (2)] and that concretes containing lightweight aggregate approximated the upper bound solution [Eq. (1)]. It can be reasoned that in the case of steel balls in a rubber matrix it is obviously more reasonable to assume equal stresses throughout the matrix and discrete phases, whereas in the case of rubber balls in a steel matrix, equal strains would seem more reasonable. Hansen applied the same analysis to cement paste containing hard grains of unhydrated cement clinker in a soft matrix of hydrated cement, which in turn contains capillary pores in a hard matrix of cement gel. By substitution into Eqs. (1) and (2),

$$\frac{1}{k_p} = \frac{v_u}{k_u} + \frac{v_h}{v_g k_g}$$

where the subscripts p, u, h, and g represent paste, unhydrated cement particles, hydrated cement, and gel, respectively. k_u is of the order of 7.5×10^6 psi and k_g of the order of 2×10^6 psi.

Maxwell [10] derived a more rigorous calculation for transport properties of a matrix containing a spherical disperse phase [Fig. 1(c)]:

$$k = \frac{v_c k_c(\frac{2}{3} + k_d/3k_c) + v_d k_d}{v_c(\frac{2}{3} + k_d/3k_c) + v_d} \tag{5}$$

where subscripts c and d represent the continuous and disperse phases. Figure 2 compares results obtained by Eqs. (1), (2), (4), and (5). Note that Eqs. (1) and (2) give the upper and lower bounds for Maxwell's and Bruggermann's results.

Since $v_c + v_d = 1$ and the modular ratio $r = k_d/k_c$, Eqs. (1) and (2) become

$$k = k_c(1 + p_u v_d)$$

and

$$k = k_c(1 + p_l v_d)$$

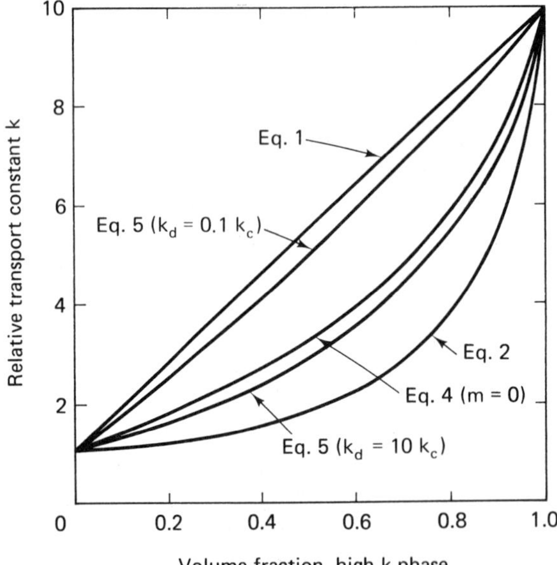

Fig. 2. Comparison of mixture laws for a two-phase composite.

where $p_u = r - 1$ and $p_l = 1 - (1/r)[1 - v_d(1 - 1/r)]$ are nondimensional constants for the upper and lower bound solutions, respectively. Ishai [11, 12] found that the elastic moduli of mortars and concretes could be expressed in the form

$$k = k_c(1 + pv_d) \tag{6}$$

where the value p depends on the elastic modular ratio k_d/k_c. Figure 3 shows that p increases with increasing modular ratio to an upper limit of about 4. p may also increase slightly with maximum particle size at a fixed modular ratio [2]. Figure 4

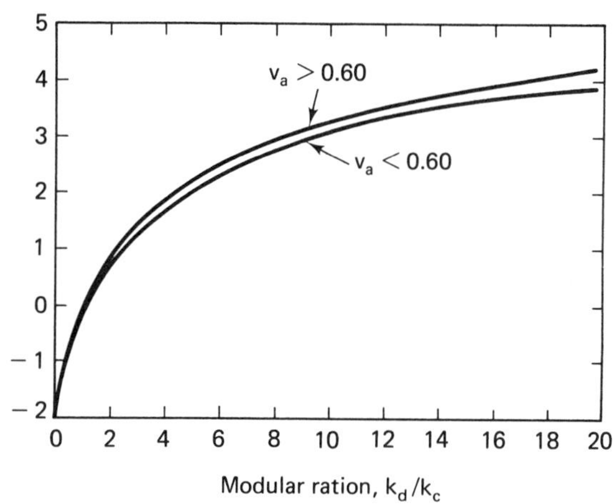

Fig. 3. Dependence of p in Eq. 6 on modular ratio, k_d/k_c. From Newman, K.: *Proceedings of the International Symposium on Theory of Arch Dams*, Southhampton, 1964. Elmsford, N.Y.: Pergamon Press, Inc., 1965, p. 683 [2].

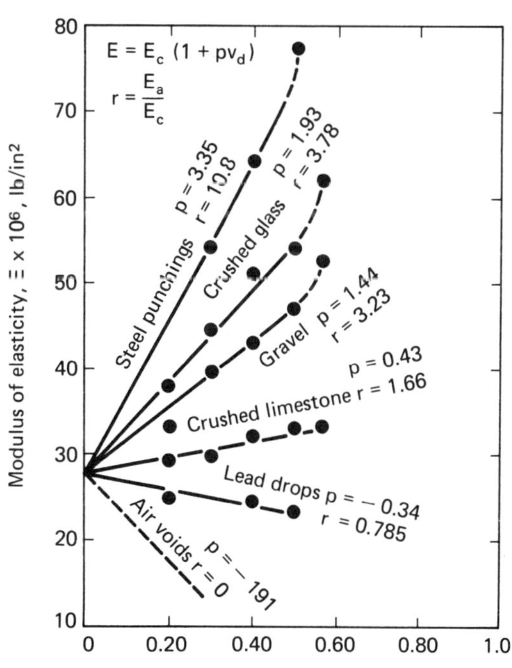

Fig. 4. Composite Young's modulus from Eq. 6 for mortar and concrete with various aggregates. From Ishai, O.: *Proc. Am. Concr. Inst.*, **59** (1962) 1363 [12].

suggests that Eq. (6) applies generally up to aggregate volume fractions of about 0.6. For an elastic material containing voids, $k_d/k_c = 0$, for which $p = -1.91$ (Fig. 3) and Eq. (6) becomes

$$k = k_c(1 - 1.91v_d)$$

The Hirsch model [Fig. 1(d)] permits representing the range of values between the upper and lower bounds, Eqs. (1) and (2), by combining these two equations with a weighting factor:

$$\frac{1}{k} = x\left(\frac{1}{v_c k_c + v_d k_d}\right) + (1 - x)\left(\frac{v_c}{k_c} + \frac{v_d}{k_d}\right) \tag{7}$$

where x and $1 - x$ represent relative proportions of material conforming to the upper and lower bound solutions [13, 14]. When there is no matrix-particle bond, no shear stresses can be transferred and $x = 0$, giving Eq. (2), which means that stresses in the matrix and particles are equal. When $x = 1$, Eq. (7) reduces to Eq. (1), corresponding to maximum bond between particles and matrix. Hirsch found values of x near 0.5 for several concrete mixtures.

A limitation of Eq. (7) is that it predicts $k = 0$ for an elastic material containing voids ($k_d = 0$). To obviate this inconsistency Counto proposed the model in Fig. 1(e), which gives the relation [15]

$$\frac{1}{k} = \frac{1 - \sqrt{v_d}}{k_c} + \frac{1}{[(1 - \sqrt{v_d})/\sqrt{v_d}]k_c + k_d} \tag{8}$$

None of the models of Eqs. (1)–(8) accounts for differences in Poisson ratios of the continuous and disperse phases. The model proposed by Hashin [16] and further developed by Hansen [17] overcomes this limitation. It consists of a spherical aggregate particle surrounded by a spherical shell of paste or mortar [Fig. 1(f)]. First the spherical particle is assumed to be a void. An internal pressure p_i and external pressure p_0 are applied, and the deformation of a unit element of the shell is expressed in terms of the Young's modulus and Poisson ratio of the matrix material and p_i and p_0. Then another expression for p_i is derived by considering compression of the particle in terms of Young's modulus of the particle. p_i is then eliminated between the two equations, and an expression is obtained for deformation of a unit element of the sphere in terms of the Young's moduli, Poisson ratios, and volume fractions of the two phases. The resulting expression can be written

$$\frac{k}{1 - 2\nu} = \left(\frac{[\nu_c k_c/(1 - 2\nu_c)]\{[1 + \nu_c/2(1 - 2\nu_c)] + \nu_d\}[k_d/(1 - 2\nu_d)]}{\{1 + [(1 + \nu_c)/2(1 - 2\nu_c)]\nu_d\}[k_c/(1 - 2\nu_c)] + [(1 + \nu_c)\nu_c/2(1 - 2\nu_d)][k_d/(1 - 2\nu_d)]}\right)\frac{k_c}{1 - 2\nu_c} \tag{9}$$

where v is the Poisson ratio, which theoretically may vary from -1.0 to $+0.5$. It is interesting that Eq. (9) gives Eqs. (1) and (2) for $\nu_c = \nu_d = \nu = -1.0$ and $+0.5$, respectively [6].

Setting $\nu_c = \nu_d = \nu = 0$ in Eq. (5) yields

$$\frac{k}{k_c} = \frac{2k_c + k_d - 2v_d(k_c - k_d)}{2k_c + k_d + v_d(k_c - k_d)} \tag{10}$$

which was derived by Maxwell in 1873 for electrical conductance, using the model of Fig. 1(f) [10]. Equation (10) has been found to apply for the electrical conductivity of emulsions even at large particle concentrations [18] and for the thermal conductivity of concretes [19, 20, 21].

For the Young's modulus of concrete, $\nu_c \sim \nu_d \sim \nu \sim 0.2$, and Eq. (9) reduces to

$$k = \frac{v_c k_c + (1 + v_d)k_d}{(1 + v_d)k_c + v_c k_c} k_c \tag{11}$$

The equation is strictly valid only for the following assumptions [20]:

1. Aggregate particles are spherical.
2. No interaction between particles (aggregate concentration small).
3. Perfect bond between particles and matrix.

The first assumption introduces negligible error for most aggregate-binder mixes. Figure 5 shows the small variation in Young's modulus for different particle shapes, assuming that $\nu_d = k_d = 0$ (voids). The model breaks down only if the disperse phase consists of columns (fiber composite) or infinite layers (laminar composite), which are we shall study in Chaps. 11 and 12.

The second assumption has been experimentally shown to remain valid to large particle concentrations for electrical and thermal conductivity [20] and probably does also for Young's modulus.

The third assumption has been found reasonable for cement mortar, where particles are small (sand size), but not for concrete, where particles are large. Figure 6 shows the correlation of Eq. (11) with experimental data for cement mortar. Hsu and Slate found that the bond between cement mortar and coarse aggregate is inferior to the bond between cement paste and sand [22]. For the limiting case of no bond, the assumption of identical stresses in the matrix and particles becomes reasonable, which leads back to Eq. (2), if the particles are also more rigid than the matrix. Hence, Young's modulus of hard aggregate concretes lies between the values calculated from Eqs. (2) and (11). If the particles are less rigid than the matrix (as with lightweight aggregates), then the bond is of little importance to composite behavior, and the particles will always follow the deformation of the matrix, for which case Eq. (11) still applies.

Whereas Eq. (11) depends on the assumption $\nu_c = \nu_d = \nu = 0.2$, somewhat more reliable values of ν for use in Eq. (9) can be obtained from

$$\nu = \nu_c(1 - v_d)^m \tag{12}$$

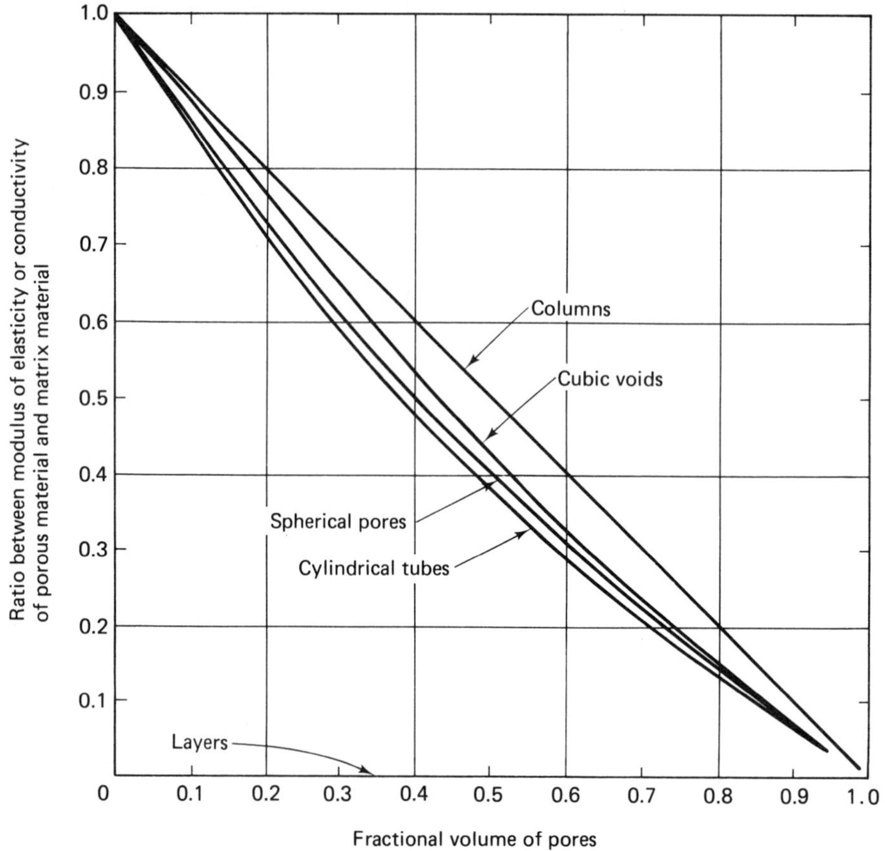

Fig. 5. Influence of pore shape on Young's modulus. From Hansen, T. C.: "Notes from a Seminar on Structure and Properties of Concrete," *Stanford University Civil Engineering Department Technical Report No. 71*, 1966 [20].

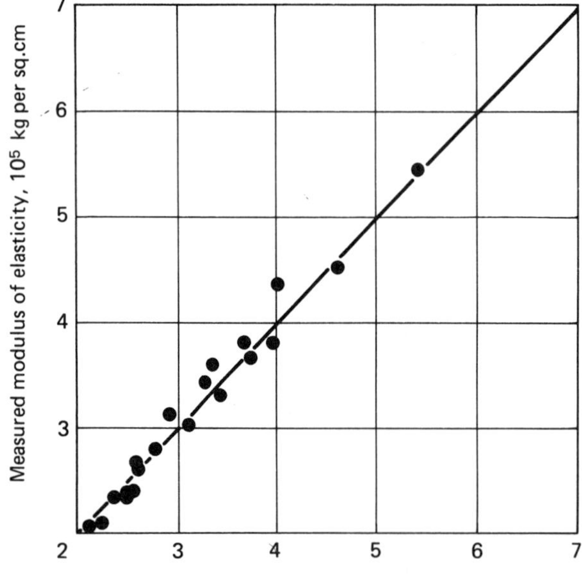

Fig. 6. Correlation of experimental Young's modulus with Eq. (11). From Hansen, T. C.: *J. Am. Concr. Inst.* **62** (Feb. 1965) 193 [17].

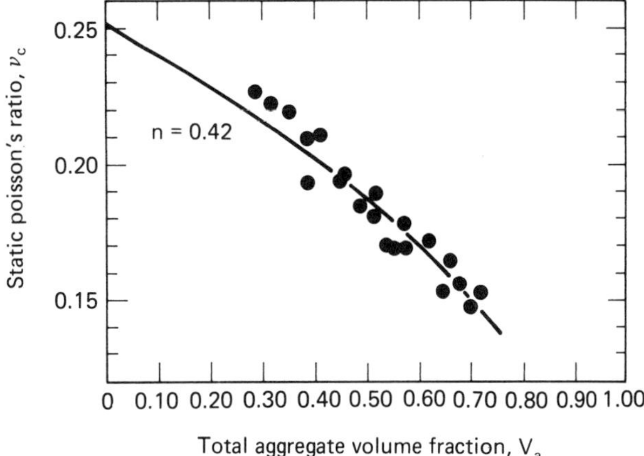

Fig. 7. Effect of volume fraction of aggregate on the Poisson ratio of mortars and concretes (see Eq. 12). From Anson, M.: *Mag. Concr. Res.* **61** (1964) 73 [23].

where m is an experimental constant. Typical data for aggregate for which $k_c \sim 10.8 \times 10^{16}$ psi and $\nu_c \sim 0.22$ are shown in Fig. 7.

For the case $k_d = 0$, as for a cement paste containing air or water voids, Eq. (9) reduces to

$$k = \frac{v_c(1 - 2\nu)}{\{1 + [(1 + \nu_c)v_d/2(1 - 2\nu_c)]\}(1 - 2\nu_c)} k_c$$

Assuming as an approximation that the Poisson ratios of both the cement gel (continuous phase) and the paste (composite) vary from 0.2 to 0.33, the following two equations are obtained for Young's modulus of the paste:

$$k = \frac{v_c k_c}{1 + v_d} \qquad \text{for } \nu_c = \nu = 0.2$$

$$k = \frac{v_c k_c}{1 + 2v_d} \qquad \text{for } \nu_c = \nu = 0.33 \tag{13}$$

Figure 8 shows the correlation between experimental data and calculations from these two equations.

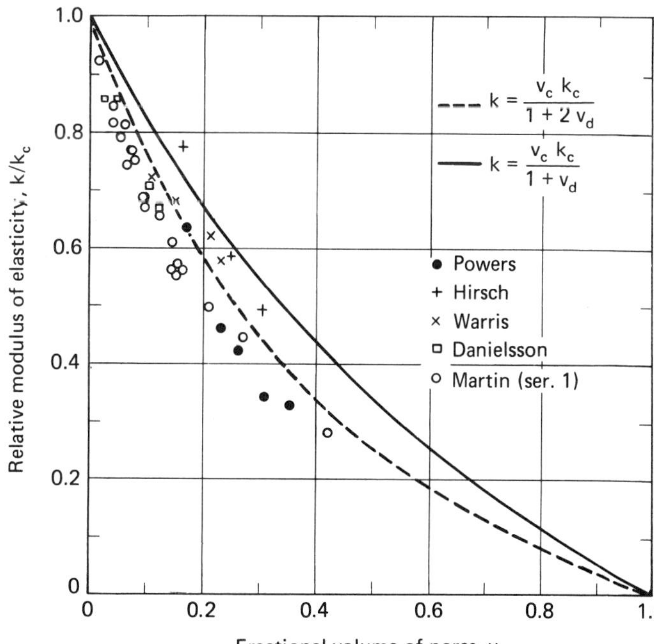

Fig. 8. Young's modulus of cement paste versus fractional volume of capillary pores. From Hansen, T. C.: *J. Am. Concr. Inst.* **62** (Feb. 1965) 193 [17].

If the pores are partially continuous, as from inadequate vibration or compaction of concrete, their influence will obviously be larger than predicted from Eq. (13). For pore continuity Hansen has proposed the following equations, which assume that the fractional volume of continuous pores is proportional to the fractional volume of total pores:

$$k = \frac{v_c^2 k_c}{1 + v_d} \qquad \text{for } \nu_c = \nu = 0.2$$

$$k = \frac{v_c^2 k_c}{1 + 2v_d} \qquad \text{for } \nu_c = \nu = 0.33$$

The following empirical equation, however, gave somewhat closer correlation with experimental results [20]:

$$k = v_c^4 k_c \tag{14}$$

Table 2 summarizes equations for Young's modulus in the forms which are particularly applicable to concretes, mortars, and cement pastes. Until experimental data are available, it must be assumed that the bulk and shear moduli, G and K, can be calculated from the usual elastic relations (see Chap. 3):

$$G = \frac{k}{2(1 + \nu)}$$

$$K = \frac{k}{3(1 - 2\nu)}$$

TABLE 2. Formulas for Young's Modulus of Concrete, Cement Mortar, and Paste

Composite	*Matrix*	*Particles*	*Condition*	*Formula for k*	*Equation*
Concrete	Cement mortar	Normal aggregate	$k_d > k_c$	$k = \dfrac{k_c k_d}{v_c k_d + v_d k_c}$	(3)
Concrete	Cement mortar	Lightweight aggregate	$k_d > k_c$	$k = \dfrac{v_c k_c + (1 + v_d)k_d}{(1 + v_d)k_c + v_c k_d} k_c$	(11)
Concrete	Cement mortar	Partially continuous voids	$k_d = 0$	$k = v_c^4 k_c$	(14)
Cement mortar	Cement paste	Sand	—	$\dfrac{1}{k} = 0.5\left(\dfrac{v_c}{k_c} + \dfrac{v_d}{k_d}\right) + 0.5\left(\dfrac{1}{v_c k_c + v_d k_d}\right)$	(7)
Cement paste	Cement gel	Air or water pores	$k_d = 0$	$k = \dfrac{v_c k_c}{1 + 2v_d}$ (for $\nu_c = \nu = 0.33$)	(13)

SOURCE: HANSEN, T. C., "Notes from a Seminar on Structure and Properties of Concrete," *Stanford University Civil Engineering Department Technical Report No. 71*, 1966 [20].

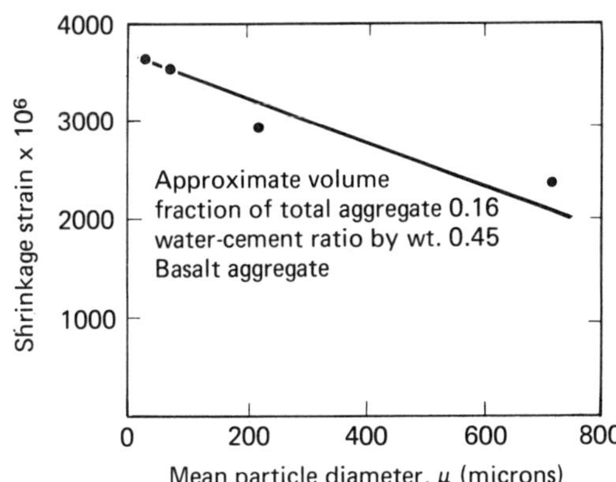

Fig. 9. Effect of mean aggregate diameter on drying shrinkage. From Alexander, K. M. and J. Wardlaw; *Proc. Am. Concr. Inst.*, **55** (1959) 303 [25].

2. *Shrinkage*

Due to the restraining effect of aggregate particles, concrete generally shrinks less than neat cement paste. For example, the shrinkage of neat cement paste is typically about 10 times the shrinkage of a concrete with 30% paste and made with high-quality, nonshrinking aggregate and about 4 times the shrinkage of mortar made with 50% paste. The difference in shrinkage between concrete and mortar is due mainly to the difference in paste content. The major factors affecting concrete shrinkage are the shrinkage of the paste and the fractional volume and shrinkage of the aggregate.

In general, concretes with water/cement ratios of 0.3 to 0.7 and aggregate volume fractions of 0.35 to 0.6 have ultimate drying shrinkage of 300 to 800 microstrain, with about 25% shrinkage occurring in 2 weeks and 75% in 1 [24].

Alexander and Wardlaw [25] found drying shrinkage to decrease with aggregate particle size in the size range from cement particles up to medium sand (Fig. 9).

Pickett [26] proposed the following relation between shrinkage strain of the concrete S, shrinkage strain of the cement paste S_p, and the aggregate volume fraction g:

$$S = S_p(1 - g)^m \tag{15}$$

where m is a constant related to the elastic properties of the paste, and aggregate and the ultimate drying shrinkage of the paste is of the order of 2,000–3,000 microstrain. Pickett suggested a value 1.7 for m of silica sand-mortar mixes (Fig. 10) and L'Hermite [27] found m to be from 1.2 to 1.7 for various aggregates.

Hansen and Neilsen [27] developed expressions for concrete shrinkage which include, in addition, the effects of modulus of elasticity and shrinkage of the aggregate, using an approach analogous to that mentioned for Hansen's derivation of Eq. (9) for the elastic modulus of concrete. If the elastic modulus of the paste is less than of the aggregate,

$$\frac{S}{S_p} = \left(1 - \frac{S_g}{S_p}\right)\frac{(E_g/E_p) - 1]g^2 - 2(E_g/E_p)g + (E_g/E_p) + 1}{(E_g/E_p) + 1} + \frac{S_g}{S_p}, \qquad E_g \geq E_p \tag{16}$$

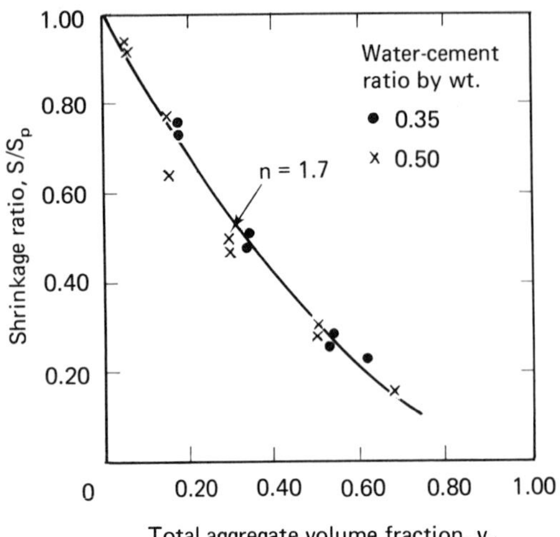

Fig. 10. Effect of aggregate volume fraction on the ratio of drying shrinkage to that of neat cement paste.: $m = 1.7$, Eq. 13. From Pickett, G.: *Proc. Am. Concr. Inst.* **52** (1956) 58 [26].

and if the paste modulus is greater,

$$\frac{S}{S_p} = \left(1 - \frac{S_g}{S_p}\right)\frac{[(E_p/E_g) + 1]g - (E/_pE_g) - 1)}{[(E_p/E_g) - 1]g - (E_p/E_g) - 1)} + \frac{S_g}{S_p}, \qquad E_g \leq E_p \tag{17}$$

where S_g = unit linear shrinkage of the aggregate,
E_g = Young's modulus of the aggregate,
E_p = Young's modulus of the paste.

Representative values for paste and aggregate are given in Table 3. The ratio between the shrinkage of the concrete and the shrinkage of its corresponding cement paste, S/S_p, is called the *shrinkage ratio.*

TABLE 3. Properties of Concrete Materials [29]

Property	*Hardened cement paste*	*Natural aggregates*	*Lightweight aggregates*
Young's modulus, psi × 10^6	1–4	5–10	0.5–1.5
Drying shrinkage, × 10^6	2,000–3,000	Negligible (with few exceptions)	
Specific creep, psi × 10^6	1–3	Negligible (with few exceptions)	High
Poisson's ratio	0.25	0.10–0.25	
Ultimate compressive strength, psi × 10^3	2–20	10–50	0.6–6
Ultimate tensile strength, psi	200–1,000	200–2,000	50–250

SOURCE: NEWMAN, K. and J. B. NEWMAN, *Structure, Solid Mechanics and Engineering Design*, M. Te'eni, ed., Proceedings of the Southampton 1970 Civil Engineering Materials Conference, Part II. New York: John Wiley & Sons, Inc. (Interscience Division), p. 963 [29].

The theory developed is applicable for various properties of all two-phase composites having roughly spherical particles as the disperse phase. For example, the

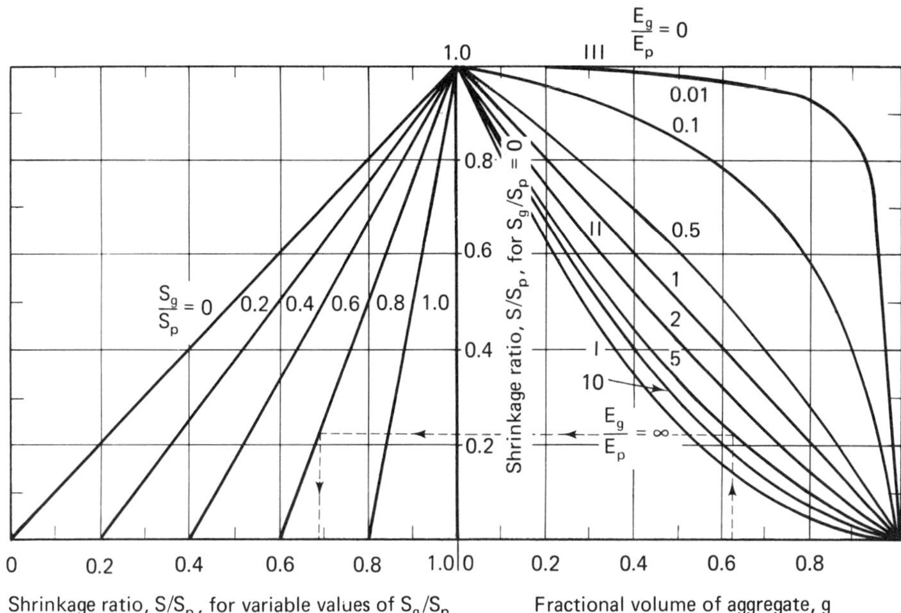

Fig. 11. Shrinkage ratio S/S_p in terms of S_g/S_p and E_g/E_p. From Hansen, T. C. and K. E. Nielsen: *J. Am. Concr. Inst.* **62** (July 1965) 783 [28].

coefficient of thermal expansion can be calculated from Eqs. (16) and (17) in the same way as the shrinkage.

Figure 11 allows solution of Eqs. (16) and (17) for E_g/E_p ratios of 0 to ∞ and and S_g/S_p ratios of 0 to 1.0. In the right half of the graph, the shrinkage ratio is plotted as a function of g for values of E_g/E_p from 0 to ∞ for zero shrinkage aggregate ($S_g/S_p = 0$).

When $E_g/E_p = \infty$, Eq. (16) gives

$$\frac{S}{S_p} = g^2 + 2g - 1 \tag{18}$$

plotted as curve I in Fig. 11. When $E_g/E_p = 1$, Eqs. (16) and (17) give

$$\frac{S}{S_p} = 1 - g \tag{19}$$

plotted as line II in Fig. 11. When $E_g/E_p = 0$, Eq. (17) gives for all values of g

$$\frac{S}{S_p} = 1 \tag{20}$$

plotted as line III. The left half of the graph is used to correct for the appropriate value of S_g/S_p.

Assuming that the cement paste does not crack around aggregate particles, Eq. (18) gives the maximum effect on concrete shrinkage of the mechanical restraint of particles. If the paste cracks, shrinkage ratios will be less than given by Eqs. (16) and (17).

If the aggregate particles are replaced by voids, Eq. (20) applies, and concrete shrinkage equals shrinkage of the corresponding cement paste for all concentrations of pores in the paste. The theory has been correlated with numerous experimental results by various investigations.

EXAMPLE 1*: Determine the shrinkage of concrete having a fractional volume of aggregate $g = 0.62$ when $E_g/E_p = 5$, $S_g/S_p = 0.6$, and the shrinkage, S_p, of the corresponding cement paste is known. Following the dashed line on the graph, $S/S_p = 0.68$. Knowing S_p, the concrete shrinkage S can be found. Note that S/S_p would have been only 0.22 if the aggregate particles had been free from shrinkage ($S_g/S_p = 0$). Even minor aggregate shrinkage significantly affects the concrete shrinkage.

Large tensile stresses which develop in the cement paste due to restraint of the aggregate particles are reduced by creep of the paste. Hansen and Neilsen suggest accounting for this effect by substituting a *modulus of deformation* of the paste, which includes creep, for the modulus of elasticity. The modulus of deformation of cement paste was found generally to be less than one third of the modulus of elasticity. Generally, the modulus of elasticity of high-quality aggregate is several times the modulus of elasticity of cement pastes and was found to be usually about 10 times the modulus of deformation of paste, including creep.

3. Creep

Aggregate-binder composites, including cement pastes, mortars, and concretes, exhibit the inelastic behavior illustrated in Chap. 3. In concretes, time-dependent deformation is complicated by the fact that hydration and drying shrinkage, both of which affect creep, proceed during the period of deformation. As long as the stress level is below that at which fracture begins, creep strain is proportional to the applied stress [27, 29, 31, 32]. This has led to the definition of *specific creep*, i.e., creep strain per unit stress. Good concretes have an ultimate specific creep of 0.4–0.8 microstrain per psi [1]. Creep rate and ultimate strain increase with the water/cement ratio [33] and decrease with the age of the concrete at loading [34] and therefore vary inversely with strength.

Although the creep behavior of concrete is not well understood, it is generally believed to be associated with the flow and diffusion of absorbed water layers within the semiamorphous entwined needles, fibrous particles, and crumbled foils which form the cement gel [20]. The secondary bonds between adjacent gel particles may be randomly broken due to thermal activation and reformed in slightly different positions under the applied load, resulting in delayed elastic deformation.

---

*By permission, from T. C. Hansen and K. E. Neilsen [28].

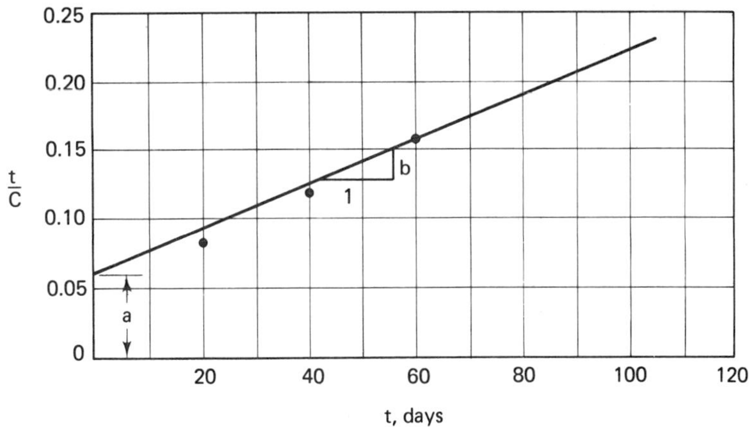

Fig. 12. Creep constants according to Ross' expression, $c = t/(a + bt)$. From Neville, A. M.: *Hardened Concrete: Physical and Mechanical Aspects.* American Concrete Institute Monograph 6, Ames, Iowa: ACI and Iowa State University Press, 1971 [35].

The fact that strain is proportional to load suggests the application of linear viscoelastic theory to concrete, i.e., generalizing Hooke's law from "instantaneous strain proportional to stress" to "strain rate and ultimate strain proportional to stress." It is convenient to express the creep-time relation as an equation, so that creep can be predicted without performing long-term tests. About a dozen equations have been suggested, most of a hyperbolic or exponential type [35]. In all such equations, certain constants must be determined experimentally; i.e., limited-time creep tests must be made using the actual mix and storage conditions. The longer the time over which creep is measured, the better the prediction.

A widely used creep equation is the one by Ross [34],

$$c = \frac{t}{a + bt} \tag{21}$$

where c is specific creep (creep strain per psi), t is time under load, and a and b are constants. A plot of t/c against t is a straight line, and the constants can be easily evaluated (Fig. 12). Another advantage of the expression is the ease with which ultimate creep can be obtained, $c_\infty = 1/b$. Also, when $t = a/b$, $c = b/2$; i.e., one half of the ultimate creep is obtained at time $t = a/b$. Note that in drawing the straight line in Fig. 12 greater weight was given to points for larger values of t because this gives a better prediction of creep values for long periods.

B. FRACTURE MECHANICS

Aggregate-binder composites may, in general, fail by either brittle or ductile fracture or by a combination of these modes in the separate phases of the composite. In bituminous concretes deformation occurs almost entirely within the continuous phase, and failure is ductile. In portland cement concretes the continuous and disperse phases both obey brittle Griffith crack theory, and the deformation behavior and crack location depend on relative elastic moduli, Poisson ratios, and strengths of the two phases. In cellular foams ($k_d = 0$) tensile failure of the cell walls may be brittle or ductile, depending on the foam material, and compressive failure may be brittle due to flexural stresses in the cell walls or ductile due to wall buckling. In this section we shall study the failure mechanisms of brittle concretes and in Chaps. 8 and 9 of bituminous mixes and cellular foams, respectively.

1. Shrinkage and Crack Initiation

In portland cement concretes both phases are brittle, each phase approximately obeys Griffith crack theory, and to study the fracture mechanics of the composite we must examine the mechanisms of *crack initiation* before loading and of *crack propagation* during loading. Cracks may occur predominantly in the continuous phase, the disperse phase, or at the interface.

In concretes, crack initiation before loading results primarily from shrinkage strains in the cement paste. Shrinkage results from the loss of water from cement gel and increases with the volume fraction of cement paste and the water/cement ratio.

The introduction of sand or aggregate into the paste reduces the drying shrinkage by (1) reducing the amount of the dimensionally unstable component and (2) producing a restraint effect which reduces volume change. Hsu found that a volume change of 0.3% in cement paste is enough to mobilize interfacial tensile stresses in the order of 1,800 psi [36]. (See Sec. D for shrinkage volume computations.) Slate and Matheus showed that volume changes even larger than 0.3% occur and that cracking before loading is probably caused, in large part, by volume changes during setting and hardening [37].

Experiments by Shah and Slate [38] with various concrete mixes suggest that the tensile and shear stresses at aggregate interfaces due to shrinkage generally increase

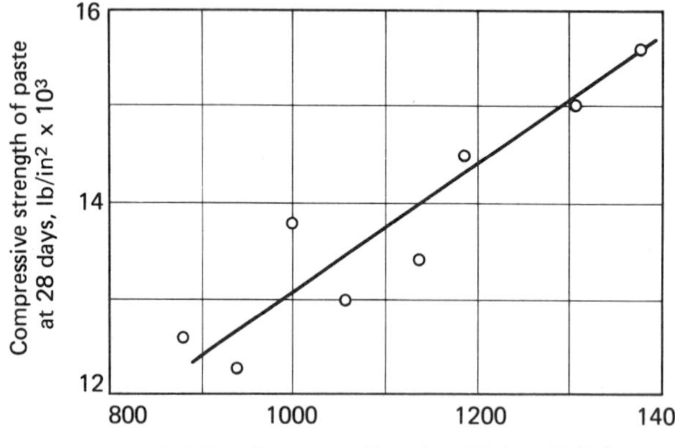

Fig. 13. Relationship between bond strength and paste strength for eight brands of ordinary portland cement at a water-cement ratio of 0.35. From Alexander, K. M., T. Wardlaw, and D. F. Gilbert; *Proceedings of the International Conference on the Structure of Concrete*, Brooks, A. E. and K. Newman, eds., London: Cement and Concrete Association, **1**, (1965) 59 [39].

with particle size, and if they exceed the paste tensile strength of the aggregate-paste bond strength, small cracks can cause a considerable reduction in shrinkage. These cracks, and their extensions upon loading, exist within the mortar and more predominantly at the interfaces between coarse aggregate and mortar. Mortar cracks tend to bridge between cracks on large aggregate particles in preference to small ones, so that a continuous crack has a minimum portion of its total length through the mortar. Bond cracks generally leave relatively clean surfaces, indicating that the actual interface is the critical region, not the nearby paste. The observations suggested that the bond between coarse aggregate and mortar was the weakest link in the heterogeneous concrete system. Paste-aggregate bond tensile strengths were found to vary from approximately 40 to 90% of the paste tensile strengths, depending on rock type, surface roughness, and water/cement ratio. Sand mortar-aggregate bond strengths were found to be only 35 to 65% of the tensile strengths of the corresponding mortars alone. Others have obtained similar results (Fig. 13). Chemical aspects of interfacial bonding were discussed in Sec. C.

2. *Loading and Crack Propagation*

Crack propagation during loading may occur at the aggregate-paste interface, in the cement paste or mortar matrix, or in the aggregate particles. The strength of regular concrete may be increased by increasing the paste strength or the bond strength. The strength of lightweight concrete can be increased by increasing the paste strength or using a stronger aggregate material.

a. DETECTION OF CRACK PROPAGATION

Crack propagation during loading can be detected by the reduction in slope of the stress-strain curve, by an increase in the Poisson ratio, by brittle lacquer and photoelastic surface coatings [40], by direct microscopic observation on the specimen surface, and by a reduction in velocity of sound through specimen (Fig. 14) [23, 41]. Crack geometries on sections cut after loading have been studied by x-ray methods and by filling the cracks with dye and observing them microscopically [39, 42].

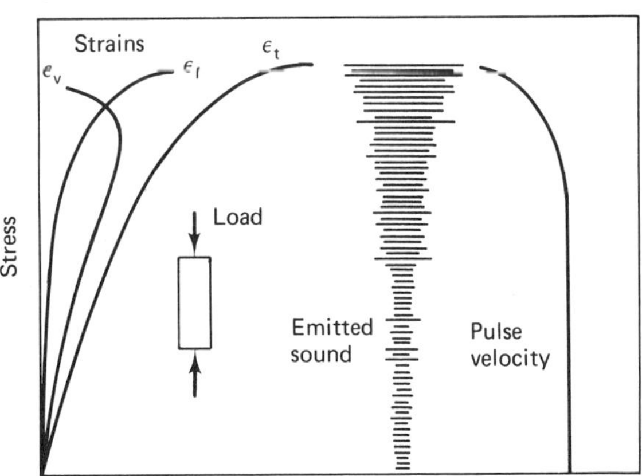

Fig. 14. Detection of microcracking on uniaxial compression by strain gage, ultrasonic pulse velocity and acoustic measurements. From Newman, K.: "Concrete Systems," in *Composite Materials*, L. Holliday, ed. Amsterdam: Elsevier Publishing Company, 1966, Chap. 8 [1].

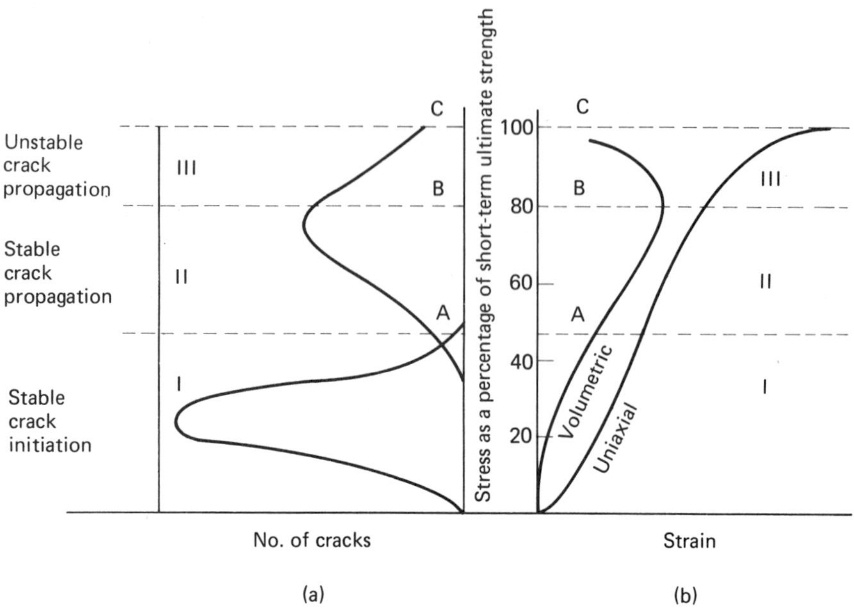

Fig. 15. Stages of concrete cracking in terms of: (a) crack initiation and propagation, (b) stress-strain. From Newman, K. and J. B. Newman in *Structure, Solid Mechanics and Engineering Design*, M. Te'eni, ed., *Proceedings of the Southampton 1970 Civil Engineering Materials Conference*, Part II. New York: John Wiley & Sons, Inc. (Interscience Division) p. 963 [29].

b. STAGES OF CRACKING

Concrete subjected to any stress state can support loads up to 40–60% of ultimate without apparent distress [29]. Above this level, noises of internal disruption can be heard until, at about 70–90% of ultimate, small cracks appear on the surface. The cracks spread and interconnect until, at ultimate load, specimens typically fracture into many pieces.

Several stages of cracking can be distinguished:

1. *Shrinkage cracking*. This occurs prior to loading due to shrinkage of cement paste between rigid aggregates [43]. Densification due to partial closing of some of the shrinkage cracks under initial loading can be observed as a small concavity near the origin of the compression stress-strain curve, with resulting increase in elastic modulus and ultrasonic pulse velocity.
2. *Crack initiation*. Under initial loading, additional microscopic cracks appear at points of high tensile strain concentration.
3. *Stable crack propagation*. As load is increased, cracks propagate, but if the stress level is maintained constant, crack propagation ceases.
4. *Unstable crack propagation*. Cracks become self-propagating under constant applied load, and failure occurs whether or not load increases. This stage occurs at 70–90% of ultimate stress and is accompanied by a dilation

of structure, indicated by a reversal in volume change behavior (Fig. 15). Below point A in Fig. 15 concrete behaves quasi-elastically, above point B it cracks self-propagate, and failure occurs at point C.

The term *discontinuity* is used to designate stress level A, at which crack propagation and noticeable nonlinearity in the uniaxial stress-strain diagram begin. In tensile stress, discontinuity is more typically as high as 70% of ultimate strength. In tension, fracture initiation is soon followed by complete failure, whereas in compression, fracturing can merely change the crack configuration and redistribute local stresses to give a more stable crack pattern and prolong failure.

In ideal brittle materials, when a crack reaches critical size, it propagates spontaneously through the material. In a heterogeneous material, such as concrete, the crack may extend into a zone of increased resistance, stop, and then extend again at a higher stress level, resulting in nonlinearity in the stress-strain curve. The nonlinearity bears no relationship to the *plasticity* of metals, which remain continuous throughout the plastic region. The stress-strain relationship for pure cement, for example, is linear to failure.

If a concrete specimen is cyclically loaded between zero load and the load at which cracks first appear, the cracks spread out and increase in number until the specimen fails in *fatigue* at a load often 35–50% of the static compressive strength [20]. If the specimen is subjected to a sustained load 20–30% higher than the load at which cracks initially develop, cracking gradually increases, and *creep* failure may eventually occur at a load 60–80% of the quick loading static compressive strength.

c. CRACK LOCATIONS

The crack patterns which develop under load depend principally on (1) the stress state of applied load and (2) the relative elastic moduli and Poisson ratios of the continuous and discrete phases. The stress state influences the overall crack pattern, and the relative moduli and Poisson ratios influence the geometry of crack development in the vicinity of individual aggregates.

Typically, cracks develop normal to the plane of maximum tensile stress, as shown for the cases of uniaxial tension, uniaxial compression, and biaxial compression in Fig. 16. This tends to give a columnar fracture pattern under uniaxial compression and a laminar fracture pattern under biaxial compression. Figure 17 shows the predominant columnar cracking of a uniaxially compressed concrete prism.

Cracking may occur

1. At the aggregate-paste interface.
2. In the cement paste or mortar matrix.
3. In the aggregate particles.

If the Young's modulus of the particle is less than that of the continuous phase, tensile stresses develop above and below the particles and compressive stresses at their sides [20, 45]. Aggregates having low moduli generally also have low strength,

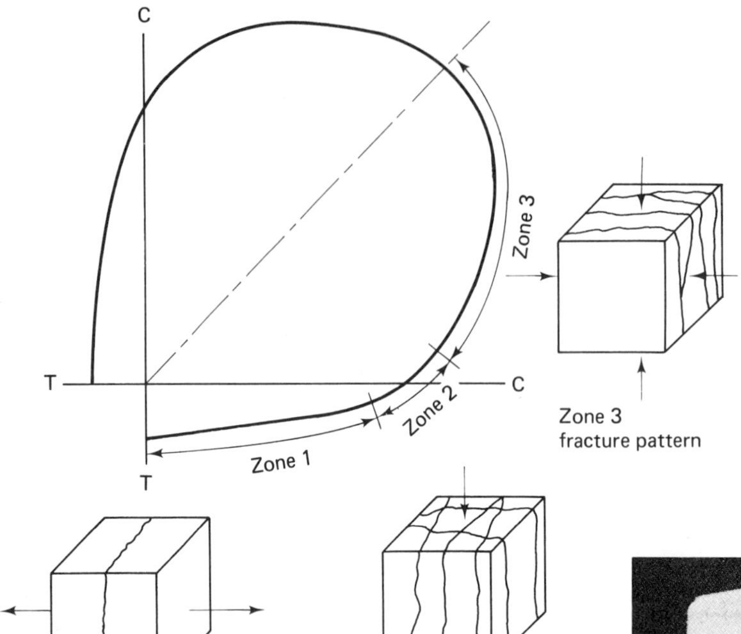

Fig. 16. Idealization of macroscopic fracture patterns. From Vile, G. W.: "Strength of Concrete Under Short-Term Static Biaxial Stress," in *The Structure of Concrete*, Brooks, A. E. and K. Newman, eds., London: Cement and Concrete Association, 1968, p. 278 [44].

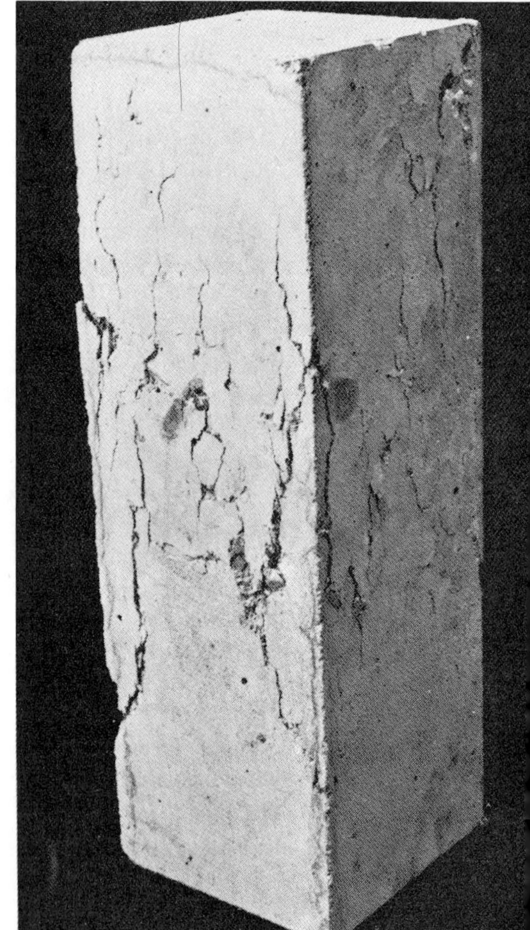

Fig. 17. Cleavage failure of uniaxially compressed concrete. From Newman, K., "Concrete Systems," in *Composite Materials*, L. Holiday, ed., Amsterdam: Elsevier Publishing Company, 1966, Chap. 8 [1].

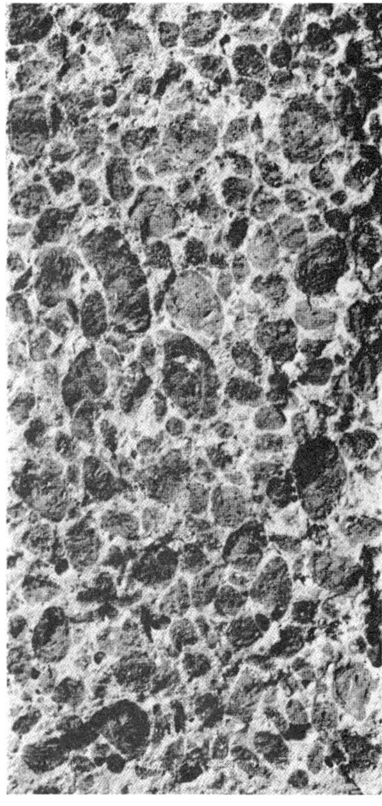

Fig. 18. Fracture of concrete made with lightweight coarse aggregate. Fracture occurs through the coarse aggregate, not along their surfaces. From Bache, H. H. and P. N-Christensen: *The Structure of Concrete*, Brooks, A. E. and K. Newman, eds., *Proceedings of an International Conference*, Imperial College. London: Cement and Concrete Association, **1** (1965) 93 [45].

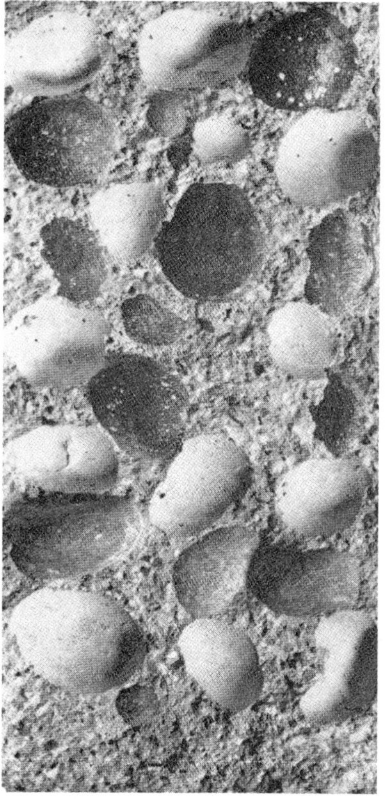

Fig. 19. Fracture of concrete made with normal aggregate. Fracture occurs along the aggregate particle surfaces. From Bache, H. H. and P. N-Christensen, eds., *The Structure of Concrete*, Brooks, A. E. and K. Newman, eds., *Proceedings of an International Conference*, Imperial College, London: Cement and Concrete Association, **1** (1965) 93 [45].

and consequently tensile failure surfaces occur in the particle parallel to the applied load (Fig. 18). The strength obviously decreases with increasing fractional volume of aggregate.

When the two phases have approximately equal moduli, cracking occurs both within and outside of the particles. If the Young's modulus of the particles is greater than that of the continuous phase, compressive stresses develop above and below the particles and tensile stresses at their sides. Aggregates having high moduli will also have high strength, and cracking will occur either in the matrix or at the sides of the larger particles, but not through them (Fig. 19). Failure again occurs on tensile or cleavage splitting surfaces parallel to the applied load or normal to the direction of minimum compressive load in triaxial loading.

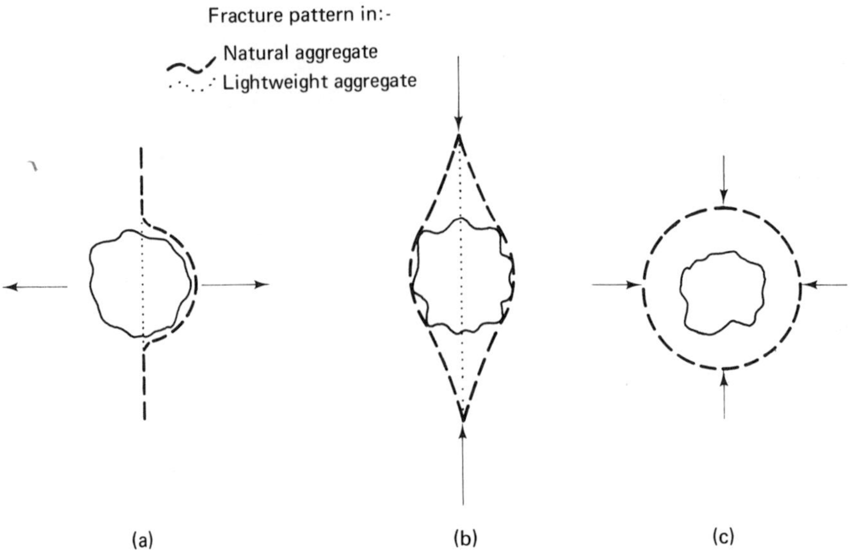

Fracture pattern in:-

Natural aggregate

Lightweight aggregate

(a) (b) (c)

Fig. 20. Fracture paths in concrete under: (a) uniaxial tension, (b) uniaxial compression, (c) biaxial compression. From Newman, K. and J. B. Newman in *Structure, Solid Mechanics and Engineering Design*, M. Te'eni, ed., *Proceedings of the Southampton 1970 Civil Engineering Materials Conference*, Part II. New York: John Wiley & Sons, Inc. (Interscience Division) p. 963 [29].

Figure 20 shows idealized cracking models for a single aggregate embedded in mortar for the case of uniaxial tension and compression and biaxial compression. In uniaxial compression, small cones of mortar tend to adhere to each end of the aggregate particle, aligned in the direction of compressive stress. In biaxial compression, the cones extend to form complete halos around the particles [29]. Figure 21 shows the conical- and halo-shaped fragments removed from specimens tested in uniaxial and biaxial compression, respectively.

3. *Simplified Strength Formulas*

Although no theory has been developed from which the strength of concrete can be computed when the paste and bond strength are known, the following empirical relationships have been proposed [39]:

$$S_c = 480 + 2.08M_p + 1.02M_b$$
$$S_f = 290 + 0.318M_p + 0.162M_b$$

where S_c = concrete compressive strength, psi,

S_f = flexural strength, psi,
M_p = paste modulus of rupture, psi,
M_b = bond modulus of rupture, psi.

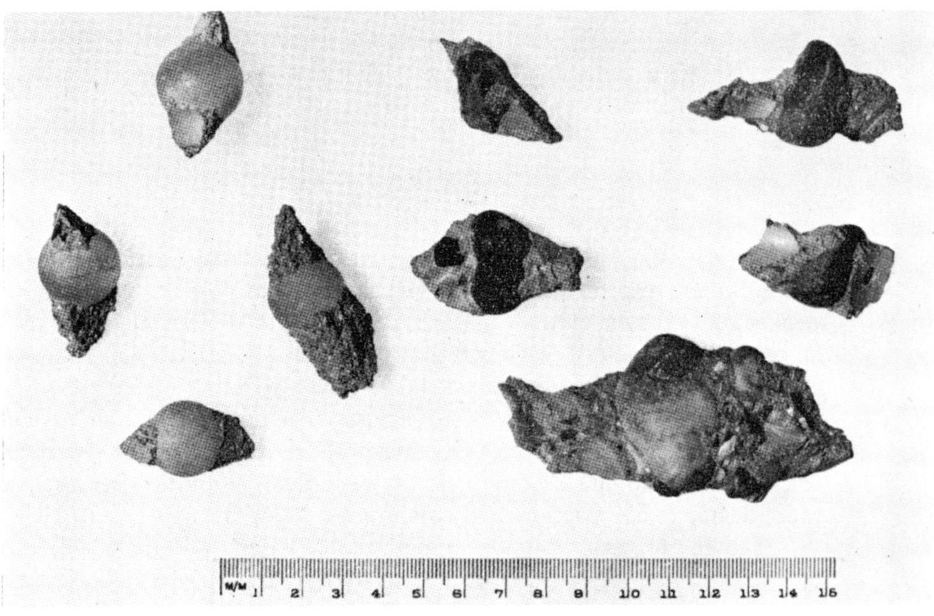

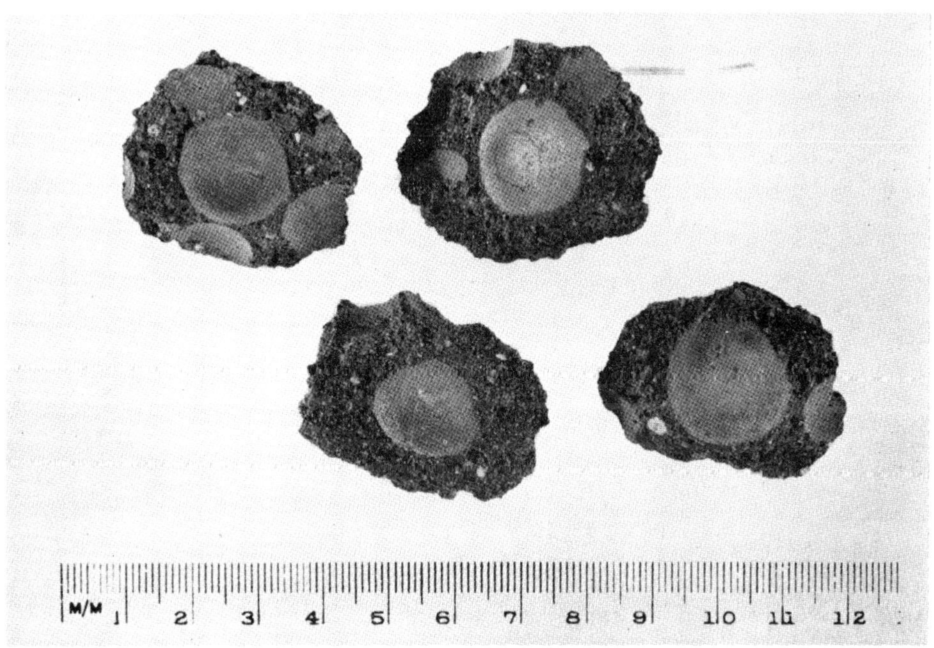

Fig. 21. Particles of coarse aggregate removed from concrete tested in: (a) uniaxial compression, showing cone formation, and (b) biaxial compression, showing 'halo' formation. From Newman, K. and J. B. Newman in *Structure, Solid Mechanics and Engineering Design*, M. Te'eni, ed., *Proceedings of the Southampton 1970 Civil Engineering Materials Conference*, Part II. New York: John Wiley & Sons, Inc. (Interscience Division), p. 963 [29].

Feret [46] was the first to define cement paste and concrete strengths in terms of the volume fractions of the constituents:

$$S_c = k\left(\frac{v_c}{v_c + v_w + v_a}\right)^2 \tag{22}$$

where k is a constant and v_c, v_w, and v_a are the volume fractions of cement, water, and air voids, respectively. Schiller [47] showed that porosity $(v_w + v_a)$ has a similar weakening effect on gypsum plasters. The strength of cement paste which contains all gel and no capillary space will be in the order of 13,000–18,500 psi [48]. Strengths of 40,000 psi have been obtained, however, by molding under pressure neat paste with a water/cement ratio as low as 0.08 [49]. The presence of a high volume fraction of unhydrated cement particles can thus produce paste strengths greatly above those of cement gels. Figure 22 shows qualitatively the relation between the water/cement ratio, degree of compaction, and compressive strength.

Hansen proposed a simple strength model based upon the compressive strength of a cube containing a spherical pore. Assuming that strength is proportional to the minimum cross-sectional area of solid material, for a cube of side 1 and pore of radius r,

$$S_c = (1 - \pi r^2)S_s$$

where S_s and S_c are the strengths of the solid material and the porous solid, respectively. Inserting the pore volume ratio, $v_p = \frac{4}{3}\pi r^3$,

$$S_c = (1 - 1.2v_p^{2/3})S_s \tag{23}$$

Wischer [50] found experimentally that for many cement pastes

$$S_c = (1 - v_p)^{2.7} \times 3100 \text{ kg/sq cm} \tag{24}$$

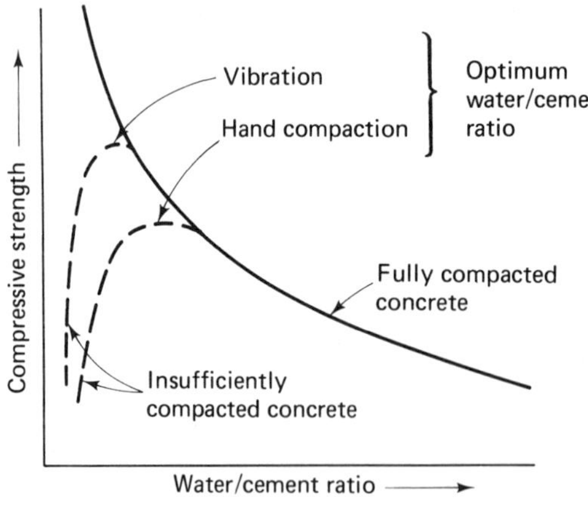

Fig. 22. Relation between water–cement ratio, degree of compaction, and compressive strength. From Hansen, T. C.: "Notes from a Seminar on Structure and Properties of Concrete," *Stanford University Civil Engineering Department Technical Report No. 71*, 1966 [20].

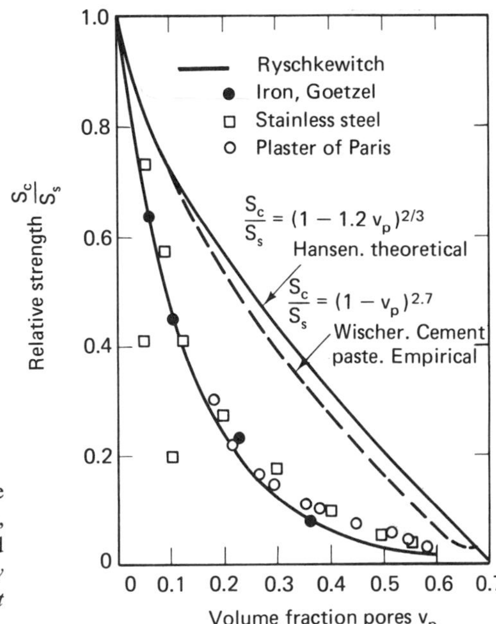

Fig. 23. Effect of porosity on compressive strength of some porous solids. From Hansen, T. C.: "Notes from a Seminar on Structure and Properties of Concrete," *Stanford University Civil Engineering Department Technical Report No. 71*, 1966 [20].

Equations (23) and (24) are plotted in Fig. 23, together with experimental strengths for other porous systems. Equation (23) may represent a reasonable upper bound for strengths of porous systems, since it does not account for partial pore continuity or buckling failure of thin walls between closely spaced pores.

One of the complicating factors in developing better theoretical relationships for concrete strength is that the water/cement ratio seems to have (1) a strong effect on mortar compressive strength, (2) a much smaller effect on the mortar tensile strength, and (3) very little effect on the coefficient of interfacial friction [38]. Hence, these three sources of strength appear to be partially independent of, and not reducible to, one another. All these strength components come into play in varying degrees under bi- or triaxially imposed stresses. This is one reason that various general strength theories for concrete, either of phenomenological origin such as octahedral stress-type theories or of more fundamental origin such as Griffith fracture mechanics, have had little success. Another reason lies in the discontinuity of the material due to the very existence of cracks.

4. Generalized Failure Criteria

Although concrete is subjected to a wide range of stress states in structures, this is rarely accounted for in design, and there is still no universally accepted criterion of failure for concrete under combined stresses.

Failure stress criteria for concretes may be related to *discontinuity*, at which the concrete can no longer withstand a fixed load without failure, or to *ultimate strength*, at which the concrete fails rapidly [2]. On a stress-strain diagram, the discontinuity point in heterogeneous brittle materials is analogous to the yield point in ductile materials. But after discontinuity, the structure is no longer continuous, and the laws of mechanics, which depend on continuity, no longer strictly apply.

Materials fail either by shear slip or by tensile separation (Sec. B in Chap. 3). Metals have lower resistance to shear than to cleavage and usually fail in shear at a stress level almost entirely dependent on deviatoric stress (the difference between maximum and minimum principal stresses) and independent of volumetric, or average principal, stress. It is usual rheologic practice to separate any stress state into its volumetric and deviatoric components and to assume that any inelastic behavior is due entirely to the deviatoric component. This assumption is valid for metals [51]. Mortars and concretes, however, have much lower resistance to tensile splitting and fail by cleavage, except at extremely high hydrostatic stress components, where a gradual transition from tensile bond failure to shear bond failure evolves. At high pressures, experiments confirm that inelastic behavior of concrete has the attribute of metal plasticity, and the structure maintains a high degree of continuity [52, 53]. At high hydrostatic pressures, therefore, discontinuity and yield criteria of all materials are approximately the same. Under more typical stress conditions, however, concrete strength depends mainly on the tensile strengths of the mortar and of the aggregate-mortar interface. The basic criterion of failure in concrete systems is therefore a limiting tensile stress which can be induced under all stress states except hydrostatic compression. In tensor terms, concrete failure depends on the magnitudes of both the deviatoric and volumetric stress tensor components.

The Griffith fracture criteria under biaxial stress (Chap. 3) does not predict the discontinuity strength of concrete in biaxial compression, mainly because it does not recognize failure in the unloaded third direction. The Coulomb-Mohr theory also

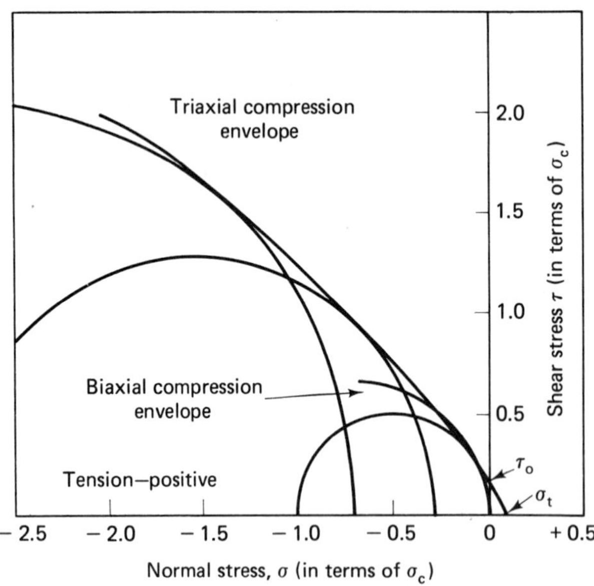

Fig. 24. Biaxial and triaxial ultimate strength Mohr envelopes for concrete in terms of ultimate uniaxial compressive strength, σ_c. From Newman, K.: *Proceedings of the International Symposium on Theory of Arch Dams*, Southampton, 1964. Elmsford, N. Y.: Pergamon Press, Inc., 1965, p. 683 [2].

cannot be applied directly to concrete for the same reason; limiting shear stress is dependent on the intermediate principal stress, and there are different envelopes for biaxial and triaxial compression (Fig. 24).

Any stress state can be expressed as a single point in three-dimensional principal stress space. If the combination of principal stresses causing discontinuity or ultimate failure are known, a three-dimensional plot of these values depicts a limiting surface of discontinuity or ultimate strength.

To understand the failure behavior of concrete it is helpful to compare the slope of this limiting surface in terms of the principal applied stresses with the shape of the limiting surface of yielding of metals. The limiting surface of yielding for metals can be represented by a cylinder whose axis lies along the line $\sigma_1 = \sigma_2 = \sigma_3$, inclined at equal angles to the principal stress axes (Fig. 10, Chap. 3) [52]. At high values of hydrostatic compression, the limiting surface of *discontinuity* for mortars and concrete is also cylindrical. However, at low hydrostatic compression, the cylinder transforms to a paraboloid and finally to a sphere (Fig. 25). The maximum tensile stress is represented by limiting planes which intersect the paraboloid to produce the elliptical surfaces shown in Fig. 25 and plotted in the σ_1-σ_3 stress plane in Fig. 26. The relative sizes of the cylinder, paraboloid, and sphere and the location of the limiting tensile stress planes depend on the specific concrete system. Most of the surface representing discontinuity can be described by an equation of the form [1]

$$\bar{\sigma} = c_1 + c_2 V_d^n$$

where $\bar{\sigma}$ is the mean normal stress, V_d is the deviatoric component of elastic strain energy (shear strain energy) and c_1, c_2, and n are constants whose values depend on the material.

Unlike discontinuity, ultimate strength of concrete cannot be expressed in terms of elastic strain energy alone, because at ultimate load large inelastic deformations have occurred and internal structure is completely disrupted. The total work of

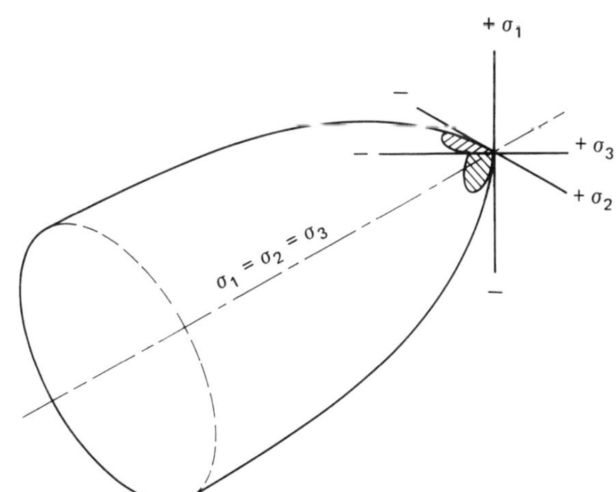

Fig. 25. Limiting surface in stress space for discontinuity of concrete. From Newman, K.: *Proceedings of the International Symposium on Theory of Arch Dams*, Southampton, 1964. Elmsford, N. Y.: Pergamon Press, Inc., 1965, p. 683 [2].

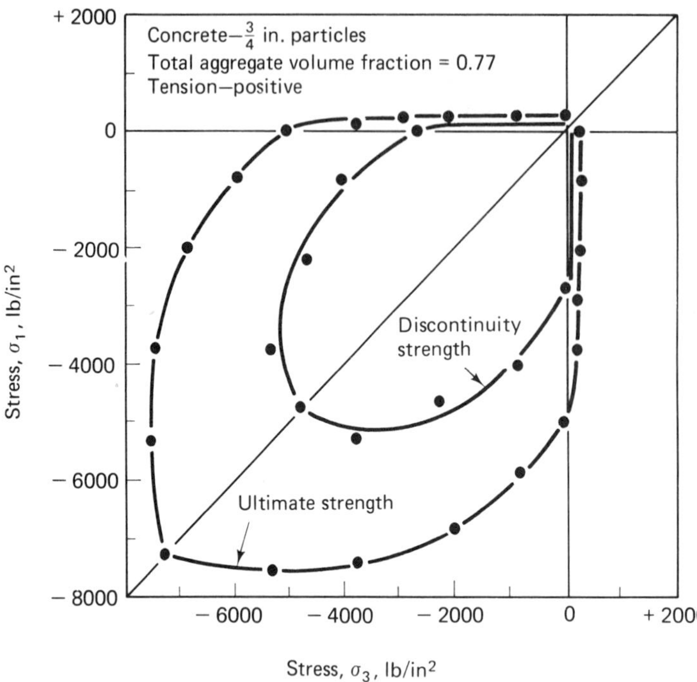

Fig. 26. The discontinuity and ultimate strengths of concrete under biaxial stress. From Vile, G. W.: Ph. D. Thesis. London: University of London, 1965 [53].

deformation given by the area under the stress-strain curve includes both elastic and inelastic components, and it has been suggested that the ultimate strengths of mortars and concretes might be of the form

$$w_\tau = f(\bar{\sigma})$$

where w_τ, the *total shear work*, is the deviatoric component of total work of deformation [1].

Newman and Newman have summarized more specific criteria for the ultimate strength surfaces in stress space for concrete. These criteria (Table 4) are based upon available experimental data and upon the assumption that all values of combined stress at failure can be related to the measured values of uniaxial tensile and compressive strengths and depend on no other mix characteristics.

The results can be represented in stress space as right sections of the volume bounded by the limiting stress surfaces, i.e., planes of constant volumetric stress which lie normal to the space diagonal axis, $\sigma_1 = \sigma_2 = \sigma_3$ (see Fig. 25). Any plane in the stress space perpendicular to the space diagonal is an octahedral plane, with equation $\sigma_1 = \sigma_2 = \sigma_3 =$ constant, and all points on this plane have the same mean normal stress, or volumetric stress, called the octahedral normal stress. Figures 27 and 28 are the plots for triaxial tension and triaxial compression, respectively. The results have been nondimensionalized with respect to uniaxial compressive strength (note the axial scales σ_1/σ_c, σ_2/σ_c, and σ_3/σ_c).

The sizes of areas enclosed by contours in triaxial tension (Fig. 27) depend on the ratio σ_c/σ_t for a particular concrete. Those in Fig. 27 are constructed upon the

TABLE 4. Ultimate Strength Criteria in Terms of Uniaxial Tensile Strength, σ_t, and Uniaxial Compressive Strength, σ_c

Stress state	*Ultimate strength, in three-dimensional stress space*
Uniaxial compression	$\sigma_1 = \sigma_c, \sigma_2 = \sigma_3 = 0$*
Uniaxial tension	$\sigma_3 = \sigma_t, \sigma_1 = \sigma_2 = 0$
Biaxial compression	$\sigma_1 = \sigma_2 = \sigma_c, \sigma_3 = 0$ (assumes that strength in equal biaxial compression is same as strength in uniaxial compression)
Biaxial tension	—
Triaxial compression ($\sigma_1 > \sigma_2 = \sigma_3$)	$\sigma_1/\sigma_c = 1 + 3.7(\sigma_3/_c)^{0.86}$
Triaxial extension ($\sigma_1 = \sigma_2 > \sigma_3$)	$\sigma_1/\sigma_c = 0.43(\sigma_3/\sigma_c - 1)$
Triaxial tension	Failure surface assumed bounded by the flat plane through the points $\sigma_1 = \sigma_t, \sigma_2 = \sigma_3 = 0$ $\sigma_2 = \sigma_t, \sigma_1 = \sigma_3 = 0$ $\sigma_3 = \sigma_t, \sigma_1 = \sigma_2 = 0$
Compression-compression-tension	Failure surface assumed bounded by a rather flat convex surface through the points $\sigma_1 = \sigma_c, \sigma_2 = \sigma_3 = 0$ $\sigma_2 = \sigma_c, \sigma_1 = \sigma_3 = 0$ $\sigma_3 = \sigma_c, \sigma_1 = \sigma_2 = 0$
Compression-tension-tension	Failure surface assumed bounded by the flat plane through the points $\sigma_1 = \sigma_c, \sigma_2 = \sigma_3 = 0$ $\sigma_2 = \sigma_t, \sigma_1 = \sigma_3 = 0$ $\sigma_3 = \sigma_t, \sigma_1 = \sigma_2 = 0$

*Compression taken as positive.

SOURCE: NEWMAN, K. and J. B. NEWMAN, *Structure, Solid Mechanics and Engineering Design*, M. Te'eni, ed., Proceedings of the Southampton 1970 Civil Engineering Materials Conference, Part II. New York: John Wiley & Sons, Inc. (Interscience Division), p. 963 [29].

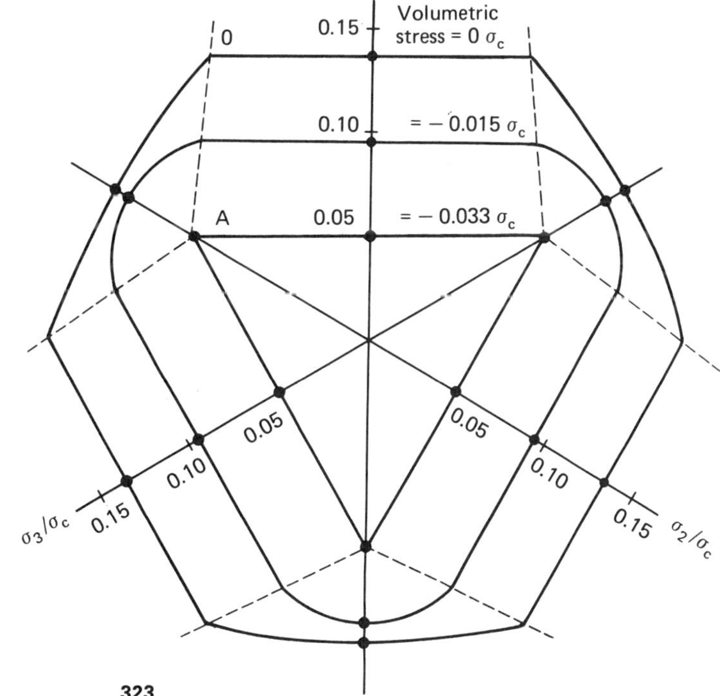

Fig. 27. Equi-volumetric stress planes through the ultimate strength failure surface for concrete in the triaxial tension region plotted in terms of the ultimate uniaxial compressive strength σ_c. From Newman, K. and J. B. Newman in *Structure, Solid Mechanics and Engineering Design*, M. Te'eni, ed., *Proceedings of the Southampton 1970 Civil Engineering Materials Conference*, Part II. New York: John Wiley & Sons, Inc. (Interscience Division), p. 963 [29].

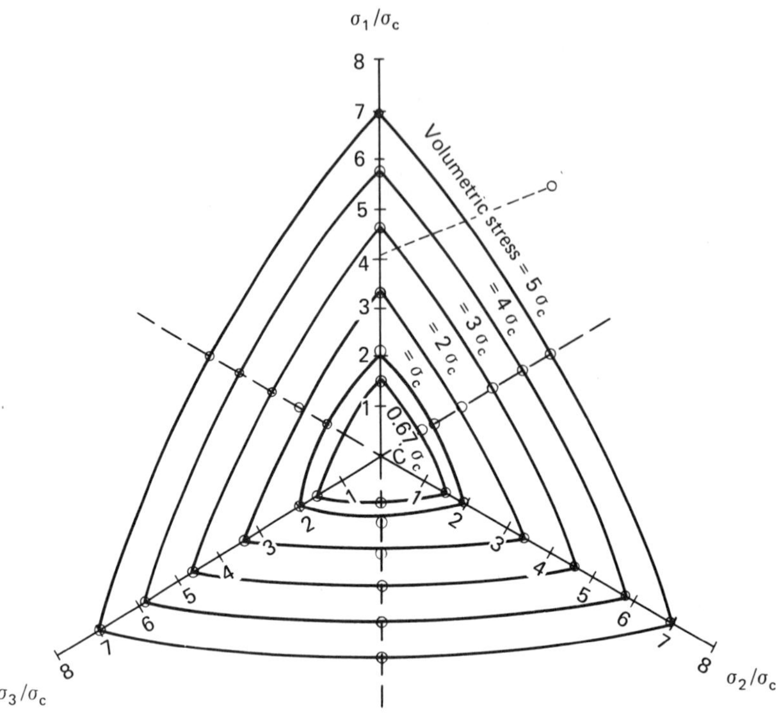

Fig. 28. Equi-volumetric stress planes through the ultimate strength failure surface for concrete in the triaxial compression region plotted in terms of the ultimate uniaxial compressive strength σ_c. From Newman, K. and J. B. Newman in *Structure, Solid Mechanics and Engineering Design*, M. Te'eni, ed., *Proceedings of the Southampton 1970 Civil Engineering Materials Conference*, Part II. New York: John Wiley & Sons, Inc. (Interscience Division), p. 963 [29].

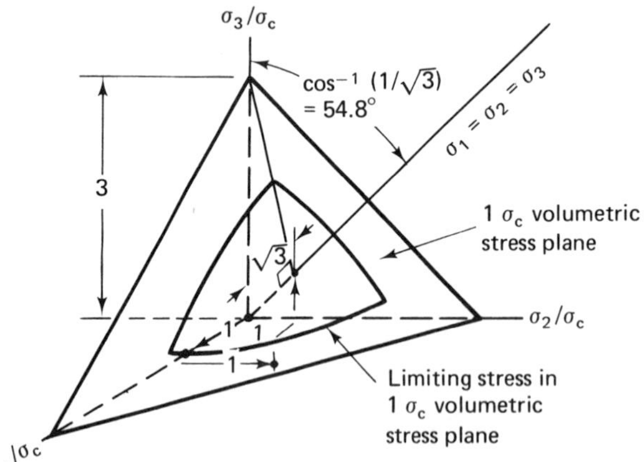

Fig. 29. Projection of volumetric (octohedral) stress plane onto principal stress axes.

basis that $\sigma_c/\sigma_t = 10$. In triaxial compression (Fig. 28) the contours are like slightly bulged equilateral triangles which become more rounded with increasing hydrostatic pressure.

The equivolumetric stress planes in Figs. 27 and 28 are planes of average principal stress, $(\sigma_1 + \sigma_2 + \sigma_3)/3$. Note also in Fig. 28 that the equivolumetric stress planes, which lie parallel to the plane of the paper in these three-dimensional stress plots, project onto the principal stress axes, σ_1/σ_c, etc., by a scale ratio of 3:1. Figure 29 illustrates this relationship.

The following example illustrates the use of Fig. 28 for determining safety against failure.

EXAMPLE 2*: A concrete has an ultimate uniaxial compressive strength of 6,000 psi. Will the following stress state, compression positive, cause failure?

$$\sigma_1 = +48{,}000 \text{ psi}; \qquad \sigma_2 = +21{,}000 \text{ psi}; \qquad \sigma_3 = +3{,}000 \text{ psi}$$

1. The nondimensionalized stresses with respect to σ_c are

$$\sigma_1 = +8.0\sigma_c; \qquad \sigma_2 = +3.5\sigma_c; \qquad \sigma_3 = +0.5\sigma_c$$

2. The volumetric stress is

$$\sigma_v = \frac{\sigma_1 + \sigma_2 + \sigma_3}{3} = \frac{12.0}{3} = +4.0\sigma_c$$

This gives the equivolumetric stress plane (Fig. 28) *within* which the deviatoric component of stress must lie if failure is not to occur.

3. The deviatoric stress components are

$$\sigma_{d1} = \sigma_1 - \sigma_v = +4.0\sigma_c$$
$$\sigma_{d2} = \sigma_2 - \sigma_v = -0.5\sigma_c$$
$$\sigma_{d3} = \sigma_3 - \sigma_v = -3.5\sigma_c$$

4. The deviatoric stress components are plotted as vectors of appropriate sign parallel to the corresponding axes (see point *P*, Fig. 28).
5. Since the deviatoric stress component lies outside the relevant equivolumetric stress plane, the state of stress will cause ultimate failure.

The solution method must be used with caution because, as with most materials, the strength of concrete is stress-path-sensitive. For example, a higher strength may be obtained by first applying hydrostatic stress, then increasing σ_1 and decreasing σ_3 to their final values, than by applying the three stresses σ_1, σ_2, σ_3 separately in succession. The results in Table 4 do not account for variations in stress path. It has been observed that discontinuity is less stress-path-dependent than ultimate strength and in that sense might be a more useful criterion for failure [29]. One disadvantage is the

*Adapted with permission from K. Newman and J. B. Newman [29].

paucity of data currently available for the discontinuity of concrete under combined stresses.

PROBLEMS

1. Estimate the Young's modulus of concrete having the following properties:

Aggregate volume fraction	0.55
Cement paste volume fraction	0.45
Aggregate modulus	7×10^6 psi
Cement modulus	2.2×10^6 psi

2. Estimate the Young's modulus of concrete having the following properties:

Cement paste	
Gel volume fraction	0.6
Air and water pore volume fraction	0.4
Gel modulus	4×10^6 psi
Cement mortar	
Cement paste volume fraction	0.5
Sand volume fraction	0.5
Sand modulus	10×10^6 psi
Concrete	
Cement mortar volume fraction	0.45
Aggregate volume fraction	0.55
Aggregate modulus (lightweight aggregate)	1×10^6 psi

   Use the computed cement paste modulus to obtain the modulus of the mortar, etc.

3. Propose and discuss one or more possible reasons drying shrinkage might decrease with aggregate particle size as shown in Fig. 9.

4. Compute the effective thermal conductivity, k (Btu/hr-in.-°F), across the width of the structural sandwich panel in the accompanying figure. The core properties are air

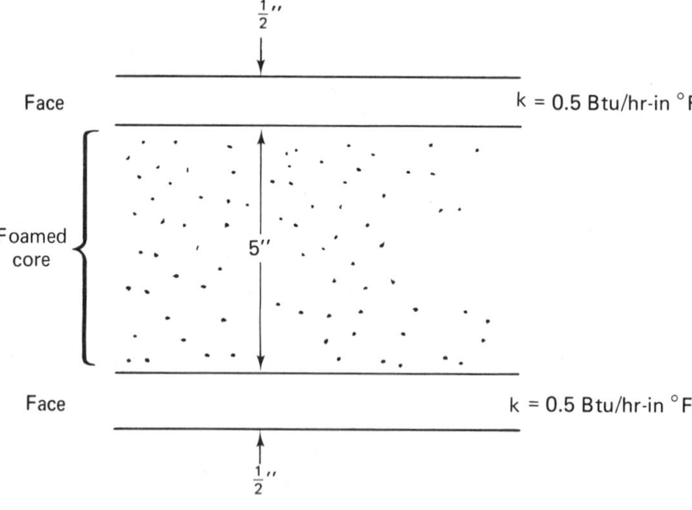

Fig. 30

voids (disperse spheres): $k = 0.01$ Btu/hr-in.-°F, volume fraction = 0.6; continuous phase: $k = 0.1$ Btu/hr-in.-°F, volume fraction = 0.4.

5. Using Hansen and Neilsen's method, compare the ultimate shrinkage of the following three concretes:

Concrete	*Aggregate fractional volume (g)*	E_p	S_p	E_g	S_g
A (natural aggregate)	0.70	2×10^6 psi	$2{,}500 \times 10^{-6}$	9×10^6 psi	~0
B (light weight aggregate)	0.70	2	2,500	1	$10{,}000 \times 10^{-6}$
C (aerated mortar)	0.70	2	2,500	0	—

6. From what you learned in Chap. 4 about cement hydration reactions, what modifications might be made in the curing of concrete to reduce long-term creep
   (a) Assuming that the load is applied 1 week after casting?
   (b) Assuming that the load is applied 4 months after casting?

7. Predict the ultimate creep of a column which will be loaded to 3,000 psi and the time at which one half of ultimate creep will be obtained, given the following laboratory test results for the same concrete mix loaded to 4,000 psi. Assume that identical curing periods and curing conditions are provided in each case before applying load and that similar environmental conditions exist during loading.

Time (Days)	*Creep (Microstrain)*
15	111
30	213
45	306
60	394

8. Why is the stress-strain relationship for pure cement linear to failure, while for concrete it is nonlinear?

9. Why do concretes in uniaxial compressive loading undergo columnar failure? Do lightweight aggregate concretes fail in the same manner? Do neat cement pastes fail likewise? Why (or why not)?

10. Enumerate differences between Hansen's assumptions in deriving Eq. (23) and what you believe to represent physical reality in porous systems, in order to explain why Eq. (23) tends to give an *upper* bound for the strengths of porous solids (see Fig. 23).

11. Determine if a concrete having an ultimate uniaxial compressive strength of 7,000 psi will fail under the following stress state:

$$\sigma_1 = +50{,}000 \text{ psi}; \qquad \sigma_2 = +22{,}000 \text{ psi}; \qquad \sigma_3 = +4{,}000 \text{ psi}$$

REFERENCES

1. NEWMAN, K., "Concrete Systems," in *Composite Materials*, L. Holliday, ed. Amsterdam: Elsevier Publishing Company, 1966, Chap. 8.
2. NEWMAN, K., *Proceedings of the International Symposium on Theory of Arch Dams*, Southampton, 1964. Elmsford, N.Y.: Pergamon Press, Inc., 1965, p. 683.
3. HASHIN, Z., *Appl. Mech. Rev.*, **17** (1964) 1.
4. BRUGGERMANN, D. A., *Phys. Z.*, **37** (1936) 906.
5. BRUGGERMANN, D. A., *Ann. Phys.*, **29** (1937) 160.
6. HANSEN, T. C., *Proceedings of the International Conference on Structure of Concrete*, Imperial College, London, 1965, p. 16.
7. PAUL, B., *Trans. AIME*, **218** (1960) 36.
8. HASHIN, Z., and S. SHTRIKMAN, *J. Mech. Phys. Solids*, **11** (1963) 127.
9. HANSEN, T. C., *Chemistry of Cement, 4th International Symposium*, Washington, D. C.: U.S. National Bureau of Standards, 1962, p. 709.
10. MAXWELL, J. C., *Treatise on Electricity and Magnetism*, Vol. 1. London: Oxford University Press, Clarendon Press Series, 1873, 365.
11. ISHAI, O., *Proc. Am. Concr. Inst.*, **58** (1961) 611.
12. ISHAI, O., *Proc. Am. Concr. Inst.*, **59** (1962) 1365.
13. HIRSCH, T. J., *Proc. Am. Concr. Inst.*, **59** (1962) 427.
14. DOUGILL, J. W., *Proc. Am. Concr. Inst.*, **59** (1962) 1363.
15. COUNTO, U. J., *Mag. Concr. Res.*, **16** (1964) 129.
16. HASHIN, Z., *J. Appl. Mech.*, **29**, No. 1 (March 1962) 143.
17. HANSEN, T. C., *J. Am. Concr. Inst.*, **62** (Feb. 1965) 193.
18. EUCKEN, A., *Beil. Torsch. Gebiete Igenieurw.* B3 (1932) 413.
19. CAMBELL-ALLEN, D., and C. P. THORNE, *Mag. Concr. Res.*, **15**, No. 43 (March 1963) 39.
20. HANSEN, T. C., "Notes from a Seminar on Structure and Properties of Concrete," *Stanford University Civil Engineering Department. Technical Report. No. 71*, 1966.
21. ALLEN, D. C., and C. P. THORNE, *Mag. Concr. Res.*, **15**, No. 43 (March 1963) 39.
22. HSU, T. C., and F. O. SLATE, *J. Am. Concr. Inst.*, **60**, No. 4 (April 1963) 465.
23. ANSON, M., *Mag. Concr. Res.*, **16** (1964) 73.
24. TROXELL, G. E., J. M. RAPHAEL, and R. E. DAVIS, *Proc. Am. Soc. Testing Mater.*, **58** (1958) 1101.
25. ALEXANDER, K. M., and J. WARDLAW, *Proc. Am. Concr. Inst.*, **55** (1959) 1303.
26. PICKETT, G., *Proc. Am. Concr. Inst.*, **52** (1956) 581.
27. L'HERMITE, R. G., *Chemistry of Cement, 4th International Symposium*, Washington, D.C.: U.S. National Bureau of Standards, 1960, 659.
28. HANSEN, T. C., and K. E. NIELSEN. *J. Am. Concr. Inst.* **62** (July 1965) 783.
29. NEWMAN, K., and J. B. NEWMAN, "Failure Theories and Design Criteria for Plain Concrete," in *Structure, Solid Mechanics and Engineering Design, Proceedings of the Southampton 1970 Civil Engineering Materials Conference*, Part II, M. Te'eni, ed. New York: John Wiley & Sons, Inc. (Interscience Division), p. 963.
30. L'HERMITE, R. G., *RILEM Bull., Paris*, 1 (1959) 21.
31. GLAMVILLE, W. H., *Bldg. Res. Stn. Tech. Paper No. 12*, London: Her Majesty's Stationery Office, 1930.
32. DAVIS, R. E., and H. E. DAVIS, *Proc. Am. Soc. Testing Mater.*, Part. II, **30** (1930) 707; *Proc. Am. Concr. Inst.*, **27** (1931) 837.

33. WAGNER, O., *Duet. Auss. Stahlbeton, Berlin,* No. 131 (1958) 74.

34. ROSS, A. D., *RILEM Bull., Paris,* **1** (1959) 55.

35. NEVILLE, A. M., *Hardened Concrete: Physical and Mechanical Aspects.* American Concrete Institute Monograph 6, Ames, Iowa: ACI and Iowa State University Press, 1971.

36. HSU, T. T., *J. Am. Concr. Inst.,* **60**, No. 3 (March 1963) 371.

37. SLATE, F. O., and R. E. MATHEUS, *J. Am. Concr. Inst.,* **64**, No. 1 (Jan. 1967) 34.

38. SHAH, S. P., and F. O. SLATE, *The Structure of Concrete*, Brooks, A. E. and K. Newman, eds., *Proceedings of an International Conference*, Imperial College, London: Cement and Concrete Association, (1965), p. 82.

39. ALEXANDER, K. M., T. WARDLAW, and D. F. GILBERT, *Proceedings of the International Conference on the Structure of Concrete,* Brooks, A. E. and K. Newman, eds., London: Cement and Concrete Association, **1** (1965), p. 59.

40. DANTU, P., *Ann. Inst. Tech. Bat. Trav. Publ.,* **11**, Ser. Essais Mesures 40, No. 121 (Jan. 1958) 54.

41. RUSCH, H., *Zement-Kalk-Gips,* Wiesbaden, **12** (1959) 1.

42. ROBINSON, G. S., Ph.D. thesis. London: University of London, 1964.

43. HSU, T. T., *Proc. Am, Concr. Inst.,* **60** (1963) 371.

44. VILE, G. W., "Strength of Concrete Under Short-Term Static Biaxial Stress," in *The Structure of Concrete,* A. E. Brooks and K. Newman, eds. London: Cement and Concrete Association, 1968, p. 278.

45. BACHE, H. H., and P. N-CHRISTENSEN, *The Structure of Concrete,* Brooks, A. E. and K. Newman, eds., *Proceedings of an International Conference,* Imperial College, Cement and Concrete Association, **1**, London (1965), p. 93.

46. FERET, R., *Bull, Soc. d'Encouragement l'Indust. Natl.,* **II** (1897) 1604.

47. SCHILLER, K. K., in *Mechanical Properties of Non-Metallic Brittle Materials,* W. H. Walton, ed. London: Butterworth & Co. (Publishers) Ltd., 1958, p. 35.

48. POWERS, T. C., *Chemistry of Cement, Proceedings of the 4th International Symposium,* Washington, D.C.: U.S. National Bureau of Standards, 1962, p. 577.

49. POWERS, T. C., *Proc. Hwy Res. Bd.,* Washington, D.C., **27** (1947) 178.

50. WISCHER, H., Compendium of Writings of the Cement Industry. Verein Deutscher Cement Works, Dusseldorf, **28**.

51. DURELLI, A. J., E. A. PHILLIPS, and C. H. TSAO, *Introduction to the Theoretical and Experimental Analysis of Stress and Strain.* New York: McGraw-Hill Book Company, 1958.

52. NADAI, A., *Theory of Flow and Fracture of Solids,* Vol. I, 2nd ed. New York: McGraw-Hill Book Company, 1950.

53. VILE, G. W., Ph.D. thesis. London: University of London, 1965.

9

BITUMINOUS MIXES

And the back-crossbar as strong as the fore,
And spring and axle and hub *encore*
And yet, *as a whole*, it will past a doubt
In another hour it will be *worn out*.

—HOLMES

In this chapter we shall summarize the general requirements for asphalt mix design and outline the two most widely used methods of mix design for paving. The properties of bitumen-aggregate interfaces and the effects of various additives to bituminous mixes will be described.

A. MIX REQUIREMENTS

The objectives of asphalt pavement design are to obtain sufficient

1. Workability.
2. Stability.
3. Durability.
4. Flexibility.
5. Skid resistance.

Mix design depends principally on aggregate gradation and asphalt content. Since optimal values of each of the above properties are normally obtained at different gradations and asphalt contents, mix design always implies a trade-off between the properties desired, as does mix design in all composites.

Workability permits efficient placement and compaction of a mix. Factors which typically contribute to workability include

1. An adequately dense aggregate grading with maximum particle size under about 1 in., to prevent segregation during handling and raveling under the roller.
2. A placing temperature at which asphalt viscosity is adequately low enough for ease of spreading, and a compaction temperature at which viscosity is high enough for firm binding of aggregate under the roller.

Stability to resist distortion under traffic loading depends on frictional interlocking among aggregate particles plus cohesiveness of the asphalt. The most stable mixes are *dense-graded angular aggregate* mixes with *just sufficient asphalt* to coat aggregate particles. Higher asphalt contents tend to lubricate more than bind, and stability is reduced.

Durability against disintegration by traffic and weathering depends on:

1. Tough aggregate particles completely coated with asphalt to protect against stripping of asphalt from aggregate by water and consequent freezing and thawing forces.
2. A minimum of air voids in the mix to retard the deterioration of asphalt by oxygen and water.

But if voids are completely filled with asphalt, stability and skid resistance are compromised. In hot weather the asphalt films thicken, reducing friction between aggregate particles, and under the compaction of traffic, asphalt is flushed to the surface and reduces skid resistance. The axiom that trouble results if voids are completely filled with asphalt is applicable to all asphalt mixes. For dense-graded mixes, air porosities of 3–5% are suggested.

Flexibility to deflect and conform to minor adjustments in base course is normally enhanced by asphalt ductility and open-graded aggregate. But ductility and open grading reduce stability, and open grading accompanied by increased air voids reduces durability.

Skid resistance is reduced primarily by bleeding or flushing, mentioned above, and aggregate polishing under traffic wear. Certain limestones which polish easily are sometimes blended with sands or natural rock asphalts to improve skid resistance.

Desirable aggregate characteristics. An ideal bituminous aggregate has

1. Strength and toughness (for resistance to degradation).
2. Angular shape (for interlocking).
3. Low porosity (resistance to freeze-thaw damage).
4. Hydrophilic surface with some roughness (for chemical and mechanical bonding to asphalt).
5. Grading appropriate to the intended use.

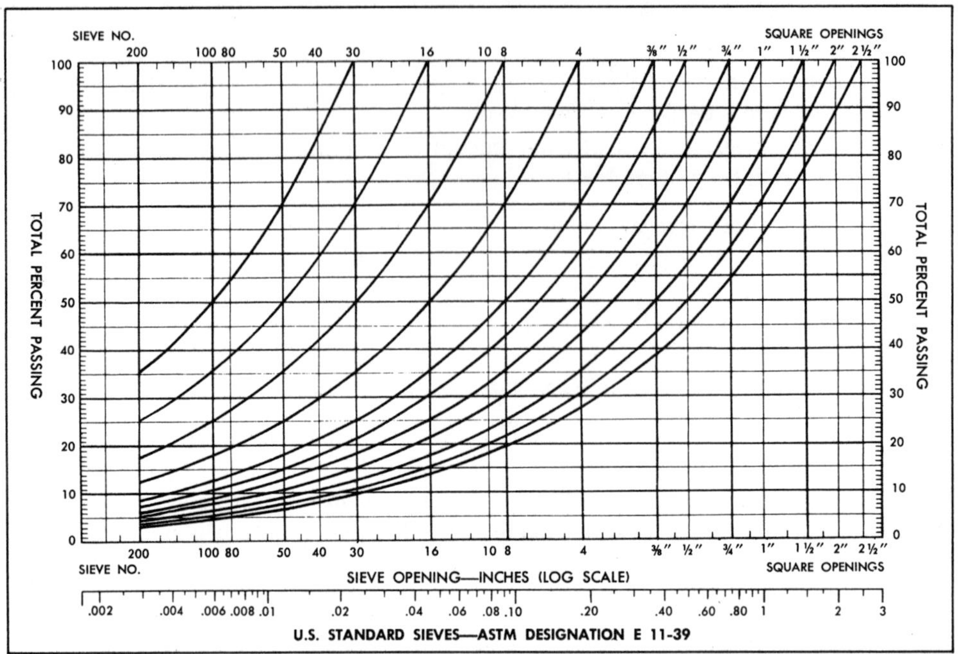

Fig. 1. Fuller maximum density curves on standard semi-log grading chart. From *Mix Design Methods for Asphalt Concrete* (Series 2: MS-2), Third Edition. College Park, Md.: The Asphalt Institute, 1969 [1].

In extremely dense-graded aggregates it is impossible to completely coat particles with sufficiently thin films to avoid particle segregation and attendant reduction of interlocking. If sufficient asphalt is used to coat and bind all particles, air void content is inadequate, or if sufficient air voids are provided, asphalt content is insufficient for optimum binding. On the other hand, extremely open-graded aggregates also have low stability due to reduced interparticle contact, and larger asphalt contents are required to maintain sufficiently low air voids for durability. Hence the densest bituminous mix gradations are generally designed to be somewhat more open-graded than indicated by Fuller's equation [Fig. 1 in this chapter and Eq. (1) in Chap. 7] or than given by a straight-line plot from the origin on the U.S. Bureau of Public Roads 0.45 power grading chart (Fig. 2). This variation from the maximum density curves is often simply obtained by slightly increasing the amounts of largest and finest aggregate to provide a degree of gap grading.

Dense-graded aggregates are most frequently used in hot mixes. The increased frequency of contact points in dense-graded materials provides greater distribution of load transfer among aggregates and decreases the probability of individual particles crushing at point loads. Open-graded aggregates (those containing much less material passing the No. 200 sieve) are more frequently used in cold plant mixes or road mixes. One-sized materials are generally used for macadams, surface treatments, and seal coats.

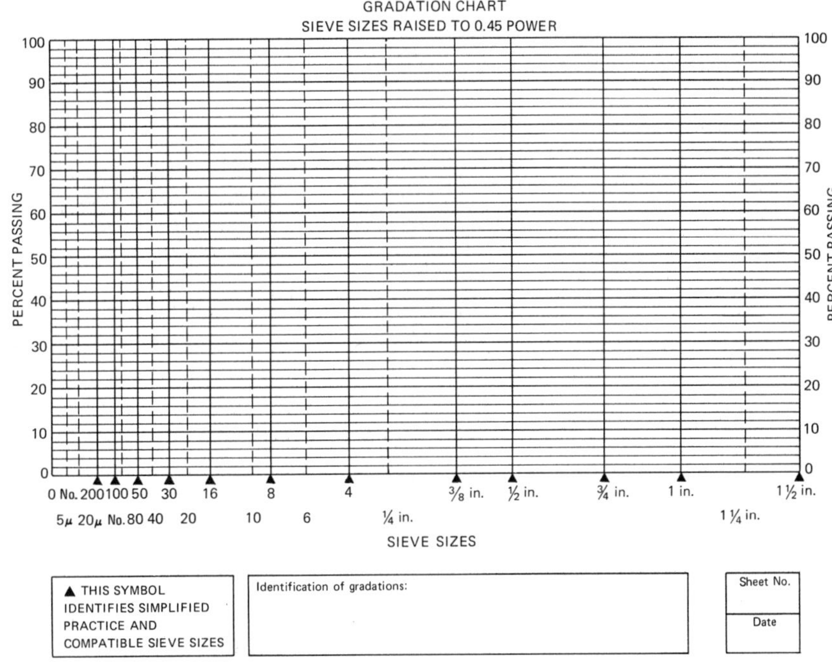

Fig. 2. U. S. Bureau of Public Roads 0.45 power grading chart. From *Mix Design Methods for Asphalt Concrete* (Series 2: MS-2), Third Edition. College Park, Md.: The Asphalt Institute, 1969 [1].

Table 1 summarizes some corrections which can be made to improve mix designs.

TABLE 1. Corrections to Asphalt Mixes

Mix deficiency	*Probable effects*	*Corrective action*
Low voids	Instability or flushing due to additional compaction under traffic	Increase amount of coarsest particles or some other specific size fraction of the aggregate
High voids	Mixes with grading curves which deviate greatly from Figs. 1 and 2 have high voids, poor workability, low stability, and reduced weather resistance, if voids include high air content	Reduce voids by adding mineral dust or by adjusting entire grading toward maximum density grading curve
Low stability		If stability cannot be sufficiently improved by adjustments in grading, crushed or more durable aggregates should be used

B. TYPES OF BITUMINOUS MIXES

Mixes may be classified according to

1. Bituminous binder used.
2. Aggregate gradation.
3. Production method.

Classes of *bituminous binders* were described in Chap.5. *Aggregates* used in asphalt mixes are designated as

Coarse aggregate	retained on No. 8 sieve
Fine aggregate	passing No. 8 sieve
Mineral dust	passing No. 200 sieve

The Asphalt Institute designates eight paving mixtures, based upon proportion of aggregate passing the No. 8 sieve. These are further divided into a total of 19 types based upon the amount of mineral dust (Table 2). The table permits selection of various asphalt mixes from coarse macadam through graded aggregate to sand and sheet mixes. Types IV are recommended for *all* pavement courses and traffic classifications. Mix type selection depends on aggregate availability, desired pavement characteristics, and past performance of similar pavements. The grading bands in Table 2

TABLE 2. Asphalt Institute Mix Compositions

Mix type	$2\frac{1}{2}$ *in.*	$1\frac{1}{2}$ *in.*	1 *in.*	$\frac{3}{4}$ *in.*	$\frac{1}{2}$ *in.*	$\frac{3}{8}$ *in.*	*No. 4*	*No. 8*	*No. 16*	*No. 30*	*No. 50*	*No. 100*	*No. 200*	*%* Asphalt*
Ia	100	35–70		0–15				0–5					0–3	3.0–4.5
IIa						100	40–85	5–20					0–4	4.0–5.0
IIb					100	70–100	20–40	5–20					0–4	4.0–5.0
IIc				100	70–100	45–75	20–40	5–20					0–4	3.0–6.0
IId			100	70–100		35–60	15–35	5–20					0–4	3.0–6.0
IIe		100	70–100	50–80		25–60	10–30	5–20					0–4	3.0–6.0
IIIa					100	75–100	35–55	20–35		10–22	6–16	4–12	2–8	3.0–6.0
IIIb				100	75–100	60–85	35–55	20–35		10–22	6–16	4–12	2–8	3.0–6.0
IIIc				100	75–100	60–85	30–50	20–35		5–20	3–12	2–8	0–4	3.0–6.0
IIId			100	75–100		45–70	30–50	20–35		5–20	3–12	2–8	0–4	3.0–6.0
IIIe		100	75–100	60–85		40–65	30–50	20–35		5–20	3–12	2–8	0–4	3.0–6.0
IVa					100	80–100	55–75	35–50		18–29	13–23	8–16	4–10	3.5–7.0
IVb				100	80–100	70–90	50–70	35–50		18–29	13–23	8–16	4–10	3.5–7.0
IVc			100	80–100		60–80	48–65	35–50		19–30	13–23	7–15	0–8	3.5–7.0
IVd		100	80–200	70–90		55–75	45–62	35–50		19–30	13–23	7–15	0–8	3.5–7.0
Va					100	85–100	65–80	50–65	37–52	25–40	18–30	10–20	3–10	4.0–7.5
Vb				100	85–100		65–80	50–65	37–52	25–40	18–30	10–20	3–10	4.0–7.5
VIa					100	85–100		65–78	50–70	35–60	25–48	15–30	6–12	4.5–8.5
VIb				100		85–100		65–80	47–68	30–55	20–40	10–25	3–8	4.5–8.5
VIIa						100	85–100	80–95	70–89	55–80	30–60	10–35	4–14	7.0–11.0
VIIIa							100	95–100	85–98	70–95	40–75	20–40	8–16	7.5–12.0

*By weight as a percentage of total mix.

are wider than allowed for good job control to permit a range of gradations to be developed within each type. Figure 3 illustrates, for example, the specification and job control grading bands and a job mix grading curve for a typical Type IVc mix.

Production methods, progressing from inexpensive to higher-quality mixes, are listed below. The lower-cost mixes usually use cutbacks or emulsions instead of hard asphalt cements which require high curing temperatures. This category includes surface treatments, penetration macadam, and road mixes. High-quality mixes include cold- and hot-laid plant mixes.

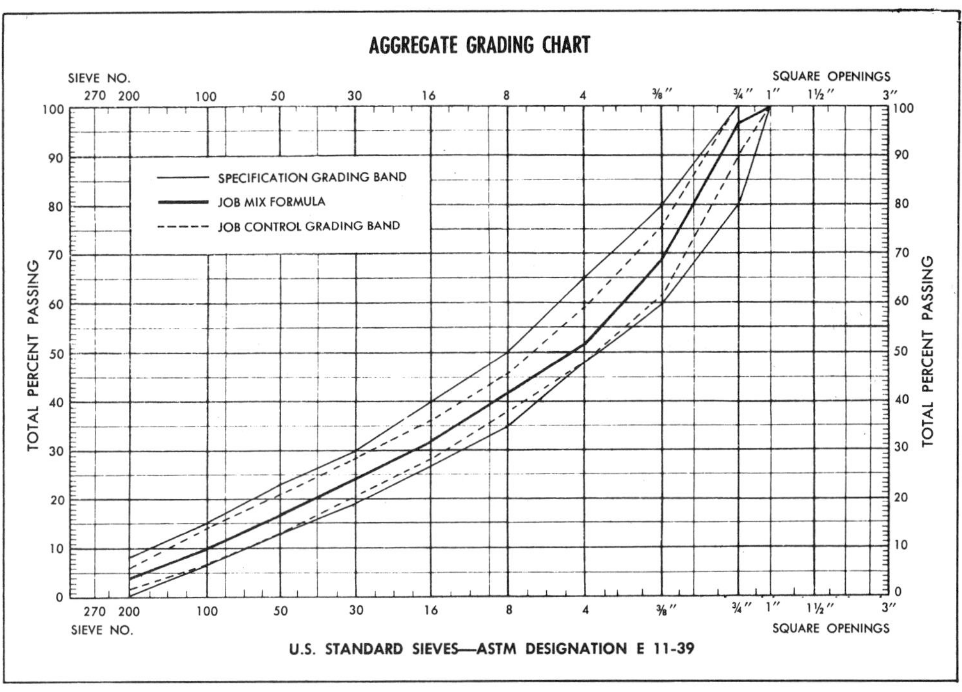

Fig. 3. Example of job-mix grading curve and job control grading band, Type IVc mix. From *Construction Specification for Asphalt Concrete* (SS-1), Fourth Edition. College Park, Md.: The Asphalt Institute, 1969 [2].

1. *Surface treatment.* A thin layer of binder covered with a single application of aggregate. Surface treatments may be used to improve abrasion resistance, skid resistance, and light reflection characteristics or to level cracked or deformed road surfaces.
2. *Penetration macadam.* A heavier-duty application, starting with a uniformly graded aggregate of about 1- to $\frac{1}{2}$-in. size, which is sprayed with asphalt, which penetrates it. A second layer of smaller aggregate which keys into the larger aggregate is placed, rolled, and sprayed. A surface course is then placed which may be either a surface treatment, previously described, or a bituminous concrete.
3. *Road mixes* generally involve placing available aggregate in a windrow and mixing asphalt and aggregate with either a road grader or traveling mix plant and then spreading and rolling. SC cutbacks and SS emulsions are often appropriate for the long mixing and placing periods required.
4. *Hot-laid plant mix* is made by mixing hot asphalt with hot aggregate (normally over 300°F) in a plant and compacting the mix at a temperature generally not exceeding 225°F. Cold-laid plant mix is similarly produced except that a cutback or emulsion is used so that mixing temperatures can be lowered and placing and compaction accomplished at ambient tempera-

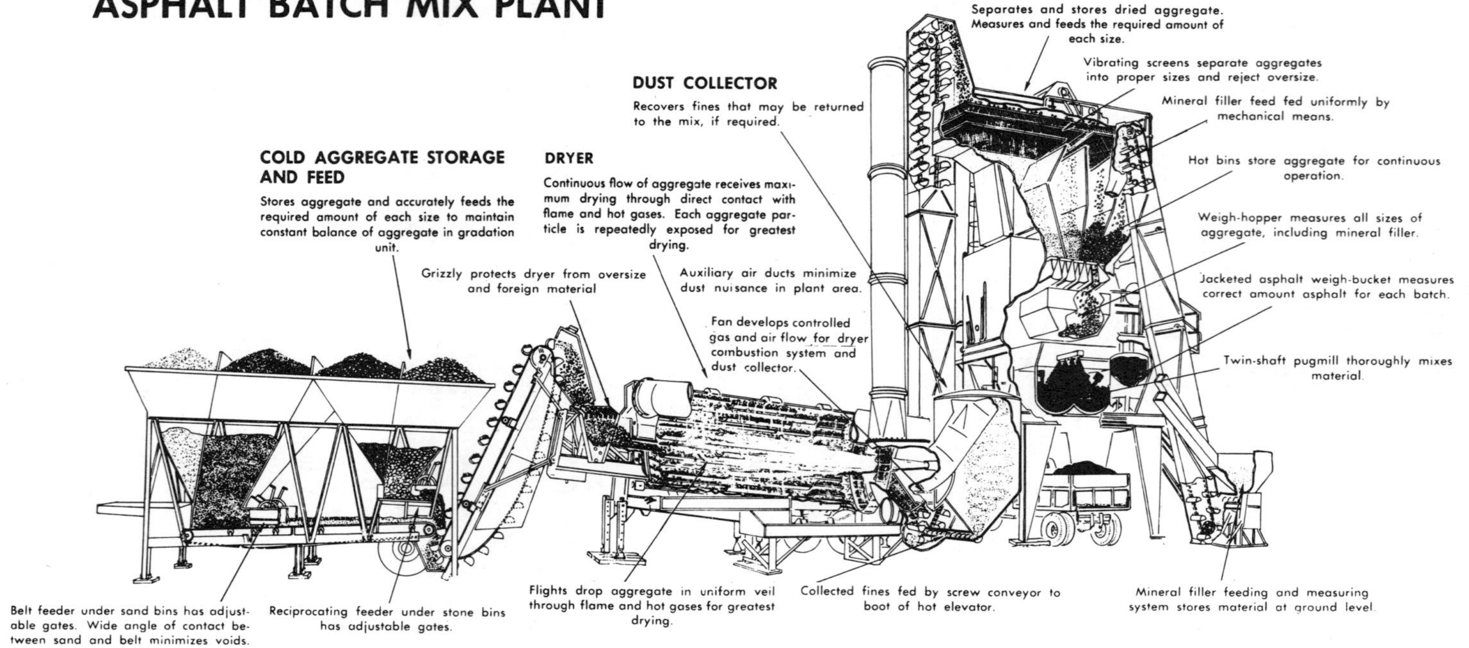

Fig. 4. Asphalt batch mix plant. From *Asphalt Plant Manual* (MS-3), Third Edition. College Park, Md.: The Asphalt Institute, 1967 [3].

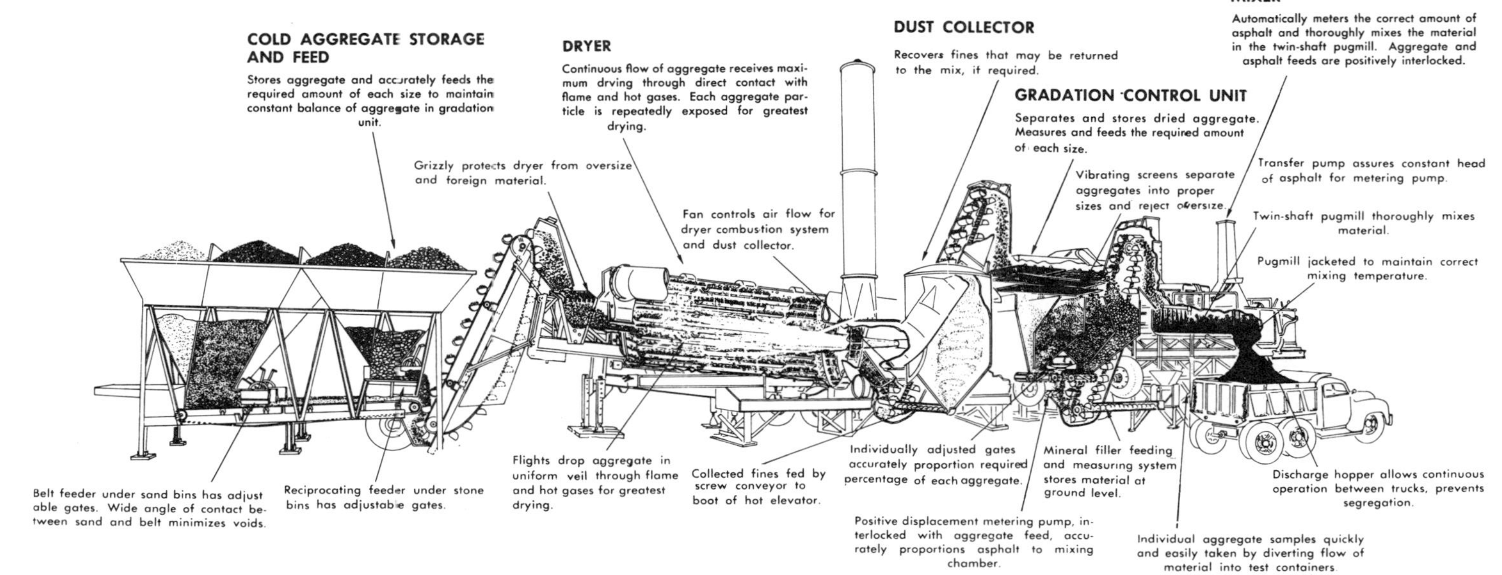

Fig. 5. Asphalt continuous mix plant. From *Asphalt Plant Manual* (MS-3), Third Edition. College Park, Md.: The Asphalt Institute, 1967 [3].

ture. Figures 4 and 5 show continuous- and batch-type mix plants. Somewhat greater mix uniformity can be obtained by batch mixing.

The *mixing temperature* for hot mixes should be the lowest temperature which will (1) provide dry aggregate and (2) provide an asphalt viscosity for suitable ease in placing the mix. At no time should the asphalt temperature exceed 350°F. The asphalt temperature rapidly adjusts to that of the stone when the two are mixed. Temperatures giving kinematic viscosities of 150–300 centistokes for dense-graded mixes are considered adequate. For open-graded mixes viscosities of 300–1,600 centistokes are used to prevent the asphalt from draining from the aggregate. Table 3 lists pugmill temperatures required to produce proper asphalt viscosities for various mixes.

TABLE 3. Suggested Mixing Temperatures

Grade of asphalt cement	*Pugmill mixing temperature of aggregate (°F)*
(For Mix Types I & II)	
40–50	225–310
60–70	225–305
85–100	225–300
120–150	225–300
200–300	225–300
(For Mix Types III–VIII)	
40–50	275–350
60–70	265–330
85–100	255–325
120–150	245–325
200–300	225–300
Grade of liquid asphalt	
MC and SC Grades	
250	135–175
800	165–205
3000	200–240
Grade of emulsified asphalt	
MS-2	50–140
SS-1	50–140
SS-1h	50–140
CMS-2	50–140
CMS-2S	50–140
CSS-1	50–140
CSS-1h	50–140

Caution: Temperature ranges indicated for MC and SC are generally above the flash points, requiring suitable equipment operated under careful supervision with regard to venting and flames or sparks.

SOURCE: *Construction Specification for Asphalt Concrete* (SS-1), 4th ed. College Park, Md.: The Asphalt Institute, 1969 [2].

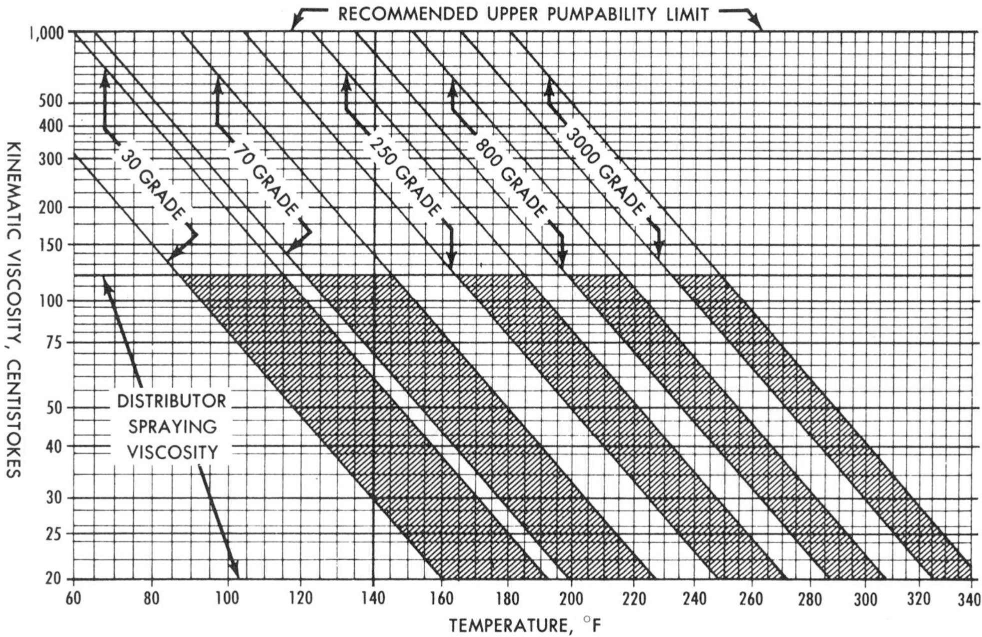

Fig. 6. Temperature-viscosity of cutbacks. From *Construction Specification for Asphalt Concrete* (SS-1), Fourth Edition. College Park, Md.: The Asphalt Institute, 1966 [2].

Asphalt road distributor spraying viscosities are normally in the range 20–120 centistokes, the higher viscosities for penetration and sealing of open surfaces and the lower viscosities for dense surfaces. Figure 6 shows temperature-viscosity relationships for cutbacks.

C. MIX DESIGN METHODS

Mix design testing is frequently used in the following stages of a project:

1. *Preliminary design and source acceptance testing*, to determine which local aggregate sources can be used to satisfy mix specifications most economically and to provide a basis for preliminary cost estimates.
2. *Job mix testing*, performed at the beginning of plant production and in conjunction with calibration of the mixing plant.
3. *Construction control testing*, for routine inspection during production.

There are two commonly used methods for testing and design of hot mixes for paving: the Marshall and Hveem methods. Table 4 shows their relative suitability for various mix types. In this section we shall outline the procedures for the two methods.

TABLE 4. Suitability of Laboratory Design Methods

	Paving mix type and description	*Marshall*	*Hveem*
I	Macadam	X	X
II	Open type	X	D
III	Coarse-graded	D	A
IV	Dense-graded	A	A
V	Fine-graded	A	A
VI	Stone sheet	A	A
VII	Sand sheet (sand asphalt)	A	A
VIII	Fine sheet (sheet asphalt)	A	A

A—Suitable
D—Doubtful
X—Unsuitable
SOURCE: *Mix Design Methods for Asphalt Concrete* (Series 2: MS-2), 3rd ed. College Park, Md.: The Asphalt Institute, 1969 [1].

1. Marshall Method

The Marshall method is the more widely used. It is applicable to hot mixes using asphalt cement and aggregate to 1-in. maximum size and is used for both design and field control. The principal feature is a *stability-flow* test on compacted specimens.

Specimens 4 in. in diameter by $2\frac{1}{2}$ in. high with various asphalt contents are compacted with a 10-1b hammer falling 18 in.; 35, 50 or 75 blows are used, depending on the design traffic. Void contents are computed, and the specimens are heated to 140°F and placed in a split breaking head for testing (Fig. 7). Stability is the maximum load in pounds which the specimen develops, and flow is the strain in $\frac{1}{100}$ in. between no load and maximum load.

Figure 8 shows typical test results. The trends in the graphs are self-evident. Notice that the unit weight and stability both decrease above an asphalt content at which asphalt begins to significantly separate aggregate particles, rather than just fill voids between particles. The percentage VMA (voids in mineral aggregate) passes through a minimum at approximately the same asphalt content, since below this level the asphalt serves primarily as a lubricant during compaction to facilitate particle orientation for greater density, while above this level the asphalt also begins to separate particles.

Optimum asphalt by the Marshall method is the average of three values:

1. Asphalt content at maximum stability.
2. Asphalt content at maximum unit weight.
3. Asphalt content at the average air void content specified in Table 5, which lists the Asphalt Institute Marshall design criteria.

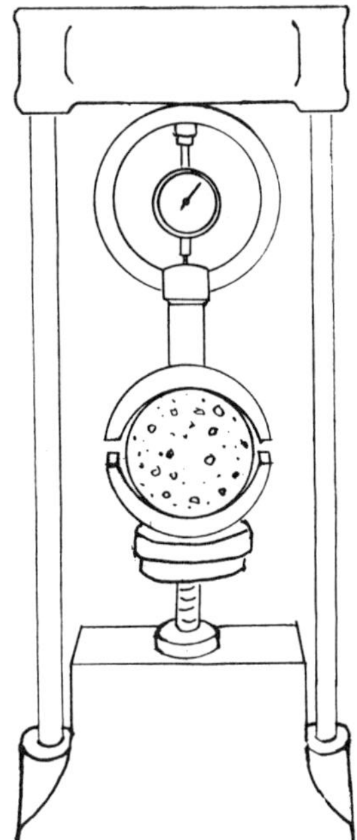

Fig. 7. Marshall test.

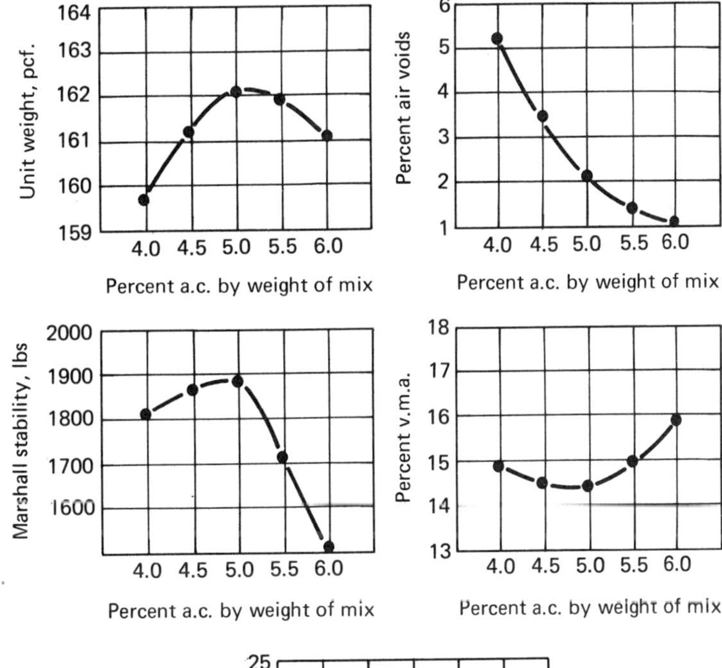

Fig. 8. Typical Marshall test results. From *Mix Design Methods for Asphalt Concrete* (Series 2: MS-2), Third Edition. College Park, Md.: The Asphalt Institute, 1966 [1].

TABLE 5. Marshall Design Criteria

Traffic category:	*Heavy*		*Medium*		*Light*	
No. of compaction blows each end of specimen:	*75*		*50*		*35*	
Test property	*Min.*	*Max.*	*Min.*	*Max.*	*Min.*	*Max.*
Stability, all mixtures	750	—	500	—	500	—
Flow, all mixtures	8	16	8	18	8	20
Percentage air voids						
Surfacing or leveling	3	5	3	5	3	5
base	3	8	3	8	3	8
Percentage voids in mineral aggregate			See Fig. 9			

Note: Compactive effort should closely approach maximum density obtained in pavement under traffic.
SOURCE: *Mix Design Methods for Asphalt Concrete* (Series 2: MS-2), 3rd ed. College Park, Md.: The Asphalt Institute, 1969 [1].

EXAMPLE 1*: Assume that the results in Fig. 8 are for a $\frac{3}{4}$-in. maximum aggregate size to be used for heavy traffic. Compute the optimum asphalt content, and determine if the mix meets the Asphalt Institute criteria.

	Percent
(a) Asphalt at maximum stability	4.8
(b) Asphalt at maximum unit weight	5.1
(c) Asphalt at 4% air voids (average of 3 and 5% range for surfacing mix, heavy traffic, in Table 5)	4.3
Optimum asphalt (average)	4.7

At 4.7% asphalt the following values are obtained from Fig. 8:

Stability	1880 lb
Flow	9
Percentage air voids	2.8
Percentage voids in mineral aggregate	14.4

Stability exceeds the minimum of 750 lb, flow is within the limiting range 8–16, and percentage voids in the mineral aggregate exceeds the minimum of 14 (Fig. 9). But percentage air voids falls below the lower limit of 3. Gradation adjustments should be made as suggested in Table 1 to provide a mix having all test valuse within allowed limits, and the most economical such mix should be identified.

Low flow values indicate brittle pavements which tend to crack early. High flow values are usually accompanied by low stabilities. Marshall mix design thus accounts for three of the basic properties discussed earlier; stability, flexibility (by flow limits), and durability (by limits on air voids).

*Examples 1 and 2, courtesy of Asphalt Institute [1].

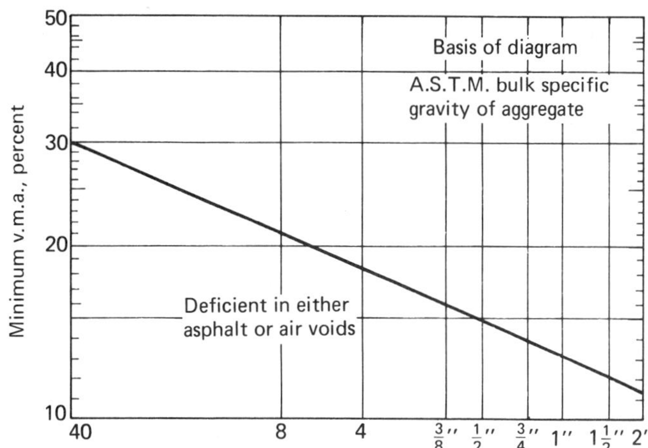

Fig. 9. Relationship between minimum VMA and maximum aggregate particle size for dense-graded mixtures. From *Mix Design Methods for Asphalt Concrete* (Series 2: MS-2), Third Edition. College Park, Md.: The Asphalt Institute, 1969 [1].

2. Hveem Method

The Hveem mix design method is applicable to paving mixes using cutbacks and emulsions as well as asphalt cements and for aggregates up to 1 in. maximum size. It is used principally for the design and field control of *dense* paving mixtures.

A unique feature of the Hveem method is a rather complicated procedure for estimating optimum asphalt content to facilitate selection of asphalt contents for the trial mixes. The procedure consists of

1. Measuring the weights of kerosene adsorbed by the aggregate fraction passing the No. 4 sieve and of SAE 10 lubricating oil adsorbed by the aggregate retained on the No. 4 sieve.
2. Estimating the aggregate surface area (square foot per pound of aggregate) by multiplying the percentage passing each sieve by a *surface area factor* and adding the products.
3. Using the above data in charts to obtain the estimated optimum asphalt content. The charts provide for approximately 5% air voids [1].

Specimens 4 in. in diameter by $2\frac{1}{2}$ in. high are then compacted with a kneading compactor using a range of asphalt contents bracketing the estimated optimum content obtained from the preceding steps. A *swell test*, *stabilometer test*, *cohesiometer test*, and *bulk density determination* are performed with the compacted specimens.

The swell test is made only on specimens prepared for that purpose. The stabilometer, cohesiometer, and bulk density measurements are each performed on the remaining specimens. Stabilometer and cohesiometer tests are performed at 140°F.

The stabilometer test uses a special triaxial test cell to measure resistance to lateral displacement under vertical load (Fig. 10). The stabilometer value is

$$S = \frac{22.2}{P_h D/(P_v - P_h) + 0.222}$$

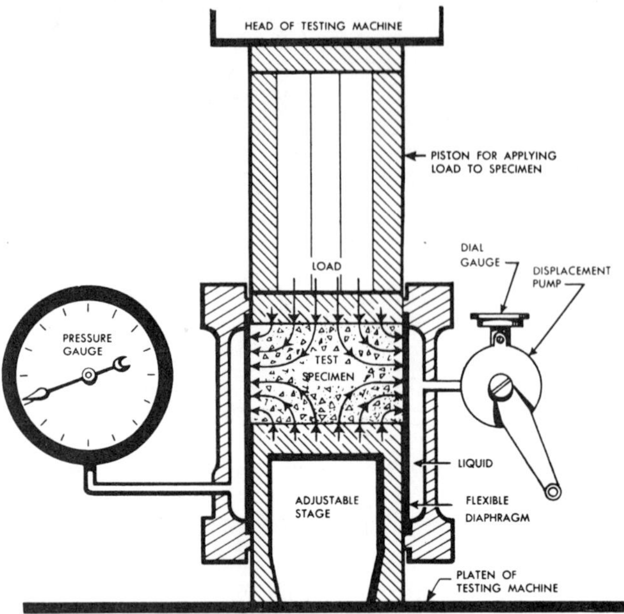

Fig. 10. Hveem stabilometer test. From *Mix Design Methods for Asphalt Concrete* (Series 2: MS-2), Third Edition. College Park, Md.: The Asphalt Institute, 1969 [1].

where S = relative stability,
D = vertical displacement,
P_v = 400 psi vertical pressure,
P_h = transmitted lateral pressure corresponding to P_v.

According to this equation, relative stability varies from 0 for a liquid having equal lateral and vertical pressures to 90 for a rigid solid which transmits zero lateral pressure.

The cohesiometer test, performed on specimens after completing the stabilometer test, is a flexural test (Fig. 11). The specimen is clamped as shown, and load is applied at a constant rate at the end of a lever arm. When the end of the arm travels $\frac{1}{2}$ in., the flow of loading shot is automatically cut off and its weight determined. The cohe-

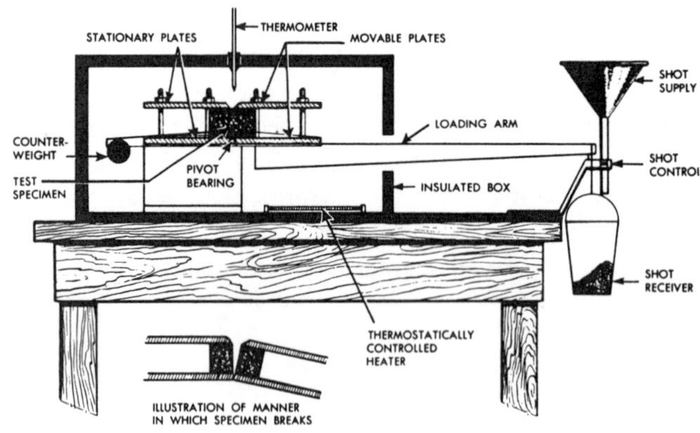

Fig. 11. Hveem cohesiometer test [1]. From *Mix Design Methods for Asphalt Concrete* (Series 2: MS-2), Third Edition. College Park, Md.: The Asphalt Institute, 1969 [1].

siometer value is

$$C = \frac{L}{W}(0.2H + 0.044H^2)$$

where C = cohesiometer value,
L = weight of shot, g,
W = width of specimen, in.,
H = height of specimen, in.

Typical test results are shown in Fig. 12. Table 6 lists commonly used Hveem design criteria.

Fig. 12. Typical Hveem test results [1]. From *Mix Design Methods for Asphalt Concrete* (Series 2: MS-2), Third Edition. College Park, Md.: The Asphalt Institute, 1969 [1].

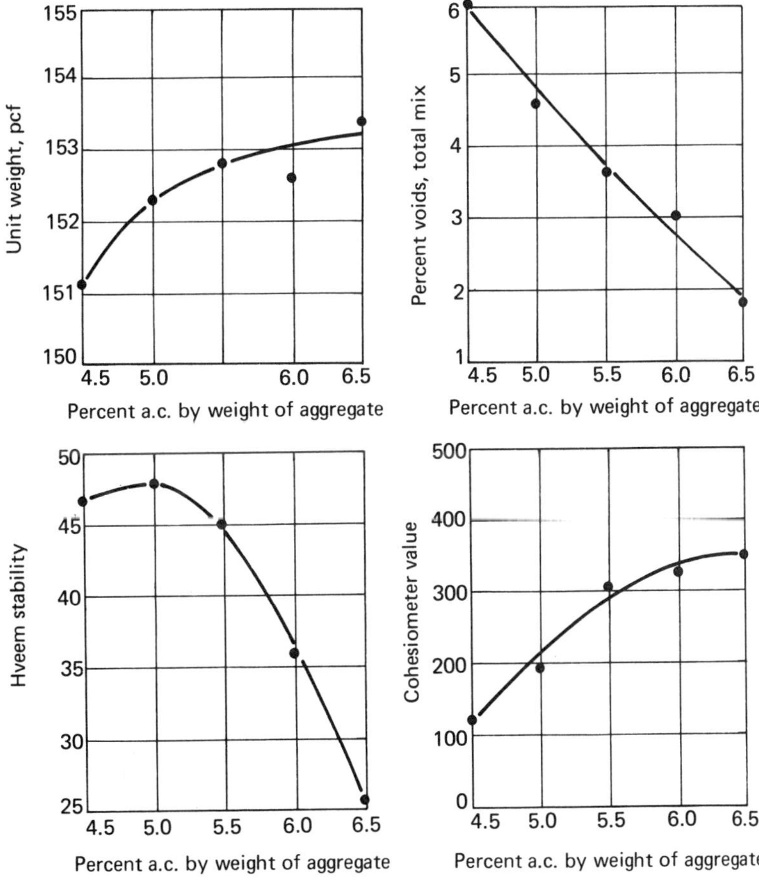

TABLE 6. Hveem Design Criteria

Traffic category: Test property	Heavy		Medium		Light	
	Min.	Max.	Min.	Max.	Min.	Max.
Stabilometer value	37	—	35	—	30	—
Cohesiometer value	50	—	50	—	50	—
Swell	Less than 0.030 in.					

Note: Although not a routine part of this design method, an effort is made to provide a minimum percentage air voids of approximately 4%.
SOURCE: *Mix Design Methods for Asphalt Concrete* (Series 2: MS-2), 3rd ed. College Park, Md.: The Asphalt Institute, 1969 [1].

Optimum asphalt content is considered to be the highest content possible without reducing stability or void content below the minimum values. Generally, air voids should approach the 4% minimum.

EXAMPLE 2: Assume that 600 g of shot are required to break a 4-in.-diameter by 2.5-in.-high specimen.

$$\text{cohesiometer value} = \tfrac{600}{4}(0.2 \times 2.5 + 0.044 \times 2.5^2) = 193$$

D. BITUMINOUS-AGGREGATE SURFACE BEHAVIOR

Two principal changes in bitumen properties account for the degradation and eventual failure of bituminous mixes:

1. Hardening (loss of ductility) due to slow volatization of the lighter fractions plus photoinitiated condensation and polymerization of the remaining components.
2. Bond separation between bitumen and aggregate.

Factors which influence the formation and performance of bitumen-aggregate bonding include mineral composition and texture of the stone surfaces; bitumen viscosity; interfacial energies between bitumen, stone, and water; aggregate coatings such as water films, dust, grease, or artificial coatings; and surface-active agents added to the asphalt. Some results obtained primarily by Thelen [4] will be described briefly.

The most usual adsorbed film is water, since water vapor is always present in air. For example, powdered quartz equilibrated in air at 25°C and 80% humidity and then dried in a combustion furnace yields an equivalent of $2\frac{1}{2}$ molecular layers of water by evaporation at 100°C and another layer at 280°C; the last layer requires over 1000°C to be completely evaporated [4]. Thus some moisture probably exists on aggregate even in hot mixes where stone is heated to over 300°F.

Spreading of asphalt or oil on a water surface occurs spontaneously to decrease the free energy of the system in accordance with the second law of thermodynamics

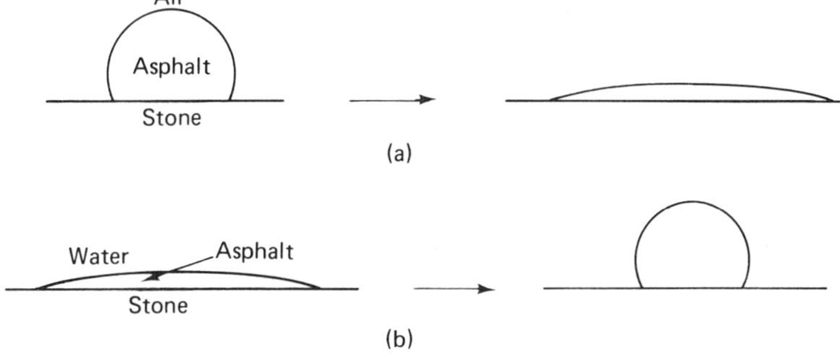

Fig. 13. (a) Spreading in air; (b) normal stripping.

[Fig. 13(a)]. The spreading rate depends primarily on viscosity. The interfacial energies are [4]

$$\gamma_{wa} = 72 \text{ ergs/cm}^2$$
$$\gamma_{ba} = 26 \text{ ergs/cm}^2$$
$$\gamma_{bw} = 30 \pm 5 \text{ ergs/cm}^2$$

where the subscripts a, b, and w stand for air, asphalt, and water, respectively. The reduction in free energy, ΔF, due to replacing the water-air interface with a water-asphalt interface is

$$\Delta F = \gamma_{wa} - (\gamma_{bw} + \gamma_{ba}) = 16 \text{ ergs/cm}^2$$

Surface energies for spreading asphalt over the water film adsorbed on rocks are slightly different:

$$\gamma_{sa} = 76 \text{ ergs/cm}^2$$
$$\gamma_{ba} = 26 \text{ ergs/cm}^2$$
$$\gamma_{bs} = 17 \pm 3 \text{ ergs/cm}^2$$

where the subscript s stands for stone. The reduction in free surface energy for spreading asphalt on stone at room temperature and humidity is approximately

$$\Delta F = 76 - (26 + 17) = 33 \text{ ergs/cm}^2$$

Stripping of asphalt from stone surfaces by water may also occur spontaneously [Fig. 13(b)]. Let

$$\gamma_{bs} = 17 \pm 3 \text{ ergs/cm}^2$$
$$\gamma_{bw} = 30 \pm 5 \text{ ergs/cm}^2$$
$$\gamma_{sw} = 0 \text{ erg/cm}^2$$

The interfacial tension between stone and water is almost zero because under usual conditions, the *stone* surface is approximately a free water surface. Then, the potential energy for stripping is

$$\Delta F = \gamma_{bs} + \gamma_{bw} - \gamma_{sw} = 47 \text{ ergs/cm}^2$$

The rates of spreading and stripping probably depend on both the asphalt viscosity and the magnitude of ΔF.

Figure 14 shows how the stripping coefficient ΔF drops to zero when stone is heated to 350–400°F (177–205°C). In this range, quartz has an adsorbed water layer on the order of $1\frac{1}{3}$ to $1\frac{1}{2}$ molecules thick. Likewise, hot mix asphalts, in which the stone is preheated, are quite resistant to stripping.

Blistering and *pitting* are mechanisms whereby bituminous coatings can be punctured and sitès created for the initiation of stripping. When the sun comes out after raindrops have formed on a pavement, the warmed asphalt tends to flow out from under a drop and spread over it [Fig. 15(a)]. As heating continues, the water blister expands and may break, leaving a pit. The potential energy for blister formation, using the previous interfacial values, is

$$\Delta F = 72 - (17 + 30) = 25 \text{ ergs/cm}^2 \quad (1)$$

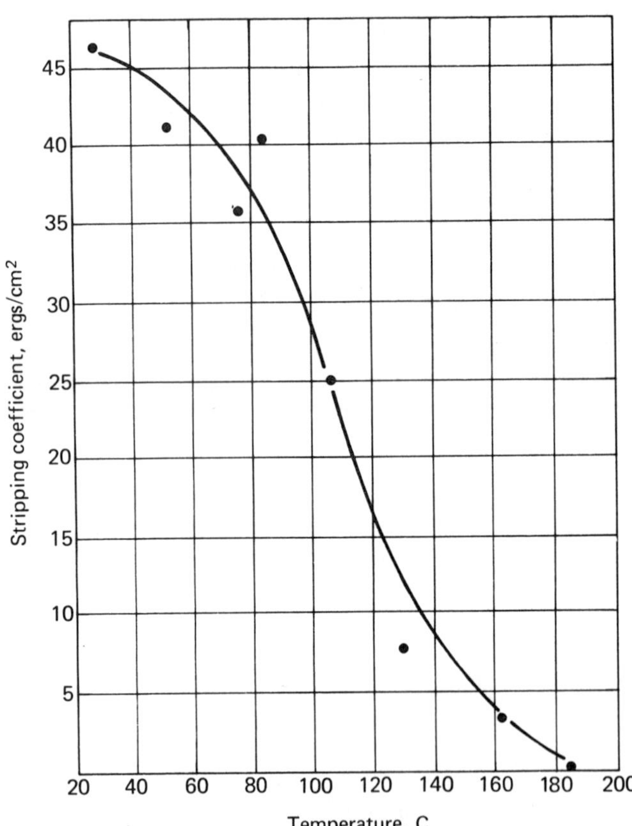

Fig. 14. Stripping coefficient of preheated quartz. From Thelen, E.: *Highway Research Board Bulletin*, Washington, D.C.: **192** (1958) 63 [4].

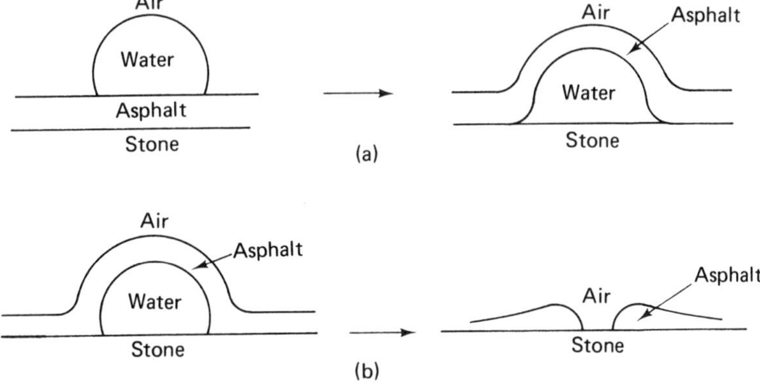

Fig. 15. (a) Blister formation; (b) pitting.

and the potential energy for pitting when the blister breaks [Fig. 15(b)] is

$$\Delta F = 17 + 30 = 47 \text{ ergs/cm}^2$$

Equation (1) shows that low interfacial tension between asphalt and water, such as provided by emulsifiers, for example, should tend to promote blister formation but retard pitting.

Diffusion of water molecules through asphalt films to aggregate surfaces provides an additional means of water damage, but diffusion is so slow it probably is much less significant than stripping or blistering. Asphalts absorb water molecules until their moisture content is in equilibrium with that of the ambient. Such absorbed water molecules may diffuse toward aggregate surfaces, for example, due to

1. *Thermal gradients.* Moisture migrates in most materials from a hot surface to a cold one. Solar heating at a pavement surface may drive absorbed water molecules toward an aggregate surface.
2. *Absorption* by hydrophilic dust particles coating the aggregate.
3. *Osmosis.* Water-soluble salts at the aggregate surface attract water to form more dilute solutions.

E. ADDITIVES

Bituminous mix additives are used for two principal purposes:

1. To improve aggregate-bitumen bonding, called adhesion or antistripping additives.
2. To improve mechanical properties of bitumens.

1. Adhesion Additives

In the previous section we indicated that if aggregate is heated sufficiently to evaporate three or more molecular layers of water, asphalt forms an almost unstrippable bond and adhesion additives seem unnecessary. If asphalt is applied to cold wet aggregate, adhesion additives may be helpful. Stripping is observed less frequently on basic stones, especially those containing some iron, than on acidic stones [5]. Also, a nonporous stone such as quartz may warrant an adhesion additive, whereas a porous stone might not. Additives may be applied directly to the stone surfaces or mixed with the asphalt. Stripping can often be drastically reduced by precoating aggregate with metal halide salts such as ferric, zinc, lead, copper, or aluminum chlorides. Additional improvements are sometimes obtained by following the salt treatment with a water-soluble soap to give a metal soap coating.

Additives mixed with the asphalt are usually primary fatty amines, fatty amino amides and imidazolines, fatty diamines, fatty acids, fatty acid-amine mixtures, and metal soaps. Fatty esters, secondary and tertiary fatty amines, and quarternary ammonium compounds are also used (see Tables 15 and 16, Chap. 2). As with emulsifiers, their selection should be based upon the acidic or basic nature of the aggregate. Table 7 shows the effects of several additives on the advancing and receding contact angles of asphalt on glass, quartz, and marble. Most additives reduce receding contact angles and have little effect on advancing contact angles. As might be expected, however, cationic additives increase the receding contact angle on marble.

TABLE 7. Influence of Some Surfactants on Asphalt-Mineral Contact Angles (degrees)

	Glass		*Quartz*		*Marble*	
Interfacially active substance	θ_r	θ_a	θ_r	θ_a	θ_r	θ_a
No addition	100	155	90	150	25	140
Montan wax (0.5%)	104	—	10	140	10	140
Oxidized paraffin wax (1%)	10–60	—	40	140	10	130
Heptadecylamine (1%)	10–60	—	55	135	65	140
Iron naphthenate (1%)	10	—	25	135	10	135
Dimethyl-sec-C_{13}-C_{18} alkyl-sulfonium methyl sulfate (1%)	—	—	35	110	60	130

γ_{bw} θ γ_{sw} γ_{as}

$$\gamma_{sw} = \gamma_{as} + \gamma_{bw} \cos \theta$$

θ_r = receding contact angle.
θ_a = advancing contact angle.
SOURCE: SAAL, R. N., "Asphalt Systems," in *Composite Materials*, L. Holliday, ed. Amsterdam: Elsevier Publishing Company, 1966, Chapter IX [6].

The *adhesion tension* of asphalt on an aggregate, Σ_b, is the maximum resistance in dynes per centimeter which asphalt is capable of withstanding when exposed to water under pressure,

$$\Sigma_b = \gamma_{bw} \cos \theta_{\text{receding}}$$

and the adhesion tension of water on the aggregate, Σ_w, is the maximum resistance which water can withstand when exposed to asphalt under pressure,

$$\Sigma_b = \gamma_{bw} \cos \theta_{\text{advancing}}$$

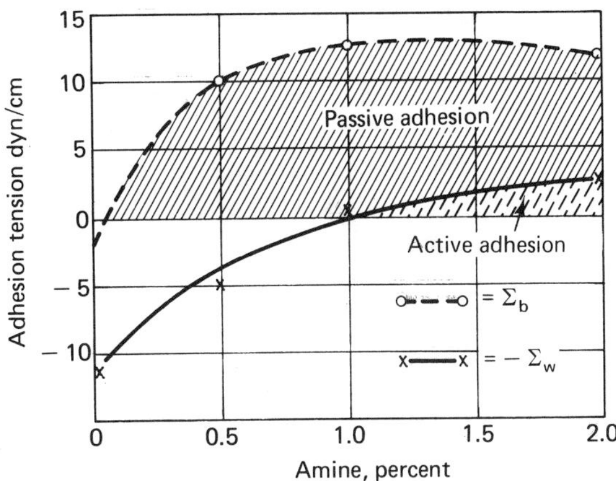

Fig. 16. Adhesion tension-concentration curves for octadecyl amine in MC cutback. From Zvejnieks, A.: *Highway Research Board Bulletin*, Washington, D. C.: **92** (1958) 26 [7].

Figure 16 shows Σ_b and Σ_w as functions of octadecyl amine additive concentration in an MC cutback. Both Σ_b and Σ_w are negative for the untreated asphalt, which means the asphalt can be replaced by water without additional forces. At an amine concentration of about 0.05%, Σ_b becomes positive, which means that water cannot replace asphalt without exterior forces. This condition is called passive adhesion. At an amine concentration of 1.0%, Σ_w becomes positive, which means that asphalt is able to displace water spontaneously from a wet aggregate surface and adhere to it. This is called active adhesion. Required concentrations of 0.5–1% to obtain active adhesion are typical of many adhesion additives.

A major disadvantage of most adhesion additives is their susceptibility to degradation after relatively short heating periods. Consequently, they are sometimes introduced with cutbacks directly at the pugmill or distributor to avoid repeated heatings.

Conservatively speaking, adhesion additives should not be used unless the need is demonstrated. They lower water-asphalt interfacial tensions and hence may promote blistering. They may also slightly increase the diffusion of water molecules through asphalt. To be effective, adhesion additives must be chemisorbed to the particular mineral surfaces present and not just reduce water-asphalt interfacial tension, as required for emulsifiers.

2. *Bulk Property Additives*

Additives used to improve bulk properties of bitumen binder phases have included mineral dusts, rubber and other high polymers, and fibers. Mineral dust, a normal component in hot mixes, increases the shear strength and reduces the temperature susceptibility of hardened binder.

Rubbers and other high polymers, typically as 0.5–3% by weight of asphalt, are in some instances found economical on bridge decks and in high-traffic areas for extending the service life of asphalts and for reducing the whipping of aggregate from paved surfaces under high-speed traffic. They may increase viscosity, softening point, and low-temperature ductility and decrease penetration and temperature susceptibility. Rubbers used include butyl, nitrile, styrene-butadiene copolymer, and neoprene. Adding in latex form to asphalt emulsions is often effective because the two emulsions can usually be formulated to be compatible. Latexes are also cheaper than most other forms of rubber. Some rubbers partially dissolve in the lower-molecular-weight fractions of asphalt, increasing their viscosity and ductility, and hence that of the whole asphalt.

The principal fiber reinforcement used in bituminous mixes is very short (and hence relatively inexpensive) asbestos fiber. Asbestos fibers provide many of the benefits of polymers at somewhat less cost and are also more effective where asphalts are subject to high temperatures and to solution by dripping oil or chemicals, as, for example, on jet ramps and taxiways.

REFERENCES

1. *Mix Design Methods for Asphalt Concrete* (Series 2: MS-2), 3rd ed. College Park, Md.: The Asphalt Institute, 1969.
2. *Construction Specification for Asphalt Concrete* (SS-1), 4th ed. College Park, Md.: The Asphalt Institute, 1969.
3. *Asphalt Plant Manual* (MS-3), 3rd ed. College Park, Md.: The Asphalt Institute, 1967.
4. THELEN, E., *Hwy Res. Bd. Bull.*, Washington, D.C.: **192** (1958) 63.
5. HUGHES, R. I., D. R. LAMB, and O. PORDES, *J. Appl. Chem.*, **10** (1960) 433.
6. SAAL, R. N., "Asphalt Systems," in *Composite Materials*, L. Holliday, ed. Amsterdam: Elsevier Publishing Company, 1966, Chap. IX.
7. ZVEJNIEKS, A., *Hwy Res. Bd. Bull.*, Washington, D.C.: **192** (1958) 26.

10

RIGID FOAMS

All at once the horse stood still,
Close by the meet'n'-house on the hill.
First a shiver, and then a thrill,
Then something decidedly like a spill,
And the parson was sitting upon a rock,
At half past nine by the meet'n'-house clock.
—HOLMES

Foams are dispersions of a large volume of gas in a small volume of liquid. They differ from gas emulsions, where the liquid volume is usually greater than that of the gas and where the bubbles tend to be spherical and their mutual interaction weak. Foam bubbles are crowded, and their shapes tend toward polyhedra separated by thin films of the liquid.

Foams of commercial value include inorganic cements, metals foamed by the introduction of metallic hydrides, and polymers. Table 1 shows properties of some polymer foams. All polymers can be expanded by a gas. Polystyrene is the principal polymer foam used in construction for thermal insulation, and rigid urethane foam is much used in the sandwich panel walls of transportation vehicles because of its low K factor.

TABLE 1. Properties of Polymer Foams*

Type	*Available form*	*Density (lb/cu ft)*	*K factor*† *at 75°F*	*Compressive strength (psi)*	*Tensile strength (psi)*	*Maximum continuous service temperature (°F)*	*Coefficient of expansion (10^{-5}) (in./in./°F)*
Acrylonitrile styrene	Expandable beads, billets	1	—	10	—	160	3
Cellulose acetate	Boards, rods	6–8	0.30	100–125	120–170	350	2
Epoxy	Blocks, spray powder	2–20	0.12	14–1,000	20–500	300	1
Methylmethacrylate styrene	Boards	1.7	0.50	10	—	175	3
Phenolic	Foam-in-place, beads	$\frac{1}{10}$–25	0.20	2–1,000	10–250	300	1
Polyethylene	Boards, rods, sheets	1.8–60	0.35	20–3,500	20–1,800	160	3
Polypropylene	Slabs, liquid	3–20	—	—	—	250	—
Polystyrene	Boards, planks, logs, beads, etc.	1–10	0.23	20–460	40–600	175	2–4
Polyvinyl chloride	Sheets, shapes, liquids, paste	3–26	0.20	10–80	200–1,200	170	5
Silicone	Powder, paste, liquid	3–16	0.28	200–900	30–400	600	—
Urea formaldehyde	Boards, shredded	$\frac{1}{3}$–1	0.18	200	120	150	—
Urethane	Boards, shapes, foam-in-place	$1\frac{1}{2}$–70	0.11	50–600	400–8,000	300	3

*Properties depend on manufacturing process, composition, and density.

†Lowest thermal conductivity listed (Btu/sq ft/hr/°F/in.). Other insulation *K* factors are

Powdered gypsum	−0.50	Rock wool	−0.32	Glass fiber	−0.23
Glass foam	−0.40	Cork	−0.28	Urethane foam (2 pcf)	
Sawdust	−0.36	Hair felt	−0.24	COr	−0.21
				fluorocarbon	−0.11

SOURCE: Reprinted from *Composite Engineering Laminates*, A. G. Dietz, ed., by permission of the MIT Press, Cambridge, Mass. [1].

Any foam geometry depends on two criteria being satisfied: The cells must fill space, and the surface energy (area) must be minimal. Ideally, three films form a junction of 120° each. Each film has a tension of 2γ/cm, where γ is the liquid surface tension and the factor 2 accounts for the two liquid-gas interfaces on each film. Three

Fig. 1. (a) Truncated octahedron; (b) minimum surface area tetrakaidecahedron; (c) pentagonal dodecahedron.

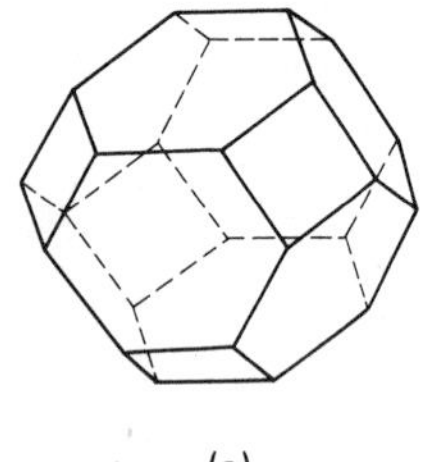

(a)

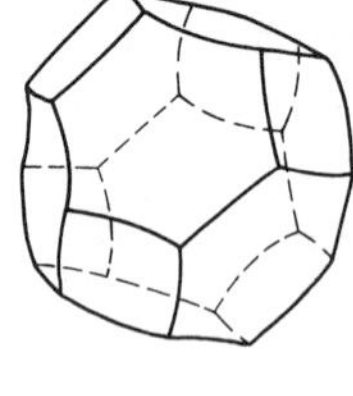

(b)

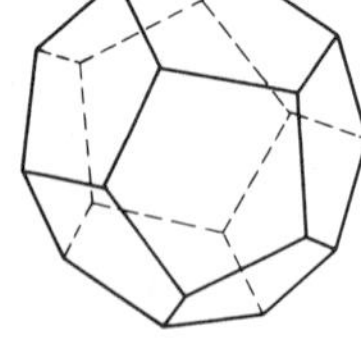

(c)

such forces can balance only if the angles between them are 120° each. The edges formed by the three films meet symmetrically in groups of four so that the angle between each pair of edges is the arc cosine $(-\frac{1}{3})$, or 109.5°.

No regular polyhedron meets these requirements. A truncated octahedron [Fig. 1(a)] satisfies the compatibility requirements but fails to meet the local geometry requirements because the angles at any vertex are 120, 120, and 90°. Kelvin showed that if the edges of the truncated octahedron are curved so each angle is 109.5°, the compatibility condition can be satisfied if the hexagonal faces have double curvature with mean zero curvature (since there is no pressure difference between adjacent cells) [2, 3]. The resulting minimum surface area tetrakaidecahedron contains eight curved hexagonal faces and six distorted, but flat, square faces [Fig. 1(b)]. It is the only configuration meeting the previous requirements.

Although no regular polyhedra meet the above requirements, the pentagonal dodecahedron (12 equilateral pentagonal faces) has face angles of 120° and almost completely fills space [4] [Fig. 1(c)]. Smith showed from Euler's law that a network of regular three-dimensional polyhedra would require each polyhedron to have 13.39 faces and each face to have 5.10 sides. The pentagonal dodecahedron approximates these values as closely as any regular polyhedron, having corner angles of 108° instead of 109.5° and relatively small mismatch in space compatibility. Thus, foam bubbles

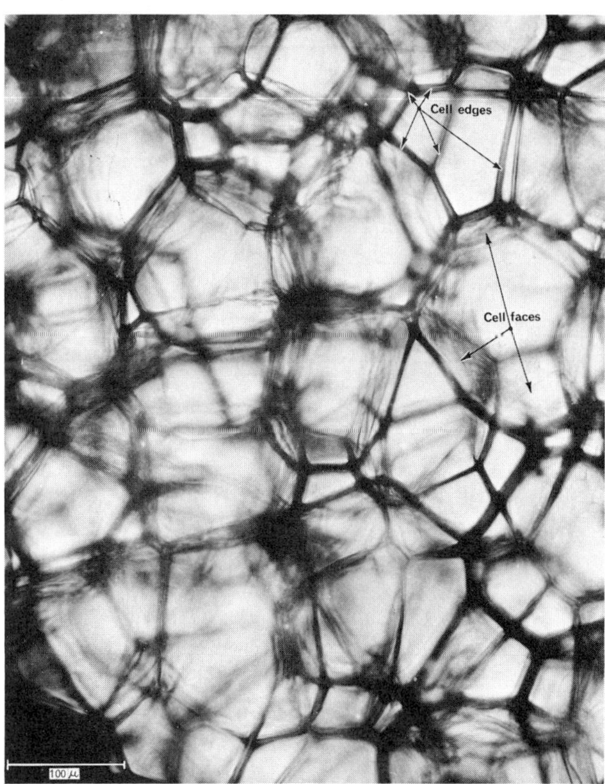

Fig. 2. Cell structure of polystyrene foam showing faces of the polyhedra (×370). From Lanceley, H. A., J. Mann, and G. Pogany: "Thermo-Plastic Systems" in *Composite Materials*, L. Holliday, ed. Amsterdam: Elsevier Publishing Company, 1966, Chap. 6 [5].

tend toward pentagonal dodecahedra, if their volumes are equal. In most foams, the bubble volumes are not equal, and their shapes form a variety of unsymmetrical polyhedra but with a predominance of four-, five-, and six-sided faces (Fig. 2). Although there is little to choose between the Kelvin model and the pentagonal dodecahedron, the dodecahedron is a simpler geometry for analytical purposes.

A. FOAM MORPHOLOGY

Rigid foams are formed by producing gas bubbles in a liquid, which then hardens by polymerization or other chemical action. As the bubbles grow in size and number the liquid between them is stretched thin. Foams have some rigidity even when their films are liquid, because the equilibrium geometry corresponds to minimum surface energy, and therefore deformation requires external work.

Foams, like other colloidal systems, can be produced by condensation or dispersion. In condensation, the gas phase is initially present as separate molecules which coalesce to form bubbles. Beer is an example, in which the solution of carbon dioxide produced by yeast becomes supersaturated when the pressure is released and the excess solute forms the disperse gas phase, or bubbles. In solid foams manufactured by condensation, the bubbles are derived from *blowing agents.* In dispersion, gas is introduced into the liquid through a small capillary to form bubbles.

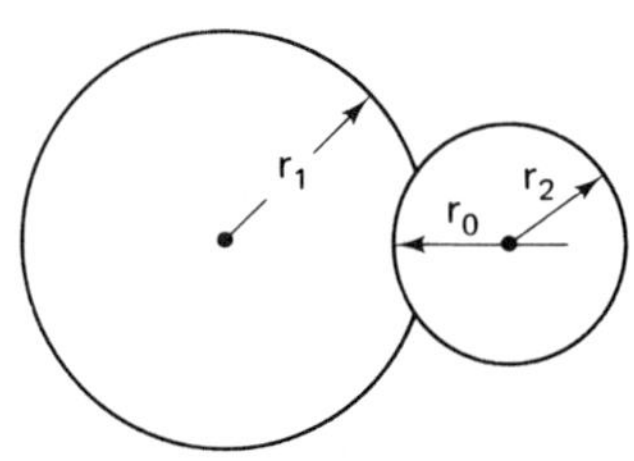

Fig. 3. Redistribution of bubble sizes.

Liquid foams are thermodynamically unstable systems, and two processes tend to reduce their surface area: redistribution of bubble sizes and film draining which leads to rupture [6]. Redistribution is caused by the difference in pressures in two adjacent bubbles of different sizes. For two bubbles of radii r_1 and r_2 sharing a partition of radius r_0 (Fig. 3), the gas pressure, from Laplace's equation, is

$$p_a + \frac{4\alpha}{r_1}$$

in the larger bubble, and

$$p_a + \frac{4\alpha}{r_2}$$

in the smaller, where p_1 is the atmospheric pressure and α the surface tension. The factor 4 accounts for each bubble having two gas-liquid interfaces, internal and external.

For equilibrium, the partition must be concave toward the smaller bubble, and its radius given by

$$\frac{4\alpha}{r_0} = \frac{4\alpha}{r_2} - \frac{4\alpha}{r_1}$$

or

$$r_0 = \frac{r_1 r_2}{r_1 - r_2}$$

Since the partition is not impermeable, gas diffuses through it from the smaller to the larger bubble (from higher to lower pressure). Thus the small bubble contracts, the large bubble expands, and the polydispersity of foam size tends to increase with age.

Film drainage of liquid from the walls may be retarded by the increasing liquid viscosity, so that final structure is determined by the capillary force causing drainage, the viscosity forces opposing drainage, and the gas pressure. If the liquid flow is not sufficiently retarded, drainage may result in rupture of the walls, leaving the edges to form an open-celled structure. Unlike aqueous foams, the cells of a rigid foam will not collapse but give a sponge-like structure of skeletal strands.

1. Density Calculations

Since the contribution of the gas phase is negligible, foam density depends on the average cell size and the thickness of films and edges between cells. The edges have approximately uniform cross sections along their lengths, thickening only where they meet. The edge is often about 2 orders of magnitude thicker than the wall, although the total wall weight is often 2–10 times the total edge weight. For a 1.25 lb/cu ft polystyrene foam, for example, the mean dimensions determined by optical microscope techniques were approximately [5]

Cell face thickness	0.44 μ
Edge thickness	2.3 μ
Edge length	38.5 μ

Assuming uniform wall thickness t and equilateral triangular edges of side b and length L, densities can be calculated from

$$d_f = d_c\left[1.38\frac{t}{L} + 0.58\left(\frac{b}{L}\right)^2\right] \quad \text{pentagonal dodecahedron}$$

$$d_f = d_c\left[1.18\frac{t}{L} + 0.46\left(\frac{b}{L}\right)^2\right] \quad \text{tetrakaidecahedron}$$

where d_f and d_c are the foam and continuous phase densities [5].

B. STRUCTURAL ANALYSIS

2. Strength Calculations

Several approximate analyses have been made for the deformation and strength characteristics of rigid foams. The one outlined here is due largely to Patel and Finnie [7].

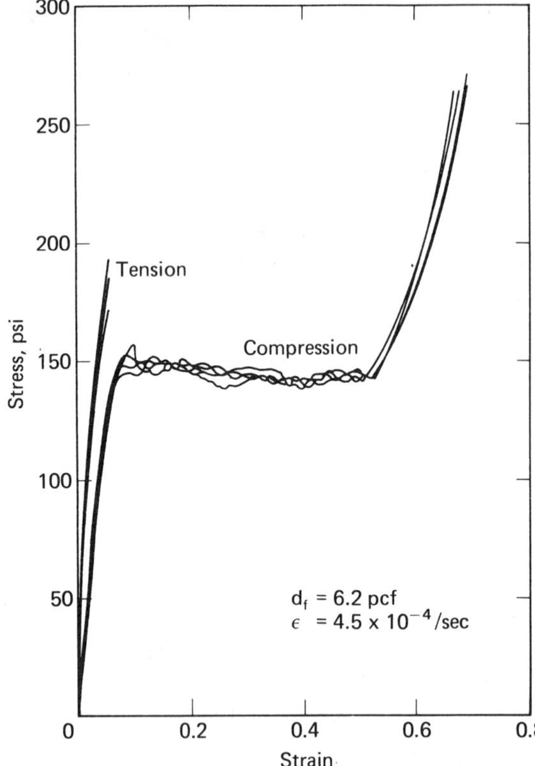

Fig. 4. Uniaxial deformation of closed-cell polyurethane foam. From Patel, M. R., and I. Finnie: *J. Mater., JMLSA*, 5, No. 4 (Dec. 1970) 909 [7].

The stress-strain diagram for a low-density closed-cell rigid polyurethane foam (Fig. 4) resembles that for wood [Fig. 1(c), Chap. 3] and is typical of foams in which cell walls and edges can fail by buckling, giving a compressive yield point followed by extensive deformation with little change in stress. The tension curves, however, undergo little deformation prior to fracture.

To compare the relative roles of edges and walls in compressive and tensile strengths, consider edges of length L and diameter d and walls of width w, length l, and thickness t. The compressive buckling strength ratio between edges and walls is

$$R_b = \left(\frac{1}{a}\right)\left(\frac{\pi d^4}{64}\right)\left(\frac{12}{wt^3}\right)\left(\frac{l^2}{L^2}\right)$$

where a is the ratio of the number of cell faces to cell edges opposing load in a given direction. A dodecahedron has 12 faces and 30 edges each shared by 2 and 3 cells, respectively, giving

$$a = \frac{12/2}{30/3} = 0.6$$

Since l cannot be smaller than the edge length L, l/L will always be greater than 1. With these values of a and l/L and observed values of d/t (50 to 100) and d/w (0.05 to 0.2) for low-density foams, R_b is of the order of 10^3 to 10^4. Similar approximations indicate the ratio of edge strength to wall strength under *tensile* loading to be of the

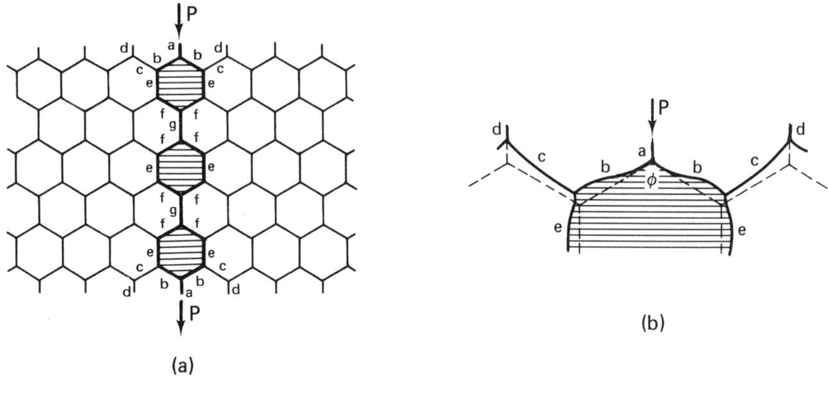

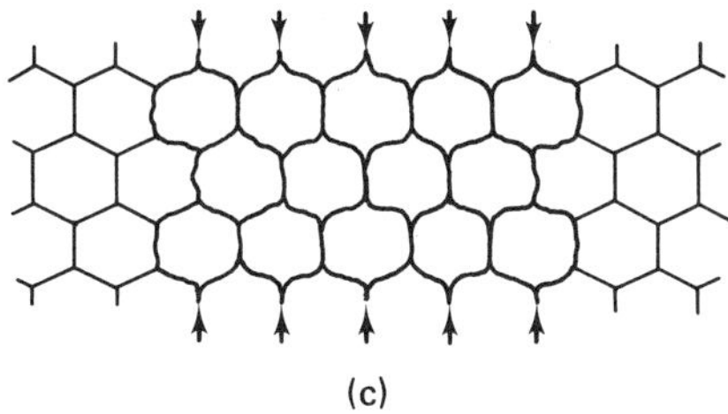

Fig. 5. Axial compression of two-dimensional model. From Patel, M. R., and I. Finnie: *J. Mater., JMLSA*, 5, No. 4 (Dec. 1970) 909 [7].

order of 10. Hence the strength contribution of the walls can be ignored in deformation which causes wall buckling, but not in wall tension. It can be shown that any edge buckling produces lateral tension in one adjacent wall and compression in the other two [8]. When the edge is in tension, the adjacent walls undergo no lateral tension, but all three stretch in the direction of applied tension.

These three-dimensional relationships can be visualized in terms of a two-dimensional hexagonal framework. In compression [Fig. 5(a)] load P applied to strut a must be supported by compressive forces P in struts b, since cell walls cannot support compression. If there is no boundary load at d, struts c do not support the horizontal components of P in struts b, and the latter tend to open up, causing bending of the struts [Fig. 5(b)] and inducing horizontal tension in the walls, indicated by the direction of crosshatching. Struts e bulge out due to rotation at the rigid junctions. The load P is transmitted principally by the heavily marked struts and crosshatched cell walls [Fig. 5(a)].

In tension (Fig. 6) struts c will again not resist the horizontal components of (tensile) force in struts b, and the unbalance will cause struts b to move in. Struts b and e will bend due to joint rigidity [Fig. 6(b)]. The cell walls will act like a set of vertically stretched strings which do not resist horizontal compression. The cell wall tensile strain is greater than strut strain, because the walls undergo strain due to

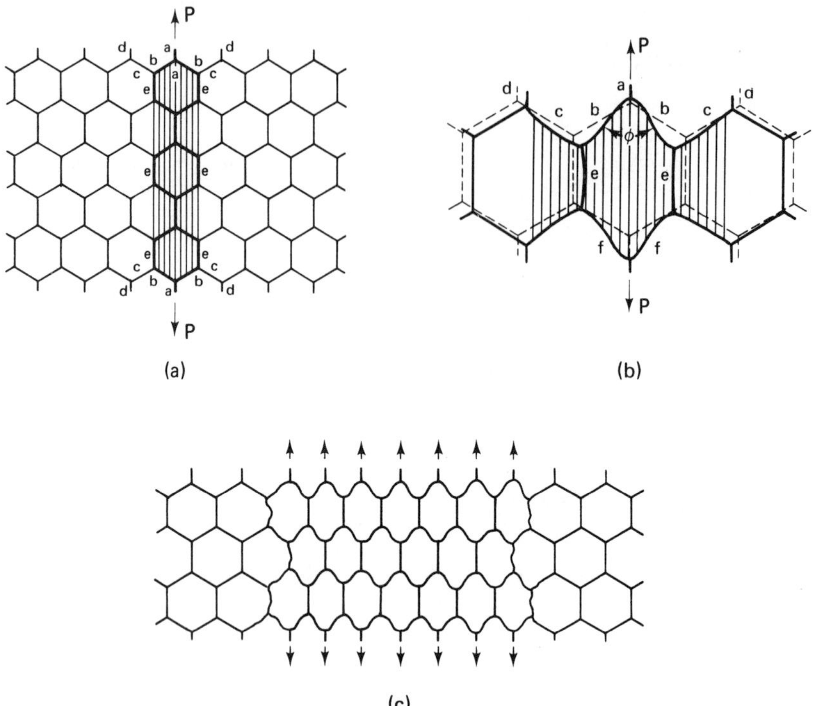

Fig. 6. Axial tension of two-dimensional model. From Patel, M. R., and I. Finnie; *J. Mater., JMLSA*, 5, No. 4 (Dec. 1970) 909 [7].

loading of the oblique struts in addition to strain due to elongation of the axial struts. Thus the cell walls will likely rupture first [8]. Since the walls support tension, there is more spread in load than under compression, but this may not affect more than one or two columns of cells.

These concepts may be generalized for approximate analysis of three-dimensional foams and foams containing nonuniform polyhedra. Because of its large slenderness ratio, an axial strut fails by elastic buckling rather than compressive yield. However, the buckling will be resisted by the attached cell walls, so that the elastic limit will be determined by simultaneous rupture of the cell wall and buckling of the attached struts.

2. *Strength-Density Relationships*

The log-log plots of strength-density and modulus-density data for various modes of foam deformation are usually approximated by straight lines. This allows representation in the form

$$\text{strength or modulus} = A(\text{density})^{B}$$

Values of exponent B for test data reported in the literature have been found to

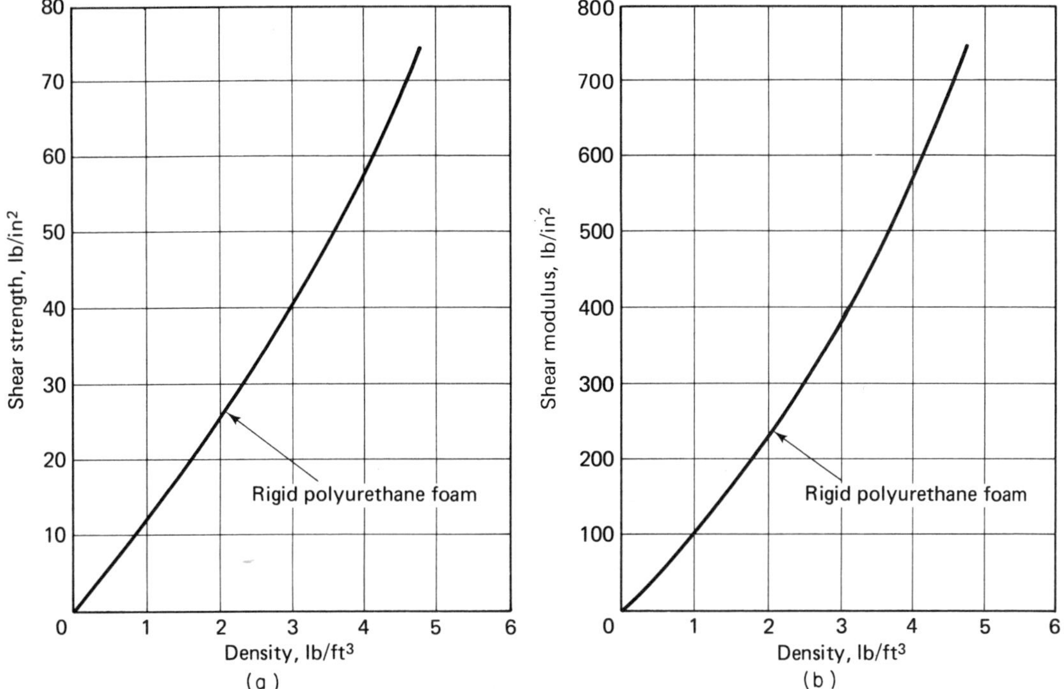

Fig. 7. (a) Shear strength, and (b) shear modulus of polyurethane foam as a function of density. From Benjamin, B.S., *Structural Design with Plastics* © 1969 by Litton Educational Publishing, Inc. New York: Van Nostrand Reinhold Company [9].

range from 1.1 to 1.8, for example, for various strength and modulus modes of polyurethane foams [7]. Since modulus is the ratio of stress to strain, the modulus-density and strength-density relations for a given deformation mode should have the same value of exponent B. Figure 7 shows typical strength- and modulus-density relationship for polyurethane foam. While these relationships can be developed from experimental data, approximate analytical expressions can also be developed, as in the following example, which relates compressive strength to density [7].

The volume of a pentagonal dodecahedron with edge length L is $5.44L^3$. Disregarding the slight geometrical incompatibility, the number of cells in a unit cube is

$$N = \frac{1}{5.44}L^3 \tag{1}$$

Since the dodecahedron has 12 faces, 30 edges, and 20 vertices, each shared by two, three, and four cells, respectively, there are $6N$ walls, $10N$ struts, and $5N$ vertices attributed to each cell. Denoting the volumes of the wall, strut, and vertex by v_w, v_s, and v_j, the foam density is

$$d_f = d_c N(5v_j + 10v_s + 6v_w)$$

where d_c is the solid phase density. Denoting K as the ratio of wall weight to total weight,

$$\begin{aligned} d_j &= d_c N(5v_j + 10v_s) + Kd_g \\ &= \frac{d_c N}{1-K}(5v_j + 10v_s) \end{aligned} \tag{2}$$

Assuming the strut junctions are cubes of side d and the struts have diameter d and length $L - d$, one obtains from Eqs. (1) and (2)

$$d_f = \frac{d_c}{1-K}\left[1.466\left(\frac{d}{L}\right)^2 - 0.524\left(\frac{d}{L}\right)^3\right] \tag{3}$$

The volume of one pentagonal cell wall [Fig. 8(a)] is $1.72(L - d)^2t$, and the total wall weight a unit cube is

$$Kd_f = 6Nd_c \times 1.72(L-d)^2t$$

or

$$\frac{t}{L-d} = \frac{0.526Kd_f}{d_c[1-(d/L)]^3} \tag{4}$$

Strut buckling is resisted by tension in an attached wall. By assuming the wall to be of uniform width [Fig. 8(b)], the tensile modulus E_t of the wall for unit strut deflection, per unit length of strut, is

$$E_t = \frac{2tE}{d_c}$$

Fig. 8. (a) Pentagonal cell wall; (b) buckling of struts supported by a cell wall. From Patel, M. R., and I. Finnie; *J. Mater., JMLSA*, 5, No. 4 (Dec. 1970) 909 [7].

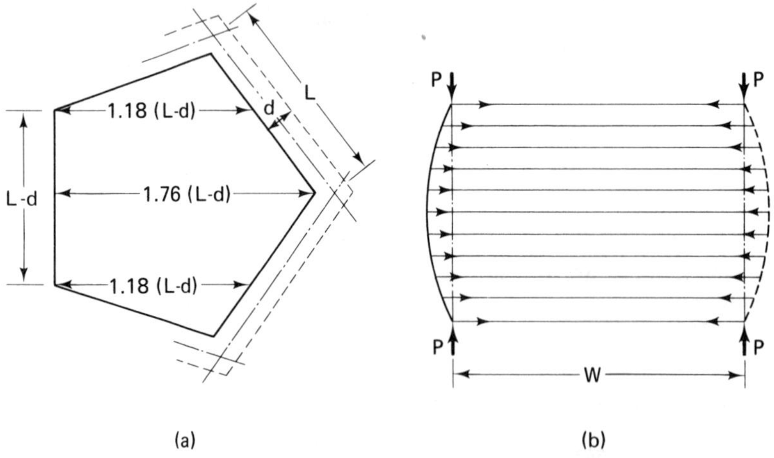

where E is Young's modulus and the factor 2 accounts for symmetrical deflection of struts on opposite sides of the wall. Assuming an approximate value $d_c = 1.5(L - d)$ for the pentagonal wall width which varies from $1.18(L - d)$ to $1.76(L - d)$ [Fig. 8(a)],

$$E_t = 1.33E\frac{t}{L-d}$$

and using Eq. (4),

$$E_t = \frac{0.7017KEd_f}{(1 - d/L)^3 d_c} \tag{5}$$

which can be considered as the reaction coefficient for a compression bar on an elastic foundation. The buckling load, from Timoshenko [10], is

$$P_b = \frac{\pi^2 EI}{L^2}\left(m^2 + \frac{E_t L^4}{m^2 \pi^4 EI}\right)$$

where I is the strut moment of inertia and m is the number of half-waves in the shape of the buckled bar. For the thin walls observed in low-density structural foams, E_t is low and there is no inflection point in the buckled bar; i.e., $m = 1$. Thus

$$P_b = \frac{\pi^2 EI}{L^2}\left(\frac{E_t L^4}{\pi^4 EI}\right)$$

and the equation for yield strength is

$$\begin{aligned}\sigma_y &= 0.75\frac{P_b}{L^2}\\ &= 0.365E\left(\frac{d}{L}\right)^4 + 0.076E_t\end{aligned} \tag{6}$$

Figure 9 compares the results from Eq. (3), (5), and (6) with some experimental results for a polymer foam.

In tensile loading no abrupt change in the slope of the stress-strain curve is observed after rupture of the cell walls. One might therefore obtain an approximate strength-density relationship by omitting the cell wall contribution to the density and strength values. In this case Eq. (3) reduces to

$$d_f = d_c\left[1.466\left(\frac{d}{L}\right)^2 - 0.524\left(\frac{d}{L}\right)^3\right] \tag{7}$$

The load in an axial strut at fracture is

$$P = (\pi d^4/4)S_s$$

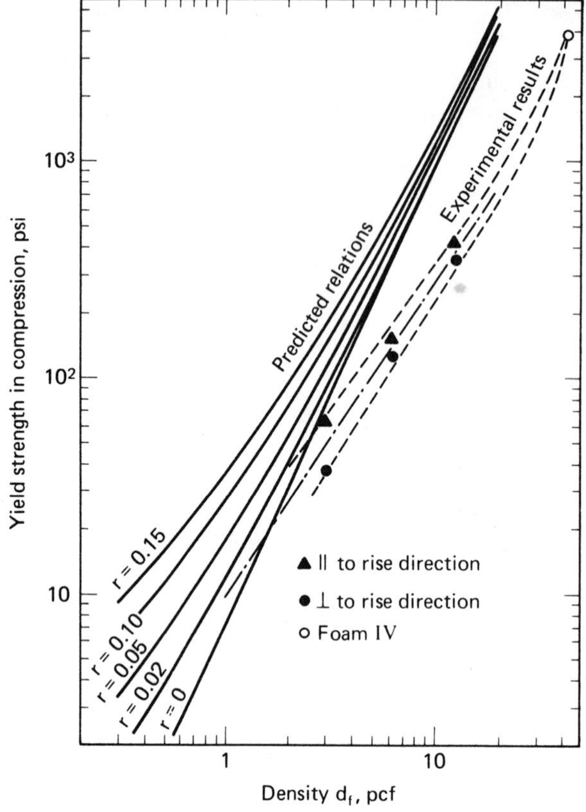

Fig. 9. Comparison of predicted and experimental yield strengths ($E = 232{,}000$ psi, $d_c =$ 77 pcf). From Patel, M. R., and I. Finnie: *J. Mater., JMLSA*, 5, No. 4 (Dec. 1970) 909 [7].

where S_s is the fracture strength of the solid phase. So the fracture strength of the foam is

$$S_c = \frac{0.75}{L^2} \cdot \frac{\pi d^2}{4} S_s$$

$$= 0.591 \left(\frac{d}{L}\right)^2 S_s \tag{8}$$

Figure 10 compares the fracture strength-density relation from Eqs. (7) and (8) with some experimental results for a polymer foam.

If d_f and K remain constant, the cell size will not affect foam strength. This can be shown analytically by examining Eqs. (3) and (6) for a fixed value of K. For example, consider the case $K = 0$, for open-celled foams. Equation (4) then becomes

$$d_f = d_c \left[1.466 \left(\frac{d}{L}\right)^2 - 0.524 \left(\frac{d}{L}\right)^3 \right]$$

or

$$d\{d_f\} = d_c \left[2.932 \left(\frac{d}{L}\right) - 1.572 \left(\frac{d}{L}\right)^2 d\left\{\frac{d}{L}\right\} \right]$$

For constant density, $d\{d_f\} = 0$, which requires $d\{d/L\} = 0$. Similarly, considering the

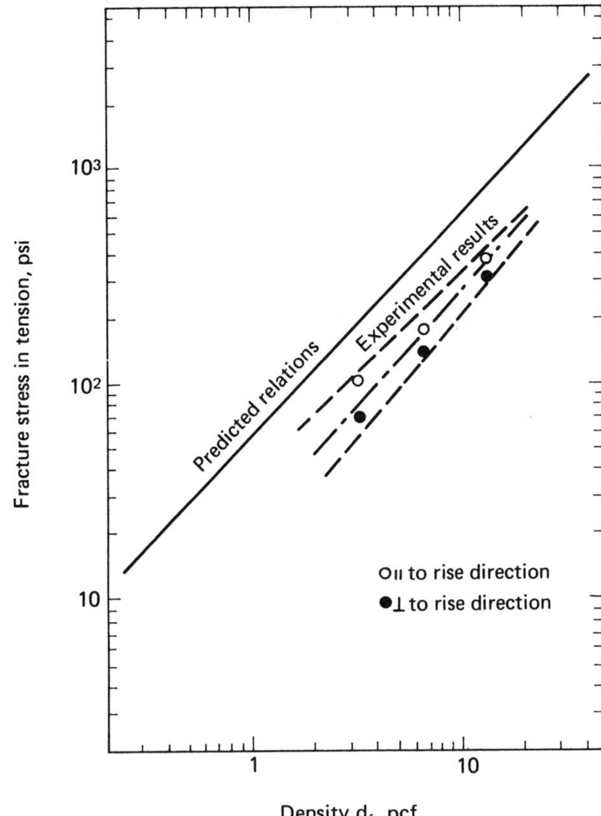

Fig. 10. Comparison of predicted and experimental tensile strengths ($S_s = 10{,}000$ psi, $d_c = 77$ pcf). From Patel, M. R., and I. Finnie: *J. Mater., JMLSA*, 5, No. 4 (Dec. 1970) 909 [7].

compressive strength for $K = 0$, Eq. (6),

$$\sigma_y = 0.365E\left(\frac{d}{L}\right)^4$$

or

$$d\sigma_y = 1.46E\left(\frac{d}{L}\right)^3 d\left\{\frac{d}{L}\right\} = 0$$

since for constant density $d\{d/L\} = 0$. The result is similar for any value of K. Tensile strength can also be shown to be independent of the cell size.

A second useful observation is that compressive yield *strain* of foams should be essentially independent of density; compressive yield occurs when the walls rupture, which requires a specific tensile strain in the walls, corresponding to a specific compressive strain in the foam.

3. Anisotropic Foams

Since structural foam cells are frequently elongated in the direction of foam rise, anisotropic strength and modulus properties are of interest. In a unit cube of such foams, more struts are oriented in a direction parallel to elongation than perpendicular to it, and one might expect decreasing tensile and compressive strengths with increasing angle θ between the cell axis and loading direction. But since compressive strength depends on buckling, the variation of strut length with ϕ may also need to be considered, giving a reverse effect upon strength.

Assume a cell to be an ellipsoid with major and minor axes a and b (Fig. 11) [7]. Then one might assume that the strut length L in any direction is proportional to the radius vector l in that direction and that the density of struts d_s oriented in any direction varies inversely with the area of a cross section of the ellipsoid normal to that direction.

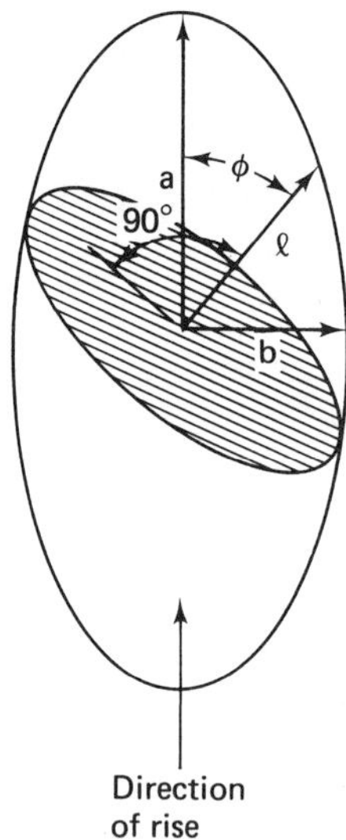

Fig. 11. Elliptical model of anisotropic cell.

$$L \propto l = \frac{a}{\sqrt{r^2 \sin^2 \phi + \cos^2 \phi}}$$

$$d_s \propto \frac{1}{A} = \frac{\sqrt{r^2 \cos^2 \phi + \sin^2 \phi}}{ab}$$

where r is the axial ratio a/b.

The strength variation with d_s alone can be represented by

$$P_\phi = C_1 \sqrt{\cos^2 \phi + \sin^2 \phi / r^2}$$

where C_1 is a constant. Normalizing with respect to the strength for $\phi = 0$,

$$\frac{P_\phi}{P_0} = \sqrt{\cos^2 \phi + \sin^2 \phi / r^2} \tag{9}$$

This relationship should approximate the tensile strength.

The strength variation determined by strut buckling would be proportional to the number of struts n and inversely proportional to the strut length squared,

$$P_\phi = C_2 \sqrt{r^2 \cos^2 \phi + \sin^2 \phi}\,(r^2 \sin^2 \phi + \cos^2 \phi)$$

which, normalized, becomes

$$P_\phi P_0 = \sqrt{\cos^2 \phi + \sin^2 \phi / r^2}(r^2 \sin^2 \phi + \cos^2 \phi) \tag{10}$$

This relationship should approximate the compressive yield strength as determined by strut buckling.

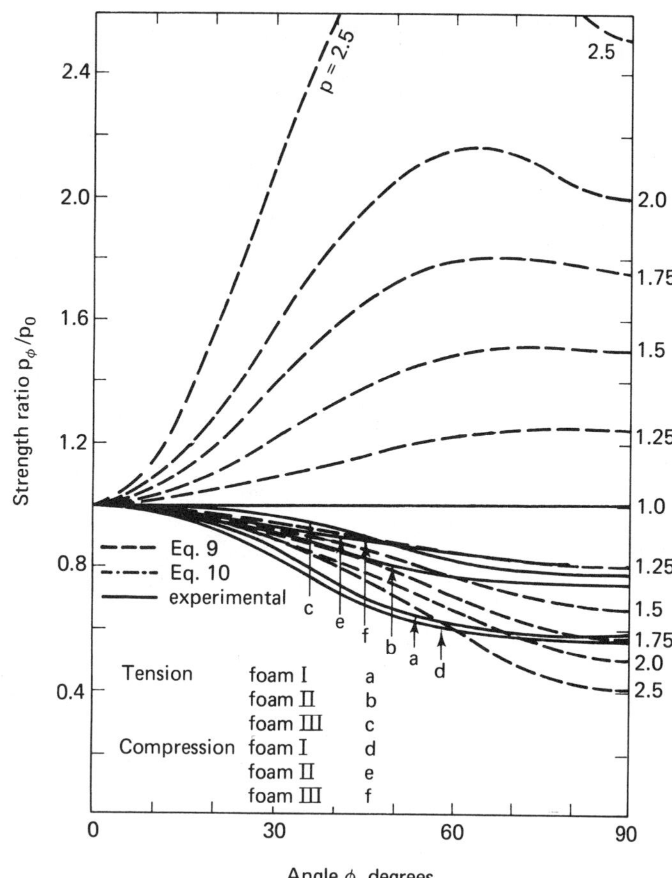

Fig. 12. Effect of anisotropy on strengths. From Patel, M. R., and I. Finnie: *J. Mater., JMLSA*, 5, No. 4 (Dec. 1970) 909 [7].

Figure 12 compares Eq. (9) and Eq. (10) with experimental strengths of several anisotropic foams having different densities. While the tensile strength trend appears to follow Eq. (9), the compression yield strengths decrease with ϕ, in contradiction to Eq. (10), indicating that strut buckling is probably not the cause of yielding. Since cell wall rupture, which must precede strut buckling, is independent of strut length, the compressive strength, like tensile strength, will depend only on the number of struts opposing deformation, and so both would follow Eq. (9).

C. TWO-COMPONENT FOAMS

Sections A and B dealt with structural foams produced by dispersing a large volume of gas in a small volme of liquid which hardens to produce approximately polyhedral-shaped voids. Foams can also be manufactured by cementing hollow spheres together with a binder. The spheres are typically of glass, ceramic, or rigid phenolic or epoxy, with diameters from 20 to 80 μ, wall thicknesses of 1 to 3% of their diameters, and weighing 9 to 24 lb/cu ft (Fig. 13 in this chapter and Fig. 7 in Chap. 1). They resemble fine-grained, easily flowing, very lightweight sands. The binder may be either an

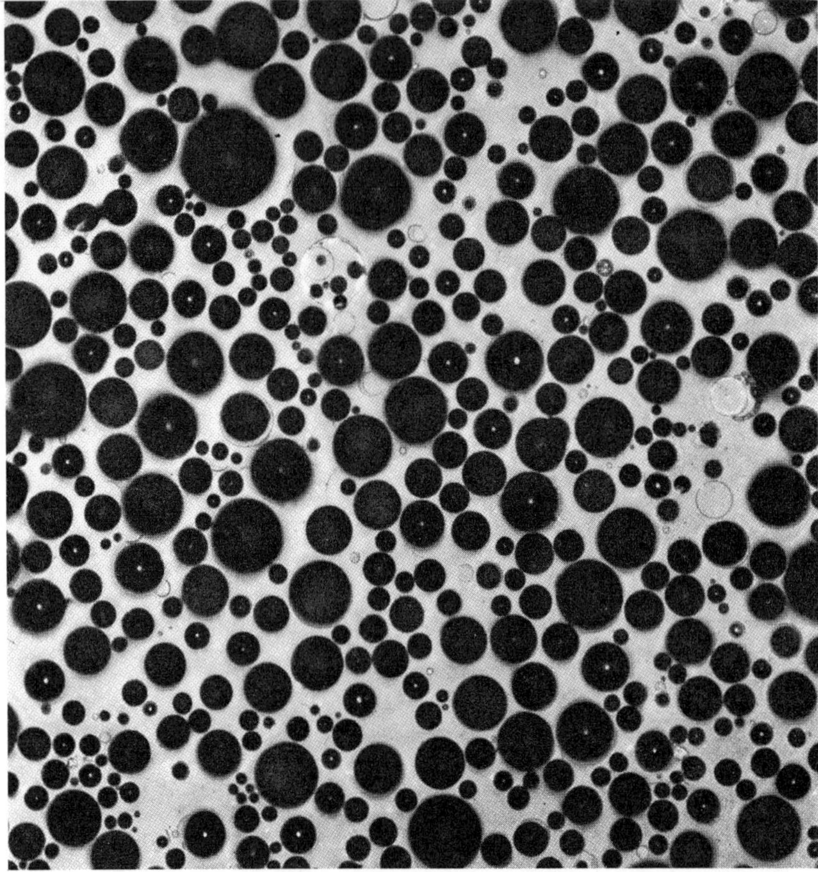

Fig. 13. Microscopic structural foam, 100x. From Deruntz. J. A., *Proceedings of the Southampton 1969 Civil Engineering Materials Conference on Structure*, M. Te'eni, ed. Part I: Solid Mechanics and Design. New York: John Wiley & Sons, Inc. (Interscience Division), 1971, p. 405 [11].

organic resin or an inorganic cement. Densities of the composite foam may range from 25 to 50 1b/cu ft, depending on the binder and bulk density and particle size gradation (void content) of the spheres. For example, polyester-glass microsphere foams with densities on the order of 40 1b/cu ft are widely used in molded simulated wood furniture manufacture. They can be sawed, drilled, nailed, and sanded like wood and require no finishing after molding. Epoxy-glass microsphere foams are used in deep-sea buoyancy applications, permitting high strength-to-weight ratios and withstanding hydrostatic pressures to depths of 40,000 ft [11]. Microspheres are used in limited applications as aggregate in the cores of sandwich panel and shell structures, although their current cost ($5–$10/1b) is high for such applications.

The stress-strain curves of microsphere foams resemble those of gas-blown foams (Fig. 4), i.e., linear elastic behavior in tension, terminating in rather abrupt brittle fracture through the matrix, and elastic-plastic behavior in compression, in which the spheres rupture due to wall buckling and the resin then simply consolidates at nearly constant stress to give a new composite of broken glass and resin. Photomicrographs show that failure tends to occur in the sphere wall approximately 45° along the meridian curve from the contact with adjacent spheres, suggesting that failure is caused by bending stress in the sphere walls due to localized loadings at the contacts or near contacts of adjacent microspheres [11].

PROBLEMS

1. Estimate the Young's modulus, the compressive yield strength, and the tensile fracture strength of a closed-cell polymer foam having the following properties:

Polymer Young's modulus	$E = 2 \times 10^5$ psi
Polymer fracture strength	$S_s = 12 \times 10^3$ psi
Polymer density	$d_c = 70$ pcf
Foam density	$d_f = 2$ pcf
Mean strut diameter	$d = 0.001$ in.
Mean strut length	$L = 0.02$ in.
Polymer weight fraction in the cell walls	$K = 0.6$

2. Assuming that foam geometry can be partially controlled by the selection of blowing agents and curing conditions, write in simplest form the objective function and constraints for determining the values of K, d, L, and d_f to minimize cost of a foam which satisfies the following constraints:

Young's modulus	$E_t \geq 8 \times 10^5$ psi
Compressive yield strength	$\sigma_y \geq 100$ psi
Tensile fracture strength	$S_c \geq 300$ psi

The properties of the solid polymer are

$$E = 3 \times 10^5 \text{ psi}$$
$$d_c = 60 \text{ pcf}$$
$$S_s = 8{,}000 \text{ psi}$$

Assume that foam cost increases with foam density.

REFERENCES

1. SCHWARTZ, R. T., and D. V. ROSATO, "Structural Sandwich Construction," in *Composite Engineering Laminates*, A. G. Dietz, ed. Cambridge, Mass.: The M.I.T. Press, 1969, Chap. 8.
2. THOMPSON, W., *Collected Mathematical and Physical Papers*, Vol. 5. New York: Cambridge University Press, 1911, p. 333.
3. SMITH, C. S., *Met. Rev.*, 9 (1964) 1.
4. BIKERMANN, J. J., *Foams and Emulsions in Chemistry and Physics of Interfaces.* Washington D.C.: American Chemical Society, 1965, p. 408.

5. LANCELEY, H. A., J. MANN, and G. POGANY, "Thermoplastic Systems," in *Composite Materials*, L. Holliday, ed. Amsterdam: Elsevier Publishing Company, 1966, Chap. 6.

6. DEVRIES, A. J., *Mededel. Rubber-Sticht.* Delft, 326 (1957) 11.

7. PATEL, M. R., and I. FINNIE, *J. Mater.*, *JMLSA*, 5, No. 4 (Dec. 1970) 909.

8. PATEL, M. R., Ph.D. thesis. Berkeley: University of California, Berkeley, 1969.

9. BENJAMIN, B. S., *Structural Design with Plastics*. New York: Van Nostrand Reinhold Company, 1969.

10. TIMOSHENKO, S., *Theory of Elastic Stability*. New York: McGraw-Hill Book Company, 1936.

11. DERUNTZ, J. A., *Proceedings of the Southampton 1969 Civil Engineering Materials Conference on Structure*, M. Te'eni, ed. Part I: Solid Mechanics and Design. New York: John Wiley & Sons, Inc. (Interscience Division), 1971, p. 405.

IV

FIBER AND LAMINATE COMPOSITES

11

FIBER-REINFORCED COMPOSITES

What do you think the parson found,
When he got up and stared around?
The poor old chaise in a heap or mound,
As if it had been to the mill and ground!
—HOLMES

Useful tensile members can be made, in principle, by joining strands of strong materials into parallel bundles. For nonbrittle materials such as most metals and polymers, this is easily accomplished by coiling individual fibers into a rope or cable. A tensile load on the rope increases the normal stress between fibers in the helix, enabling stress transmission between the fibers by sliding friction. Since the strongest materials are all very brittle (Sec. B in Chap. 3), they cannot be used in ropes because the coiling and surface abrasion between fibers would cause extensive cracking. They can, however, be bound together in a deformable matrix which has a shear modulus several times smaller than the Young's modulus of the fibers.

The matrix serves several functions: (1) It protects the surface of the fiber against brittle fracture initiated by abrasion or atmospheric corrosion. (2) It transfers the longitudinal load between fibers by virtue of its shear strength and its adhesion to the fibers (this function corresponds to the interfiber friction in a rope). (3) It deflects cracks which may occur in fibers in a longitudinal direction (Fig. 1) and transfers the resulting load increase over appreciable lengths of the adjacent fibers, as well as redistributing the load to both segments of the broken fiber. This contrasts to the effect of a Griffith crack in a monolithic brittle material, which results in total failure at low stress levels. Depending on the fiber material, these mechanical properties of the matrix can be provided by polymers or by metals such as aluminum or copper.

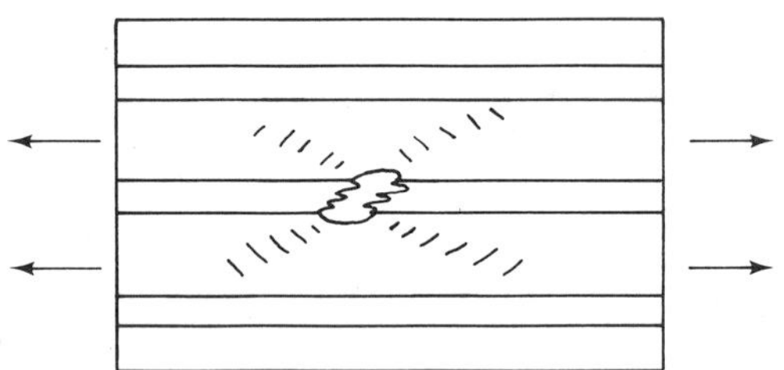

Fig. 1. Crack deflection in a weak matrix.

The beauty of fiber-reinforced composites is that they provide high fracture toughness at high-strength levels, a combination impossible to obtain with a single material. The strong but brittle fibers provide high strength, while plastic flow at the tip of a crack in the matrix reduces stress concentration, absorbs energy, and provides fracture toughness. Since the matrix deflects cracks parallel to the fibers, the fibers will not all fail in one plane. To extend a crack through the material, it is therefore necessary to pull the fibers out of the matrix as they break. This *pull-out* work provides a major portion of the fracture toughness.

In this chapter, we shall first study composites of brittle fibers in a ductile matrix, such as glass fiber-reinforced plastics, and then ductile fibers in a brittle matrix, such as polymer or metal fiber-reinforced concretes.

A. BRITTLE FIBERS IN A DUCTILE MATRIX

Some manufacturing materials and processes for fiber-reinforced plastics will be described before studying the mechanics of brittle fiber-ductile matrix systems.

1. Manufacturing Processes

a. MATERIALS

Typical fiber materials include whiskers which can be grown as filamentary crystals, extruded or drawn glass and ceramic fibers, and high-strength metal wires. Table 1 shows properties of some typical fibers in these three classes. From this table, one can calculate the strength per unit weight and stiffness per unit weight, important properties in many applications of fiber-reinforced composites. Although whiskers have the greatest strengths, their slow growth rate and the current difficulties of cleaning, sorting, and final fabrication place their costs in the hundreds of dollars per pound, severely restricting their uses. Some progress has been made in the direct fabrication of fiber-reinforced metals by the growth and arrangement of fibers using controlled phase transformations, both in the solid state and from the melt. For example, needle-like crystals have been formed by precipitation in the systems aluminum-magnesium-silicon and aluminum-magnesium-copper [1, 2]. From strength theory (Sec. B in Chap. 3), the highest-strength fibers should be anticipated from covalently bonded solids of low atomic weight, such as carbides, borides, silicides, nitrides, and oxides of the trivalent and transition metals.

Figure 2 shows approximately the theoretically potential strength-density ratios of several fiber-reinforced composites.

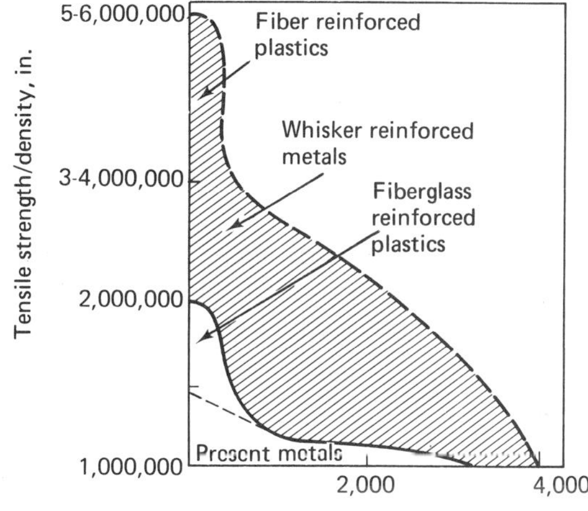

Fig. 2. Potential strength-density ratios of fiber-reinforced materials. Reproduced by permission, from Rosen, B. W., *Fiber Composite Materials*, Metals Park, Ohio: American Society of Metals, 1965 [2].

(1) Glass-reinforced plastics. The largest tonnage of fiber-reinforced material now manufactured is fiber glass. The glass is normally coated with a polymeric or metallic bonding agent which protects it from atmospheric corrosion prior to use and improves bonding to the specific matrix material [3]. Glass is comparatively inexpensive, strong, and adaptable to a wide variety of composite manufacturing techniques. Glass is not very stiff, however (Table 1), and bridges and airplane wings could not be made of glass-reinforced composites because they would bend too much under load as the fibers stretch. Useful temperature ranges are 200–300°C for most polymer matrices, and glass itself softens appreciably above 400°C. Thus, the main

TABLE 1. Properties of Fiber Materials

Class	*Material*	*Tensile strength (psi × 10⁶)*	*Young's modulus (psi × 10⁶)*	*Specific gravity*	*Typical diameter (mm)*	*Melting temperature (°C)*
Whisker	Graphite	3.0	98	2.2	—	3000*
	Al_2O_3	2.2	76	4.0	3–10	2050
	Iron	1.8	28	7.8	—	1540
	Si_3N_4	2.0	55	3.1	3–10	1900*
	SiC	3.0	100	3.2	1–3	2600
	Si	1.1	26	2.3	—	1450
	AlN	—	50	3.3	—	2450
	Boron	—	64	2.3	10–30	—
Glass, ceramic, polymer fibers	Asbestos	0.85	27	2.5	—	500†
	Drawn silica	0.86	10.5	2.5	35	1700
	Boron glass	0.35	55	2.3	10	—
	High-tenacity Nylon 66	0.12	0.7	1.1	2–10	—
Strong metal wires	Carbon steel	0.57	30	7.8	25–100	—
	Molybdenum	0.30	53	10.3	25	2610
	Tungsten	0.42	50	19.3	50	3380

*Sublimation temperatures.
†Loses water.
SOURCE: KELLY, A., and G. R. DAVIES. *Metal. Rev.*, **10** (1965) 1 [1].

advantage of glass-reinforced plastics in structural applications is their strength and toughness combined with light weight. Principal structural limitations are the low stiffness, uncertain weathering durability, and low resistance to fire and heat distortion.

Table 2 and Fig. 3 illustrate the commercial forms of glass fiber reinforcements.

Polyester resins account for about 85% of resin materials used with glass fiber. Other thermosetting resins include epoxies, phenolics, silicones, melamines, acrylics, and polyester resins modified with acrylics. Some thermoplastic resins used with glass reinforcement include polystyrene, nylon, polycarbonates, and acetals.

Inorganic fillers such as clay, talc, calcium carbonate, and calcium silicate may be used to reduce cost or improve surface appearance, strength, fire resistance, or moldability of fiber-reinforced plastics. Other additions may include coloring pigments, mold release compounds such as zinc stearate, and fire-retardant additives such as antimony oxide or chlorinated waxes and ultraviolet absorbers.

(2) Jute fibers. Jute reinforcement can be used either alone or with glass fiber. Unlike glass fiber, however, jute is available only as woven fabric or cloth. Jute is also a springy fiber, and surface fibers in the cloth tend to keep adjacent reinforcement layers apart, leaving resin-rich intervening bands. This can be overcome by pressing or flattening the cloth, called *calendering*.

TABLE 2. Commercial Forms of Glass Fiber Reinforcements

Form	*Description*	*Suitable processes**	*Typical applications*
Yarn	Twisted strands	Weaving, filament winding	Aircraft, marine, electrical laminates
Rovings	Continuous strands	Filament winding, continous panel, spray-up, pultrusion, preforming for matched die molding	Pipes, auto bodies, rocket motor cases
Woven fabrics	Woven cloths from glass fiber yarns	Hand lay-up, vacuum bag, autoclave, high-pressure laminating	Aircraft, marine, electrical flat sheet and tubing
Woven rovings	Woven strands, coarser and heavier than fabrics	Hand lay-up	Marine, large containers
Chopped strands	Strands cut $\frac{1}{4}$ to 2 in. long	Premix molding, wet slurry performing	Appliance, electrical, ordinance components
Reinforcing mats	Chopped or continuous strands in random matting	Matched die molding, hand lay-up, centrifugal casting	Sheets, auto and truck bodies
Surfacing mats	Random mat of small fiber size	Matched die molding, hand lay-up, filament winding	For smooth surfaces, auto bodies, special housings

*Processes are described in the following section.

(3) Asbestos fibers. Asbestos fibers of the amphibole group, which include crocidolite (blue) and amosite (gray-brown) asbestos, are widely used as reinforcements. Asbestos fibers are used with both thermosets such as phenolics, epoxies, and polyesters and with thermoplastics such as nylon, vinyls, and polypropylenes. Asbestos and glass fibers sometimes find complementary uses. For example, high-strength glass fiber-wound epoxy pipe is sometimes combined with an asbestos-phenolic liner to provide broad chemical resistance. Some advantages of asbestos include

1. High Young's modulus.
2. The best chemical resistance of any inorganic fiber available in commercial quantities.
3. High surface area and excellent fiber-resin bond. Both crocidolite and amosite fibers are compatible with polyester, epoxide, phenolic, and silicone resins, without fiber surface treatments.
4. High temperature resistance—effective in improving fire retardance and reducing heat distortion.
5. Available in various forms (mat, tape, and loose fiber) and can be used in combination with glass and a variety of other fibers.

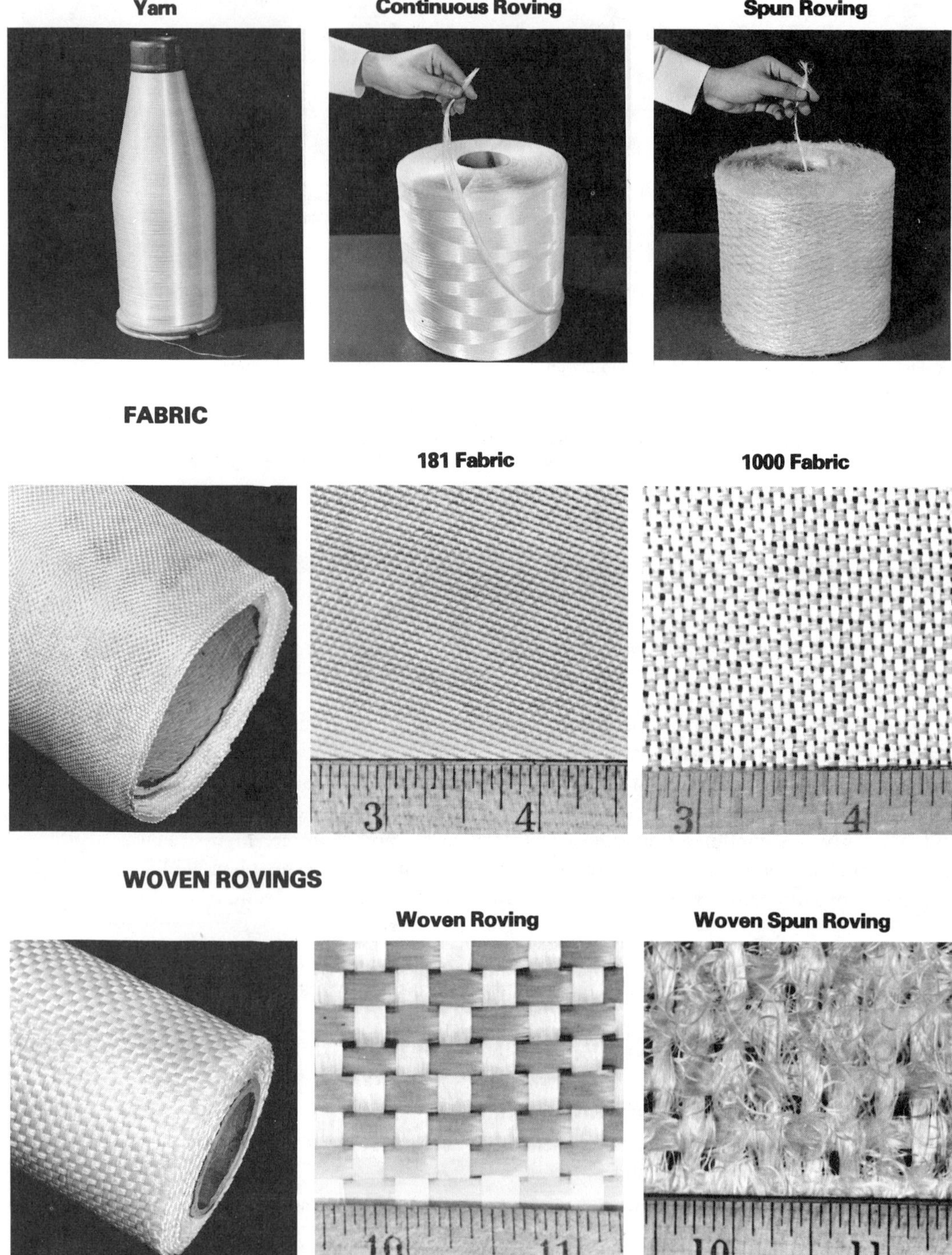

Fig. 3. Commercial forms of glass fiber reinforcements. (*Courtesy of Owens-Corning Fiberglas Corporation.*)

CHOPPED STRANDS

REINFORCING MATS

Chopped Strand Mat

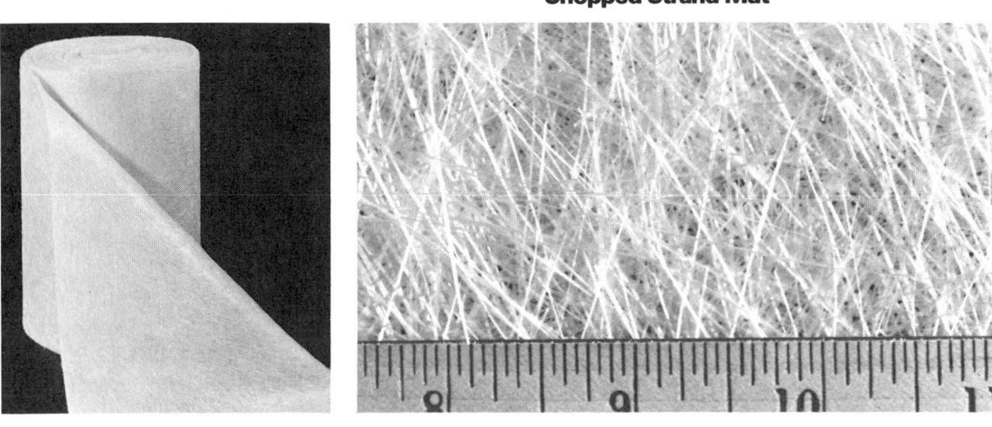

SURFACING MAT

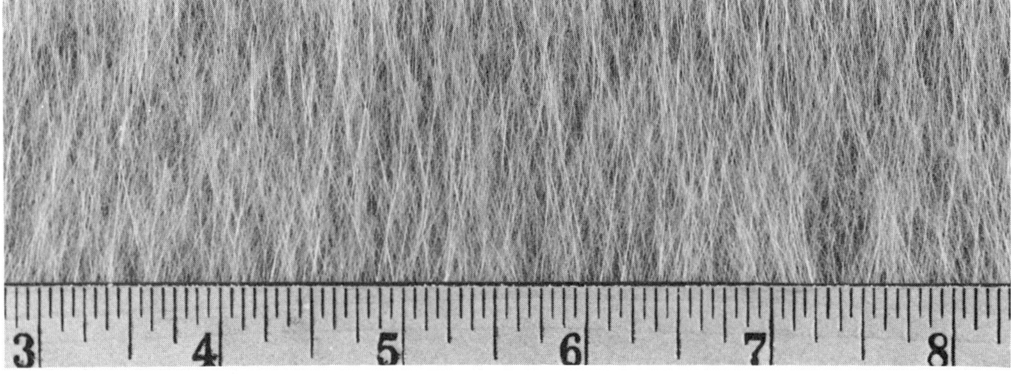

Fig. 3. *Continued*

b. MOLDING METHODS

Molding methods for glass fiber-reinforced plastics are classed as open and closed mold processes.

Open molds are single-cavity, either male or female, and are used with little or no pressure. The principal characteristics of open mold process are [4]

1. Low investment due to need for only one mold and lack of other equipment. Due to low pressures, mold can be of plaster or reinforced plastics.
2. Relatively high labor cost.
3. Rapid mold production and easy design changes.

The principal characteristics of the molded object are

1. Only one side is finished.
2. Complex shapes and large objects may be formed.

Closed molds are two-piece male and female molds, usually made of metal. Principal characteristics of closed-mold processes are

1. Low labor cost.
2. Efficient use of raw materials.
3. Highest production rate.
4. Higher equipment and mold cost.

The principal characteristics of the molded object are

1. Two sides finished.
2. Excellent duplication of parts.

Major types of the two molding processes are

Open Mold

1. Hand lay-up
2. Spray-up
3. Encapsulation
4. Filament winding
5. Centrifugal casting
6. Continuous pultrusion

Closed Mold

1. Matched-die molding
   a. Preform
   b. Mat
   c. Fabric
   d. Premix/molding compound

B. Injection molding

C. Continuous laminating

In open mold methods, glass fiber and resin are placed in the mold manually and worked in with squeegees or rollers to remove air. For high-quality surfaces, gel coat is sprayed on the mold prior to lay-up. Open mold processes are shown in Table 3.*

---

*Tables 3–7 abstracted with permission from Owens-Corning *Fiberglas* [4].

TABLE 3. Open Mold Process

	 CONTACT MOLDING	 VACUUM BAG	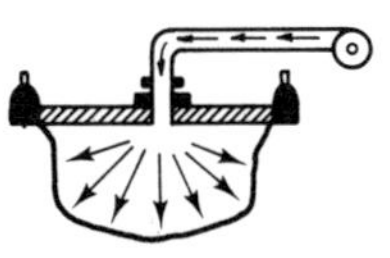 PRESSURE BAG	 AUTOCLAVE
Description	Resin is in contact with air. Lay-up normally cures at room temperature. Heat may accelerate cure. A smoother exposed side may be achieved by wiping on cellophane.	Cellophane or polyvinyl acetate is placed over lay-up. Joints are sealed with plastic; vacuum is drawn. Resultant atmospheric pressure eliminates voids and forces out entrapped air and excess resin.	Tailored bag—normally rubber sheeting—is placed against lay-up. Air or steam pressure up to 50 psi is applied between pressure plate and bag.	Modification of the pressure bag method: after lay-up, entire assembly is placed in steam autoclave at 50 to 100 psi. Additional pressure achieves higher glass loadings and improved removal of air.
Advantages	1. Simplest process. 2. Low cost molds 3. No size restrictions 4. Max. design flexibility 5. Min. equipment needed 6. Gel-coats possible	1. Higher glass loading 2. Better unfinished side 3. Less air and voids 4. Better adhesion in sandwich constructions possible	1. Higher glass loading 2. Dense, void-free moldings 3. Undercuts possible 4. Cores and inserts used	1. Undercuts possible 2. 65 per cent glass loadings 3. Dense, void-free moldings 4. Cores and inserts used
Limitations	1. Labor per unit is high 2. One finished surface 3. Quality depends on operator	1. More labor 2. Surface next to bag not as good as surface next to mold 3. Quality depends on operator	1. Only female molds 2. More labor 3. Surface next to bag not as good as surface next to mold 4. Quality depends on operator	1. Extra labor to load autoclave 2. Autoclave is expensive 3. Size of autoclave limits size of parts that can be made 4. Quality depends on operator

	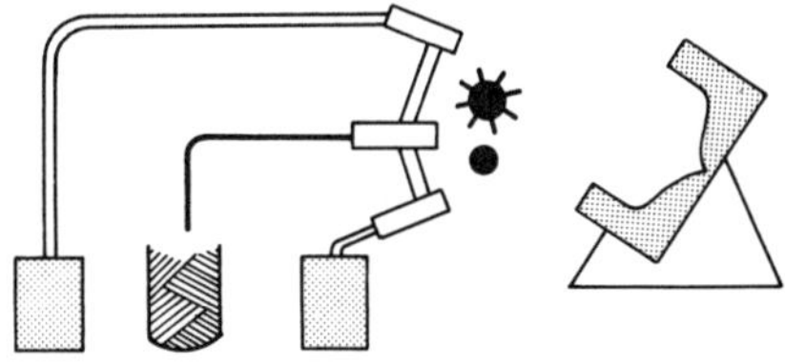 SPRAY-UP	ENCAPSULATION
Description	Roving is fed through a chopper and ejected into a resin stream, which is directed at the mold by either of two spray systems: (1) A gun carries resin premixed with catalyst, another gun carries resin premixed with accelerator. (2) Ingredients are fed into a single run mixing chamber ahead of the spray nozzle. By either method the resin mix precoats the strands and the merged spray is directed into the mold by the operator. The glass-resin mix is rolled by hand to remove air, lay down the fibers, and smooth the surface. Curing is similar to hand lay-up.	Short chopped strands are combined with catalyzed resin and pured into open molds. Cure is at room temperature. A post-cure of 30 minutes at 200°F is normal.
Advantages	1. Inexpensive equipment 2. Uses lowest cost form of reinforcement (roving). 3. Complex shapes with reverse curves can be formed with minimum waste 4. Labor for complex shapes is less than with hand lay-up molding 5. Permits on-site fabrication	1. Simple process, low tooling cost 2. Can be automated 3. High materials utilization 4. Inserts of any size, shape, material or number may be used

TABLE 3. *Continued*

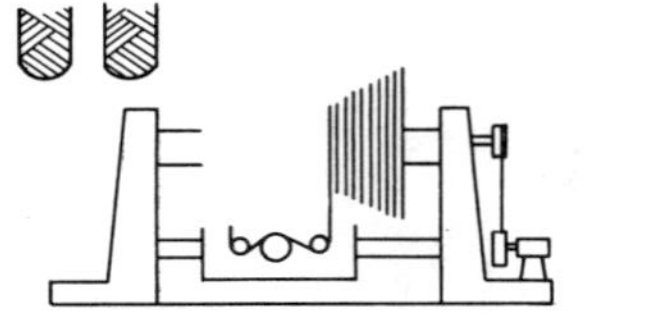 FILAMENT WINDING	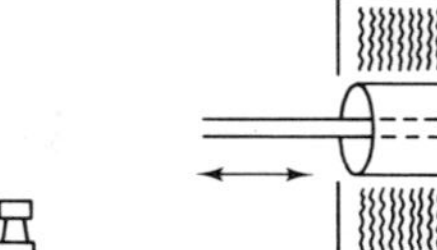 CENTRIFUGAL CASTING	CONTINUOUS PULTRUSION
Uses continuous reinforcement to achieve efficient utilization of glass fiber strength. Roving or single strands are fed from a creel through a bath of resin and wound on a mandrel. Preimpregnated roving is also used. Special lathes lay down glass in a predetermined pattern to give max. strength in the directions required. When the right number of layers have been applied, the wound mandrel is cured at room temperature or in an oven.	Round objects such as pipe can be formed using the centrifugal casting process. Chopped strand mat is positioned inside a hollow mandrel. The assembly is then placed in an oven and rotated. Resin mix is distributed uniformly throughout the glass reinforcement. Centrifugal action forces glass and resin against walls of rotating mandrel prior to and during the cure. To accelerate cure, hot air is passed through the oven.	Continuous strand—roving or other forms of reinforcement—is impregnated in a resin bath and drawn through a die which sets the shape of the stock and controls the resin content. Final cure is effected in an oven through which the stock is drawn by a suitable pulling device.
1. Highest strength to weight ratio 2. Highest degree of control over uniformity and orientation 3. Can be automated at high volumes 4. Fusible or collapsible mandrels can produce complex shapes 5. Integral closures may be wound in 6. Uses lowest cost reinforcement	1. Minimum labor involved 2. Can be automated at high volumes 3. Low tooling cost; simple mandrel 4. Gives good inside-outside surfaces 5. Low material waste 6. Uniform void-free wall thickness 7. External threads possible 8. Make cylinders up to 20-ft length	1. Continuous operation 2. Adaptable to small cross-sectional areas 3. Unidirectional high strength parts
1. Shape is restricted to surface of revolution, i.e., round, oval, taper, some squares 2. In high pressure uses (150 to 10,000 psi) inner liners are needed to prevent leakage	1. Shape limited to cylinders without taper and with uniform thickness 2. High equipment cost	Usually limited to cross-sections from $\frac{1}{8}$ to 6 inch diameter

TABLE 4. Preforming Methods

			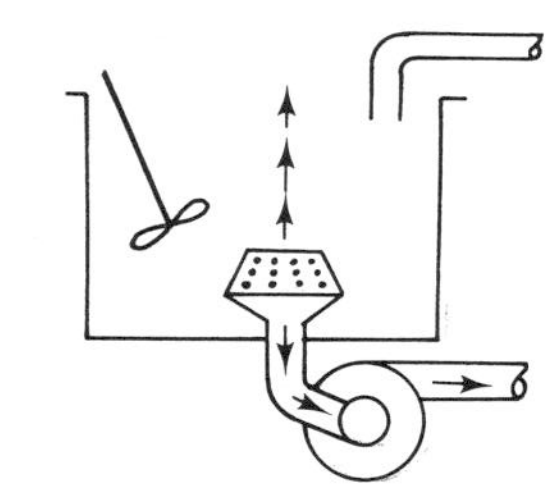
	DIRECTED FIBER	PLENUM CHAMBER	WATER SLURRY
Description	Roving is cut into 1 to 2 inch lengths of chopped strand which are blown through a flexible hose onto a rotating preform screen. Suction holds them in place while a binder is sprayed on the preform and cured in an oven. The operator controls both deposition of chopped strands and binder.	Roving is fed into a cutter on top of plenum chamber. Chopped strands are directed onto a spinning fiber distributor to separate chopped strands and distribute strands uniformly in plenum chamber. Falling strands are sucked onto preform screen. Resinous binder is sprayed on. Preform is positioned in a curing oven. New screen is indexed in plenum chamber for repeat cycle.	Chopped strands are pre-impregnated with pigmented polyester resin and blended with cellulosic fiber in a water slurry. Water is exhausted through a contoured, perforated screen and glass fibers and cellulosic material are deposited on the surface. The wet preform is transferred to an oven where hot air is sucked through the preform. When dry, the preform is sufficiently strong to be handled and molded.

In closed mold processes, mat or fabric is tailored for the mold by textile cutting techniques. When the product shape is too complex to use mat or fabric, a *preform*, or mat of chopped strands bonded together in the shape of the end product, is used. Preforming may be done in the air or in water slurry. Air methods include the directed fiber and plenum chamber processes (Table 4).

The mass production method for manufacturing glass fiber-reinforced plastics is matched die molding. Mat, fabric, or preform reinforcing is combined with a resin mix at the press just prior to or just after placing in the mold. Heated metal molds form and cure the part at 100 to 300 psi and temperatures from 225 to 300°F. Cure cycles range from less than 1 to 5 min, depending on product size and shape.

High-pressure laminating, a variation, uses molding pressures to 3,000 psi and curing temperatures to 350°F. Reinforcing mat or fabric is impregnated with resign in this way. Resins are usually phenolics, melamines, silicones, or epoxies. Table 5 shows the principal closed mold processes.

Table 6 shows the product limitations and characteristics attainable by several of the open and closed mold processes, and Table 7 lists current manufacturing cost ranges, including materials, labor, and processing but not tooling.

c. FIBER-MATRIX INTERFACE

Failure of reinforced plastics occurs usually in shear at or near the fiber-matrix boundary. Much research is therefore directed at understanding the forces at this boundary and developing methods to improve the stress transfer across it.

Regarding glass behavior near an interface, cation activity studies and surface energy considerations suggest that the chemical composition of glass fiber at its surface differs from average glass composition [5]. The more polarizable ions shift toward the surface. The remaining residual surface forces are satisfied by adsorption of water, giving SiOH, AlOH, etc., groups, and possibly by absorption of oxygen [6].

Regarding resin behavior near the interface, a variety of phenomena is either known, or suspected, to occur, including direct chemical bonding of certain resins to the glass surface, van der Waal's bonding, and entropic segregation [7]. Entropic segregation is believed to occur with the shorter chain molecules concentrating near the surface, because longer molecules suffer a relatively greater loss in entropy as they are brought toward a surface. Such segregation could enhance the short-range mobility of the resin in the vicinity of the interface and possibly lower the resin strength [7]. The central structural concept underlying all three mechanisms is the loss of conformational freedom of the polymer molecules in the vicinity of the surface, so that the interphase region develops properties that differ from bulk resins. A comprehensive survey and bibliography on glass-resin interfaces has been prepared by Eakins [8].

Surface treatments. To improve resin-fiber bonding in glass-reinforced systems, a variety of chemical surface treatments for the glass is used. The industrial terms for these surface treatments are

TABLE 5. Closed Mold Processes

	PREMIX/MOLDING COMPOUND	INJECTION MOLDING	CONTINUOUS LAMINATING
Description	Prior to molding, glass reinforcement, usually chopped spun roving, is thoroughly mixed with resin, pigment, filler, and catalyst. The premixed material can be extruded into a rope-like form for easy handling or may be used in bulk form. The premix is formed into accurately weighed charges and placed in the mold cavity under heat and pressure. Amount of pressure varies from 100 to 1500 psi. Length of cycle depends on cure temperature, resin, and wall thickness. Cure temperatures range from 225° F to 300° F. Time varies from 30 seconds to 5 minutes.	For use with thermoplastic materials. The glass and resin molding compound is introduced into a heating chamber where it softens. This mass is then injected into a mold cavity that is kept at a temperature below the softening point of the resin. The part then cools and solidifies.	Fabric or mat is passed through a resin dip and brought together between cellophane covering sheets; the lay-up is passed through a heating zone and the resin is cured. Laminate thickness and resin content are controlled by squeeze rolls as the various plies are brought together.
Advantages	1. Unlimited part and cross-sectional configuration 2. Can mold inserts and attachments 3. Low cost mass-production using low cost molding materials—140/lb and up. High performance molding compounds are usually more expensive 4. Parts weighing 200 pounds have been made 5. Molding to close tolerances possible 6. Direct labor is low 7. Molded in holes and threads	1. Automated process used on high production runs 2. Low direct labor 3. High reproducibility 4. Complex details easily molded 5. Very small, delicate precision parts possible	1. No limit to length of panels produced 2. Automated process 3. Low tooling cost 4. Wide variety of surface textures 5. Many different shapes 6. Uniform wall thickness
Limitations	1. Strengths generally lower than preform 2. Knit lines may cause weakness	1. Part size limited 2. High priced molds	1. $\frac{3}{32}$-in max. wall thickness 2. Uneconomical for short runs

TABLE 6. Product Limitations and Characteristics Attainable by Open and Closed Mold Processes

	Spray-up, Hand lay-up (contact)	Hand lay-up (Pressure Bag)	Filament Winding	Continuous Pultrusion	Matched-die molding Premix, Molding compound	Matched-die molding Preform, Mat
Minimum Inside Radius (inches)	¼	½	⅛	NA	1/32	⅛
Molded-in Holes	Large	Large	NR	NA	Yes	Yes
Trimmed-in Mold	No	No	Yes	Yes	Yes	Yes
Built-in Cores	Yes	Yes	Yes	NA	Yes	Yes
Undercuts	Yes	Yes	No	No	Yes	No
Minimum Draft Recommended (degrees)	0°	5°	2°-3°	NA	1°	1°
Minimum Practical Thickness (inches)	.060	.060	.010	.037	.060	.030
Maximum Practical Thickness (inches)	.500	1	3	1	1	.250
Normal Thickness Variation (inches)	±.020	±.020	±.010	±.005	±.002	±.008
Maximum Thickness Buildup	As Desired	As Desired	As Desired	NA	As Desired	2 to 1 Max.
Corrugated Sections	Yes	Yes	Circumferential Only	In Longitudinal Direction	Yes	Yes
Metal Inserts	Yes	Yes	Yes	No	Yes	Yes
Surfacing Mat	Yes	Yes	Yes	Yes	No	Yes
Maximum Size Part to Date (Sq. ft.)	3000	2000	1500	NA	25	200
Limiting Size Factor	Mold Size	Bag Size	Lathe Bed Length & Swing	Pull Capacity	Press cap., Flow	Press Dimens.
Metal Edge Stiffness	Yes	NR	Yes	No	Yes	Yes
Bosses	Yes	NR	No	No	Yes	Yes
Fins	Yes	Yes	No	NR	Yes	NR
Molded in Labels	Yes	Yes	Yes	Yes	No	Yes
Raised Numbers	Yes	Yes	No	No	Yes	Yes
Gel Coat Surface	Yes	Yes	Yes	No	No	Yes
Shape Limitations	None	Flexibility of Bag	Surface of Revolution	Constant Cross Section	Moldable	Moldable
Translucency	Yes	Yes	Yes	Yes	No	Yes
Finished Surfaces	One	One	One	Two	Two	Two
Strength Orientation	Random	Ply Orient.	Depend on Wind	Directional	Random	Random
Typical Glass Ldg. % by Wt.	20-30	45-65	75-90	25-70	10-35	25-40

NR—Not Recommended NA—Not Applicable

Size. A treatment applied immediately after fibers are formed.

Temporary sizes, used for producing *yarns* and *fabrics*, simply lubricate and protect strands from abrading one another until they can be spun into yarn or woven into fabric. The size is then dissolved and washed away, and a permanent finish is applied to improve resin bonding. Temporary sizes contain such ingredients as dextrinized starch gum, hydrogenated vegetable oil, cationic lubricant, or gelatin.

TABLE 7. Typical Manufacturing Costs, 1972 (Less Tooling)*

Process	*Material*	*Product cost (cents/lb)*
Matched die	Premix-polyester	30–45
Matched die	Preform or mat-polyester	42–75
Spray-up	Roving-polyester	50–100
Contact (hand lay-up)	Mat-polyester	75–200
Pultrusion	Roving-polyester	45–75
Filament winding	Roving-epoxy	55–150
Contact (pressure bag)	Fabric-epoxy	200 and up

*Tooling costs cover preform screens (where used), molds, cooling fixtures, and assembly jigs.

Compatible sizes, used for *roving*, *mat*, and *chopped strand*, are not removed and serve the dual function of protecting strands from one another and coupling with the plastic resin. Compatible sizes are categorized by their coupling agent, which is usually a silane or a chrome complex. The silanes have hydrolyzable groups, such as methoxy or ethoxy, and can be dispersed in water. The functional groups include amino, vinyl, glycidoxy, and methacryloxy. It is still a moot point whether the molecular structures do in fact react with glass on one end and resin on the other. But in practice, the effect of "bridging" between resin and glass is obtained by using these agents [9]. Chrome sizes can reduce the static electricity on glass which causes processing problems in such operations as preforming and spray-up.

Binder. A treatment to hold strands of fiberglass together in *mat* or *preform* during manufacturing of the molded object. High- and low-solubility binders are available for varying needs. Polyvinyl acetate is commonly used because of its good adhesive properties and compatibility with a variety of resins.

Finish (or *coupling* agent, or *keying* agent). A treatment applied after fibers are fabricated into *yarn* or *woven fabric* and cleaned, to make the yarn or fabric compatible with molding resins.

As with compatible sizes, keying agents contain an organic group which, at least theoretically, can react with the resin during its polymerization and alkoxide groups or chloride ions which can react with OH groups in the glass surface. Examples are [5]

Formula	Name
$CH_2{=}CH{-}Si(OC_2H_5)_3$	vinyltriethoxysilane
$CH_2{=}\underset{\displaystyle CH_3}{\underset{\vert}{C}}H{-}COO{-}(CH_2)_3{-}Si(OCH_3)_3$	γ-methacryloxypropyl-trimethoxysilane
$CH_2{=}CH{-}SiCl_3$	vinyltrichlorosilane
$CH_2{=}\underset{\displaystyle CH_3}{\underset{\vert}{C}}{-}COOCr_2(OH)Cl_4$	methacrylatochromic chloride (suggested formula)
$NH_2{-}(CH_2)_3{-}Si(OC_2H_5)_3$	γ-aminopropyltriethoxysilane

In most cases, keying agents are applied to glass fibers from aqueous solutions. The silane part of organosilanes is believed to react with Si—OH groups on the glass fiber surface to form Si—O—Si bonds. The organic part of the silane is expected to react with or be compatible with the specific functionality of the resin being used. For example, the silane used with unsaturated polyesters would have *double bond* functionality, silanes with epoxies would have amino or hydroxy functionality, etc.

Keying agents have been found to be deposited on fibers as discrete globules on the order of a micron apart, rather than as continuous films, leaving substantial portions of glass potentially exposed to attack by corrosive solutions, such as water and acids on E glass [5].

We shall now obtain some of the useful relationships which govern fiber-reinforced composite behavior [10].

2. Fiber-Matrix Mechanics

a. FIBER-MATRIX STRESS TRANSFER

Consider a tensile test in which a rigid cylindrical fiber is pulled from a soft matrix [Fig. 4(a)]. Equating axial forces and neglecting the adhesion force at the fiber end,

$$P_x = \pi r^2 \sigma_x = \int_0^x 2\pi r \tau_x dx \tag{1}$$

or

$$\sigma_x = \frac{2}{r} \int_0^x \tau_x \, dx \tag{1a}$$

where P_x = fiber tensile force at distance x from fiber end,
r = fiber radius,
σ_x = fiber stress at distance x,
τ_x = interfacial shear stress at x.

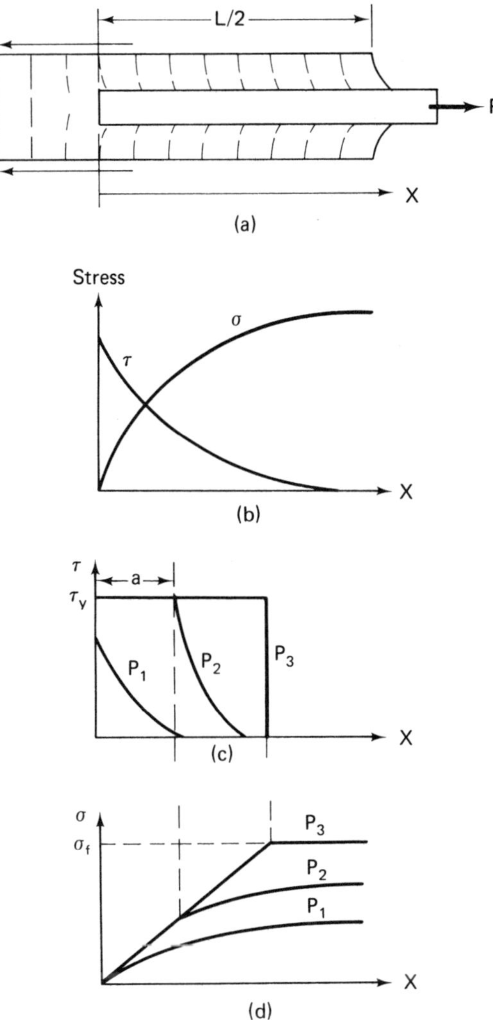

Fig. 4. Fiber-matrix stress transfer.

Typical variations of fiber stress σ and shear stress τ when both fiber and matrix deform elastically are shown in Fig. 4(b). τ is maximum at the fiber end and minimum at the fiber center, while σ is maximum at the center. The larger the ratio of matrix shear modulus to fiber Young's modulus, G_M/E_f, the more rapidly the fiber stress increases with distance from the fiber end. Specific formulations of Eq. (1) usually neglect the stress concentrating effects of fiber ends and the effect on the stress in one fiber of a nearby fiber end [2, 11]. Shear stress will actually be concentrated at the fiber end in proportion to the difference in elastic moduli of the fiber and matrix and inversely to the square root of the minimum radius of curvature of the fiber end [1].

The matrix near fiber ends will usually begin to fail at stress levels much below the ultimate strength of the composite. A metal matrix will flow plastically; a polymer matrix usually fails in shear at the matrix-fiber interface.

In the case of metal matrices, the yield strain of the matrix is usually lower than the failure strain of the fiber. As load is applied to the composite, differential displacements between matrix and fiber produce shear stresses in the matrix near the fiber ends. At small load P_1 [Fig. 4(c)], τ is governed by elastic behavior of the matrix. At greater load, plastic flow occurs, and the matrix shear stress reaches a limiting value $\tau_y = \sigma_y/2$, where the subscript y designates matrix yield stresses. Figures 4(c) and 4(d) show the stress conditions after the matrix yields but before maximum fiber stress, σ_f, is reached (curves P_2) and when maximum fiber stress is reached (curves P_3). At load P_2, for example, the matrix deforms plastically in the range $0 < x < a$, resulting in constant τ in this range and therefore a linear increase in fiber stress σ, in accordance with Eq. (1),

$$\sigma_x = \frac{2\tau_y x}{r} \qquad 0 < x < a \tag{2}$$

The matrix may also yield plastically in the range $x > a$, but the difference between longitudinal displacements in matrix and fiber are small enough to produce a shear stress at the interface less than τ_y. Increasing the load P increases plastic flow length a and fiber stress σ.

If the fiber failure strain is much greater than the matrix yield strain, we can neglect the fiber stress increase in the region $x > a$ and say that if the value of σ_x in Eq. (2) reaches σ_f before a equals $L/2$ [Fig. 4(a)] the fiber will break somewhere in the range $(L/2) - a$. If a reaches $L/2$ before the fiber stress reaches σ_f, composite failure will occur by plastic flow of the matrix. There is thus a *critical fiber length*,

$$l_c = \frac{r\sigma_f}{\tau_y} \tag{3}$$

which must be exceeded if the fiber is to be broken, or, alternatively, a critical aspect ratio,

$$\frac{l_c}{d} = \frac{\sigma_f}{2\tau_y} \tag{3a}$$

where d is the fiber diameter. If this aspect ratio is not exceeded, the matrix will merely continue to flow and load the fiber to a maximum stress,

$$\sigma = \frac{2\tau_y L}{d}$$

at its midpoint. Polymer matrices, which fail in shear between fiber and matrix rather than plastic flow, yield stress diagrams similar to those of metals and will be considered shortly.

b. DEFORMATION

Tensile stress-strain curves parallel to the fiber axis are quite similar for both continuous* and discontinuous fibers, provided the discontinuous fibers exceed the critical length. There may be up to four regions of the stress-strain curve [1]: (1) Matrix and fibers both elongate elastically; (2) matrix flows plastically, and fibers extend elastically; (3) both matrix and fibers deform plastically; and (4) the fibers fail. Fiber failure may lead either to immediate composite failure or to continuing plastic flow of the matrix containing broken fibers, at a reduced stress level.

In region (1), Young's modulus of a *continuous* fiber composite is approximately

$$E_c = E_f v_f + E_m v_m \tag{4}$$

where E is Young's modulus, v is volume fraction, and the subscripts c, f, and m designate composite, fiber, and matrix, respectively. This mixture law applies only when the Poisson ratios of the two phases are equal. For other cases it is actually a theoretical lower bound, since for a unit strain ϵ under uniaxial load, the deformation energy of a unit volume, $\frac{1}{2}E_c\epsilon^2$, exceeds that of the sum of fibers and matrix, free of any mutual interaction, $\frac{1}{2}(E_f v_f + E_m v_m)\epsilon^2$ [12]. Exact calculations of E_c are difficult and have been considered by Hashin and Rosen [13] and Hill [12]. Since the elastic modulus of the fibers is often greater than 10^7 psi and that of a strong polymer is less than 5×10^5 psi, E_c is usually close to $E_f v_f$.

In region (2) Young's modulus of the composite at failure is

$$E_c = E_f v_f + \left(\frac{d\sigma_m}{d\epsilon_m}\right)_s v_m$$

where $(d\sigma_m/d\epsilon_m)_s$ is the stress-strain curve slope of the matrix at the failure strain of the fibers. The approximation $E_c = E_f v_f$ is again valid, for the same reason. In region (2) the composite behavior is quasi-elastic. When tensile load is removed, the specimen returns almost to its initial length, at first with both components contracting elastically and finally with the matrix deforming plastically in compression [1]. A small permanent set remains, with the fibers in tension and the matrix in compression.

Region (3) is, of course, absent for composites containing brittle fibers. Metal wires are the only high-strength fibers which flow plastically at room temperature.

When a composite contains *discontinuous* fibers, the fiber contribution to the

*Fibers which extend the full length of test specimen.

modulus is less than $E_f v_f$ because fiber stress builds up from the ends. The average stress is $(1/L)\int_0^L \sigma_x\,dx$. If τ is constant, as represented by Eq. (2), the stress builds up linearly in a distance,

$$a = E_f \frac{\epsilon r}{2\tau y} \tag{5}$$

and the average stress $\bar{\sigma}_f$ in a fiber stretched to the breaking strain over its midsection (Fig. 5) is

$$\begin{aligned}\bar{\sigma}_f &= \frac{\sigma_f}{L}(L - l_c) + \frac{\sigma_f}{L}\frac{l_c}{2} \\ &= \sigma_f\left(1 - \frac{l_c}{2L}\right)\end{aligned} \tag{6}$$

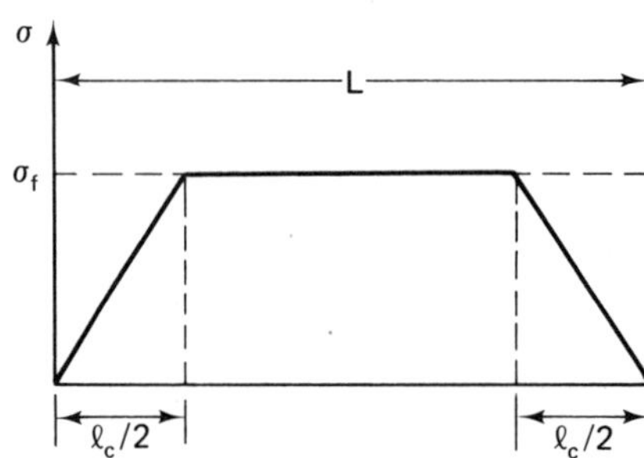

Fig. 5. Linear stress distribution along a fiber, length L.

The composite Young's modulus will be less than for a continuous fiber composite because of the load reduction at the fiber ends. The ratio of modulus of a discontinuous composite E_{dc} to that of a continuous one E_{cc} is

$$\frac{E_{dc}}{E_{cc}} = \frac{1}{E_f \epsilon L}\int_0^L \sigma\,dx \tag{7}$$

Here the contribution of the matrix, E_m, is neglected, and the integral is evaluated at the same strain ϵ for both composites. From Fig. 5 and Eq. (6),

$$E_{dc} = E_{cc}\left(1 - \frac{r}{2\tau_y L}E_f\epsilon\right) \tag{7a}$$

and, therefore, the modulus of a discontinuous composite decreases with increasing strain.

When the matrix is a *polymer*, the matrix stress-strain curve is nonlinear, and there is no sharp division between elastic and plastic behavior, as with metals. τ is therefore not constant, σ_x builds up nonlinearly, and the average stress may be written

$$\bar{\sigma}_f = \sigma_f\left[1 - (1 - \beta)\frac{l_c}{L}\right] \tag{8}$$

where $\beta\sigma_f$ is the average fiber stress within a distance $l_c/2$ from either end. Particular solutions have been obtained by Cox [14] and Dow [11]. Both obtained

$$P = E_f A_f \epsilon\left[1 - \frac{\cosh\beta(L/2 - x)}{\cosh\beta(L/2)}\right] \qquad 0 < x < \frac{L}{2}$$

where $E_f A_f \epsilon$ is the load borne by the fiber subject to strain ϵ, equal to that of the composite. The value of β depends on the elastic constants of matrix and fibers and on the fiber diameter and fiber spacing. The solutions are not exact in that stress concentrations due to the fiber end shapes and fiber-fiber interactions are not considered.

Shear failure of polymer matrices near the fiber ends can often be expected after the composite has undergone only a small portion of its ultimate strain [1]. Further straining of the composite increases the fiber length over which matrix shear failure occurs and hence decreases the composite modulus with increasing strain. Outwater [15] visualized this phenomenon as a shear zone traveling in from the fiber end, detaching matrix from fiber similar to the detachment of a rubber sheet from a rigid spot of glue as the sheet is stretched. The rubber sheet detaches from the edge toward the center of the glue spot without significantly altering the strain pattern in the rubber. There is experimental evidence that this may occur; for example, the elastic modulus of some discontinuous glass fiber composites permanently decreases after small strains. The stress buildup at the fiber ends is believed to be linear, as for a plastic matrix, but the limiting shear stress is determined by sliding friction between fiber and matrix, rather than by plastic yield in the matrix. Neglecting any shear in the matrix, Eq. (2) is then written

$$\sigma_x = \frac{2\mu p x}{r} \tag{9}$$

where μ is the coefficient of sliding friction and p is the radial pressure between fiber and matrix. The modulus and critical aspect ratio for discontinuous fibers in polymer matrices can be obtained by replacing τ_y in Eqs. (3) and (7) by μp,

$$\frac{l_c}{d} = \frac{\sigma_f}{2\mu p} \tag{10}$$

and

$$E_{dc} = E_{cc}\left(1 - \frac{r}{2\mu p L} E_f \epsilon\right) \tag{11}$$

The dependence on μ in Eqs. (9)–(11) implies a sensitivity to chemical environment, and particularly to moisture, at the fiber-matrix interface. Most polymer matrices shrink on curing, although the volume change of the composite may be slight at small volume fractions of the matrix. Outwater proposed that the radial pressure due to shrinkage would be approximately that exerted by a thin-walled tube of thickness $t/2$, internal diameter d, and hoop stress σ_y, i.e.,

$$p = \frac{t\sigma_y}{d}$$

where t is the minimum distance between adjacent fiber surfaces, d their diameter, and σ_y the tensile yield stress of the matrix. Interfacial pressures of $\sim$ 700 psi have been measured for glass in a polyester [16]. Taking $\mu = 0.4$, μp is then 280 psi, much less than the limiting shear value of work-hardened metal matrices ($\sim$ 20,000 psi). Critical aspect ratios are thus predicted to be one to two orders of magnitude greater for polymers than for metal matrices [1, 17]. Although the load transfer is formally similar to the case of metals, an essential difference is that polymers do not flow plastically around the ends of fibers but break away and slide over them, leaving voids at the fiber ends. Polymer matrices might also be expected to perform poorly in fatigue, because the sliding friction between polymer and fibers under cyclic loading would generate heat, and polymers are poor heat conductors.

c. STRENGTH

The tensile strength of a composite containing more than a certain volume fraction, $v_{\min}$, of *continuous* fibers is given by the simple mixture law [18]

$$\sigma_c = \sigma_f v_f + \sigma'_m(1 - v_f), \qquad v_f > v_{\min} \tag{12}$$

where σ_f is the fiber tensile strength and σ'_m the stress borne by the matrix at the ultimate tensile strain of the fibers. σ'_m is approximately the value determined from a tensile test on the matrix material alone. Since very strong fibers strain elastically to failure, the failure strain is quite small, and σ'_m is very small with respect to σ_f. The validity of Eq. (12) has been established for a variety of fiber-reinforced composites [19, 20, 21]. For brittle circular fibers the experimental strengths decrease above about $v_f = 80\%$ due to fiber-fiber contact, although the upper limit for v_f is $\pi/2\sqrt{3} = 89.8\%$ [1].

If the matrix has a larger failure strain than the fibers and can also work-harden, the composite will have greater strength than the matrix alone only if

$$\sigma_c > \sigma_u \tag{13}$$

where σ_u is the tensile strength of the matrix. From Eqs. (12) and (13), the *critical volume fraction*, v_{crit}, which must be exceeded for fiber strengthening, is

$$v_{crit} = \frac{\sigma_u - \sigma'_m}{\sigma_f - \sigma'_m} \simeq \frac{\sigma_w}{\sigma_f} \tag{14}$$

where $\sigma_w = \sigma_u - \sigma'_m$, the stress due to work hardening of the matrix after the fibers fail. When the work-hardening stress σ_w is small, v_{crit} is small and reinforcement will be obtained even at very low fiber concentrations.

Also, if the matrix can work-harden, the strength will be given by Eq. (12) only at high enough fiber concentrations that

$$\sigma_c \geq \sigma_u(1 - v_f) \tag{15}$$

and Eq. (12) applies only if $v_f > v_{min}$, where

$$v_{min} = \frac{\sigma_u - \sigma'_m}{\sigma_f + (\sigma_u - \sigma'_m)} = \frac{\sigma_w}{\sigma_f + \sigma_w} \tag{16}$$

For $v_f < v_{min}$ the strength is $\sigma_c = \sigma_u(1 - v_f)$. (17)

v_{crit} is always greater than v_{min}, and their relationship is shown in the stress-strain diagrams [Fig. 6(a)] and the composite strength-fiber volume diagrams [Fig. 6(b)]. Ductile fibers, such as metals, fail by necking. If, as the fibers elongate, the increase in stress carried by the matrix exceeds the reduction in stress carried by the fibers due to necking, the composite strength is increased even at very small fiber volume fractions. This is indicated by the curve for ductile fibers in Fig. 6(b) [1, 22].

Fig. 6. (a) Stress-strain diagrams for composite, and for matrix only; (b) strength-volume fraction diagrams for ductile and brittle continuous fibers.

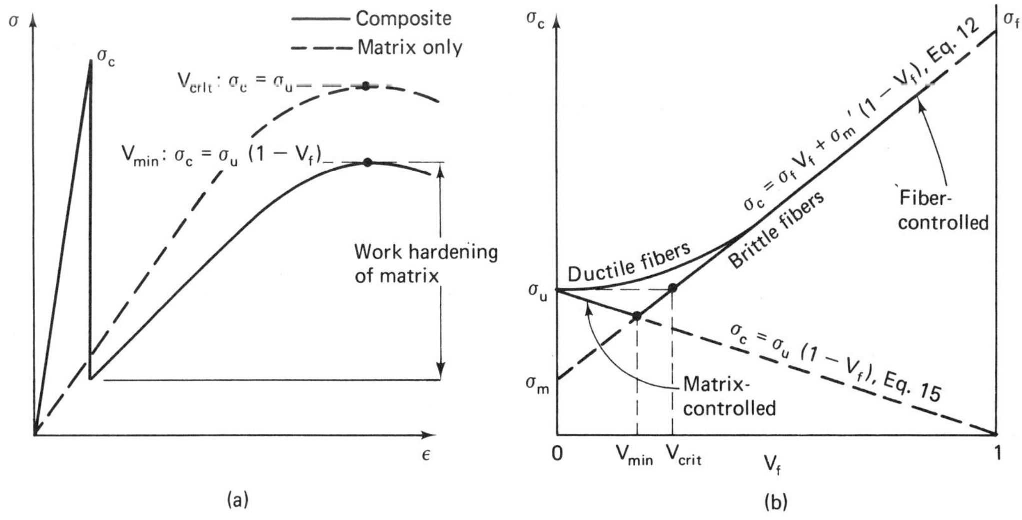

In *discontinuous* fibers the tensile stress builds up from the fiber ends, and if the fiber distribution is truly random, the composite strength is given by replacing σ_f in Eq. (12) by the average fiber stress $\bar{\sigma}_f$,

$$\sigma_c = \bar{\sigma}_f v_f + \sigma'_m(1 - v_f), \qquad v_f > v_{\min} \tag{18}$$

where

$$\bar{\sigma}_f = \sigma_f\left(1 - \frac{l_c}{2L}\right) \tag{6}$$

if the stress buildup from the fiber ends is linear (see Fig. 5), or

$$\bar{\sigma}_f = \sigma_f\left[1 - (1 - \beta)\frac{lc}{L}\right] \tag{8}$$

if nonlinear, where $\beta\sigma_f$ is the average fiber stress in the distance $l_c/2$ from either end. Substituting Eq. (8) into (18) and letting $\alpha = L/l_c$,

$$\sigma_c = \sigma_f v_f\left(1 - \frac{1 - \beta}{\alpha}\right) + \sigma'_m(1 - v_f), \qquad v_f > v_{\min} \tag{19}$$

The validity of Eq. (19) has been established experimentally for various fiber composites up to about $v_f = 0.6$, beyond which fiber-fiber interaction becomes noticeable [1, 22].

Comparison of Eqs. (12) and (19) shows that discontinuous fibers strengthen a composite less than continuous ones. Assuming $\beta = 0.5$, the ratio of strengths is

$$\frac{\sigma_{c_{\text{disc}}}}{\sigma_{c_{\text{cont}}}} = 1 - \frac{1}{2\alpha[1 + (\sigma'_m/\sigma_f)(1/v_f - 1)]}$$

Figure 7 shows this ratio as a function of α for the limiting value $v_f = 1$. For $v_f < 1$, the strength ratio is greater, so that feasible ratios occur in the crosshatched portion of the figure. As shown, for $\alpha > \sim 5$ discontinuous fiber composites are almost as strong as those containing continuous fibers.

The critical volume fraction for increased strength with discontinuous fibers can be obtained by substituting $\bar{\sigma}_f$ from Eq. (8) for σ_f in Eq. (14):

$$v_{\text{crit}} = \frac{\sigma_u - \sigma'_m}{\sigma_f[1 - (1 - \beta)/\alpha] - \sigma'_m}; \qquad \alpha \geq 1 \tag{20}$$

Equation (19) gives the composite strength only if the fiber volume exceeds a value $v_{\min}$, which can be derived in a way similar to the corresponding value for continuous fiber composites. In this case, *immediate* failure of the composite occurs if

$$\sigma_c \geq \sigma_u(1 - v_f) + \frac{\beta}{\alpha}\sigma_f v_f \tag{21}$$

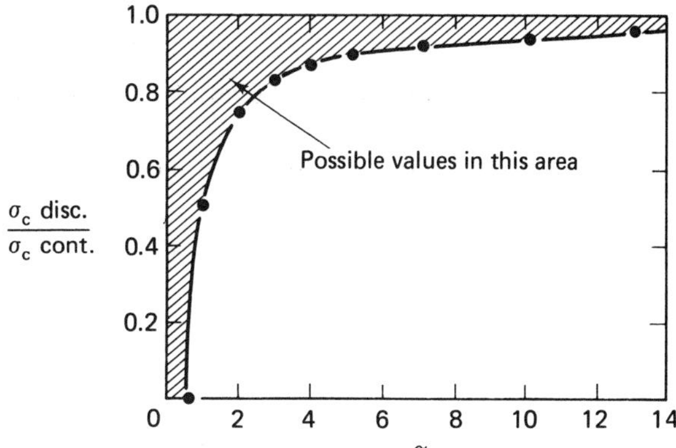

Fig. 7. Strength ratio between composites containing discontinuous and continuous fibers. From Kelley, A., and G. R. Davies. Metal. Rev., 10 (1965) 1 [1].

where the last term accounts for the fact that fibers which have ends within $l_c/2$ of the cross section at which the first fiber fails will slip out rather than break. The proportion of these fibers is $1/\alpha$. From Eqs. (19) and (21),

$$\sigma_f v_f\left(1 - \frac{1-\beta}{\alpha}\right) + \sigma'_m(1 - v_f) > \sigma_u(1 - v_f) + \frac{\beta}{\alpha}\sigma_f v_f$$

from which

$$v_{\min} = \frac{\sigma_w}{\sigma_f(1 - 1/\alpha) + \sigma_w} \tag{22}$$

At volume fractions less than $v_{\min}$ composite strength is given by the equality in Eq. (21).

For discontinuous fibers of length $L < l_c$, $v_{\min}$ [Eq. (22)] cannot be reached, and all failure will be due to matrix flow. In this case,

$$\sigma_c = \sigma_u(1 - v_f) + \bar{\sigma}'_f v_f$$

where $\bar{\sigma}'_f$ is the average fiber stress. If the fiber stress builds up linearly,

$$\sigma_c = \sigma_u(1 - v_f) + \frac{\tau L}{d} v_f$$

where $\tau = \tau_y$ for metal matrices and $\tau = \mu p$, according to Outwater, for polymer matrices.

Values of l_c, for use in design and for computing the work of fracture, can be experimentally determined by several methods. In metallic matrices where transfer length is determined by the shear strength of the matrix rather than of the interface, values of l_c/d are given within a factor of 2 simply by Eq. (3). A value of l_c/d can also be found by observing the fracture surface of a composite containing $v_f > v_{\text{crit}}$ of uniform fibers. All fibers having ends within a distance $l_c/2$ of the fracture surface will pull out of the matrix rather than break. The critical length can be found from

$$\frac{n_f}{n_p} = \frac{L}{l_c} - 1 \tag{23}$$

where n_f and n_p are the numbers of fibers which fractured and pulled out, respectively [10, 22].

Alternatively, from Eq. (19),

$$\frac{d\sigma_c}{dv_f} = \sigma_f\left(1 - \frac{1-\beta}{\alpha}\right) - \sigma'_m$$

$$= (\sigma_f - \sigma'_m) - \sigma_f\left(\frac{1-\beta}{\alpha}\right); \qquad L > l_c$$

Assuming $\beta = \frac{1}{2}$,

$$\frac{d\sigma_c}{dv_f} = (\sigma_f - \sigma'_m) - \frac{\sigma_f}{2} \cdot \frac{l_c}{L}; \qquad L > l_c$$

or

$$l_c = \frac{2L(\sigma_f - \sigma_{m'} - d\sigma_c/dv_f)}{\sigma_f} \tag{24}$$

The value of l_c may be computed from measured values of σ_c for various volume fractions v_f of a given fiber length, L.

EXAMPLE 1: From the following data, determine the critical fiber length:

$$\sigma_f = 6 \times 10^5 \text{ psi}, \qquad \sigma_{m'} = 18 \times 10^3 \text{ psi}, \qquad L = 0.5 \text{ in.}$$

v_f	*Measured* σ_c
0.40	60×10^3 psi
0.45	80×10^3 psi

Solution: $d\sigma_c/dv_f \sim \Delta\sigma_c/\Delta v_f = (80 - 60)10^3/(0.45 - 0.40) = 4 \times 10^5$.
$l_c = 2(0.5)(6 \times 10^5 - 18 \times 10^3 - 4 \times 10^5)/6 \times 10^5 \sim 0.33 \text{ in.} < 0.5 \text{ in.}$ **OK.**

In general, some compromise between strength and toughness will be required for specific applications. High strength requires high v_f and short transfer length. But high toughness requires large l_c, and L should be just less than l_c to ensure fiber slipping rather than fiber fracture. Large values of l_c can be obtained by making either matrix shear strength or interfacial shear strength as low as possible.

d. COMPRESSIVE FAILURE

Compressive failure has been studied by Rosen [2] and Scheurch [23]. Both considered failure of a composite consisting of a set of parallel plates under edge loading, analyzed two limiting modes of buckling failure by energy methods, and obtained similar results. Compressive strengths were found to be governed by instability modes analogous to the buckling of a column on an elastic foundation. The solution outlined here follows that of Rosen [2].

Consider parallel fibers represented as plates of thickness h and length L, each subjected to compressive load P and separated by matrix of dimension $2c$. Two limiting buckling modes may be visualized: Adjacent fibers buckle in opposite directions [Fig. 8(a)] or in phase [Fig. 8(b)]. The first mode is called the extension mode, because the matrix material extends or compresses in the direction normal to the fibers. The second mode is the shear mode, because matrix deformation is primarily in shear on planes normal to the fibers. The fibers are relatively stiff compared with the binder, and fiber shear deformation is neglected.

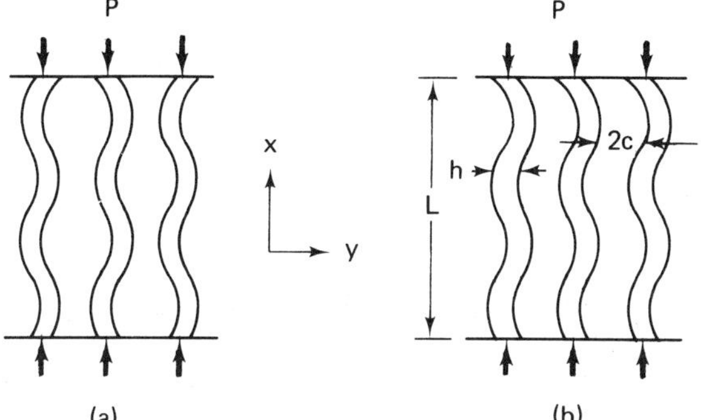

Fig. 8. Buckling failure of a fiber composite. (a) Extension mode; (b) shear-mode. From Rosen, B. W., *Fiber Composite Materials.* Metals Park, Ohio: American Society for Metals, 1965 [2].

The condition for instability is given by equating the internal strain energy for the fiber V_f and binder V_m to the work W done by the external loads during buckling,

$$V_f + V_m = W \tag{25}$$

The procedure is to compute the stresses corresponding to the two limiting buckling modes and to take the lower as the critical buckling stress. Any combination of the two modes will have a buckling stress larger than the smaller of the individual modes.

Assume each fiber to buckle into a sinusoidal pattern expressed by the following series in Ψ, the displacement in the y direction:

$$\psi = \sum_n a_n \sin\frac{n\pi x}{L} \tag{26}$$

For the extension mode the transverse strain in the binder is independent of y, so that

$$\epsilon_y = \frac{\psi}{c} \tag{27}$$

and

$$\sigma_y = \frac{E_m \psi}{c} \tag{28}$$

Assuming axial and shear strain energy to be negligible compared with transverse strain energy,

$$V_m = \frac{1}{2}\int_v \sigma_y \epsilon_y \, dv \tag{29}$$

where v is the matrix volume. Substituting Eqs. (26)–(28) into Eq. (29),

$$V_m = \frac{E_f L}{2c} \sum_n a_n^2 \tag{30}$$

and for a beam on elastic foundation [24],

$$V_f = \left(\pi^4 \frac{E_f h^3}{48L^3}\right) \sum_n n^4 a_n^2 \tag{31}$$

and

$$W = \frac{P\pi^2}{4L} \sum_n n^2 a_n^2 \tag{32}$$

Substituting Eqs. (30)–(32) into Eq. (25),

$$p = \frac{\pi^2 E_f h^3}{12L^2} \cdot \frac{\sum_n n^4 a_n^2 + (24L^4 E_m/\pi^4 ch^3 E_f) \sum_n a_n^2}{\sum_n n^2 a_n^2} \tag{33}$$

Equation (33) is minimum for some particular value of h, say m; hence the critical fiber stress is

$$\sigma_{f_{\text{crit}}} = \frac{\pi^2 E_f h^2}{12L^2}\left[m^2 + \frac{24L^4 E_m}{\pi^2 ch^3 E_f}\left(\frac{1}{m^2}\right)\right] \tag{34}$$

Since m is large, it can be treated as a continuous variable, and Eq. (34) is minimized by setting $\partial\sigma_{f_{\text{crit}}}/\partial(\mathrm{m}^2) = 0$, which yields

$$\sigma_{f_{\text{crit}}} = 2\left[\frac{RE_m E_f}{3(1-R)}\right]^{1/2} \tag{35}$$

where

$$R = \frac{h}{h + 2c}$$

The composite stress is

$$\begin{aligned} \sigma_c &= R\sigma_{f_{\text{crit}}} \\ &= 2R\left[\frac{RE_m E_f}{3(1-R)}\right]^{1/2} \end{aligned} \tag{36}$$

and the strain at critical stress can be obtained from Eq. (35),

$$\epsilon_{\text{crit}} = \frac{\sigma_{f_{\text{crit}}}}{E_f} = 2\left[\frac{R}{3(1-R)}\right]^{1/2}\left(\frac{E_m}{E_f}\right)^{1/2}$$

The *shear instability* mode [Fig. 8(b)] is evaluated similarly. In the binder,

$$\gamma_{xy} = \frac{\partial u_y}{\partial_x} + \frac{\partial u_x}{\partial_y} \tag{37}$$

Since the transverse displacement is independent of the y coordinate,

$$\left.\frac{\partial u_y}{\partial_x}\right|_b = \left.\frac{du_y}{d_x}\right|_f \tag{38}$$

Since the shear strain is independent of y,

$$\frac{du_x}{d_y} = (\tfrac{1}{2}c)[u_x(c) - u_x(-c)] \tag{39}$$

Since the fiber shear deformation is negligible,

$$u_x(c) = \frac{h}{2}\left.\frac{du_y}{dx}\right|_f \tag{40}$$

Substituting Eq. (39) into (38),

$$\frac{du_x}{d_y} = \frac{h}{2c}\left.\frac{du_y}{dx}\right|_f \tag{41}$$

Substituting Eqs. (41) and (38) into Eq. (37),

$$\gamma_{xy} = \left[1 + \left(\frac{h}{2c}\right)\right]\left.\frac{du_y}{dx}\right|_f \tag{42}$$

and

$$\tau_{xy} = G_m \gamma_{xy} \tag{43}$$

Assuming axial strain energy to be negligible compared with transverse shear strain energy,

$$V_m = \frac{1}{2}\int_v \tau_{xy}\gamma_{xy}\, dv \tag{44}$$

Substituting Eqs. (26), (42), and (43) into Eq. (44),

$$V_m = G_m c\left(1 + \frac{h}{2c}\right)^2 \left(\frac{\pi^2}{2L}\right) \sum_n a_n^2 n^2 \tag{45}$$

Using Eq. (45) in place of Eq. (30) in Eq. (25) and proceeding as for the extension mode,

$$\sigma_{f_{crit}} = \frac{G_m}{R(1-R)} + \frac{\pi^2 E_f}{12}\left(\frac{mh}{L}\right)^2 \tag{46}$$

Since L/m is the buckle wavelength, the second term in Eq. (46) is small for wavelengths large compared with the fiber diameter, and the buckling stress is approximately

$$\sigma_{f_{\text{crit}}} \frac{G_m}{R(1-R)}$$

and the composite stress and critical strain are

$$\sigma_c = \frac{G_m}{1-R} \tag{47}$$

and

$$\epsilon_{\text{crit}} = \frac{1}{R(1-R)} \frac{G_m}{E_f}$$

The lower of the values given by Eqs. (36) and (47) is the better estimate of compressive strength. Equations (36) and (47) are shown in Fig. 9 for E glass fibers in an epoxy matrix. At low fiber volume fraction the extension mode is the limiting stress,

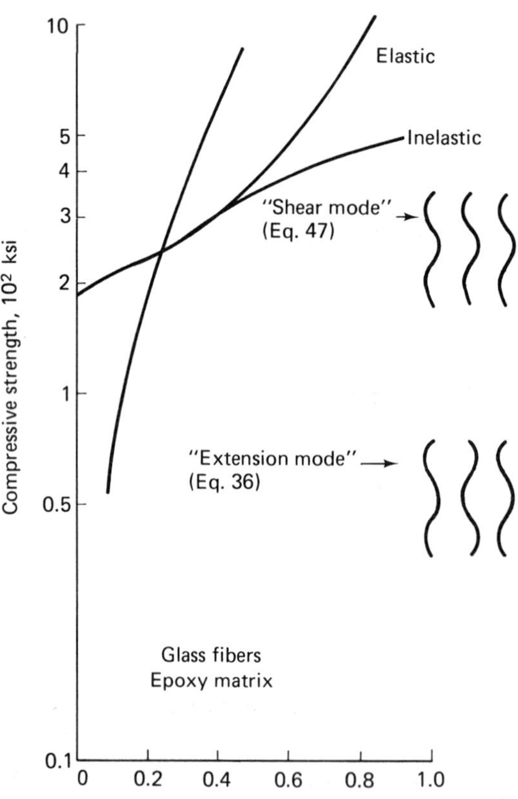

Fig. 9. Compressive strength of glass reinforced epoxy composites. Reproduced by permission from Rosen, B. W., *Fiber Composite Materials.* Metals Park, Ohio: American Society for Metals, 1965 [2].

while at high volume fraction the shear mode limits. To achieve a strength of 500 ksi of this composite would require axial strains greater than 5%, or beyond the proportional limit for shear stiffness of the binder. The approximate "inelastic" branch was therefore added by replacing the binder modulus in Eq. (47) by a modulus which varies linearly from the elastic value of the epoxy at 1% strain to zero at 5% strain [2]. The nature of failure modes and magnitudes of failure stresses have been supported qualitatively by experimental results [2]. It is apparent that at high fiber volume fractions the matrix shear stiffness G_m is the limiting factor in compressive strength.

e. ORIENTATION EFFECTS

(1) Strength; uniaxial fibers. When a tensile load is applied at an angle ϕ [Fig. 10(a)] to the direction of fiber alignment, three stresses are important [10]: σ_c, required for tensile failure of the fibers and matrix and given by either Eq. (12) or (19); τ_u, for shear failure of the matrix or of the fiber-matrix interface; and σ_u, for tensile failure normal to the fibers. σ_u will depend on either the tensile strength of the matrix, in which case

$$\sigma_u = \sigma_{\text{ult}}(1 - v_f)$$

where σ_{ult} is the ultimate tensile strength of the matrix, or on the tensile strength of the fiber-matrix interface, in which case

$$\sigma_u = \sigma_i v_f$$

where σ_i is the tensile strength of the fiber-matrix interface.

Failure of the composite by fiber fracture [Fig. 10(b)] requires a tensile stress,

$$\sigma = \sigma_c \sec^2 \phi \tag{48}$$

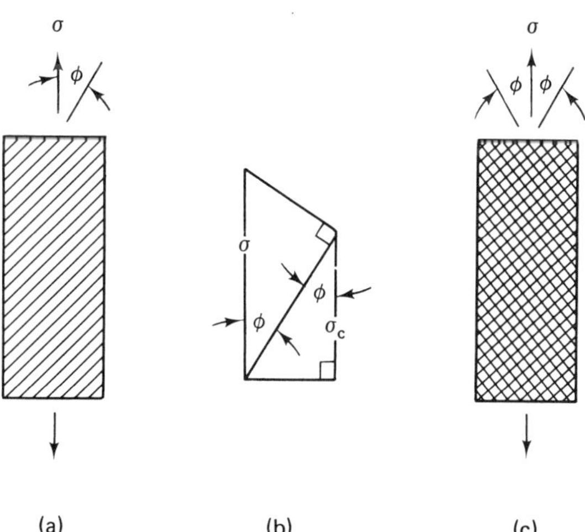

Fig. 10. Orientation effects. (a) Single fiber orientation; (b) stress diagram for failure by fiber fracture; (c) double orientation.

Failure by shear parallel to the fibers requires a tensile stress,

$$\sigma = 2\tau_u \operatorname{cosec} 2\phi \tag{49}$$

By analogy to thin brazed joints, the appropriate value of τ_u will be greater than the value measured in a bulk specimen of the matrix, due to interference from the fibers. Metallic matrices undergo considerable failure strain in shear, during which ϕ decreases as the matrix shears. The angle ϕ in Eq. (49) must therefore be the value at which ultimate shear stress is reached. If τ_u is the shear strength of the fiber-matrix interface, ϕ undergoes little change before failure. Failure by matrix or matrix-interface tension normal to the fibers requires a tensile stress,

$$\sigma = \tau_u \operatorname{cosec}^2 \phi \tag{50}$$

The failure mode requiring the lowest stress, from Eqs. (48)–(50), is assumed to be the one which occurs.

Figure 11 shows how σ depends on ϕ for the three failure modes for uniaxial silica fibers in an aluminum matrix [25]. The observed failure modes agree with those predicted, and the equations are quite representative, if the expected change of ϕ is accounted for when Eq. (49) controls.

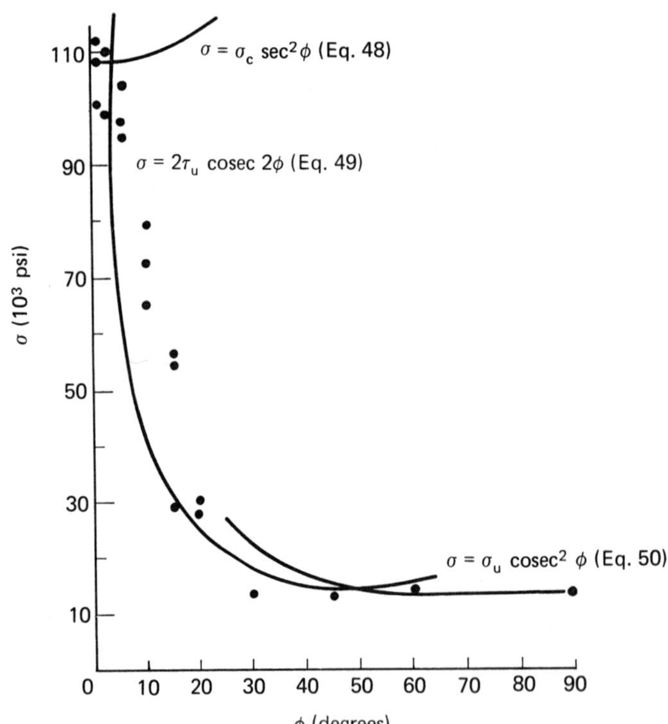

Fig. 11. Effect of stress orientation angle ϕ on measured and computed tensile strengths σ for 50% volume fraction silica fibers in an aluminum matrix. From Jackson, P. W., and D. Cratchley, *J. Mech. Phys. Solids*, **14** (1966) 49 [25].

A critical angle ϕ_{crit} above which strength falls rapidly can be obtained by equating Eqs. (48) and (49),

$$\phi_{crit} = \tan^{-1}\frac{\tau_u}{\sigma_c}$$

For the composite shown, $\phi_{crit} \sim 3.5°$, indicating marked anisotropy. Anisotropy can be reduced by laminating together sheets with fibers in different directions, as in plywood manufacture. Figure 12 shows results for two-layer laminates of 50% silica fibers in an aluminum matrix corresponding to Fig. 11 and for 66% E glass fiber in an epoxy resin. For the aluminum matrix, large tensile strength is maintained to about $\phi = 25°$, because at smaller angles, assuming perfect bonding between lamina, shear parallel to the fibers in one layer must be accompanied by a component of the stress represented by Eq. (48) in the adjacent layer. The reduction in anisotropy due to lamination is similar for the resin matrix, but not so pronounced because τ_u is much less than for metals.

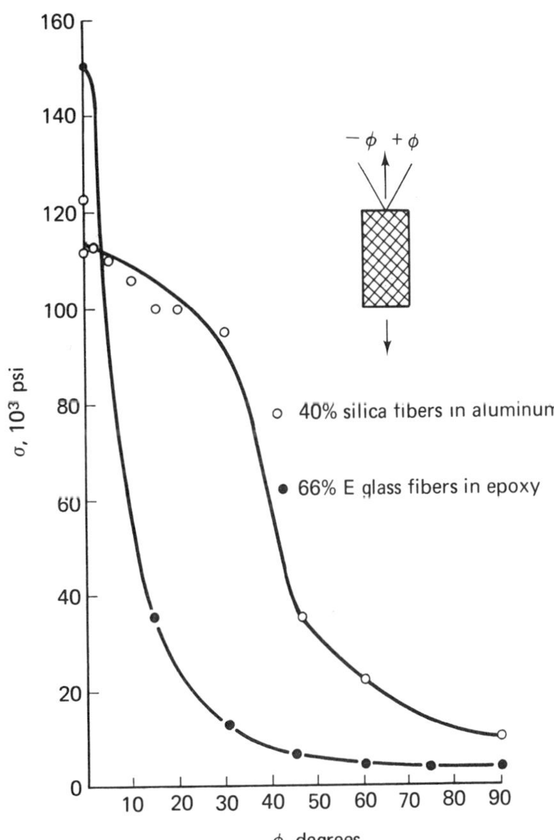

Fig. 12. Effect of two-dimensional lamination on tensile strengths. From Kelly, A., and G. R. Davies, *Metal. Rev.*, **10** (1965) 1 [1].

(2) Deformation. Deformation behavior is frequently an important design criteria for fiber-reinforced composites. We shall study deformation behavior for three cases: uniaxial fibers, uniform fiber orientation in a plane, and multiaxial fiber orientation in a plane.

Uniaxial fibers. For isotropic materials, elastic properties such as Young's and shear moduli and the Poisson ratio are independent of orientation with respect to applied stress and are given by Eqs. (1) and (2) of Chap. 3. For orthotropic materials such as uniaxial fiber composites, these elastic properties vary with orientation. From classical elastic theory, the angular dependence of Young's modulus for uniaxial long fibers is [26, 27, 28]

$$\frac{E_L}{E_1} = \cos^4 \phi + \frac{E_L}{E_T} \sin^4 \phi + \frac{1}{4}\left(\frac{E_L}{G_{LT}} - 2\nu_{LT}\right)\sin^2 2\phi \tag{51}$$

where E_L, E_T, and E_1 are Young's moduli in the direction parallel to the fibers (longitudinal direction), perpendicular to the fibers (transverse direction), and at an angle ϕ to the fiber orientation, respectively. G_{LT} is the shear modulus for these directions, and ν_{LT} is the Poisson ratio for a transverse strain caused by a longitudinal stress. Equation (51) accounts for the effect of shear distortion when a normal load is applied to a specimen at angles other than 0° or 90° to the fibers. E_L, E_T, G_{LT}, and ν_{LT} can be experimentally determined or estimated from the following theoretical equations, assuming zero contiguity among fibers and perfect adhesion between fibers and matrix [29, 30]:

$$E_L = E_m v_m + E_f v_f \tag{52}$$

$$E_T = 2[1 - \nu_f + (\nu_f - 1)\nu_m]\left[\frac{M_f(2M_m + G_m - G_m/M_f - M_m)v_m}{(2M_m + G_m) + 2(M_f - M_m)v_m}\right] \tag{53}$$

$$G_{LT} = G_m \frac{2G_f - (G_f - G_m)v_m}{2G_m + (G_f - G_m)v_m} \tag{54}$$

$$\nu_{LT} = \frac{M_f \nu_f (2M_m + G_m)v_f + M_m \nu_m (2M_f + G_m)v_m}{M_f(2M_m + G_m)v_f - G_m(M_f - M_m)v_m} \tag{55}$$

where the Ms are areal moduli given by $M_m = E_m/2(1 - \nu_m)$ and $M_f = E_f/2(1 - \nu_f)$. Subscripts m and f refer to the matrix and fiber phases, and v_m and v_f are the corresponding volume fractions.

A stress σ in the 1 direction causes orthogonal strains

$$\epsilon_1 = \frac{\sigma_1}{E_1} \tag{56}$$

and

$$\epsilon_2 = -\nu_{12}\epsilon_1 \tag{57}$$

in the 1 and 2 directions, respectively, in which

$$\nu_{12} = \frac{E_1}{E_L}\left[\nu_{LT} - \frac{1}{4}\left(1 + 2\nu_{LT} + \frac{E_L}{E_T} - \frac{E_L}{G_{LT}}\right)\sin^2 2\phi\right] \tag{58}$$

Unlike isotropic materials, stress σ_1 causes shear distortion, except at 0° and 90°, given by

$$\gamma_{12} = \frac{-m_1\sigma_1}{E_L} \tag{59}$$

in which

$$m_1 = \sin 2\phi\left[\nu_{LT} + \frac{E_L}{E_T} - \frac{1}{2}\frac{E_L}{G_{LT}} - \cos^2\phi\left(1 + 2\nu_{LT} + \frac{E_L}{E_T} - \frac{E_L}{G_{LT}}\right)\right] \tag{60}$$

The shear stress-strain relationship is

$$\gamma_{12}\frac{\tau_{12}}{G_{12}} \tag{61}$$

where

$$\frac{G_{LT}}{G_{12}} = \frac{G_{LT}}{E_L}\left[\left(1 + 2\nu_{LT} + \frac{E_L}{E_T}\right) - \left(1 + 2\nu_{LT} + \frac{E_L}{E_T} - \frac{E_L}{G_{LT}}\right)\cos^2 2\phi\right] \tag{62}$$

Unlike isotropic material, shear stress τ_{12} causes longitudinal strains,

$$\epsilon_1 = \frac{-m_1\tau_{12}}{E_L} \tag{63}$$

and

$$\epsilon_2 = \frac{-m_2\tau_{12}}{E_L} \tag{64}$$

in which

$$m_2 = \sin 2\phi\left[\nu_{LT} + \frac{E_L}{E_T} - \frac{1}{2}\frac{E_L}{E_{LT}} - \sin^2\phi\left(1 + 2\nu_{LT} + \frac{E_L}{E_T} - \frac{E_L}{G_{LT}}\right)\right] \tag{65}$$

The principal Poisson ratios are related:

$$\frac{\nu_{LT}}{\nu_{TL}} = \frac{E_L}{E_T} \tag{66}$$

Figure 13, from Dietz [31], illustrates the relationships for Eqs. (51), (58), and (62) as functions of ϕ for the following specific values corresponding, for example, to

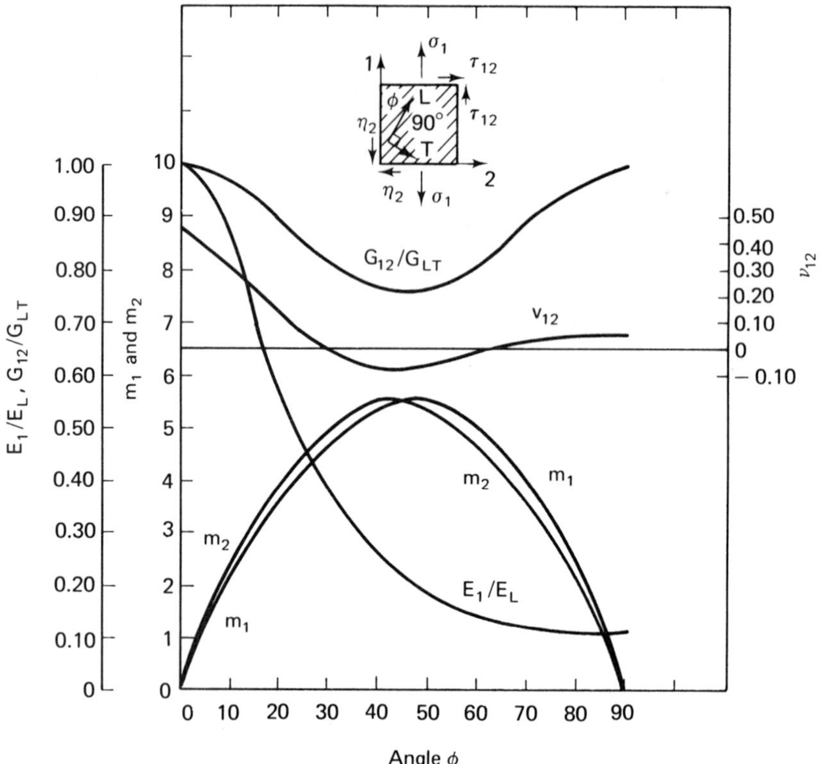

Fig. 13. Elastic constants of a specific orthotropic material. $\phi = 0°$ represents longitudinal direction. From Dietz, A. G. H., "Fiber Reinforced Composite Materials Engineering Analysis" in *Environmental Effects on Polymeric Materials*, Vol. II: Materials, D. V. Rosato and R. T. Schwartz, eds. New York: John Wiley & Sons, Inc. (Interscience Division) 1968, Sec. 15 [31].

an intermediate polyester resin fiber glass:

$$E_L = 5 \times 10^6 \text{ psi}, \qquad \nu_{LT} = v_{0°} = 0.450, \qquad G_{LT} = 5.5 \times 10^5 \text{ psi}$$
$$E_T = 5 \times 10^5 \text{ psi}, \qquad \nu_{TL} = v_{90°} = 0.045$$

Factors m_1 and m_2 in Fig. 13 account for direct and shear strains caused by shear and direct stresses, respectively.

EXAMPLE 2: Find the strains in the 1 and 2 directions on the plate at the top of Fig. 13 if $\sigma_1 = 15{,}000$ psi, $\tau_{12} = 5{,}000$ psi, and $\phi = 25°$.

Solution: From Eqs. (51), (62), (58), (60), and (65),

$$\frac{E_L}{E_1} = 2.101 \quad \text{or} \quad E_1 = 2.28 \times 10^6 \text{ psi}$$

$$\frac{G_{LT}}{G_{12}} = 1.18 \quad \text{or} \quad G_{12} = 4.66 \times 10^5 \text{ psi}$$

$$\nu_{12} = 0.0173, m_1 = 2.76, \qquad m_2 = 4.14$$

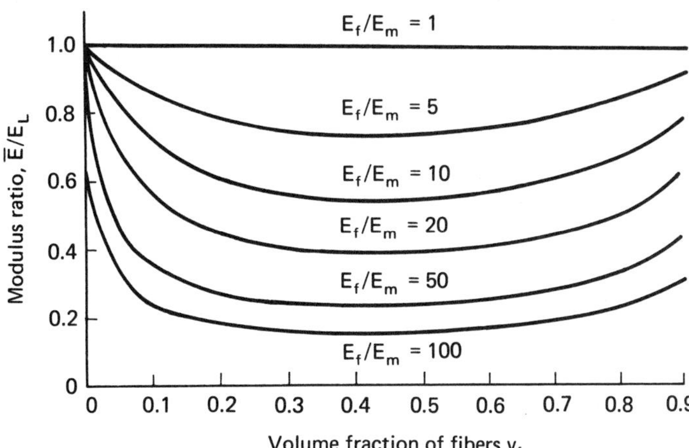

Fig. 14. Effect of volume fraction and modular ratio on Young's modulus when fibers are oriented uniformly in a plane. From Nielson, L. E., and P. E. Chen, *J. Mater.* 3, No. 2 (June 1968) 352 [32].

From Eqs. (56), (57), and (59), the strains caused by σ_1 are

$$\epsilon_1 = \frac{15{,}000}{2.28 \times 10^6} = 6.58 \times 10^{-3}$$

$$\epsilon_2 = -0.0173(6.58 \times 10^{-3}) = -1.14 \times 10^{-5}$$

$$\gamma_{12} = \frac{-2.76\,(15{,}000)}{5 \times 10^6} = -8.28 \times 10^{-3}$$

From Eqs. (61), (63), and (64), the strains caused by τ_{12} are

$$\gamma_{12} = \frac{5{,}000}{4.66 \times 10^5} = 1.07 \times 10^{-3}$$

$$\epsilon_1 = \frac{-2.76(5{,}000)}{5 \times 10^6} = 2.76 \times 10^{-3}$$

$$\epsilon_2 = \frac{-4.14(5{,}000)}{5 \times 10^6} = -4.14 \times 10^{-3}$$

Total strains are therefore

$$\gamma_{12} = 2.40 \times 10^{-3}$$

$$\epsilon_1 = 9.34 \times 10^{-3}$$

$$\epsilon_2 = -4.15 \times 10^{-3}$$

Uniform fiber orientation in a plane. Neilson and Chen [32] solved the case of uniform fiber orientation in all directions in a plane by averaging Eq. (51) over all values of ϕ:

$$\bar{E} = \frac{\int_0^{\pi/2} E_\phi \, d\phi}{\int_0^{\pi/2} d\phi} \tag{67}$$

Figure 14 shows values of $\bar{E}$ from Eq. (67) for typical polymer values of $G_m = 1.45 \times 10^5$ psi, $E_m = 3.92 \times 10^5$ psi, and $\nu = 0.35$. The result is a reasonable approx-

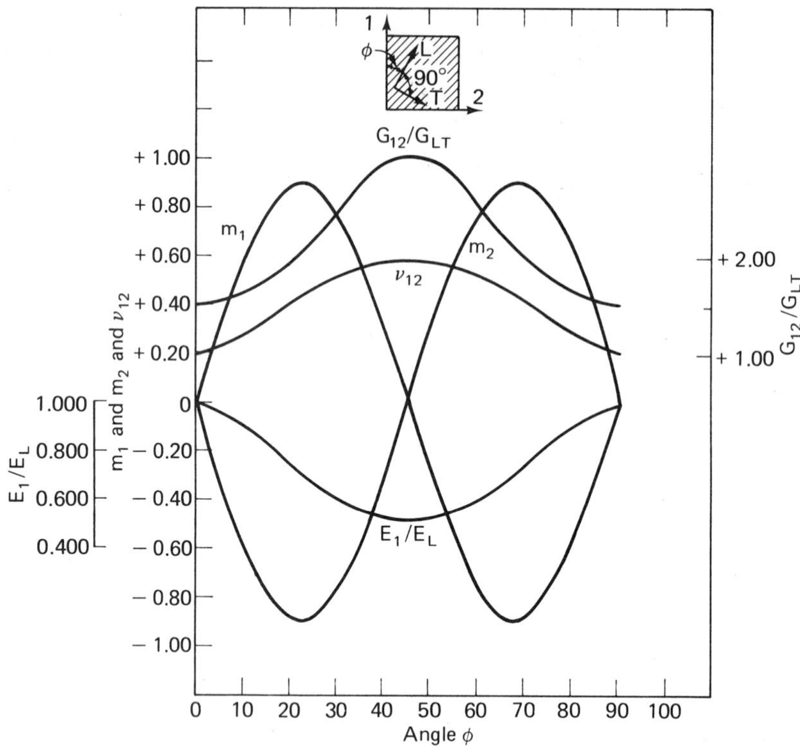

Fig. 15. Elastic Constants of a specific balanced orthotropic material. From Dietz, A. G. H., "Fiber Reinforced Composite Materials Engineering Analysis," in *Environmental Effects on Polymeric Materials*, Vol. II: Materials, D. V. Rosato and R. T. Schwartz, eds. New York: John Wiley & Sons, Inc. (Interscience Division) 1968, Sec. 15 [31].

imation for any fiber composite having the same E_f/E_m ratio, since slight changes in Poisson's ratio would not affect the results appreciably. For most composites of practical interest, $\bar{E}/E_L$ varies from about 0.15 to 0.70.

Multiaxial fiber orientation in a plane. In addition to uniaxial and uniform orientation in a plane, fibers may be oriented along some small number of discrete planar axes. If the longitudinal and transverse properties are equal ($E_L = E_T$ and $\nu_{LT} = \nu_{TL}$), the orthotropic material is said to be balanced. An example is a 90° fiber weave with equal strand density in both directions. In this case, properties are symmetrical about $\phi = 45°$, as shown in Fig. 15, for the specific values $E_L = E_T = 3 \times 10^6$ psi, $G_{LT} = 5 \times 10^5$ psi, and $\nu_{LT} = \nu_{TL} = 0.20$. Problems involving Fig. 15 can be solved in a manner analogous to Example 2.

If one is dealing with a system having high τ, so that anisotropy is significantly reduced with a laminate having two or more fiber orientations, reasonable approximations for the strength and for an upper bound on Young's modulus can be obtained by considering only the single deformation and failure mode represented by Eq. (51). Krenchel [33] did this and compared the results with experimental data for glass fiber composites. If lateral contraction due to the Poisson ratio is ignored and the fibers are sufficiently long to ignore end effects, the reinforcement efficiency, χ, is

$$\chi = \epsilon a \cos^4 \phi \tag{68}$$

where a is the volume portion of the fiber group at angle ϕ from the direction of tensile load. Table 8 gives χ values for several fiber orientations.

TABLE 8. Reinforcement Efficiency, χ, for Several Fiber Orientations; Poisson Effect Ignored

Fiber orientation	*Stress direction*	*Reinforcement efficiency,* χ
1. All fibers parallel	1. Parallel to fibers	1
	2. Perpendicular to fibers	0
2. Fibers in two directions, proportions a_1 and a_2 perpendicular to one another	1. Parallel to direction of a_1 fiber (a_2 fiber)	a_1 (a_2)
	2. Angle $\pi/4$ to fiber direction	$\frac{1}{4}$
3. Four equal layers or groups of fibers at $\pi/4$ to one another	1. Parallel to any one fiber group or layer	$\frac{3}{8}$
	2. Angle $\pi/8$ to any one fiber group or layer	$\frac{3}{8}$
4. Fibers uniformly distributed in plane	Any (in plane)	$\frac{3}{8}$
5. Fibers uniformly distributed in three dimensions in space	Any	$\frac{1}{5}$

SOURCE: KRENCHEL, H., *Fiber Reinforcement*. Copenhagen: Akademisk Forlag, 1964 [33].

Usually lateral contraction cannot be ignored, and shear must be considered, as well as axial loading. For an applied force along the x axis and fibers lying at $+\phi$ and $-\phi$ to the x axis, the fiber force components parallel to the x and y axes, respectively, are

$$P_x = E_f v_f(\epsilon_x \cos^4 \phi + \epsilon_y \sin^2 \phi \cos^2 \phi) \tag{69}$$

$$P_y = E_f v_f(\epsilon_x \sin^2 \phi \cos^2 \phi + \epsilon_y \cos^4 \phi) \tag{70}$$

Since internal and external forces must balance in the x and y directions,

$$E_m k(\epsilon_x + \nu_m \epsilon_y) + \sum P_x = \sigma \tag{71}$$

$$E_m k(\nu_m \epsilon_x + \epsilon_y) + \sum P_y = 0 \tag{72}$$

where $k = v_f/(1 - \nu_m^2)$. The Poisson ratio ν for the composite is

$$\nu = \frac{\epsilon_y}{\epsilon_x} \tag{73}$$

and the composite Young's modulus E is

$$E = \frac{\sigma}{\epsilon_x} \tag{74}$$

In an analogous manner, the shear modulus G is found to be

$$G = (1 - v_f)G_m + v_f E_f \sum a \sin^2 \phi \cos^2 \phi \tag{75}$$

Table 9 summarizes the equations for several fiber orientations. References [34] and [35] contain additional useful data on fiber orientation effects.

TABLE 9. Effect of Fiber Orientation on Composite Properties Including the Poisson Effect [33]

	$\nu =$ [*Eqs.* (73),(69)–(72)]	$\sigma =$ [*Eqs.* (71), (69)]	$E =$ [*Eqs.* (74), (69), (71), (72)]	$G =$ [*Eq.* (75)]
1. Reinforcement with parallel fibers				
a. Force parallel to fibers	ν_m	$E_m k(\epsilon_x + \nu_m \epsilon_y) + E_f v_f \epsilon_x$	$E_m(1 - v_f) + E_f v_f$	$(1 - v_f)G_m$
b. Force perpendicular to fibers	$\dfrac{k\nu_m}{k + n v_f}$	$E_m k(\epsilon_x + \nu_m \epsilon_y)$	$E_m k(1 - \nu_m \nu)$	$(1 - v_f)G_m$
2. Two equal groups of fibers at right angles				
a. Force parallel to one fiber group	$\dfrac{k\nu_m}{k + \frac{1}{2} n v_f}$	$E_m k(\epsilon_x + \nu_m \epsilon_y) + \frac{1}{2} E_f v_f \epsilon_x$	$E_m k(1 - \nu_m \nu) + \frac{1}{2} E_f v_f$	$(1 - v_f)G_m$
b. Force at half-angle between adjacent groups	$\dfrac{k\nu_m + \frac{1}{4} n v_f}{k + \frac{1}{4} n v_f}$	$E_m k(\epsilon_x + \nu_m \epsilon_y) + \frac{1}{4} E_f v_f (\epsilon_x + \epsilon_y)$	$E_m k(1 - \nu_m \nu) + \frac{1}{4} E_f v_f (1 - \nu_m)$	$(1 - v_f)G_m + \frac{1}{4} v_f E_f$
3. Four or more equal fiber groups uniformly oriented, force parallel to or at half-angle between adjacent groups	$\dfrac{k\nu_m + \frac{1}{8} n v_f}{k + \frac{3}{8} n v_f}$	$E_m k(\epsilon_x + \nu_m \epsilon_y) + E_f v_f (\frac{3}{8} \epsilon_x + \frac{1}{8} \epsilon_y)$	$E_m k(1 - \nu_m \nu) + E_f v_f (\frac{3}{8} - \frac{1}{8} \nu)$	$(1 - v_f)G_m + \frac{1}{8} v_f E_f$
4. Fibers randomly orientated				

SOURCE: KRENCHEL, H., *Fiber Reinforcement.* Copenhagen: Akademisk Forlag, 1964 [33].

B. DUCTILE FIBERS IN A BRITTLE MATRIX

Two techniques are especially useful for improving the performance of brittle nonhomogeneous materials such as concrete. One is the impregnation of voids and microcracks with more ductile materials such as polymers; the other is fiber reinforcement. Void filling primarily improves the precracking characteristics and strength of concrete, and fiber reinforcement primarily the postcracking characteristics and work of fracture. But both techniques can offer some improvement in both ranges of strain.

By filling the concrete microcracks resulting from curing shrinkage with a ductile polymer, fracture energy under external load is absorbed by the ductile phase, and crack propagation through the brittle phase is retarded, giving increased strength. Similar end results can be obtained by mixing certain emulsion latexes in the fresh concrete or by various additives which improve the bond between cement paste and aggregate (Sec. F in Chap. 7).

Alternatively, fibers have been used to reinforce brittle materials since antiquity; straw for sun-baked brick, horse hair for plasters, and more recently asbestos and other fibers for cements and concretes. Unless the fiber has greater tensile modulus than the brittle matrix and greater length than the critical length for the particular mixture, it cannot appreciably improve strength but can significantly increase impact resistance and work of fracture.

The combination of crack filling with a ductile material and fiber reinforcement can be used simultaneously to tailor-make concretes to specified strength and ductility requirements. In this section we shall study only the second technique, fiber reinforcement of brittle materials.

1. Manufacturing Processes

Fiber-reinforced cements and concretes can be manufactured by a variety of processes, many adapted from other industries [36]. For example,

1. The dewatering by vacuum of asbestos cement board (from papermaking).
2. The simultaneous spray-up of a low-water-content portland cement slurry and chopped glass fibers from separate nozzles and then removal of excess water by applying vacuum to a porous mold (from fiber-reinforced plastics) [37].

Some practical advantages of fiber-reinforced concretes are

1. Increased resistance to cracking and spalling.
2. Increased resistance to thermal shock.
3. Less maintenance and longer life under exposure to abrasion and climatic weathering.
4. Material properties which permit design in thinner sections, at potentially faster production rates.

Some potential advantages in terms of specific applications of fiber-reinforced concretes include

Application	*Advantages*
Roadways and bridge decks	Improve fatigue strength under wheel loads, reduce penetration of deicing agents due to improved crack control, permit thinner bridge slabs having reduced dead weight.
Building frames	Permit alternative code design criteria allowing full utilization of tensile steel strength* and the use of plastic in lieu of elastic design methods, since fiber reinforcement may allow the development of plastic hinges in otherwise brittle concrete. Permit an increase in allowable working

*The use of less than balanced steel reinforcement is adopted to ensure ductile failure. The ACI Code, for example, limits tensile reinforcement in singly reinforced beams to 75% of the balanced ratio.

	stress of concrete due to reduced statistical scatter of strengths of ductile (fiber-reinforced) concrete compared with brittle (without fiber) concrete. Allow a reduction of secondary reinforcing such as shrinkage steel, stirrups in beams, and ties or spiral reinforcing in columns.
Pipes	Lighter weight designs, reduction in reinforcing, improved resistance to damage during handling, greater ease in making service connections to existing pipe; i.e., openings can be made without shattering the pipe or having to cut through heavy reinforcing steel.
Waterfront and marine structures	Improve impact and postcracking strength caused by wave and ship forces. Improve sulfate resistance due to improved crack control.
Linings for cement kilns, open hearth furnaces, etc.	Reduction in spalling due to severe thermal gradients and time-varying temperature.

a. MATERIALS

Table 10 compares the characteristics of several classes of fibers for concretes. Fiber dispersion always requires a longer mixing time than for mixing without fiber, although the addition of small amounts of air entrainer improve dispersion, especially of organic fibers. The proportion of finest aggregate is usually reduced slightly to provide space for the fibers, and either the water/cement ratio or both water and cement contents may have to be increased to maintain required workability.

TABLE 10. Properties of Various Fibers Used with Concrete

	Mechanical properties			*Chemical properties*		*Mixing properties*		*Cost/strength ratio*
Fiber	*Tensile strength* (10^3 *psi*)	*Young's modulus* (10^6 *psi*)	*Elongation at fracture* (%)	*Fiber-matrix shear strength* (*psi*)	*Fiber corrosion in portland cement*	*Balling and segregation*	*Required w/c ratio*	[*\$/cu in./psi = \$/lb-in.* ($\times 10^{-2}$)]
Metallic								
Steel	40–600	29	0.5–35.0	Medium	Mild	Severe	Medium	0.07–0.35*
Mineral								
Asbestos fiber	80–140	10	1.5–3.5	High	Negligible	Medium	High	0.04–0.30
Glass fiber	150–500	12–20	0.6	~300 [38]	Severe	Medium	Medium	0.07–0.42
Organic								
Nylon	110–120	0.6	16–20	Low	Negligible	Mild	Low	0.31–0.45
Polypropylene	80–110	0.5	25	Low	Negligible	Mild	Low	0.19–0.26
Polyethylene	100	0.2–0.6	10	Low	Negligible	Mild	Low	0.18–0.28

*14 to 50 cents/lb for 0.015- to 0.040-in. diameters. Prices increase significantly for finer diameters and higher carbon contents.

(1) ***Metal fibers.*** Steel fibers have a Young's modulus 10 times that of concrete, have reasonably good bond, and have high elongation at fracture. With about 4% by volume, flexural strength can be doubled and compressive strength increased by about 50% [39].

The micro steel wires used in radial belted tires are available in diameters as low as 0.003 in. and have strengths as high as 500,000 psi. However, the wire cost increases significantly for finer diameters, and it is probable that fine steel wire having high carbon content may not survive corrosion in cement pastes.

A major limitation in the use of steel fibers is segregation and balling during mixing, even with wire lengths considerably below the critical length. Balling is related to three factors: the aspect ratio, the volume percentage of fibers, and their rigidity (rigid fibers are more troublesome than flexible ones). Typically, to reduce balling and segregation to within tolerable limits during mixing, steel fibers significantly shorter than the critical length must be used, giving low fiber tensile efficiency. As tensile reinforcement, improvement by fibers is therefore inferior to improvement by the same volume of conventional reinforcing rod, because fibers (1) more easily pull out, (2) are randomly oriented, and (3) are dispersed throughout the cross section of the beam.

Characteristically a plain mortar with a slump of 7 in. will have a slump of only 1 in. with 2.5 volume % fibers having an aspect ratio of 90. Mixing is very difficult with fibers having aspect ratios greater than 100 or fiber volumes larger than 4%. Mixing problems are generally increased, and improvements due to fiber addition decrease with increasing aggregate size [40].

(2) ***Mineral fibers.*** Asbestos cement normally contains 6–18% fibers by volume. Since asbestos fibers absorb large quantities of water, a large water/cement ratio is required. Because readily available asbestos supplies are being depleted and because of the lung cancer, asbestosis, associated with asbestos, substitutes are being sought.

Glass-reinforced concrete is less brittle than asbestos cement and more fire-resistant than glass-reinforced plastics. The strength, stiffness, and price of glass-reinforced cement lie between the corresponding values for asbestos cement and glass-reinforced plastics.

The main problem with commercially available glass fibers (E glass) is that they are chemically attacked by the high alkaline environment of cement paste and as a result lose their strength with time.

The alkali corrosion problem of glass is solved in several ways:

1. Preparation of more alkali-resistant glasses, either with special compositions of the bulk glass or by subjecting conventional A or E glass to surface diffusion treatments. Alkali-resistant glasses currently cost over $1/1b compared with less than 50 cents/lb for more ordinary forms of glass.
2. Application of various protective coatings to the fiber. Organic coatings such as polyvinyl chloride and polyvinyl acetate protect fibers from alkali corrosion but also reduce bond shear strength between cement and fibers. An aqueous solution of a polyvalent metal and a nonmetallic base has been

claimed to increase bonding between fiber and cement by forming a water-insoluble layer on the fiber surface [41].

3. The use of less alkaline cements such as high alumina cements, or the use of additives to neutralize portland cement. Although ordinary glass can be combined with high alumina cements without severe corrosion, these cements are on the order of three times as expensive as portland cements. For use with portland cements, cation exchangers such as zeolite clays or sulfonated polyvinyl-aryl compounds which combine with the harmful Ca^{+2}, Na^{+}, or K^{+} ions have been proposed [42]. The use of fluosilicate salts such as $MgSiF_6$ has been proposed not only for neutralizing alkali in the cement paste but also for strengthening the glass fiber by etching reaction [43].

(3) Organic fibers. Some organic fibers are also susceptible to deterioration by alkalis in cement (cotton, rayon, polyester, etc.). Those which are not include nylon, polypropylene, polyethylene, and Saran. Nylon mixtures suffer somewhat more of a viscosity effect than corresponding mixtures with polypropylene, polyethylene, and Saran. In some applications it is desirable to remove excess water after molding by pressing or vacuum treatment.

Because of their considerably lower Young's moduli, most organic fibers do not significantly increase the strength of portland cement, and often they reduce it [39]. The main function of organic fibers is to improve the impact resistance or ductility of concrete and hold the structure together past the limit strain of concrete, taken as 0.003 in the ACI Code [44]. To achieve this, fibers need not have high modulus. Increases in impact resistance of 10 to 25 times have been observed with nylon and polypropylene fibers. Although organic fibers are more easily mixed with concrete than metal fibers, they have poorer bond. Critical aspect ratios as high as 765 (1.3-in. length for 15 denier fibers) have been measured for nylon fibers in concrete [45].

b. CEMENT-FIBER BOND

Little is known about the nature of bonding and bond mechanisms between cements and various fibers. Adhesion between glass fiber mats and cement has been measured using an electromechanical adhesion meter originally developed for measuring the adhesion of insulating layers of varnish. For four cement types used with alkaline (A) and nonalkaline (E) glass, it was concluded that [46]

1. Adhesion decreased with increasing water/cement ratios above a ratio of about 0.3.
2. High alumina cement adhered best to E glass.
3. Pozzolanic portland cement adhered best to A glass. Fiber glass-portland cement shear strengths on the order of 300 psi have been measured in single fiber pull-out tests [38]. This compares favorably with measured values of shear strength between siliceous aggregate and portland cement [47].

2. Fiber-Matrix Mechanics

a. FIBER-MATRIX STRESS TRANSFER

Although the mechanics for fiber-matrix stress transfer are well established and correlated with a plethora of experimental data for brittle fibers in ductile matrices such as soft metals [1] and polymers [15], experimentally documented expressions for stress transfer in the reverse case, i.e., ductile fibers in a brittle matrix, are rare [38]. The low moduli of elasticity and high breaking strain of ductile matrices permit efficient elastic transfer of load to the fibers and reliable models of behavior. These conditions do not exist in brittle matrices such as concrete, and in addition the matrix itself is not homogeneous.

Although the limited amount of fiber-reinforced concrete research to date has been nearly all experimental, even an approximate treatment of the problem in terms of classical mechanics is useful in determining which factors are important in the strength of fiber-reinforced concretes. This section will parallel the sequence in Sec. A for brittle fibers in a ductile matrix.

Recall that in ductile matrices the matrix flows around discontinuous fibers and the fiber tension builds up linearly (in metal matrices) or nonlinearly (in polymer matrices) [Fig. 16(a)] as a function of the shear distribution near the fiber ends in

Fig. 16. Fiber-matrix shear and fiber tension distributions along a fiber. (a) Typical of brittle fiber in metal matrix (uniform shear, linear tension increase from fiber end); (b) typical of brittle fiber in polymer matrix (nonuniform shear, debonding at fiber end and nonlinear tension increase), (c) possible stress distribution for ductile fiber in brittle matrix.

l_c = critical fiber length.

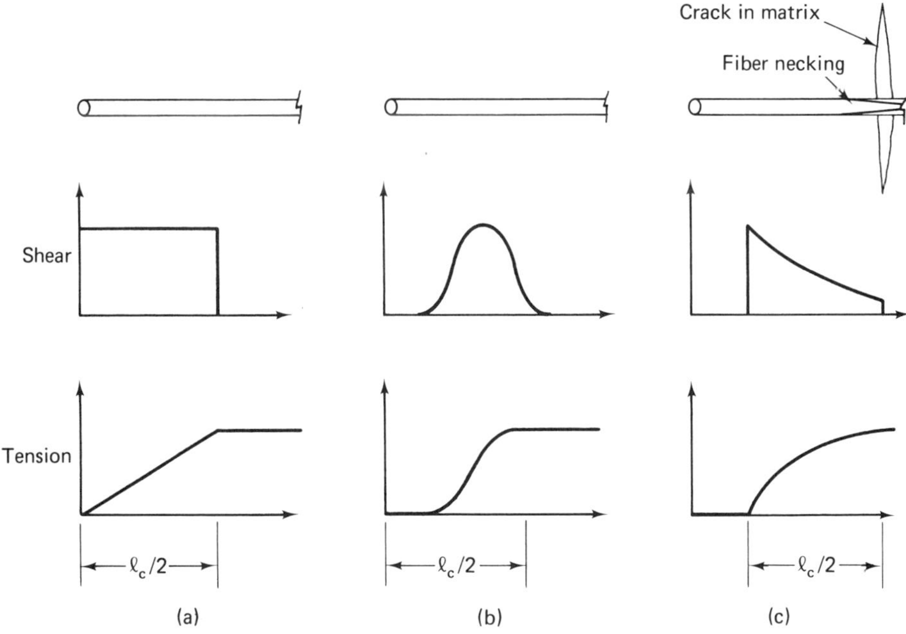

these two matrix materials. In addition, in the case of polymer matrices, the ductile matrix may tear away from the ends of a fiber at high tensile strain values [Fig. 16(b)] analogous to the tearing of a soft rubber sheet from the edges of a spot of rigid glue placed on it as the sheet is stretched [15].

In the case of fibers in a brittle matrix, different stress-transfer mechanisms can be hypothesized. For example, one might use the physical analogies of pulling a helical spring from inside a rigid pipe in which it has been snugly coiled or of pulling an earthworm from its hole. As tension is applied, the fiber elongates and contracts away from its rigid hole, undergoing maximum loss of interfacial shear at the cross section of highest tensile loading. The shear and tensile stress distributions along the length of a ductile fiber resisting a perpendicular tensile crack in a rigid matrix at the fiber midpoint may, for example, resemble Fig. 16(c), where tension is, of course, the integral of shear along the fiber and the area under the shear function equals the fiber tensile load at the crack. The unstressed portion of fiber at the left in Fig. 16(c) would serve no purpose in the composite system under the assumed load and crack configuration.

Assumptions: The following material property assumptions will be used,

1. Matrix undergoes brittle failure.
2. Fiber undergoes either brittle or ductile failure.
3. Tensile failure strain is lower for the matrix than for the fiber.
4. The limiting bond stress between matrix and fiber is the interfacial shear strength, rather than shear strength within either the matrix or fiber phases.

If we assume interfacial shear strength, τ, to obey the Coulomb equation, then

$$\tau = c + n \tan \phi$$

where c is static cohesion between fiber and matrix, n is the interfacial radial pressure, and ϕ the friction angle. The radial pressure can be caused, for example, by the curing shrinkage of cement paste, analogous to the radial pressure on fibers caused by the curing shrinkage of polymer matrices [see Eq. (9)]. The fiber tensile stress in the vicinity of a matrix crack is then, from Eq. (1a),

$$\begin{aligned}\sigma_x &= \sigma_{x0} - \frac{2}{r}\int_0^x \tau_x \, dx \\ &= \sigma_{x0} - \frac{2}{r}\int_0^x (c + n_x \tan \phi)\, dx, \qquad \sigma_x \geq 0 \qquad (76)\end{aligned}$$

where σ_{x0} is the fiber tensile stress at the matrix crack, x is the distance along the fiber away from the crack, and c is a constant but equal to zero where fiber necking is sufficient to eliminate contact between the fiber and its cavity. The radial pressure is

$$n_x = n_{x0} - \sigma_x \nu \qquad (77)$$

where n_{x0} is the radial pressure when $\sigma_x = 0$, and ν is the fiber Poisson ratio, assumed to remain constant throughout both elastic and plastic deformation of the fiber. Substituting Eq. (77) into Eq. (76),

$$\sigma_x = \sigma_{x0} - \frac{2cx}{r} - \frac{2xn_{x0} \tan \phi}{r} + \frac{2\nu \tan \phi}{r} \int_0^x \sigma_x \, dx, \qquad \sigma_x \geq 0 \tag{78}$$

This is an integral equation of the Volterra type [48]; that is, the wanted function appears under the integral sign, and one of the limits is the independent variable. Differentiating with respect to x,

$$\sigma_x = \sigma_{x0} + \frac{c + n_{x0} \tan \phi}{\nu \tan \phi}(e^{-2x\nu \tan \phi/r} - 1), \qquad \sigma_x \geq 0 \tag{79}$$

The fiber critical length, l_c (neglecting the crack width), can be obtained by setting $\sigma_{x0} = \sigma_f$, the fiber tensile strength, and solving for the value of $x(= l_c/2)$, which makes $\sigma_x = 0$ in Eq. (79):

$$l_c = \frac{r}{\nu \tan \phi} ln \left(\frac{\sigma_f \nu \tan \phi}{c + n_{x0} + \tan \phi} \right) \tag{80}$$

This corresponds to Eq. (3) for the critical length of fibers in a ductile matrix. Equation (80) represents a lower bound critical length, or a critical length in the case of only a single matrix crack at the fiber midpoint. For cases of more extensive cracking, the critical length is greater.

b. DEFORMATION

As in ductile matrices, the tensile stress-strain curves parallel to the fiber axis should be quite similar for both continuous and discontinuous fibers, provided the discontinuous fibers exceed the critical length. There may be two regions of stress-strain behavior in fiber-reinforced brittle matrices: (1) below the strain level at which the matrix begins to crack and (2) above this level.

In region 1, both fiber and matrix are assumed to behave elastically, and the simple mixture law applies, as for the corresponding strain region in ductile matrices. The lower bound of the composite Young's modulus is again

$$E_c = E_f v_f + e_m v_m \tag{4}$$

In region 2, the matrix progressively cracks until, in the limit, the entire tensile load may be carried by the fibers. For this limiting condition,

$$E_c = \frac{d\sigma_f}{d\epsilon_f} v_f \tag{81}$$

where $d\sigma_f/d\epsilon_f$ is the stress-strain curve slope of the fibers, whether deforming elastically or plastically. Equations (4) and (81) are the lower and upper bound limits of deformation behavior for the region of matrix cracking, where the matrix contribution decreases with increasing strain, in the manner of a brittle coating on a fiber which undergoes progressive cracking as the fiber is stretched. In region 2, composite behavior is quasi-elastic due to progressive matrix cracking. These two regions of stress state exist simultaneously in adjacent micro-areas of the composite, i.e., quasi-elastic where the matrix is cracked and the fibers are in a local state of pull-out and elastic at locations between cracks.

Discontinuous fibers. As in ductile matrices, when the composite contains discontinuous fibers the fiber contribution to the modulus in region 2 is less than $(d\sigma_f/d\epsilon_f)v_f$ because the fiber stress decreases away from the matrix cracks [Fig. 16(c)]. Unlike the case of ductile matrices, the average fiber stress depends on the locations and sizes of matrix cracks along the fiber and therefore is indeterminant, except in a statistical sense [49] (Fig. 17). A unique relationship between the Young's moduli for composites having continuous and discontinuous fibers, E_{dc}/E_{cc}, therefore cannot be developed for brittle matrices, as for ductile matrices [see Eq. (6a)].

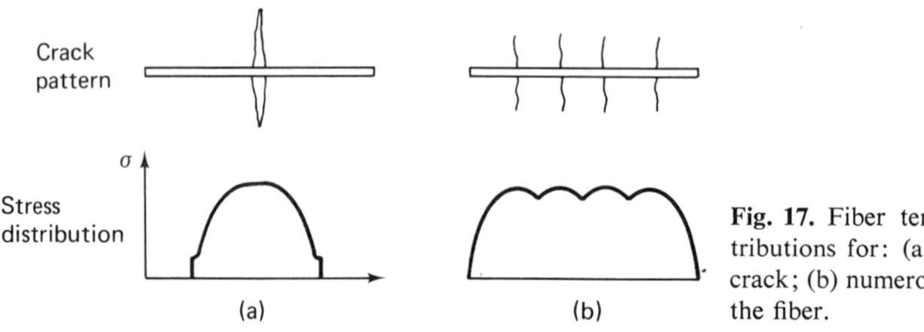

Fig. 17. Fiber tensile stress distributions for: (a) Single matrix crack; (b) numerous cracks along the fiber.

Figure 18 shows representative flexural load-deflection results for cements with glass, steel, and nylon reinforcement. Fiber lengths of 1 to 3 in. were used in amounts limited by viscosity during mixing [50]. Beams containing glass (curve *c*) broke at low deflections. Beams containing steel wire (curve *e*) maintained high modulus over a large deflection range, presumably due to the high modulus of the steel, and then deformed at almost constant load as the wires pulled out. Beams containing nylon (curves *f* and *g*) had much lower moduli but returned almost to their original shapes after being deformed well beyond peak load, leaving only a few visible cracks.

c. STRENGTH

Given the same material property assumptions for which the equations of deformation were developed, fiber-reinforced brittle matrices can fail in the following modes:

1. Matrix fails; then fibers fail in (a) tension or (b) bond at a lower composite stress level.
2. Matrix fails; then fibers fail in tension at a higher composite stress level.
3. Matrix fails; then fibers fail in bond at a higher composite stress level.

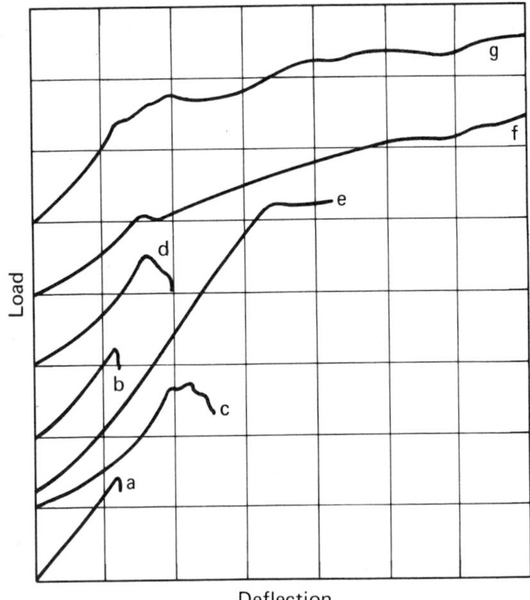

Fig. 18. Representative flexural load-deflection curves. (a) Cement mortar; (b) neat cement; (c) neat cement with resin-coated fiber glass; (d) with glass flake; (e) with 1 in. steel wire; (f) with 1 in. nylon fibers; and (g) with 3 in. nylon fibers. From Goldfine, S., Mod. Plastics, 42 (April, 1965) 156 [50].

For mode 1, the simple mixture law analogous to Eqs. (4) and (12) applies:

$$\sigma_c = \sigma_m(1 - v_f) + \sigma'_f v_f, \qquad v_f < v_{\text{crit}} \tag{82}$$

where σ'_f is the fiber stress at ultimate matrix strain, assuming that fiber and matrix undergo equal strains. It is apparent from Eq. (82) that if $\sigma'_f < \sigma_m$, mode 1 failure will occur at a lower value of σ_c for *all fiber volumes* than for the matrix alone. If the condition $v_m > v_{\text{crit}}$ is not satisfied, failure occurs by either mode 2 or mode 3.

For mode 2, where load is carried entirely by the fibers,

$$\sigma_c = \sigma_f v_f, \qquad v_f > v_{\text{crit}} \tag{83}$$

It is apparent from Eq. (83) that in mode 2 failure the minimum volume fraction, v_{min}, which must be exceeded for fiber strengthening is

$$v_{\text{min}} = \frac{\sigma_m}{\sigma_f} \tag{84}$$

For mode 3, which can occur only with discontinuous fibers, the strength depends on the nature of cracking and is indeterminant, for the same reasons that region 2 deformation with discontinuous fibers was found to be indeterminant. Progressive failure by a combination of modes 2 and 3 might also occur, resulting, for example, from flawed fibers [38].

Critical volume fraction. For continuous fibers which have a larger failure strain than the matrix and which can also work-harden, there exists a critical fiber volume, v_{crit}, below which mode 1 failure occurs and above which mode 2 failure occurs.

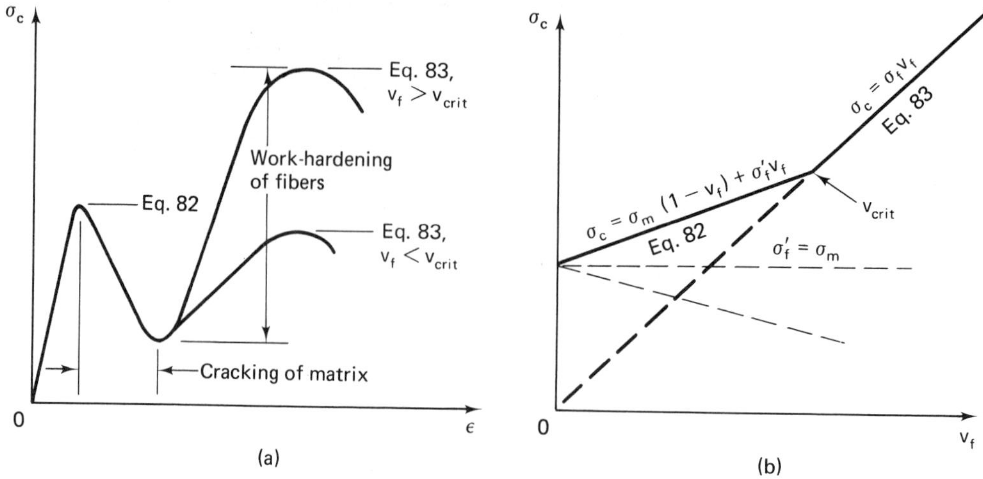

Fig. 19. Fiber-reinforced brittle matrix composite behavior. (a) Stress-strain diagram; (b) strength-volume fraction diagram.

Equating Eqs. (82) and (83),

$$v_{\text{ctit}} = \frac{\sigma_m}{\sigma_w' + \sigma_m} \tag{85}$$

where $\sigma_w' = \sigma_f - \sigma_f'$, the stress due to work hardening of the fibers after the matrix fails.

Figure 19 shows graphically the relationship between failure modes 1 and 2, corresponding to Fig. 6 for the ductile matrix case. It is apparent from Fig. 19(b) that if $\sigma_f' < \sigma_m$, $v_f > v_{\text{crit}}$ must be satisfied, to obtain any strengthening benefit from the fiber.

Discontinuous fibers. In discontinuous fiber brittle composites, failure may occur also by fiber pull-out (mode 3). Expressions for strength then become indeterminant, as well as the expressions for critical fiber length, critical fiber volume ratio, and the work of fracture due to fiber pull-out.

Experiments have shown that the work of fracture of plain concrete can be increased by up to 100 times with 2% by volume of steel wire and up to 40 times with glass fiber [51]. Pull-out tests on individual wires embedded in concrete for 7 weeks indicated the following length-average dynamic bond strengths:

low carbon steel wire in wet-cured concrete ~ 1.5 MNm^{-2}
in dry-cured concrete ~ 0.7 MNm^{-2}
high carbon steel wire (brass-plated) wet cured ~ 0.5 MNm^{-2}
dry cured ~ 0.5 MNm^{-2}
[6.9 kilonewtons/m^2 (kNm^{-2}) = 1 psi]

The results confirmed that the chemical reaction which enhances bonding between cement and steel is intensified in the presence of moisture and inhibited by the brass coating.

d. COMPRESSIVE FAILURE

In brittle matrices, compressive failure results in tensile splitting on the maximum principal stress plane, tensile stresses being positive (see Sec. B in Chap. 8). Therefore, fibers should be aligned in the direction of maximum principal stress, or normal to the direction of maximum compressive stress, rather than parallel to it, as for ductile matrices (Sec. A2d). Compressive failure of a specimen typically results from the lateral instability of the columnar elements separated by tension cracks. With increasing strain, each element is subjected to lateral pressure from its adjoining elements. Since the outer elements have no lateral support on the outside, they yield first, and cracks progressively open up from the outside inward. An example of lateral support in a brittle material is the spiral reinforcement of columns, which after the ultimate load is reached keeps the core from disintegrating even though the outer shell has spalled off.

e. ORIENTATION EFFECTS

In the absence of reported experimental results on fiber orientation effects in brittle matrices, those expressions in Sec. A2e for ductile matrices which are not dependent in their derivation on matrix properties might also be applied tentatively to brittle composites. An expression for the work of fracture of randomly oriented fiber-reinforced concretes has been developed by Iyengar [45].

f. RESEARCH AREAS

Various schemes can be postulated for improving the efficiency of fiber-reinforced concretes, for example,

1. The use of fibers with enlarged or bulbous ends, which, by providing a shorter critical length, would permit the use of shorter fibers, reducing the viscosity and balling problems during mixing and the required water/cement ratio for workability.
2. The use of polymeric or water-soluble polyelectrolyte fibers, which might be irreversibly posttensioned after the concrete has hardened, for example, by dehydration, by radiation, or by chelation with calcium or other metal ions in the cement paste.

C. DESIGN OPTIMIZATION

In Sec. C in Chap. 1, it was noted that effective use of composite materials requires that material design be included as an integral part of structural design. The following example illustrates the simultaneous optimization of material and structural properties.

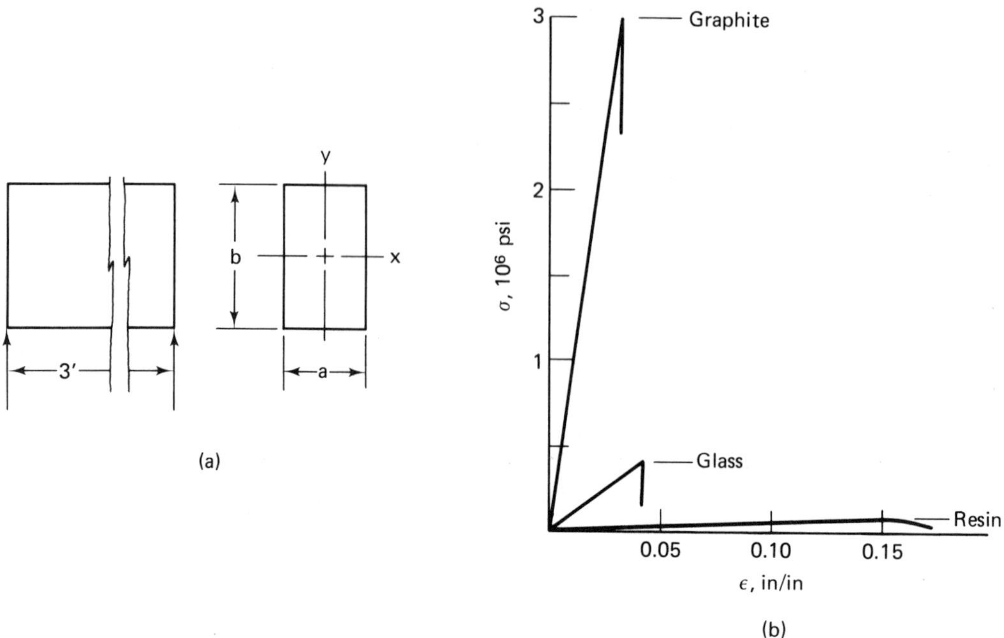

Fig. 20. (a) Beam dimensions; (b) stress-strain characteristics.

EXAMPLE 3: A uniaxial continuous fiber-reinforced rectangular beam is to be designed using resin and a combination of glass and graphite fibers. The graphite fibers, which have higher strength and modulus but higher cost, may be used to keep the cross-sectional dimensions of the beam within prescribed limits.

The beam will be simply supported on a 3-ft span and must resist, at separate times, center span loads of 500 lb in the x direction and 1,000 lb in the y direction [Fig. 20(a)]. The allowable midspan deflections are 0.30 in. in the x direction and 0.20 in. in the y direction. The allowable cross-section dimensions are $a < 1.0$ in. and $b < 2.0$ in.

Figure 20(b) shows the stress-strain characteristics of the component materials. Their relative costs per unit volume are 1, 2.5, and 200 for resin, glass, and graphite, respectively. A minimum of 20% resin by volume must be used for adequate bonding and separation of fibers.

Determine the optimum volume fractions of the two fibers if the beam is to have a uniform composite cross section.

Solution: Ignoring, for simplicity of illustration, the possibility of beam buckling and shear failure, we recognize two failure mode constraints: flexural strength and deflection, in both the x and y directions. We also observe that, depending on the relative proportions of glass and graphite fibers and assuming strain compatibility of the three components, the composite may reach its maximum flexural strength

a. When the fiber having the lowest failure strain (graphite) breaks, in which case

$$\sigma_c = \sigma_r v_r + \sigma_{gl} v_{gl} + \sigma_{gr} v_{gr} \tag{86}$$

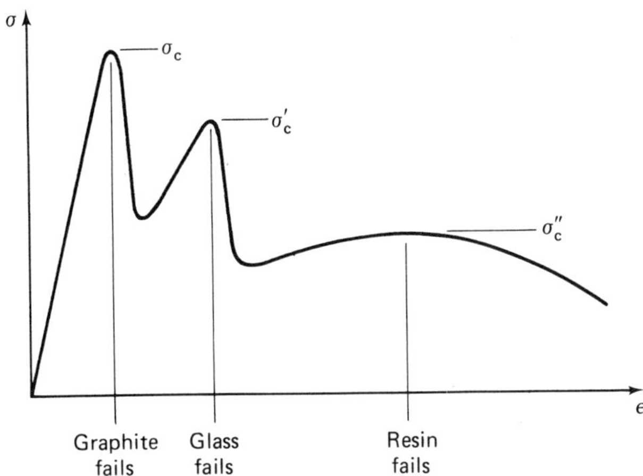

Fig. 21. Stress-strain diagram for two-fiber composite.

b. At the failure strain of the remaining fiber, for which

$$\sigma'_c = \sigma'_r v_r + \sigma'_{gl} v_{gl} + \sigma'_{gr} v_{gr} \tag{87}$$

or

c. At the failure strain of the resin, for which

$$\sigma''_c = \sigma''_r v_r + \sigma''_{gl} v_{gl} + \sigma''_{gl} v_{gr} \tag{88}$$

where σ = strength; v = volume fraction; subscripts $r, gl,$ and gr = resin, glass, and graphite, respectively; and superscripts (none), $'$, and $''$ indicate component average stresses at values of composite strain corresponding to failure of the graphite, glass, and resin, respectively. For example, σ''_{gl} represents the statistical length-average stress borne by fractured glass fibers of less than critical length at the strain level which causes failure of the resin. Figure 21 is a hypothetical composite stress-strain diagram illustrating the strength peaks at the corresponding three values of strain. The peak rounding is due to the statistical distribution of actual fiber breaking strains and the slipping of broken fibers having less than critical length.

The beam cost is $abl[c_{gl}v_{gl} + c_{gr}v_{gr} + c_r(1 - v_{gl} - v_{gr})]$, where abl is the beam volume and c is cost per unit volume.

The beam moments of inertia about the y and x axes are $ba^3/12$ and $ab^3/12$, respectively.

Assuming the simple mixture law to hold for small deflections, the composite Young's modulus, E_c, is

$$E_c = E_{gl}v_{gl} + E_{gr}v_{gr} + E_r(1 - v_{gl} - v_{gl}) \tag{89}$$

in which the values $E_{gr} = 100 \times 10^6$, $E_{gl} = 10 \times 10^6$, and $E_r = 0.5 \times 10^6$ psi are the slopes in Fig. 20(b).

The deflections for midpoint loadings of the simply supported beam are

$$\Delta_x = \frac{500(3 \times 12)^3}{48E_c(a^3b)/12} = \frac{5.83}{a^3b(9.5v_{gl} + 99.5v_{gr} + 0.5)} \tag{90}$$

and

$$\Delta_y = \frac{1{,}000(3 \times 12)^3}{48E_c(ab^3)/12} = \frac{11.66}{ab^3(9.5v_{gl} + 99.5v_{gr} + 0.5)} \tag{91}$$

in the x and y directions, respectively.

The corresponding maximum bending stresses are

$$\sigma_{cx} = \frac{3(500)(3 \times 12)}{2a^2b} = \frac{27{,}000}{a^2b} \tag{92}$$

and

$$\sigma_{cy} = \frac{3(1{,}000)(3 \times 12)}{2ab^2} = \frac{54{,}000}{ab^2} \tag{93}$$

The optimization problem can now be written as follows:

$$\text{minimize} \quad Z = ab[c_{gl}v_{gl} + c_{gl}v_{gr} + c_r(1 - v_{gl} - v_{gr})]$$

or

$$\begin{aligned}\text{minimize} \quad Z &= ab(2.5v_{gl} + 200v_{gr} - v_{gl} - v_{gr})^* \\ &= ab(1.5v_{gl} + 199v_{gr})\end{aligned} \tag{94}$$

subject to

$$\left.\begin{aligned}&\frac{5.83}{a^3b(9.5v_{gl} + 99.5v_{gr} + 0.5)} \leq 0.30 \quad (95)\\ &\frac{11.66}{ab^3(9.5v_{gl} + 99.5v_{gr} + 0.5)} \leq 0.20 \quad (96)\end{aligned}\right\} \text{deflection}$$

$$\left.\begin{aligned}&\frac{27{,}000}{a^2b} \leq \max(\sigma_c, \sigma'_c, \sigma''_c) \quad (97)\\ &\frac{54{,}000}{ab^2} \leq \max(\sigma_c, \sigma'_c, \sigma''_c) \quad (98)\end{aligned}\right\} \text{flexural strength}$$

$$\left.\begin{aligned}&0 \leq a \leq 1.0 \quad (99)\\ &0 \leq b \leq 2.0 \quad (100)\end{aligned}\right\} \text{dimensions}$$

This is a nonlinear optimization problem, since the objective function and constraints both contain products of variables. There are four design variables—a, b, v_{gl}, and v_{gr}—and six constraints. Nonlinear mathematical techniques for the solution are readily available [52] but beyond our scope. A rational solution will be presented.

Investigating first the possibility of using the cheapest component, resin alone, for a beam of maximum allowable dimensions,

$$\sigma_{cx} = \frac{27{,}000}{2} > 12{,}000 \text{ psi [resin strength, from Fig. 20(b)]}$$

eliminating resin alone as a material candidate.

*The constant 1 may be omitted from the optimization.

We next investigate the possibility of using the two cheapest components, resin and glass, for a beam of maximum dimensions and maximum glass loading of $v_{gl} = 80\%$. From Eqs. (86) and (92),

$$\sigma_c = 5{,}000(0.20) + 400{,}000(0.80) = 321{,}000 > \frac{27{,}000}{2} \text{ psi} \quad \textbf{OK.}$$

And by Eq. (90),

$$\Delta_x = \frac{5.83}{2}[9.5(0.8) + 0.5] = 0.36 < 0.30 \text{ in.} \quad \textbf{Not allowed.}$$

Thus for the glass-resin composite, optimal design is controlled by deflection, not by strength. This will also necessarily be true for composites containing both fibers, because $E_{gr}/E_{gl} \sim \sigma_{gr}/\sigma_{gl}$ (100/10 ~ 3/0.4) and the strength with glass fibers was satisfactory by a wide margin (321,000 ≫ 27,000/2). Further, the fact that Young's modulus/cost ratio of glass ($10 \times 10^6/2.5$) is greater than of graphite ($100 \times 10^6/200$) suggests that optimal design will incorporate the maximum fiber loading (80%), using just enough graphite to satisfy the governing deflection constraint. The limiting value of E_c from Eqs. (90) and (91), after substituting $\Delta x = 0.3$ in. and $\Delta y = 0.2$ in., is $E_c = 9.74 \times 10^6$ psi, in Eq. (90). Substituting this value into Eq. (89) and letting $v_{gl} + v_{gr} = 0.80$ yield

$$v_{gl} = 0.782 \quad \text{and} \quad v_{gr} = 0.018$$

Had it not been apparent by inspection that one of the two deflection constraints would control the design, the use of triangular property maps (see Sec. B in Chap. 7) (Fig. 22) combined with a plot of moments of inertia I_x and I_y as functions of beam dimensions a and b (Fig. 23) would provide sufficient representation for a graphical solution. The unlined areas in Figs. 22a, b, c represent the constraint $v_r \geq 20\%$. Strengths in Fig. 22(b) are plotted in three zones, corresponding to the three strength equations. For example, σ_c at the point P can be obtained by computing σ_c by each of the three equations for corresponding volume fractions v_{gl} and v_{gr} and plotting the maximum of these three values. Since strengths vary linearly with composition, Fig. 22(b) may be prepared by computing strengths at only three points from each of the Eqs. (86)–(88).

Fig. 22. Property maps for two-fiber composite. (a) Young's modulus, $\times 10^6$ psi; (b) flexural strength, $\times 10^6$ psi; (c) relative cost per unit volume.

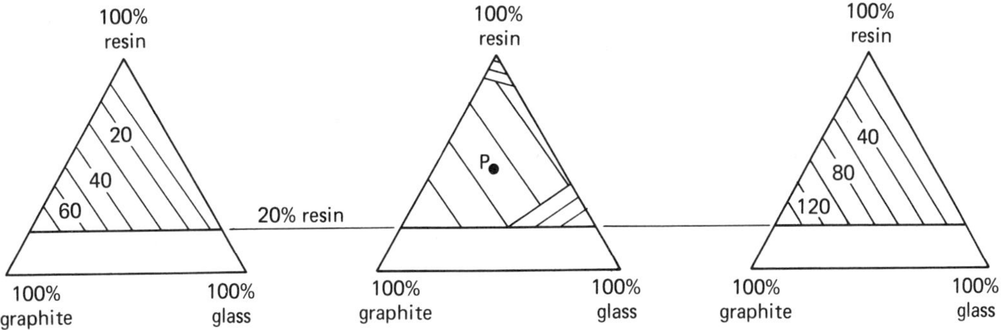

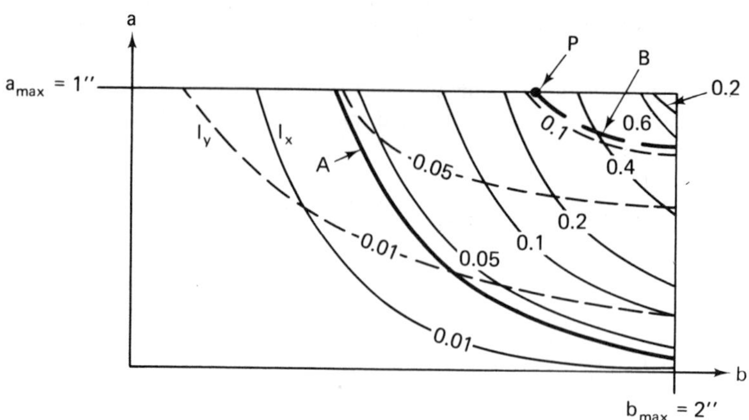

Fig. 23. I_x and I_y as functions of beam dimensions a and b. Curves A and B represent deflection constraints for x and y axis loadings, using an arbitrary value $E_c = 40 \times 10^6$ psi in Eqs. (90) and (91). Point P represents the beam cross section of minimum area for the arbitrary constraints.

McCullough and Peterson [53] discuss a similar example. More complex structural elements would require more involved treatments.

PROBLEMS

1. What is the critical length of an asbestos fiber having a tensile strength of 3×10^5 psi and a diameter of 10^{-5} in. embedded in phenolic having a shear yield strength of 18,000 psi? What is the average stress in a fiber 0.5 in. long stressed to its breaking strength at midsection, assuming stress buildup from the fiber ends to be linear?
2. 0.01-in.-diameter fibers 4 in. long are to be used for axial reinforcement in a tension specimen. If the maximum fiber stress is 300,000 psi and the maximum bond (shear) stress between fibers and matrix is 500 psi, which failure mode would you expect, bond or fiber tension? Assume that bond stress is constant (tensile stress builds up linearly from the fiber ends). What is the critical length/diameter ratio for fibers in this system?
3. What is the minimum volume fraction of a uniaxially reinforced continuous glass fiber-epoxy composite if the fiber tensile strength is 4×10^5 psi, the matrix tensile strength is 10,000 psi, and the stress borne by the matrix at fiber failure is 8,000 psi? What fiber volume fraction is required for the composite strength to exceed that of the matrix alone? What is the composite tensile strength for a fiber volume fraction of 0.50?
4. What is the composite tensile strength using the data in Problem 3 except that the fibers are 0.5 in. long and 0.008 in. in diameter and the matrix shear strength is 5,000 psi? What are the corresponding critical and minimum fiber volume fractions?
5. Refer to the fiber and matrix stress-strain curves in Fig. 24 (indicate equation numbers from the text, where used).
   (a) For continuous uniaxial fibers, what fiber volume fraction must be exceeded for the composite to be stronger in tension than the matrix alone?
   (b) For a fiber diameter of 0.01 in. and a metal matrix (linear stress increase at ends)

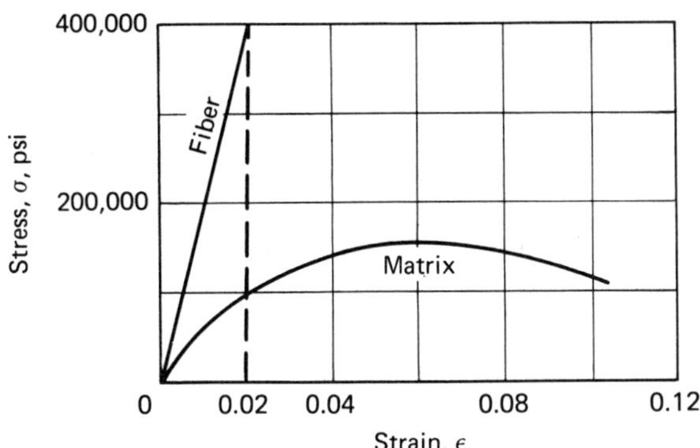

Fig. 24.

having a shear yield stress of 20,000 psi, what fiber length must be exceeded so that some fibers break rather than pull out of the matrix?

(c) If the fibers are 0.5 in. long and the fiber volume fraction is 0.6, what is the Young's modulus of the composite at failure strain?

(d) What is the composite strength?

(e) If the fibers were continuous but the matrix metal was foamed so that the volume fraction of air to total matrix = 0.6, what is the Young's modulus at failure strain? Assume the foam bubbles to be much smaller than the fiber diameter.

6. Compute the work of fracture required to pull a single 0.005-in.-diameter fiber from a matrix in which it is embedded to a depth of 0.5 in. Assume that the yield shear stress of the matrix is 20,000 psi, independent of shear strain.

7. What is the critical volume ratio (glass to total volume) for the fiber glass-reinforced polymer whose stress-strain characteristics are shown in Fig. 25?

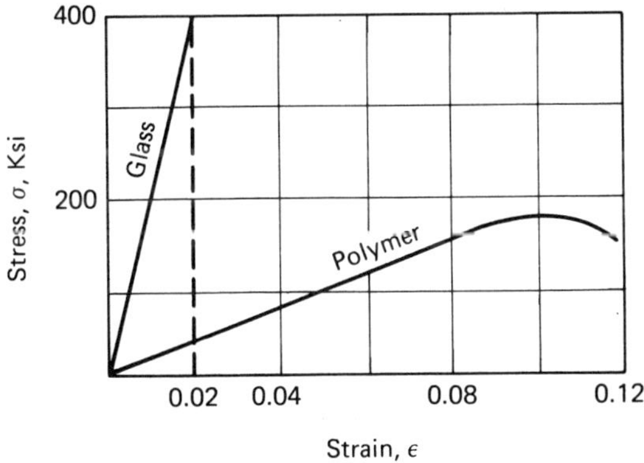

Fig. 25.

8. Uniaxial continuous fibers of two species are used to reinforce a matrix. Tensile stress-strain curves for the three components of the composite are shown in Fig. 26. The volume ratios are $v_{f1} = 0.3$, $v_{f2} = 0.4$, and $v_m = 0.3$. Using the simple mixture law, compute the Young's modulus of the composite at a strain of 0.01.

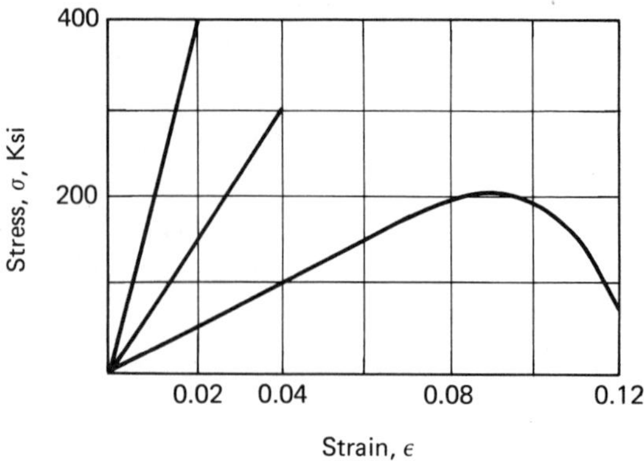

Fig. 26.

9. If in a glass fiber-polyester composite the volume cost ratio of glass to polyester is 2.5, what is the cheapest fiber volume ratio to carry tensile load? Assume continuous fibers having 4×10^5 psi tensile strength, a stress borne by the matrix at fiber failure of 8,000 psi, and an ultimate matrix strength of 13,000 psi.

10. Formulate the objective function and constraints for a problem similar to Problem 9 with the allowable strain of the tensile member introduced as a constraint.

11. The fracture surface of a tensile test specimen having 1-in.-long nylon fibers in portland cement mortar contained 72 fibers which pulled out and 54 fibers which fractured. What is the critical fiber length?

12. Determine the critical fiber length in a uniaxially reinforced silica fiber-aluminum composite from the following tensile test results:

Fiber Volume Fraction	*Composite Tensile Strength (psi)*
0.30	50×10^3
0.35	70×10^3

The fiber strength is 5×10^5 psi, the fiber length is 0.6 in., and the stress borne by the matrix at fiber failure can be taken as 16,000 psi.

13. Assuming that the ratio $h/(h + 2c)$ in Fig. 8 can be represented as the fiber volume fraction, determine the fiber volume fraction at which the strengths by shear and extension modes of compressive failure are equal. Assume Young's moduli of 10×10^6 psi and 3.5×10^5 psi for the fiber and matrix, respectively, and a matrix shear modulus of 1.2×10^5 psi.

14. For a fiber-matrix unit cost ratio of 2.5 and the moduli of Problem 13, determine the cheapest fiber volume ratio to produce a composite to carry a compressive load. Assume that the fiber volume ratio can be represented as $h/(h + 2c)$.

15. Formulate the objective function and constraints for a problem similar to Problem 14 with the allowable strain of the compressive member introduced as a constraint.

16. Determine the diagonal tensile strength at $\phi = 8°$ of a uniaxially reinforced continuous fiber composite having the following properties:

$$V_f = 0.40 \qquad \sigma_u = 14{,}000 \text{ psi}$$
$$\sigma_f = 6 \times 10^5 \text{ psi} \qquad \tau_u = 6{,}000 \text{ psi}$$
$$\sigma'_m = 9{,}000 \text{ psi}$$

Assume brittle fibers (see Fig. 6).

17. Determine the composite Young's modulus in the direction of tensile load for a fiber-reinforced composite having the following properties:

Two layers of fiber reinforcement at $\phi = \pm\pi/4$
$v_f = 0.50$
$E_f = 12 \times 10^6$ psi
$E_m = 3 \times 10^5$ psi
$\nu_m = 0.38$

18. Compute the tensile work of fracture of a 1-sq-in.-cross-section rod reinforced with randomly distributed uniaxial fibers of length 1 in. and diameter 0.005 in. if the yield shear stress between matrix and fibers is 20,000 psi, independent of shear strain. The fiber strength is $\sigma_f = 300{,}000$ psi, and the fiber volume ratio is $v_f = 0.70$.

REFERENCES

1. KELLY, A., and G. R. DAVIES. *Metal. Rev.*, 10 (1965) 1.
2. ROSEN, B. W., *Fiber Composite Materials.* Metals Park, Ohio: American Society for Metals, 1965.
3. ARRIDGE, R. G., A. A. BAKER, and D. CRATCHLEY, *J. Sci. Instruments*, 41 (1964).
4. *Fiberglas.* Toledo, Ohio: Owens-Corning Corporation, 1974.
5. LOEWENSTEIN, K. L., "Glass Systems," in *Composite Materials*, L. Holliday, ed. Amsterdam: Elsevier Publishing Company, 1966, Chap. 5.
6. ANDERSON, S., and D. D. KIMPTON, *J. Am. Ceram. Soc.*, 43 (1960) 484.
7. PETERSON, J. M., *Bull. Am. Phys. Soc.*, XV (March 1970) 329.
8. EAKINS, W. J., *Plastic Report 18.* Dover, N.J.: Plastics Technical Evaluation Center, Picatinny Arsenal, Nov. 1964.
9. LOWRIE, R. E., "Glass Fibers for High-Strength Composites," in *Modern Composite Materials*, L. J. Broutman and R. H. Krock, eds. Reading, Mass.: Addison-Wesley Publishing Company, Inc., 1967, Chap. 11.
10. KELLY, A., *Strong Solids.* New York: Oxford University Press, 1966.
11. DOW, N. F., *General Electric Co. Rpt. R635D61.* Schenectady, N.Y.: General Electric Co., 1963.
12. HILL, R. J., *Mech. Phys. Solids*, 12 (1964) 199, 213; 13 (1965) 189.
13. HASHIN, Z., and B. W. ROSEN, *J. Appl. Mech.*, 31 (1964) 223.
14. COX, H. L., *Brit. J. Appl. Phys.*, 3 (1952) 72.
15. OUTWATER, J. O., JR., *Mod. Plastics*, 33 (1956) 156.
16. DANIEL, I. M., and A. J. DURELLI, *Paper 19A.* 16th Conference of the Society of the Plastics Industry, 1961.

17. PARRATT, N. J., *Rubber Plastics Age*, 41, No. 3 (1960) 263.

18. MCDANIELS, D. L., R. W. JECH, and J. W. WEETON, *Metal Prog.*, 78, No. 6 (1960) 118.

19. CRATCHLEY, D., *Powder Met.*, 11 (1963) 59.

20. TYSON, W. R., Ph.D. thesis. University of Cambridge, 1964.

21. PIEHLER, H. R., *Trans. Met. Soc. AIME*, 233 (1965) 12.

22. KELLY, A., and W. R. TYSON, *J. Mech. Phys. Solids*, 13 (1965) 329.

23. SCHEURCH, H., *NASA Report CR-202*, Washington D.C.: Natl. Aeron. & Space Admin., 1965.

24. TIMOSHENKO, S., *Theory of Elastic Stability*. New York: McGraw-Hill Book Company, 1936.

25. JACKSON, P. W., and D. CRATCHLEY, *J. Mech. Phys. Solids*, 14 (1966) 49.

26. SHAFFER, B. W., *Soc. Plastics Engrs. Trans.*, 4 (1964) 267.

27. TSAI, S. W., *Proceedings of the 6th Annual Symposium of Filament Structure Technology*, Albuquerque, N.M., 1965.

28. STAVSKY, Y. and N. J. HOFF, "Mechanics of Composite Structures" (Chap. 1), in *Composite Engineering Laminates*, A. G. Dietz, ed. New York: John Wiley & Sons, Inc. (1969), p. 1.

29. HERMANS, J. J., *Koninklijke Nederlandse Akademic van Wetenschappen, Amsterdam, Proc. Series B*, 70, No. 1 (1967) 1.

30. TSAI, S. W., *Natl. Aeron. and Space Admin. Rpt. CR71*, Washington, D.C.: Natl. Aeron. & Space Admin., 1964.

31. DIETZ, A. G. H., "Fiber Reinforced Composite Materials Engineering Analysis," in *Environmental Effects on Polymeric Materials*, Vol. II: Materials, D. V. Rosato and R. T. Schwartz, eds. New York: John Wiley & Sons, Inc. (Interscience Division), 1968, Sec. 15.

32. NEILSON, L. E., and P. E. CHEN, *J. Mater.*, 3, No. 2 (June 1968) 352.

33. KRENCHEL, H., *Fiber Reinforcement*. Copenhagen: Akademisk Forlag, 1964.

34. LUBIN, GEORGE, ed., *Handbook of Fiberglass and Advanced Plastics Composites*. New York; Van Nostrand Reinhold Company, 1969.

35. MCCULLOUGH, R. L., *Concepts of Fiber-Resin Composites*. New York: Marcel Dekker, Inc., 1971.

36. ALLEN, H. G., *J. Composite Mater.*, 5 (April 1971) 194.

37. BIRYUKOVICH, D. L., et al., *Budiveli'nik, Kiev* (1964), CERA Translation No. 12.

38. MARIES, A., and A. C. TSEUNG, *Proceedings of the Southampton 1969 Civil Engineering Materials Conference on Structure, Solid Mechanics and Engineering Design*, Part II. New York: John Wiley & Sons, Inc. (Interscience Division), 1970.

39. MONFORE, G. E., *J. Port. Cem. Assoc. Res. Develop. Labs* (Sept. 1968) 43.

40. SHAH, S. P., *Proceedings of the Conference on New Materials in Concrete Construction*. Chicago Circle: University of Illinois, Dec. 1971, p. 1-IV.

41. SHANNON, R. F., and R. H. MITCHELL, Coating of Glass Fibers for Use in Pressure-Molded Cementitious Products. U.S. Patent 2,793,130, 1957 (to Owens-Corning Fiberglas Corp.).

42. SHANNON, R. E., Glass-Reinforced Cementitious Articles. U.S. Patent 3,147,127, 1964 (to Owens-Corning Fiberglas Corp.).

43. PROCTOR, B. A., *Phys. Chem. Glasses*, 3 (1962) 7.

44. "American Concrete Institute Building Code Requirements for Reinforced Concrete," *ACI Standard 318–71*. Detroit: American Concrete Institute (1971).

45. IYENGAR, M. K., "Organic Fiber-Reinforced Concrete," Ph.D. thesis. Newark, Del.: University of Delaware, 1973.

46. POSHCHENKO, A. A., V. P. SERBIN, and I. I. KLETCHENKOV, "The Adhesion Between Hardened Cement Paste and Glass Fiber," *Izv. Stroit. Arkhitckt.* (*Novosibirsk*), 8, No. 4 (1965) 89.

47. TAYLOR, M. A., and B. B. BROMS, *J. Am. Concr. Inst.*, 61 (1964) 939.

48. HILDEBRAND, F. B., *Advanced Calculus for Applications*. Englewood Cliffs, N.J.: Prentice-Hall, Inc., 1962.

49. NAAMAN, A. E., A. S. ARGON, and F. MOAVENZADEH, *J. Cem. Concr. Res.*, 3 (1973) 397.

50. GOLDFINE, S., *Mod. Plastics*, XLII (April 1965) 156.

51. HARRIS, B., J. VARLOW, and C. D. ELLIS, *J. Cem. Concr. Res.*, 2 (1972) 447.

52. STARK, R. M., and R. L. NICHOLLS, *Mathematical Foundations for Design: Civil Engineering Systems*. New York: McGraw-Hill Book Company, 1972.

53. MCCULLOUGH, R. L., and J. M. PETERSON, "Principles of Materials Design for Multicomponent Composite Systems," Seattle: *Boeing Scientific Research Laboratories Doc. D1-82-0941*, July 1971.

12

STRUCTURAL LAMINATES

You see, of course, if you're not a dunce,
How it went to pieces all at once,
All at once, and nothing first,
Just as bubbles do when they burst.

—HOLMES

Laminar composites provide a means to obtain desired properties such as strength or rigidity in given directions while maintaining economy of materials. Examples of laminates in construction include laminated timber beams and plywood, glass fiber laminates for concrete forms, car and truck bodies, cargo containers and boat hulls, and an endless variety of structural sandwich panels and shells. A very challenging laminate design problem is the determination of optimal cord directions and ply thicknesses in rubber tires.

In the Sec. A we shall develop skills for the analysis and design of laminates, and in Sec. B we shall deal with sandwich panels, a specific class of laminates used extensively in building construction.

A. LAMINAR COMPOSITES

In this section we shall first extend the method for deformation of uniaxial and multiaxial fiber orientations from the previous chapter to the stress and strain analyses of some simple laminates under specialized loading conditions (Sec. A1). Then after summarizing Hooke's law and the stress and strain transformations basic to all problems in solid mechanics, a more general method will be developed for laminate analysis (Sec. A2).

In contrast to the micromechanics approach of the previous chapter, the macromechanics approach used here ignores fiber-resin interaction and considers the lamina as homogeneous media. The principal references are Dietz [1], Ashton et al. [2], and Stausky and Hoff [3].

1. Simplified Analysis for Special Laminar and Loading Conditions

Consider a symmetrical laminate having a and b layers with principal longitudinal and transverse directions L_a and T_a and L_b and T_b, respectively, and total thicknesses of the a layers, b layers, and total laminate being t_a, t_b, and t, respectively (Fig. 1). External stresses σ_1, σ_2, and τ_{12} are applied in the 1 and 2 directions, as shown. The 1 direction makes angles α and β with the L_a and L_b directions, respectively.

To obtain laminar stresses in the individual layers σ_{1a}, σ_{2a}, τ_{12a} and σ_{1b}, σ_{2b}, τ_{12b} we observe that the sums of internal stresses in the 1 and 2 directions must equal the external stresses in these directions and that strains in a symmetrical laminate with

Fig. 1. Symmetrical laminate with different orthotropic layers a and b oriented at angles α and β to applied stresses σ_1, and σ_2 and τ_{12}.

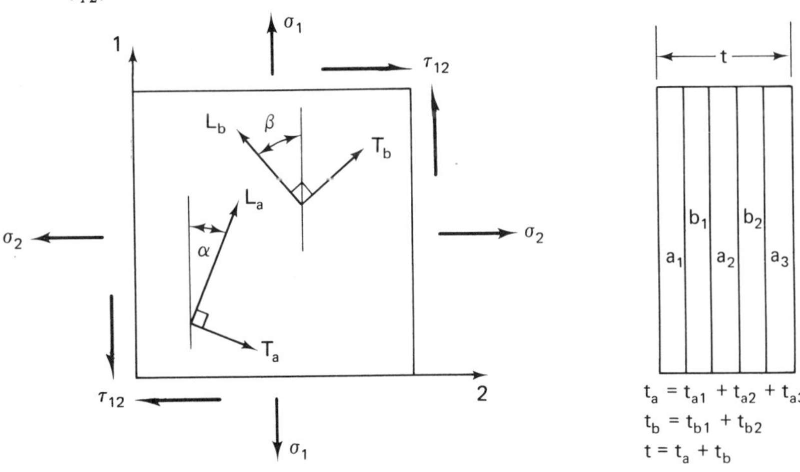

perfect bonding must be the same in all layers. Therefore,

$$\sigma_{1a}t_a + \sigma_{1b}t_b = \sigma_1 t \tag{1}$$

$$\sigma_{2a}t_a + \sigma_{2b}t_b = \sigma_2 t \tag{2}$$

$$\tau_{12a}t_a + \tau_{12b}t_b = \tau_{12} t \tag{3}$$

and

$$\epsilon_{1a} = \epsilon_{1b} = \epsilon_1, \qquad \epsilon_{2a} = \epsilon_{2b} = \epsilon_2, \qquad \gamma_{12a} = \gamma_{12b} = \gamma_{12} \tag{4}$$

The strains may be expressed in terms of stresses as

$$\epsilon_1 = \begin{cases} \epsilon_{1a} = \dfrac{\sigma_{1a}}{E_{1a}} - \nu_{21a}\dfrac{\sigma_{2a}}{E_{2a}} - m_{1a}\dfrac{\tau_{12a}}{E_{La}} \\[2ex] \epsilon_{1b} = \dfrac{\sigma_{1b}}{E_{1b}} - \nu_{21b}\dfrac{\sigma_{2b}}{E_{2b}} - m_{1b}\dfrac{\tau_{12b}}{E_{Lb}} \end{cases} \tag{5}$$

$$\epsilon_2 = \begin{cases} \epsilon_{2a} = -\nu_{12a}\dfrac{E_{1a}}{\sigma_{1a}} + \dfrac{\sigma_{2a}}{E_{2a}} - m_{2a}\dfrac{\tau_{12a}}{E_{La}} \\[2ex] \epsilon_{2b} = -\nu_{12b}\dfrac{\sigma_{1b}}{E_{1b}} + \dfrac{\sigma_{2b}}{E_{2b}} - m_{2b}\dfrac{\tau_{12b}}{E_{Lb}} \end{cases} \tag{6}$$

$$\gamma_{12} = \begin{cases} \gamma_{12a} = -m_{1a}\dfrac{\sigma_{1a}}{E_{La}} - m_{2a}\dfrac{\sigma_{2a}}{E_{La}} + \dfrac{\tau_{12a}}{G_{12a}} \\[2ex] \gamma_{12b} = -m_{1b}\dfrac{\sigma_{1b}}{E_{Lb}} - m_{2b}\dfrac{\sigma_{2b}}{E_{Lb}} + \dfrac{\tau_{12b}}{G_{12b}} \end{cases} \tag{7}$$

where m_1 and m_2 for any layer are given by Eqs. (60) and (65) in Chap. 11. Solution of Eqs. (1)–(7) yields the following simultaneous equations:

$$A_{11}\sigma_{1a} + A_{12}\sigma_{2a} + A_{13}\tau_{12a} = \frac{t}{t_a t_b}\left(\frac{\sigma_1}{E_{1b}} - \nu_{21b}\frac{\sigma_2}{E_{2b}} - m_{1b}\frac{\tau_{12}}{E_{Lb}}\right) \tag{8}$$

$$A_{21}\sigma_{1a} + A_{22}\sigma_{2a} + A_{23}\sigma_{12a} = \frac{t}{t_a t_b}\left(-\nu_{12b}\frac{\sigma_1}{E_{1b}} + \frac{\sigma_2}{E_{2b}} - m_{2b}\frac{\tau_{12}}{E_{Lb}}\right) \tag{9}$$

$$A_{31}\sigma_{1a} + A_{32}\sigma_{2a} + A_{33}\tau_{12a} = \frac{t}{t_a t_b}\left(-m_{1b}\frac{\sigma_1}{E_{Lb}} - m_{2b}\frac{\sigma_2}{E_{Lb}} + \frac{\tau_{12}}{G_{12b}}\right) \tag{10}$$

in which

$$A_{11} = \frac{1}{E_{1a}t_a} + \frac{1}{E_{1b}t_b} \qquad A_{12}{}^* = -\frac{\nu_{21a}}{E_{2a}t_a} - \frac{\nu_{21b}}{E_{2b}t_b} \qquad A_{13} = -\frac{m_{1a}}{E_{La}t_a} - \frac{m_{1b}}{E_{Lb}t_b}$$

$$A_{21}{}^* = -\frac{\nu_{12a}}{E_{1a}t_a} - \frac{\nu_{12b}}{E_{1b}t_b} \qquad A_{22} = \frac{1}{E_{2a}t_a} + \frac{1}{E_{2b}t_b} \qquad A_{23} = -\frac{m_{2a}}{E_{La}t_a} - \frac{m_{2b}}{E_{Lb}t_b}$$

$$A_{31} = A_{13} \qquad A_{32} = A_{23} \qquad A_{33} = \frac{1}{G_{12a}t_a} + \frac{1}{G_{12b}t_b}$$

---

*$A_{21} = A_{12}$, numerically.

The following example illustrating applications of these expressions is adopted with permission from A. G. H. Dietz [1].

EXAMPLE 1: Consider a 9.6-in.-ID cylindrical pressure vessel having an 0.20-in.-thick wall subjected to 800 psi of internal pressure. The circumferential and longitudinal wall stresses, σ_1 and σ_2, are

$$\sigma_1 = \frac{pr}{t} = 19{,}200 \text{ psi}$$

$$\sigma_2 = \frac{pr}{2t} = 9{,}600 \text{ psi}$$

Determine the circumferential and longitudinal strains ϵ_1, ϵ_2, and γ_{12} for each of the three two-ply laminate constructions shown in Fig. 2(b). In each case $t_a = t_b = 0.10$ in., $\nu_{LTa} = \nu_{TLa} = \nu_{LTb} = \nu_{TLb} = 0.20$, $G_{LTa} = G_{TLa} = G_{LTb} = G_{TLb} = 0.5 \times 10^6$ psi and $E_{La} = E_{Lb} = E_{Ta} = E_{Tb} = 3 \times 10^6$ psi (balanced lamina having the characteristics shown in Fig. 15 of Chap. 11).

Referring to Fig. 2, for case 1,

$$E_{1a} = E_{1b} = E_{2a} = E_{2b} = 3 \times 10^6 \text{ psi}$$

$$\nu_{12a} = \nu_{21a} = \nu_{12b} = \nu_{21b} = 0.20$$

$$m_{1a} = m_{2a} = m_{1b} = m_{2b} = 0$$

$$A_{11} = A_{22}, \qquad A_{12} = A_{21}$$

$$A_{13} = A_{31} = A_{23} = A_{32} = 0$$

Equations (8)–(10) therefore become

$$A_{11}\sigma_{1a} + A_{12}\sigma_{2a} + 0 = \frac{t}{t_a t_b}\left(\frac{\sigma_1}{E_{1b}} - \nu_{21b}\frac{\sigma_2}{E_{2b}} + 0\right)$$

$$A_{21}\sigma_{1a} + A_{22}\sigma_{2a} + 0 = \frac{t}{t_a t_b}\left(-\nu_{12}\frac{\sigma_1}{E_{1b}} + \frac{\sigma_2}{E_{2b}} + 0\right)$$

$$A_{33}\sigma_{12a} = 0$$

and their solution is

$$\sigma_{1a} = \sigma_{1b} = \sigma_1 = 19{,}200 \text{ psi}$$

$$\sigma_{2a} = \sigma_{2b} = \sigma_2 = 9{,}600 \text{ psi}$$

$$\tau_{12a} = \tau_{12b} = 0$$

This result, with internal direct stresses equal to applied stresses and shear stresses equal to zero, might have been expected because of laminar symmetry with respect to the 1 and 2 directions.

The same result is obtained in case 2, where $m_1 = m_2 = 0$ at 45° and shear stresses are zero.

In case 3, from Eqs. (51), (62), (58), (60), and (65), respectively, of Chap. 11,

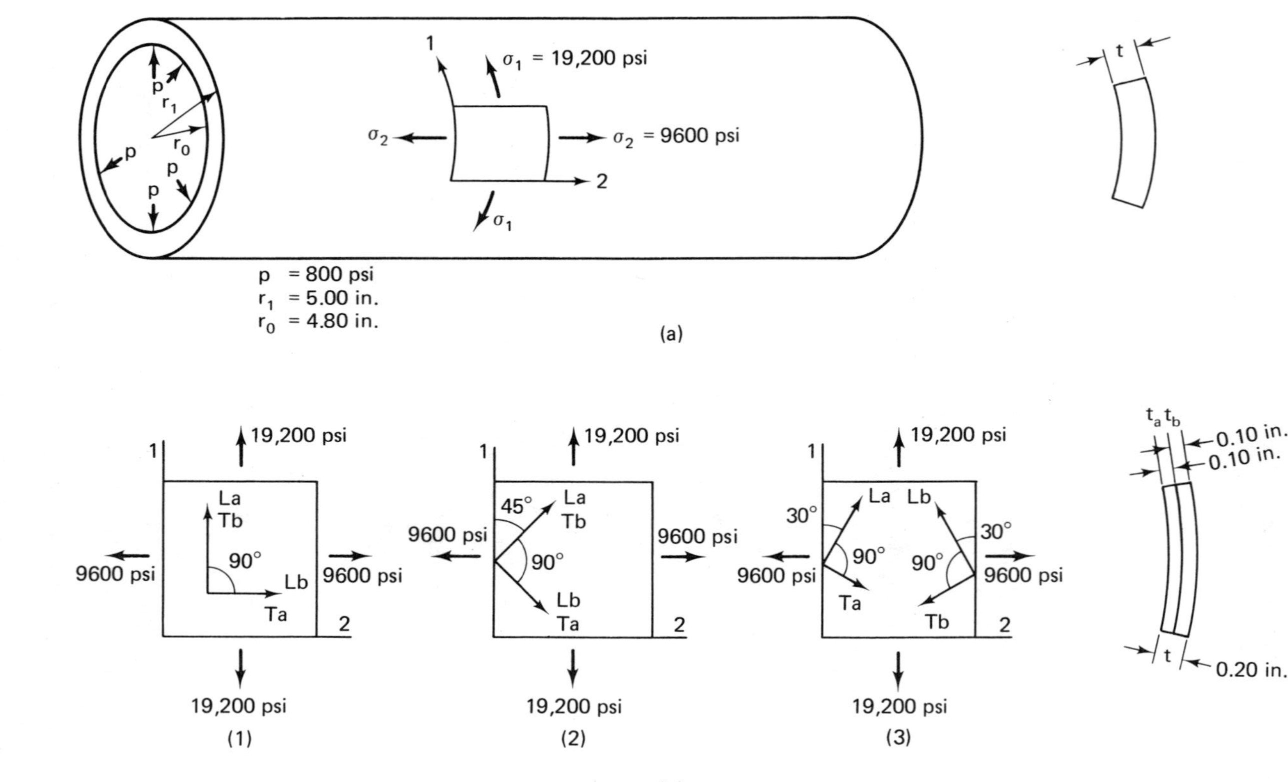

Fig. 2. Fiber glass-reinforced thin wall cylinder. (a) Applied stresses; (b) lamina orientation. From Dietz, A. G. H., "Fiber Reinforced Composite Materials Engineering Analysis," in *Environmental Effects on Polymeric Materials*, Vol. II: Materials, D. V. Rosato and R. T. Schwartz, eds. New York: John Wiley & Sons, Inc. (Interscience Division) 1968, Sec. 15 [1].

$$E_{1a} = E_{1b} = E_{2a} = E_{2b} = E_{30^\circ} = E_{60^\circ} = 0.597 \times 3 \times 10^6 = 1.78 \times 10^6 \text{ psi}$$

$$G_{12a} = G_{12b} = 1.82 \times 0.5 \times 10^6 = 0.91 \times 10^6 \text{ psi}$$

$$\nu_{12a} = \nu_{12b} = \nu_{30^\circ} = \nu_{60^\circ} = \nu_{21a} = \nu_{21b} = 0.523$$

$$m_{1a} = m_{30^\circ} = 0.775, \qquad m_{1b} = -m_{1a} = -0.775$$

The values of m_{1b} and m_{2b} are negative because the 30° angle of the longitudinal direction L_b of layer b is measured in the negative direction, whereas it is positive for the a layer.

Equations (8)–(10) become

$$A_{11}\sigma_{1a} + A_{12}\sigma_{2a} + 0 = \frac{t}{t_a t_b}\left(\frac{\sigma_1}{E_{1b}} - \nu_{21b}\frac{\sigma_2}{E_{2b}} + 0\right)$$

$$A_{21}\sigma_{1a} + A_{22}\sigma_{2a} + 0 = \frac{t}{t_a t_b}\left(-\nu_{12b}\frac{\sigma_1}{E_{1b}} + \frac{\sigma_2}{E_{2b}} + 0\right)$$

$$0 + 0 + A_{33}\tau_{12a} = \frac{t}{t_a t_b}\left(-m_{1b}\frac{\sigma_1}{E_{Lb}} - m_{2b}\frac{\sigma_2}{E_{Lb}} + 0\right)$$

The first two of these equations are identical to those for cases 1 and 2:

$$\sigma_{1a} = \sigma_{1b} = \sigma_1 = 19{,}200 \text{ psi}$$

$$\sigma_{2a} = \sigma_{2b} = \sigma_2 = 9{,}600 \text{ psi}$$

The third equation, however, does not equal zero, and its solution, together with Eq. (5), gives

$$\tau_{12a} = 6{,}750 \text{ psi}$$

$$\tau_{12b} = -6{,}750 \text{ psi}$$

Thus, in case 3 shear stresses are developed in the cylinder wall even without externally applied shear stresses. The difference between shear stresses in the lamina must be carried by shear in the adhesive bond between them; i.e.,

$$6{,}750 - (-6{,}750) = 13{,}500 \text{ psi}$$

In all three cases internal direct stresses equaled applied stresses, due to the symmetric orientation of the laminar axes with respect to the applied stress directions. In the first two cases, the chosen orientations resulted also in zero shear stresses. In more general cases the direct stresses in the individual layers do not equal the applied direct stresses, nor are they equal in the various layers. In addition, bending and twisting deformations may result from direct applied stresses, and normal deformations may result from bending and twisting loads.

For more general problems in laminate analysis, it is useful to employ the matrix solutions developed in the following sections. The few necessary matrix operations are described in Appendix C.

2. General Analysis

Definitions. Material *symmetry* is denoted by the invariance of properties such as stiffness or thermal conductivity under a specified transformation of coordinates. *Isotropic* means that material properties at a point in a body are not a function of orientation. In isotropic materials all planes which pass through a point are planes of material property symmetry. *Transverse isotropy* indicates invariance of properties under a transformation of rotation in a given plane. *Orthotropic* materials have only three mutually perpendicular planes of material property symmetry passing through any point. *Aeolotropic* materials have only one plane of material property symmetry, and *anisotropic* materials have no planes of material property symmetry.

Each symmetry class is a more generalized case of the previous class and requires more coefficients to completely describe the material properties at a point. For stiffness (Young's modulus) the following numbers of coefficients are required:

Symmetry Class	*Number of Coefficients to Describe Stiffness*
Isotropic	2
Transverse isotropic	5
Orthotropic	9
Aeolotropic	13
Anisotropic	21

This will be illustrated shortly.

A material of any symmetry class may be either *homogeneous* or *heterogeneous*. Properties do not vary with position (as contrasted with orientation) in homogeneous materials but do in heterogeneous materials.

Hooke's law. Hooke's law (Sec. A in Chap. 3) is the relationship between stress and strain. For a homogeneous isotropic material in one-dimensional stress, Hooke's law is

$$\sigma = E\epsilon \tag{11}$$

Therefore, one elastic constant, E, is required to write Hooke's law for one-dimensional stress in isotropic materials. In two- and three-dimensional stress a second elastic constant, Poisson's ratio, ν, is required to define the stress-strain relationship. For example, in two-dimensional (plane) stress, Eq. (6) of Chap. 3 reduces to

$$\begin{aligned} \sigma^{1} &= (\epsilon_1 + \nu\epsilon_2)\frac{E}{1-\nu^2} \\ \sigma_2 &= (\epsilon_2 + \nu\epsilon_1)\frac{E}{1-\nu^2} \\ \tau_{12} &= \gamma_{12}G \end{aligned} \tag{12}$$

The third elastic constant, shear modulus G, is related to the other two by

$$G = \frac{E}{2(1 + \nu)} \tag{13}$$

Therefore two independent constants are required to write Hooke's law for two- or three-dimensional stress in isotropic materials.

An orthotropic material (three planes of symmetry) subjected to plane stress requires four independent elastic constants. This is the most frequently encountered case in the study of laminated composites, including plywoods and fiber-reinforced laminates. It requires nine constants in three-dimensional stress.

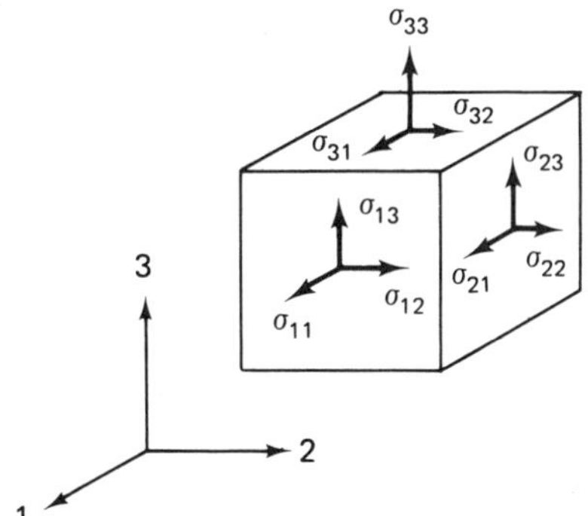

Fig. 3. Three dimensional stress.

An anistropic material requires 6 elastic constants for plane stress and 21 constants for three-dimensional stress. The generalized Hooke's relation for three-dimensional stress in an anistropic material (Fig. 3) is

$$\begin{bmatrix} e_{11} \\ e_{22} \\ e_{33} \\ e_{23} \\ e_{31} \\ e_{12} \\ e_{32} \\ e_{13} \\ e_{21} \end{bmatrix} = \begin{bmatrix} F_{1111} & F_{1122} & F_{1133} & F_{1123} & F_{1131} & F_{1112} & F_{1132} & F_{1113} & F_{1121} \\ F_{2211} & F_{2222} & F_{2233} & F_{2223} & F_{2231} & F_{2212} & F_{2232} & F_{2213} & F_{2221} \\ - & & & & & & & & - \\ - & & & & & & & & - \\ - & & & & & & & & - \\ - & & & & & & & & - \\ - & & & & & & & & - \\ - & & & & & & & & - \\ F_{2111} & F_{2122} & F_{2133} & F_{2123} & F_{2131} & F_{2112} & F_{2132} & F_{2113} & F_{2121} \end{bmatrix} \begin{bmatrix} \sigma_{11} \\ \sigma_{22} \\ \sigma_{33} \\ \sigma_{23} \\ \sigma_{31} \\ \sigma_{12} \\ \sigma_{32} \\ \sigma_{13} \\ \sigma_{21} \end{bmatrix} \tag{14}$$

The **F*** matrix is the flexibility matrix, which gives strain-stress relations for the material. The inverse of the flexibility matrix, the stiffness matrix, **S**, will later be used

*Boldface letters indicate a matrix.

for obtaining stress-strain relations. It has been shown [4] that the stress, strain, flexibility, and stiffness matrices must be symmetrical; i.e.,

$$\sigma_{12} = \sigma_{21}, \qquad \sigma_{23} = \sigma_{32}, \qquad \sigma_{13} = \sigma_{31}$$
$$\epsilon_{12} = \epsilon_{21}, \qquad \epsilon_{23} = \epsilon_{32}, \qquad \epsilon_{13} = \epsilon_{31}$$
$$F_{1122} = F_{2211}, \qquad F_{1121} = F_{2111}, \qquad \text{etc.}$$

These symmetries permit the reduction of Eq. (14) to only 6 independent stresses and strains, and 36 components of the **F** matrix of which only 21 are independent. It can also be shown, from equilibrium considerations, that $\sigma_{12} = \tau_{12}$, $\sigma_{13} = \tau_{13}$, $\epsilon_{12} = \gamma_{12}$, and $\epsilon_{13} = \gamma_{13}$. After substituting a contracted notation to shorten the subscripts, the three-dimensional stress-strain relation for an anisotropic material becomes

$$\begin{bmatrix} \epsilon_1 \\ \epsilon_2 \\ \epsilon_3 \\ \gamma_{23} \\ \gamma_{13} \\ \gamma_{12} \end{bmatrix} = \begin{bmatrix} F_{11} & F_{12} & F_{13} & F_{14} & F_{15} & F_{16} \\ F_{12} & F_{22} & F_{23} & F_{24} & F_{25} & F_{26} \\ - & & & & & - \\ - & & & & & - \\ - & & & & & - \\ F_{16} & F_{26} & F_{36} & F_{46} & F_{56} & F_{66} \end{bmatrix} \begin{bmatrix} \sigma_1 \\ \sigma_2 \\ \sigma_3 \\ \tau_{23} \\ \tau_{13} \\ \tau_{12} \end{bmatrix}$$

The corresponding relation for an orthotropic material in three-dimensional stress reduces to

$$\begin{bmatrix} \epsilon_1 \\ \epsilon_2 \\ \epsilon_3 \\ \gamma_{23} \\ \gamma_{13} \\ \gamma_{12} \end{bmatrix} = \begin{bmatrix} F_{11} & F_{12} & F_{13} & 0 & 0 & 0 \\ F_{12} & F_{22} & F_{23} & 0 & 0 & 0 \\ F_{13} & F_{23} & F_{33} & 0 & 0 & 0 \\ 0 & 0 & 0 & F_{44} & 0 & 0 \\ 0 & 0 & 0 & 0 & F_{55} & 0 \\ 0 & 0 & 0 & 0 & 0 & F_{66} \end{bmatrix} \begin{bmatrix} \sigma_1 \\ \sigma_2 \\ \sigma_3 \\ \tau_{23} \\ \tau_{13} \\ \tau_{12} \end{bmatrix} \qquad (15)$$

requiring nine independent elastic constants.

In two-dimensional stress we assume that $\sigma_3 = \sigma_{23} = \tau_{13} = 0$, and substitution into Eq. (15) yields $\gamma_{23} = \gamma_{13} = 0$ and $\epsilon_3 = F_{13}\sigma_1 + F_{23}\sigma_2$. Thus ϵ_3 is not independent but a function of the other normal strains and may be dropped. Equation (15) then reduces to

$$\begin{bmatrix} \epsilon_1 \\ \epsilon_2 \\ 0 \\ 0 \\ 0 \\ \gamma_{12} \end{bmatrix} = \begin{bmatrix} F_{11} & F_{12} & F_{13} & 0 & 0 & 0 \\ F_{12} & F_{22} & F_{23} & 0 & 0 & 0 \\ F_{13} & F_{23} & F_{33} & 0 & 0 & 0 \\ 0 & 0 & 0 & F_{44} & 0 & 0 \\ 0 & 0 & 0 & 0 & F_{55} & 0 \\ 0 & 0 & 0 & 0 & 0 & F_{66} \end{bmatrix} \begin{bmatrix} \sigma_1 \\ \sigma_2 \\ 0 \\ 0 \\ 0 \\ \tau_{12} \end{bmatrix}$$

or more compactly as

$$\begin{bmatrix} \epsilon_1 \\ \epsilon_2 \\ \gamma_{12} \end{bmatrix} = \begin{bmatrix} F_{11} & F_{12} & 0 \\ F_{12} & F_{22} & 0 \\ 0 & 0 & F_{66} \end{bmatrix} \begin{bmatrix} \sigma_1 \\ \sigma_2 \\ \tau_{12} \end{bmatrix} \tag{16}$$

the stress-strain relation for an orthotropic material in plane stress.

Stress and strain transformations. In laminated composites the material property axes of individual lamina are laid at various orientations, and it is useful to transform their stresses and strains to a single coordinate set for the entire laminate. This requires use of the stress and strain transformations, which are given graphically by Mohr's circle.

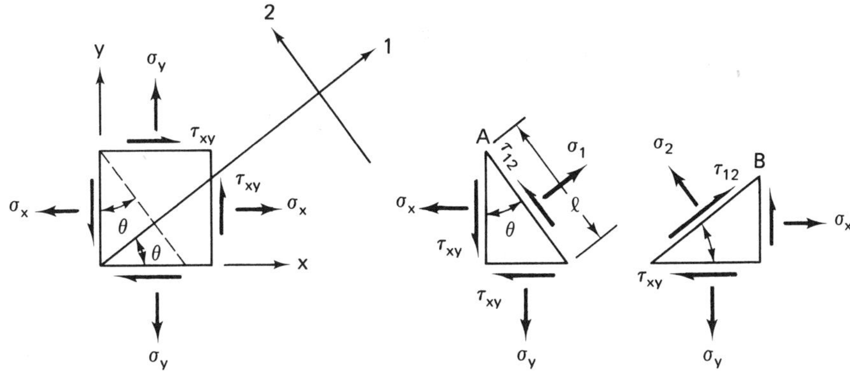

Fig. 4. Stress transformation.

Consider the transformation of two-dimensional *stress* from the *x-y* coordinates through angle θ to the 1–2 coordinates (Fig. 4). Equating forces in the 1 direction on free body *A*,

$$\sigma_1 l - \sigma_x \cos\theta(l\cos\theta) - \sigma_y \sin\theta(l\sin\theta) - \tau_{xy}\cos\theta(l\sin\theta) - \tau_{xy}\sin\theta(l\cos\theta) = 0$$

from which

$$\sigma_1 = \sigma_x \cos^2\theta + \sigma_y \sin^2\theta + 2\tau_{xy}\sin\theta\cos\theta \tag{17}$$

Equating forces in the 2 direction on free body *A* in similar fashion gives

$$\tau_{12} = (\sigma_y - \sigma_x)\sin\theta\cos\theta + \tau_{xy}(\cos^2\theta - \sin^2\theta) \tag{18}$$

and equating forces in the 2 direction on body *B* gives

$$\sigma_2 = \sigma_x \sin^2\theta + \sigma_y \cos^2\theta - 2\tau_{xy}\sin\theta\cos\theta \tag{19}$$

Equating forces in the 1 direction on body *B* gives Eq. (17).

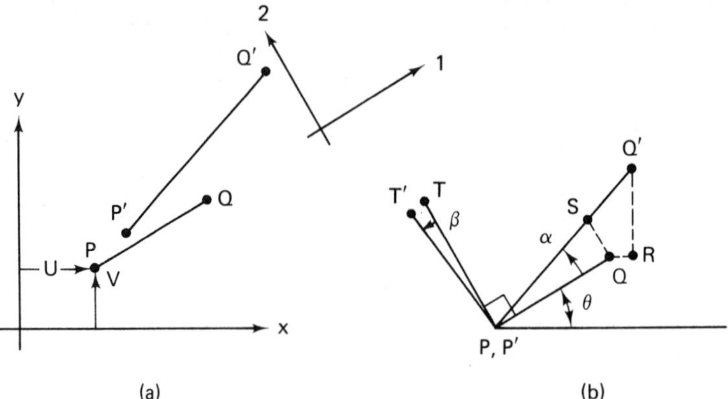

Fig. 5. Strain transformation.

To obtain the *strain* transformation relations, consider an infinitesimal line segment PQ deformed and displaced to final position $P'Q'$, in Fig. 5(a). Since we are presently concerned only with strain, and not displacement, imagine translating line $P'Q'$ back so that P' coincides with P [Fig. 5(b)]. Then the x-y components of the deformation QQ' are

$$QR = \frac{\Delta u}{\Delta x}\Delta x + \frac{\Delta u}{\Delta y}\Delta y$$

$$Q'R = \frac{\Delta v}{\Delta y}\Delta x + \frac{\Delta v}{\Delta y}\Delta y$$

or

$$QR = \frac{\partial u}{\partial x}dx + \frac{\partial u}{\partial y}dy$$

$$Q'R = \frac{\partial v}{\partial x}dx + \frac{\partial v}{\partial y}dy \tag{20}$$

since the displacements u and v are functions of x and y and PQ is infinitesimal so that Δx and $\Delta y \longrightarrow 0$.

The 1 and 2 components of the deformation QQ' are obtained by elementary trigonometry:

$$QS = Q'R\cos\theta - QR\sin\theta$$

$$SQ' = QR\cos\theta + Q'R\sin\theta \tag{21}$$

where the small angle α is ignored in comparison with θ. The normal strain of segment PQ in the 1 direction is

$$\epsilon_1 = \frac{SQ'}{PQ} \tag{22}$$

Substituting Eqs. (20) and (21) into Eq. (22),

$$\epsilon_1 = \left(\frac{\partial u}{\partial x}\frac{dx}{ds} + \frac{\partial u}{\partial y}\frac{dy}{ds}\right)\cos\theta + \left(\frac{\partial v}{\partial x}\frac{dx}{ds} + \frac{\partial v}{\partial y}\frac{dy}{ds}\right)\sin\theta$$

or

$$\epsilon_1 = \left(\frac{\partial u}{\partial x}\right)\cos^2\theta + \left(\frac{\partial u}{\partial y} + \frac{\partial v}{\partial x}\right)\sin\theta\cos\theta + \left(\frac{\partial v}{\partial y}\right)\sin^2\theta \tag{23}$$

The terms in parentheses are the normal and shear strains referred to the x-y axes, which were derived in Eqs. (21) and (22). Thus

$$\epsilon_1 = \epsilon_x \cos^2\theta + \epsilon_y \sin^2\theta + \gamma_{xy}\sin\theta\cos\theta \tag{24}$$

The 2 component of normal strain can be obtained by substituting $\theta + 90°$ for $90°$ in Eq. (24):

$$\epsilon_2 = \epsilon_x \sin^2\theta + \epsilon_y \cos^2\theta - \gamma_{xy}\sin\theta\cos\theta \tag{25}$$

To transform the shear strain, the angular change from the right angle TPQ to $T'P'Q'$ is required. Assuming small angular rotations,

$$\alpha = \frac{QS}{PQ}$$

Substituting from Eqs. (20) and (21),

$$\alpha = \left(\frac{\partial v}{\partial x}\frac{dx}{ds} + \frac{\partial v}{\partial y}\frac{dy}{ds}\right)\cos\theta - \left(\frac{\partial u}{\partial x}\frac{dx}{ds} + \frac{\partial u}{\partial y}\frac{dy}{ds}\right)\sin\theta$$

and substituting from Eq. (23),

$$\alpha = \frac{\partial v}{\partial x}\cos^2\theta + \left(\frac{\partial v}{\partial y} - \frac{\partial u}{\partial x}\right)\sin\theta\cos\theta - \frac{\partial u}{\partial y}\sin^2\theta \tag{26}$$

The value of β, representing the rotation of segment PT, is obtained by replacing θ with $\theta + 90°$ in Eq. (26):

$$\beta = \frac{\partial v}{\partial x}\sin^2\theta - \left(\frac{\partial v}{\partial y} - \frac{\partial u}{\partial x}\right)\sin\theta\cos\theta - \frac{\partial u}{\partial y}\cos^2\theta$$

The shear strain referred to the 1–2 axes is then

$$\gamma_{12} = \alpha - \beta = \left(\frac{\partial v}{\partial y} - \frac{\partial u}{\partial x}\right)2\sin\theta\cos\theta + \left(\frac{\partial v}{\partial x} + \frac{\partial u}{\partial y}\right)(\cos^2\theta - \sin^2\theta)$$

or

$$\frac{\gamma_{12}}{2} = -\epsilon_x\sin\theta\cos\theta + \epsilon_y\sin\theta\cos\theta + \frac{\gamma_{xy}}{2}(\cos^2\theta - \sin^2\theta) \tag{27}$$

The stress transformation equations (17)–(19) in matrix form are

$$\begin{bmatrix}\sigma_1\\ \sigma_2\\ \tau_{12}\end{bmatrix} = \begin{bmatrix}\cos^2\theta & \sin^2\theta & 2\sin\theta\cos\theta\\ \sin^2\theta & \cos^2\theta & -2\sin\theta\cos\theta\\ -\sin\theta\cos\theta & \sin\theta\cos\theta & \cos^2\theta - \sin^2\theta\end{bmatrix}\begin{bmatrix}\sigma_x\\ \tau_{xy}\\ \tau_{xy}\end{bmatrix} = \mathbf{T}\begin{bmatrix}\sigma_x\\ \sigma_y\\ \tau_{xy}\end{bmatrix} \tag{28}$$

The strain transformation equations (24), (25), and (27) in matrix form are

$$\begin{bmatrix} \epsilon_1 \\ \epsilon_2 \\ \gamma_{12}/2 \end{bmatrix} = \begin{bmatrix} \cos^2\theta & \sin^2\theta & 2\sin\theta\cos\theta \\ \sin^2\theta & \cos^2\theta & -2\sin\theta\cos\theta \\ -\sin\theta\cos\theta & \sin\theta\cos\theta & \cos^2\theta - \sin^2\theta \end{bmatrix} \begin{bmatrix} \epsilon_x \\ \epsilon_y \\ \gamma_{xy}/2 \end{bmatrix} = \mathbf{T} \begin{bmatrix} \epsilon_x \\ \epsilon_y \\ \gamma_{xy}/2 \end{bmatrix} \tag{29}$$

where **T** is the transformation matrix, a tensor of second order (see Appendix C). Equations (28) and (29) are used to transform any two-dimensional stress or strain state from one coordinate set to another and are independent of the material symmetry.

a. LAMINAR CONSTITUTIVE EQUATIONS

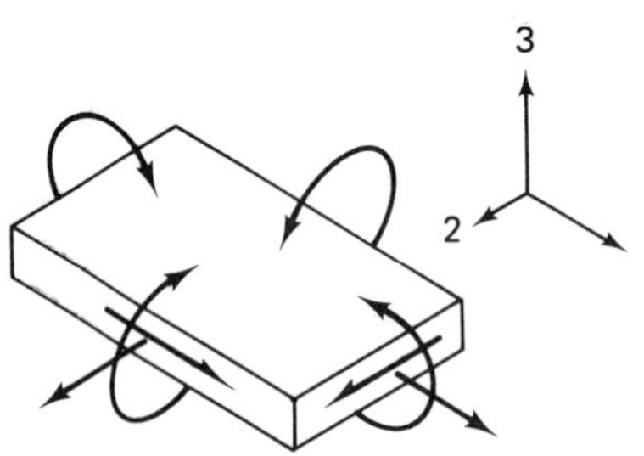

Fig. 6. Plane stress state.

The term *constitutive* refers to the stress-strain relationships of a material which describe its mechanical constitution. In plates, certain of the preceding Hooke's law stresses can be neglected. For example, in bending, the stress perpendicular to the plate, σ_3, is assumed negligible compared with the normal stresses σ_1 and σ_2 induced by bending. Also, by assuming that any perpendicular to the plate midplane remains perpendicular after deformation, shear strains γ_{13} and γ_{23} and normal strain ϵ_3 become zero. Consequently shear stresses γ_{13} and γ_{23} are also zero. These are the Kirchhoff assumptions for plate deformation. The remaining stresses are shown in Fig. 6.

For a laminated composite formed of homogeneous *isotropic* lamina in plane stress the stress-strain relationship in the kth plate may be written as Eq. (12) in matrix form:

$$\begin{bmatrix} \sigma_1 \\ \sigma_2 \\ \tau_{12} \end{bmatrix}_k = \begin{bmatrix} S_{11} & S_{12} & 0 \\ S_{12} & S_{22} & 0 \\ 0 & 0 & S_{66} \end{bmatrix}_k \begin{bmatrix} \epsilon_1 \\ \epsilon_2 \\ \gamma_{12} \end{bmatrix}_k$$

where the stiffness matrix components are

$$S_{11} = S_{22} = \frac{E}{1 - \nu^2}$$

$$S_{12} = \frac{E\nu}{1 - \nu^2}$$

$$S_{66} = \frac{E}{2(1 - \nu)} = G$$

The two independent elastic constants are E and ν.

For a laminated composite formed of homogeneous *orthotropic* lamina having their principal material axes coincident with the laminate reference axis (Fig. 7) the strain-stress relationship in the kth plate is given by Eq. (16),

$$\begin{bmatrix} \epsilon_1 \\ \epsilon_2 \\ \gamma_{12} \end{bmatrix}_k = \begin{bmatrix} F_{11} & F_{12} & 0 \\ F_{12} & F_{22} & 0 \\ 0 & 0 & F_{66} \end{bmatrix}_k \begin{bmatrix} \sigma_1 \\ \sigma_2 \\ \tau_{12} \end{bmatrix}_k$$

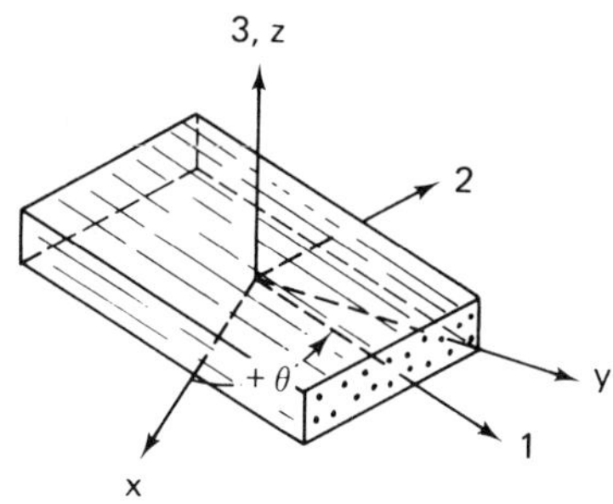

Fig. 7. Orthotropic lamina.

where the flexibility matrix components are

$$F_{11} = \frac{1}{E_{11}}$$

$$F_{22} = \frac{1}{E_{22}}$$

$$F_{12} = -\frac{\nu_{12}}{E_{11}} = -\frac{\nu_{21}}{E_{22}}$$

$$F_{66} = \frac{1}{G_{12}}$$

The four independent elastic constants are the Young's moduli E_{11} and E_{22} in the 1 and 2 directions, the shear modulus G_{12}, and the major Poisson ratio, ν_{12}. The fifth constant ν_{21} may be expressed as a function of the other constants due to the requirement of symmetry of the stiffness matrix,

$$\nu_{21}E_{11} = \nu_{12}E_{22} \tag{30}$$

Since the stiffness matrix is the inverse of the flexibility matrix, $\mathbf{S} = \mathbf{F}^{-1}$ (Appendix E), the stress-strain relations for a composite of orthotropic lamina become

$$\begin{bmatrix} \sigma_1 \\ \sigma_2 \\ \tau_{12} \end{bmatrix}_k = \begin{bmatrix} S_{11} & S_{12} & 0 \\ S_{12} & S_{22} & 0 \\ 0 & 0 & 2S_{66} \end{bmatrix}_k \begin{bmatrix} \epsilon_1 \\ \epsilon_2 \\ \gamma_{12}/2 \end{bmatrix}_k \tag{31}$$

where

$$S_{11} = \frac{E_{11}}{1 - \nu_{12}\nu_{21}}$$

$$S_{22} = \frac{E_{22}}{1 - \nu_{12}\nu_{21}}$$

$$S_{12} = \frac{\nu_{21}E_{11}}{1 - \nu_{12}\nu_{21}} = \frac{\nu_{12}E_{22}}{1 - \nu_{12}\nu_{21}}$$

$$S_{66} = G_{12}$$

Where the lamina principal axes (1,2) are not coincident with the laminate reference axes (x, y) the stress-strain relations for each plate must be transformed to the reference axes by use of Eq. (28) or (29) to determine the laminate stress-strain relations. Or if the equations are inverted (Appendix C),

$$\begin{bmatrix} \sigma_x \\ \sigma_y \\ \tau_{xy} \end{bmatrix}_k = \mathbf{T}^{-1} \begin{bmatrix} \sigma_1 \\ \sigma_2 \\ \tau_{12} \end{bmatrix}_k \quad \text{or} \quad \begin{bmatrix} \epsilon_x \\ \epsilon_y \\ \gamma_{xy}/2 \end{bmatrix}_k = \mathbf{T}^{-1} \begin{bmatrix} \epsilon_1 \\ \epsilon_2 \\ \gamma_{12}/2 \end{bmatrix}_k \tag{33}$$

It can be shown that the inverse of the **T** matrix is obtained simply by reversing the sign of θ in the **T** matrix.

From Eqs. (29), (31), and (33),

$$\begin{bmatrix} \sigma_x \\ \sigma_y \\ \tau_{xy} \end{bmatrix}_k = \mathbf{T}^{-1}\mathbf{S}_k\mathbf{T} \begin{bmatrix} \epsilon_x \\ \epsilon_y \\ \gamma_{xy}/2 \end{bmatrix}_k$$

The resulting matrix multiplication yields

$$\begin{bmatrix} \sigma_x \\ \sigma_y \\ \tau_{xy} \end{bmatrix}_k = \begin{bmatrix} \bar{S}_{11} & \bar{S}_{12} & \bar{S}_{16} \\ \bar{S}_{12} & \bar{S}_{22} & \bar{S}_{26} \\ \bar{S}_{16} & \bar{S}_{26} & \bar{S}_{66} \end{bmatrix}_k \begin{bmatrix} \epsilon_x \\ \epsilon_y \\ \gamma_{xy} \end{bmatrix}_k \tag{34}$$

where the stiffness matrix components, now referred to the laminate reference axes, are

$$\begin{aligned}
\bar{S}_{11} &= S_{11} \cos^4 \theta + 2(S_{12} + 2S_{66}) \sin^2 \theta \cos^2 \theta + S_{22} \sin^4 \theta \\
\bar{S}_{12} &= (S_{11} + S_{22} - 4S_{66}) \sin^2 \theta \cos^2 \theta + S_{12}(\sin^4 \theta + \cos^4 \theta) \\
\bar{S}_{16} &= (S_{11} - S_{12} - 2S_{66}) \sin \theta \cos^3 \theta + (S_{12} - S_{22} + 2S_{66}) \sin^3 \theta \cos \theta \\
\bar{S}_{22} &= S_{11} \sin^4 \theta + 2(S_{12} + 2S_{66}) \sin^2 \theta \cos^2 \theta + S_{22} \cos^4 \theta \\
\bar{S}_{26} &= (S_{11} - S_{12} - 2S_{66}) \sin^3 \theta \cos \theta + (S_{12} - S_{22} + 2S_{66}) \sin \theta \cos^3 \theta \\
\bar{S}_{66} &= (S_{11} + S_{22} - 2S_{12} - 2S_{66}) \sin^2 \theta \cos^2 \theta + S_{66}(\sin^4 \theta + \cos^4 \theta)
\end{aligned} \tag{35}$$

Although it appears that six elastic constants are now required, $\bar{S}_{16}$ and $\bar{S}_{26}$ are linear combinations of the remaining four constants and are not independent. Notice that the rotations θ are with respect to the composite reference axes, not with respect to the material axes of the individual lamina.

EXAMPLE 2: Determine the laminar stresses $(\sigma_1, \sigma_2, \tau_{12})$ and strains $(\epsilon_1, \epsilon_2, \gamma_{12})$ due to an applied tensile stress $\sigma_x = 10$ psi at an angle $\theta = 20°$ to the principal material axis of an orthotropic lamina. The laminar constants are

$$E_{11} = 20 \times 10^6$$
$$E_{22} = 4 \times 10^6$$
$$G_{12} = 1.0 \times 10^6$$
$$\nu_{12} = 0.3$$

From Eq. (30), $\nu_{21} = 0.06$. From Eq. (31), the lamina stiffness matrix is

$$\mathbf{S} = \begin{bmatrix} 20.37 & 1.22 & 0 \\ 1.22 & 4.07 & 0 \\ 0 & 0 & 1.00 \end{bmatrix} \times 10^6$$

From Eq. (35),

$$\bar{S}_{11} = 20.37 \cos^4 20 + 2(1.22 + 2 \times 1) \sin^2 20 \cos^2 20 + 4.07 \sin^4 20 = 16.60$$

and

$$\bar{S}_{12} = 3.08, \quad \bar{S}_{16} = 4.84, \quad \bar{S}_{22} = 4.12, \quad \bar{S}_{26} = 0.40, \quad \bar{S}_{66} = 2.86$$

From Eq. (34),

$$\begin{bmatrix} 10 \\ 0 \\ 0 \end{bmatrix} = \begin{bmatrix} 16.6 & 3.08 & 4.84 \\ 3.08 & 4.12 & 0.40 \\ 4.84 & 0.40 & 2.86 \end{bmatrix} \begin{bmatrix} \epsilon_x \\ \epsilon_y \\ \gamma_{xy} \end{bmatrix} \times 10^3$$

which yields $\epsilon_x = 0.00143$, $\epsilon_y = -0.000846$, and $\gamma_{xy} = -0.00230$ in./in. Then from Eqs. (28) and (29),

$$\sigma_1 = 8.83, \quad \sigma_2 = 1.17, \quad \tau_{12} = -3.21 \text{ psi}$$

and

$$\epsilon_1 = 0.000424, \quad \epsilon_2 = 0.000161, \quad \gamma_{12} = -0.00323 \text{ in./in.}$$

b. LAMINATES

The preceding equations for behavior of a single lamina provide the basis for computing stresses and strains in laminate composites. In addition, it is useful to define the strain-displacement relationship for the laminate and the stress and moment resultants at the laminate midplane.

(1) Strain-displacement relationship. Consider a point, O, on the midplane of a laminate, which undergoes displacement components u_0, v_0 in the x and z directions due to loading (Fig. 8). If we assume that the normal to the midplane AOB remains straight and normal after loading and that the length AOB remains unchanged, then the x and z displacement components of point P on the normal AOB are

$$u_p = u_0 - z_p \frac{\partial w}{\partial x} \tag{36}$$

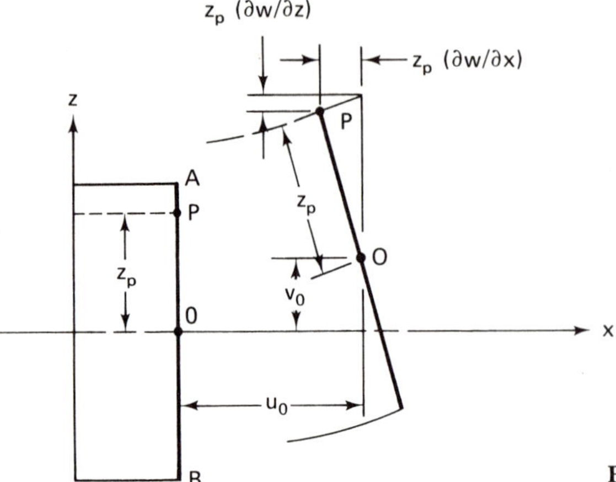

Fig. 8. Laminate deformation.

and

$$v_p = v_0 - z_p \frac{\partial w}{\partial z} \tag{37}$$

where the distances and slopes are indicated in Fig. 8.

The strains ϵ_x, ϵ_y, γ_{xy} at any point in the deformed laminate can now be determined in terms of the displacements. The strains ϵ_z, γ_{xz}, γ_{yz} are considered negligible, according to our assumption that the midplane normal remains straight, normal, and unchanged in length after deformation. Substituting Eq. (36) into Eq. (8) of Chap. 3,

$$\epsilon_x = \frac{\partial u}{\partial x} = \frac{\partial}{\partial x}\left(u_0 - z_p \frac{\partial w}{\partial x}\right) = \frac{\partial u_0}{\partial x} - z_p \frac{\partial^2 w}{\partial x^2} \tag{38}$$

Similarly,

$$\epsilon_y = \frac{\partial v_0}{\partial y} - z_p \frac{\partial^2 w}{\partial y^2} \tag{39}$$

Since shear strain is defined as angular change [Eq. (9) of Chap. 3],

$$\begin{aligned}\gamma_{xy} &= \frac{\partial u}{\partial y} + \frac{\partial v}{\partial x} = \frac{\partial}{\partial y}\left(u_0 - z_p \frac{\partial w}{\partial x}\right) + \frac{\partial}{\partial x}\left(v_0 - z_p \frac{\partial w}{\partial y}\right) \\ &= \frac{\partial u_0}{\partial y} + \frac{\partial v_0}{\partial x} - 2z_p \frac{\partial^2 w}{\partial x\, \partial y}\end{aligned} \tag{40}$$

Equations (38)–(40) can be written in terms of the midplane strains ϵ_x^0, ϵ_y^0, γ_{xy}^0 and midplane curvatures k_x, k_y, k_{xy} defined as

$$\begin{aligned} \epsilon_x^0 &= \frac{\partial u_0}{\partial x} & k_x &= -\frac{\partial^2 w}{\partial x^2} \\ \epsilon_y^0 &= \frac{\partial v_0}{\partial y} & k_y &= -\frac{\partial^2 w}{\partial y^2} \\ \gamma_{xy}^0 &= \frac{\partial u_0}{\partial y} + \frac{\partial v_0}{\partial x} & k_{xy} &= -\frac{2\partial^2 w}{\partial x\, \partial y} \end{aligned} \tag{41}$$

Substituting identities (41) into Eqs. (38)–(40),

$$\begin{bmatrix} \epsilon_x \\ \epsilon_y \\ \gamma_{xy} \end{bmatrix} = \begin{bmatrix} \epsilon_x^0 \\ \epsilon_y^0 \\ \gamma_{xy}^0 \end{bmatrix} + z \begin{bmatrix} k_x \\ k_y \\ k_{xy} \end{bmatrix} \tag{42}$$

which is the strain-displacement relationship for small deformation of a laminate. Combining Eqs. (34) and (42),

$$\begin{bmatrix} \sigma_x \\ \sigma_y \\ \tau_{xy} \end{bmatrix}_k = \begin{bmatrix} \bar{S}_{11} & \bar{S}_{12} & \bar{S}_{16} \\ \bar{S}_{12} & \bar{S}_{22} & \bar{S}_{26} \\ \bar{S}_{16} & \bar{S}_{26} & \bar{S}_{66} \end{bmatrix}_k \begin{bmatrix} \epsilon_x^0 \\ \epsilon_y^0 \\ \gamma_{xy}^0 \end{bmatrix} + z \begin{bmatrix} \bar{S}_{11} & \bar{S}_{12} & \bar{S}_{16} \\ \bar{S}_{12} & \bar{S}_{22} & \bar{S}_{26} \\ \bar{S}_{16} & \bar{S}_{26} & \bar{S}_{66} \end{bmatrix}_k \begin{bmatrix} k_x \\ k_y \\ k_{xy} \end{bmatrix} \tag{43}$$

or

$$\boldsymbol{\sigma}_k = \bar{\mathbf{S}}_k \boldsymbol{\epsilon}^0 + z\bar{\mathbf{S}}_k \mathbf{k}$$

which can be used to compute the stress in a lamina if the laminate midplane strain and curvature are known.

(2) Stress and moment resultants. Since the stresses vary from layer to layer in a laminate, it is helpful to employ stress and moment resultants acting on the midplane which are equivalent to the actual stresses distributed across the thickness of the laminate. The resultant stresses and moments are defined in units of forces per unit length of midplane and moment per unit length of midplane, respectively. For the plate element in Fig. 9, the stress resultants are

$$N_x = \int_{-h/2}^{h/2} \sigma_x \, dz, \qquad N_y = \int_{-h/2}^{h/2} \sigma_y \, dz, \qquad N_{xy} = \int_{-h/2}^{h/2} \tau_{xy} \, dz \tag{44}$$

and the moment resultants are

$$M_x = \int_{-h/2}^{h/2} \sigma_x z \, dz, \qquad M_y = \int_{-h/2}^{h/2} \sigma_y z \, dz, \qquad M_{xy} = \int_{-h/2}^{h/2} \tau_{xy} z \, dz \tag{45}$$

The double arrows in Fig. 9 indicate positive moment vectors, according to the right-

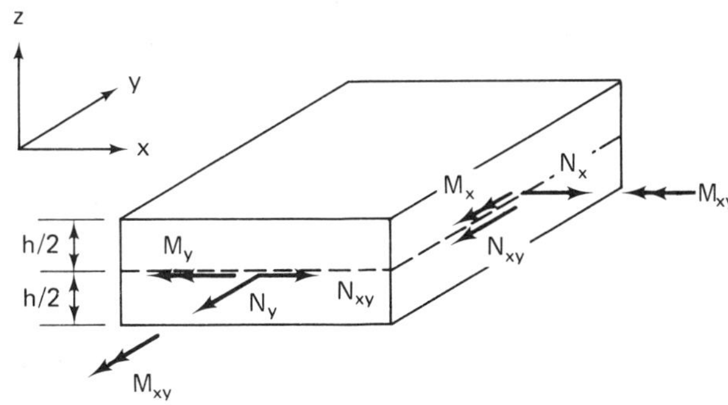

Fig. 9. Stress and moment resultants on plate element.

hand rule. In these equations only the stresses $\sigma_x, \sigma_y, \tau_{xy}$ have been considered, according to our previous practice for plane stress of a plate.

(3) Laminate constitutive equations. The laminate constitutive (stress-strain) equations are obtained by combining Eqs. (43)–(45). Writing Eq. (44) in matrix form,

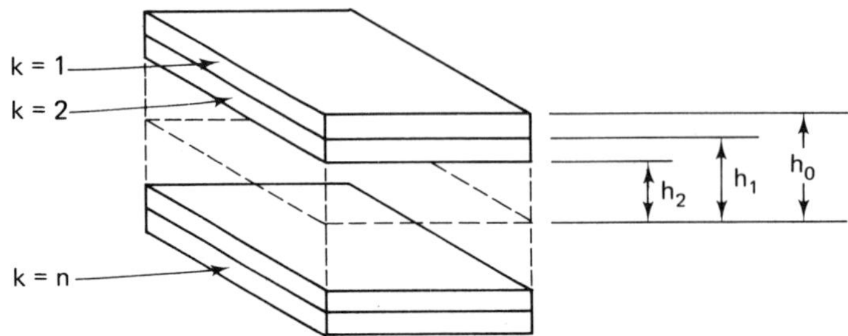

Fig. 10. Lamina notation.

separating the continuous integral into an integral over each of n lamina (Fig. 10), and substituting into Eq. (43) yields

$$\begin{bmatrix} N_x \\ N_y \\ N_{xy} \end{bmatrix} = \sum_{k=1}^{n} \int_{h_{k-1}}^{h_k} \begin{bmatrix} \sigma_x \\ \sigma_y \\ \tau_{xy} \end{bmatrix}_k dz$$

$$= \sum_{k=1}^{n} \left\{ \int_{h_{k-1}}^{h_k} \begin{bmatrix} \bar{S}_{11} & \bar{S}_{12} & \bar{S}_{16} \\ \bar{S}_{12} & \bar{S}_{22} & \bar{S}_{26} \\ \bar{S}_{16} & \bar{S}_{26} & \bar{S}_{66} \end{bmatrix}_k \begin{bmatrix} \epsilon_x^0 \\ \epsilon_y^0 \\ \gamma_{xy}^0 \end{bmatrix} dz + \int_{h_{k-1}}^{h_k} \begin{bmatrix} \bar{S}_{11} & \bar{S}_{12} & \bar{S}_{16} \\ \bar{S}_{12} & \bar{S}_{22} & \bar{S}_{26} \\ \bar{S}_{16} & \bar{S}_{26} & \bar{S}_{66} \end{bmatrix}_k \begin{bmatrix} k_x \\ k_y \\ k_{xy} \end{bmatrix} z\, dz \right\} \tag{46}$$

Since $\boldsymbol{\epsilon}^0$ and $\mathbf{k}$ are not functions of z, and $\mathbf{S}$ within any (h_{k-1} to h_k) layer is not a function of z, Eq. (46) can be written

$$\begin{bmatrix} N_x \\ N_y \\ N_{xy} \end{bmatrix} = \sum_{k=1}^{n} \left\{ \begin{bmatrix} \bar{S}_{11} & \bar{S}_{12} & \bar{S}_{16} \\ \bar{S}_{12} & \bar{S}_{22} & \bar{S}_{26} \\ \bar{S}_{16} & \bar{S}_{26} & \bar{S}_{66} \end{bmatrix}_k \begin{bmatrix} \epsilon_x^0 \\ \epsilon_y^0 \\ \gamma_{xy}^0 \end{bmatrix} \int_{h_{k-1}}^{h_k} dz + \begin{bmatrix} \bar{S}_{11} & \bar{S}_{12} & \bar{S}_{16} \\ \bar{S}_{12} & \bar{S}_{22} & \bar{S}_{26} \\ \bar{S}_{16} & \bar{S}_{26} & \bar{S}_{66} \end{bmatrix}_k \begin{bmatrix} k_x \\ k_y \\ k_{xy} \end{bmatrix} \int_{h_{k-1}}^{h_k} z\, dz \right\} \tag{47}$$

and since $\boldsymbol{\epsilon}^0$ and $\mathbf{k}$ are not functions of k, Eq. (47) reduces to

$$\begin{bmatrix} N_x \\ N_y \\ N_{xy} \end{bmatrix} = \begin{bmatrix} A_{11} & A_{12} & A_{16} \\ A_{12} & A_{22} & A_{26} \\ A_{16} & A_{26} & A_{66} \end{bmatrix} \begin{bmatrix} \epsilon_x^0 \\ \epsilon_y^0 \\ \gamma_{xy}^0 \end{bmatrix} + \begin{bmatrix} B_{11} & B_{12} & B_{16} \\ B_{12} & B_{22} & B_{26} \\ B_{16} & B_{26} & B_{66} \end{bmatrix} \begin{bmatrix} k_x \\ k_y \\ k_{xy} \end{bmatrix} \tag{48}$$

or

$$\mathbf{N} = \mathbf{A}\boldsymbol{\epsilon}^0 + \mathbf{B}\mathbf{k}$$

where

$$A_{ij} = \sum_{k=1}^{n} (\bar{S}_{ij})_k (h_k - h_{k-1}) \tag{49}$$

$$B_{ij} = \frac{1}{2} \sum_{k=1}^{n} (\bar{S}_{ij})_k (h_k^2 - h_{k-1}^2) \tag{50}$$

Equation (48) gives the midplane stress resultants in terms of the midplane strains and laminate curvatures.

The moment resultants can be obtained similarly by writing Eq. (45) in matrix form, separating into integrals over each of the n lamina, substituting into Eq. (43), and simplifying to yield

$$\begin{bmatrix} M_x \\ M_y \\ M_{xy} \end{bmatrix} = \begin{bmatrix} B_{11} & B_{12} & B_{16} \\ B_{12} & B_{22} & B_{26} \\ B_{16} & B_{26} & B_{66} \end{bmatrix} \begin{bmatrix} \epsilon_x^0 \\ \epsilon_y^0 \\ \gamma_{xy}^0 \end{bmatrix} + \begin{bmatrix} D_{11} & D_{12} & D_{16} \\ D_{12} & D_{22} & D_{26} \\ D_{16} & D_{26} & D_{66} \end{bmatrix} \begin{bmatrix} k_x \\ k_y \\ k_{xy} \end{bmatrix} \tag{51}$$

or

$$\mathbf{M} = \mathbf{B}\boldsymbol{\epsilon}^0 + \mathbf{D}\mathbf{k}$$

where

$$D_{ij} = \frac{1}{3} \sum_{k=1}^{n} (\bar{S}_{ij})_k (h_k^3 - h_{k-1}^3) \tag{52}$$

Notice in Eq. (48) that for a general laminate normal stress resultants may be induced by curvature as well as by midplane strain and in Eq. (51) that moment resultants can be induced by midplane strain as well as by curvature. This coupling between bending and stretching, or cross-elasticity effect, distinguishes the behavior of generally laminated composites from that of homogeneous orthotropic or isotropic plates.

Combining Eqs. (48) and (51),

$$\begin{bmatrix} N_x \\ N_y \\ N_{xy} \\ \hline M_x \\ M_y \\ M_{xy} \end{bmatrix} = \left[\begin{array}{ccc|ccc} A_{11} & A_{12} & A_{16} & B_{11} & B_{12} & B_{16} \\ A_{12} & A_{22} & A_{26} & B_{12} & B_{22} & B_{26} \\ A_{16} & A_{26} & A_{66} & B_{16} & B_{26} & B_{66} \\ \hline B_{11} & B_{12} & B_{16} & D_{11} & D_{12} & D_{16} \\ B_{12} & B_{22} & B_{26} & D_{12} & D_{22} & D_{26} \\ B_{16} & B_{26} & B_{66} & D_{16} & D_{26} & D_{66} \end{array}\right] \begin{bmatrix} \epsilon_x^0 \\ \epsilon_y^0 \\ \gamma_{xy}^0 \\ \hline k_x \\ k_y \\ k_{xy} \end{bmatrix} \tag{53}$$

or

$$\begin{bmatrix} N \\ \hline M \end{bmatrix} = \left[\begin{array}{c|c} A & B \\ \hline B & D \end{array}\right] \begin{bmatrix} \epsilon^0 \\ \hline k \end{bmatrix}$$

which is the total laminate constitutive equation. The quantities A, B, and D are the elastic areas, the elastic statical moments, and the elastic moments of inertia, respectively.

Equation (53) can be rearranged in other useful forms by partially or totally inverting the matrices. For example, solving Eq. (48) for the midplane strain gives

$$\boldsymbol{\epsilon}^0 = \mathbf{A}^{-1}\mathbf{N} - \mathbf{A}^{-1}\mathbf{B}\mathbf{k} \tag{54}$$

Substituting Eq. (54) into Eq. (51) and rearranging,

$$\mathbf{M} = \mathbf{B}\mathbf{A}^{-1}\mathbf{N} + (-\mathbf{B}\mathbf{A}^{-1}\mathbf{B} + \mathbf{D})\mathbf{k} \tag{55}$$

Solving Eq. (55) for **k**,

$$\mathbf{k} = (\mathbf{D} - \mathbf{B}\mathbf{A}^{-1}\mathbf{B})^{-1}\mathbf{M} - (\mathbf{D} - \mathbf{B}\mathbf{A}^{-1}\mathbf{B})^{-1}\mathbf{B}\mathbf{A}^{-1}\mathbf{N} \tag{56}$$

Substituting Eq. (56) into Eq. (54),

$$\boldsymbol{\epsilon}^0 = -\mathbf{A}^{-1}\mathbf{B}(\mathbf{D} - \mathbf{B}\mathbf{A}^{-1}\mathbf{B})^{-1}\mathbf{M} + \{\mathbf{A}^{-1} + \mathbf{A}^{-1}\mathbf{B}(\mathbf{D} - \mathbf{B}\mathbf{A}^{-1}\mathbf{B})^{-1}\mathbf{B}\mathbf{A}^{-1}\}\mathbf{N} \tag{57}$$

Combining Eqs. (56) and (57) in matrix form yields the completely inverted form of Eq. (53):

$$\begin{bmatrix} \epsilon^0 \\ \hline k \end{bmatrix} = \left[\begin{array}{c|c} A' & B' \\ \hline C' & D' \end{array}\right] \begin{bmatrix} N \\ \hline M \end{bmatrix} \tag{58}$$

where

$$\begin{aligned} A' &= \mathbf{A}^{-1} + \mathbf{A}^{-1}\mathbf{B}(\mathbf{D} - \mathbf{B}\mathbf{A}^{-1}\mathbf{B})^{-1}\mathbf{B}\mathbf{A}^{-1} \\ B' &= -\mathbf{A}^{-1}\mathbf{B}(\mathbf{D} - \mathbf{B}\mathbf{A}^{-1}\mathbf{B})^{-1} \\ C' &= -(\mathbf{D} - \mathbf{B}\mathbf{A}^{-1}\mathbf{B})^{-1}\mathbf{B}\mathbf{A}^{-1} \\ D' &= (\mathbf{D} - \mathbf{B}\mathbf{A}^{-1}\mathbf{B})^{-1} \end{aligned} \tag{59}$$

EXAMPLE 3: Determine the midplane strains $\epsilon_x^0, \epsilon_y^0, \gamma_{xy}^0$ and curvatures k_x, k_y, k_{xy} produced by an axial tensile stress resultant $N_x = 10$ ksi on a two-ply laminate having the following properties:

Lamina	*a*	*b*
Thickness	0.2 in.	0.3 in.
θ	0	$+20°$
E_{11}	30×10^6	20×10^6
E_{22}	5×10^6	4×10^6
G_{12}	1×10^6	1×10^6
ν_{12}	0.25	0.3

Observe that lamina a has $\theta = 0$ and that lamina b is identical to that used in Example 2. We first obtain $\bar{\mathbf{S}}$ matrices as in Example 2. From Eq. (30),

$$\nu_{a21} = 0.0417 \qquad \nu_{b21} = 0.0600$$

From Eq. (32),

$$\mathbf{S}_a = \begin{bmatrix} 30.32 & 1.263 & 0 \\ 1.263 & 5.053 & 0 \\ 0 & 0 & 1 \end{bmatrix} \times 10^6 \qquad \mathbf{S}_b = \begin{bmatrix} 20.37 & 1.222 & 0 \\ 1.222 & 4.073 & 0 \\ 0 & 0 & 1 \end{bmatrix} \times 10^6$$

From Eq. (35),

$$\bar{\mathbf{S}}_a = \begin{bmatrix} 30.32 & 1.263 & 0 \\ 1.263 & 5.053 & 0 \\ 0 & 0 & 1.0 \end{bmatrix} \times 10^6 \qquad \bar{\mathbf{S}}_b = \begin{bmatrix} 16.60 & 3.081 & 4.834 \\ 3.081 & 4.120 & 0.403 \\ 4.834 & 0.403 & 2.859 \end{bmatrix} \times 10^6$$

The thickness dimensions (Fig. 10) are $h_0 = -0.25$, $h_1 = -0.05$, $h_2 = 0.25$. From Eqs. (49), (50), and (52),

$$\mathbf{A} = \begin{bmatrix} 11.04 & 1.18 & 1.45 \\ 1.18 & 2.25 & 0.1209 \\ 1.45 & 0.1209 & 1.06 \end{bmatrix} \times 10^6 \qquad \mathbf{B} = \begin{bmatrix} -0.4115 & 0.0545 & 0.1450 \\ 0.0545 & -0.0280 & 0.0121 \\ 0.1450 & 0.0121 & 0.0558 \end{bmatrix} \times 10^6$$

$$\mathbf{D} = \begin{bmatrix} 0.2438 & 0.0227 & 0.0254 \\ 0.0227 & 0.0477 & 0.0021 \\ 0.0254 & 0.0021 & 0.0202 \end{bmatrix} \times 10^6$$

From Eq. (59),

$$\mathbf{A}' = \begin{bmatrix} 0.1717 & -0.0841 & -0.2429 \\ -0.0905 & 0.4965 & 0.0925 \\ -0.2298 & 0.0632 & 1.486 \end{bmatrix} \times 10^{-6}$$

$$\mathbf{B}' = \begin{bmatrix} 0.6334 & 0.1071 & -1.289 \\ -0.4377 & 0.1843 & 0.6065 \\ -1.267 & -1.126 & -0.8797 \end{bmatrix} \times 10^{-6}$$

$$\mathbf{C}' = \begin{bmatrix} 0.5889 & -0.3192 & -1.306 \\ -0.4375 & 0.5587 & 0.5950 \\ -1.289 & 0.6065 & -0.8797 \end{bmatrix} \times 10^{-6}$$

$$\mathbf{D}' = \begin{bmatrix} 7.374 & 3.737 & -9.670 \\ -3.737 & 23.24 & 3.426 \\ -9.670 & 3.426 & 72.66 \end{bmatrix} \times 10^{-6}$$

Combining the above results in Eq. (58) and solving,

$$\epsilon_x^0 = 0.00172 \text{ in./in.} \qquad k_x = 0.00589 \text{ in./in.}$$

$$\epsilon_y^0 = -0.000904 \text{ in./in.} \qquad k_y = -0.00438 \text{ in./in.}$$

$$\gamma_{xy}^0 = -0.00230 \text{ in./in.} \qquad k_{xy} = -0.0129 \text{ in./in.}$$

This example shows that the states of stress and strain in a general laminated composite are quite complex. They cannot, as yet, be calculated by shortcut methods with reliable accuracy. The bending-stretching coupling caused by the **B** matrix is due to the heterogeneity in stacking of the lamina and can also exist even in the case of isotropic lamina.

(4) Simplifications due to symmetry. Major simplifications in the constitutive equations can result if **B** is zero and if the 16, 26 terms are zero. Since the B_{ij} are even functions of h_k, **B** is zero for laminates with midplane symmetry. Midplane symmetric laminates are the most commonly constructed to eliminate undesirable warping due to in-plane loads and to differential thermal expansions. Laminate orthotropy with respect to the applied loads reduces the A_{16} and A_{26} terms to zero. According to Eq. (49) the only way to obtain zero A_{ij} terms is to have some $\bar{S}_{ij}$ positive and some negative. According to Eq. (35) the $\bar{S}_{16}$ and $\bar{S}_{26}$ for $+\theta$ are equal and opposite in sign to the corresponding terms for $-\theta$. Thus if for every lamina at $+\theta$ there is a corresponding one at $-\theta$ (orthotropy with respect to applied forces), $A_{16} = A_{26} = 0$.

For laminates with midplane subjected to membrane stresses but no bending moments, Eq. (58) reduces to

$$\begin{aligned} \boldsymbol{\epsilon}_0 &= \mathbf{A'N} = \mathbf{A}^{-1}\mathbf{N} \\ \mathbf{k} &= \mathbf{D'M} = 0 \end{aligned} \tag{60}$$

EXAMPLE 4: Determine the lamina stresses and strains produced by an axial tensile stress resultant $N_x = 20$ ksi on a three-ply laminate having the following properties:

Lamina	*a*	*b*	*c*
Thickness	0.2 in.	0.3 in.	0.2 in.
θ	0	$+20°$	0
E_{11}	30×10^6	20×10^6	30×10^6
E_{22}	5×10^6	4×10^6	5×10^6
G_{12}	1×10^6	1×10^6	1×10^6
ν_{12}	0.25	0.3	0.25

We observe midplane symmetry and that laminae *a* and *b* are identical to those in Example 3. From Eq. (49),

$$\mathbf{A} = \begin{bmatrix} 17.11 & 1.430 & 1.450 \\ 1.430 & 3.257 & 0.1209 \\ 1.450 & 0.1209 & 1.258 \end{bmatrix} \times 10^6$$

$$\mathbf{A}^{-1} = \begin{bmatrix} 0.0670 & -0.0266 & -0.0747 \\ -0.0266 & 0.3187 & 0.000087 \\ -0.0747 & 0.000087 & 0.8810 \end{bmatrix} \times 10^{-6} \qquad \text{(see Appendix C)}$$

From Eq. (60),

$$\begin{bmatrix} \epsilon_x \\ \epsilon_y \\ \gamma_{xy} \end{bmatrix} = A^{-1} \begin{bmatrix} 20{,}000 \\ 0 \\ 0 \end{bmatrix} = \begin{bmatrix} 0.00134 \\ 0.000533 \\ 0.00149 \end{bmatrix} \begin{matrix} \text{in./in. strains in} \\ \text{all three laminae} \end{matrix}$$

From Eq. (34),

$$\begin{bmatrix}\sigma_x\\ \sigma_y\\ \tau_{xy}\end{bmatrix}_{0^\circ} = \begin{bmatrix}\sigma_1\\ \sigma_2\\ \tau_{12}\end{bmatrix}_{0^\circ} = \begin{bmatrix}41{,}303\\ 4{,}386\\ 1{,}493\end{bmatrix} \text{ psi stresses in laminae } a \text{ and } c$$

$$\begin{bmatrix}\sigma_x\\ \sigma_y\\ \gamma_{xy}\end{bmatrix}_{+20^\circ} = \begin{bmatrix}31{,}111\\ 6{,}927\\ 10{,}963\end{bmatrix}_{+20^\circ} \text{ psi stress in lamina } b$$

The stresses and strains in the natural coordinate system of each lamina can be determined from the transformation equations (28) and (29):

$$\begin{bmatrix}\sigma_1\\ \sigma_2\\ \tau_{12}\end{bmatrix}_{+20^\circ} = \begin{bmatrix}35{,}329\\ 2{,}709\\ 626\end{bmatrix}_{+20^\circ} \text{ psi principal stresses in lamina } b$$

$$\begin{bmatrix}\epsilon_1\\ \epsilon_2\\ \gamma_{12}\end{bmatrix}_{+20^\circ} = \begin{bmatrix}0.00221\\ -0.00033\\ 0.00088\end{bmatrix}_{+20^\circ} \text{ in./in. principal strains in lamina } b$$

No transformation of stresses and strains are required for the 0° laminae, since their axes coincide with the laminate axis. The resulting stresses and strains appear in Fig. 11.

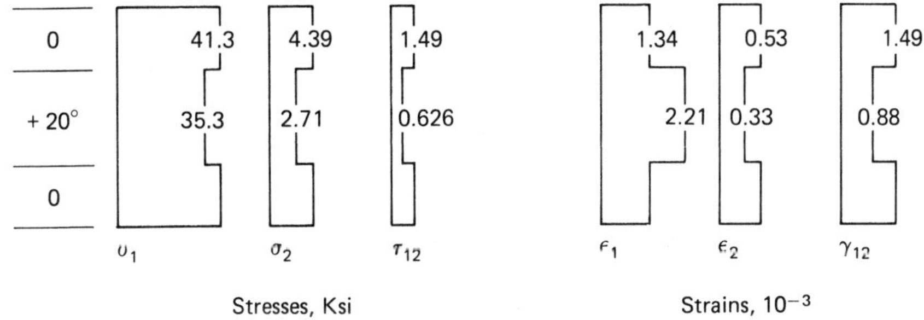

Fig. 11. Laminae stresses and strains in the 1-2 coordinate systems.

B. SANDWICH PANELS

Sandwich panels are a particular class of laminates having two thin strong sheets of dense material separated by a thick core of low-density material which is less stiff and strong. This arrangement combines stiffness with lightness, because the stiff faces are at a maximum distances from the neutral axis, similar to the flanges of an I-beam. The faces carry most of the axial loading and transverse bending stresses. The core carries most of the shear and also stabilizes the thin compressive faces against buckling. The bond between core and face must resist shear and transverse tensile stress resulting from the wrinkling tendency of the compressive face.

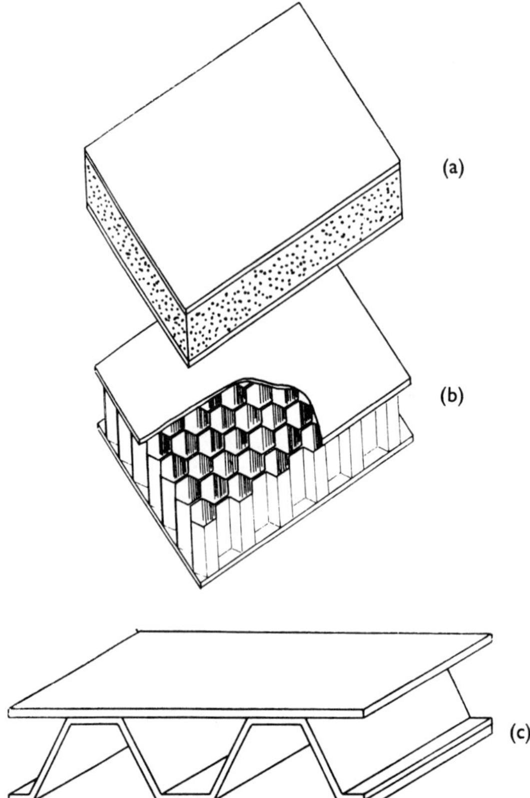

Fig. 12. Structural core geometries. (a) Foamed; (b) honeycomb; (c) corrugated. From Allen, H. G., *Analysis and Design of Structural Sandwich Panels.* Elmsford, N.Y.: Pergamon Press, Inc., 1969 [5].

Sandwich panels of a wide variety are used extensively for interior and exterior walls and roofs of buildings (see Chap. 1). Typical face materials include aluminum, fiber-reinforced plastics, plywood, gypsum board, asbestos cement, and concrete. Typical core materials include foamed polymers and inorganic cements (Chap. 10) and various structural corrugated shapes (Figs. 12–14). Corrugated and honeycomb cores are made in a wide variety of materials: paper, cotton, glass cloth, aluminum foil, and combinations with plastic foam. Plastic impregnates such as phenolics, nylon-phenolics, or polyesters are used with the first three to provide sufficient rigidity.

The two common methods of manufacturing honeycomb core are by corrugation [Fig. 15(a)] and by expansion [Fig. 15(b)]. In the corrugation method sheet material is first corrugated into half-hexagons and then stacked and bonded. In the expansion method, glue lines are rolled onto flat sheets which are then stacked, bonded to each other, and then expanded to form the honeycomb. The glue lines are stag-

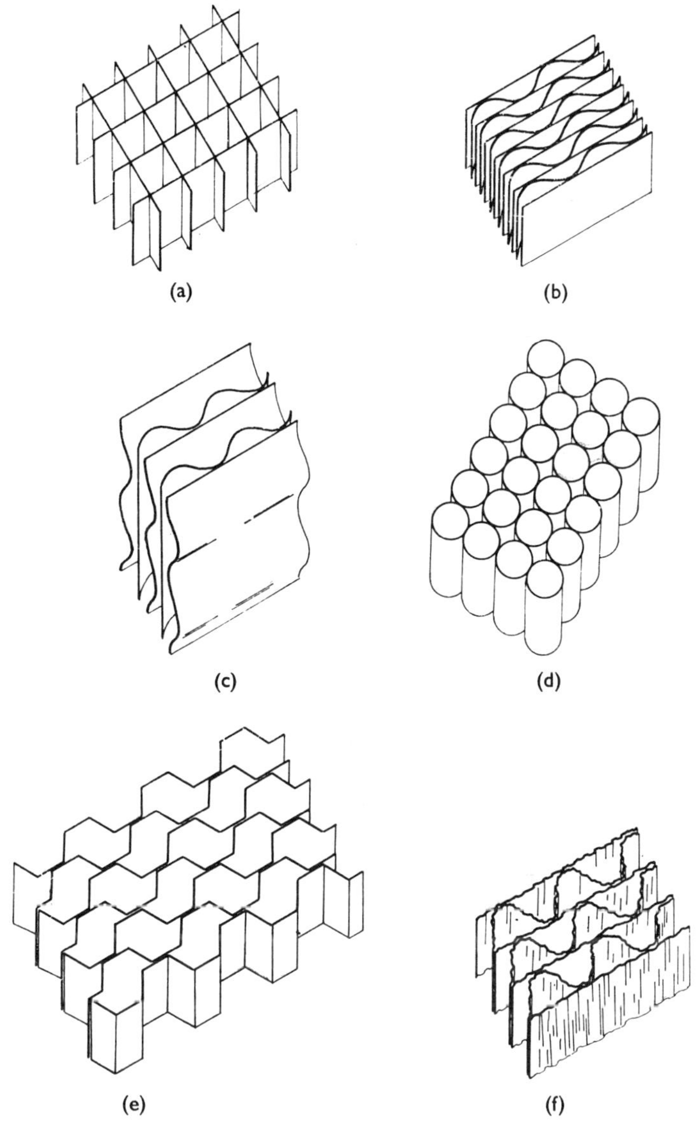

Fig. 13. Varieties of "honeycomb" core. Shapes (e) and (f) can be deformed for use between curved face sheets. From Allen, H. G., *Analysis and Design of Structural Sandwich Panels*. Elmsford, N.Y.: Pergamon Press, Inc., 1969 [5].

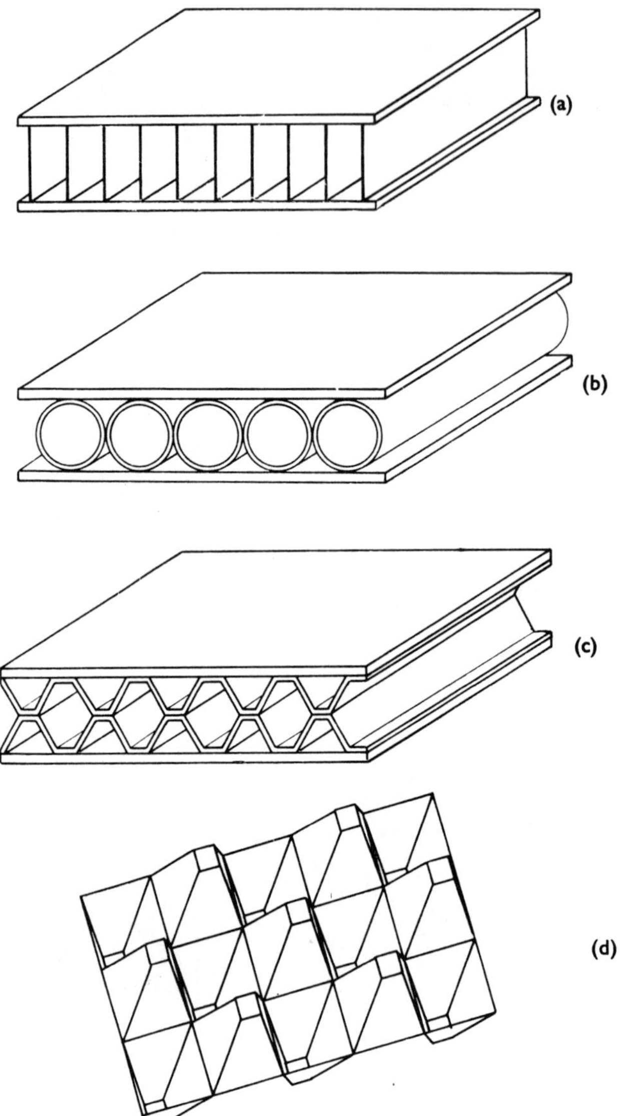

Fig. 14. Varieties of "corrugated" core. From Allen, H. G., *Analysis and Design of Structural Sandwich Panels.* Elmsford, N.Y.: Pergamon Press, Inc., 1969 [5].

gered on alternate sheets, and the cell size of the finished honeycomb is controlled by the distance between glue lines. Electrical heating elements may be used above and below the stack to cure the glue.

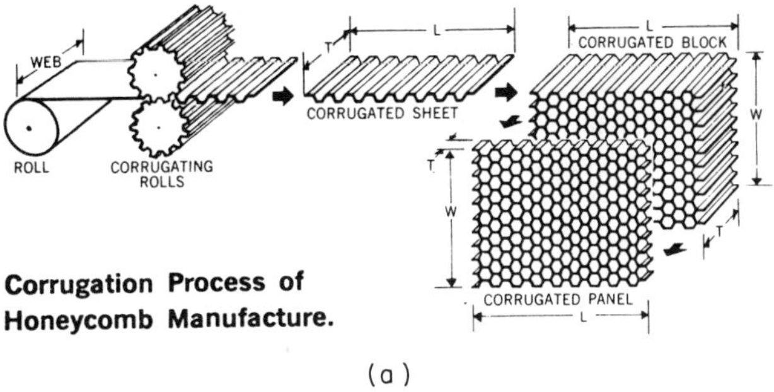

(a)

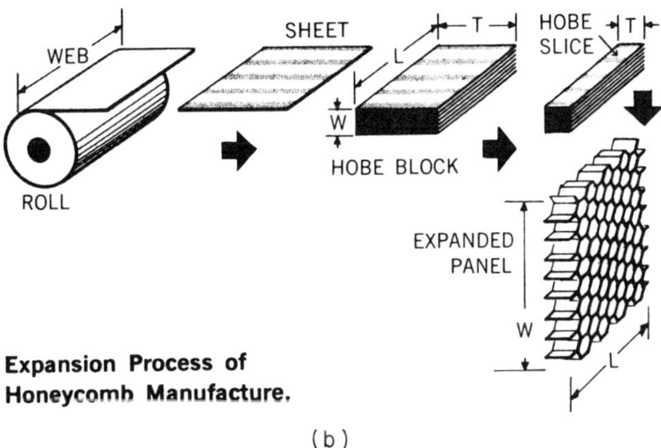

(b)

Fig. 15. Honeycomb manufacture. (a) Corrugation process; (b) expansion process. (*Courtesy of Hexcel Corporation.*)

1. Structural Analysis

The theory of sandwich construction has an extensive literature which has been summarized by Plantema [6]. The usual loading conditions are beam loading, as in floor and roof elements, and edgewise compression loading, as in load-bearing wall elements.

A sandwich panel under *beam loading* may fail in any of six ways:

1. Tensile or compressive failure of faces [Eq. (62)].
2. Shear failure of core [Eq. (63)].

3. Adhesive failure between core and face [Eq. (63)].
4. Local compressive failure of core, due to a concentrated load on the face [Eq. (64)].
5. Excessive panel deflection due to inadequate flexural stiffness of facings and shear stiffness of core [Eq. (65)].
6. Wrinkling of the compressive face [Eq. (66)].

The pertinent equations appear in column (3) of Table 1 and may be compared with the corresponding equations for isotropic rectangular beams [column (1)], and the generalized equations for laminated beams [column (2)].

TABLE 1. Equations for Beam Loading

Property	*(1)* *Rectangular isotropic beam*	*(2)* *General laminated beam*	*(3)* *Sandwich beam with faces of equal thickness*
Neutral axis location, $\bar{x}$	$= \frac{d}{2}$	$= \frac{\sum E_i A_i x_i}{\sum E_i A_i}$ (61)	$= \frac{d}{2}$
Flexural rigidity, D	$= E\frac{bd^3}{12}$	$= \sum E_i I_i$ (62)	$= E_f \frac{bf^3}{6} + E_f \frac{bfs^2}{2} + E_c \frac{bc^3}{12}$ (65)
Bending stress, σ	$= \frac{6M}{bd^2}$ (at outermost fiber)	$= \frac{ME_x x}{D}$ (63) (at any fiber)	$\sigma_f = \frac{M}{D} E_f \left(\frac{c+f}{2}\right)$ (66) and $\sigma_c = \frac{M}{D} E_c \left(\frac{c}{2}\right)$
Shear stress, τ	$= \frac{VQ}{bI} = \frac{3V}{2bd}$ (at neutral axis)	$= \frac{VQ'}{bD}$ (64) (at any fiber)	$\tau_c = \frac{V}{D}\left(E_f \frac{fs}{2} + E_c \frac{c^2}{8}\right)$ (67)
Compressive stress in core, σ_n, under point load P			$= 0.0582P\left[\frac{E_f f^3(1-\nu_c^2)}{E_c(1-\nu_f^2)}\right]^{2/3}$ (68)
Deflection, Δ			$= \frac{k_b PL^3}{D} + \frac{k_s PL}{N}$ (69)
Wrinkling of compressive face, σ_w			$= k_w(E_f E_c G_c)^{1/3}$ (70)

The following notation is used in Table 1:

A_i = cross-sectional area of lamina i
b = beam width
c = core thickness
d = beam depth
D = flexural stiffness of faces
E = Young's modulus
E_c = Young's modulus of core
E_f = Young's modulus of faces
E_i = Young's modulus of lamina i
E_x = Young's modulus of lamina at distance x from neutral axis
f = face thickness
G_c = shear modulus of core

I = second moment
I_i = second moment of lamina i about neutral axis
k_b, k_s = constants dependent on the beam loading
k_w = theoretical or empirical buckling coefficient
L = beam length
M = bending moment
N = shear stiffness of core
P = total load
Q = first moment about the neutral axis of the portion of the cross section between the neutral axis and the top or bottom of the cross section
Q' = weighted first moment, E_iA_ix about the neutral axis of the portion of the cross section between the horizontal plane in question and the outer edge (top or bottom) of the cross section
s = distance between centers of two faces
V = beam shear load
x = distance from neutral axis to any point
x_i = distance from some datum, such as bottom of cross section, to center of lamina i
ν_c = Poisson ratio of core
ν_f = Poisson ratio of faces
σ_c = bending stress in core, at extreme fiber
σ_f = bending stress in faces
τ_c = maximum (centroidal) shear stress in core

For isotropic materials the neutral axis of a rectangular cross section is at middepth, and the familiar formulas appear in column (1) of Table 1. For laminated beams the neutral axis is not necessarily at middepth of a rectangular section and must first be found [Eq. (61)]. The maximum bending stress does not necessarily occur at the outermost fiber, as in isotropic materials.

In Eq. (65) the second term, representing face stiffness due to bending about the panel centroid, invariably dominates. The first and third terms are each less than 1% of the second term when $d/f > 5.77$ and when

$$\frac{E_f}{E_c}\frac{f}{c}\left(\frac{d}{c}\right)^2 > 16.7, \text{ respectively}$$

For these conditions, Eq. (65) may be written

$$D \sim E_f\frac{bfs^2}{2} \tag{65a}$$

The variation in core shear stress [Eq. (67)] across the depth of core is less than 1% if $(E_f/E_c)(fs/c^2) > 25$. For this condition, it is permissible to set $E_c = 0$ in Eq. (67), which then reduces to

$$\tau_c = \frac{V}{D}\frac{E_f fs}{2} \tag{67a}$$

A frequently useful approximation is the concept of an antiplane core, in which the Young's modulus parallel to the faces is zero but the shear modulus perpendicular to the faces is finite. A honeycomb core is an approximation to an antiplane core.

Equation (58), for the core compressive stress under a point surface load, is empirical, based upon the assumption that the facing is an infinitely large plate on an elastic foundation [7].

The deflection of a sandwich beam with antiplane core and thin faces may be approximated by superposing the bending deflection due to the faces acting alone and the shear deflection due to the core acting alone. The two terms in Eq. (69) account for bending and shear deflections, respectively. The shear stiffness is $N = G_c bs^2/c$. If $d/c \sim 1$, this may be written simply as

$$N = G_c bs \tag{71}$$

Values of k_b and k_s in Eq. (69) for various loading conditions appear in Table 2.

TABLE 2. k_b and k_s Values to Use in Eq. (69) for Various Loading and Support Conditions (Courtesy of Hexcel Corporation)

MANNER OF LOADING AND END SUPPORT	Point of Deflection	K_B	K_S
UNIFORM LOAD SIMPLY SUPPORTED (P/a; span a)	MIDSPAN	$\frac{5}{384}$	$\frac{1}{8}$
UNIFORM LOAD FIXED ENDS (P/a; span a)	MIDSPAN	$\frac{1}{384}$	$\frac{1}{8}$
POINT LOAD AT MIDSPAN SIMPLY SUPPORTED (P; $a/2$)	MIDSPAN	$\frac{1}{48}$	$\frac{1}{4}$
POINT LOAD AT MIDSPAN FIXED ENDS (P; $a/2$)	MIDSPAN	$\frac{1}{192}$	$\frac{1}{4}$
POINT LOADS AT QUARTER SPAN SIMPLY SUPPORTED ($P/2$, $P/2$; $a/4$, $a/4$)	MIDSPAN	$\frac{11}{768}$	$\frac{1}{8}$
POINT LOADS AT QUARTER SPAN SIMPLY SUPPORTED ($P/2$, $P/2$; $a/4$, $a/4$)	LOAD POINT	$\frac{1}{96}$	$\frac{1}{8}$
UNIFORM LOAD CANTILEVER (P/a; span a)	FREE END	$\frac{1}{8}$	$\frac{1}{2}$
POINT LOAD AT FREE END CANTILEVER (P; span a)	FREE END	$\frac{1}{3}$	1

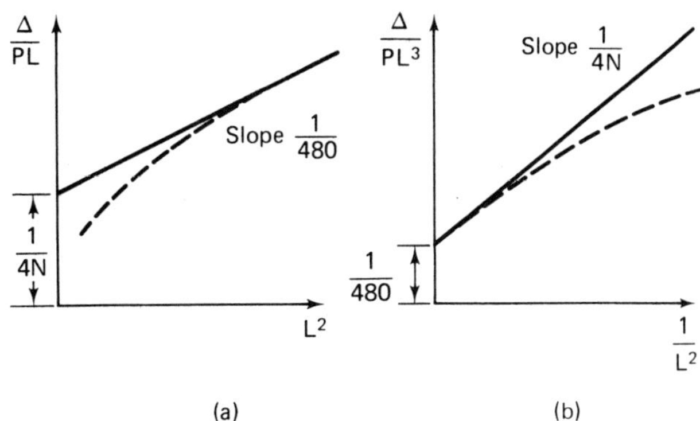

Fig. 16. Experimental determination of stiffnesses D and N. From Allen, H. G., *Analysis and Design of Structural Sandwich Panels*. Elmsford, N.Y.: Pergamon Press, (1969) [5].

It is always helpful to verify the flexural and shear stiffnesses, D and N, of new sandwich constructions experimentally. The three-point load test is commonly used, for which, from Table 2 and Eq. (69),

$$\Delta = \frac{PL^3}{48D} + \frac{PL}{4N} \tag{69a}$$

Values of D and N may be obtained in the following manner [5].

Equation (69a) is recast in either of two forms:

$$\frac{\Delta}{PL} = \frac{L^2}{48D} + \frac{1}{4N}$$

or

$$\frac{\Delta}{PL^3} = \frac{1}{48D} + \frac{1}{4N} \cdot \frac{1}{L^2}$$

The first equation represents a straight line in a plot of Δ/PL against L^2 [solid line, Fig. 16(a)], and the second in a plot of Δ/PL^3 against $1/L^2$ [solid line, Fig. 16(b)]. If Δ/P is measured for several spans, the straight line may be plotted in either figure and the stiffnesses D and N obtained from the slope and intercept of the vertical axis. It is best to average the results of several loads at each span length to obtain the corresponding value of Δ/P. Deflection measurements should be made on the underside of the beam under the load and on top of the beam over the supports to avoid any error due to core crushing at the load and reaction points.

Sandwiches with thick faces and very weak cores, i.e., plywood faces on polystyrene foam core, give curved data plots [dashed lines, Fig. (16)], due to the apparent increase in shear stiffness caused by the bending stiffness of the faces.

A sandwich panel under *edgewise compressive loading* may also fail in any of six ways (Fig. 17):

1. Compressive failure [Eq. (72)].
2. General buckling [Eqs. (73a) and (73b)].
3. Face wrinkling [Eq. (70)].
4. Face dimpling [Eq. (74)].

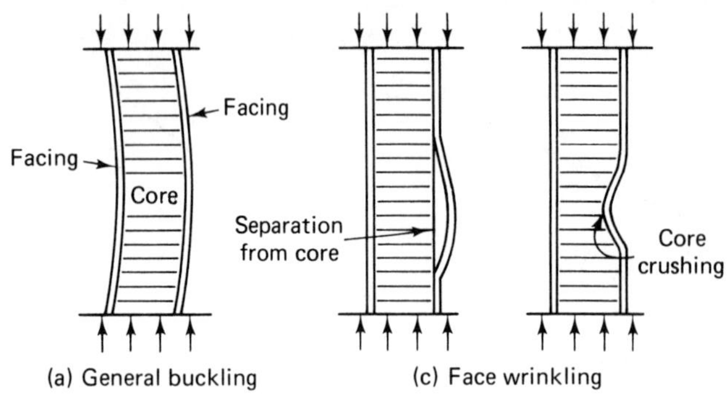

(a) General buckling (c) Face wrinkling

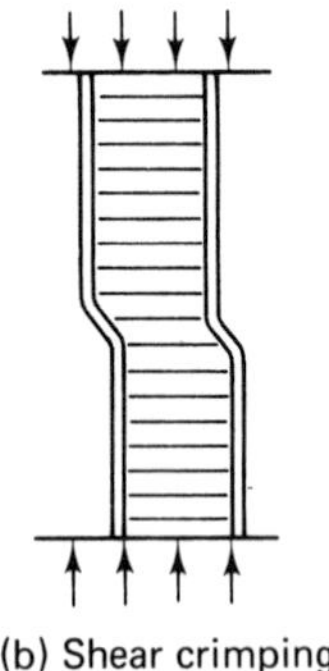

(b) Shear crimping

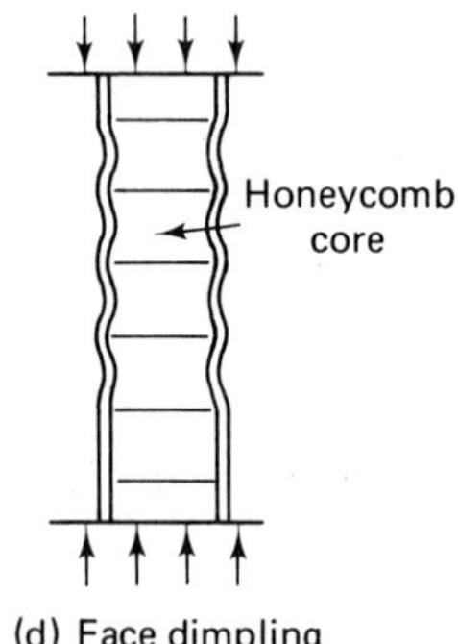

(d) Face dimpling

Fig. 17. Some sandwich panel failure modes under edgewise compressive load. From Nordby, G. M., and W. C. Crissman, "Strength Properties and Relationships Associated with Various Types of Fiberglass-Reinforced Facing Sandwich Structure." Army Fort Eustis Report 65–15 (AD621522) ASTIA. Springfield, Virginia: U. S. Army, Aug. 1965 [8]

5. Excessive compressive deformation [Eq. (75)].
6. Shear crimping.

The pertinent equations are in Table 3.

TABLE 3. Equations for Edgewise Compressive Loading; Sandwich Panel Having Equal Faces

Edge compressive load, P, per inch width of panel	$= 2f\sigma'_f + c\sigma'_c$	(72)
Critical buckling load, P_{cr}, on panel		
a. Panel ends simply supported	$= \dfrac{\pi^2 D}{L^2[1 + (\pi^2 D/L^2 N)]}$	(73a)
b. Panel ends fixed	$= \dfrac{4\pi^2 D}{b^2[1 + (\pi^2 D/b^2 N)]^2}$	(73b)
Face wrinkling stress, σ_w	$= k_w(E_f E_c G_c)^{1/3}$	(70)
Face dimpling stress, σ_d	$= k_d E_f \dfrac{f^2}{a^2}$	(74)
Compressive unit strain, e	$= \dfrac{P/(c + 2f)}{2fE_f + cE_c}$	(75)

The following additional notation is used in Table 3:

a = spacing between points of honeycomb or corrugated core support for the facings
k_d = theoretical or experimental dimpling coefficient
σ'_c = compressive stress in core
σ'_f = compressive stress in faces

The core usually carries so little compressive load that the last term in Eq. (72) can be ignored. Equations (73a) and (73b) are variations of the Euler equations which account for the low stiffness of the core. The second terms in the denominators represent shear deformation of the cores. Equation (73b) is valid provided that the length L of the panel is at least as great as the width b and provided that the second term in the brackets of the denominator is not greater than unity.

Shear crimping [Fig. 17(b)] is caused by low shear stiffness and is not as predictable as general buckling. Face wrinkling for compressive loading [Fig. 17(c)], as well as for beam loading, has been analyzed, and general design parameters are described in the literature [5]. Face dimpling [Fig. 17(d)] normally is a problem only with nonhomogeneous cores such as honeycomb.

Three additional failure modes, impact resistance, dent resistance, and creep under load, are difficult to predict analytically and are usually evaluated experimentally.

2. Design Optimization

Structural sandwich design consists of determining an economical combination of the thicknesses and material properties of the faces and core in order to carry the shear, bending, axial, and other stresses induced by loads. The design problem is simplified to the extent that any of these design parameters may be fixed in advance. Nonstructural requirements such as fire resistance and acoustical and thermal transmissibility may also be specified.

Panels may be designed for minimum weight or minimum construction cost. In the aircraft industry, minimum weight is important because operating costs are critically sensitive to weight. In the building industry, construction cost is more important because transportation and handling costs are typically incurred only once—at the time of erection.

Design optimization examples will be illustrated using both the minimum-weight and minimum-cost criteria.

a. MINIMUM-WEIGHT CRITERION

The following optimum ratios between face and core weights for the various panel failure modes were originally developed by E. W. Kuenzi of the U.S. Forest Products Laboratory [9].

(1) Bending stiffness. The flexural rigidity, D, of a sandwich having identical thin faces and a core of negligible bending stiffness was given by Eq. (65a). The rigidity for the case of *nonidentical* thin faces can be obtained from mechanics as

$$D = \frac{E_1 f_1}{\lambda_1} s^2 \frac{\beta}{1+\beta} \tag{76}$$

where E_1, E_2 and f_1, f_2 are the Young's moduli and thicknesses of faces 1 and 2, respectively, λ_1 is one minus the square of the Poisson's ratio, and

$$\beta = \frac{E_2 f_2 \lambda_1}{E_1 f_1 \lambda_2} \tag{77}$$

The sandwich weight is

$$W = w_1 f_1 + w_2 f_2 + w_c c \tag{78}$$

where w_1, w_2, w_c are the densities of the faces and core, respectively. The weight of any adhesive between faces and core is assumed constant and therefore omitted from the calculation.

From Eq. (76),

$$f_1 = \frac{D\lambda_1(1 + \beta)}{E_1 s^2 \beta}$$

and from Eq. (77),

$$f_2 = \beta f_1 \frac{E_1 \lambda_2}{E_2 \lambda_1}$$

Substituting these expressions into Eq. (78) and minimizing the sandwich weight with respect to s,

$$s^3 = \frac{2D\lambda_1}{E_1 w_c}\left(w_1 + \beta \frac{E_1 \lambda_2}{E_2 \lambda_1} w_2\right)\frac{1 + \beta}{\beta} \tag{79}$$

Substituting D from Eq. (76) into Eq. (79),

$$\frac{f_1}{s} = \frac{w_c}{2[w_1 + \beta(E_1\lambda_2/E_2\lambda_1)w_2]} \tag{80}$$

Substituting β from Eq. (77) into Eq. (80) and rewriting,

$$\frac{w_c s}{2(w_1 f_1 + w_2 f_2)} = 1 \tag{81}$$

Since the core weight is $W_c \simeq w_c s$ and the facing weight is $W_f = w_1 f_1 + w_2 f_2$, Eq. (81) yields

$$W_c = 2W_f \tag{82}$$

Thus for a minimum-weight sandwich of specified stiffness, D, the core must weigh twice as much as the faces, independent of the core and face materials used. The same result is obtained by minimizing sandwich weight with respect to f_1 or f_2 rather than with respect to s.

If sandwich weight is minimized with respect to β instead of with respect to s, one obtains

$$\beta = \left(\frac{w_1 E_2 \lambda_1}{w_2 E_1 \lambda_2}\right)^{1/2} \tag{83}$$

The optimal values of s, f_1, and f_2 can then be found from Eqs. (79), (80), and (77) respectively after substituting Eq. (83) in each. For many combinations of face and core materials, $\beta \simeq 1$.

In design, the optimum weight ratios can often be only approximated, depending on available thicknesses of core and face materials.

(2) Bending strength. For a sandwich of thin equal faces on a core of negligible bending stiffness, Eq. (66) reduces to

$$\sigma_f = \frac{M}{fs} \tag{84}$$

Following the procedure for bending stiffness, Eq. (84) is solved for f, substituted into Eq. (78), and minimized to give

$$s^2 = 2\frac{w_f}{w_c}\frac{M}{\sigma_f} \tag{85}$$

where w_f is the density of both faces. Substituting the value of M from Eq. (84) into Eq. (85),

$$\frac{f}{s} = \frac{w_c}{2w_f} \tag{86}$$

and substituting this result into Eq. (78) gives

$$W_c = W_f \tag{87}$$

Thus, for a minimum-weight sandwich of specified bending strength the weights of core and faces must be equal.

If face wrinkling or dimpling controls, σ_w from Eq. (70) or σ_d from Eq. (74), respectively, should be substituted for σ_f in Eqs. (84) and (85).

Wrinkling stress [Eq. (70)], depends on core and face properties. If it is assumed that core elastic properties are related to face properties in proportion to densities as follows,

$$E_c = k_1\frac{w_c}{w_f}E_f, \qquad G_c = k_2\frac{w_c}{w_f}E_f$$

then Eq. (70) becomes

$$\sigma_w = KE_F\left(\frac{w_c}{w_f}\right)^{2/3} \tag{88}$$

where $K = k_w(k_1/k_2)^{1/3}$. Substituting Eq. (88) into Eq. (85) yields

$$s^2 = \frac{2M}{KE_f(w_c/w_f)^{5/3}}$$

and as before,

$$\frac{f}{s} = \frac{w_c}{2w_f} \tag{86}$$

Therefore, the minimum-weight sandwich for which bending moment resistance is governed by wrinkling of the compression face must also have a core weight equal to the weight of faces.

For dimpling stress [Eq. (74)], if one assumes an effective core density which is related to face density and which is inversely proportional to the spacing between points of face support, a, then

$$a = \frac{k_1 w_f}{w_c}$$

and Eq. (74) becomes

$$\sigma_d = KE_f f^2 \left(\frac{w_c}{w_f}\right)^2$$

where $K = k_d/k_1^2$. Proceeding as before and minimizing,

$$3W_c = W_f \tag{89}$$

(3) Buckling under compressive edge load. The sandwich buckling is determined by bending stiffness, D, and shear stiffness, N [Eqs. (73a) and (73b)]. For the simply supported case, one of the variables f or s may be eliminated by solving Eq. (73a) and then substituting into Eq. (78) and minimizing by the Lagrange multiplier technique [10]. The final result is

$$(1 - Q)W_c = 2W_f \tag{90}$$

where Q is a function of the core shear modulus. If the core shear modulus is large, $Q \simeq 0$, and the core should weigh twice as much as the faces. This was the result obtained for prescribed bending stiffness [Eq. (82)] and should be expected since for large core shear modulus buckling depends on bending stiffness [see Eq. (73a)]. The effect of $Q \neq 0$ reduces the ratio of core weight to face weight [3].

The results are summarized in Table 4.

TABLE 4. Optimum Core-Face Weight and Thickness Ratios for Minimum-Weight Sandwich Panels

Criteria	*Optimum* W_c/W_f	*Optimum* s/f
Bending stiffness	2	$1 + 4w_f/w_c$
Bending strength		
a. σ_f (normal face stress governs)	1	$1 + 2w_f/w_c$
b. σ_w (wrinkling governs)	1	$1 + 2w_f/w_c$
c. σ_d (dimpling governs)	$\frac{1}{3}$	$1 + \frac{2}{3}(w_f/w_c)$
Buckling under compressive load	2*	$1 + 4w_f/w_c$*

*Approximate.

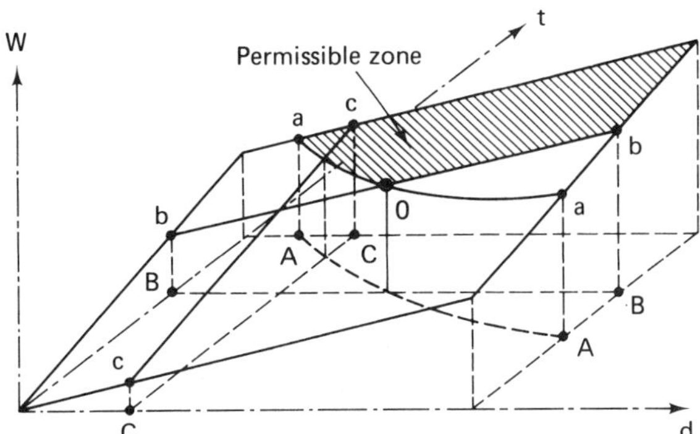

Fig. 18. Graphical representation of panel design optimization with respect to weight. From Allen, H. G., *Analysis and Design of Structural Sandwich Panels*, Elmsford, N. Y.: Pergamon Press, (1969) [5].

Any of the criteria in Table 4 may govern. Therefore, sandwich design involves nonlinear optimization in which Eq. (78) is the objective function and any of Eqs. (65) to (75) may be constraints [10]. The problem has been illustrated graphically by Allen [5]. In Fig. 18, the sandwich weight [Eq. (78)] is represented vertically above the horizontal axes s and f. Curves A, B, and C in the horizontal plane represent any of the constraint equations (65) to (75) and curves a, b, and c represent their vertical projections onto the surface representing sandwich weight. The feasible design region of this surface is shaded, and the optimal design, point O, represents minimum weight on the shaded surface.

(4) Design examples. A common design procedure is to minimize weight for the particular failure mode believed most likely to govern and then to check safety with respect to the remaining failure modes. Alternatively, if all design parameters (thicknesses and properties of core and faces) are fixed except one, which is to be solved for, we may solve the equation representing each failure mode for the single design parameter and take its limiting value. A third procedure, which is more efficient when several or all design parameters are to be found and/or when the probable limiting failure mode is uncertain, is to resort to one of several methods of nonlinear optimization with inequality constraints [10]. The first two procedures are illustrated in Examples 5–7 and the third procedure in Example 8.

EXAMPLE 5: Find the core and face thicknesses, c and f, for a minimum-weight sandwich beam given the following data:

Beam load and support conditions: uniformly loaded, simply supported
Face properties: σ'_f, E_f, k_w, w_f
Core properties: τ'_c, E_c, G_c, w_c

The primes indicate allowable stresses. Assume faces of equal thickness and end supports of such design that core crushing above the supports does not occur.

In the following solution sequence we assume that flexural stress in the face sheets governs the design and then check safety with respect to the remaining failure modes.

Solution:

(a) Determine maximum bending moment and shear, M and V.
(b) Calculate an approximate value for the product fs from Eq. (84), $fs = M/\sigma_f$.
(c) Select a trial starting value of s/f from the weight-minimizing relations in Table 4. For example, if it is anticipated that either normal face stress or wrinkling will govern the design, set $s/f = 1 + 2w_f/w_c$.

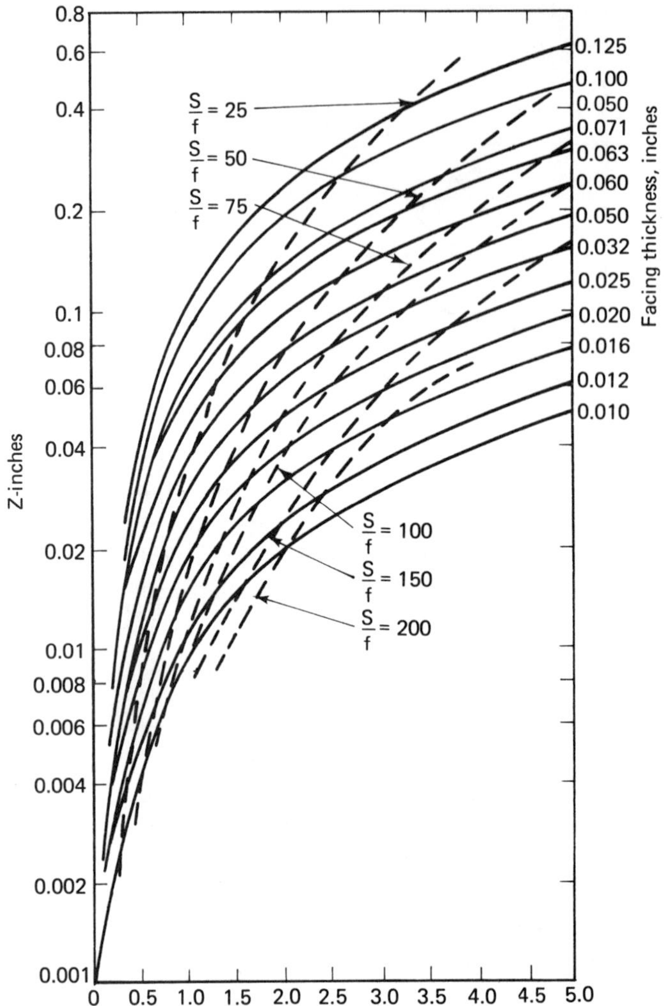

Fig. 19. Graphical solution for s from values of fs and s/f for the weight minimizing relation $s/f = 1 + 2w_f/w_c$. (*Courtesy of Hexcel Corporation.*)

(d) From Fig. 19 find the value of s associated with the value of fs determined in step (b) and with the value of s/f determined in step (c). If metal faces are to be used, move horizontally on the chart to the nearest standard sheet guage.

(e) Check the face stress [Eq. (66)] with the values of s and f thus determined.

(f) Determine the core thickness ($c = s - f$) and check core shear stress [Eq. (67)].

(g) Check safety with respect to remaining failure modes as needed: deflection, using k_b and k_s values from Table 2 [Eq. (69)], wrinkling [Eq. (70)], and possibly dimpling [Eq. (74)].

(h) If any of the failure modes in steps (f) or (g) are found to be limiting, return to step (c), using the corresponding value of s/f from Table 4. Note that step (d) must then be accomplished by solving the appropriate equations, because Fig. 18 can be used only for the minimizing weight relation $s/f = 1 + 2w_f/w_c$.

EXAMPLE 6: Find the core thickness, c, for a uniformly loaded, simply supported minimum-weight sandwich beam, given all core properties and the properties and thicknesses of identical faces. Again assume no core crushing above the supports.

Solution: Since only one design parameter, c, is required, we can easily solve each of the failure mode equations (66), (67), and (69) for c and take the limiting value.

Assume that $c \simeq s$. Then substituting Eq. (65a) into Eq. (66),

$$c_1 = \frac{M}{\sigma_1 f}$$

where σ_1 is the lesser of the allowable ultimate face strength and the wrinkling strength, σ_w, from Eq. (70). Substituting Eq. (65a) into Eq. (67a),

$$c_2 = \frac{V}{\tau'_c}$$

Substituting D from Eq. (65a), N from Eq. (71), and the appropriate values of k_b and k_s from Table 2 into Eq. (69),

$$c_3 = \frac{PL^2}{16G_c\Delta'}\left\{1 + \sqrt{1 + \frac{20}{3f}\frac{G_c^2\Delta'}{E_f P}}\right\}$$

where Δ' is the allowable midspan deflection. The design core thickness is the maximum of the values c_1, c_2, and c_3.

It is apparent that any solution method which provides a minimum-weight design can also provide a minimum-cost design, and one such method for a specific set of panel properties is summarized in the next section.

b. MINIMUM-COST CRITERION

The cost-minimizing formulas [Eqs. (91)–(94)] were developed for the following panel properties [11]:

1. Uniformly loaded, simply supported panel beam with specified maximum midspan deflection.
2. No material in the panel passes the rupture point at the specified deflection.
3. Core contributes to panel stiffness only by resisting shear, not by flexural compression or tension.
4. Costs per unit weight of core and per unit volume of faces are constant.
5. All nonstructural requirements, such as fire resistance and core thermal conductivity, are ignored.

The formulas are

$$P_{ct} = \frac{2.38 \times 10^8(c_1 + c_2t_2/t_1)(1 + E_2t_2/E_1t_1)E_s^3t_1}{E_2t_2c_cp_c}\left(\frac{Y}{LW}\right)^2 \quad (91)$$

$$\frac{t_2}{t_1} = \sqrt{\frac{E_1c_1}{E_2c_2}} \quad (92)$$

$$t_1 = \frac{5.05 \times 10^{-7} L c_c p_c}{E_s(c_1 + c_2 t_2/t_1)} \frac{LW}{Y} \frac{1 - N_{ct}}{N_{ct}} \frac{1 - 2N_{ct}}{2 - 3N_{ct}} \tag{93}$$

$$t_c = \frac{\dfrac{8.7 \times 10^{-4} L}{E_s} \pm \sqrt{\dfrac{75.7 \times 10^{-8} L^2}{E_s^2} + \dfrac{Y}{LW} \dfrac{(36 \times 10^{-5} L^3)(1 + E_2 t_2/E_1 t_1)}{E_2 t_2}}}{2Y/LW} \tag{94}$$

where P_{ct} = core cost factor,
c_1, c_2 = costs of faces 1 and 2, cents/cu in.
c_c = core cost, cents/lb
E_1, E_2 = Young's moduli of faces, psi
E_s = shear modulus of core, psi
p_c = core density, pcf
Y = allowable length, in.
L = span length, in.
W = panel width, in.
t_1, t_2 = face thicknesses, in.
N_{ct} = optimal cost ratio, core material cost per total material cost
t_c = core thickness, in.

The formulas incorporate three rules concerning core density, core cost, and face density:

1. The optimum core material is that having the lowest density which provides adequate shear strength.
2. If the shear modulus of the core is great so that shear deformation is negligible, the core should cost just twice as much as the faces. The relative core cost increases with increasing shear modulus.
3. The ratio of face thickness varies inversely with the square root of the products of face Young's modulus and costs per unit volume, as expressed by Eq. (92).

Solutions are obtained by solving Eq. (91) for P_{ct} and obtaining the corresponding value of N_{ct} from Fig. 20 and then solving Eqs. (92)–(94) for t_1, t_2, and t_c. The

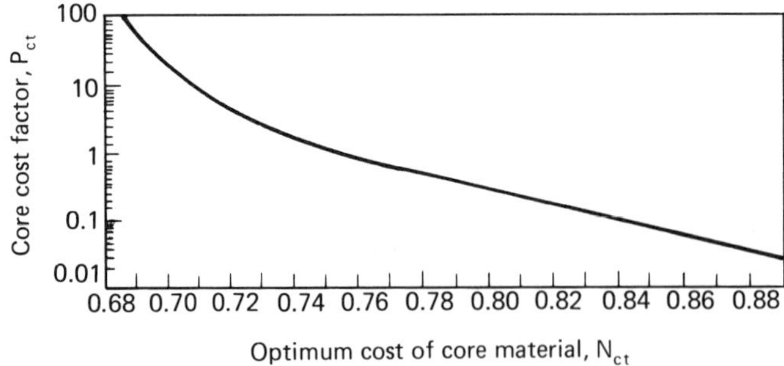

Fig. 20. Graph to obtain optimal cost ratio, N_{ct}. From Strausky, Y. and N. J. Hoff, "Mechanics of Composite Structures," in *Composite Engineering Laminates*, A. G. Dietz, ed. Cambridge, Mass.: The MIT Press, 1969, Chap. 1. [3].

following example, reprinted by permission from Schwartz and Rosato [12], applies these equations to a panel having aluminum and steel faces over a urethane foam core.

EXAMPLE 7: The panel measures 240 in. long (L) by 48 in. wide (W) and the allowable deflection (y) is 1 in. The aluminum (face 1) has a modulus of elasticity (E_1) of 10^7 psi, and its cost (c_1) is 40 cents/lb $\times$ 0.099 lb/cu in., or 3.96 cents/cu in. Steel (face 2) has a modulus (E_2) of 3×10^7 psi, and its cost (c_2) is 11 cents/lb $\times$ 0.283 lb/cu in. or 3.11 cents/cu in. The polyurethane core material has a shear modulus (E_s) of 500 psi, a density (p_c) of 3 pcf, and a cost (c_c) of 45 cents/lb.

By substituting these figures in the equations, values obtained are

(a) Relative thickness of the faces: The thickness ratio of the steel to aluminum face as calculated from Eq. (92) is 0.65.
(b) Relative cost of the core: The core cost factor calculated from Eq. (91) is 5.6. Using this value in conjunction with the appropriate curve, the optimum cost of the core material (N_{ct}—or the cost of the core divided by the combined cost of the core and faces) is 0.72. This means that the cost of the core should constitute 72% of the total cost of core and facing materials.
(c) Thickness of the faces: The thickness of the aluminum face as calculated from N_{ct} and Eq. (93) is 0.035 in. Then, from Eq. (92), the steel face is 0.023 in.
(d) Thickness of the core: From Eq. (94), the thickness of the core is 7.2 in.

The next example does not satisfy the specific assumptions required for Eqs. (91)–(94) and illustrates a more general cost minimization prodedure.

EXAMPLE 8*: Precast autoclaved concrete roof sandwich panels are to have concrete compressive faces, foamed concrete cores, and asbestos fiber-reinforced tensile faces. Find the thicknesses of the two faces and core and the *core density factor* which minimize first cost, given the following data:

Load and supports:

$$100 \text{ psf}; \qquad L = 20 \text{ ft, simply supported}$$

Face properties:

$$\sigma'_{f1} = 5{,}000 \text{ psi} \qquad E_{f1} = 40 \times 10^6 \text{ psi}$$

$$\sigma'_{f2} = \min \begin{cases} 3{,}000 \text{ psi} \\ 20{,}000\, c^2 \text{ psi}\dagger \end{cases} \qquad E_{f2} = 8 \times 10^6 \text{ psi}$$

Core properties:

$$\tau'_c = 100 d_0^{\frac{3}{2}} \text{ psi} \qquad G_c = 1.3 \times 10^6 d_0 \text{ psi}$$

Subscripts 1 and 2 represent tensile and compressive faces, respectively, and d_0 is the core density factor (solids volume per total core volume). Design against failure by bending stresses σ_{f1} and σ_{f2} and core shear stress τ_c, with a maximum midspan deflection of 0.5 in.

*Adapted with permission from Stark and Nicholls [10].

†Allowable stress for buckling of compressive face.

The tension face costs five times as much per unit volume as the compressive face, and the core costs the same per unit *weight* as the compressive face.

Solution: Assume that $c \simeq s$. Then substituting Eq. (65a) into Eq. (66),

$$\sigma_{f1} = \frac{M}{cf_1} \tag{95}$$

and

$$\sigma_{f2} = \frac{M}{cf_2} \tag{96}$$

Substituting Eq. (65a) into Eq. (67a),

$$\tau_c = \frac{V}{c} \tag{97}$$

Substituting N from Eq. (71) and values of k_b and k_s from Table 2 into Eq. (69),

$$\Delta = \frac{5PL^3}{384D} + \frac{PL}{8G_c c} \tag{98}$$

The problem becomes

$$\text{minimize } C = c_1 + c_2 + c_c \tag{99}$$

where the notation indicates costs of panel, tension face, compression face, and core, respectively.

For a 1-in.-wide beam, $V = 83.2$ lb, and $M = 5{,}000$ lb-in.

From Eq. (95),

$$f_1 = \frac{M}{c\sigma_{f1}}$$

and

$$c_1 \geq \left\{5\left(\frac{M}{c\sigma_{f1}}\right) = 5\left[\frac{5{,}000}{5{,}000c}\right] = \frac{5}{c}\right\} \tag{100}$$

From Eq. (96),

$$c = \frac{M}{c\sigma_{f2}}$$

and

$$c_2 \geq \left\{1\frac{M}{c\sigma_{f2}} = \max\left[\frac{5{,}000}{3{,}000c}, \frac{5{,}000}{20{,}000f_2^2c}\right]\right\} \tag{101}$$

From Eq. (97),

$$d_0 \geq \left[\sqrt{\frac{0.832}{c}} = \frac{0.912}{c^{0.5}}\right]$$

and

$$c_c \geq [1(cd_0) = 0.912c^{0.5}] \tag{102}$$

From Eq. (98),

$$\frac{29.9 \times 10^6}{D} + \frac{3.84 \times 10^{-3}c}{[c + (f_1 + f_2)/2]^2 d_0} \leq 0.5 \tag{103}$$

in which D, using Eq. (62), is

$$D = \sum E_i I_i = 40 \times 10^6 f_1 \bar{x}^2 + 8 \times 10^6 f_2 \left[c + \frac{f_1 + f_2}{2} - \bar{x}\right]^2 \tag{104}$$

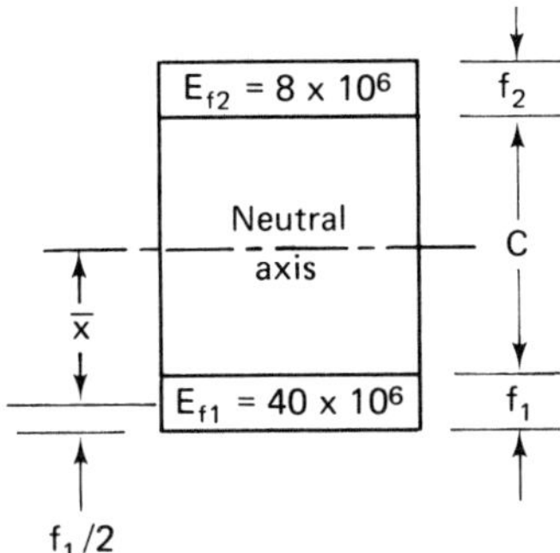

We neglect the moments of inertia of the compressive and tensile faces about their own centroids. From Eq. (61),

$$\bar{x} = \frac{cf_2 + (f_2/2)(f_1 + f_2)}{5f_1 + f_2} \tag{105}$$

Since Eq. (103) is cumbersome, we assume that one or more of the stress constraints [Eqs. (100)–(102)] are binding and test the resulting solution for deflection (103).

Because no lower bound for d_0 was specified and since d_0 does not appear in Eq. (100) or (101), Eq. (102) binds. Also, since face stresses are not affected by the location of the neutral axis [due to the assumption of thin faces in Eqs. (95) and (96)], Eqs. (100) and (101) also bind (an understressed face would be reduced in thickness to develop maximum stress).

As a first trial, let $c_c = 1.66/c$ (assuming compression, not buckling controls). Substituting into Eq. (99) gives

$$C = \frac{5}{c} + \frac{1.66}{c} + 0.912c^{0.5} = \frac{6.66}{c} + 0.912c^{0.5}$$

Setting $\partial C/\partial c$ to zero yields $c = \underline{6.00 \text{ in.}}$

Substituting into Eq. (101), $f_2 = 5{,}000/6(3{,}000) = \underline{0.277 \text{ in.}}$, and the allowable buckling stress $= 40{,}000(0.277)^2 = 3{,}080 > 3{,}000$ psi. Buckling does not control, and the assumption that $c_c = 1.66/c$ is sound.

From Eq. (102),

$$d_0 = \frac{0.912}{6^{0.5}} = \underline{0.420.}$$

Finally,

$$f_1 = \frac{5{,}000}{6(5{,}000)} = \underline{0.166 \text{ in.}}$$

(a) Three-pinned portal frame

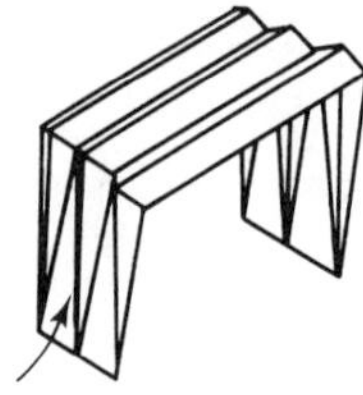

(b) Two-pinned portal frame

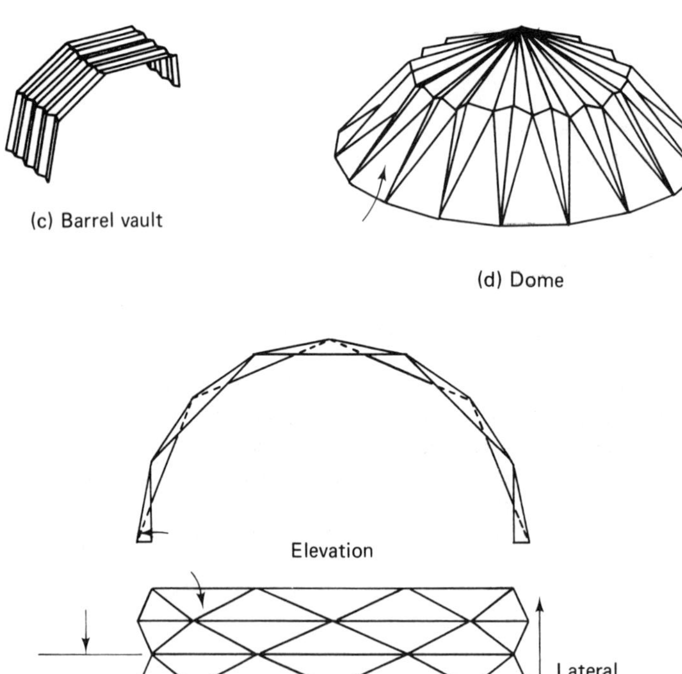

(c) Barrel vault

(d) Dome

(e) Barrel vault

Fig. 21. Sandwich panel folded plate structures. From Benjamin, B. S., *Structural Design with Plastics.* New York: Van Nostrand Reinhold Company (Litton Educational Division), 1969 [14].

Using Eqs. (103)–(105) to check deflection,

$$\bar{x} = 1.56, \qquad D = 60.9 \times 10^6$$

and

$$\frac{29.9 \times 10^6}{60.9 \times 10^6} + \frac{3.84 \times 10^{-3}(6)}{6.22(0.420)} = 0.495 < 0.5 \text{ in.} \quad \textbf{Deflection OK.}$$

Additional structural efficiency can be gained by providing a core material whose density, and hence strength, increases in proportion to bending stress from the neutral axis to the faces. One technique for providing such a core consists of lining the insides of the faces with open-textured glass or acrylic fibers then injecting urethane foam into the core cavity, providing a foam density which varies from as low as 1.5 pcf in the center of the core to as high as 28 pcf at the faces where the foam has penetrated the fibers [13]. Nonuniform core sandwich panels have been used in curtain walls, sanitary fixtures, transportation vehicles, and many other applications.

Structural efficiency on a larger scale can be gained by combining sandwich panels into various forms of folded plate structures [14]. Figures 21(a) and (b) show two- and three-pinned folded plate portal frames. The sandwich panels must be end- and edge-shaped in order to mate each other. The triangular panels marked A may be omitted for openings between columns. Figure 21(e) shows a folded plate barrel vault which can be built with just a single shape of element, resembling a rhomboid folded along its longer diagonal [14].

Ashton and Whitney [15], Calcote [16], and Lubin [17] are excellent references for more advanced problems in the analysis and design of sandwich panels.

PROBLEMS

1. A laboratory scale sandwich panel 3 in. wide having $\frac{1}{16}$-in. fiber-reinforced faces on a $1\frac{1}{2}$-in.-thick foamed core deflected the following amounts under three-point load tests:

Span Length (Simply Supported)	*Midspan Load*	*Midspan Deflection*
1 ft	100 lb	0.008 in.
2	100	0.026
3	100	0.060

Determine the required face thickness of a full-scale panel in order to limit midspan deflection to 0.2 in. for a 12-ft simply supported span carrying 80-psf uniform load and having similar face and core materials with a 5-in.-thick core.

2. Find the core and identical face thicknesses, c and f, to minimize the weight of a sandwich beam, given the following data:

Load and supports: 40 psf uniform load; $L = 12$ ft, simply supported.

Face properties: $\lambda'_f = 1{,}200$ psi, $E_f = 10^6$ psi, $k_w = 0.5$, $w_f = 168$ pcf.
Core properties: $\sigma'_c = 100$ psi, $E_c = 8{,}000$ psi, $\tau'_c = 70$ psi, $G_c = 2{,}000$ psi, $w_c = 4$ pcf.

Maximum allowable midspan deflection = 0.2 in. Assume a foam core so that dimpling failure can be ignored and wide or yielding supports so that core crushing over the supports can be ignored.

3. Find the required foamed polymer core density, w_c, for an aluminum skin sandwich panel having the following properties:

Load and supports: 20 psf uniform load; $L = 10$ ft, simply supported.
Face properties: $\sigma'_f = 35{,}000$ psi, $E_f = 10^7$ psi, $f = 0.08$ in., $k_w = 0.5$.
Core thickness: $c = 5.92$ in.

The core properties are plotted as a function of core density in Fig. 22. The allowable midspan deflection is 0.3 in. Assume the supports to be sufficiently wide and/or yielding to prevent core crushing above the supports.

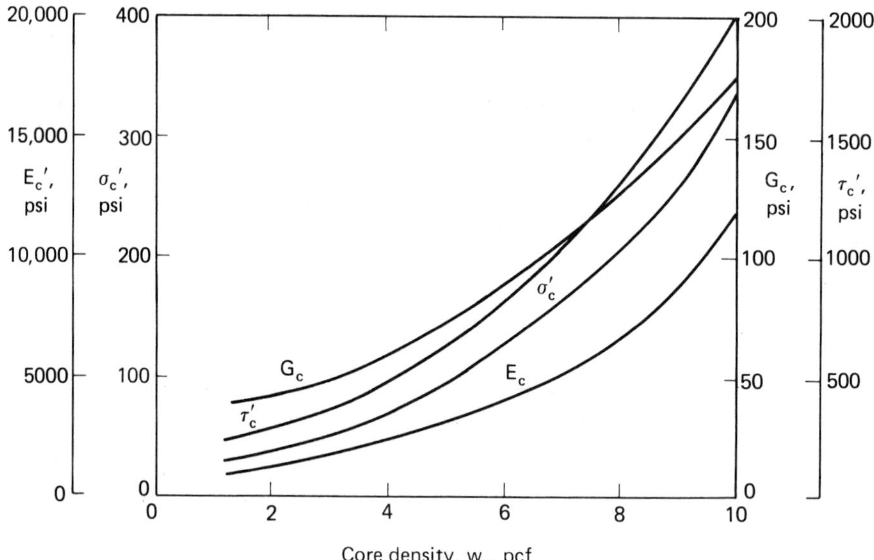

Fig. 22.

4. Write the equations for determining the required core thickness, c, of a pin-ended sandwich strut, given the load, core properties, and properties and thickness of identical faces.

5. A sandwich panel having fiber-reinforced faces on a foamed core is to be designed for minimum cost. Determine the optimum thicknesses of the core and each face, the optimum core density, and the fiber volume ratio in the tensile face, given

Load and supports: 80 psf uniform load; $L = 15$ ft, simply supported.
Face properties: $\sigma'_{f1} = k_1 v_{f1}$ psi, $E_{f1} = k_2 v_{f1}$ psi, $c_{f1} = \$k_3 v_{f1}$/cu in. (tensile face), $\sigma'_{f2} = k_4 v_{f2}$ psi, $E_{f2} = k_5 v_{f2}$ psi, $c_{f2} = \$k_6 v_{f2}$/cu in. (compressive face).
Core properties: $G_c = k_7 d_0$ psi, $c_c = \$k_8 d_0$/cu in.
Maximum allowable midspan deflection: 0.5 in.

Subscripts 1 and 2 refer to tensile and compressive faces, respectively, v_f is the fiber-volume ratio, and d_0 is the core density factor (solids volume per core volume). Although there are six potential failure modes for a sandwich panel beam, assume that only face-bending stresses and midspan deflection may govern.

REFERENCES

1. DIETZ, A. G. H., "Fiber Reinforced Composite Materials Engineering Analysis," in *Environmental Effects on Polymeric Materials*, Vol. II: Materials, D. V. Rosato and R. T. Schwartz, eds. New York: John Wiley & Sons, Inc. (Interscience Division), 1968, Sec. 15 II.
2. ASHTON, J. E., J. C. HALPIN, and P. H. PETIT, *Primer on Composite Materials: Analysis*. Westport, Conn.: Technomic Publishing Co., 1969.
3. STAUSKY, Y., and N. J. HOFF, "Mechanics of Composite Structures," in *Composite Engineering Laminates*, A. G. H. Dietz, ed. Cambridge, Mass.: The M.I.T. Press, 1969, Chap. 1.
4. TSAI, S. W., "Mechanics of Composite Materials," Part II, *U.S. Air Force Material Laboratories Technical Report 66–149*, Dayton, Ohio: U.S. Air Force Material Laboratories, 1966.
5. ALLEN, H. G., *Analysis and Design of Structural Sandwich Panels*. Elmsford, N.Y.: Pergamon Press, Inc., 1969.
6. PLANTEMA, F. J., "Sandwich Construction," Vol. 3 of *Airplane, Missile and Spacecraft Structures*, N. J. Hoff, ed. New York: John Wiley & Sons Inc., 1966.
7. FENTON, J. B., *Automotive Design Eng.* (Nov. 1962).
8. NORDBY, G. M., and W. C. CRISSMAN, "Strength Properties and Relationships Associated with Various Types of Fiberglass-Reinforced Facing Sandwich Structure," *Army Ft. Eustis Report* **65–15** (*AD621522*) *ASTIA*. Springfield, Va.: U.S. Army Aug., 1965.
9. KUENZI, E. W., "Minimum Weight Structural Sandwich," *Forest Prod. Lab. Rept. 086*. Madison, Wisc.: U.S. Department of Agriculture Forest Service, Forest Products Lab, Jan., 1965.
10. STARK, R. M., and R. L. NICHOLLS, *Mathematical Foundations for Design: Civil Engineering Systems*. New York: McGraw-Hill Book Company, 1972.
11. HARTSOCK, J. A., *Mater. Design Eng.* (March 1966) 72.
12. SCHWARTZ, R. T., and D. V. ROSATO, "Structural Sandwich Construction," in *Composite Engineering Laminates*, A. G. H. Dietz, ed. Cambridge, Mass.: The M.I.T. Press, 1969, Chap. 8.
13. SIREN, R. L., *J. Soc. Plastics Engrs.*, 21 (Nov. 1965) 1290.
14. BENJAMIN, B. S., *Structural Design with Plastics*. New York: Van Nostrand Reinhold Company, 1969.
15. ASHTON, J. E., and J. M. WHITNEY, *Theory of Laminated Plates*. Westport, Conn.: Technomic Publishing Co., 1970.

16. CALCOTE, L. R., *Analysis of Laminated Composite Structures.* New York: Van Nostrand Reinhold Company, 1969.

17. LUBIN, G., ed., *Handbook of Fiberglass and Advanced Plastics Composites.* New York: Van Nostrand Reinhold Company, 1969.

V

PHYSICAL METHODS

13

EXPERIMENTAL OBSERVATION OF COMPOSITES

End of the wonderful one-hoss shay.
Logic is logic. That's all I say.
—HOLMES

Analytical techniques for studying material properties can be classed as chemical and physical. Chemical methods, such as titration, ion exchange, pH, and chromatography, measure the presence and quantities of certain atoms, ions, functional groups, or molecular species but generally give only indirect clues as to their structural arrangement. On the other hand, physical methods such as optical and electron microscopy, x-ray and electron diffraction, spectroscopy, and thermal analysis generally provide information on structural arrangement, from which a partial knowledge of chemical composition can sometimes also be inferred. The two classes of methods are thus complementary.

In this chapter we shall introduce several physical methods which have been widely used for quality control of construction materials such as asphalts, cements, glasses, ceramics, and metals and for research and development work with these materials. Three especially useful groups of physical methods are

1. *Optical* methods, using either visible light or electron optics.
2. *Diffraction* methods, using an x-ray, electron, or neutron beam.
3. *Spectrographic* methods, i.e., recording a characteristic of the electromagnetic spectrum—as in emission, atomic absorption, or infrared spectroscopy and in x-ray fluorescence or electron probe analysis.

The physical methods described in the chapter include electron microscopy, x-ray diffraction, infrared spectroscopy, and differential thermal analysis. The petrographic microscope, while also widely used in the study, development, and commercial control of construction composites, is omitted because of the numerous excellent guides to its use [1–5].

A. ELECTRON MICROSCOPY

In the electron microscope, an electron beam performs a function similar to that of light in the light microscope. Since the electron beam has a wavelength in the order of 0.05 A, compared with 4,000–7,000 A for visible light, the resolving power is much greater. Resolution of spacings as close as 2–10 A is possible, and magnifications as high as 100,000. Magnetic or electrostatic lenses in an electron microscope perform a function similar to that of glass lenses in a light microscope. As in light microscopy, the beam may be transmitted through the sample or reflected from it. Reflection techniques permit focusing an electron beam on an area as small as 100 A in diameter and scanning across the surface of the sample. The reflected beam can be observed directly, or the secondary emission of x-rays from the sample generated by the incident beam can be analyzed to identify chemical composition within the small area under focus.

2. Apparatus

Figure 1 shows one variety of commerical electron microscope, and Fig. 2 illustrates the similarity in optics with the light microscope. Note that the arrangement of lenses and the ray paths are identical in the two systems.

The radiation source in an electron microscope is an electron gun, which typically consists of a heated tungsten cathode filament about 0.1 mm in diameter and an accelerating anode. The important factors in gun design are to obtain high electron emission in a concentrated beam (for electrical and thermal efficiency) and to emit electrons at constant velocity (for optical quality). Operating voltages are normally 50–100 kv.

The beam of electrons is focused onto the sample by one or two condenser lenses, and the electrons scattered by the sample are focused with an objective and one or more projector lenses to produce a highly magnified image on a fluorescent screen.

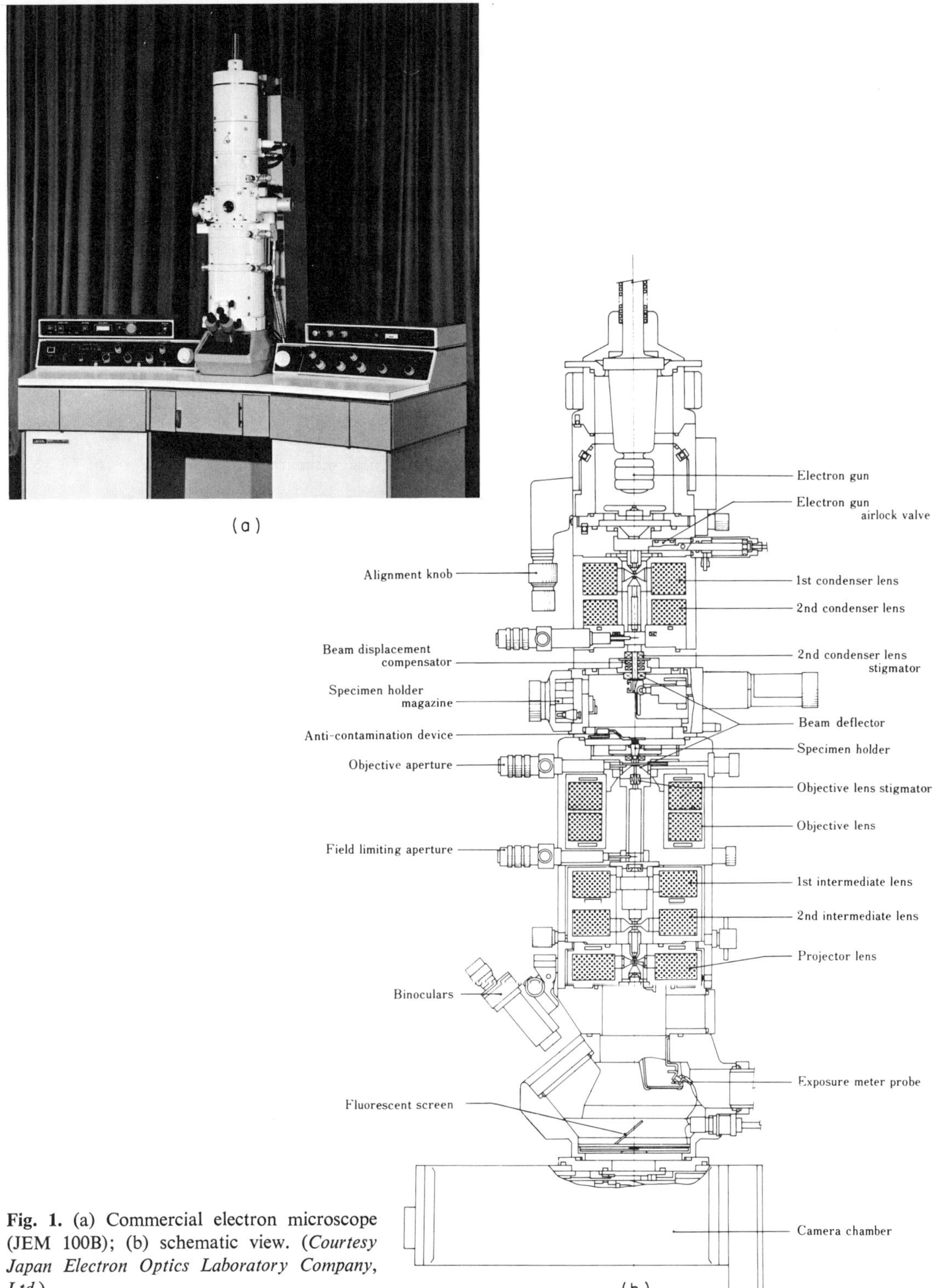

Fig. 1. (a) Commercial electron microscope (JEM 100B); (b) schematic view. (*Courtesy Japan Electron Optics Laboratory Company, Ltd.*)

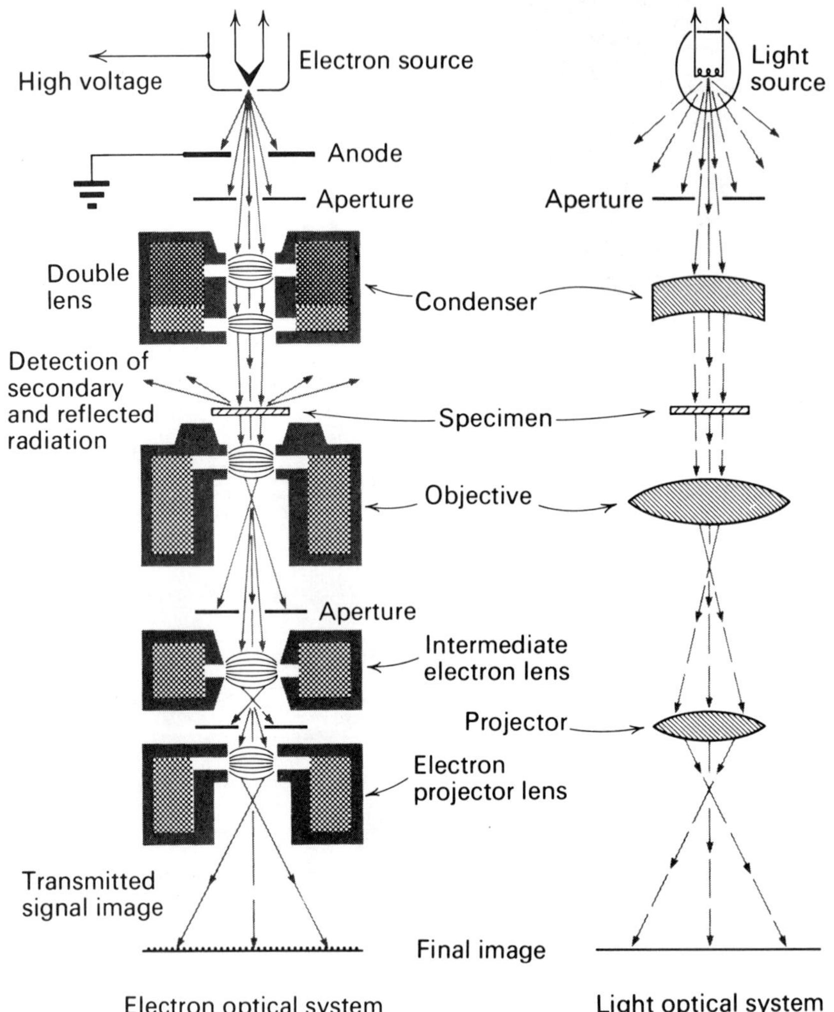

Fig. 2. Comparison of optics in typical light and electron microscopes. From Murr, L. E., *Electron Optical Applications in Materials Science*. New York: McGraw-Hill Book Company, 1970 [6].

The image may be recorded by direct exposure of photographic film to the electron beam. Magnifying power is normally controlled by regulating current to the intermediate electron lens.

The fact that electron optical systems are characterized by small aperture angles permits sharp images to be obtained for considerable distances above and below the object plane, i.e., a large depth of focus. This permits good images of rough surfaces with the electron microscope. It also allows pairs of images to be recorded with the sample tilted approximately 10° between exposures, so that the resulting prints can be mounted and viewed stereoscopically. Commercial goniometric stages are available which permit both rotation and tilting. Sample heating and cooling devices are also available. A heat-resistant substrate such as silica or alumina film on a platinum grid is sometimes used to study thermal transformations of materials [7].

2. Electron Optics

The ray paths of electrons in an electric or magnetic field are identical to the paths of light where glass lenses are the refractive medium. In an electric or magnetic field, the refractive index at each point depends on electrical field strength, just as

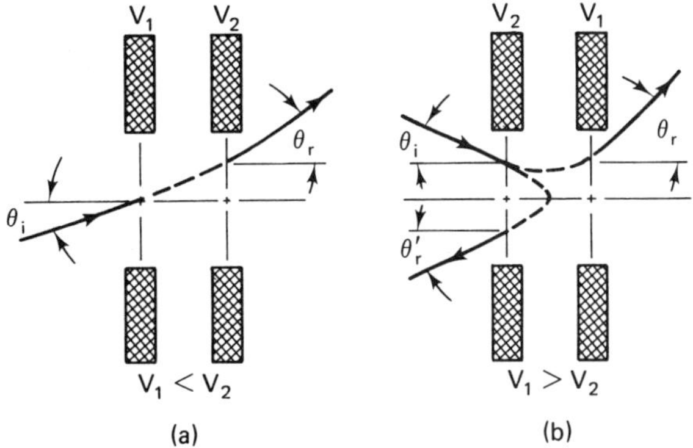

Fig. 3. Electron (a) refraction, and (b) reflection due to change in field strength.

light refraction depends on optical density. The velocity change and refraction of an electron beam passing from high potential V_1 to low potential V_1 [Fig. 3(a)] is defined by Snell's law, and as with light, when ϕ_i is large, the beam is reflected [Fig. 3(b)]:

$$\frac{\sin \theta_r}{\sin \theta_i} = \left(\frac{V_1}{V_2}\right)^{1/2}, \qquad \theta_i < \sin^{-1}\left(\frac{V_1}{V_2}\right)^{1/2} \qquad \text{refraction}$$

$$\theta'_r = \theta_i, \qquad \theta_i > \sin^{-1}\left(\frac{V_1}{V_2}\right)^{1/2} \qquad \text{reflection}$$

The deflection of the electron beam can be accomplished by either an electrostatic lens [Fig. 4(a), or an electromagnetic lens [Fig. 4(b)]. The magnifying power of the electrostatic lens depends on the voltage applied across the plates.

Electromagnetic lenses have largely superseded electrostatic lenses because of their greater stability in focusing electron beams. The magnifying power of an electromagnetic lens is proportional to the number of turns, N, of the coil, the current, I, and the extent of the field region. From the standpoint of microscope construction, it is desirable to concentrate the field into as short a region as possible. This is accomplished by surrounding the lens coil with an annularly slotted high-susceptibility soft iron casing and using a central high-susceptibility pole piece.

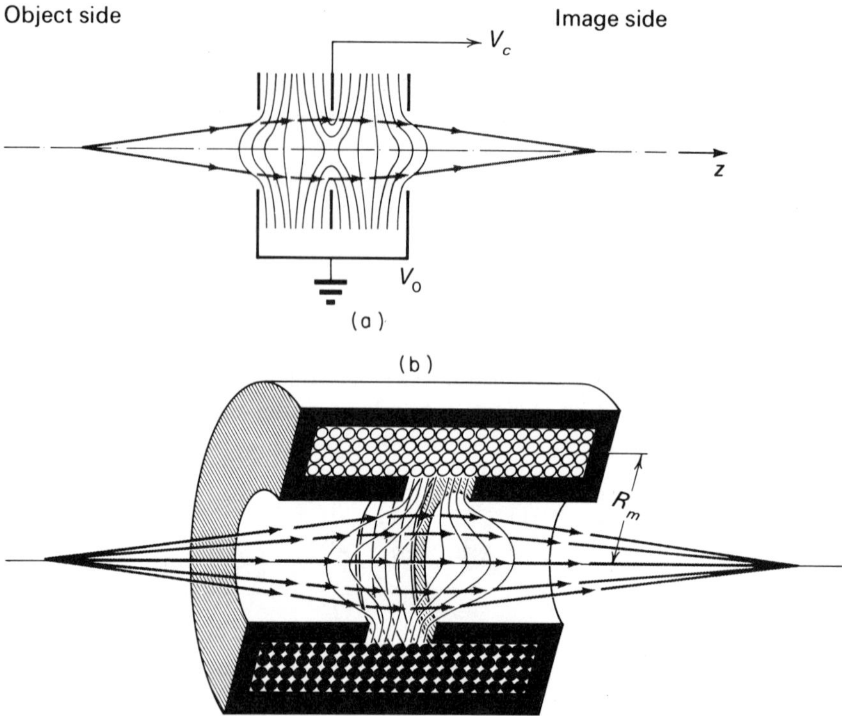

Fig. 4. Electron lenses. (a) Electrostatic; (b) electromagnetic. From Murr, L. E., *Electron Optical Applications in Materials Science*. New York: McGraw-Hill Book Company, 1970 [6].

3. Modes of Use

By employing minor instrumental modifications, electron optical systems can be used to study materials by a variety of modes. These include

1. *Electron microscopy*, in which an image is obtained of the sample, or of a replica made from its surface, by either transmission or reflection techniques.
2. *Electron microprobe analysis*, in which the electron beam is focused to a sharp point on the sample, and either the absorption or backscatter of electrons is monitored, or the secondary emissions of either electrons or x-rays are recorded and analyzed [8]. Variations of electron probe analysis include the direct imaging of the sample by exposure to x-rays instead of to an electron beam (x-ray microscopy) and the analysis of x-ray absorption [9] or of secondary x-rays generated in the sample by exposure to an incident x-ray beam (x-ray fluorescence).
3. *Electron diffraction*, with which crystallographic determinations can be made by techniques, similar in some ways to x-ray diffraction techniques (see Sec. B).

The electron probe techniques are useful for chemical analyses of surface areas as small as 100 A, which is approximately the limit of focal sharpness due to various types of aberration. Electron diffraction is useful for both chemical and structural determinations over areas down to about the same size. Electron probe analysis and diffraction are analytical techniques and require use of the sample material directly, whereas microscopy is an imaging technique and may use either the sample or a replica of its surface. With some instruments it is possible to perform analyses by several of the modes on the same sample. The electron probe can be used as a scanning instrument, with which information is obtained by scanning the sharply focused electron beam across the sample surface. Several of these techniques will be briefly described.

While an electron image of a sample can show its geometry and topography, it contains no information about chemical composition. Figure 5 depicts the interaction of an electron beam with a sample, including those secondary emissions which can be analyzed for the determination of chemical content. The magnitude of transmitted radiation depends of course on the sample thickness and incident electron velocity.

The electron microprobe analyzer uses the same electron source, condenser, and objective lenses as shown in Fig. 2 for the microscope. The objective lens, however, is used to focus the electron beam onto the sample, rather than for imaging the sample detail [10, 11, 12]. The sample is placed at the focal point of the objective lens

Fig. 5. Interaction of an electron beam with a solid. From Murr, L. E., *Electron Optical Applications in Materials Science*. New York: McGraw-Hill Book Company, 1970 [6].

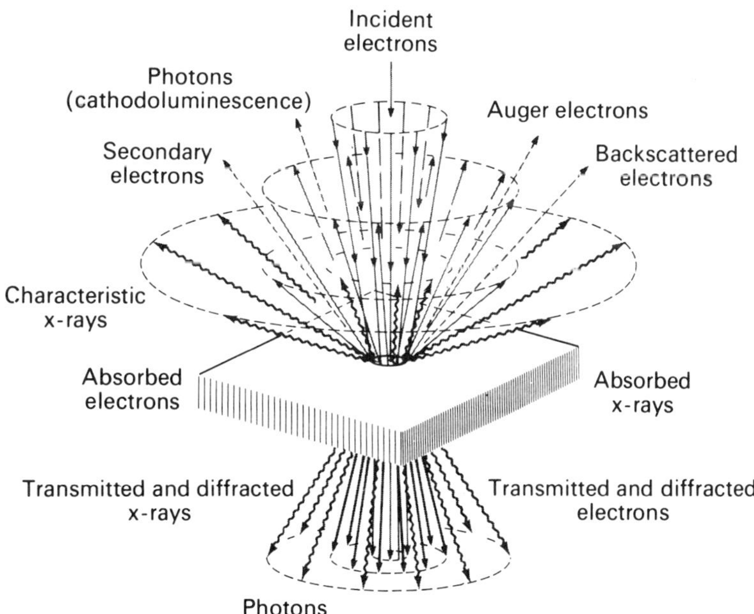

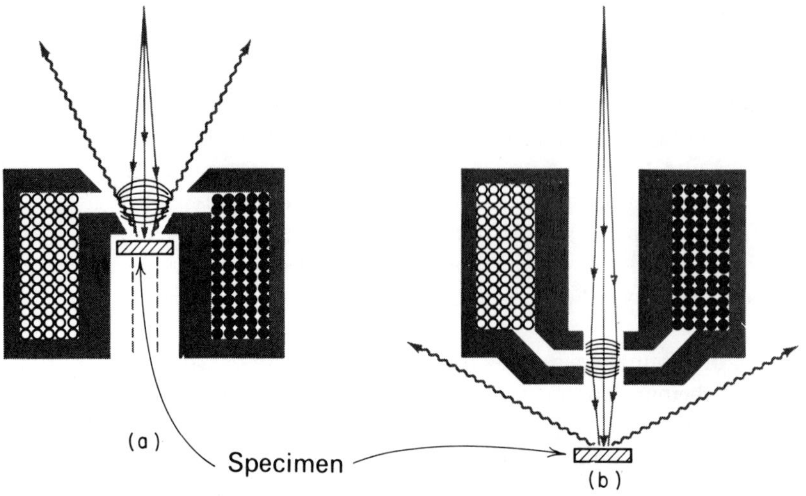

Fig. 6. Objective lens arrangements for electron probes. (a) Sample within lens; (b) lens inverted. From Murr, L. E., *Electron Optical Applications in Materials Science*. New York: McGraw-Hill Book Company, 1970 [6].

either by mounting the sample within the lens [Fig. 6(a)] or by inverting the lens [Fig. 6(b)].

In *x-ray emission microprobe analysis*, an analyzing crystal (monochrometer) is inserted between the emission point on the sample and a detector (Fig. 7). The

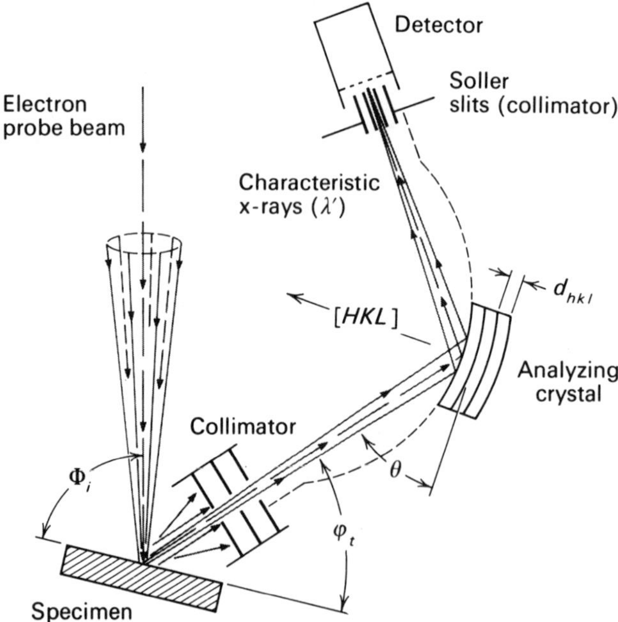

Fig. 7. Filtering and detection system for x-ray emission microprobe analysis. From Murr, L. E., *Electron Optical Applications in Materials Science*. New York: McGraw-Hill Book Company, 1970 [6].

analyzing crystal serves to filter out all x-rays except those which satisfy Bragg's law,

$$n\lambda = 2d \sin \theta \tag{1}$$

By collimating the beam incident to the analyzing crystal, all but one specific wavelength can be excluded from the detector. The necessary θ angle for detecting the x-ray wavelength of a specific element can be positioned mechanically on a goniometer as in x-ray diffractometry [13]. The detector is identical to standard x-ray detectors [13] and may be of the Geiger counter, proportional counter, scintillation counter, or photomultiplier types. Table 1 lists d spacings for some common analyzing crystals.

TABLE 1. Common Analyzing Crystals

Crystal designation	*Reflection plane (hkl)*	$2d_{hkl}$ *(A)*
Muscovite (mica)	002	19.840
Ammonium dihydrogen phosphate (ADP)	101	10.640
Low quartz (SiO_2)	101	6.687
Calcite ($CaCO_3$)	104	6.071
Sodium chloride (NaCl)	200	5.641
Lithium fluoride (LiF)	200	4.027
Topaz	303	2.712

SOURCE: MURR, L. E., *Electron Optical Applications in Materials Science*. New York: McGraw-Hill Book Company, 1970.

Elements are identified using Moseley's law,

$$\frac{1}{\lambda} = C(z - \sigma)^2 \tag{2}$$

where λ is the wavelength of x-ray emission; C is a constant associated with electron jumps to a specific electron shell, as the K or L shells; and σ is a nuclear screening constant, having values for K emission spectra of about 3 to 5 for atomic numbers from 30 to 90, respectively. Equation (2) is plotted in Fig. 8 for K_α and L_α emissions. Thus, specific elements in the reaction zone can be identified by measuring wavelengths of the emitted x-rays and comparing them with the data in Fig. 8. Quantitative estimates can be obtained by using the relationship

$$\frac{I_A}{I'_A} = C_A g(E) \tag{3}$$

where I_A is the measured intensity of, say, K_α, x-radiation from element A in the sample being investigated; I'_A is the measured intensity of similar wavelength from a reference sample of pure element A; C_A is the concentration of element A in the sample being investigated; and $g(E)$ is an energy loss ratio for corrections due to mass absorption, which may be approximately evaluated by preparing mixtures of known ratios of the detected substances. The intensities observed at the detector will of course depend on the intensity of the incident electron beam and the angles of incidence θ_i and takeoff θ_t (Fig. 5). These three parameters must therefore be kept constant when using Eq. (3) for quantitative analysis.

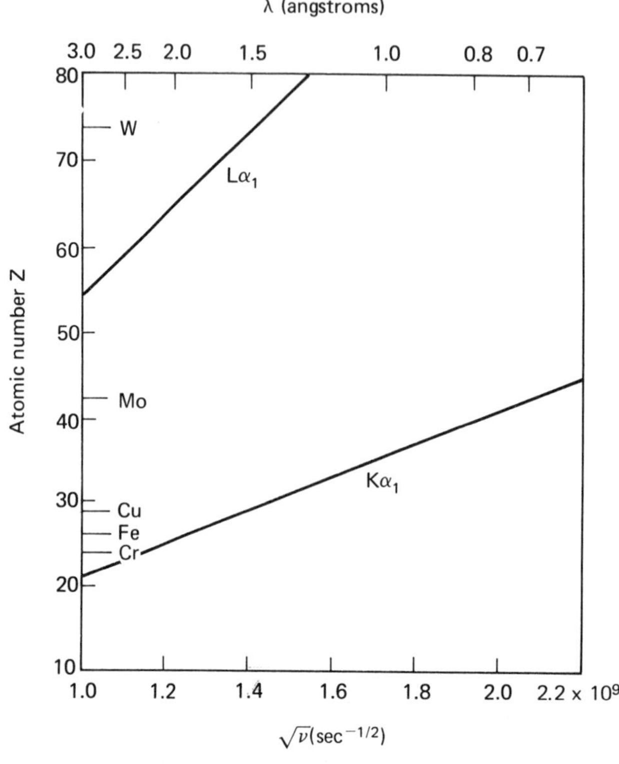

Fig. 8. Atomic number versus wavelength for K_α and L_α x-ray emission. From Cullity, B. D., *Elements of X-Ray Diffraction.* Reading, Mass.: Addison-Wesley Publishing Company, Inc., 1956 [13].

By the addition of a set of deflection coils in which the current can be varied to control the location of focus upon the sample, the instrument can be converted to a *scanning electron microprobe.* Figure 9 illustrates the additional design features. As the probe beam is scanned across the specimen the detector output is amplified and used to bias a cathode-ray tube synchronized with the probe scanning frequency, and the screen image varies in brightness with the x-ray emission intensity. A quantitative microanalysis of any specific element across the scanned area of the specimen is observed directly by setting the goniometer to the θ angle to satisfy Eq. (1) for the emission wavelength of the desired element. Figures 10(b) to 10(f) show such scans obtained for various elements from the polished surface of a cement clinker.

Less powerful but still useful techniques include scanning and recording the intensity of either backscattered, secondary, or absorbed electrons (see Fig. 5). This can be done with the instrumental arrangement shown in Fig. 11. Figure 10(a) is an absorbed electron image of a cement clinker. The cathode-ray signal was modulated so that high atomic number elements appear bright. Mean atomic numbers for clinker compounds are CaO (14.0), C_4AF (13.2), C_3S (12.7), C_2S (12.3), and C_3A (12.2). According to Gillott [14], the dark rounded areas at the bottom of Fig. 10(a) are C_2S

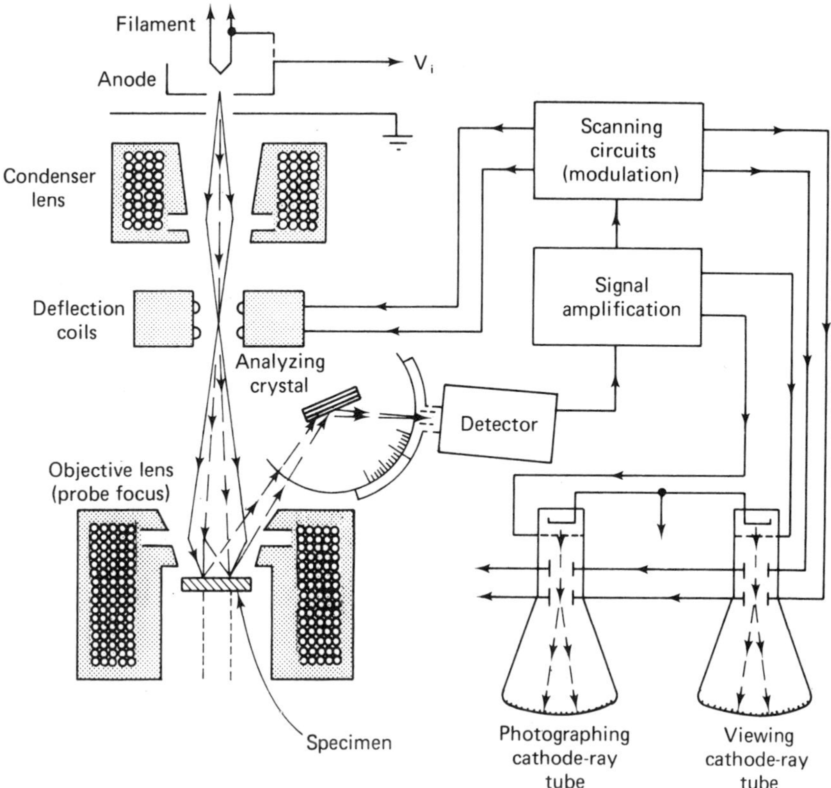

Fig. 9. Scanning electron probe and x-ray microanalyzer. From Murr, L. E., *Electron Optical Applications in Materials Science*. New York: McGraw-Hill Book Company, 1970 [6].

and the lighter areas at the top are C_3S. The interstitial material, rich in ferrite, appears white, and there are indications of free CaO (white inclusions in C_2S) and of C_3A [very dark areas corresponding to the dark areas in Fig. 10(b)].

The development of the *scanning electron microscope* has closely paralleled that of the scanning microprobe, and the design features are identical to those shown in Fig. 11 for the monitoring of secondary electron emission. The separation of second-

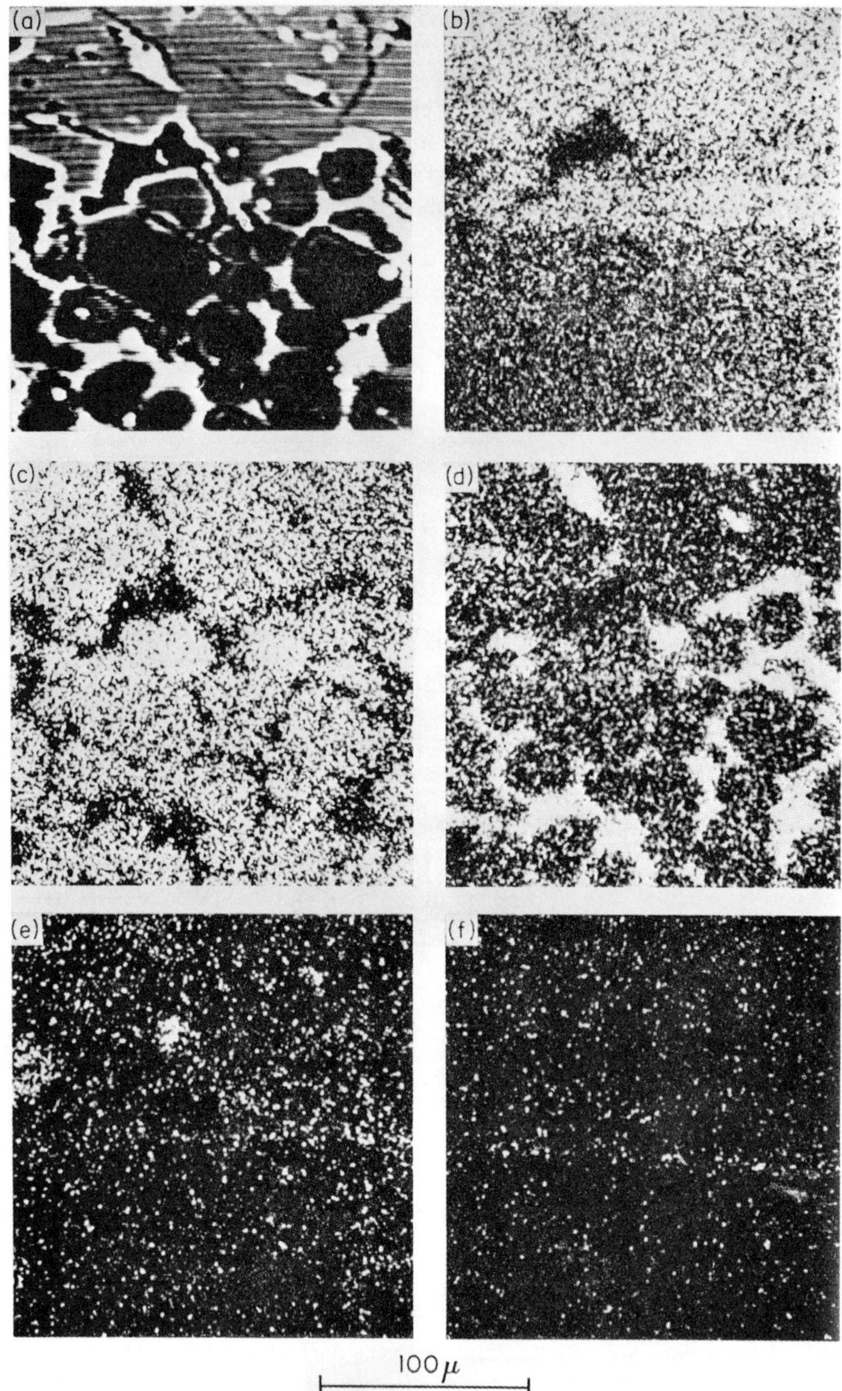

Fig. 10. Electron probe results for a polished surface of cement clinker. (*Courtesy Japan Electron Optics Laboratory Company, Ltd.*) (a) Absorbed electron image, (b)-(f) characteristic x-ray images for (b) CaK_α, (c) SiK_α, (d) FeK_α, (e) AlK_α, (f) TiK_α.

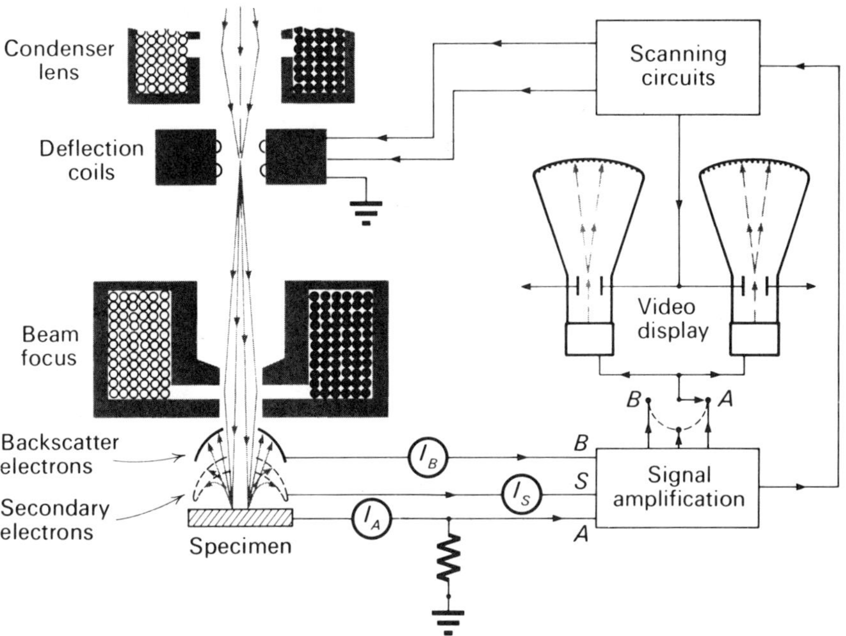

Fig. 11. Specimen scanning and measurement of backscatter, secondary, and absorbed electrons. From Murr, L. E., *Electron Optical Applications in Materials Science*. New York: McGraw-Hill Book Company, 1970 [6].

ary and backscatter electron detection is made possible by the fact that secondary electrons, which result from atomic excitation by the high-energy primary beam, have much lower energy and a wider range of energy values than do backscattered electrons, which result from inelastic collisions. The resolution of the scanning electron microscope is limited to about 100 A [15] as compared with 2–10 A for electron transmission microscopy, but the large field depth obtainable with the electron optical system often permits better imaging of rough surfaces than does the optical microscope, even at magnifications at which detail can be resolved optically.

Several additional techniques, *x-ray fluorescence*, *absorption microanalysis*, and *x-ray microscopy*, depend on excitation of the sample with an x-ray beam, instead of an electron beam. Characteristic x-radiation generated in the sample due to x-ray excitation is called x-ray fluorescence. The simplest method consists of placing a metal target at the focal point of the electron beam just above the sample (Fig. 12). The resulting fine x-ray beam then falls on the sample, or secondary target. The incident and transmitted x-ray intensities (Fig. 12) are related by

$$I_t/I_0 = e^{-(\mu/\rho)_{\text{eff}} m} \tag{4}$$

where I_t = transmitted x-ray intensity,
I_0 = incident intensity, detected with the specimen removed,
$(\mu/\rho)_{\text{eff}}$ = composite mass absorption coefficient (μ = absorption coefficient, ρ = density),

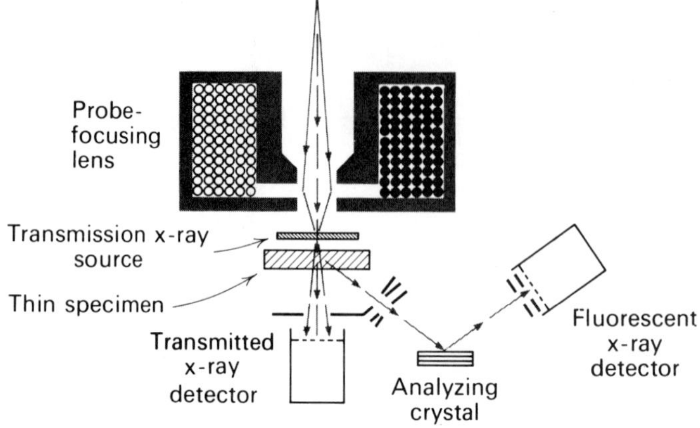

Fig. 12. Microanalysis by x-ray fluorescence and absorption. Adapted from Murr, L. E., *Electron Optical Applications in Materials Science*. New York: McGraw-Hill Book Company, 1970 [6].

m = mass per unit area of absorber.

For an absorbing sample containing n elements of concentrations C_i,

$$\left(\frac{\mu}{\rho}\right)_{\text{eff}} = \sum_{i=1}^{n} C_i \left(\frac{\mu}{\rho}\right)_i$$

where absorptions for individual elements are given by

$$\mu = Kz^{3.3}\lambda^3$$

in which K is a constant, Z is the atomic number, and λ the wavelength of incident x-radiation. Substitution into Eq. (4) allows the ratio I_t/I_0 to be related to the element identity and concentration.

By measuring transmitted intensities for two different x-ray wavelengths (by using two target elements) one can eliminate the unknown background element mass and solve for the mass of the desired element. Rewriting Eq. (4),

$$I_{t1} = I_{01}e^{-[(\mu_1/\rho)_x m_x + (\mu_1/\rho)' m']}$$

and

$$I_{t2} = I_{02}e^{-[(\mu_2/\rho)_x m_x + (\mu_2/\rho)' m']}$$

where I_{t1} and I_{t2} are the transmitted intensities for the two incident wavelengths and m' and m_x are the background element mass and the mass of the desired element, respectively. Combining and solving for m_x,

$$m_x = \frac{ln(I_{01}/I_{t1}) - (\mu_1/\mu_2)' \, ln(I_{02}/I_{t2})}{(\mu_1/\rho)_x - (\mu_2/\rho)_x(\mu_1/\mu_2)'}$$

The intensity of the detected signal is increased slightly by the presence of fluorescent x-rays, in addition to those which are transmitted, and correction for this effect can be made for greater accuracy. Cullity [13] provides additional details.

In x-ray microscopy, the transmitted x-ray detector in Fig. 12 is simply replaced with a photographic film for the observation of structural detail in the specimen by direct imaging of transmitted x-rays.

Electron diffraction is analogous in many ways to x-ray diffraction (Sec. B). Just as a beam of x-rays has a dual wave-particle character, a stream of particles has certain wave motion properties. Particles such as electrons and neutrons are diffracted from crystals just as x-rays are, and the diffraction directions obey Bragg's law, Eq. (1). The difference between x-ray, and neutron diffraction are such that the three methods supplement one another, each giving certain information which the others cannot.

For electron diffraction, the wavelength λ to use in Bragg's law is given by the theory of wave mechanics as

$$\lambda = \frac{h}{mv} \tag{5}$$

where h is Planck's constant (see Sec. C) and m and v are the particle mass and velocity (mv = particle momentum). Combining Eq. (5) and Eq. (7) of Sec. B,

$$\lambda = \left(\frac{150}{v}\right)^{0.5}$$

where V is the applied voltage in practical units. Some major differences between electron and x-ray diffraction are that electron wavelengths are shorter (normally in the order of 0.5 A, compared with 1.5 A for x-rays), an electron beam can be scattered by the nucleus of an atom as well as its electrons, and electrons are much less penetrating than x-rays. Because of the latter property, electron diffraction is successfully used to study thin surface coatings such as metal films deposited by evaporation, electrodeposits, and oxide films on metals.

4. Sample Preparation Techniques

Sample preparation often presents considerable difficulty, but several excellent references are available [16]. For transmission microscopy the sample should be thin enough to transmit the electron beam, or within 1,000 to 2,000 A for many materials under normal operating voltages of 50–100 kv. A few materials can be cut to the required thinness with a diamond knife mounted in an ultramicrotome [17]. A thin metal layer must be evaporated onto nonconducting samples to dissipate the charge produced by the incident electron beam. The sample should not be damaged by the high vacuum or by heating due to bombardment with the electrons. Special sample preparation techniques have been developed to minimize these effects for materials such as clays, where loss of water of hydration may alter both the shapes of individual particles and the structure of the entire mass [18–22].

Thin plastic or evaporated films supported on metal grids with 2- or 3-mm openings are used for mounting particulate samples or replicas made from the

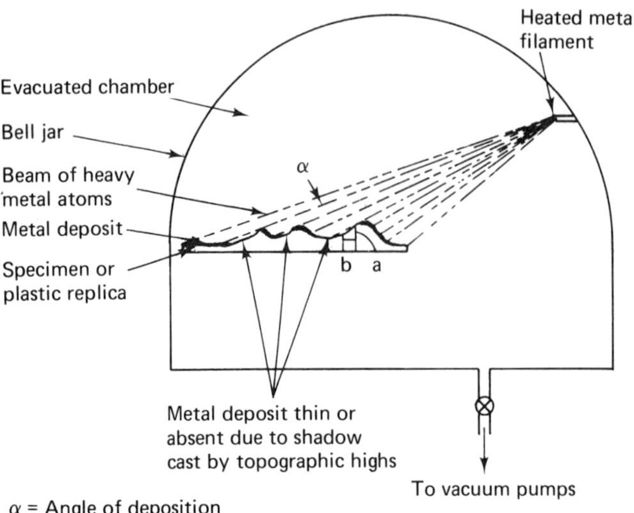

Fig. 13. Vacuum chamber for replication and shadow casting. From Gillott, J. E., *Clay in Engineering Geology*. Amsterdam, Elsevier Publishing Company, 1968 [14].

surfaces of bulk materials. A replica is a thin film which is transparent to electrons and which reproduces the surface details of the sample. The replica is usually plastic or a thin evaporated layer of silica, carbon, or carbon-platinum. Carbon is especially easily prepared and handled for evaporation techniques. Even difficult materials such as bulk clays have been successfully replicated [23].

Either particulate samples or replicas can be shadow-cast with a heavy metal such as gold, chromium, or platinum [14] to improve contrast and to obtain the heights of surface features from measurements of shadow length and shadowing angle. The shadow-casting is done in a separate vacuum chamber before mounting the sample and sample holder in the microscope (Fig. 13). Carbon and carbon-platinum replication and shadow-casting can be accomplished by passing a 30–50-amp current at 20–30 v in bursts of 1 sec or less through two pointed carbon or carbon-platinum rods held in contact by a light spring [24, 25]. Usually a few experiments are necessary to find the optimum shadowing angle and mass of metal to use [7].

Whereas replicas are useful for microscopic studies, all the analytical modes, i.e., with the microprobe, require use of the sample material itself. Reliable methods of dispersing and mounting in grids without alteration of the material are essential. One technique involves sprinkling of the powder on a glass slide and dispersing by vibration before evaporation of a layer of carbon as a support. Hardened cement pastes have reportedly been dispersed by crushing, grinding, and lathe cutting with little distortion of the gel structure [7].

B. X-RAY DIFFRACTION

Most of our knowledge of the arrangement of atoms in crystals comes from x-ray and electron diffraction studies. This is possible because of the repetitive pattern in crystalline materials, which allows them to act as three-dimensional diffraction gratings. X-rays are more commonly used than electrons because of their greater penetrating power. Electrons, because they are charged particles, do not penetrate deeply and are consequently useful for studying surface phenomena. Neutron diffraction also complements x-ray diffraction in useful ways, which will be described. All three techniques date from 1912 when von Laue predicted and showed experimentally that crystalline solids would diffract x-rays.

X-rays have frequency and wavelength and can be treated as waves or as photons of energy obeying the laws of quantum theory. As with other electromagnetic radiation, they can be transmitted, reflected, refracted, and diffracted. Their particular advantage for studying solids is their short wavelength. The x-ray spectrum extends from the gamma ray to the ultraviolet regions, or approximately 0.1 to 1,000 A (Fig. 14). The short wavelengths around 1–2 A are the most useful for diffraction work.

X-rays are scattered by collisions with electrons in a crystal. Diffraction, or constructive interference, occurs only if the difference in the distance traveled by two diffracted waves which are initially in phase is an integral number of wavelengths, so that they are again in phase. All x-ray diffraction depends on the Bragg law,

$$n\lambda = 2d \sin \theta \tag{6}$$

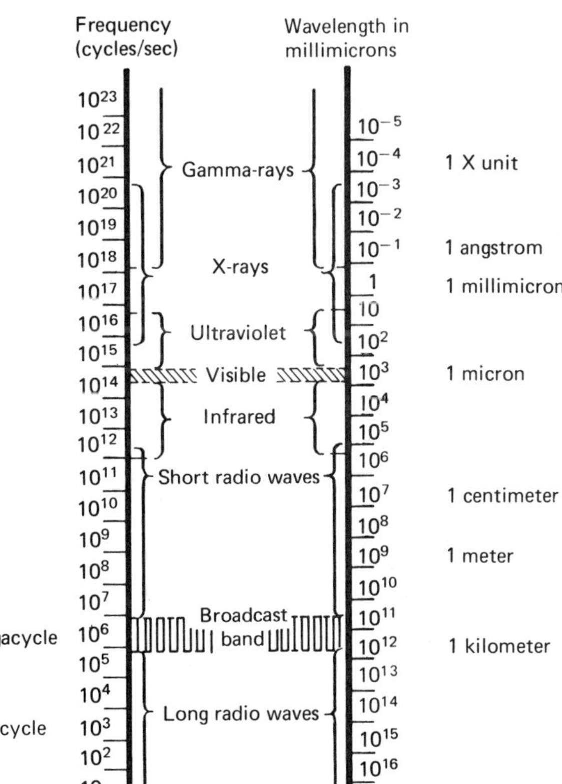

Fig. 14. The electromagnetic spectrum. Boundaries between regions are somewhat arbitrary. From Sears, F. W., *Optics*, Third edition, Reading, Mass.: Addison-Wesley Publishing Company, 1949 [26].

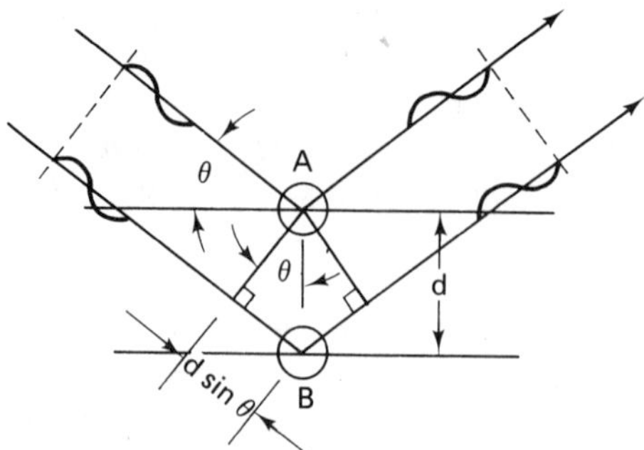

Fig. 15. Bragg law diffraction.

where n is an integer, λ the wavelength of radiation used, d the spacing between two crystal planes, and θ the angle of incidence. In Fig. 15, A and B represent atoms (electron swarms) on two crystal planes. The total path difference between the two rays is $2d \sin \theta$. Physically, Eq. (6) states that λ, d, and θ must have values which yield integral values of n in order to produce an intensity peak. The observed θ values of intensity peaks for radiation of known λ can be used in Eq. (6) to compute the interplanar spacings d.

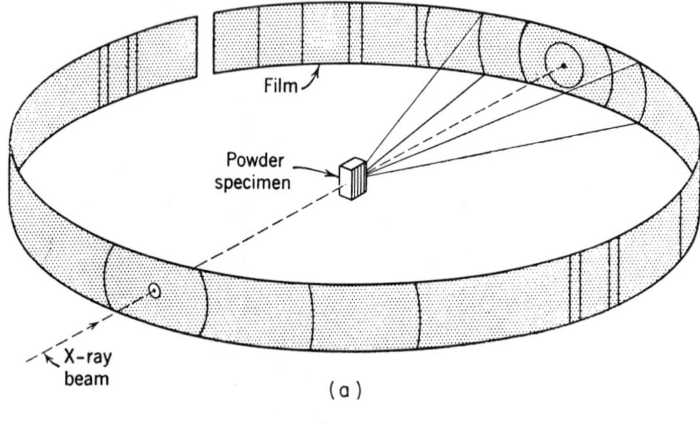

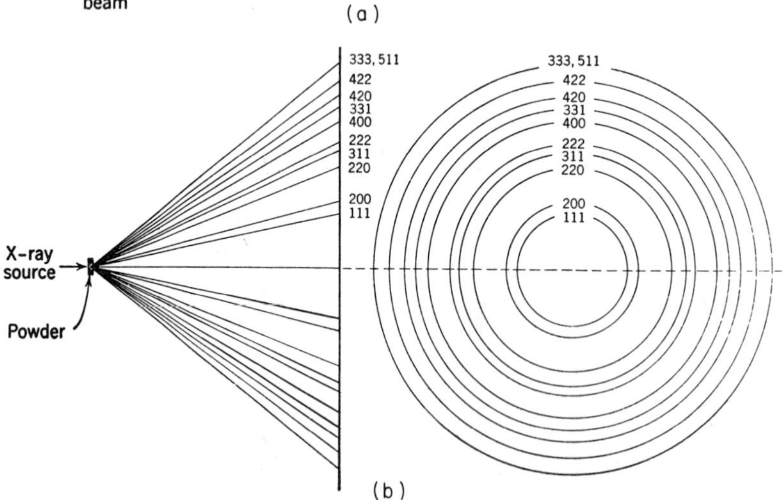

Fig. 16. Diffraction pattern from a powdered specimen. (a) On a circular strip film. From Daniels, F., and A. Alberty, *Physical Chemistry*, Third Edition. New York: John Wiley & Sons, Inc., 1966 [27]. (b) On flat film. From Bragg, W. L., *The Crystalline State*, Vol. I: A General Survey. London: G. Bell & Sons, Ltd. 1949 [28].

1. Apparatus and Techniques

There are basically two methods of using x-ray diffraction: the Laue method and the Hull-Scherrer-Debye method. In the Laue method a *single* crystal of the material is rotated while being exposed to *white* x-radiation of wavelengths about 0.5–2.5 A. Each diffracting set of planes selects the correct wavelength from the white radiation and produces a spot on a photographic film. The crystal structure is computed from the locations of the spots. In the Hull-Scherrer-Debye method a finely powdered specimen containing *many* crystals randomly oriented is rotated while being exposed to *monochromatic* x-radiation, the theory being that all possible diffraction planes will be available to allow diffraction with the monochromatic radiation. The resulting diffraction appears as cones radiating from the specimen (Fig. 16) instead of discrete rays, as in the Laue method. Since powdered specimens are more useful for most applications, we shall omit further consideration of the Laue technique.

The basic features of a powder diffractometer are shown in Fig. 17. The powder specimen is mounted with a flat surface coincident with axis P, perpendicular to the page, about which it can be rotated. X-rays from line source A, also perpendicular to the page, pass through collimating slit B, are selectively diffracted by sample C, and pass through collimating slit D and filter E and then into counter F. An increase in the widths of the collimating slits increases the intensity of any diffraction line, but at some loss of resolution. The filter serves to remove undesired wavelengths. Components D, E, and F are mounted on a carriage which also rotates about axis P and is geared to the specimen holder G so that rotation of the specimen through θ is accompanied by rotation of the counter through 2θ, maintaining equal instrumental angles of incidence and reflection on the specimen. The counter may be driven at

Fig. 17. X-ray diffractometer (schematic).

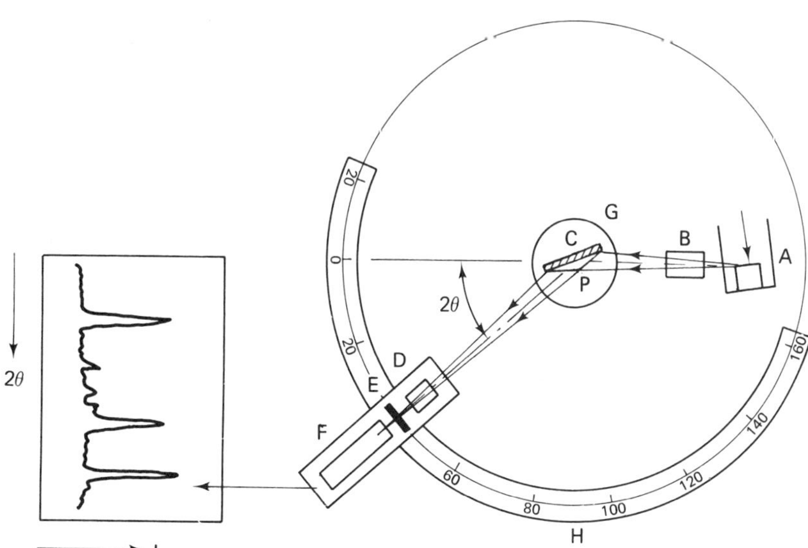

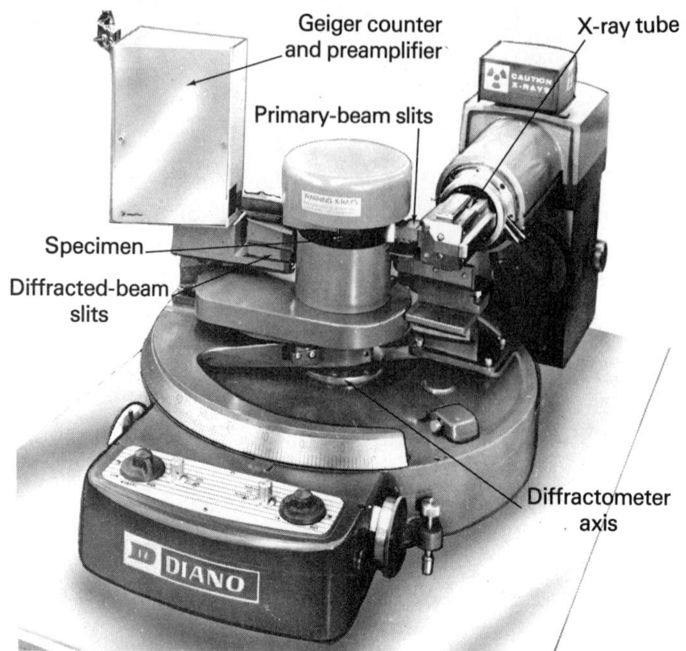

Fig. 18. A commercial diffractometer. (*Courtesy of Diano Corporation.*)

constant angular velocity or rotated by hand to any 2θ value indicated on scale H. The amplified output of the counter, usually of the proportional counter or Geiger counter types, can be used for recording diffracted x-ray intensity, I, as a function of 2θ. One type of commercial diffractometer is shown in Fig. 18. Since diffraction lines are recorded sequentially, it is essential to have good voltage stabilization for maintaining constant intensity of the incident beam. Diffraction patterns at high or low temperature are obtained by surrounding the specimen with an appropriate heating or cooling unit.

Powder specimens of most materials should be ground to 10 μ or less for reproducible relative line intensities. Powders can be enclosed in a fine capillary tube of nondiffracting material such as glass or plastic, cemented in a thin layer to a glass slide with a nondiffracting cement such as collodion or shellac, or compacted dry into a recess in a glass or plastic plate. In each case care should be taken to produce as smooth a surface as possible without causing preferential orientation of grains at the surface by the smoothing or compacting process. Relative peak intensities can be altered both by nonrandom crystal orientation and by excessive x-ray absorption at low θ angles due to surface roughness.

2. Production of X-Rays

X-rays are produced when a charged particle is rapidly decelerated. In an x-ray tube this is accomplished by applying a high d-c voltage across two metal electrodes so that electrons from a heated cathode filament strike the anode, or target, with very high velocity. X-rays radiate in all directions from the point of impact. The rate of electron bombardment can be varied by varying the filament temperature, and the impacting velocity by varying the voltage across the electrodes. The kinetic energy, KE, of an electron striking the target is

$$\text{KE} = ev = \tfrac{1}{2}mv^2 \tag{7}$$

where e is the electron charge (4.8×10^{-10} esu), ν is the voltage (in esu) across the electrodes, m is the electron mass (9.11×10^{-28} g), and v is its impact velocity. The velocity is about one-third of the speed of light at a typical operating level of $\nu =$ 30,000 v (practical units).* About 1% of the kinetic energy is converted to x-radiation, the remainder being lost in longer wavelength radiation, mostly in the form of heat. A wide spectrum of x-radiation is obtained, with the intensity-wavelength relationship typically varying with applied tube voltage as shown in Fig. 19. The impinging electrons dislodge electrons from the target material, and the subsequent return of electrons to recreate a stable electron configuration gives rise to x-radiation.

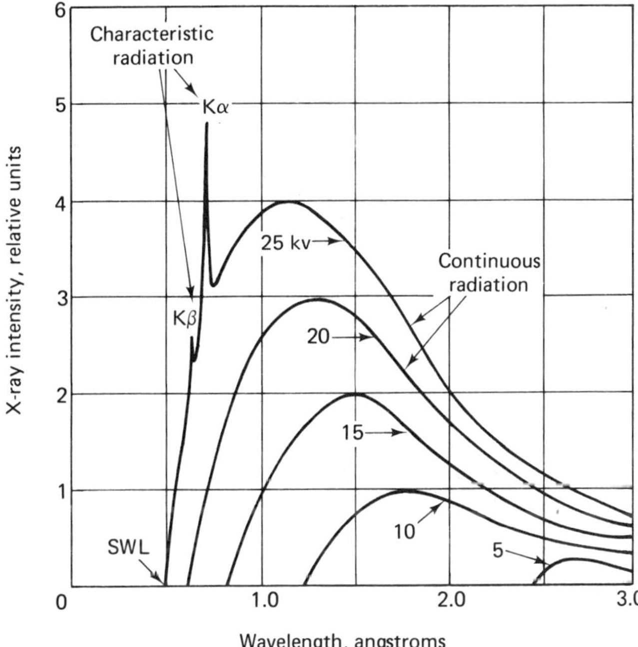

Fig. 19. X-ray spectrum of molybdenum as a function of applied voltage (schematic). From Cullity, B. D., *Elements of X-Ray Diffraction.* Reading, Mass.: Addison-Wesley Publishing Company, Inc., 1956 [13].

Some electrons are stopped and give up their total energy on one impact; others lose it through several collisions. Electrons stopped in one collision give rise to radiation of maximum energy, i.e., minimum wavelength, which can be calculated from

$$\lambda_{\text{swl}} = \frac{hc}{ev} = \frac{12{,}400}{v}$$

*1 V (esu) = 300 V (practical units).

where λ_{sw1} is the short wavelength limit in angstroms (Fig. 19), h is Planck's constant (6.62×10^{-27} erg-sec), and c is the velocity of light. Electrons stopped by a series of collisions give rise to continuous or white radiation, the wavelength depending on the energy lost in each collision. The curves in Fig. 19 become higher and shift to the left with increasing applied voltage because the number of photons per second and the average energy per photon both increase.

Whereas valency electrons can be dislodged easily and give rise to very long wavelength radiation, considerably more energy is required to remove electrons from the inner shells. When the tube voltage is raised above a certain level, short-intensity peaks appear at specific wavelengths (Fig. 19), characteristic of the target metal—and therefore called characteristic lines.

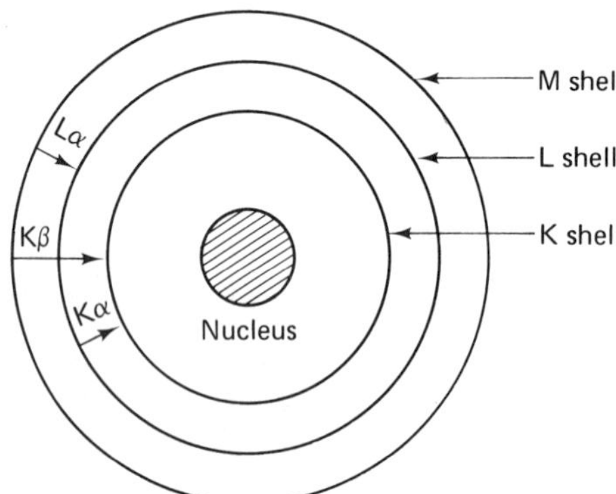

Fig. 20. Atom electronic transitions.

If one of the bombarding electrons has sufficient energy to dislodge an electron from the K shell, an outer electron immediately falls into the resulting vacancy, emitting energy in the process (Fig. 20). The energy emitted is the characteristic K-radiation. Since the K shell vacancy may be filled by an electron from any of the outer shells, there are a series of K lines. The K_α and K_β lines, for example, result from filling the K shell vacancy by an electron from the L and M shells, respectively. Since it is more probable that the K vacancy will be filled by an L electron, the K_α line is stronger than the K_β line. Each of these is actually a multiple line. For example, the K_α doublet consists of $K_{\alpha 1}$ and $K_{\alpha 2}$ lines, since the two electrons having opposite spin in the K shell have slightly different energy levels. But this small energy difference is hardly resolvable and not important for x-ray diffraction. In fact, for most elements the K_α doublet is less than 0.001 A wide at its half-peak value, providing an extremely accurate measuring scale for determining interplanar spacings from the Bragg equation. L, M, and higher characteristic lines originate in the same way as do the K lines, each vacancy being filled by an electron from some outer shell. Ordinarily only the K lines are used for x-ray diffraction, the longer wavelengths being too easily absorbed.

The binding energy of K electrons is not the same for all elements, and characteristic wavelength depends on the target metal selected. Moseley found a linear relation between the atomic number Z and the square root of the characteristic line frequency ν:

$$\sqrt{\nu} = C(Z - \sigma)$$

where C and σ are constants. Figure 8 showed how characteristic frequency varies with atomic number.

When monochromatic K_α radiation is desired, it is necessary to filter out the K_β and continuous radiation (Fig. 19) to reduce background noise. This can be accomplished by inserting in the x-ray beam a thin metal foil or powdered film having an atomic number one or two less than that of the target metal. Such a metal will absorb the K_β component much more than the K_α component because of the abrupt change in absorption coefficient between these two wavelengths. Figure 21 illustrates filtering of radiation from a copper target ($Z = 29$) with a nickel filter ($Z = 28$).

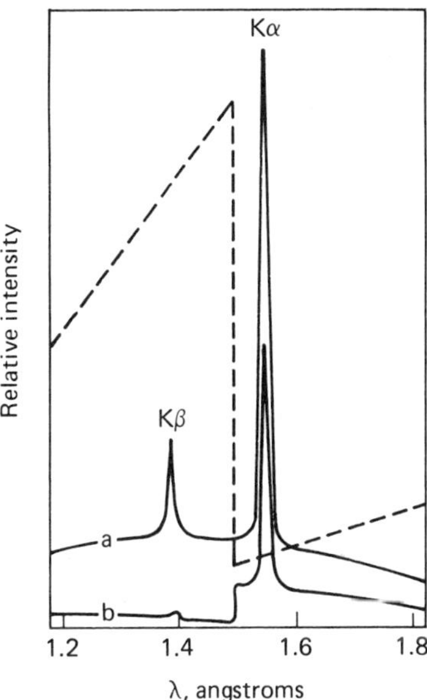

Fig. 21. Copper radiation spectra. (a) Before; and (b) after passage through a nickel filter. Dashed line represents the mass absorption of nickel (schematic).

3. Directions and Intensities of Diffracted Beams

To interpret diffraction data, either for the determination of crystal structure or for the analysis of mixtures, we must understand how both the *direction* and the *intensity* of diffracted beams depend on crystal structure. Diffraction directions depend solely on the shape and size of the unit cell. Conversely, only the shape and size of an unknown crystal can be determined from its diffraction directions. The intensities of diffracted beams will provide clues to the positions of atoms within the cell. All that is needed to determine unit cell shape and size is a relation obtained by combining the Bragg law with the appropriate plane-spacing equation from Appendix B. For example, combining

$$\lambda = 2d \sin \theta \tag{6}$$

and

$$d = \left[\frac{a^2}{h^2 + k^2 + l^2}\right]^{0.5} \tag{8}$$

gives

$$\sin^2 \theta = \frac{\lambda^2}{4a^2}(h^2 + k^2 + l^2) \tag{9}$$

for the cubic crystal system. The corresponding equation for the tetragonal system is

$$\sin^2 \theta = \frac{\lambda^2}{4}\left(\frac{h^2 + k^2}{a^2} + \frac{l^2}{c^2}\right)$$

and so on for the remaining crystal systems. Subsequent examples will illustrate the use of these equations.

The diffracted intensity depends primarily on the *atomic scattering factor*, f, of each atom and the *position* of each atom in the unit cell. The atomic scattering factor is the ratio of the x-ray scattering power of an atom compared with the scattering power of an electron. It measures the efficiency of cooperation in scattering by the electrons in a single atom. The second important factor, the *structure factor*, is a measure of the efficiency of scattering by various atoms in the unit cell. Its value depends on the positions of atoms, since each atom will scatter radiation with an amplitude proportional to its value of f but with a phase dependent on its position in the cell. Hence the computation of structure factor requires the vector addition of sine waves all of the same frequency but of varying amplitudes and phases. These relationships can be derived by starting with the scattering by a single electron, then by an atom, and finally by a unit cell.

An *electron* which has been set into oscillation by an x-ray beam scatters x-rays in all directions, and the intensity of the scattered beam depends on the angle of scattering. The intensity I of the beam scattered by a single electron of charge e and mass m, at a distance r from the electron, was found by J. J. Thompson to be

$$I = I_0 \frac{e^4}{r^2 m^2 c^4} \sin^2 \alpha \tag{10}$$

where I_0 is the incident beam intensity, c the velocity of light, and α the angle between the scattering direction and the direction in which the electron is accelerated. Suppose that we wish to find the intensity of scattered radiation at point P from the incident x-ray beam xO striking an electron at O (Fig. 22). Since the incident

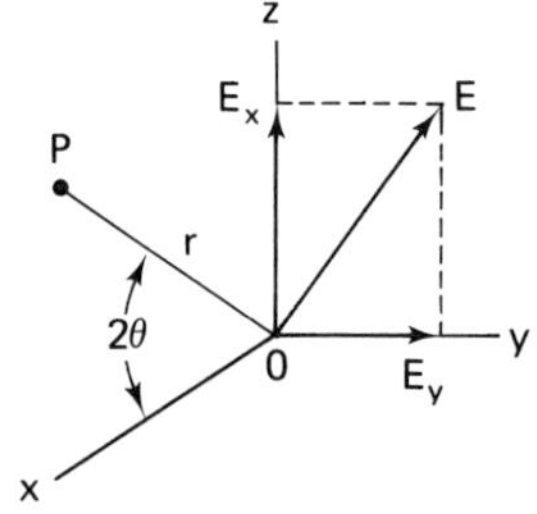

Fig. 22. X-ray scattering by an electron.

beam is unpolarized, its electric vector **E** has a random direction in the *yz* plane. Assuming a resolution into plane-polarized components in the *yz* plane having electric vectors $\mathbf{E}_y$ and $\mathbf{E}_z$,

$$\mathbf{E}^2 = \mathbf{E}_y^2 + \mathbf{E}_z^2$$

Since the direction of **E** is random, $\mathbf{E}_y = \mathbf{E}_z$ and

$$\mathbf{E}_y^2 = \mathbf{E}_z^2 = \frac{\mathbf{E}^2}{2}$$

Since **E** measures wave amplitude and the intensity of a wave is proportional to its amplitude squared, the intensity of these two components of the incident beam is

$$I_{0y} = I_{0z} = I_{0/2}$$

The *y* component of the incident beam accelerates the electron in the *y* direction and produces a scattered beam intensity at *P*, by Eq. (10), of

$$I_{P_y} = I_{0y}\frac{e^4}{r^2m^2c^4}$$

because $\alpha = \angle yOP = \pi/2$. Likewise for the *z* component,

$$I_{P_z} = I_{0z}\frac{e^4}{r^2m^2c^4}\cos^2 2\theta$$

because $\alpha = \pi/2 - 2\theta$. The total scattered intensity at *P* is the sum of the two components

$$\begin{aligned} I_P &= I_{P_y} + I_{P_z} \\ &= I_0\frac{e^4}{r^2m^2c^4}\left(\frac{1 + \cos^2 2\theta}{2}\right) \end{aligned} \tag{11}$$

This is the Thompson equation for scattered x-ray intensity (erg/sq cm/sec) by a single electron. Fortunately, all terms except $1 + \cos^2 2\theta$ are constant for a single experiment. This variable term, called the *polarization factor*, arises because the incident beam is *unpolarized.*

When an *atom* is exposed to x-radiation each electron scatters radiation in accordance with Eq. (11). This equation shows that the radiation scattered by the nucleus will be negligible. Although the nucleus carries a charge, its mass is so large it cannot be made to oscillate by exposure to x-rays. The atomic scattering factor, *f*, is defined as

$$f = \frac{\text{amplitude of the wave scattered by an atom}}{\text{amplitude of the wave scattered by one electron}} \tag{12}$$

The amplitude of the wave scattered by an atom depends on the phase relationships

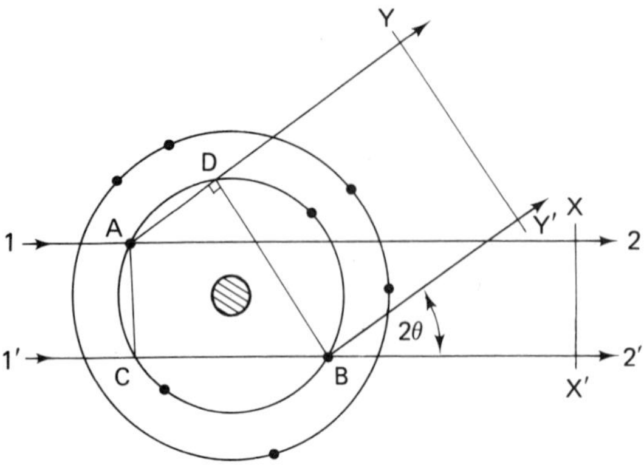

Fig. 23. X-ray scattering by an atom.

between waves scattered by the individual electrons. Consider the two electrons A and B in Fig. 23. For $\theta = 0$, the waves scattered by A and B are exactly in phase, because the path lengths 1-2 and 1′-2′ are equal. For $\theta \neq 0$, the two waves are out of phase by the distance $CB - AD$. It is apparent then that $f = Z$, the valence, for any atom when $\theta = 0$ and that f decreases as θ increases, the angular phase shift being greater for shorter wavelength radiation. The variation of f with $\sin\theta/\lambda$ has been calculated for various elements [13] and is shown for copper in Fig. 24.

Similar phase differences arise when a *unit cell* is exposed to radiation. The waves scattered by atoms in the cell are not in phase, except when $\theta = 0$, and we must determine how the phase differences depend on their three-dimensional arrangement in the cell. In addition, the *amplitudes* of scattered radiation will vary with the atomic numbers of various elements in the unit cell. In analogy to the atomic scattering factor f, the structure factor $|F|$ is defined as

$$|F| = \frac{\text{amplitude of the wave scattered by all atoms in a unit cell}}{\text{amplitude of the wave scattered by one electron}} \tag{13}$$

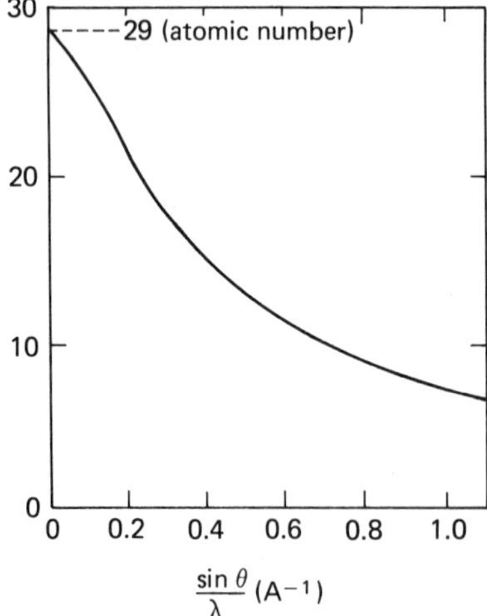

Fig. 24. Atomic scattering factor of copper.

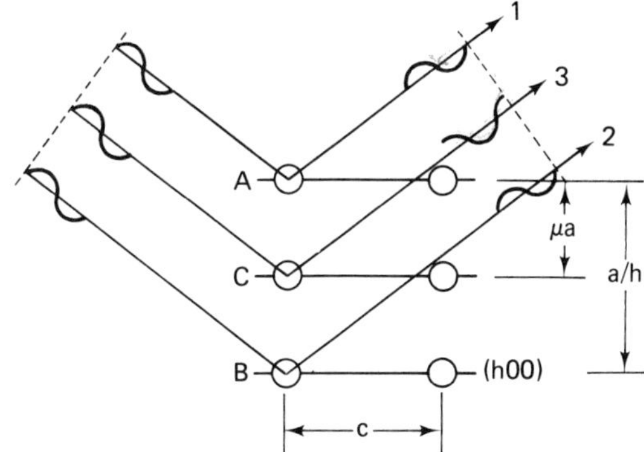

Fig. 25. Reduction of diffraction intensity caused by an out-of-phase component.

To derive F, consider the reduction in Bragg diffraction intensity from the ($h00$) planes (planes A and B in the unit cell, Fig. 25) caused by the out-of-phase diffraction from the C plane located at arbitrary distance ua, where u is a fractional coordinate. From the definition of Miller indices (Sec. B in Chap. 2), the distance AB is

$$AB = d_{h00} \frac{a}{h}$$

By Bragg's law, the first-order phase shift between waves 1 and 2 is 2π radians, and by similar triangles the phase shift ϕ between waves 1 and 3 is

$$\phi = \frac{ua}{a/h}(2\pi) = 2\pi hu$$

The derivation can be extended to three dimensions to obtain

$$\phi = 2\pi(hu + kv + lw) \tag{14}$$

where v and w are additional fractional coordinates. The relation is applicable to a unit cell of any shape. If atoms A and C are of different kinds, the two waves will

Fig. 26. (a) Sine wave addition; (b) wave vector in the complex plane.

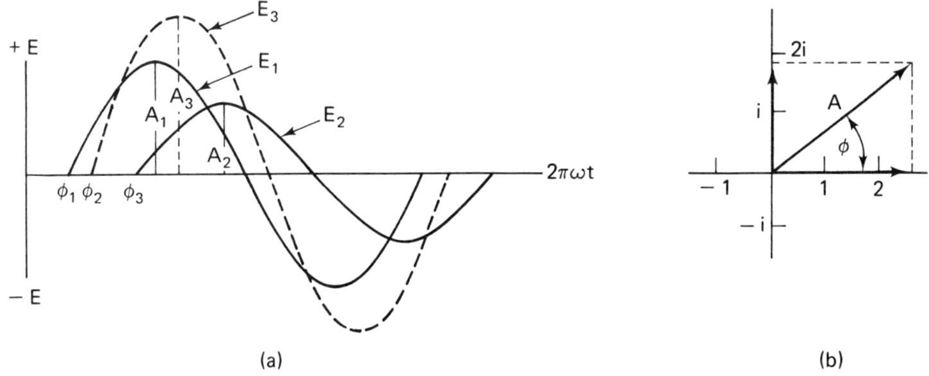

differ in amplitude as well as phase, and to obtain F we must sum sine waves of the same frequency but varying amplitudes and phases. This can be done in several ways. A convenient method is to express the waves as complex exponential functions. The sum of two sine waves of the same frequency [Fig. 26(a)] is a sine wave (dashed), also of the same frequency. A sine-wave vector can be represented in the complex plane as shown in Fig. 26(b), where A is the maximum amplitude and ϕ the phase angle. The vector drawn to any point in the complex plane represents the complex number $a + bi$, where a is a real value plotted on the abscissa and bi is an imaginary value ($i = \sqrt{-1}$) plotted on the ordinate. The vector representation for the sine wave is therefore $A \cos \phi + i \sin \phi$. Writing the power series expansion for $e^{i\phi}$, $\cos \phi$ and $\sin \phi$ yields

$$e^{i\phi} = \cos \phi + i \sin \phi \tag{15}$$

or

$$Ae^{i\phi} = A \cos \phi + Ai \sin \phi$$

The expression on the left is a complex exponential function. Using Eqs. (7) and (14), the contribution to scattered waves by a single atom in a unit cell can be expressed in complex exponential form as

$$Ae^{i\phi} = fe^{2\pi i(hu+kv+lw)}$$

and the structure factor F for a given hkl reflection is obtained by summing the contributions of all atoms

$$F_{hkl} = \sum_{n=1}^{N} f_n e^{2\pi i(hu_n+kv_n+lw_n)} \tag{16}$$

where N is the number of atoms in the unit cell. F is a complex number expressing both amplitude and phase of the resultant wave, and its absolute value $|F|$ gives the resultant wave amplitude in terms of the amplitude contributed by a single electron [Eq. (13)]. Since the intensity of a diffracted beam is proportional to its amplitude squared, Eq. (16) enables the calculation of the intensity of any hkl reflection from a knowledge of the atom positions.

The use of Eq. (16) will now be illustrated for several simple unit cells. But first, several relationships for the integer values of n arise frequently enough to be worth noting:

$$\begin{aligned} e^{n\pi i} &= (-1)^n \\ e^{n\pi i} &= e^{-n\pi i} \\ e^{ix} + e^{-ix} &= 2 \cos x \end{aligned} \tag{17}$$

These relationships can be verified by using Eq. (15).

For the simplest cell having one atom at the origin, i.e., fractional coordinate 000,

$$F = fe^{2\pi i(0)} = f \quad \text{and} \quad F^2 = f^2$$

For a body-centered cell having two atoms of the same kind at 000 and $\frac{1}{2}\frac{1}{2}\frac{1}{2}$,

$$F = fe^{2\pi i(0)} + fe^{2\pi i(h/2+k/2+l/2)} = f[1 + e^{\pi i(h+k+l)}]$$

$$F = 2f \quad \text{and} \quad F^2 = 4f^2 \qquad \text{when } (h + k + l) \text{ is even}$$

$$F = 0 \quad \text{and} \quad F^2 = 0 \qquad \text{when } (h + k + l) \text{ is odd}$$

For a face-centered cell having four atoms of the same kind at 000, $\frac{1}{2}\frac{1}{2}0$, $\frac{1}{2}0\frac{1}{2}$, and $0\frac{1}{2}\frac{1}{2}$,

$$F = fe^{2\pi i(0)} + fe^{2\pi i(h/2+k/2)} + fe^{2\pi i(h/2+l/2)} + fe^{2\pi i(k/2+l/2)}$$

$$= f[1 + e^{\pi i(h+k)} + e^{\pi i(h+l)} + e^{\pi i(k+l)}] \tag{18}$$

If h, k, and l are unmixed, then the three sums $h + k$, $h + l$, and $k + l$ are even integers, and each term in Eq. (18) has the value 1:

$$F = 4f \quad \text{and} \quad F^2 = 16f^2 \qquad \text{(unmixed indices)}$$

If h, k, and l are mixed, the sum of the three exponentials is -1. For example, for the (120) plane, $F = f(1 + 1 - 1 - 1)$, or

$$F = 0 \quad \text{and} \quad F^2 = 0 \qquad \text{(mixed indices)}$$

Reflections would therefore occur on the (111) and (220) planes but not on the (110) and (210) planes.

Such relationships may be summarized as in Fig. 27. Since the structure factor is independent of the shape and size of the unit cell the relationships in Fig. 27 also apply to simple, body-centered, and face-centered crystals of systems other than cubic. However, any additional atoms in the cell may partially or completely cancel some of the predicted peaks. For example, diamond is a face-centered cell in which additional atoms cancel the 200, 222, 420, etc., reflections.

The analytical procedure for determining crystal structure is to index the observed diffraction peaks and then determine the crystal structure from a knowledge of which planes do and do not produce peaks and from the relative intensities of the observed peaks. The Bragg law, Eq. (6), gives the directions of diffracted beams and the structure factor, Eq. (16), their intensities.

The structure factor is one of six factors which affect the relative intensities of diffraction peaks from a powder specimen:

1. Polarization factor.
2. Structure factor.
3. Multiplicity factor.
4. Absorption factor.
5. Temperature factor.
6. Lorentz factor.

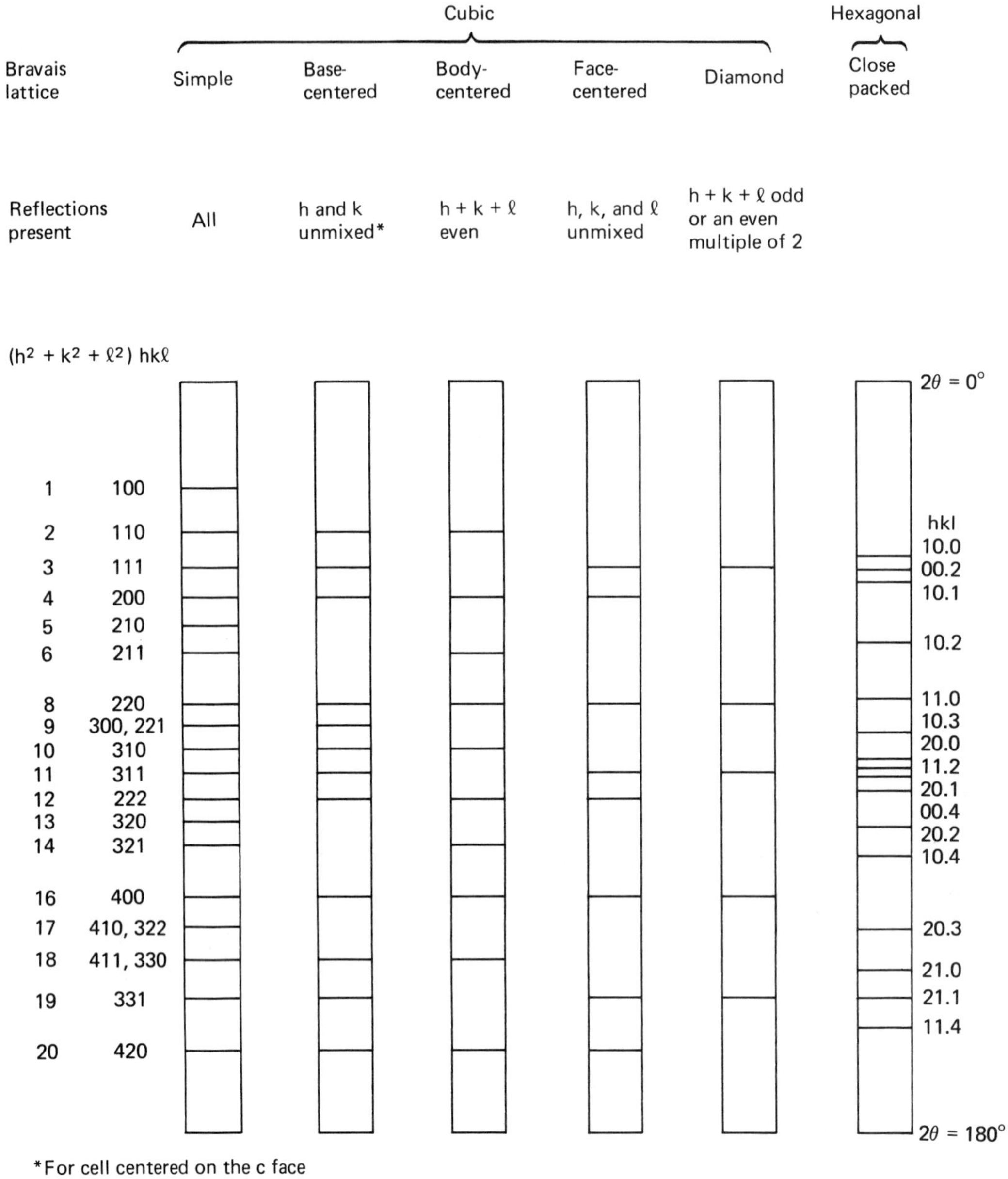

	Cubic					Hexagonal
Bravais lattice	Simple	Base-centered	Body-centered	Face-centered	Diamond	Close packed
Reflections present	All	h and k unmixed*	h + k + ℓ even	h, k, and ℓ unmixed	h + k + ℓ odd or an even multiple of 2	

$(h^2 + k^2 + \ell^2)$	$hk\ell$
1	100
2	110
3	111
4	200
5	210
6	211
8	220
9	300, 221
10	310
11	311
12	222
13	320
14	321
16	400
17	410, 322
18	411, 330
19	331
20	420

$2\theta = 0°$

hkl: 10.0, 00.2, 10.1, 10.2, 11.0, 10.3, 20.0, 11.2, 20.1, 00.4, 20.2, 10.4, 20.3, 21.0, 21.1, 11.4

$2\theta = 180°$

*For cell centered on the c face

Fig. 27. Diffraction patterns for simple lattices. 2θ values correspond to CuK_α radiation with lattice parameter $a = 3.50$ A: $c/a = 1.633$ for hexagonal close-packed crystals.

The first two have already been derived, and the remaining four will be briefly described.

The multiplicity factor accounts for the fact that planes having different indices often have equal spacings, and therefore the probability of a crystal being oriented so as to diffract a given cone (Fig. 16) depends on the number of such planes having equal spacings. In a cubic cell, for example, the 100, 010, and 001 planes all have equal spacing, giving a multiplicity factor of 3.* Likewise the 111, $11\bar{1}$, $1\bar{1}\bar{1}$, and $1\bar{1}1$ planes have equal spacing, giving a multiplicity factor of 4. Therefore the intensity of the 111 plane peak would be four-thirds that of the 100 planes, other things being equal.

The absorption factor accounts for the loss of intensity of the beam in traversing the sample. Fortunately, for a flat plate powder specimen making equal angles with the incident and diffracted beams, the absorption factor is independent of the angle θ and can be ignored.

The temperature factor accounts for the decrease in intensity of a diffracted beam due to the smearing out of lattice planes caused by thermal agitation of individual atoms. Atoms undergo some thermal vibration about their positions even at zero absolute temperature, and at room temperature the amplitudes typically vary from 2 to 8% of the mean atom spacings, depending on the elastic constants and melting temperature of the crystal.

The Lorentz factor is really a combination of three geometrical factors each of which accounts for a variation of integrated diffraction intensity as a function of the θ angle. The relative integrated intensity, or area under the curve of intensity versus (as a function of) 2θ, is used instead of the peak height, since the former is characteristic of the specimen, while the latter is influenced by instrumental adjustments. The three geometrical factors account for properties such as the diffraction intensity at angles slightly off the Bragg angle, where diffracted rays are only partially out of phase, and for the number of randomly oriented particles occurring near enough to each Bragg angle to contribute to the corresponding intensity peak. The Lorentz factor, as derived in texts on x-ray diffraction [13], is

$$\frac{1}{4\sin^2\theta\cos\theta}$$

This factor is usually combined with the polarization factor, in Eq. (11), and the coefficient $\frac{1}{8}$ is omitted to give

$$\text{Lorentz} - \text{polarization factor} = \frac{1+\cos^2 2\theta}{\sin^2\theta\cos\theta}$$

The overall effect of these two factors, as shown in Fig. 28, is to decrease the intensities of reflections at intermediate angles compared with those in forward and backward directions.

---

*Six instead of 3, if we include the corresponding planes with index signs reversed.

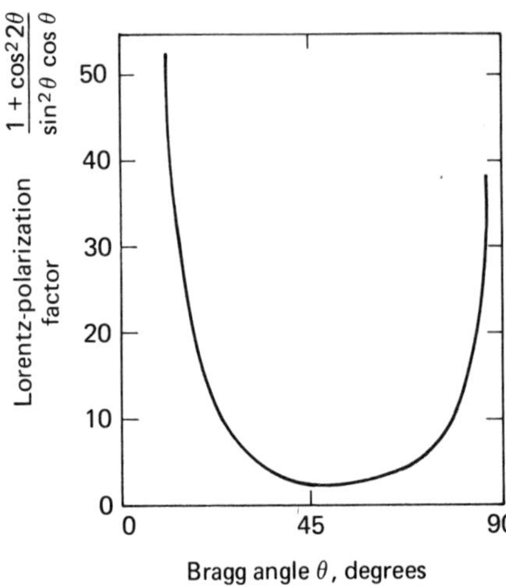

Fig. 28. Lorentz polarization factor. From Cullity, B. D., *Elements of X-Ray Diffraction.* Reading, Mass.: Addison-Wesley Publishing Company, Inc., 1956 [13].

We can now combine all the intensity factors into a single equation for powder diffraction work, omitting the terms which are constant for all lines of the pattern:

$$I = |F|^2 p \frac{1 + \cos^2 2\theta}{\sin^2 \theta \cos \theta} \tag{19}$$

where I is the relative integrated intensity (arbitrary units), F is the structure factor [Eq. (23)], p is the multiplicity factor, and θ is the Bragg angle. Omitted terms include the intensity of the incident beam and the charge and mass of an electron from the Thompson equation, among others.

4. Determination of Crystal Structure

Although the structures of several thousand inorganic and organic crystals have been determined and catalogued, new substances are constantly being synthesized, and the structures of many old ones are still unknown. The structural determination of crystals of high symmetry, namely cubic, can often be made from diffraction patterns in a very few hours, whereas highly complex substances, as even the simpler proteins, may require months of effort by several crystallographers. Let us consider only the most basic principles and see how they can be used to solve some of the simplest structures from powder patterns.

The previous section indicated that the shape and size of the unit cell determines the diffraction line positions and that the arrangement of atoms within the unit cell determines their relative intensities. Although the diffraction pattern can be predicted directly from a knowledge of the structure, the reverse problem, calculating the structure directly from the observed pattern, has never been solved. Since all we can measure is intensity, we can determine amplitude but not phase, which means we cannot compute the structure factor but only its absolute value $|F|$. We must therefore use trial and error: assume a structure, calculate its diffraction pattern, compare it with the observed one, and repeat until the two agree. The process involves three steps [13]:

1. Determine the unit cell size and shape from the θ values of the diffraction lines. One first assumes which of the seven crystal systems the structure belongs to and then assigns Miller indices to each line based upon this assumption. If Miller indices cannot be correctly assigned, an alternative crystal system is assumed. This step is called *indexing the pattern*. Once the shape of the unit cell is thus determined, its size is obtained from the positions and Miller indices of the diffraction lines.
2. Determine the number of each atom species per unit cell from the cell size and the measured density and chemical composition of the specimen.
3. Determine the positions of atoms in the cell from the relative intensities of diffraction lines.

The third step is usually the most difficult. The three steps are illustrated in the following example.

EXAMPLE 1: Figure 29 is a diffraction pattern obtained with Cu K_α radiation ($\lambda = 1.54$ A) for a specimen whose chemical composition and density were determined to be CaF_2 and 3.08 g/cc. Determine the crystal structure (crystal class, Bravais lattice, unit cell parameters, and positions of atoms within the cell).

For a trial crystal system we select the cubic unit cell first, since it is easiest to analyze and easiest to eliminate if our selection is wrong. We would then proceed progressively to more complex systems having lower symmetry. The following results are illustrated in Table 2.

TABLE 2. Trial Solution for CaF_2 Structure

(1) θ	(2) *Relative measured intensity*	(3) $\sin\theta$	(4) d	(5) $(h^2+k^2+l^2)$	(6) hkl	(7) $\frac{\sin\theta}{\lambda}$	(8) $f_{Ca^{++}}$	(9) f_{F^-}	(10) $F_{(hkl)}$	(11) $\lvert F\rvert^2$	(12) P	(13) $\frac{1+\cos^2 2\theta}{\sin^2\theta\cos\theta}$	(14) $I_{calc} \times 10^3$	(15) *Relative calculated intensity*
14.1	98	0.244	3.16	3	111	0.158	15.2	7.5	$4f_{Ca} + 0f_F$	60.8^2	4	30.87	456	84
23.5	100	0.399	1.93	8	220	0.259	12.5	5.6	$4f_{Ca} + 8f_F$	94.8^2	6	10.05	542	100
27.8	31	0.466	1.65	11	311	0.302	11.5	4.8	$4f_{Ca} + 0f_F$	46.0^2	12	6.86	174	32
34.3	10	0.562	1.37	16	400	0.365	10.1	4.0	$4f_{Ca} + 8f_F$	72.4^2	3	4.32	71.4	13
37.9	9	0.613	1.25	19	331	0.397	9.3	3.5	$4f_{Ca} + 0f_F$	37.2^2	12	3.56	59.0	11
43.7	16	0.690	1.12	24	422	0.447	8.7	3.2	$4f_{Ca} + 8f_F$	60.4^2	12	2.90	127	23
47.0	5	0.730	1.05	27	{511 333	0.474	8.4	2.4	$4f_{Ca} + 0f_F$	33.6^2	12 + 4 = 16	2.76	49.7	9

Columns (1) and (2): From diffraction pattern. For an initial trial, approximate values for column (2) may be obtained by measuring relative peak heights, instead of peak areas, and assigning the highest peak a value of 100.

Column (3): From column (1).

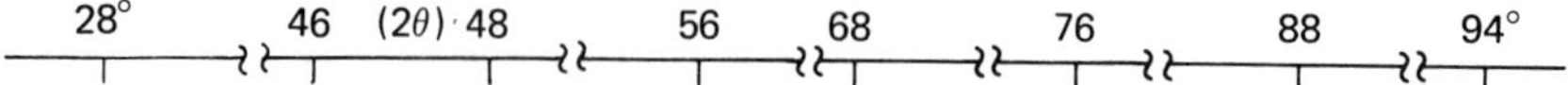

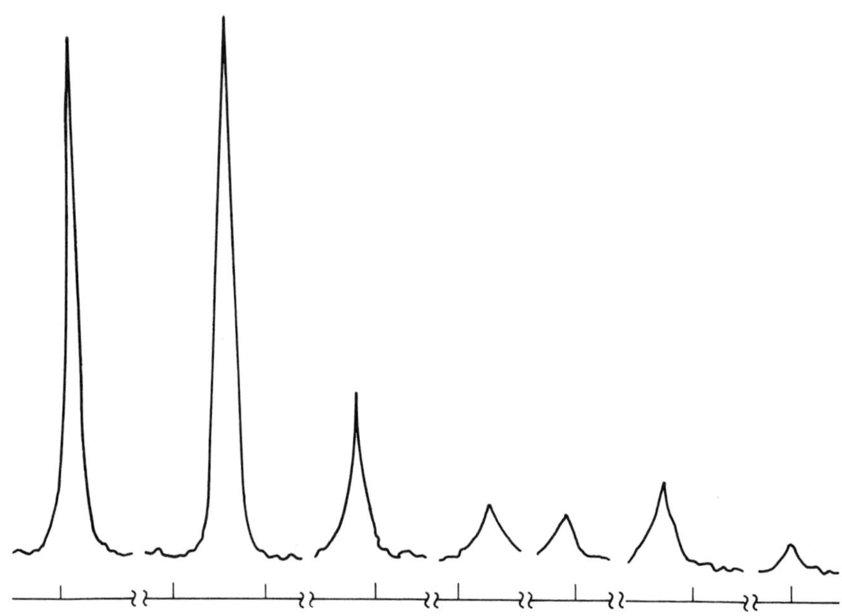

Fig. 29. Diffraction pattern for calcium fluoride.

Column (4)*:* From the Bragg law, $\lambda = 2d \sin \theta$.

Columns (5) *and* (6)*:* Having initially assumed a cubic unit cell, we attempt to index the pattern by finding values of *hkl* which satisfy the indexing equation for the particular crystal class:

$$\text{cubic:} \quad \frac{\sin^2 \theta}{(h^2 + k^2 + \pi l^2)} = \frac{\lambda^2}{4a^2} \tag{9}$$

Since $(h^2 + k^2 + l^2)$ is always integral and $\lambda^2/4a^2$ is a constant for any one pattern, we must find a set of integers giving $(h^2 + k^2 + l^2)$ values which will yield a constant quotient when divided one by one into the observed $\sin^2 \theta$ values. This can be done by marking with a pencil the first 5 or 6 d values from column (4) on the D scale of a slide rule, inverting the slide so that the B scale touches the D scale, and then searching for a single setting of the slide which makes the pencil marks on the D scale coincide with integers on the B scale. The slide inversion squares the sine term in Eq. (9), and the integer matching satisfies the requirement of a constant quotient. Because of systematic errors which have not been described, the integer matching is never exact but is usually close enough to select the proper integer values of $(h^2 + k^2 + l^2)$ if the B scale is shifted very slightly from line to line to compensate for the systematic errors. If a set of integers satisyfing Eq. (9) cannot be found, the substance is not cubic and other crystal systems must be explored. The integers in column (5) can be observed on the B scale and the hkl values in column (6) obtained by inspection. Note that $(h^2 + k^2 + l^2) = 27$ can be obtained from two sets of hkl values.

We can now determine the Bravais lattice of the specimen by comparing the $(h^2 + k^2 + l^2)$ values in column (5) with those in Fig. 27. We observe that the integers (3, 8, 11, 16, . . .) match the diamond cubic structure. We could almost have deduced this by comparing the relative spacings between peaks in Fig. 27 with the spacings in Fig. 29. There is an almost regular spacing of lines in both simple and body-centered cubic, but the former contains almost twice as many lines. The face-centered cubic has a repetition of two closely spaced lines followed by a single line, the diamond pattern contains fewer lines, whereas hexagonal structures and those of lower symmetry have larger numbers of irregularly spaced lines. We can also compute the single lattice parameter, a, required to define the cubic unit cell, by inserting into Eq. (8) any pair of corresponding values from columns (4) and (5) in Table 2. We select $(h^2 + k^2 + l^2) = 16$, since it is a perfect square, and obtain

$$a = d(h^2 + k^2 + l^2)^{0.5} = 1.37(4) = 5.48 \text{ A}$$

A slightly more accurate value of a could be obtained by selecting one of the higher-order reflections, due to the systematic errors mentioned previously.

Having indexed the pattern, we determine the weight W, and hence the number of molecules N, per unit cell (step 2) from our knowledge of its volume V and density ρ:

$$W = 6.03 \times 10^{23} \rho V = 6.03 \times 10^{23}(3.08)(5.48 \times 10^{-8})^3 = 312$$

and

$$N = \frac{312}{40 + 2(19)} = 4 \text{ molecules of } CaF_2 \text{ per unit cell}$$

where 6.03×10^{23} is Avogadro's number, 10^{-8} converts angstroms to centimeters, and the atomic weights of Ca and F are 40 and 19, respectively. Hence there are 12 atoms per unit cell (4 Ca + 8 F), whereas the diamond cubic structure has only 8 atoms (Fig. 30). This means that we must determine the locations of the remaining 4 atoms so that the resulting relative intensities for our hypothesized structure agree with the observed intensities [column (2) of Table 2]. This leads us to step 3, the determination of atom positions.

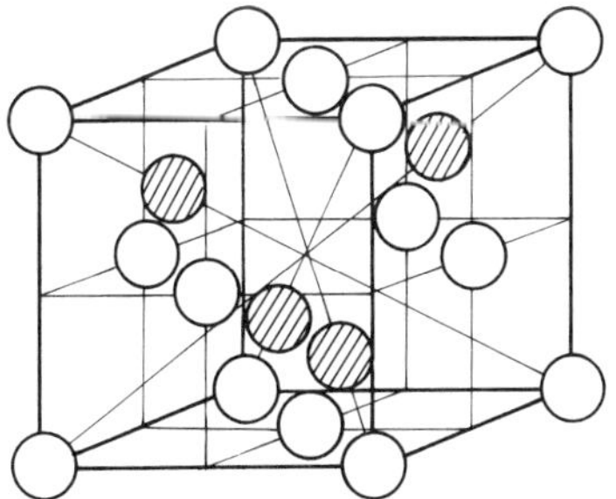

Fig. 30. Diamond cubic structure. Unshaded spheres form face-centered cubic structure. Shaded spheres occupy tetrahedral locations.

A variety of aids are available for hypothesizing a plausible trial structure, including the use of Pauling's coordination rules (Sec. B in Chap. 2), the use of three-dimensional structural models, a knowledge of atomic spacings obtained by other means such as band spectra (Sec. C), and, above all, an accumulated knowledge of previously solved structures of similar compounds. For CaF_2 we might hypothesize a trial structure in which calciums occupy the face positions and fluorines the internal positions of a diamond cubic cell, with the extra four fluorines located to occupy the largest internal void spaces remaining in the diamond structure, i.e.,

4 Ca^{++} at 000, $\frac{1}{2}\frac{1}{2}0$, $\frac{1}{2}0\frac{1}{2}$, $0\frac{1}{2}\frac{1}{2}$	(face positions)
4 F^- at $\frac{1}{4}\frac{1}{4}\frac{1}{4}$, $\frac{3}{4}\frac{1}{4}\frac{3}{4}$, $\frac{3}{4}\frac{3}{4}\frac{1}{4}$, $\frac{1}{4}\frac{3}{4}\frac{3}{4}$,	(diamond internal positions)
4 F^- at $\frac{1}{4}\frac{1}{4}\frac{3}{4}$, $\frac{3}{4}\frac{1}{4}\frac{1}{4}$, $\frac{3}{4}\frac{3}{4}\frac{3}{4}$, $\frac{1}{4}\frac{3}{4}\frac{1}{4}$	("void" positions)

The ion positions may be written in summary form as

4 Ca^{++} at 000 + face-centering translations.
8 F^- at $\frac{1}{4}\frac{1}{4}\frac{1}{4}$ + internal quarter point translations.

We then apply Eqs. (16) and (19) to determine the relative intensities of our hypothesized structure by the steps tabulated in columns (7)–(15) of Table 2.

Column (7): Required to determine f_{Ca} and f_F.

Columns (8) and (9): From tabulated atomic scattering factors, such as Appendix 8 of Cullity [13].

Column (10): From Eq. (16),

$$F_{(111)} = f_{Ca}[e^0 + e^{2\pi i(1)} + e^{2\pi i(1)} + e^{2\pi i(1)}] + f_F[e^{2\pi i(3/4)} + e^{2\pi i(7/4)} + \cdots + e^{2\pi i(5/4)}]$$

Using Eq. (17),

$$F_{111} = f_{Ca}(1 + 1 + 1 + 1) + f_F(-i - i - i - i + i + i + i + i) = 4f_{Ca}$$

and the same for all lines having all odd indices. In a similar manner,

$$F_{220} = 4f_{Ca} + 8f_F$$

and the same for all lines having all even indices.

Column (11): From columns (8), (9), and (10).

Column (12): The multiplicity factor can be obtained from Table 7 in Appendix B or by enumerating all planes of equal spacing by permuting the appropriate index integers and signs, i.e.,

111 planes		*220 planes*				*311 planes*
111	= 1	220	202	022	= 3	3
$\bar{1}11, 1\bar{1}1, 11\bar{1}$	= 3	$\bar{2}20$ $2\bar{2}0$	$\bar{2}02$ $20\bar{2}$	$0\bar{2}2$ $02\bar{2}$	= 6	9
$\bar{1}\bar{1}1, 1\bar{1}\bar{1}, \bar{1}1\bar{1}$	= 3	$\bar{2}\bar{2}0$	$\bar{2}0\bar{2}$	$0\bar{2}\bar{2}$	= 3	9
$\bar{1}\bar{1}\bar{1}$	= 1				12	3
	8					24
(4 nonparallel planes)		(6 nonparallel planes)				(12 nonparallel planes)

Column (13): The Lorentz-polarization factor, from Fig. 28.

Column (14): Integrated intensity, the product of columns (11), (12), and (13), from Eq. (19).

Column (15): Relative intensities, by setting the largest intensity in column (14) to 100.

The agreement here between the calculated integrated intensities and observed (peak height) intensities is satisfactory. Comparisons should be made between lines which are not too far apart, since we omitted the absorption and temperature factors from the calculation, and the intensity corrections due to these factors become significant only over broad ranges of 2θ. Note that the complete structural determination has involved two stages of trial and error, first in the assumption of crystal class [columns (3)–(6)] and second in the assumption of atom positions [columns (10)–(15)].

After the structure appears to be in accord with diffraction data, it may be advisable to check for possible errors by computing interatomic distances. For the hypothesized CaF_2 structure, the closest Ca-F spacing, by the Pythagorean theorem, is $\sqrt{3}(a/4) = 2.37$ A. The distances of closest approach in the pure elements [29] is 2.36 A for Ca and 2.72 A for F, giving an average of 2.54 A, as compared with 2.37 A. The difference is reasonable, due to the addition and subtraction of electrons to the shells in forming the ionic bond. As a general rule, the interatomic distance of a compound is less than the average distance of closest approach of the pure elements due to the attractive force between unlike atoms.

More sophisticated techniques are required for indexing noncubic crystals. Whereas the cubic system has only one unknown parameter, a, noncubic systems have two to six unknown parameters. Graphical methods can be used for the systems which contain no more than two unknown parameters, i.e., tetragonal, hexagonal, and rhombohedral. The plane-spacing equation for the *tetragonal* system (Appendix B)

$$\frac{1}{d^2} = \frac{h^2 + k^2}{a^2} + \frac{l^2}{c^2} \tag{20}$$

may be rewritten

$$\frac{1}{d^2} = \frac{1}{a^2}\left[(h^2 + k^2) + \frac{l^2}{(c/a)^2}\right]$$

or

$$2 \log d = 2 \log a - \log\left[(h^2 + k^2) + \frac{l^2}{(c/a)^2}\right] \tag{21}$$

where the dimensions a and c are the unknown unit cell parameters. The difference between the $(2 \log d)$ values for any two sets of planes depends only on the c/a ratio, instead of their individual values; i.e.,

$$2 \log d_1 - 2 \log d_2 = -\log\left[(h_1^2 + k_1^2) + \frac{l_1^2}{(c/a)^2}\right] + \log\left[(h_2^2 + k_2^2) + \frac{l_2^2}{(c/a)^2}\right] \tag{22}$$

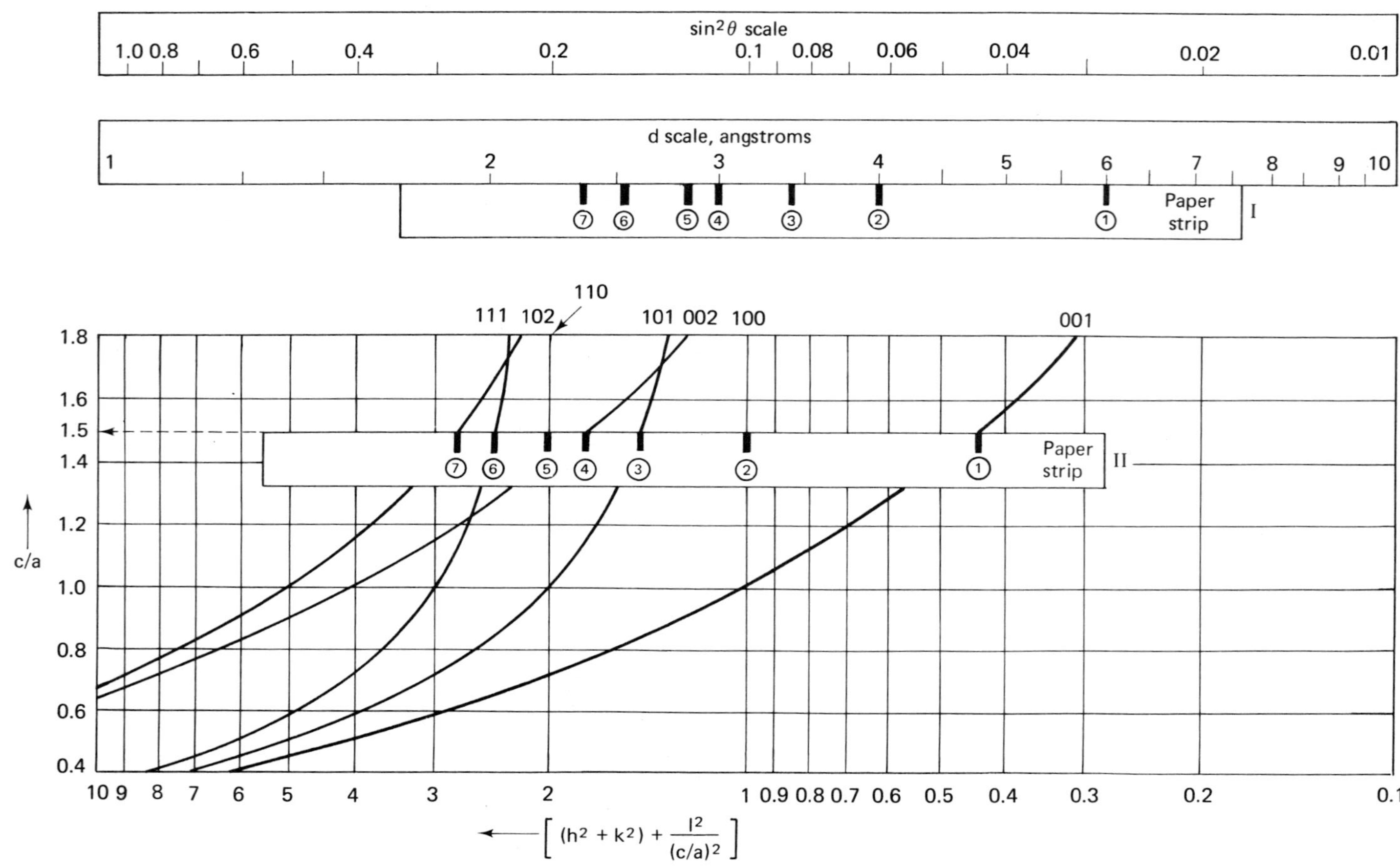

Fig. 31. Partial Hull–Davey chart for simple tetragonal lattices. From Cullity, B. D., *Elements of X-Ray Diffraction.* Reading, Mass.: Addison-Wesley Publishing Company, Inc., 1956 [13].

where subscripts 1 and 2 denote the two sets of planes. This reduction of variables provides the basis for the Hull-Davey chart (Fig. 31). Each curve on the chart represents one or more sets of *hkl* planes which have equal *d* spacings. When $l = 0$ the curve is a straight line parallel to the c/a axis. A single-range logarithmic *d* scale is constructed which spans two ranges of the $[(h^2 + k^2) + l^2/(c/a)^2]$ scale and increases in the opposite direction, since the coefficient of log *d* in Eq. (22) is -2 times the coefficient of $\log[(h^2 + k^2) + l^2/(c/a)^2]$. Thus the *d* values of any two planes are separated by the same horizontal distance as the two corresponding curves on the chart for a given c/a ratio. The chart and scale are used for indexing by marking the *d* values computed from Bragg's law for each observed diffraction line along the edge of a strip of paper laid against the *d* scale and then sliding the strip both horizontally and vertically (maintaining horizontal alignment) on the chart until a position is found where each mark on the strip coincides with a line on the chart. Vertical and horizontal movements correspond to trying various c/a and a values, respectively. When a fit is obtained, the c/a value is read from the abscissa scale and the Miller indices from the corresponding curves. The d, h, k, and l values of the two highest angle lines can then be used to solve Eq. (20) simultaneously for a and c. The chart can also be used to index cubic cells since a tetragonal cell with $c/a = 1$ is cubic. Some Hull-Davey charts use a $\sin^2 \theta$ scale instead of a *d* scale, obtained by combining Eq. (22) with the Bragg law, Eq. (6). The $\sin^2 \theta$ scale, shown at the top of Fig. 31, eliminates the computation of *d* for each line from Bragg's law. Figure 31 includes only the curves which were found to match the prominent lines of a particular diffraction pattern. A complete Hull-Davey chart for body-centered tetragonal lattices is shown in Fig. 32. An alternative construction, the Bunn chart, is based upon slightly different functions of *hkl* and c/a but employs a similar solution method.

Hexagonal crystals can be indexed by an identical procedure. The equation analogous to Eq. (21) for constructing the Hull-Davey chart is

$$2 \log d = 2 \log a - \log\left[\frac{4}{3}(h^2 + hk + k^2) + \frac{l^2}{(c/a)^2}\right]$$

and the indices and values of a and c are obtained in the same way. The plane spacing equation (Appendix B) analogous to Eq. (20) is

$$\frac{1}{d^2} = \frac{4}{3}\left(\frac{h^2 + hk + k^2}{a^2}\right) + \frac{l^2}{c^2}$$

Rhombohedral crystals, with parameters a and α, can be referred to hexagonal axes [13] and therefore also be indexed with a hexagonal Hull-Davey or Bunn chart. Thus any two-parameter crystal can be indexed graphically.

The remaining systems, orthorhombic (parameters a, b, c), monoclinic (parameters a, b, c, β), and triclinic (parameters a, b, c, α, β, γ), require more sophisticated indexing methods, because of the large numbers of variables. Such methods usually depend on the examination of single crystals instead of powder diffraction.

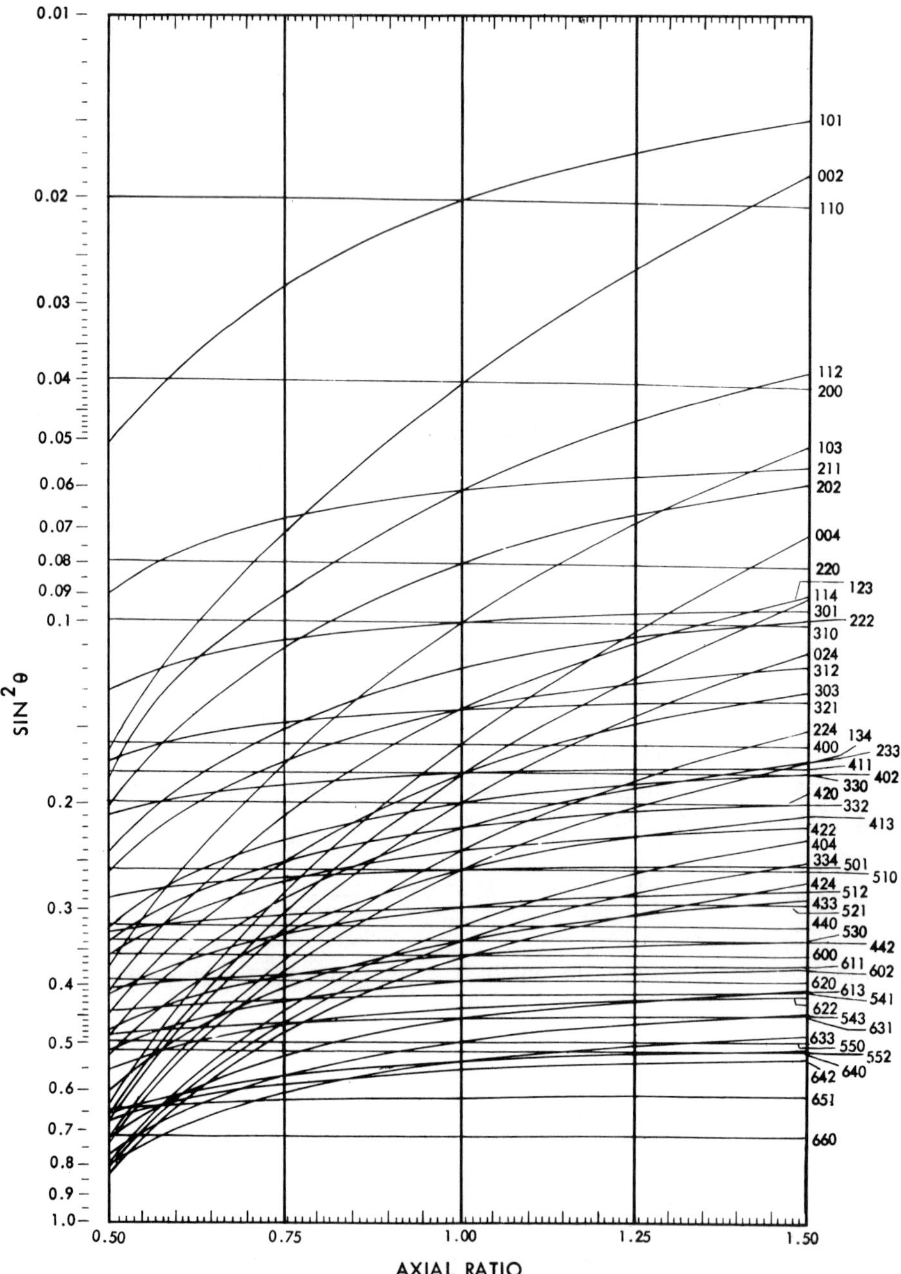

Fig. 32. Hull–Davey chart for body-centered tetragonal lattices. From Cullity, B. D., *Elements of X-Ray Diffraction*. Reading, Mass.: Addison-Wesley Publishing Company, Inc., 1956 [13].

The complete structural determination of the more complex systems almost always depends on the use of space group theory and Fourier series. Space group theory permits the listing of all possible arrangements of n atoms in a unit cell of given symmetry, such as hexagonal. It can thereby greatly aid in the selection of trial structures. These methods are described in crystallography texts [30].

5. Qualitative and Quantitative Analysis Methods

Qualitative x-ray analysis is based upon identifying the diffraction pattern of each crystalline substance present and quantitative analysis upon comparing the intensities of their diffraction lines.

For *qualitative* analyses, the ASTM diffraction data index book and card file [31] with listings for over 7,000 substances are now almost universally used. The index book contains an alphabetical index with d values and relative intensities of the three strongest lines of each substance and a numerical index in which substances are grouped according to the d values of their first, second, and third strongest lines in hierarchial order. Each card in the card file contains a serial number to which it can be referred from the index book and the d values and relative intensities of all major

Fig. 33. Diffraction pattern of an aggregate sample.

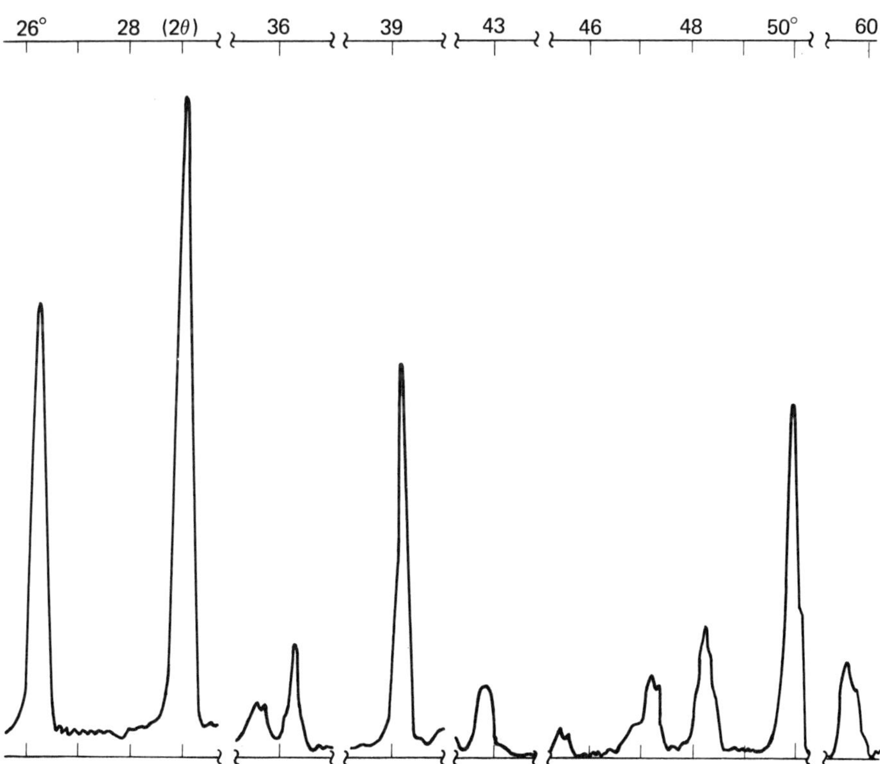

lines. After tabulating the observed d values and relative intensities, an unknown can be identified by locating the proper d_1, d_2, and d_3 values in the numerical index, checking their relative intensities, and then confirming the remaining d and relative intensity values from the card file. Consider the following example:

EXAMPLE 2: A concrete aggregate sample submitted for x-ray analysis gave the diffraction pattern in Fig. 33, using Cu K_α radiation ($\lambda = 1.54$ A). It is suspected that the rock may contain varying amounts of any of the minerals shown in Table 3, whose three most intense lines were obtained from ASTM x-ray data [31]. If the specific gravity of the rock is found to be 2.68, can you estimate approximately the weight percentage of each mineral identified in the x-ray diffraction pattern?

TABLE 3. Anticipated Minerals, Line Intensities, and Specific Gravities

Mineral	*d Spacings for the three most intense lines (A)*	*Specific gravity*
Biotite	10.7, 3.37, 2.22	3.03
Calcite	3.04, 2.29, 2.10	2.72
Feldspar	3.29, 2.97, 3.99	2.72
α-Quartz	3.34, 4.26, 1.82	2.64

Computing d values for the major observed peaks,

θ	$I_{relative}$	d (1.54/2 sin θ)
13.1	68	3.40
14.5	100	3.07
18.1	17	2.48
19.6	63	2.30
21.4	10	2.11
23.6	12	1.92
24.1	21	1.89
25.0	56	1.83
29.8	15	1.55

From a matching of d values for only the three major peaks of each mineral, it appears that the sample contains only the two minerals calcite and α-quartz, in which case the weight percentages can be obtained from specific gravity values. Letting x be the percent by weight of calcite,

$$2.68 = x(2.72) + (1 - x)2.64$$

from which $x = 50\%$.

Quantitative analysis depends on the fact that the line intensities of a given phase depend on the concentration of that phase in the mixture. Two of the most important quantitative methods are the *direct comparison method* and the *internal standard method.* Both methods depend on finding the ratio of intensity of a diffraction line of the phase in question to the intensity of some standard *reference line.* In the direct comparison method the reference line is a line from another phase in the mixture; in the internal standard method it is a line from an additional substance mixed with the specimen. It is essential that integrated intensity, not peak intensity, be measured in both methods since instrumental variations and variations in microstrain and grain size can profoundly affect peak intensity but will not affect integrated intensity.

The direct comparison method depends on the relationship

$$\frac{I_1}{I_2} = \frac{R_1 c_1}{R_2 c_2} \tag{23}$$

in which

$$R = I\frac{e^{-2M}}{V^2}$$

where subscripts 1 and 2 represent two crystalline phases in the mixture, c is the volume fraction of each phase, V the unit cell volume, I the integrated intensity from Eq. (19), and e^{-2M} the temperature factor, which is derived in x-ray diffraction texts [13]. From Eq. (23) the ratio c_1/c_2 can be obtained by measuring I_1/I_2 and calculating R_1/R_2. Independent checks can be obtained by using several pairs of lines. The same method can be used for n phases and the individual volume fractions obtained from the additional relationship

$$c_1 + c_2 + \cdots + c_n = 1$$

The internal standard method depends on the use of a calibration curve obtained by mixing various proportions of one of the phases identified in the mixture with some standard substance and plotting the ratio of the intensity of one line of the phase to one line of the standard as a function of the concentration ratio of the two substances. The specimen, which may contain any number of phases, is then mixed in any fixed proportion with the standard substance, and the intensity ratio of the same two lines is observed in its diffraction pattern. This intensity ratio is then used with the calibration chart to determine the concentration ratio of that phase to the standard in the composite material (specimen plus standard) and thereby obtain the concentration of that particular phase in the undiluted specimen. The procedure can be repeated for each crystalline phase identified in the specimen, i.e., requiring a single diffraction pattern for the composite material plus a calibration chart for each crystalline phase in the sample. The method depends on the fact that line intensities of the standard and the phase being measured are reduced in equal proportions by the absorption of the mixture. A desirable standard is one having strong diffraction lines near those of the phases being measured but not superposed on them.

C. INFRARED SPECTROSCOPY

Spectroscopy is used routinely for both the identification and the chemical structural analysis of construction materials such as bitumens, pigments, and synthetic polymeters.

Spectroscopy is the study of interractions between matter and electromagnetic radiation. Atoms in molecules and crystals continuously vibrate with natural frequencies in the range of 10^{13} to 10^{14} cycles per second, which is the frequency of infrared radiation. Vibrations which are accompanied by a change in dipole moment cause absorption of infrared radiation. Several vibration modes may occur for a particular atomic group, each at a particular frequency which is normally independent of the other modes. If the amount of radiation absorbed by a substance is plotted against the incident wavelength, the resulting graph reflects the presence of specific chemical bonds and can therefore be used for structural identification. Whereas spectra associated with atoms are caused by electrons moving from one electronic energy level to another, interatomic spectra are usually characterized by either bond stretching or bending vibration modes. In addition, to these fundamental absorption bands, there are multiples of the fundamental frequencies (overtones) and frequencies which are the sum and difference of two or more fundamental frequencies, called combination lines. Wavelength is conventionally expressed in microns (μ) and frequency in wave number —the number of cycles per centimeter, with units of cm^{-1}.

Infrared spectroscopy is probably the most powerful single technique available to qualitatively identify organic materials and to determine molecular structure. Mass spectrometry gives the molecular weight and formula, and nuclear magnetic resonance the number and type of protons, but only infrared indicates in a direct manner the presence of key functional groups. A match between the infrared spectra of an unknown sample and a reference sample is a simple, and almost positive, method of identification. Currently, over 130,000 reference spectra are available, most of which are available in indexed form. Spectroscopic theory will be explained in terms of very simple molecules, and then extensions will be briefly indicated to use with construction composites.

1. Theory

The observed absorption frequencies for molecular rotation and for bond stretching and bending are related to the atomic masses, bond distances, bond strengths, and bond angles. Let us consider some relationships for a simple diatomic molecule and then briefly inspect their extension to more complicated molecules.

Both the emission and absorption of electromagnetic radiation by a molecule is described by

$$\epsilon = h\nu = h\frac{c}{\lambda} \tag{24}$$

where ϵ = energy of the quantum of radiation emitted or absorbed,
h = Planck's constant = 6.62×10^{-27} erg-sec,
ν = frequency of emitted or absorbed radiation, cycles per second,
c = speed of light = 3.00×10^{10} cm/sec,
λ = wavelength of radiation, cm.

For a diatomic molecule, Eq. (24) may be written approximately as

$$\epsilon = h\nu = \Delta E_e + \Delta E_v + \Delta E_r \tag{25}$$

where ΔE_e, ΔE_v, and ΔE_r are energy changes in the molecule due to electron transition, stretching, and rotation modes, respectively. The kinetic energy of translation of a molecule is normally small and has been neglected. For most molecules, $\Delta E_e \gg \Delta E_v \gg \Delta E_r$, so that electronic, vibrational, and rotational spectra are observed in different regions of the spectrum. Pure rotational spectra involve only changes from one rotational energy level to another and are found in the far infrared ($\lambda \simeq 0.003$–0.06 cm) or in the microwave region ($\lambda \simeq 0.06$–30 cm). Vibration-rotation spectra involve changes in both vibrational and rotational energy levels and are observed in the near infrared ($\lambda \simeq 8{,}000$–$300{,}000$ A). Electronic spectra involve changes in electronic, vibrational, and rotational energy levels and are found in the visible ($\lambda \simeq 4{,}000$–$8{,}000$ A) or ultraviolet ($\lambda \simeq 1{,}500$–$4{,}000$ A) regions.

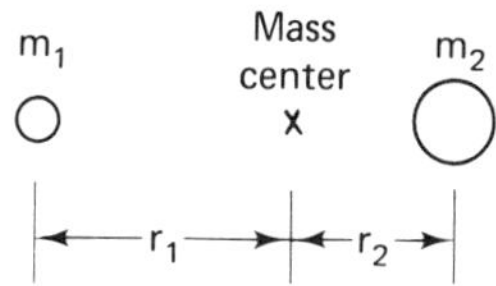

Fig. 34. Representation of a diatomic molecule.

We shall first compute the bond length of a diatomic molecule from its rotation mode frequency and then the bond force constant from its vibration mode frequency.

The moment of inertia I of a diatomic molecule (Fig. 34) is

$$I = m_1 r_1^2 + m_2 r_2^2 \tag{26}$$

where m_1 and m_2 are the masses of the two atoms and r_1 and r_2 their respective distances to the mass center. Taking first moments about the centers of mass 1 and mass 2,

$$r_1 = \frac{m_2 r}{m_1 + m_2} \quad \text{and} \quad r_2 = \frac{m_1 r}{m_1 + m_2} \tag{27}$$

respectively, where the atomic spacing is $r = r_1 + r_2$. Substituting into Eq. (26),

$$I = m_1 \left(\frac{m_2 r}{m_1 + m_2}\right)^2 = m_2 \left(\frac{m_1 r}{m_1 + m_2}\right)^2$$

which reduces to

$$I = \frac{m_1 m_2}{m_1 + m_2} r^2 \tag{28}$$

The rotational energies of a diatomic molecule are quantized, and it can be shown, by solving the Schrödinger wave equation for a diatomic molecule (Sec. A in Chap. 2) that the allowed rotational energy is

$$E_r = J(J + 1)\frac{h^2}{8\pi^2 I}$$

where J, the rotational quantum number, is an integer and E is the allowed energy of a rotating diatomic molecule. The energy difference for a transition from one rotational level to the next is therefore

$$\Delta E_r = \frac{h^2}{8\pi^2 I}[J(J+1) - (J-1)J] = \frac{h^2}{4\pi^2 I}J \tag{29}$$

Equation (29) is more frequently expressed in terms of wave numbers. From Eq. (25),

$$\Delta E_r = \frac{hc}{\lambda} \quad \text{or} \quad \frac{1}{\lambda} = \bar{\nu} = \frac{\Delta E_r}{hc} \tag{30}$$

where $\bar{\nu}$ is the wave number. Substituting Eq. (29) in Eq. (30),

$$\bar{\nu} = \frac{h}{4\pi^2 Ic}J$$

Since J is an integer, the pure rotational spectra of a diatomic molecule will consist of a number of equally spaced lines. The difference in wave numbers of two adjacent lines is therefore

$$\bar{\nu}_1 - \bar{\nu}_2 = \frac{h}{4\pi^2 Ic}(J_1 - J_2) = \frac{h}{4\pi^2 Ic} \tag{31}$$

EXAMPLE 3: Calculate the interatomic distance for HCl. The pure rotation spectrum consists of almost equally spaced lines 20.7 cm^{-1} apart [32].

From Eq. (31),

$$20.7 = \frac{6.62 \times 10^{-27}}{4(3.14)^2(3 \times 10^{10})I}, \qquad \text{from which } I = 2.70 \times 10^{-40} \text{ g-cm}^2$$

Since the masses of H and Cl are 1.008 and 35.45, we have from Eq. (28)

$$2.70 \times 10^{-40} = \frac{(1.008)(35.45)}{(1.008 + 35.45)(6.02 \times 10^{23})} r^2$$

from which $r = 1.29 \times 10^{-8}$ cm. We have neglected the existence of isotopes of chlorine in HCl.

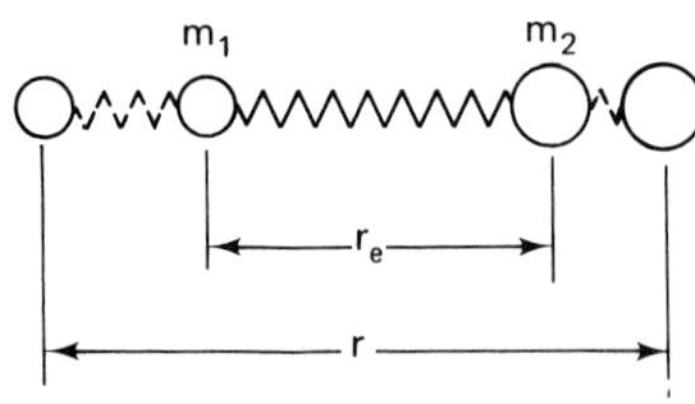

Fig. 35. Diatomic molecule as a simple harmonic oscillator.

Next compute the bond force constant from the vibration frequency mode. A diatomic molecule can be considered as two vibrating masses, m_1 and m_2, connected by a spring with equilibrium length r_e (Fig. 35). To a first approximation, the system can be treated as a simple harmonic oscillator, i.e., one in which the resisting force is proportional to the displacement. The restoring force, f, is then

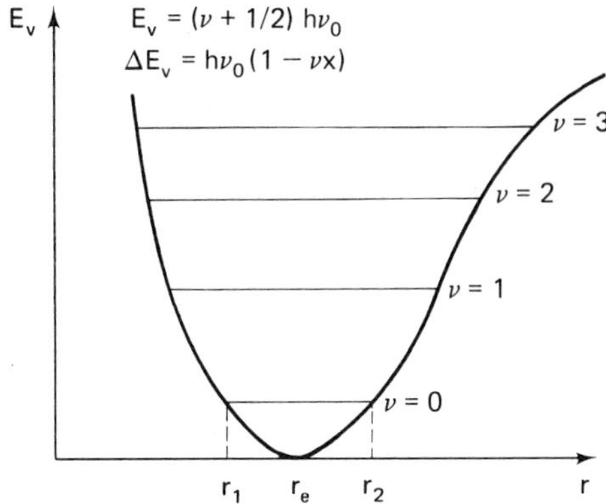

Fig. 36. Vibrational energy levels in a diatomic molecule.

$$f = k(r - r_e)$$

where k is the force constant and $r - r_e$ is the displacement from the equilibrium internuclear distance. Since force is related to potential energy by $f = dV/dr$, we have $dV/dr = k(r - r_e)$, and $V = \frac{1}{2}k(r - r_e)^2$, a familiar result from statics. A plot of V versus r is parabolic and approximates the bottom of the curve in Fig. 36. If the Schrödinger wave equation is solved for a simple harmonic oscillator, it can be shown that the allowed vibrational energies, E_v, are

$$E_v = (v + \tfrac{1}{2})h\nu_0 \tag{32}$$

where v is the vibrational quantum number (0, 1, 2, 3, . . .) and ν_0 the fundamental vibration frequency. If $v = 0$, $E_v = \frac{1}{2}h\nu_0$. This is called the zero-point energy, the vibrational energy at zero temperature (Fig. 36). Note that at zero temperature vibration still occurs between internuclear distances r_1 and r_2. The fundamental frequency of a harmonic oscillator can be derived from Hooke's law and is

$$\nu_0 = \frac{1}{2\pi}\sqrt{k\left(\frac{m_1 + m_2}{m_1 m_2}\right)} \tag{33}$$

From Eq. (32), the energy difference for a transition from one vibrational level to the next is

$$\Delta E_v = h\nu_0[v + 1 + \tfrac{1}{2} - (v + \tfrac{1}{2})] = h\nu_0 \tag{34}$$

Expressing Eq. (34) in terms of the wave number,

$$\bar{\nu}\frac{\Delta E_v}{hc} = \frac{h\nu_0}{hc} = \frac{\nu_0}{c} \tag{35}$$

Substituting Eq. (33) into Eq. (35),

$$\bar{\nu} = \frac{1}{2\pi c}\sqrt{k\left(\frac{m_1 + m_2}{m_1 m_2}\right)} \tag{36}$$

EXAMPLE 4: Calculate the force constant for HBr if its pure vibration wave number (with zero rotational energy) is found spectrographically to be 2560 cm^{-1} [32].

The masses of H and Br are 1.008 and 79.909. From Eq. (36),

$$2560 = \frac{1}{2(3.14)(3.00 \times 10^{10})}\left(\frac{1.008 + 79.909}{1.008 \times 79.909}\right)^{1/2} k^{1/2}$$

$$k = 3.84 \times 10^5 \text{ dynes/cm}$$

Force constants of the hydrogen halides are on the order HF > HCl > HBr > HI.

As Fig. 36 shows, actual molecules are not strict harmonic oscillators, their potential energy curves are not truly parabolic, and the energy levels are unequally spaced. The energy difference between two adjacent states is

$$\Delta E_v = h\nu_0(1 - vx)$$

where x is an anharmonicity constant with a value generally less than 0.1.

Thus far we have considered only diatomic molecules. The determination of fundamental frequencies of a polyatomic molecule or a crystal structure is generally much more difficult. Some simplification is possible where there are molecular or ionic groups in a crystal with bond forces much stronger than those linking them to their immediate neighbors. The fundamental frequencies are then shifted only slightly by the presence of the neighboring groups. This is often the case for a water molecule in a crystal lattice. It has nine degrees of freedom: six translational and rotational oscillations of the molecule as a whole and three vibration modes. The translation and rotation modes all lie at frequencies below 900 cm^{-1}. The vibrations can be resolved into a set of modes in which each atom moves, to a first approximation, in simple harmonic motion. Linear and nonlinear molecules of n atoms have $3n - 5$ and $3n - 6$ fundamental modes, respectively. For example, the three modes of water correspond approximately to in-phase and out-of-phase stretching of the O—H bonds and bending of the H—O—H bond angle (Fig. 37). The vibrations are generally

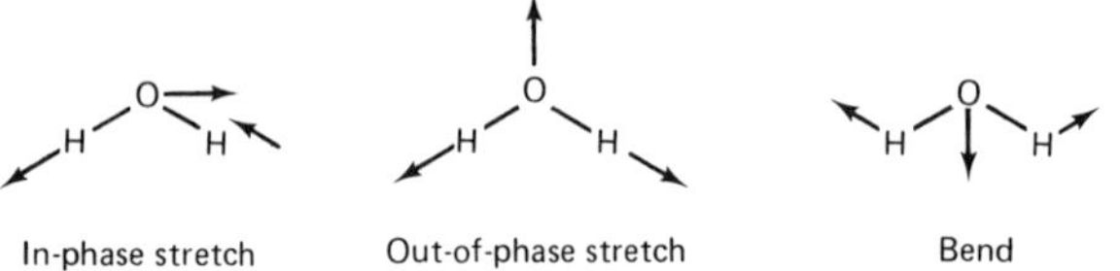

Fig. 37. Vibrations of a water molecule.

displaced by hydrogen bonding in the crystal but are usually recognizable as such. Bonds corresponding to each fundamental vibration mode are not always observed. For example, carbon dioxide undergoes no change of dipole movement during symmetric stretching and therefore does not absorb radiation (Fig. 38). Also the two bending motions occur in perpendicular planes but are otherwise identical and therefore have the same frequency and are said to be degenerate. Such symmetry considerations often simplify the interpretation of spectra for complex molecules.

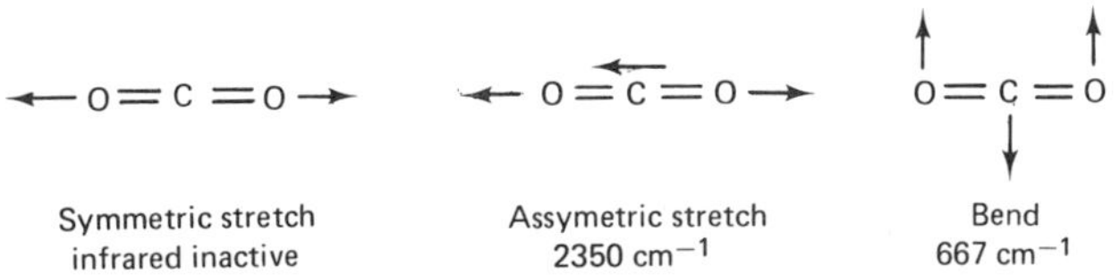

$\leftarrow O=C=O \rightarrow$ Symmetric stretch infrared inactive

$\leftarrow O=\overleftarrow{C}=O \rightarrow$ Assymetric stretch 2350 cm^{-1}

$O=C=O$ (bend arrows) Bend 667 cm^{-1}

Fig. 38. Vibrations of carbon dioxide.

In addition to its uses for identification and structural interpretation, infrared is used for quantitative analysis. Quantitative analysis is based upon the Beer-Lambert law, which states that the amount of monochromatic light absorbed is proportional to the concentration of absorbing molecules and the thickness of the sample. Accuracies with double-beam optical null instruments on the order of 1–3% are routinely obtained, and somewhat better accuracies are obtained with single-beam instruments [33]. *Difference spectroscopy* is a useful quantitative technique in which a standard sample is placed in the reference beam and the difference between the absorption of the standard and test samples is measured.

2. Apparatus

Because the entire infrared spectrum cannot be covered by a single instrument, it is divided into the near-infrared (13,000–4,000 cm^{-1}), mid-infrared (4,000–200 cm^{-1}) and far-infrared (200–10 cm^{-1}) regions. Spectrophotometers for all three regions have a radiation source, an optical system to isolate radiation approximating a single frequency, a sample holder, a detector, and a recording system. A sodium chloride prism useful for the range 2–15 μ was the usual optical element in instruments until the 1960s. Filter-grating and prism-grating instruments are now used which offer

TABLE 4. Spectrophotometers for the Near-, Mid-, and Far-Infrared Regions

		Near-infrared		*Mid-infrared*		*Far-infrared*	
Wavelength, μ:	0.8		2.5–3		50		1,000
Wave number cm^{-1}:	12,500		4,000		200		10
Radiation source		Tungsten filament lamp		Globar, Nerst glower, or Nichrome wire coil		High-pressure mercury arc lamp	
Optics		One or two quartz prisms or prism-grating double monochromator		Two to four plane diffraction gratings with either a fore-prism monochromator or infrared filters to remove unwanted radiation		Double-beam grating for use to 700 μ; interferometric Michelson spectro-photometers for use to 100 μ (no prisms, grating, or slits)	

Remarks: Near-infrared: Useful for quantitative analysis based upon O—H, N—H, and C—H groups. Intensities of the two beams are compared electronically and the ratio recorded. Instruments can generally take a different source and detector for use in the ultraviolet and visible regions.

Mid-infrared: Most are double-beam instruments using the optical null principle.

Far-infrared: This region is useful to study low-energy vibrations of organic and inorganic molecules. Interferometric instruments have a better signal-to-noise ratio than conventional spectrophotometers.

SOURCE: WHETSEL, K. B., *Chem. Eng. News*, **46** (Feb. 5, 1968) 82 [33].

higher resolution to permit separation of closely spaced bands and higher scanning speeds for a given resolution and noise level with more accurate measurement of band positions and intensities. Table 4 summarizes some properties of spectrophotometers for the three regions.

Since the mid-1960s several types of rapid-scan spectrophotometers have become available, with scan times in the range of microseconds to seconds, instead of the 10–15 min typical of conventional instruments. Rapid-scan photometers have been used to study fast reactions, to detect transient species generated in flash photolysis, and to measure spectra of chromatographic fractions as they elute from a column.

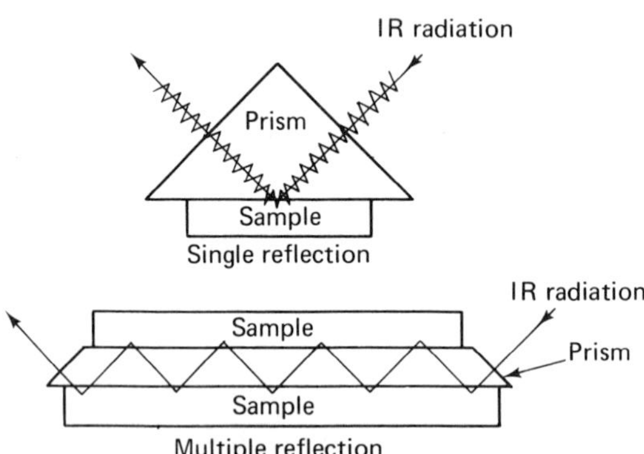

Fig. 39. Attenuated total reflectance technique. Reprinted with permission from Whetsel, K. B., *Chemical and Engineering News* (Feb. 5, 1968), 82. Copyright by the American Chemical Society.

The *attenuated total reflectance* technique permits the infrared analysis of opaque materials such as leather, rubber, certain polymers, and pressed powders which cannot be prepared in sections thin enough to allow transmission. Figure 39 illustrates the optics of this method. The ray will be reflected at the interface between prism and sample at angles of incidence greater than the critical angle. Since the incident beam penetrates the sample to a depth of a few micrometers, the reflected beam bears a resemblance to the absorption spectrum obtained by normal methods. Multiple reflections are sometimes used to increase the absorption peak height for samples that do not give intimate contact with the prism surface. The use of multiple reflections is analogous to increasing the path length in a transmission cell. The method has also been used to analyze thin surface coatings on papers and metals [33].

3. Identification of Minerals

Applications of infrared spectroscopy to inorganic chemistry are less numerous than to organic chemistry, but their number is steadily increasing. It is now possible to study frequencies in the range below 400 cm^{-1}, where many inorganic spectra appear. Infrared analysis of minerals is especially helpful when there is ambiguity in the interpretation of x-ray diffraction data. For example, the presence of chlorite is often uncertain in samples containing kaolin. The distinction is easy with infrared because kaolin minerals show a bond at 3,698 $\pm$ 2 cm^{-1}, which is not shown by chlorite [34]. Infrared provides direct information on the nature of water in minerals, it provides information on amorphous components in crystalline materials not obtainable from x-ray diffraction data, and it is probably the simplest method to detect carbonate in silicates. Infrared has given information on bond types and orientations and on the nature of isomorphous substitutions in crystal lattice [34]. Mineralogical changes induced by heat, acid treatment, clay-organic reactions, and ion exchange reactions have also been studied by infrared analysis [34].

All silicates show strong absorption lines in the range 1,100–900 cm^{-1} associated with Si—O stretching vibrations. Unusual features in spectra outside of this range assist in the identification of components in mixtures, for example, the bonds shown by α-dicalcium silicate hydrate (1288 cm^{-1}) and xonotlite (1200 cm^{-1}). Si—O bending vibrations occur at lower frequencies and vary more in frequency for different silicates. In layer-structure silicates such as the clay minerals, infrared spectral assignments are complicated by the fact that frequencies are sometimes significantly shifted, not only by the influence of nearest neighbor groups but also by ions in the second coordination sphere [34].

Figure 40 illustrates general characteristics of absorption spectra for two groups of clay minerals. The absorption at 2.75 μ has been shown to be due to unbonded OH and absorptions from 2.7 to 3.2 μ and 6.15 to 7.55 μ to be due to bonded OH [35]. The absorption at about 6.1 μ observed in some clay minerals is believed due to absorbed water and at 9 μ to the Si—O bond. The origins of some absorption bands in clay minerals are still not known.

The mid-infrared region has been the most useful one for mineralogical work, and infrared analysis of minerals is usually done with powdered samples. Although the weaker hydroxyl absorption bands and combination frequencies in the 10,000–1300-cm^{-1} range can be examined in thin section, the high-frequency stretching bands of silicates would require sample thicknesses in the order of 1 μ, or beyond thin-section capability. The particle size of the powders should be smaller than the minimum wavelength of radiation used, because the scattering of radiation by larger particles varies sharply in the region of absorption and distorts the absorption band shapes. Larger particles also cause loss of detail in the absorption bands. Early work with clays and cement components was done in an oil medium, which had the disadvantage of adding its own spectra to that of the minerals [35]. Dry powders are sometimes deposited from isopropyl alcohol onto alkali halide disks or from water onto silver chloride [36].

The method of attenuated total reflection, described previously, has been used successfully on large mineral crystals [36]. Since reflected intensities are a function of refractive indices as well as of absorption coefficients, reflection spectra are not identical with absorption spectra. Reflected spectra obtained with polarized radiation can provide useful information on the directions of dipole changes associated with absorption bands [36].

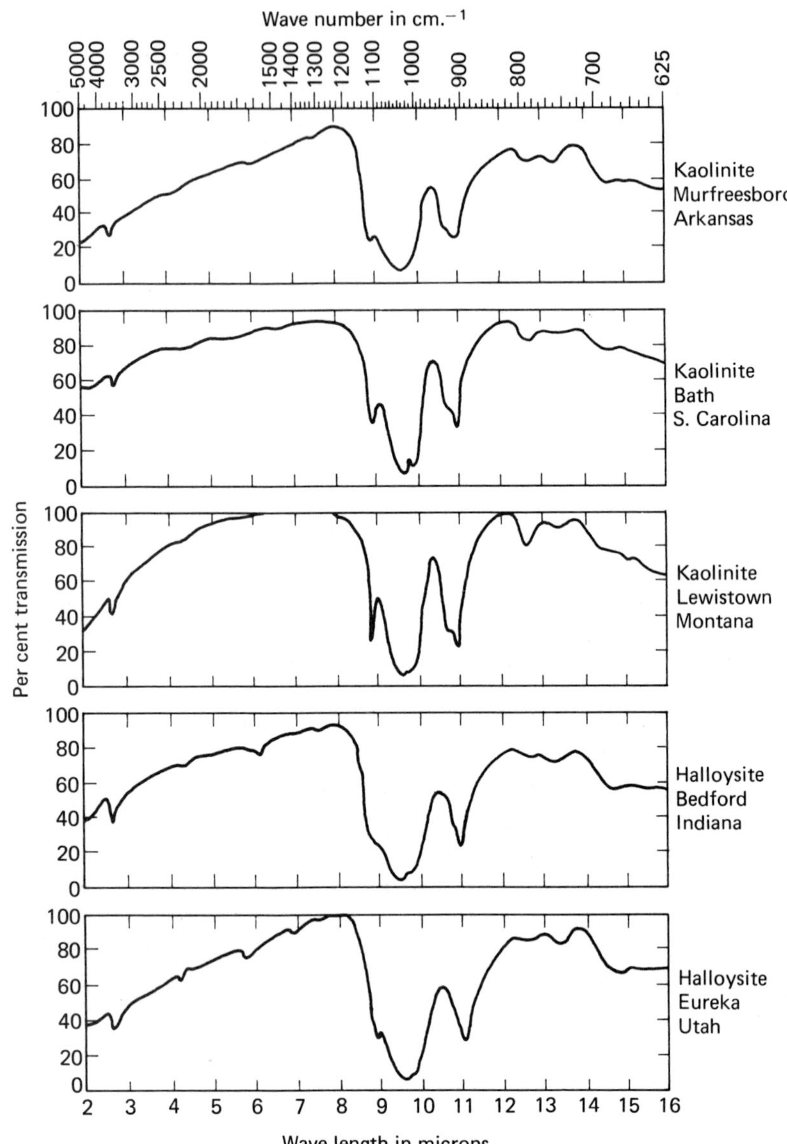

Fig. 40. Absorption spectra for (a) kaolinite minerals; (b) montmorillonite minerals. From Adler, et al., Report 8. *Am. Petrol. Inst. Proj. 49*. New York: Columbia University, 1950 [37], as presented by Grim, R. E., *Clay Mineralogy*, Second Edition. New York: McGraw-Hill Book Company, 1968 [35].

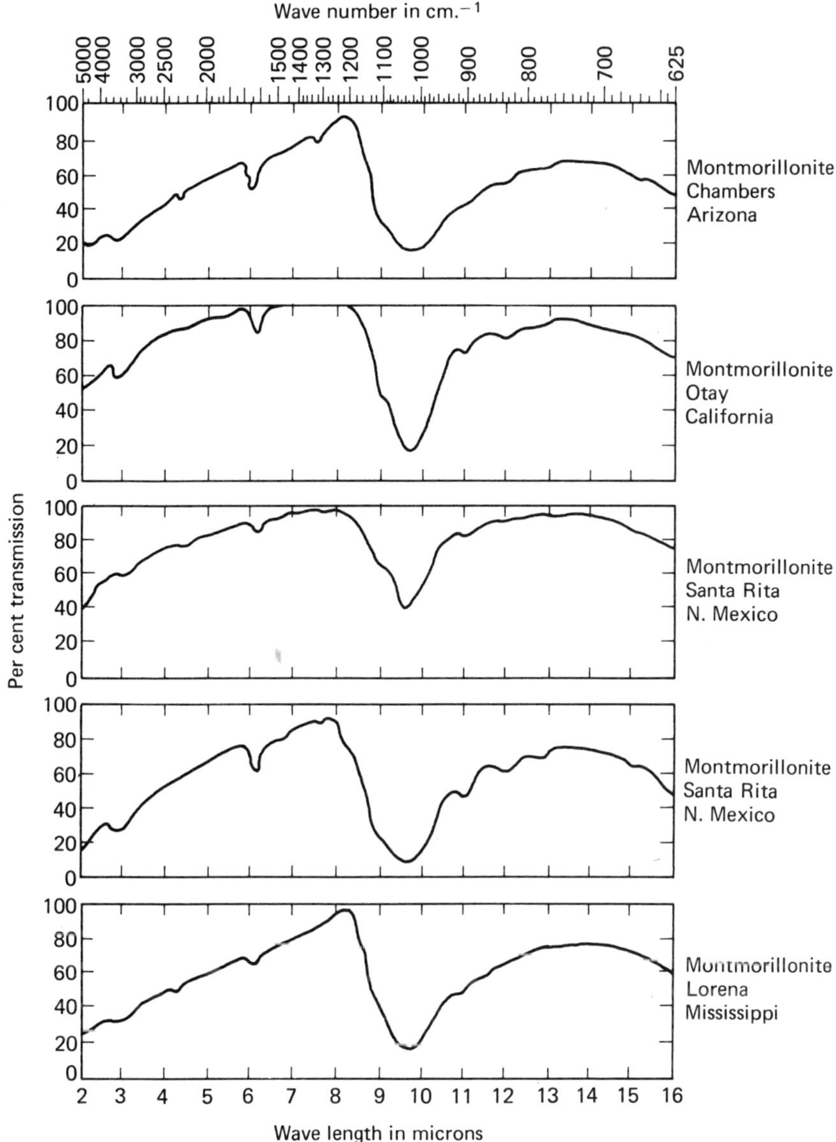

Fig. 40. *Continued*

D. DIFFERENTIAL THERMAL ANALYSIS

1. Principles of Use

Physical and chemical changes such as dehydration, decarbonation, oxidation, crystalline transition, decomposition, and lattice destruction are usually accompanied by the absorption or liberation of heat. Differential thermal analysis permits the identification of compounds by observing the temperatures at which they undergo endothermic or exthermic reactions due to these changes while being slowly heated. Observations of the relative heats of reaction allow approximate quantitative analyses. The method is useful for all materials which exhibit thermal reactions, whether crystalline or amorphous. It is therefore frequently used as a complementary technique to x-ray diffraction. DTA is widely applied in cement chemistry and clay mineralogy and also finds application in such fields as catalysis, coal chemistry, and polymer, soap, and lubricating systems.

The sample and a thermally inert reference material, such as calcined alumina, are heated side by side at a constant rate, commonly 10°C/min, and the temperature difference between the two materials is recorded as a function of the furnace temperature. Temperature differences are registered by use of thermocouples in the sample and reference materials and a very sensitive galvanometer. When the sample undergoes a reaction which absorbs heat its temperature drops below that of the inert reference material and an endothermic peak is recorded—and similarly for an exothermic reaction. Exothermic reactions are customarily plotted as upward peaks and endothermic reactions as downward peaks. The curves obtained are really temperature difference curves, rather than a differential, and the name of the technique is somewhat a misnomer.

DTA is simpler and quicker to perform than x-ray diffraction but suffers from several experimental limitations, of which the overlapping of adjacent peaks is perhaps the greatest. A technique often useful for quantitative analysis is to add to the reference sample different amounts of one component already identified in the test sample until the peaks for that component disappear, indicating approximately equal amounts of that component in both samples. This so-called double differential technique sometimes allows quantitative identifications to less than 1%, or a sensitivity similar to that for x-ray diffraction.

An additional technique is to record the weight change of the sample simultaneously with the temperature difference curve. Such *thermogravimetric* measurements require a sensitive balance and a means for independently suspending the sample within the furnace [38]. Reactions which involve only energy changes, such as crystal inversion, crystallization, fusion, and certain types of solid-state reactions are detected by DTA but not by thermogravimetry. Thermogravimetry has special value in clay work where heating is often accompanied by evolution of gases, such as water from gypsum and hydrated silica, aluminate, and iron compounds or CO_2 from carbonate minerals.

The generalized peak shown in Fig. 41 might result from an exothermic reaction which begins at a temperature corresponding to point A and ends at point D. The vertex C simply represents a point at which the rate of heat loss from the sample to its surroundings equals the rate of heat liberated by the reaction. The maximum reaction rate may occur at some point B, between A and C. The temperature A, at which the reaction starts, is the most useful one, but the curve usually departs from the

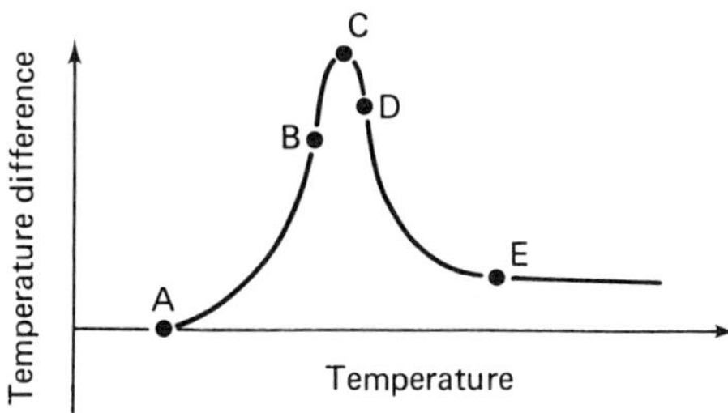

Fig. 41. Generalized exothermic peak.

base line so gradually that point C can be determined more reliably, and this peak temperature is the one usually quoted. Some factors which affect the peak temperature include the heating rate, the particle size and pretreatment of the sample, the dilution and packing of the sample, the design and material use for the sample holder, and the furnace atmosphere. For example, a lower heating rate gives smaller, sharper peaks at lower temperatures (point C moves toward point A in Fig. 41). Therefore, rigid standardization of experimental detail is necessary to minimize variations.

In some instances the base line drifts after a transformation due to a change in thermal properties of the sample material. Either an inversion or a reaction can produce a sharp change in the specific heat of the sample, and when a gas takes part in a reaction there is also a change in bulk density. Base line drifts due to such causes may be as much a characteristic of the sample as are the peak temperatures.

Several approximate theories have been proposed for quantitative determination based upon peak area measurement. Speil [39] and Yagfarov and Berg [40] proposed theories based upon a constant heating rate which lead to the identical result

$$M = k\int_{t_1}^{t_2} \Delta T \, dt$$

where M is the sample mass, k a constant whose value depends on the sample geometry, ΔT the temperature difference, and t_1 and t_2 the times for beginning and completion of the peak; i.e., the integral represents peak area on a ΔT versus time plot. There have been several refinements which account for the effects of additional experimental variables [41]. It has been observed that the peak area is proportional to the reaction heat per unit volume of the sample, sample holder, and furnace wall but independent of heating rate (if the rate is constant), the reaction rate, and the specific heat of the sample [42].

2. Apparatus

Basic DTA equipment consists of a furnace, a temperature regulator, a sample block, thermocouples, and a temperature-recording system (Figs. 42 and 43). The symmetrical and easily reproducible positioning of the sample holder in the furnace is especially important, as well as an accurately controlled heating rate. Best results are

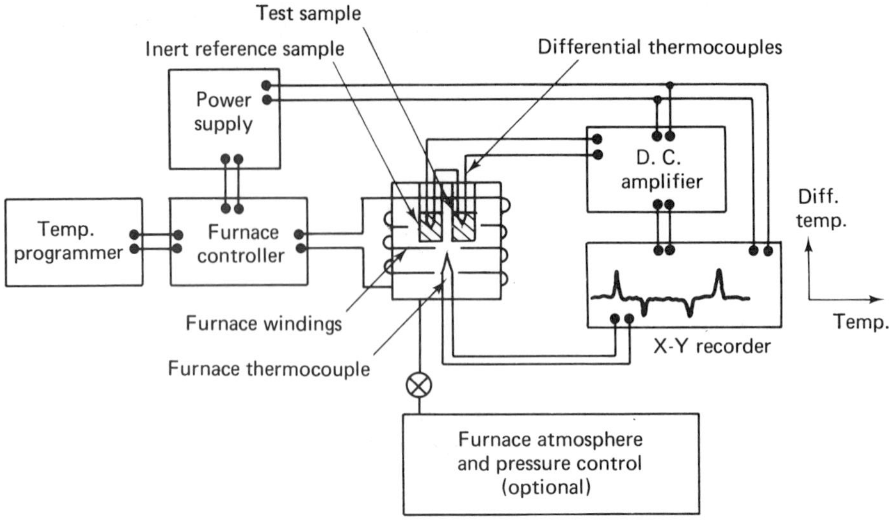

Fig. 42. DTA components.

Fig. 43. Commercial differential thermal apparatus. (*Courtesy of E. I. DuPont de Nemours Co.*)

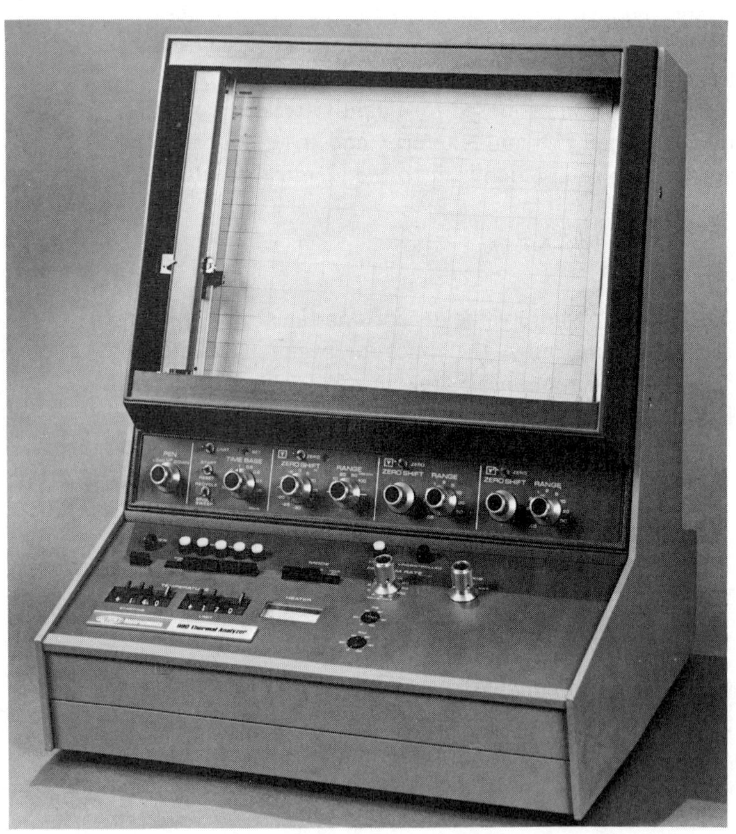

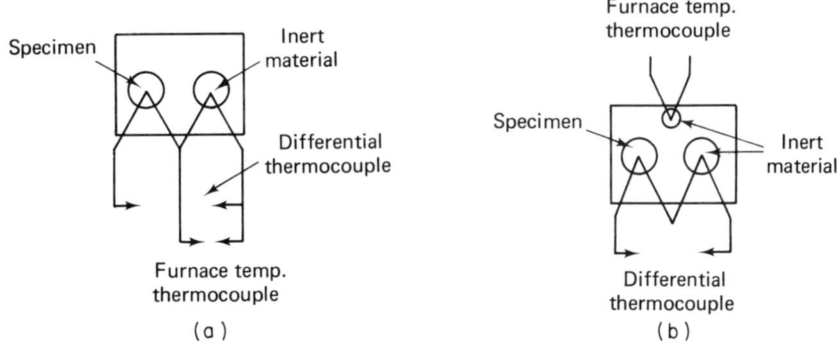

Fig. 44. Temperature measuring arrangements. (a) Two thermocouples; (b) three thermocouples.

obtained if the sample is highly diluted and if the materials in the test and reference samples have similar specific heats and thermal conductivities. Calcined alumina is the most commonly used inert material, although calcined kaolinite is often used for clay investigations because its thermal properties are closer to those of clays [43].

Figure 44 shows *temperature-measuring arrangements* using two and three thermocouples. Usually chromel/alumel or platinum/platinum-rhodium thermocouples are used. The former are very sensitive, cheap, and easily constructed but can only be used to about 1100°C; the latter can be used to about 1700°C but are more expensive and less sensitive [44]. To minimize heat conduction into and out of a sample, thermocouple wires should be thin and should make a long path through the sample. Since thin wires break and corrode more easily, a compromise is often reached at about 0.2–0.4-mm diameter [42]. Junctions with copper leads require a temperature equalizer, and for high sensitivity work the copper leads should be of screened wire suitably earthed [44].

The *sample holder* is usually of nickel, stainless steel, or ceramic [45]. Metal sample holders reduce base line drift, but the peaks tend to be small because of rapid heat transfer through the sample holder. In general, metal holders permit greater accuracy because they are more homogeneous and can be machined more accurately. Ceramic holders give larger peaks because of their lower thermal conductivity, thus permitting the same sensitivity with a smaller sample or a less sensitive recording device. Ceramic holders are also preferred in controlled atmosphere work where reactive gases are used [44].

Temperature regulation is usually accomplished by a motor-driven autotransformer which controls voltage to the furnace wires, which are usually made of nichrome (to 1000°C) or platinum alloys (to 1750°C) [44]. Atmospheric control in the furnace is sometimes used to suppress an undesired reaction, to enhance the desired reaction, or to resolve overlapping peaks by varying the pressure of a gaseous reaction product [46–49]. Studies have also been performed in which water evolution or gaseous decomposition products were continuously analyzed and recorded during heating [50].

Sample particle size can affect the thermal curves of most materials, and particles passing a 100 mesh sieve are usually used. Uniform packing of the specimen into the sample well is also important. The temperatures of the low-temperature dehydration peaks in clays, in particular, are sensitive to environmental changes. Clay samples are often saturated with the same cation (e.g., Ca^{+2} or Mg^{+2}) and then equilibrated for 4 days over a saturated solution of $Mg(NO_3)_2 \cdot 6\ H_2O$ to bring them to constant moisture content.

3. *Identification of Minerals*

Much of the data on thermal peak temperatures and the associated reactions for specific materials are now available in the form of tables and punched-card indices [51]. We shall look briefly at thermal curves of typical materials from two groups: the clay minerals and portland cement components.

The major clay mineral thermal reactions occur in three ranges: a 50–200°C range of endothermic peaks due to the loss of water, a 450–700°C range of endothermic peaks due to the loss of OH ions from the aluminositicate structure, and an 800–1400°C range containing endothermic peaks due to the breakdown of crystal structure and one or more exothermic peaks due to the crystallization of new phases [52]. The sizes of the low-temperature peaks reflect the amount of absorbed water, and the presence of more than one peak indicates water that is bound with different energies. The peak temperatures for OH ion loss are influenced by the mineral crystallinity. Well-crystallized kaolinite has a peak at about 600°C, chlorite at 550–650°C, illite at 500–550°C, and vermiculite a broad peak at about 550°C. The montmorillonite peak is at about 700°C, though some expanding lattice minerals have peaks at 500–600°C [34]. Thermograms of some principal clay minerals are shown in Fig. 45 and for a mixture of kaolinite and montmorillonite in Fig. 46.

DTA has been useful in cement research to study the kiln reactions during manufacture and to study the hydration reactions during use. The thermal reactions of binary, ternary, and quaternary systems involving CaO, SiO_2, Fe_2O_3, and Al_2O_3 have been especially helpful in studying the chemistry of clinkering processes [53, 54]. In some instances, samples have been withdrawn at the temperature of thermal inflections and examined by x-ray diffraction. Thermograms for some raw materials of portland cement manufacture are shown in Fig. 47, for a binary mixture in Fig. 48, and for a ternary mixture in Fig. 49. In Fig. 48 the exotherm at 950–1000°C is attributed to the simultaneous formation of CA and $C_{12}A_7$. The endotherm at 1170°C represents a change from γ- to α-Al_2O_3, and at 1300°C melting begins [55]. In Fig. 47 the exotherm at 980°C results from the formation of C_4AF, and the endotherm at 1160°C for the lower Al_2O_3 contents reflects the formation of C_2F.

The following results by Greene [56] as quoted by Ramachandran et al. [57] illustrate an application of DTA for the study of portland cement hydration reactions:

> Typical thermograms of portland cement hydrated for various periods are shown in Fig. 50. The unhydrated cement exhibits two endothermic peaks at 140 and 170°C

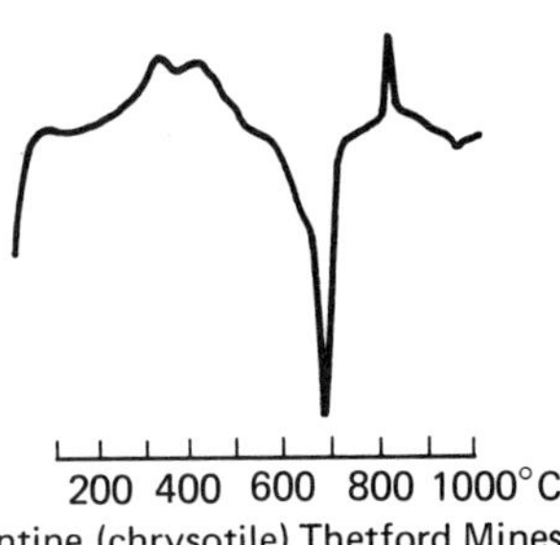

Serpentine (chrysotile) Thetford Mines, Quebec, Canada

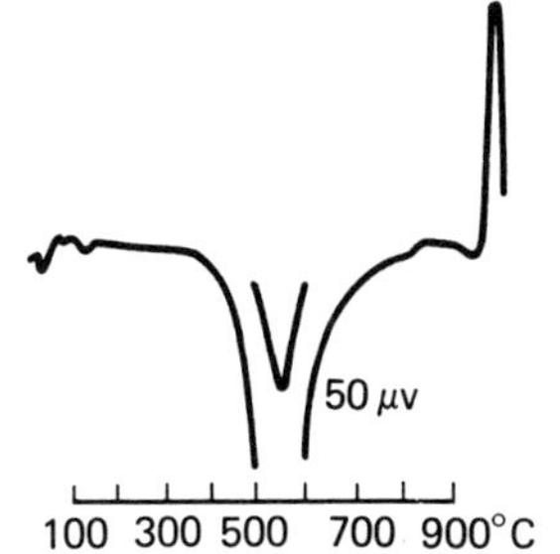

Kaolinite Macon, Georgia, U.S.A.

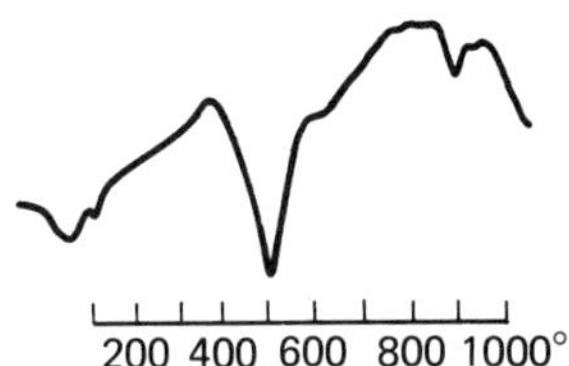

Illite Fithian, Illinois, U.S.A.

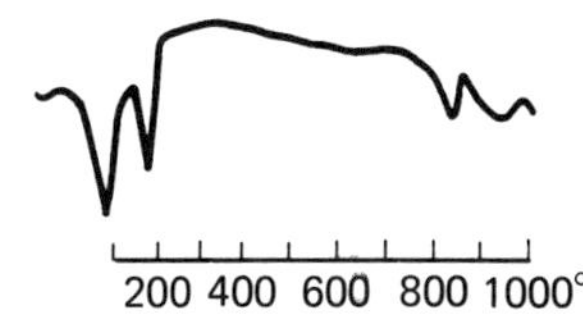

Vermiculite Montana, U.S.A.

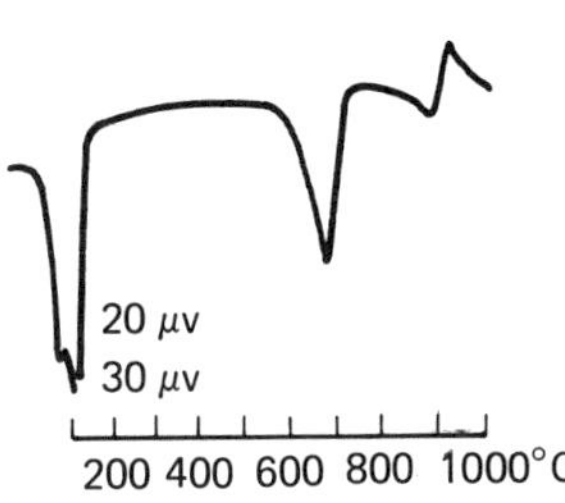

Montmorillonite (Bentonite) Upton, Wyoming, U.S.A.

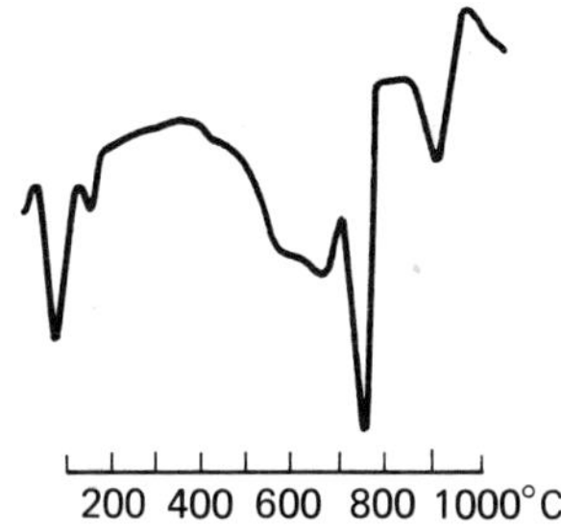

Metabentonite Tazewell, Virginia, U.S.A.

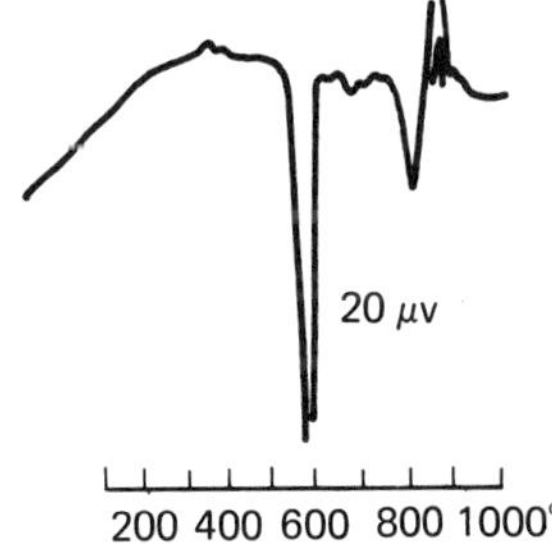

Clinochlore - 14 Å (Tem Press Synthetic)

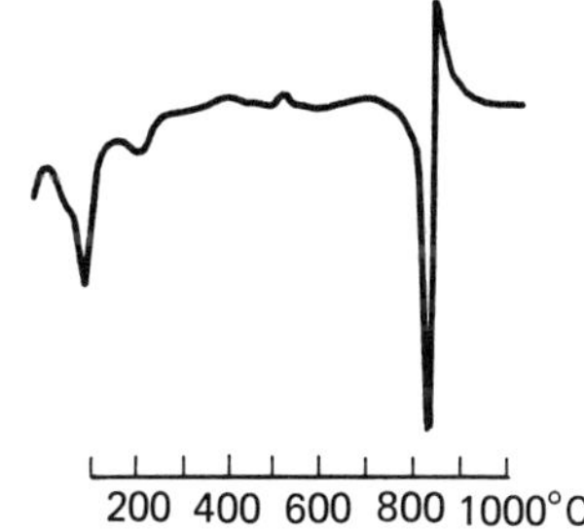

Clinochlore - 7 Å (Tem Press Synthetic)

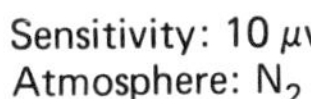

Fig. 45. Thermograms of some principal clay minerals. From Gillott, J. E., *Clay in Engineering Geology*. Amsterdam: Elsevier Publishing Company, 1968 [34].

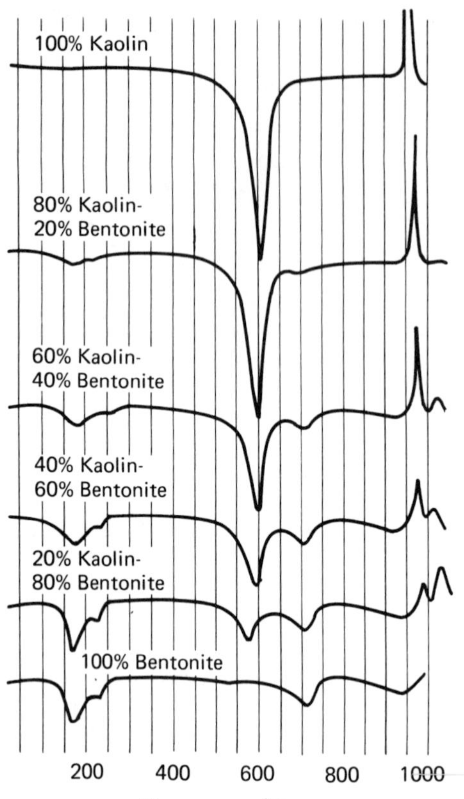

Fig. 46. Thermograms of kaolin-bentonite mixtures. From Spiel, S., *Tech. Pap. Bur. Min.* Washington, D. C.: No. 664 (1945) 1, [39].

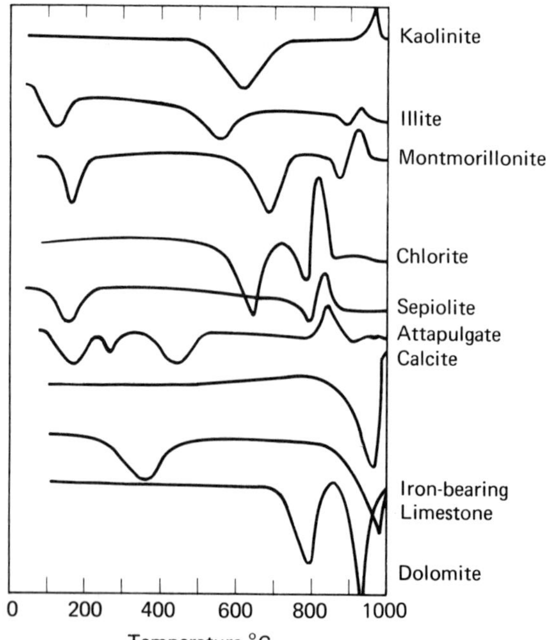

Fig. 47. Thermograms of some raw materials in portland cement manufacture. From Ramachandran, V. S., R. F. Feldman, and P. J. Sereda, *Hwy. Res. Bd. Record*, No. 62, (1964) 40 [57].

Fig. 48. Thermograms of CaO-Al_2O_3 mixtures. From Barta, R., *Differential Thermal Analysis*. Chemie, 1958, p. 257 [55].

CaO Al2O3
% %
19 81
25 75
30 70
36 64
42 58
48 52
55 45
62 38
68 32
1000
1200
1400
Temperature °C

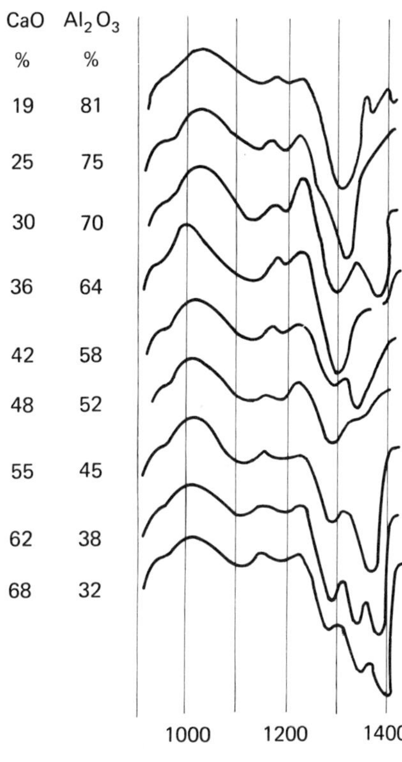

CaO Al2O3 Fe2O3
% % %
25 40 35
35 35 30
45 22 35
1000
1200
1400
Temperature °C

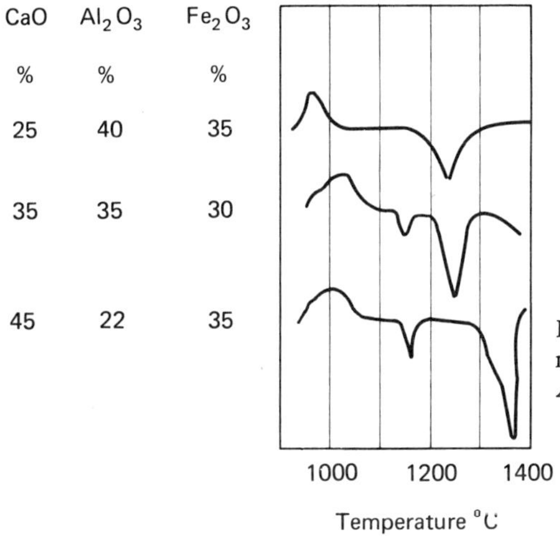

Fig. 49. Thermograms of CaO-Al_2O_3-Fe_2O_3 mixtures. From Barta, R., *Differential Thermal Analysis*. Chemie, 1958, p. 257 [55].

Unhydrated
5 minutes
1 hour
4 hours
24 hours
7 days
200
400
600
800
Temperature °C

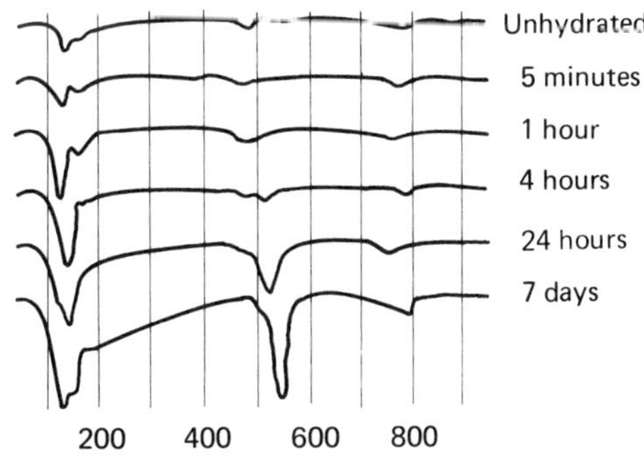

Fig. 50. Thermograms of portland cement hydrated for various lengths of time.

due to stepwise dehydration of gypsum. The endothermal dent below 500°C is attributed to $Ca(OH)_2$ formed during exposure to air. The broad endothermic effect in the range 700 to 800°C is caused by the decomposition of $CaCO_3$, also formed by exposure to air. Five minutes after hydration, an endothermic peak appears at 130°C due to the formation of high-sulfate calcium sulfoaluminate, $3\,CaO \cdot Al_2O_3 \cdot 3\,CaSO_4 . 31\,H_2O$. The decrease in gypsum content is evident from the reduction of the gypsum peaks. An hour after hydration, the intensities of the gypsum peaks decrease further and the peak due to sulfoaluminate becomes more pronounced. At 4 hr, an additional endothermic peak above 500°C is observed. The peak below 500°C is believed to be due to chemisorbed water on the surface of free lime particles, and that above 500°C to the more coarsely crystalline $Ca(OH)_2$ formed by crystallization through solution. After 24 hr of hydration, the double peak due to gypsum disappears and a small endothermic shoulder appears on the low temperature flank of the sulfoaluminate peak due to calcium silicate hydrate. After 7 days this endothermic peak increases. The appearance of low-sulfate calcium sulfoaluminate, $3\,CaO \cdot Al_2O_3 \cdot CaSO_4 \cdot 12\,H_2O$, or to a solid solution of this compound with tetracalcium aluminate hydrate.

References 58 and 59 describe additional applications of DTA to construction materials.

PROBLEMS

1. A Cu K_∞ (1.54-A) x-ray beam strikes a simple cubic crystal having a lattice parameter of 2.75 A. What angle must the beam make with the 111 planes for diffraction to occur?
2. Compute the diffraction angles θ for the following planes with Cu K_α radiation:
   (a) The 211 plane in a cubic crystal having $a = 3.5$ A.
   (b) The 011 plane in a tetragonal crystal having $a = 8.6$ and $c = 6.1$ A.
   (c) The 103 plane in a hexagonal crystal having $a = 2.3$ and $c = 3.6$ A.
3. The unit cell of a certain tetragonal crystal has four atoms of the same kind at $0\frac{1}{2}\frac{1}{4}$, $0\frac{1}{2}\frac{3}{4}$, $\frac{1}{2}0\frac{1}{4}$, $\frac{1}{2}0\frac{3}{4}$. What are the values of F^2 for the 002, 001, 100, and 111 planes?

*4. A Debye-Scherrer pattern of tungsten (BCC) is made with Cu K_∞ radiation. The first four lines have the following θ values and corresponding atomic scattering factors:

θ	f_w
20.3°	58.1
29.2	51.6
36.7	46.8
43.6	43.4

Determine the Miller indices of these lines and calculate their relative integrated intensities.

*5. The powder pattern of aluminum, made with Cu K_α radiation, contains eight lines whose $\sin^2\theta$ values are 0.1118, 0.1487, 0.294, 0.403, 0.439, 0.583, 0.691, and 0.727. Index these lines and calculate the lattice parameter. Try a cubic structure first.

---

*From *Elements of X-Ray Diffraction*, by B. D. Cullity, 1956, by permission of Addison-Wesley Publishing Company, Inc., Reading, Mass.

6. The powder pattern, using Cu K_α radiation ($\lambda = 1.54$ A), of a body-centered tetragonal material contains the following $\sin^2 \theta$ values:

0.0292	0.0933
0.0320	0.115
0.0520	0.128
0.0638	0.135
0.0849	0.156

Using the Hull-Davey chart (Fig. 32), determine the Miller indices of the indicated lines and the cell parameters a and c.

REFERENCES

1. MUIR, I. D., "Microscopy: Transmitted Light," in *Physical Methods in Determinative Mineralogy*, J. Zussman, ed. New York: Academic Press, Inc., 1967, Chap. 2.
2. MIDGLEY, H. G., and H. F. TAYLOR, "Optical Microscopy," in *The Chemistry of Cements*, Vol. II, H. W. Taylor, ed. New York: Academic Press, Inc., 1964, Chap. 20.
3. HARTSHORNE, N. H., and A. STUART, *Practical Optical Crystallography*. London: Edward Arnold, & Co., 1964.
4. BLOSS, F. D., *An Introduction to the Methods of Optical Crystallography*. New York: Holt, Reinhart and Winston, Inc., 1961.
5. WAHLSTROM, E. E., *Optical Crystallography*, 3rd ed. New York: John Wiley & Sons, Inc., 1962.
6. MURR, L. E., *Electron Optical Applications in Materials Science*. New York: McGraw-Hill Book Company, 1970.
7. GARD, J. A., "Electron Microscopy and Diffraction," in *The Chemistry of Cements*, Vol. II, H. F. Taylor, ed. New York: Academic Press, Inc., 1964, Chap. 21.
8. CASTAING, R., *Proceedings of the 3rd International Conference on Electron Microscopy*, London. London: Royal Microscopical Society, 1954, p. 300.
9. LONG, J. V., and J. D. MCCONNELL, *Mineral. Mag.*, **32** (1959) 117.
10. CASTAING, R., *Adv. Electron. Phys.*, **13** (1961) 317.
11. HEINRICH, K. F., *Bibliography on Eelctron Probe X-Ray Analysis and Related Subjects*. New York: John Wiley & Sons, Inc., 1965.
12. WITTRY, D. B., *ASTM Spec. Tech. Publ.*, **349** (1964) 128.
13. CULLITY, B. D., *Elements of X-Ray Diffraction*. Reading, Mass.: Addison-Wesley Publishing Company, Inc., 1956.
14. GILLOTT, J. E., *Clay in Engineering Geology*. Amsterdam: Elsevier Publishing Company, 1968.
15. OATLEY, C. W., W. C. NIXON, and R. F. PEASE, *Adv. Electron. Phys.*, **21** (1965) 181.
16. HAINE, M. E., and V. E. COSSLETT, *The Electron Microscope, the Present State of the Art*. New York: Interscience Publishers, 1961.
17. MASER, M., R. V. RICE, and H. P. KLUG, *Am. Mineralogist*, **45** (1960) 680.
18. DRUMMOND, D. G., ed., "The Practice of Electron Microscopy," *J. Royal Microscop. Soc.*, **70** (1950), 158 pp.
19. FISCHER, R. B., *Applied Electron Microscopy*. Bloomington: Indiana University Press, 1953.
20. HALL, C. E., *Introduction to Electron Microscopy*. New York: McGraw-Hill Book Company, 1953.

21. KAY, D., ed., *Techniques for Electron Microscopy*. 2nd ed. Oxford: Blackwell Scientific Publications Ltd., 1965.

22. BRYDON, J. E., H. M. RICE, and G. C. SCOTT, *Can. J. Soil Sci.*, **43** (1963) 404.

23. COMER, J. J., and J. W. TURLEY, *J. Appl. Phys.*, **26** (1955) 346.

24. BRADLEY, D. E., *Brit. J. Appl. Phys.*, **5** (1954) 65.

25. BRADLEY, D. E., *Brit. J. Appl. Phys.*, **10** (1959) 198.

26. SEARS, F. W., *Optics*. 3rd ed. Reading, Mass.: Addison-Wesley Publishing Company, Inc., 1949.

27. DANIELS, F., AND A. ALBERTY, *Physical Chemistry*, 3rd ed. New York: John Wiley & Sons, Inc., 1966.

28. BRAGG, W. L., *The Crystalline State*, Vol. I: A General Survey. London: G. Bell & Sons, Ltd., 1949.

29. PAULING, L., *The Nature of the Chemical Bond*. 3rd ed. Ithaca, N.Y.: Cornell University Press, 1960.

30. LIPSON, H., and W. COCHRAN, *The Crystalline State*, Vol. III: The Determination of Crystal Structures. London: G. Bell & Sons, Ltd., 1953.

31. *The X-Ray Powder Data File*. Philadelphia: American Society of Testing Materials, 1968.

32. KITTSLEY, S. L., *Physical Chemistry*. New York: Barnes & Noble, Inc., College Outline Series, 1969.

33. WHETSEL, K. B., *Chem. Eng. News*, **46** (Feb. 5, 1968) 82.

34. GILLOTT, J. E., *Clay in Engineering Geology*. Amsterdam: Elsevier Publishing Company, 1968.

35. GRIM, R. E., *Clay Mineralogy*. 2nd ed. New York: McGraw-Hill Book Company, 1968.

36. FARMER, V. C., "Infra-red Spectroscopy of Silicates and Related Compounds," in *The Chemistry of Cements*, Vol. II, H. F. Taylor, ed. Academic Press, Inc., 1964, Chap. 23.

37. ADLER, H., E. E. BRAY, N. P. STEVENS, J. M. HUNT, W. D. KELLER, E. E. PICKETT, and P. F. KERR, *Report 8*, *Am. Petrol. Inst. Proj. 49*. New York: Columbia University, 1950.

38. MACKENZIE, R. C., "Differential Thermal Analysis," in *The Chemistry of Cements*, Vol. II, H. F. Taylor, ed. New York: Academic Press, Inc., 1964, Chap. 22.

39. SPEIL, S., *Tech. Pap. Bur. Min.*, *Washington*, No. 664 (1945) 1.

40. YAGFAROV, M. S., and L. G. BERG, *Izv. Kazan. Fil. Akad. Nauk SSSR*, *Ser. Khim.*, 31, No. 3 (1958), p. 211.

41. MACKENZIE, R. C., and B. D. MITCHELL, *Analyst*, 87 (1962) 420.

42. SEWELL, E. C., and D. B. HONEYBORNE, Theory and Quantitative Use in *The Differential Thermal Investigation of Clays*, R. C. Mackenzie, ed. London: Mineralogical Society, 1957, Chap. 3.

43. MACKENZIE, R. C., and B. D. MITCHELL, Apparatus and Technique for Differential Thermal Analysis in *The Differential Thermal Investigation of Clays*, R. C. Mackenzie, ed. London: Mineralogical Society, 1957, Chap. 2.

44. MITCHELL, B. D., and R. C. MACKENZIE, *Clay Mineral Bull.*, **4** (1959) 31.

45. WEBB, T. L., *Nature*, London, **174** (1954) 686.

46. MACKENZIE, R. C., ed., *The Differential Thermal Investigation of Clays*. London: Mineralogical Society, 1957.

47. WEBB, T. L., D.Sc. thesis. Pretoria, 1958.

48. PAPAILHAU, J., *Bull. Soc. Franc. Mineral.*, **82** (1959) 367.

49. LODDING, W., and L. HAMMELL, *Analyst. Chem.*, **32** (1960) 657.

50. AYRES, W. M., and E. M. BENS, *Analyst. Chem.*, **33** (1961) 568.

51. MACKENZIE, R. C., *The Scifax Differential Thermal Analysis Data Index*. London: Cleaver-Hume Press Ltd., 1962.

52. WAHL, F. M., and R. E. GRIM, *Proc. Natl. Conf. Clay Mineral.*, **12** (1963) 69.

53. JEFFERY, J. W., *Proc. 3rd Intl. Symp. Chem. Cements*, London (1952) 30.

54. NURSE, R. W., *Proc. 3rd Intl. Symp. Chem. Cements*, London (1952) 56.

55. BARTA, R., *Differential Thermal Analysis*. Chemie, 1958, p. 257.

56. GREENE, K. T., *Proc. 4th Intl. Symp. Chem. Cement*, Washington, D. C. (1960) 359.

57. RAMACHANDRAN, V. S., R. F. FELDMAN, and P. J. SEREDA, *Hwy Res. Bd. Record*, No. 62 (1964) 40.

58. RAMACHANDRAN, V. S., *Applications of Differential Thermal Analysis in Cement Chemistry*. New York: Chemical Publishing Company, Inc., 1968.

59. MACKENZIE, R. C., ed., *Differential Thermal Analysis*. New York: Academic Press, Inc., 1970.

APPENDICES

APPENDIX A

Properties of Elements

TABLE 1. Periodic Table of the Elements

IA	IIA	IIIB	IVB	VB	VIB	VIIB	VIII	VIII	VIII	IB	IIB	IIIA	IVA	VA	VIA	VIIA	
								1 H 1.0080									2 He 4.003
3 Li 6.940	4 Be 9.013											5 B 10.82	6 C 12.010	7 N 14.008	8 O 16.0000	9 F 19.00	10 Ne 20.183
11 Na 22.997	12 Mg 24.32											13 Al 26.98	14 Si 28.09	15 P 30.975	16 S 32.066	17 Cl 35.457	18 A 39.944
19 K 39.100	20 Ca 40.08	21 Sc 44.96	22 Ti 47.90	23 V 50.95	24 Cr 52.01	25 Mn 54.93	26 Fe 55.85	27 Co 58.94	28 Ni 58.69	29 Cu 63.54	30 Zn 65.38	31 Ga 69.72	32 Ge 72.60	33 As 74.91	34 Se 78.96	35 Br 79.916	36 Kr 83.8
37 Rb 85.48	38 Sr 87.63	39 Y 88.92	40 Zr 91.22	41 Nb 92.91	42 Mo 95.95	43 Tc (99)	44 Ru 101.7	45 Rh 102.91	46 Pd 106.7	47 Ag 107.880	48 Cd 112.41	49 In 114.76	50 Sn 118.70	51 Sb 121.76	52 Te 127.61	53 I 126.91	54 Xe 131.3
55 Cs 132.91	56 Ba 137.36	57–71 Rare Earths	72 Hf 178.6	73 Ta 180.88	74 W 183.92	75 Re 186.31	76 Os 190.2	77 Ir 193.1	78 Pt 195.23	79 Au 197.2	80 Hg 200.61	81 Tl 204.39	82 Pb 207.21	83 Bi 209.00	84 Po (210)	85 At (210)	86 Rn 222
87 Fr (223)	88 Ra 226.05	89– Acti- nides															

57 La 138.92	58 Ce 140.13	59 Pr 140.92	60 Nd 144.27	61 Pm (145)	62 Sm 150.43	63 Eu 152.0	64 Gd 156.9	65 Tb 159.2	66 Dy 162.46	67 Ho 164.94	68 Er 167.2	69 Tm 169.4	70 Yb 173.04	71 Lu 174.99	Rare earths (Lanthanide series)
89 Ac (227)	90 Th 232.12	91 Pa 231	92 U 238.07	93 Np (237)	94 Pu (242)	95 Am (243)	96 Cm (245)	97 Bk (249)	98 Cf (249)	99 E (253)	100 Fm (254)	101 Mv (256)			Actinide series

SOURCE: ARMSTRONG, R. L. and J. D. KING, *The Electromagnetic Interaction*, Englewood Cliffs, N. J.: Prentice-Hall, Inc., 1973, p. 469.

TABLE 2. Atomic Numbers and Weights

Element	Symbol	Atomic Number	Atomic Weight	Element	Symbol	Atomic Number	Atomic Weight
Actinium	Ac	89	227	Mercury	Hg	80	200.61
Aluminum	Al	13	26.98	Molybdenum	Mo	42	95.95
Americium	Am	95	[243]	Neodymium	Nd	60	144.27
Antimony	Sb	51	121.76	Neon	Ne	10	20.183
Argon	Ar	18	39.944	Neptunium	Np	93	[237]
Arsenic	As	33	74.91	Nickel	Ni	28	58.71
Astatine	At	85	[210]	Niobium	Nb	41	92.91
Barium	Ba	56	137.36	Nitrogen	N	7	14.008
Berkelium	Bk	97	[249]	Osmium	Os	76	190.2
Beryllium	Be	4	9.013	Oxygen	O	8	16
Bismuth	Bi	83	209.00	Palladium	Pd	46	106.4
Boron	B	5	10.82	Phosphorus	P	15	30.975
Bromine	Br	35	79.916	Platinum	Pt	78	195.09
Cadmium	Cd	48	112.41	Plutonium	Pu	94	[242]
Calcium	Ca	20	40.08	Polonium	Po	84	210
Californium	Cf	98	[249]	Potassium	K	19	39.100
Carbon	C	6	12.011	Praseodymium	Pr	59	140.92
Cerium	Ce	58	140.13	Promethium	Pm	61	[145]
Cesium	Cs	55	132.91	Protactinium	Pa	91	231
Chlorine	Cl	17	35.457	Radium	Ra	88	226.05
Chromium	Cr	24	52.01	Radon	Rn	86	222
Cobalt	Co	27	58.94	Rhenium	Re	75	186.22
Columbium (see Niobium)				Rhodium	Rh	45	102.91
				Rubidium	Rb	37	85.48
Copper	Cu	29	63.54	Ruthenium	Ru	44	101.1
Curium	Cm	96	[245]	Samarium	Sm	62	150.35
Dysprosium	Dy	66	162.46	Scandium	Sc	21	44.96
Erbium	Er	68	167.2	Selenium	Se	34	78.96
Europium	Eu	63	152.0	Silicon	Si	14	28.09
Fluorine	F	9	19.00	Silver	Ag	47	107.880
Francium	Fr	87	[223]	Sodium	Na	11	22.991
Gadolinium	Gd	64	157.26	Strontium	Sr	38	87.63
Gallium	Ga	31	69.72	Sulfur	S	16	32.066
Germanium	Ge	32	72.60	Tantalum	Ta	73	180.95
Gold	Au	79	197.0	Technetium	Tc	43	[99]
Hafnium	Hf	72	178.50	Tellurium	Te	52	127.61
Helium	He	2	4.003	Terbium	Tb	65	158.93
Holmium	Ho	67	164.94	Thallium	Tl	81	204.39
Hydrogen	H	1	1.0080	Thorium	Th	90	232.05
Indium	In	49	114.82	Thulium	Tm	69	168.94
Iodine	I	53	126.91	Tin	Sn	50	118.70
Iridium	Ir	77	192.2	Titanium	Ti	22	47.90
Iron	Fe	26	55.85	Tungsten	W	74	183.86
Krypton	Kr	36	83.80	Uranium	U	92	238.07
Lanthanum	La	57	138.92	Vanadium	V	23	50.95
Lead	Pb	82	207.21	Xenon	Xe	54	131.30
Lithium	Li	3	6.940	Ytterbium	Yb	70	173.04
Lutetium	Lu	71	174.99	Yttrium	Y	39	88.92
Magnesium	Mg	12	24.32	Zinc	Zn	30	65.38
Manganese	Mn	25	54.94	Zirconium	Zr	40	91.22
Mendelevium	Mv	101	[256]				

A value given in brackets denotes the mass number of the most stable known isotope.

SOURCE: SORUM, C. H., *How to Solve General Chemistry Problems*, 2nd ed., Englewood Cliffs, N. J.: Prentice-Hall, Inc., 1958.

TABLE 3. Electron Configurations of Atoms in Their Normal States*

Atomic Number	Element	K	L		M			N				O				P			Q
		1s	2s	2p	3s	3p	3d	4s	4p	4d	4f	5s	5p	5d	5f	6s	6p	6d	7s
1	H	1																	
2	He	2																	
3	Li	2	1																
4	Be	2	2																
5	B	2	2	1															
6	C	2	2	2															
7	N	2	2	3															
8	O	2	2	4															
9	F	2	2	5															
10	Ne	2	2	6															
11	Na				1														
12	Mg				2														
13	Al				2	1													
14	Si		10		2	2													
15	P		Neon core		2	3													
16	S				2	4													
17	Cl				2	5													
18	A				2	6													
19	K							1											
20	Ca							2											
21	Sc						1	2											
22	Ti						2	2											
23	V						3	2											
24	Cr						5	1											
25	Mn						5	2											
26	Fe						6	2											
27	Co			18			7	2											
28	Ni			Argon core			8	2											
29	Cu						10	1											
30	Zn						10	2											
31	Ga						10	2	1										
32	Ge						10	2	2										
33	As						10	2	3										
34	Se						10	2	4										
35	Br						10	2	5										
36	Kr						10	2	6										
37	Rb											1							
38	Sr											2							
39	Y									1		2							
40	Zr									2		2							
41	Cb									4		1							
42	Mo									5		1							
43	Tc					36				5		2							
44	Ru					Krypton core				7		1							
45	Rh									8		1							
46	Pd									10									
47	Ag									10		1							
48	Cd									10		2							
49	In									10		2	1						
50	Sn									10		2	2						

*From F. Daniels-A. Alberty, *Physical Chemistry*. New York: John Wiley & Sons, Inc., 1955.

APPENDIX B

X-ray Diffraction Data

The value of d, the distance between adjacent planes in the set (hkl), may be found from Table 4.

TABLE 4. Plane Spacings

Cubic:	$\frac{1}{d^2} = \frac{h^2 + k^2 + l^2}{a^2}$
Tetragonal:	$\frac{1}{d^2} = \frac{h^2 + k^2}{a^2} + \frac{l^2}{c^2}$
Hexagonal:	$\frac{1}{d^2} = \frac{4}{3}\left(\frac{h^2 + hk + k^2}{a^2}\right) + \frac{l^2}{c^2}$
Rhombohedral:	$\frac{1}{d^2} = \frac{(h^2 + k^2 + l^2)\sin^2\alpha + 2(hk + kl + hl)(\cos^2\alpha - \cos\alpha)}{a^2(1 - 3\cos^2\alpha + 2\cos^3\alpha)}$
Orthorhombic:	$\frac{1}{d^2} = \frac{h^2}{a^2} + \frac{k^2}{b^2} + \frac{l^2}{c^2}$
Monoclinic:	$\frac{1}{d^2} = \frac{1}{\sin^2\beta}\left(\frac{h^2}{a^2} + \frac{k^2\sin^2\beta}{b^2} + \frac{l^2}{c^2} - \frac{2hl\cos\beta}{ac}\right)$
Triclinic:	$\frac{1}{d^2} = \frac{1}{V^2}(S_{11}h^2 + S_{22}k^2 + S_{33}l^2 + 2S_{12}hk + 2S_{23}kl + 2S_{13}hl)$*

*In the equation for triclinic crystals, V = volume of unit cell, $S_{11} = b^2c^2\sin^2\alpha$, $S_{22} = a^2c^2\sin^2\beta$, $S_{33} = a^2b^2\sin^2\gamma$, $S_{12} = abc^2(\cos\alpha\cos\beta - \cos\gamma)$, $S_{23} = a^2bc(\cos\beta\cos\gamma - \cos\alpha)$, and $S_{13} = ab^2c(\cos\gamma\cos\alpha - \cos\beta)$.

The unit cell volumes are given by the equations in Table 5.

TABLE 5. Unit Cell Volumes

Cubic:	$V = a^3$
Tetragonal:	$V = a^2c$
Hexagonal:	$V = \frac{\sqrt{3}\,a^2c}{2} = 0.866a^2c$
Rhombohedral:	$V = a^3\sqrt{1 - 3\cos^2\alpha + 2\cos^3\alpha}$
Orthorhombic:	$V = abc$
Monoclinic:	$V = abc\sin\beta$
Triclinic:	$V = abc\sqrt{1 - \cos^2\alpha - \cos^2\beta - \cos^2\gamma + 2\cos\alpha\cos\beta\cos\gamma}$

TABLE 6. Quadratic Forms of Miller Indices

Cubic					Hexagonal	
	hkl					
$h^2 + k^2 + l^2$	*Simple*	*Face-centered*	*Body-centered*	*Diamond*	$h^2 + hk + k^2$	*hk*
1	100				1	10
2	110	...	110		2	
3	111	111	...	111	3	11
4	200	200	200		4	20
5	210				5	
6	211	...	211		6	
7					7	21
8	220	220	220	220	8	
9	300, 221				9	30
10	310	...	310		10	
11	311	311	...	311	11	
12	222	222	222		12	22
13	320				13	31
14	321	...	321		14	
15					15	
16	400	400	400	400	16	40
17	410, 322				17	
18	411, 330	...	411, 330		18	
19	331	331	...	331	19	32
20	420	420	420		20	
21	421				21	41
22	332	...	332		22	
23					23	
24	422	422	422	422	24	
25	500, 430				25	50
26	510, 431	...	510, 431		26	
27	511, 333	511, 333	...	511, 333	27	33
28					28	42
29	520, 432				29	
30	521	...	521		30	
31					31	51
32	440	440	440	440	32	
33	522, 441				33	
34	530, 433	...	530, 433		34	
35	531	531	...	531	35	
36	600, 442	600, 442	600, 442		36	60
37	610				37	43
38	611, 532	...	611, 532		38	
39					39	52
40	620	620	620	620	40	
41	621, 540, 443				41	
42	541	...	541		42	
43	533	533	...	533	43	61
44	622	622	622		44	
45	630, 542				45	
46	631	...	631		46	
47					47	
48	444	444	444	444	48	44
49	700, 632				49	70, 53

TABLE 7. Multiplicity Factors for Powder Photographs

Cubic:	$\frac{hkl}{48^*}$	$\frac{hhl}{24}$	$\frac{0kl}{24^*}$	$\frac{0kk}{12}$	$\frac{hhh}{8}$	$\frac{00l}{6}$	
Hexagonal and rhombohedral:	$\frac{hk\cdot l}{24^*}$	$\frac{hh\cdot l}{12^*}$	$\frac{0k\cdot l}{12^*}$	$\frac{hk\cdot 0}{12^*}$	$\frac{hh\cdot 0}{6}$	$\frac{0k\cdot 0}{6}$	$\frac{00\cdot l}{2}$
Tetragonal:	$\frac{hkl}{16^*}$	$\frac{hhl}{8}$	$\frac{0kl}{8}$	$\frac{hk0}{8^*}$	$\frac{hh0}{4}$	$\frac{0k0}{4}$	$\frac{00l}{2}$
Orthorhombic:	$\frac{hkl}{8}$	$\frac{0kl}{4}$	$\frac{h0l}{4}$	$\frac{hk0}{4}$	$\frac{h00}{2}$	$\frac{0k0}{2}$	$\frac{00l}{2}$
Monoclinic:	$\frac{hkl}{4}$	$\frac{h0l}{2}$	$\frac{0k0}{2}$				
Triclinic:	$\frac{hkl}{2}$						

The designation *hhl* means that two Miller indices are identical, *hhh* that three are idential, etc. Multiplicity factors are for all permutations of the indicated Miller indices. For example, in cubic, the factor 12 applies to the planes 011, 101, 110, 022, 202, 220, etc.

*These are the usual multiplicity factors. In some crystals, planes having these indices comprise two forms with the same spacing but a different structure factor, and the multiplicity factor for each form is half the value given above. In some cubic crystals, for example, the (123) plane belongs to one form and has a certain structure factor, while the (321) plane belongs to another form and has a different structure factor. There are 48/2=24 planes in the first form and 24 planes in the second. (See Cullity, B. D., *Elements of X-Ray Diffraction*. Reading, Mass.: Addison-Wesley Publishing Company, Inc., 1956.)

APPENDIX C

Matrices and Tensors

Matrix and tensor operations provide a systematic shorthand for numerous problems in mathematics. The following brief summary is useful, particularly for applications in laminar composites (Chap. 12).

a. MATRIX ALBEGRA

A matrix is a rectangular array of i rows and j columns of quantities such as

$$\mathbf{A} = \mathbf{A}_{ij} = \begin{bmatrix} A_{11} & A_{12} & A_{13} \\ A_{21} & A_{22} & A_{23} \end{bmatrix}, \qquad i = 1, 2;\ j = 1, 2, 3$$

A row matrix can be expressed as

$$\mathbf{A}_i = [A_{i1} \quad A_{i2} \quad A_{i3}], \qquad j = 1, 2, 3$$

and a column matrix as

$$\mathbf{A}_j = \begin{bmatrix} A_{1j} \\ A_{2j} \end{bmatrix}, \qquad i = 1, 2$$

Row and column matrices are called vectors. For example, a force vector can be expressed as

$$\mathbf{F} = [F_x \quad F_y \quad F_z]$$

Two matrices can be *added* or *subtracted* only if they have the same number of columns and rows. The sum is obtained by adding corresponding ij elements in each matrix, and the difference is obtained by subtracting corresponding elements; i.e.,

$$\begin{bmatrix} 1 & 5 \\ 4 & 2 \\ 3 & 1 \end{bmatrix} + \begin{bmatrix} 3 & 2 \\ 1 & 4 \\ 1 & 6 \end{bmatrix} = \begin{bmatrix} 4 & 7 \\ 5 & 6 \\ 4 & 7 \end{bmatrix}$$

and

$$\begin{bmatrix} 2 & 5 \\ 4 & 2 \\ 3 & 1 \end{bmatrix} - \begin{bmatrix} 3 & 2 \\ 1 & 4 \\ 1 & 6 \end{bmatrix} = \begin{bmatrix} -1 & 3 \\ 3 & -2 \\ 2 & -5 \end{bmatrix}$$

The *product* of a scalar s and a matrix $\mathbf{A}$ is obtained by multiplying each element of $\mathbf{A}$ by the scalar; i.e., if

$$\mathbf{A} = \begin{bmatrix} A_{11} & A_{12} \\ A_{21} & A_{22} \end{bmatrix}$$

then

$$s\mathbf{A} = \begin{bmatrix} sA_{11} & sA_{12} \\ sA_{21} & sA_{22} \end{bmatrix}$$

The product of two matrices $\mathbf{A}$ and $\mathbf{B}$ is written $\mathbf{AB} = \mathbf{C}$, where the elements of the resulting matrix $\mathbf{C}$ are, by definition,

$$a_{ik}b_{kj} = c_{ij}$$

$\mathbf{A}$ is called the premultiplier and $\mathbf{B}$ the postmultiplier; i.e., $\mathbf{A}$ is postmultiplied by $\mathbf{B}$. Two matrices can be multiplied only if the number of columns in the premultiplier equals the number of rows in the postmultiplier. The resulting matrix will have the same number of rows as the premultiplier and the same number of columns as the postmultiplier. In general, $\mathbf{AB} \neq \mathbf{BA}$. For example,

$$\begin{bmatrix} 1 & 5 \\ 4 & 2 \\ 3 & 1 \end{bmatrix} \times \begin{bmatrix} 3 & 1 & 1 \\ 2 & 4 & 6 \end{bmatrix} = \begin{bmatrix} 1\times3+5\times2 & 1\times1+5\times4 & 1\times1+5\times6 \\ 4\times3+2\times2 & 4\times1+2\times4 & 4\times1+2\times6 \\ 3\times3+1\times2 & 3\times1+1\times4 & 3\times1+1\times6 \end{bmatrix}$$

$$= \begin{bmatrix} 13 & 21 & 31 \\ 16 & 12 & 16 \\ 11 & 7 & 9 \end{bmatrix}$$

but

$$\begin{bmatrix} 3 & 1 & 1 \\ 2 & 4 & 6 \end{bmatrix} \times \begin{bmatrix} 1 & 5 \\ 4 & 2 \\ 3 & 1 \end{bmatrix} = \begin{bmatrix} 3\times1+1\times4+1\times3 & 3\times5+1\times2+1\times1 \\ 2\times1+4\times4+6\times3 & 2\times5+4\times2+6\times1 \end{bmatrix}$$

$$= \begin{bmatrix} 10 & 18 \\ 36 & 24 \end{bmatrix}$$

The *transpose* of a matrix **A**, designated **A**′, is obtained by interchanging the rows and columns of matrix **A**.

A *unit* matrix is a square matrix in which the diagonal elements are unity and the remaining elements are zero; i.e.,

$$\mathbf{I} = \begin{bmatrix} 1 & 0 & 0 \\ 0 & 1 & 0 \\ 0 & 0 & 1 \end{bmatrix}$$

The *cofactor* of an $m \times m$ square matrix is obtained by replacing each element of the matrix by its cofactor. The cofactor of an element is the product of the determinant of the matrix having $m - 1$ rows and $m - 1$ columns, obtained by removing the ith row and jth column, by the term $(-1)^{i+j}$. For example, for the matrix

$$\mathbf{A} = \begin{bmatrix} 2 & 3 & 1 \\ 1 & 5 & 3 \\ 4 & 2 & 1 \end{bmatrix} \tag{1}$$

we obtain

$$\text{Co } A_{11} = \begin{bmatrix} 5 & 3 \\ 2 & 1 \end{bmatrix} (-1)^{(1+1)} = (5 - 6)(-1)^2 = -1$$

$$\text{Co } A_{12} = \begin{bmatrix} 1 & 3 \\ 4 & 1 \end{bmatrix} (-1)^{(1+2)} = (1 - 12)(-1)^2 = 11$$

$$\text{Co } A_{13} = \begin{bmatrix} 1 & 5 \\ 4 & 2 \end{bmatrix} (-1)^{(1+3)} = (2-20)(-1)^4 = -18$$

$$\text{Co } A_{21} = \begin{bmatrix} 3 & 1 \\ 2 & 1 \end{bmatrix} (-2)^{(2+1)} = (3-2)(-1)^3 = 1$$

and the cofactor matrix of **A** is

$$\text{Co } \mathbf{A} = \begin{bmatrix} -1 & 11 & -18 \\ -1 & -2 & 8 \\ 4 & -5 & 7 \end{bmatrix}$$

The transpose, unit matrix, and cofactor are used to obtain the widely useful *inverse* matrix. For certain square matrices **A**, the inverse matric $\mathbf{A}^{-1}$ is defined by

$$\mathbf{AA}^{-1} = \mathbf{I}$$

where **I** is the unit matrix and the inverse matrix $\mathbf{A}^{-1}$ is

$$\mathbf{A}^{-1} = \frac{(\text{Co }\mathbf{A})'}{|A|}$$

e.g., the inverse matrix equals the transpose of the cofactor matrix divided by the determinant. For example, the inverse of the matrix in Eq. (1) is

$$\mathbf{A}^{-1} = \frac{\begin{bmatrix} -1 & -1 & 4 \\ 11 & -2 & -5 \\ -18 & 8 & 7 \end{bmatrix}}{\begin{bmatrix} 2 & 3 & 1 \\ 1 & 5 & 3 \\ 4 & 2 & 1 \end{bmatrix}}$$

Evaluating the determinant,

$$\begin{matrix} 2 & 3 & 1 \\ 1 & 5 & 3 \\ 4 & 2 & 1 \\ 2 & 3 & 1 \\ 1 & 5 & 3 \end{matrix} \qquad \begin{aligned} &(2\times5\times1 + 1\times2\times1 + 4\times3\times3) \\ &\quad - (1\times5\times4 + 3\times2\times2 + 1\times3\times1) = 13 \end{aligned}$$

Therefore,

$$\mathbf{A}^{-1} = \begin{bmatrix} -\frac{1}{13} & -\frac{1}{13} & \frac{4}{13} \\ \frac{11}{13} & -\frac{2}{13} & -\frac{5}{13} \\ -\frac{18}{13} & \frac{8}{13} & \frac{7}{13} \end{bmatrix}$$

It may be verified by matrix multiplication that

$$\mathbf{AA}^{-1} = \begin{bmatrix} 1 & 0 & 0 \\ 0 & 1 & 0 \\ 0 & 0 & 1 \end{bmatrix}$$

b. TENSORS

A tensor is a physical entity which obeys specific laws of transformation. The transformation most used in applied mechanics is the simple rotation of axes transformation. There are different orders of tensors, each having its own transformation

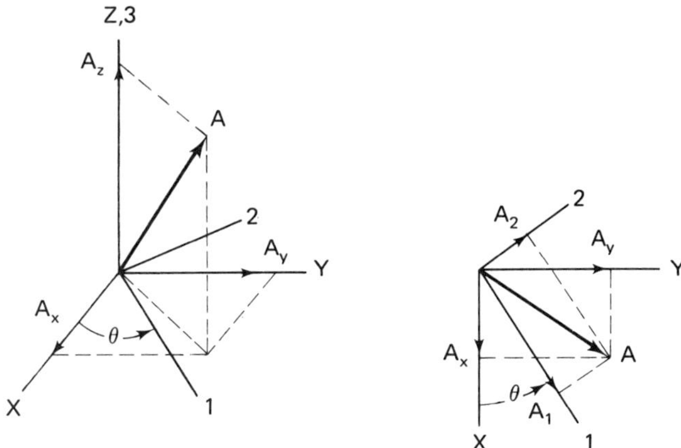

Fig. 1. Coordinate transformation for a vector.

relations. Some familiar tensors are work and density (zero-order tensors, or scalars), force and velocity (first-order tensors, or vectors), and stress and strain (second-order tensors, or just *tensors*). Although tensors are written in the form of matrices, a matrix is not a tensor unless it obeys the tensor transformation laws.

No transformation law is required for a *zero-order tensor* (scalar), because it has magnitude but no direction. In other words, under a rotation of axes a scalar quantity does not change.

The components of a *first-order tensor* (vector) do change under a rotation of axes. For example, if we wish to transform the components of a vector A from the coordinate set (x,y,z) to the set (1,2,3) by rotation about the $(z,3)$ axis (Fig. 1), the following transformation matrix can be used:

$$\mathbf{T} = \begin{bmatrix} \cos & \sin & 0 \\ -\sin & \cos & 0 \\ 0 & 0 & 1 \end{bmatrix} \tag{2}$$

and the transformation would be

$$\begin{bmatrix} A_1 \\ A_2 \\ A_3 \end{bmatrix} = \begin{bmatrix} \cos & \sin & 0 \\ -\sin & \cos & 0 \\ 0 & 0 & 1 \end{bmatrix} \begin{bmatrix} A_x \\ A_y \\ A_z \end{bmatrix}$$

Thus, Eq. (2) is the transformation law for rotating a first-order tensor about the z or 3 axis.

A symmetric* *second-order tensor* is an array of elements,

$$\begin{bmatrix} a_{11} & a_{12} & a_{13} \\ a_{21} & a_{22} & a_{23} \\ a_{31} & a_{32} & a_{33} \end{bmatrix}$$

**Symmetric* refers to the condition $a_{12} = a_{21}$, etc. Since nonsymmetric tensors are less common in engineering, we shall not consider them.

which transforms with a rotation of axes according to the following equations. For the diagonal terms,

$$a_{kk} = l^2 a_{11} + m^2 a_{22} + n^2 a_{33} + 2lma_{12} + 2lna_{13} + 2mna_{23}$$

For the off-diagonal terms,

$$a_{kj} = ll' a_{11} + mm' a_{22} + nn' a_{33} + (lm' + ml')a_{12} + (ln' + nl')a_{13} + (mn' + nm')a_{23}$$

where l, m, n are direction cosines for the k direction and l', m', n' are direction cosines for the j direction.

The graphic form of these transformations in only two dimensions is the familiar Mohr's circle for stress or strain. The transformation relation in two dimensions may be written

$$\mathbf{T} = \begin{bmatrix} \cos^2\theta & \sin^2\theta & 2\sin\theta\cos\theta \\ \sin^2\theta & \cos^2\theta & -2\sin\theta\cos\theta \\ -\sin\theta\cos\theta & \sin\theta\cos\theta & (\cos^2\theta - \sin^2\theta) \end{bmatrix} \tag{3}$$

This transformation would be used to transform plane stress and strain from axes (x,y) to axes (1,2) through rotation θ about the (2,3) axis; i.e.,

$$\begin{bmatrix} \sigma_1 \\ \sigma_2 \\ \tau_{12} \end{bmatrix} = \mathbf{T} \begin{bmatrix} \sigma_x \\ \sigma_y \\ \tau_{xy} \end{bmatrix}$$

and

$$\begin{bmatrix} \epsilon_1 \\ \epsilon_2 \\ \gamma_{12}/2 \end{bmatrix} = \mathbf{T} \begin{bmatrix} \epsilon_x \\ \epsilon_y \\ \gamma_{xy}/2 \end{bmatrix}$$

A three-dimensional stress tensor is usually expressed in the following form:

$$\begin{bmatrix} \sigma_{xx} & \tau_{xy} & \tau_{xz} \\ \tau_{yx} & \sigma_{yy} & \tau_{yz} \\ \tau_{zx} & \tau_{zy} & \sigma_{zz} \end{bmatrix} \tag{4}$$

Notice in this array that the first subscript identifies a row, while the second subscript identifies a column. The normal stresses form the diagonal elements.

Some important properties of the stress tensor are

1. It is symmetric; i.e., shear stresses with reversed indices are equal, so that the array in Eq. (4) is symmetric about the diagonal.
2. The average of the normal stresses forms a scalar field; i.e., the sum of a

set of orthogonal normal stresses at a point is independent of the set of directions chosen. Actually, this is true for any second-order tensor. The sum of diagonal terms is called a second-order tensor inavrient, because it is independent of the orientation of the reference axes. In elasticity, the average normal stress is called the *bulk stress*, and in viscous flow, simply the *pressure*.

Certain material properties, such as the thermal expansion coefficient, α, are also second-order tensors unless the material is isotropic. If it is isotropic, these properties are scalars, the same in all directions, and no transformation is required. For an *orthotropic* material the transformation from the (x,y) to the $(1,2)$ plane is

$$\begin{bmatrix} \alpha_1 \\ \alpha_2 \\ 0 \end{bmatrix} = \mathbf{T} \begin{bmatrix} \alpha_x \\ \alpha_y \\ \alpha_{xy}/2 \end{bmatrix}$$

The Hooke's law stiffnesses of an anisotropic material are components of a *fourth-order tensor*. The transformation relations for a specially orthotropic material (Sec. B2 in Chap. 2) for rotation about the $(z,3)$ axis is

$$\begin{bmatrix} A_{11} \\ A_{22} \\ A_{12} \\ A_{66} \\ A_{16} \\ A_{26} \end{bmatrix} = \begin{bmatrix} m^4 \\ n^4 \\ m^2n^2 \\ m^2n^2 \\ -m^3n \\ -mn^3 \end{bmatrix} \begin{bmatrix} n^4 & 2m^2n^2 & 4m^2n^2 \\ m^4 & 2m^2n^2 & 4m^2n^2 \\ m^2n^2 & (m^4 + n^4) & -4m^2n^2 \\ m^2n^2 & -2m^2n^2 & (m^2 - n^2)^2 \\ mn^3 & (m^3n - mn^3) & 2(m^3n - mn^3) \\ m^3n & (mn^3 - m^3n) & 2(mn^3 - m^3n) \end{bmatrix} \begin{bmatrix} a_{11} \\ a_{22} \\ a_{12} \\ a_{66} \\ 0 \\ 0 \end{bmatrix} \tag{5}$$

where $m = \cos\theta$ and $n = \sin\theta$. Equation (5) relates rotation with respect to the material coordinate system rather than with respect to the x, y coordinates as in Eqs. (2) and (3).

Equations (2), (3), and (5) show the increasing complexity of the transformation relation with increasing order of tensor. At any given point in a tensor field only one value is needed to define a scalar, three values to define a vector, nine to define a second-order tensor, etc. These requirements are independent of the coordinate system (orthogonal, cylindrical, etc.) used.

APPENDIX D

Equilibrium Equations for Plates

Consider an element of a plate having stress resultants acting at the midplane [Fig. 2(a)], moment resultants [Fig. 2(b)], and transverse shear stress resultants [Fig.

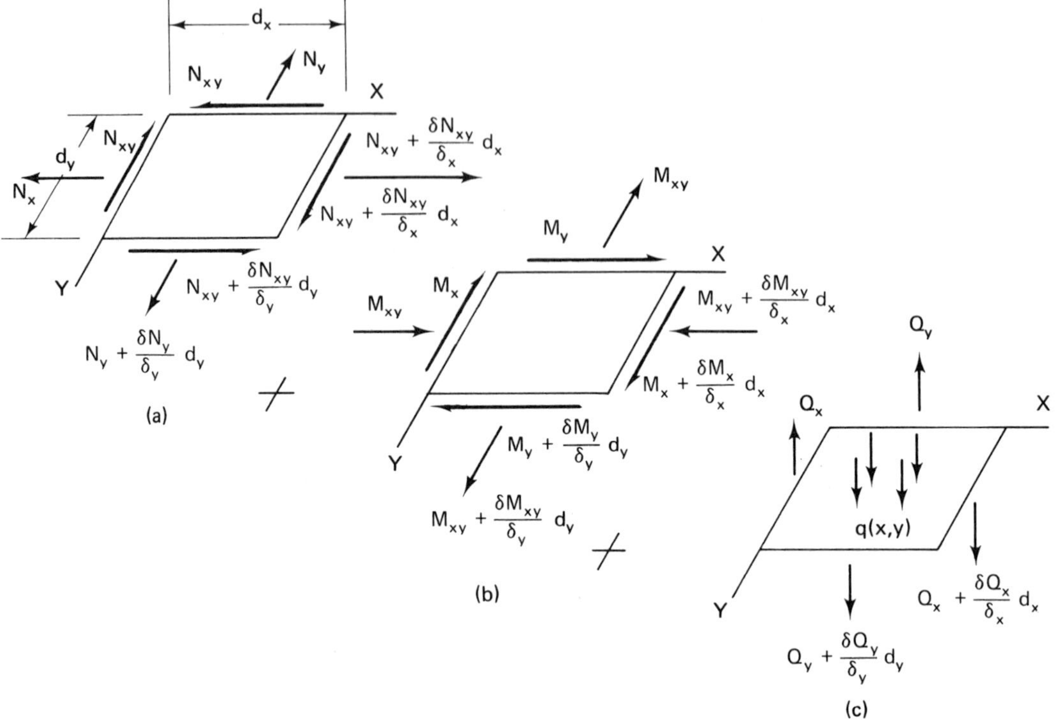

Fig. 2. Midplane stress and moment resultants.

2(c), resulting from the distributed normal load $q(x,y)$. The transverse shear stress resultants, in a manner analogous to the definitions of N_x, N_y, and N_{xy}, are the integrals of the transverse shears across the plate thickness, h:

$$\left.\begin{aligned} Q_x &= \int_{-h/2}^{+h/2} \tau_{xz}\, ds \\ Q_y &= \int_{-h/2}^{+h/2} \tau_{yz}\, dz \end{aligned}\right\} \qquad \text{lb/ft}$$

Equilibrium requires a balance of forces in the x, y, and z directions and a balance of moments about the x and y axes.

Summing x direction forces,

$$N_x\, dy + \frac{\partial N_x}{\partial x} dy + N_{xy}\, dx + \frac{\partial N_{xy}}{\partial y} dx\, dy - N_x\, dy - N_{xy}\, dx = 0$$

or

$$\frac{\partial N_x}{\partial y} + \frac{\partial N_{xy}}{\partial x} = 0 \tag{6}$$

Summing y direction forces,

$$N_y\,dx + \frac{\partial N_y}{\partial y}dy\,dx + N_{xy}\,dy + \frac{\partial N_{xy}}{\partial x}dx\,dy - N_y\,dx - N_{xy}\,dy = 0$$

or

$$\frac{\partial N_y}{\partial y} + \frac{\partial N_{xy}}{\partial x} = 0 \tag{7}$$

Summing z direction forces,

$$Q_x\,dy + \frac{\partial Q_x}{\partial x}dx\,dy + Q_y\,dx + \frac{\partial Q_y}{\partial y}dy\,dx - Q_x\,dy - Q_y\,dx + q(x,y)\,dx\,dy = 0$$

or

$$\frac{\partial Q_x}{\partial x} + \frac{\partial Q_y}{\partial y} + q(x, y) = 0 \tag{8}$$

Summing moments about the x axis,

$$-M_y\,dx - \frac{\partial M_y}{\partial y}\,dy\,dx - M_{xy}\,dy - \frac{\partial M_{xy}}{\partial x}\,dx\,dy + Q_y\,dx\,dy + \frac{\partial Q_y}{\partial y}\,dy\,dx\,dy$$
$$+ q(x,y)\,dx\,dy\,(dy/2) + Q_x\,dy\,(dy/2) + \frac{\partial Q_x}{\partial x}\,dx\,dy\,(dy/2)$$
$$+ M_y\,dx + M_{xy}\,dy - Q_x\,dy\,(dy/2) = 0$$

Dropping the higher-order terms, which vanish for arbitrarily small dx and dy,

$$-\frac{\partial M_y}{\partial y} - \frac{\partial M_{xy}}{\partial x} + Q_y = 0$$

or

$$Q_y = \frac{\partial M_y}{\partial y} + \frac{\partial M_{xy}}{\partial x} \tag{9}$$

Summing moments about the y axis,

$$M_x\,dy + \frac{\partial M_x}{\partial x}\,dx\,dy + M_{xy}\,dx + \frac{\partial M_{xy}}{\partial y}\,dy\,dx - Q_x\,dy\,dx - \frac{\partial Q_x}{\partial x}\,dx\,dy\,dx$$
$$- Q_y\,dx\,(dx/2) - \frac{\partial Q_y}{\partial y}\,dy\,dx\,(dx/2) + q(x,y)\,dx\,dy\,(dx/2) - M_x\,dy$$
$$- M_{xy}\,dx + Q_y\,dx\,(dx/2) = 0$$

and dropping the higher-order terms,

$$\frac{\partial M_x}{\partial x} + \frac{\partial M_{xy}}{\partial y} = Q_x \tag{10}$$

Equations (6)–(10) are the equilibrium equations for a plate element. The last three can be combined to eliminate the transverse shears and yield a single moment equilibrium equation:

$$\frac{\partial}{\partial x}\left(\frac{\partial M_x}{\partial x} + \frac{\partial M_{xy}}{\partial y}\right) + \frac{\partial}{\partial y}\left(\frac{\partial M_y}{\partial y} + \frac{\partial M_{xy}}{\partial x}\right) = -q(x, y)$$

or

$$\frac{\partial^2 M_x}{\partial x^2} + 2\frac{\partial^2 M_{xy}}{\partial x\,\partial y} + \frac{\partial^2 M_y}{\partial y^2} = -q(x, y) \tag{11}$$

APPENDIX E

Standard Tests for Construction Materials

Frequently used American Society for Testing Materials tests are listed alphabetically in groups corresponding to the topics of Chaps. 4–12, followed by a final group of tests suitable for all materials. Many of the test methods are also described by the American Association of State Highway Officials, the Federal Test Method Standard, and the Underwriter Laboratories.

CHAP. 4: INORGANIC CEMENTS

Hydraulic and Portland Cements	*ASTM Designation*
Air Content of Hydraulic Cement Mortar	C185
Alkalinity and Free Alkali in Portland Cement	C150
Autoclave Expansion of Portland Cement	C151
Bleeding of Cement Pastes and Mortars	C243
Blended Hydraulic Cements, Specs. for	C595
Calcium Sulfate in Hydrated Portland Cement Mortar	C265
Chemical Analysis of Hydraulic Cement	C114
Chemical Resistance of Mortars	C267
Compressive Strength of Hydraulic Cement Mortars	C109, C349
Drying Skrinkage of Mortar Containing Portland Cement	C596
False Set of Portland Cement	C359, C451
Fineness of Hydraulic and Portland Cements	C115, C184, C204, C430
Flexural Strength of Hydraulic Cement Mortars	C348
Heat of Hydration of Hydraulic Cement	C186
Natural Cement, Specs. for	C10
Optimum SO_3 in Portland Cement	C563
Portland Cement, Specs. for	C150
Potential Expansion of Portland Cement Mortars Exposed to Sulfate	C452
Tensile Strength of Hydraulic Cement Mortars	C190
Time of Setting of Hydraulic Cement by Vicat Needle	C191

Fly Ash	
Fly Ash and Pozzolans for Use with Lime, Specs. for	C593
Fly Ash and Raw or Calcined Natural Pozzolans for Use in Portland Cement Concrete	C618
Fly Ash for Use as an Admixture in Portland Cement Concrete	C311
Lime	
Hydrated Lime for Structural Purposes, Specs. for	C259
Limestone, Quicklime, and Hydrated Lime, Chemical Analysis	C25
Quicklime and Hydrated Lime for Silica Brick Manufacture, Specs. for	C49

CHAP. 5: ORGANIC CEMENTS

Bituminous Materials	*ASTM Designation*
Accelerated Weathering Tests for Bituminous Materials	D529
Effect of Heat and Air on Asphaltic Materials	D1754
Failure End Point in Accelerated and Outdoor Weathering of Bituminous Materials	D1670
Asphalt Cement	
Asphalt Cements Used in Pavement Construction, Specs. for	D946
Ductility	D113
Flash Point	D92
Penetration	D5
Softening Point	D36
Solubility	D4
Specific Gravity	D70
Thin Film Over Test	D1754
Viscosity, Kinematic	D2170
Rapid and Medium Curing Cutbacks	
Cutback Asphalts, Spec. for	D2028
Distillation	D402
Flash Point	D1310
Specific Gravity	D70
Viscosity, Kinematic	D2170
Water in Asphalt	D95
Slow Curing Cutback	
Asphalt Residue of 100 Penetration	D243
Distillation	D402
Ductility	D113
Flash Point	D92

Float Test	D139
Solubility	D4
Specific Gravity	D70
Viscosity, Kinematic	D2170
Water in Asphalt	D95

Emulsified Asphalt

Cationic Emulsified Asphalt, Specs. for	D2397
Cement Mixing	D244
Coating Test	D244–61T
Demulsibility	D244
Emulsified Asphalt, Specs. for	D977
Films Deposited from Bituminous Emulsions	D466
Oil Distillate	D244
Particle Charge Test	D244
pH Test	E70
Residue from Distillation	D244
Settlement	D244
Sieve Test	D244
Specific Gravity	D70
Viscosity	D244

Synthetic Polymers

Abrasion Wear	D1242
Accelerated Service Tests (temperature and humidity extremes)	D756
Accelerated Weathering Test; Carbon Arc without Filters	G23
Acetone Extraction Test for Degree of Cure of Phenolics	D494
Bearing Strength	D953
Brittleness Temperature of Plastics by Impact	D746
Compressive Properties of Rigid Plastics	D695
Deflection Temperature Under Load	D648
Deformation Under Load	D621
Epoxy Resin, Spec. for	D1763
Flammability of Plastics 0.05 in. and Under in Thickness	D568
Flammability of Plastics Over 0.05 in. in Thickness	D635
Flexural Fatigue	D671
Flexural Properties of Plastics	D790
Flow Temperature Test for Thermo-plastic Molding Materials	D569
Gloss	D523
Indentation Hardness of Plastics by Durometer	D1706
Index of Refraction	D542
Izod Impact Strength	D256
Joint Sealing Compounds, Waterstops and Gaskets	C443, C509, C510, D545, D994, D1190, D1191, D1752

Light Diffusion	E166, E167
Linear Thermal Expansion (fused-quartz tube method)	D696
Luminous Transmittance and Haze of Transparent Plastics	D1003
Mar Resistance	D673
Optical Uniformity and Distortion	D637
PVC Films and Sheets	D1593, D1755, D1927, D2123
Resistance of Plastics to Artificial Weathering, Using Fluorescent Sunlamp and Fog Chamber	D1501
Resistance of Plastics to Chemical Reagents	D543
Rockwell Indentation Hardness	DD785
Shear Strength	D2345
Short-Time Stability at Elevated Temperature of Plastics Containing Chlorine	D793
Specific Gravity by Displacement of Water	D792
Surface Abrasion	D1044
Tear Resistance of Film and Sheeting	D1004
Tensile Properties of Plastics	D638, D2290
Tensile Properties of Thin Plastic Sheets and Films	D882
Tensile Strength of Molded Electrical Insulating Materials	D651
Volatile Loss	D1203
Warpage of Sheet Plastics	D1181
Water Absorption of Plastics	D570
Water Vapor Permeability	E96

CHAP. 6: AGGREGATES

Coarse Aggregates	*ASTM Designation*
Coarse Aggregates for Bituminous Paving Mixtures	D692
Resistance to Abrasion of Coarse Aggregates, by Los Angeles Machine	C131 C535
Sieve Analysis of Fine and Coarse Aggregates	C136
Specific Gravity and Absorption of Coarse Aggregates	C127
Concrete Aggregates	
Concrete Aggregates, Specs. for	C88
Petrographic Examination of Aggregates	C295
Potential Alkali Reactivity of Carbonate Rocks (rock cylinder method)	C586
Potential Reactivity of Aggregates (chemical method)	C289
Soundness of Aggregates by Use of Sodium Sulfate or Magnesium Sulfate	C33
Fine Aggregates	
Effect of Organic Impurities in Fine Aggregate	C40

on Mortar Strength	C87
Fine Aggregate for Bituminous Paving Mixtures	D1073
Organic Impurities in Sands	C40
Specific Gravity and Absorption of Fine Aggregate	C128
Surface Moisture in Fine Aggregate	C70

Lightweight Aggregate

Lightweight Aggregates for Concrete Masonry Units	C331
Lightweight Aggregates for Insulating Concrete	C332
Lightweight Aggregates for Structural Concrete	C330

CHAP. 6: FIBERS

Asbestos Fiber	*ASTM Designation*
Asbestos Roving, Specs. and Tests for	D375
Asbestos Yarns, Specs. and Tests for	D299
Heat of Aging of Asbestos Textiles	D1573
Method of Sampling Asbestos Fiber for Testing	D2590
Glass Fiber	
Acetone Extraction and Ignition of Strands, Yarns, and Roving	D2587
Glass Fiber Strands, Yarns, and Rovings	D2343
Glass Fabrics, Specs. and Tests for	D579
Glass Yarns, Specs. and Tests for	D578
Testing and Tolerances for Woven Glass Tapes	D580
Woven Glass Fabric, Cleaned and After-finished with Amino-saline Type Finishes	D2408
Woven Glass Fabric, Cleaned and After-finished with Chrome Complexes	D2410
Woven Glass Fabric, Cleaned and After-finished with Vinyl-silane Type Finishes	D2409
Woven Roving Glass Fabric for Polyester Glass Laminates	D2510

CHAPS. 7 AND 8: PORTLAND CEMENT CONCRETE

Concrete	*ASTM Designation*
Abrasion Resistance of Concrete	C418
Ball Penetration in Fresh Portland Cement Concrete	C360
Bleeding of Concrete	C232
Cement Content of Hardened Portland Cement Concrete	C85
Compressive Strength of Cylindrical Concrete Specimens	C39
Concrete Made by Volumetric Batching and Continuous	

Mixing, Specs. for	C685
Creep of Concrete in Compression	C512
Flexural Strength of Concrete	C78, C293
Fundamental, Transverse, Longitudinal, and Torsional Frequencies of Concrete Specimens	C215
Length Change of Cement Mortar and Concrete	C157
Making and Accelerated Curing of Concrete Compression Test Specimens	C31
Packaged Dry Combined Materials for Mortar and Concrete, Specs. for	C387
Potential Alkali Reactivity of Cement-Aggregate Combinations	C227
Pulse Velocity through Concrete	C597
Ready-mixed Concrete, Specs. for	C94
Resistance of Concrete to Rapid Freezing and Thawing in Water	C290
Scaling Resistance of Concrete Surfaces Exposed to Deicing Chemicals	C672
Slump of Portland Cement Concrete	C143
Specific Gravity, Absorption, and Voids in Hardened Concrete	C642
Concrete Admixtures	
Air-Entraining Admixtures for Concrete, Specs. and Testing	C233, C260
Chemical Admixtures for Concrete, Specs. for	C494
Effectiveness of Mineral Admixtures in Preventing Excessive Expansion of Concrete Due to the Alkali-Aggregate Reaction	C441
Fly Ash for Use as an Admixture in Portland Cement Concrete	C311
Concrete, Air Content	
Air, Content of Freshly Mixed Concrete	C173, C231
Microscopical Determination of Air Void Content—in Hardened Concrete	C457
Concrete Blocks	
Drying Shrinkage of Concrete Block	C426
Concrete Masonry	
Bond Strength of Mortar to Masonry Units	E149
Hollow Load-Bearing Concrete Masonry Units	C90
Masonry Cement, Specs. for	C91
Solid Load-bearing Concrete Masonry Units	C145
Concrete Pipe	
Concrete Pipe for Irrigation or Drainage, Specs. for	C118
Concrete Sewer, Storm Drain, and Culvert Pipe, Specs. for	C14

Physical Properties of Concrete Pipe or Tile	C497
Porous Concrete Pipe, Specs. for	C654
Reinforced Concrete Low-Head Pressure Pipe, Specs. for	C361

CHAP. 9: BITUMINOUS MIXES

Bituminous Paving Mixtures	*ASTM Designation*
Bituminous Mining Plant Inspection	D290
Bulk Specific Gravity of Compacted Bituminous Mixtures	D1188, D2726
Coating and Resistance to Stripping of Bituminous-Aggregate Mixtures Containing Adhesion Promoting Agents	D2727
Coating and Stripping of Bitumen-Aggregate Mixtures	D1664
Compressive Strength of Bituminous Mixtures	D1074
Crushed Stone, Crushed Slag, and Crushed Gravel for Macadam Base and Surface Courses	D693, D694
Degree of Particle Coating of Bituminous-Aggregate Mixtures	D2489
Density of Bituminous Concrete in Place by Nuclear Method	D2950
Effect of Water on Bituminous Mixtures	D1075
Hot-Mixed, Hot-Laid Asphalt, Emulsified Asphalt, and Tar Paving Mixtures, Specs. for	D1663, D1753, D2629
Mineral Filler for Bituminous Paving Mixtures, Specs. for	D242
Quantitative Extraction of Bitumen for Bituminous Paving Mixtures	D2172
Requirements for Mixing Plants for Hot-Mixed, Hot-Laid Bituminous Paving Mixtures, Specs. for	D995
Resistance to Plastic Flow of Bituminous Mixtures Using Marshall Apparatus	D1559
Resistance to Plastic Flow of Fine-Aggregate Bituminous Mixtures by Means of the Hubbard-Field Apparatus	D1138

CHAP. 10: RIGID FOAMS

Cellular Plastics	*ASTM Designation*
Compressive Strength of Rigid Cellular Plastics	D1621
Rigid Urethane Foam, Specs. for	C591, D1623
Water Absorption of Rigid Cellular Plastics	D2842
Lightweight Concrete	
Compressive Strength of Lightweight Insulating Concrete	C495

CHAP. 11: FIBER-REINFORCED COMPOSITES

Asbestos Cements	*ASTM Designation*
Asbestos-Cement Fiberboard Insulating Planels	C551

Asbestos-Cement Pipes:	
Linings for	C541
Methods of Testing	C500
Nonpressure Sewer	C428
Pressure	C296
Asbestos-Cement Sheets, Shingles, Siding	C220 to C223, C459
Bitumized Fiber Pipe	
Homogeneous Bitumized Fiber Pipe	D2314
Laminated-Wall Bitumized Fiber Perforated Pipe	D2418
Fiber-Reinforced Plastics	
Acetone Extraction and Ignition of Strands, Yarns, and Roving for Reinforced Plastics	D2587
Apparent Horizontal Shear Strength of Reinforced Plastics by Short-Beam Method	D2344
Hydrostatic Compressive Strength of Glass-Reinforced Plastic Cylinders	D2586
Interlaminar Shear Strength of Structural Reinforced Plastics at Elevated Temperatures	D2733
Preparation and Tensile Testing of Filament-Wound Pressure Vessels	D2587
Tensile Properties of Oriented Fiber Composites	D3039

CHAP. 12: STRUCTURAL LAMINATES

Sandwich Panels	*ASTM Designation*
Delamination Strength of Honeycomb Type Core Material	C363
Density of Core Materials for Structural Sandwich Constructions	C271
Edgewise Compressive Strength of Flat Sandwich Constructions	C364
Flatwise Compressive Strength of Sandwich Cores	C365
Flexural and Flexural-Creep of Sandwich Constructions	C480, C481
Shear Fatigue of Sandwich Core Materials	C394
Shear Test in Flatwise Plane of Sandwich Constructions	C273
Water Absorption of Core Materials for Structural Sandwich Constructions	C272
Corrugated Panels	
Bearing Load of Corrugated Plastic Panels (for rivet and bolt fastenings)	D1602
Diffuse Light Transmission Factor of Reinforced Plastic Panels	D1494
Fiberglass-Reinforced Sheets	C581, C582
Glass-Fiber-Reinforced Thermoset Corrugated Structural Plastic Panels	D1919

Transverse Load of Corrugated Reinforced Plastic Panels	D1502

GENERAL TEST METHODS

Flammability Tests	*ASTM Designation*
Density of Smoke from Burning Plastics	D2843
Fire Tests of Building Construction and Materials	E119
Fire Tests of Roof Coverings	E108
Fire Tests of Window Assemblies	E163
Flammability of Plastic Sheeting and Cellular Plastics	D1692
Flammability of Plastics, Self-Extinguishing Types	D757
Flammability of Plastics Using the Oxygen Index Method	D2863
Flammability of Self-Supporting Plastics	D635
Ignition Properties of Plastics	D1929
Surface Burning Characteristics of Building Materials	E84
Surface Flammability of Building Materials, Tunnel Furnace Method	E286
Surface Flammability of Materials Using a Radiant Heat Energy Source	E162
Nondestructive Tests	
Evaluating Performance Characteristics of Pulse-Echo Ultrasonic Testing Systems	E317
Immersed Ultrasonic Testing by Reflection Method, Using Pulsed Longitudinal Waves	E214
Ultrasonic Testing by Resonance and Reflection Methods	E113, E114
Transmission Tests	
Laboratory Measurement of Airborne Sound Transmission Loss in Building Partitions	E90
Measurement of Airborne Sound Insulation in Buildings	E336
Sound Transmission Class, Classification for Determination of	E413
Thermal Conductivity of Materials, Guarded Hot Plate Test	C177
Water Vapor Transmission of Thick Materials	C355

INDEX